FUTURE TRENDS FOR TOP MATERIALS

Authored By

AF886445

Mário J. F. Calvete

Department of Chemistry, Faculty of Science and Technology, University of Coimbra
3004-535 Coimbra, Portugal

Bentham Science Publishers Ltd.
Executive Suite Y - 2
PO Box 7917, Saif Zone
Sharjah, U.A.E.
Email: subscriptions@benthamscience.org

BENTHAM SCIENCE

CONTENTS

FOREWORD

Chemistry is central to science, engineering and medicine, and has been at the forefront of the development of advanced materials for applications ranging from personal electronics, lighting and displays, solar energy conversion, energy storage, through environmental remediation, to clinical diagnosis, and advanced therapies for the treatment of cancer.

In this book, *Future Trends for Top Materials*, my colleague Mário J. F. Calvete presents an extensive review of the chemistry of some of the most important materials for these applications. These include the pyrrole-based systems phthalocyanines, porphyrins, porphyrinoids, boron dipyrromethene (bodipy) dyes, together with polymethine (cyanine) systems, and the carbon allotropes of fullerenes, carbon nanotubes and graphene. Further, from our understanding of mechanisms, stereochemistry and chemical bonding, it is now possible to design molecular architecture with specific materials functions. Dendrimers, metal-organic and covalent- organic frameworks are discussed within this perspective. The author also considers relevant aspects of metal and semiconductor based nanoparticles, including superparamagnetic systems and quantum dots. In the final chapter, he summarizes possible achievements based on relevancy of the most important materials.

All of the systems discussed are treated in terms of their synthesis and functionalization, before discussing their most important applications (real and potential). Examples are given in catalysis, photovoltaics, sensing, imaging, nonlinear optics, optoelectronics, biomedicine… This book provides both a good starting point for those entering the field, and a presentation of the state-of-the-art in for more experienced researchers.

The author is to be commended for both the breadth and depth of the work described, and for the extensive bibliography. This can form the basis for a course in advanced materials chemistry, but also acts as an excellent entry for those starting research in this area.

The author, Mário Calvete, has extensive experience in many of the topics discussed, and is to be commended for this excellent publication.

Hugh D. Burrows
Chemistry Department
University of Coimbra, Portugal

PREFACE

The idea behind this book arose from a necessity typical from Organic Chemists. Exploration of search engines in diverse topics usually leads to isolated results on each certain thematic issue. Hence, a directed and simple approach seemed necessary. After a careful compilation, discussion of synthetic methodology, historical perspective and relevance is provided for each of the families of compounds here described. The most representative and notable examples are mentioned/scrutinized at historical level, either regarding synthetic aspects or applications (which can be across any field, from medical, biological or materials applications, among others).

This contribution focuses on defined compounds (or types of compounds) which are sure candidates to occupy a prominent place in synthesis (and in diverse applications) in the forthcoming years, for a large number of readers, who by any motivation want to explore new compounds for any given application. In this book, general and established synthetic methodologies for several compounds are also given, in an attempted way to provide straightforward approaches for researchers who intend to start new topics, as a hands on reference guide for scientists working in the fields of chemistry, physics, biology, biomedicine, materials science, polymer science, nanotechnology or supramolecular science.

This eBook is divided in 9 chapters, in following order: phthalocyanines, polymethines, porphyrins, BODIPYs, dendrimers, carbon allotropes, organic frameworks, nanoparticles and future prospects. Each chapter covers detailed synthetic aspects of the most established preparation routes for each type of compounds, while giving a historical perspective, with selected information on actual and outstanding applications of each material, unraveling what likely might be the future for each one. I would like to refer to the last chapter, which deals with the synthesis and functionalization of several types of inorganic nanoparticles, but emphasizing the organic functionalization of such inorganic materials.

Bottom-line, it is intended to provide a clear concept on each section as a "*since when/how/why/what's next*" overall idea, targeted for researchers either from industries or universities, who by any motivation want to explore new compounds for any given application, providing general and established synthetic approaches, or to readers who would like to know more in a concise manner, without spending unnecessary efforts on isolated thematic search through available search engines.

CONFLICT OF INTEREST

The author confirms that this eBook content has no conflict of interest.

ACKNOWLEDGEMENTS

Declared none.

Mário J. F. Calvete
Department of Chemistry
University of Coimbra
Coimbra, Portugal
Email: mcalvete@qui.uc.pt

BIOGRAPHY

Mário José Ferreira Calvete is of Portuguese nationality and he was born in Portugal in 1975. He received his Industrial Chemistry diploma from the University of Coimbra in 2000 and his Ph.D. in Natural Sciences – Organic Chemistry in 2004 from the Eberhard Karls University of Tübingen, Germany, under the guidance of Prof. Dr. H. C. Michael Hanack on the synthesis and optical applications of phthalocyanines derivatives.

After a period in University/Industry working on Materials Science field, under the supervision of Prof. Dr. H. C. Michael Hanack, he returned to Portugal in 2007 for a postdoctoral position at the University of Aveiro with Prof. Dr. José Cavaleiro, working on the field on porphyrin-phthalocyanine heteroarrays. In 2010 he was appointed an Auxiliary Researcher at the University of Coimbra, currently being also Auxiliary Professor at the University of Coimbra.

His main research interest focus on the design and photophysics of near infrared absorbing/emitting materials and its conjugation with supports (of organic or inorganic nature) for application/study in homogeneous/heterogeneous catalysis and theranostics.

FUTURE TRENDS FOR TOP MATERIALS

CHAPTER 1

Phthalocyanines as Top Materials

Abstract: Phthalocyanines are colored aromatic macrocycles, composed of four iminoisoindoline units, giving origin to a highly conjugated planar macrocycle, known for its remarkable stability. Their remarkable features have rendered them wide attention, which continues to live on nowadays. Herein is presented a conspectus on the field of phthalocyanines, where emphasis is set on their preparation methods, along with more detailed examples on several promising applications with 418 references provided. Some synthetic details on the preparation of state-of-the-art compounds are given as well.

Keywords: Biomedicine dyes, Energy materials, Phthalocyanines, Subphthalocyanines.

INTRODUCTION

Phthalocyanines are intensely blue-green colored aromatic macrocyclic compounds, consisted by an 18 π-electron system, composed of four iminoiso indoline units, conferring a remarkable stability for an organic compound. Nevertheless, phthalocyanines offer great structural flexibility, being capable of forming coordination metal complexes with the majority of periodic table elements. Phthalocyanine structures are directly associated to the naturally existing porphyrins, with the differences being the four benzo-subunits and the nitrogen atoms at each of the four *meso* positions, reason why phthalocyanines are occasionally referred to as tetrabenzotetraazaporphyrins.

Other quite known related macrocycles include the tetraazaporphyrins, more known as porphyrazines (without the four benzo-subunits at each of the four *meso* positions) and naphthalocyanines (with additional four benzo-subunits at each of the four *meso* positions) (Fig. **1**).

Less known, but also important are the subphthalocyanines, which are the lower homologues of phthalocyanines (having three *N*-fused iminoisoindoline units

Mário J. F. Calvete

bearing a boron core). These compounds possess delocalized 14 π-electronic system, being considered aromatic compounds, although they possess nonplanar cone-shaped structure.

Nomenclature in phthalocyanines follows adapted IUPAC nomenclature codes. A system of abbreviations is necessary to avoid the long-winded nature of Pc nomenclature demanded by the IUPAC system. Fig. (**2**) shows the accepted numbering system of the Pc ring, where sixteen possible substitution sites in the macrocycle appear.

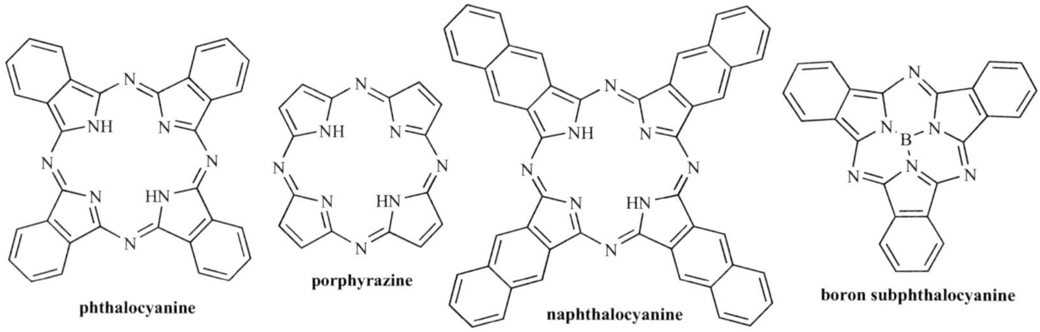

Fig. (1). Phthalocyanine derivatives.

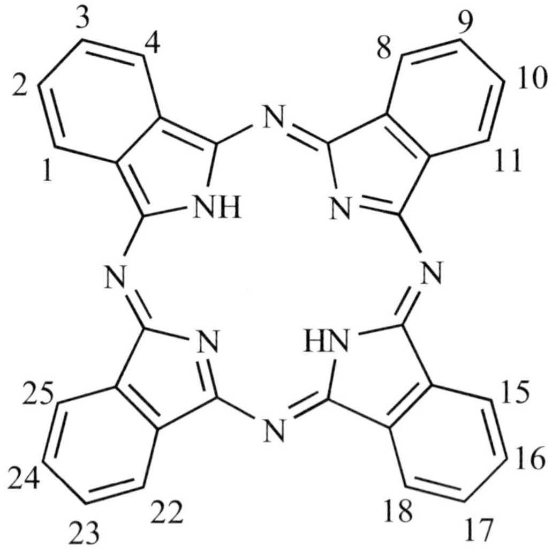

Fig. (2). Phthalocyanine ring numbering.

As many great findings in History, phthalocyanines were discovered fortuitously. Braun and Tcherniac observed a dark, insoluble material during the preparation of 2-cyanobenzamide from phthalimide and acetic acid in 1907 [1]. Nevertheless, less importance was given to such findings at the time. Later, in 1927, a remarkably stable blue material, ascribed as copper phthalocyanine, was synthesized by de Diesbach and von der Wied in 23% yield, as a side reaction of 1,2-dibromobenzene with copper(I) cyanide in pyridine (Rosenmund-von Braun reaction) [2]. Almost coincidentally, the preparation of a blue-green material occurred during the industrial synthesis of phthalimide from phthalic anhydride, at the Grangemouth plant of Scottish Dyes Ltd. in the year 1928. Dandridge and Dunsworth, two laboratory employees analyzed the side-reaction meticulously and discovered that it was prone to occur inside older iron reactors, which could release some iron splinters from reactor cracks during production. Initial studies on these iron containing pigments reveled an outstandingly stable insoluble dye, whose preparation and properties gave rise to a patented, granted in 1929 [3]. One year earlier, Imperial Chemistry Industries (ICI) had acquired Scottish Dyes Ltd., and the curiosity of the new owners provoked a decisive chain of events, by sending some samples to Professor Jocelyn Thorpe and Imperial College, London, who in turn gave the investigation to newly appointed Reginald P. Linstead (later entitled Sir) who, through collaboration with ICI, managed to meticulously explore most aspects of this new substance. Indeed, the term phthalocyanine was the first used by him [4], as a derivatization from the Greek words *phthal* (rock oil) and *cyanine* (blue).

In following years phthalocyanine structure was clarified, along with procedures for the preparation of several metal phthalocyanines and their metal-free complexes [5 - 10]. Research continued until it was interrupted by World War II, including seminal outcomes reporting the first synthesis of naphthalocyanines [11] and accomplishment of two dozen different metals inside phthalocyanine cavities [12]. In the subsequent years after World War II, research hasted and great developments were reported on the family of phthalocyanines. For instance, in 1972, subphthalocyanines were accidentally discovered by Ossko and Meller [13], when attempting to synthesize boron containing phthalocyanines *via* cyclotetramerization reaction of boron trichloride with phthalonitrile in chloronaphthalene at 200 °C. Unexpectedly, a pink colored product was isolated the structure of which was determined later by Kietaibl using X-ray crystallography [14].

Through direct retrieval from a search engine (Web of Knowledge^{SM}-Thomson Reuters®) the hits obtained (approximate value, including patents) on phthalocyanine related reports since its discovery, appear as follows: 1907-1939 (40 reports); 1940-1959 (110 reports); 1960-1979 (2800 reports); 1980-1989 (5500 reports); 1990-1999 (13000 reports); 2000-2009 (23000 reports) and finally from 2010-2013 (12300 reports).

SYNTHESIS OF PHTHALOCYANINE DERIVATIVES

In general, phthalocyanine synthesis proceeds from single step cyclotetramerization reactions, normally using benzoic acid derivatives, *e.g.* phthalimide, phthalic anhydride, *o*-cyanobenzamide, phthalonitrile (the most used) or isoindolinediimine [15 - 19] (Scheme **1**).

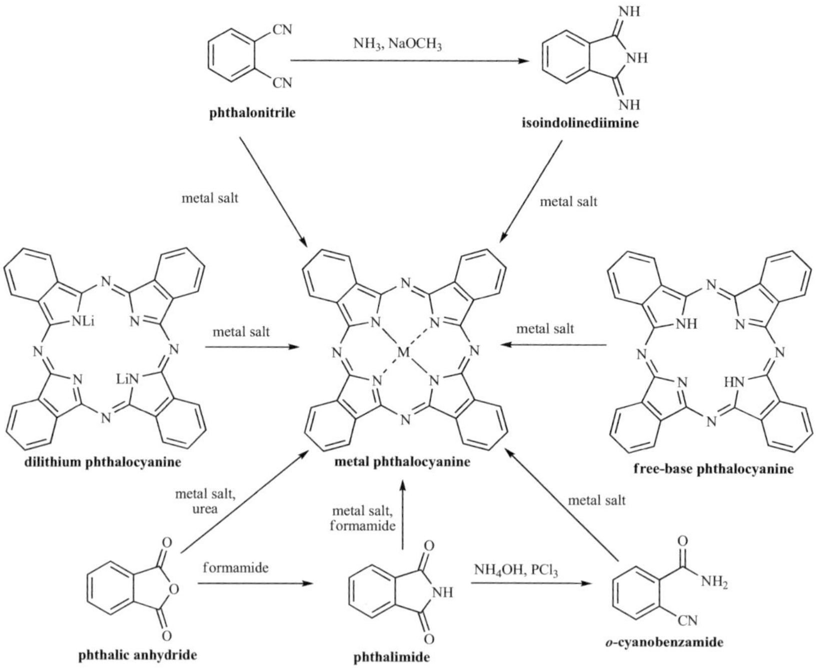

Scheme (1). Common synthetic pathways for phthalocyanine synthesis.

A direct and convenient method for the preparation of metal containing phthalocyanines is heating phthalonitrile with a metal salt in the presence of high boiling solvents, such as quinoline, dimethylformamide, 1-chloronaphthalene, DMAE. Dimethylaminoethanol, octan-1-ol, hexan-1-ol pentan-1-ol. In addition, some hindered bases such as 1,8-diazabicycloundec-7-ene and 1,5-diazabicyclo(4.3.0) non-5-ene can also be used to aid cyclotetramerization, giving good results in many cases.

Phthalimide, phthalic anhydride and phthalic acid precursors can also be applied successfully for metal phthalocyanine synthesis, particularly phthalic anhydride, as most commercial processes are based on this compound. Such reactions are normally carried out in melted urea, acting as a convenient nitrogen source (ammonia), with ammonium molybdate as catalyst, transforming the anhydride in phthalimide *in situ*. Occasionally, low reactive phthalonitriles can also be transformed in diiminoisoindoline, prepared from the reaction of phthalonitrile and ammonia, using sodium methoxide as catalyst.

The low solubility in most solvents presented by phthalocyanines is a challenging issue in any application, which is caused by the strong interaction of the π-electrons of their extended aromatic system and appropriately called π-stacking. Preventing this strong interaction increases solubility and can be facilitated by substituents in three ways: i) axial substituents, by modulating the valence of the central metal; ii) peripheral substituents that are sufficiently space demanding; iii) non-peripheral substituents. The first approach, which deals with central metal coordination with additional axial ligands (axial substitution) was firstly developed by Kenney group, in 1962 [20]. Generally, covalently bound axial ligands require the central metal ion to be at least in a +3 or +4 oxidation state, *e.g.* Al^{3+}, Ga^{3+}, In^{3+}, Si^{4+}, Ge^{4+} and Sn^{4+}, since phthalocyanine itself has a -2 oxidation state and can only be attached after cyclotetramerization. In addition, suitable ligands such as pyridines can also form stable coordination bonds with many central metal ions, accounting for the enhanced solubility of metal phthalocyanines in pyridine and quinoline, firstly reported by Cariati [21].

Albeit the many studies committed to the explanation of the mechanism of phthalocyanine formation, the exact and unambiguous course of the reaction still remains unclear. Bottom-line, the mechanism of phthalocyanine formation can be considered as a stepwise polymerization of the appropriate precursor followed by coordination of the central metal ion and ring closure originating the macrocyclic molecule [22 - 29]. An interface of two factors might be the driving force for phthalocyanine macrocycle formation: the template effect of the central metal ion/resulting stabilization the complex and thermodynamic stabilization/ aromaticity involved in the cyclotetramerization/consequent ring closure.

Phthalocyanine solubility in general organic solvents can also be improved by appending long chain or bulky substituents at the macrocycle periphery (2 and/or 3 positions), referred as peripheral substitution, or at non peripheral positions (1 and/or 4 positions) (Fig. **3**). Non-peripheral substituents are most effective in order to diminish π-stacking. Due to mutual sterical hindrance, the phthalocyanine is distorted out of the plane, but to the expense of increased strain in the molecule. In contrast, peripheral substituents have to be highly space demanding in order to prevent π-stacking. Nevertheless, out-of-plane distortion is minimal and resonance stabilization is maintained.

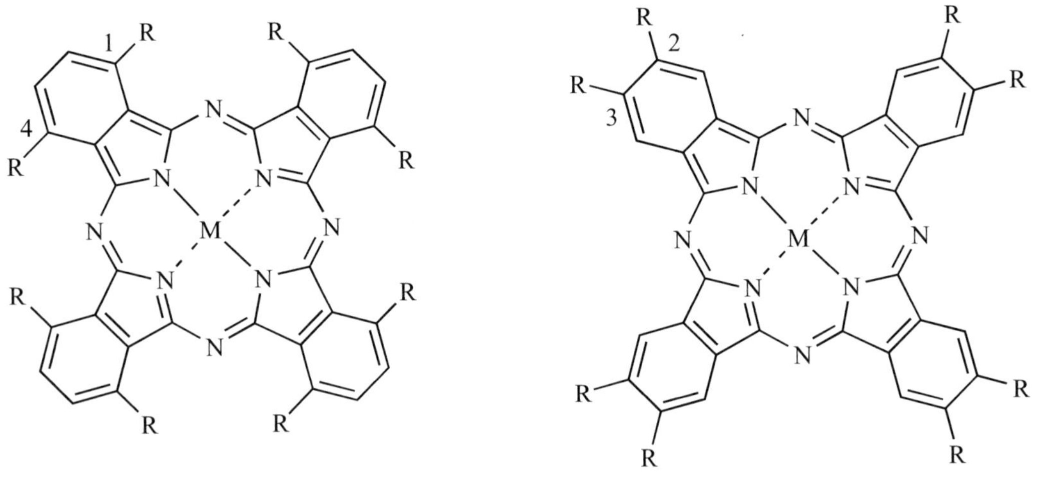

Fig. (3). Non peripherally and peripherally substituted phthalocyanines.

Substitution at the 1, 2, 3 or 4 positions of a phthalonitrile yields tetrasubstituted phthalocyanines, while substituting at the 1,2, 2,3, 1,3 or 1,4 positions gives octasubstituted phthalocyanines. Moreover, simultaneous substitution at all phthalonitrile positions is also possible, giving rise to hexadecasubstituted phthalocyanines. In general, the peripherally substituted phthalocyanines studied in more detail are the tetra- and octa- substituted ones [15].

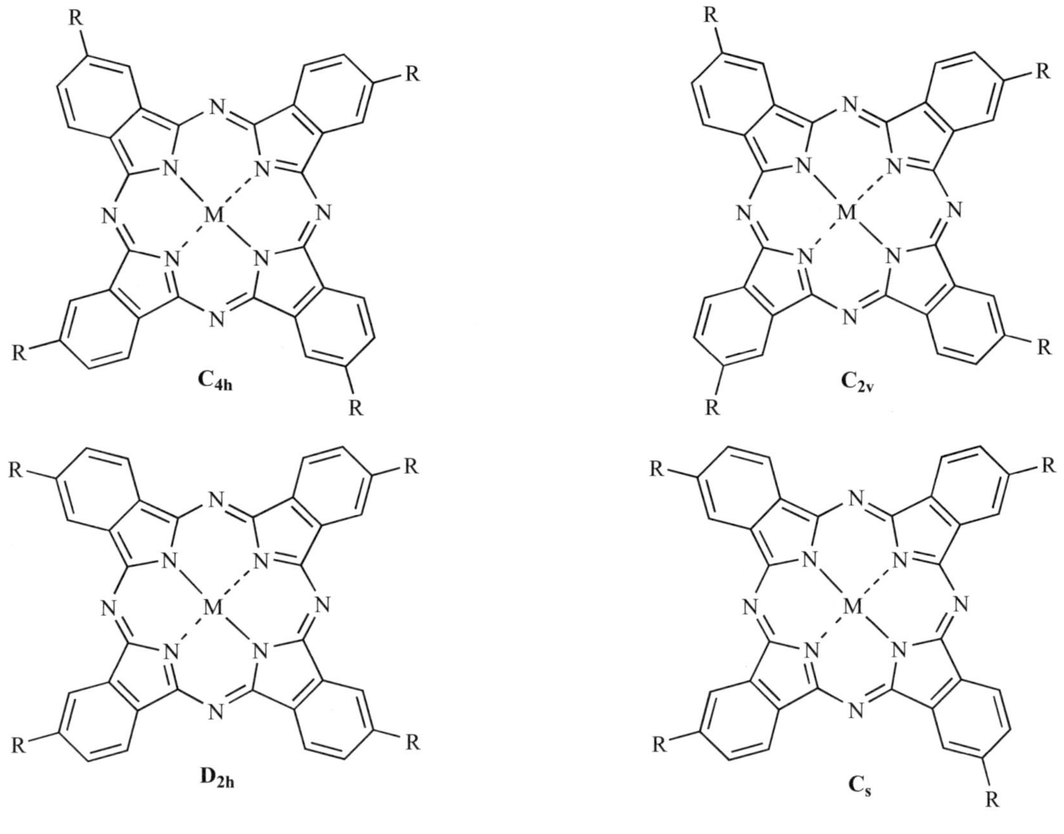

Fig. (4). Tetra substituted phthalocyanine's isomers.

The solubility of tetrasubstituted phthalocyanines is higher than that of octasubstituted phthalocyanines, due to the fact that tetrasubstituted phthal ocyanines appear as a mixture of isomers (Fig. **4**), consequently leading to a lower degree of order in the solid state, when compared to symmetrically octa- or even hexadecasubstituted phthalocyanines.

Furthermore, the less symmetrical isomers have a higher dipole moment derived from the more unsymmetrical disaposition of the substituents at the phthalocyanine ring periphery [30, 31]. This was proven by Hanack group, after complete separation of the four structural isomers of nickel tetra (2-ethylhexyloxy) phthalocyanine, for the first time by HPLC [32], and later of a family of 2,9,(10),16,(17),23,(24)-tetrasubstituted phthalocyanines [31].

Summing up, the introduction of substituents at the phthalocyanine macrocycle is of utmost importance for most concrete applications. Although modulation of the phthalocyanine is possible, the relative stability of the macrocycle hampers reactivity towards many synthons, mainly originating mixtures with several degrees of substitution. Therefore, prior phthalonitrile (or any other phthalocyanine precursor) functionalization followed by cyclotetramerization is the main approach, leading to an exact number of substituents on the desired phthalocyanine (metallated or free-base), of which is important to refer the comprehensive reviews of Nemykin on the synthesis of substituted phthalonitriles and their transformation in phthalocyanines [33, 34].

Modulation of the benzene rings in phthalocyanines in a non symmetrical manner (*i.e.,* differentially substituted) is sometimes necessary, in order to introduce specific functionalities for the most varied applications.

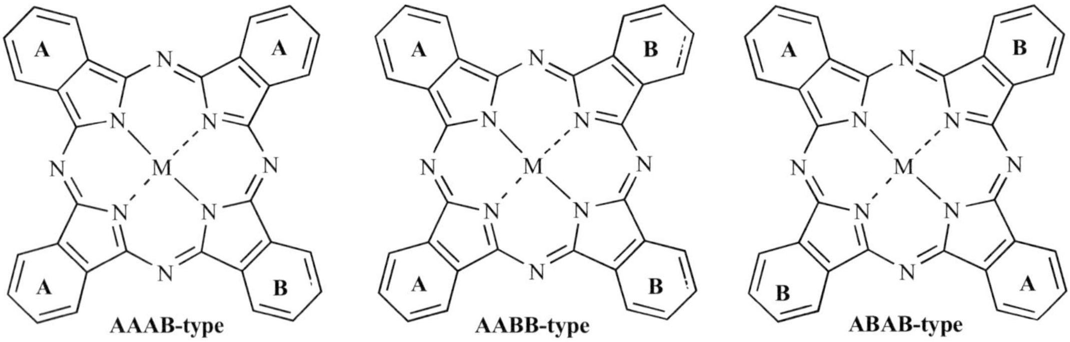

Fig. (5). Unsymmetric substitution patterns on phthalocyanines.

The chemistry of unsymmetrically substituted phthalocyanines had its beginnings with Linstead in 1955 [35], followed by work of Leznoff in 1982 [36], making a

28 year hiatus between the two first reports on unsymmetrical phthalocyanines. However, after that, the number of reports on the preparation of unsymmetrical phthalocyanines increased nearly exponentially, including several important reviews on the topic [37 - 43].

The main possible types of unsymmetrical phthalocyanines are A_3B, ABAB and AABB types, as represented in Fig. (**5**). Most of preparations rely on the use of statistical cross condensation reactions between two different phthalonitriles, varying the ratios between them, favoring a specified statistical product, according to Table **1**, in a model developed by Cook [44]; however not considering template, steric and electronic effects. In addition, it does not consider that the particle reservoir decreases during the reaction. However, this can be neglected, to a good approximation of the large particle number.

Table 1. Predicted relative yields from statistical condensation of different phthalonitriles.

A:B	AAAA	AAAB	ABAB	AABB	ABBB	BBBB
1:1	6.25	25	12.5	25	25	6.25
3:1	31.6	42.2	7.0	14.1	4.7	0.4
9:1	65.6	29.2	1.6	3.2	0.4	0.01
Permutations	1	4	2	4	4	1

Strategies for obtaining unsymmetrical A_3B-type phthalocyanines are by far the most developed ones, of which the above mentioned methodology is the most used. Other less used approaches include the subphthalocyanine methodology (Scheme **2**), firstly developed by Kobayashi [45], where a subphthalocyanine was expanded to an unsymmetrical phthalocyanine, by reacting it with a diiminoisoindoline derived compound. This method revealed to be quite selective and efficient in several cases, allowing as well to prepare phthalocyanines otherwise inaccessible by other methods [46, 47].

Another method for A_3B unsymmetrical phthalocyanines is based on solid phase synthesis, a method developed by Lever [36], involving the connection of a substituted phthalonitrile to a polymeric support. This polymer-linked precursor might then be reacted with another differently substituted phthalonitrile, originating a polymer-linked A_3B phthalocyanine, which can then be selectively unattached from the polymer.

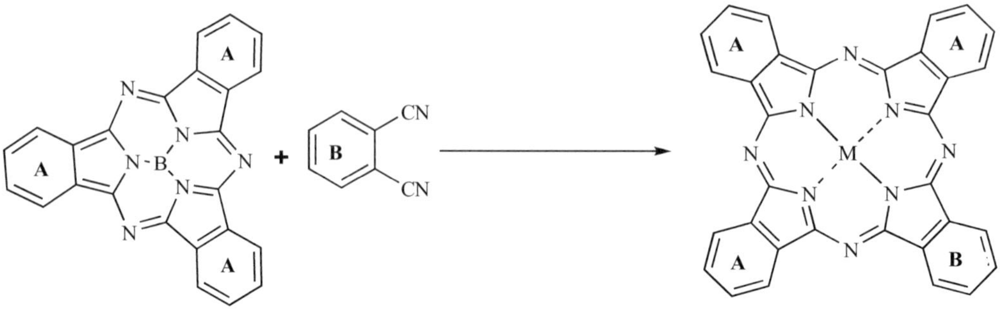

Scheme (2). Synthesis of unsymetrically substituted phthalocyanine from subphthalocyanine.

ABAB and AABB phthalocyanines can seldom be isolated from statistical mixtures, due to their similar polarity and solubility. Nevertheless, some methods have been developed to prepare this type of phthalocyanines. For instance, Young developed a method [48], where 1,3-diiminoisoindoline and 1,3,3-trichloroisoindolenine derivatives where reacted in presence of a reducing agent and base (Scheme **3**).

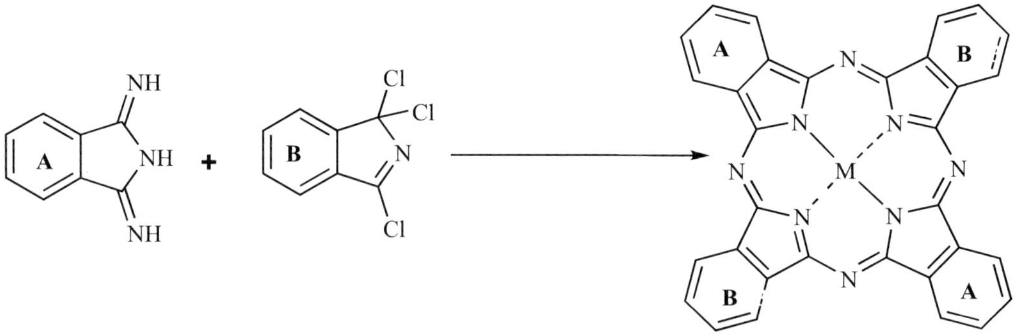

Scheme (3). Synthesis of ABAB substituted phthalocyanine.

The authors alleged that the only formed derivative was the ABAB phthalocyanine, in 50% yields; however, Hanack tried such a procedure but the obtained yields were lower than those achieved by Young (15-25%) [49].

Nevertheless, this route affords high selectivity for the desired compounds.

Additionally, AABB phthalocyanines are even more complicated to prepare. In this case the synthesis of this type of phthalocyanines occurs through an intermediate designated "half-phthalocyanine", which is a diiminoisoindoline dimer (AA), further reacted with a B type phthalonitrile, yielding an AABB phthalocyanine. This intermediate was first reported by Oliver [26] and later followed by Leznoff [50] and Kobayashi [51 - 54].

APPLICATIONS OF PHTHALOCYANINES

The list of phthalocyanine applications is enormous. A simple search on the number of patents issued throughout the years reveals more than 15000 patents worldwide where phthalocyanines are involved directly. The approximate production of phthalocyanines derivatives passes 60000 tons per year. Initially after their discovery, phthalocyanines were mostly used as dyes in the textile, paper and other industries due to their chemical, photochemical, and thermal stabilities [15, 18, 55, 56]. Phthalocyanines are widely used in inks (ballpoint pens, printing inks, *etc.*), coloring for plastics and metal surfaces as well as dyestuff for jeans and other clothing. Interestingly, copper phthalocyanine was certified as a food dyestuff in Germany and for coloring contact lenses in the United States [57].

In more recent decades the chemistry of substituted and unsubstituted phthalocyanines has experienced a remarkable expansion and, in addition to traditional applications, substituted have found potential applications in many other fields. It is worth mention both the Porphyrin Handbook, published between 1999 and 2003, with 20 volumes and 122 chapters, 24 of them exclusively concerning phthalocyanines and their properties [58]. Later, this successful series of books were complemented with the acclaimed Handbook of Porphyrin Science, still unfinished, containing so far 35 volumes and 180 chapters, having 21 chapters concerning phthalocyanines [59]. Both series are edited by the eminent professors Kevin M. Smith, Roger Guilard and Karl M. Kadish.

Like mentioned above, phthalocyanines have more recently found applications in catalysis [60 - 86], chemical sensors [87 - 111], as Langmuir-Blodgett films [112 -

126], semiconductors [30, 92, 127 - 150], materials for ink-jet printing [55, 151], electrophotographic applications [152 - 157], optical data storage [158 - 166], liquid crystals [167 - 189], light-harvesting modules for dye-sensitized solar cells and organic photovoltaics [139, 190 - 203], nonlinear optics [184, 204 - 222], photosensitizers for photodynamic cancer therapy [223 - 237], antibacterial composites [238 - 256] and theranostics [257 - 261].

Applications in Catalysis

Catalysis is a highly significant metallophthalocyanine applications. For example, the Merox process involves the catalytic oxidation of mercaptans using sulfonated cobalt phthalocyanine (petrol sweetening), (Fig. **6**, 1) to remove most of sulfur from petrol [262].

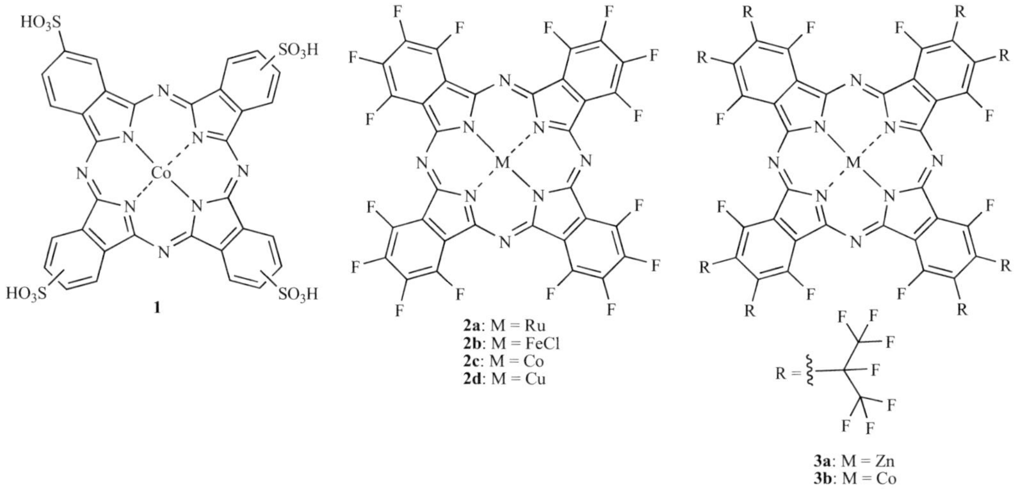

Fig. (6). Phthalocyanines for application in catalysis.

Another well documented field is research on the degradation of pollutants using catalytic amounts of metal phthalocyanines, including in photocatalytic and electrocatalytic processes [66, 67, 69, 263].

Though centered in oxidation, other processes including several C−C bond formation reactions, preparation of nitrogen-containing compounds and reduction reactions and can be proficiently catalyzed by metal phthalocyanines. They own several advantages over other types of processes, *e.g.* noble metals. Their reactivity can be modulated by appropriate structural modifications, either by changing the nature of the central metal or by modifying the structure of the macrocyclic ligand.

The peripheral substituents are crucial in the catalytic chemistry of metallopht halocyanines [34]. Changing the electron donating or electron-withdrawing substituent's character allows obtaining the desired solubility, tuning catalytic properties and stability of phthalocyanine complexes.

For example, appending fluorinated substituents to metallophthalocyanine molecules increases their stability against radical, electrophilic or nucleophilic attacks [264]. For instance metallated hexadecafluorinated phthalocyanines (with ruthenium (**2a**)) [265], iron (**2b**) [266], cobalt (**2c**) [267] and copper (**2d**) [267] as central metals have been prepared and used in a variety of catalytic reactions (Fig. **6**). In addition, all-fluorinated phthalocyanines bearing perfluoroalkyl groups such as **3a** [268] and **3b** [269] have also been used in catalytic reactions.

Immobilization by encapsulation (a method named ship-in-a-bottle synthesis) has performed for the immobilization of several phthalocyanine derivatives, like for instance unsubstituted iron(III) phthalocyanine **4** [270] or tetrasubstituted zinc phthalocyanine **5** [271] derivatives (Fig. **7**) and used for photocatalytic pesticide degradation. Other authors described the mesoporous silica covalent anchoring of cobalt(II), manganese(III) and iron(III) phthalocyanines (**6a-c**) [64, 272, 273]. Also copper hexadecafluorinated phthalocyanine **7** was encapsulated inside a mesoporous material (Fig. **7**) [274], through an innovative approach, which consisted in reacting hexadecafluorinated copper phthalocyanine with 3-aminopropy-ltrimethoxysilane, followed by posterior tetraethylorthosilicate co-condensation yielding an hexagonally ordered material, having high surface area.

Applications as Langmuir-Blodgett Films

Organic thin films formed from conjugated organic molecules are of particular interest, since they may lead to the preparation of inorganic-organic heterostructures with differentiated properties, including ultra-fast optical response, chemical sensing, reactivity and biocompatibility [275]. Reactivity of reactive centers with analytes is ruled by thickness and film characteristics. For instance, ultrathin films are satisfactorily used in molecular separation, reception and discrimination.

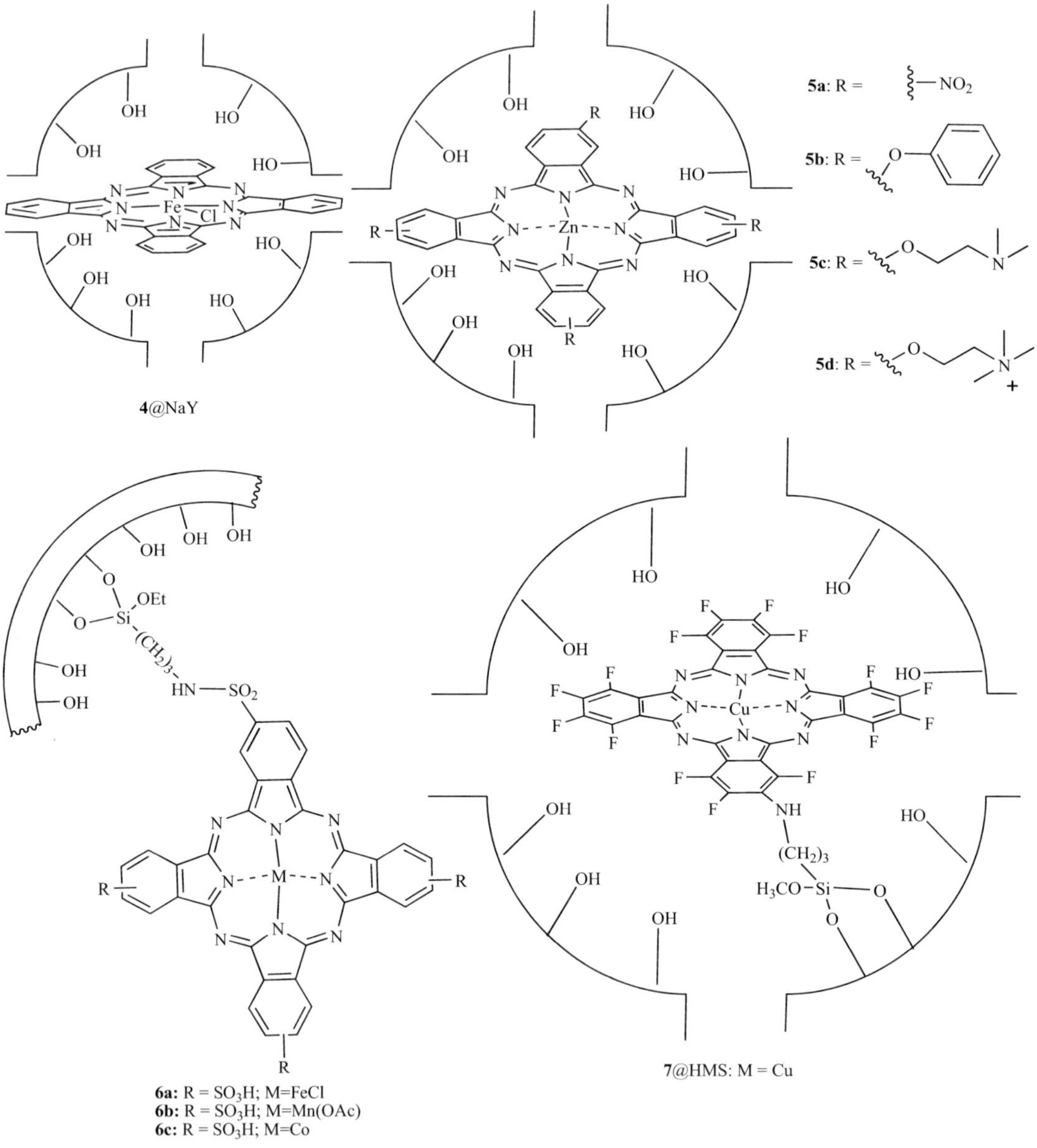

Fig. (7). Immobilized phthalocyanines for application in catalysis.

In this regard, phthalocyanines conform to the requisite of thermal and mechanical stability, necessary for Langmuir-Blodgett based functional devices. The use of the Langmuir-Blodgett method is essential, since some applications, such as photovoltaic cells, optical storage, electrochromic displays, color filter dyes and electro-luminescence [276, 277], require layers of molecular dimensions.

The first reports on Langmuir-Blodgett phthalocyanines applications emerged in the 1930s, unsuccessfully using unsubstituted iron and magnesium phthalocyanines [278]. The revival of interest in Langmuir-Blodgett films of phthalocyanines begun in the early 1980s [120, 121], by Robert's group. Their promising results instigated research on deposition studies of substituted phthalocyanines soluble in organic solvents, particularly the tris-*N*-isopropyl - aminomethyl derivative, **4a**, and the tetra-*tert*-butyl derivatives **4b** (Fig. **8**) [120, 121].

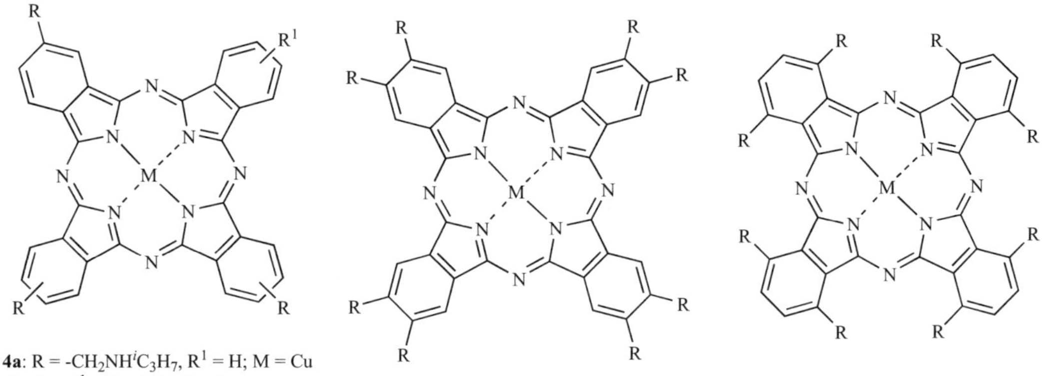

4a: R = $-CH_2NH^iC_3H_7$, R^1 = H; M = Cu
4b: R = R^1 = ^{t}Bu; M = Cu, Co
4c: R = R^1 = *p*-cumylphenoxy; M = Cu
4d: R = R^1 = CO_2R^2 (R^2 = alkyl); M = Cu
4e: R = R^1 = $CO_2NHC_{18}H_{37}$; M = Ni

5a: R = $-CH_2OC_{12}H_{25}$; M = Cu

6a: R = -OAlkyl; M = several metals
6b: R = -Alkyl; M = several metals

Fig. (8). Phthalocyanines for application in Langmuir-Blodgett films.

Research on both substituted and unsubstituted phthalocyanines has continued, aiming to produce highly ordered films for particular applications. Reports on films of non substituted compounds included annealing a film of free base phthalocyanine at 300 °C [279, 280], or even zinc phthalocyanine [281]. However, research on films of substituted phthalocyanine derivatives has been much more intense, for instance in various metallated tetrasubstituted phthalocyanines. These include aryloxy groups **4c** [282], alkoxy groups [283], esters **4d** [284, 285] and amides **4e** [286, 287] (Fig. **8**).

Usually, tetrasubstituted phthalocyanines form materials that are isomeric mixtures, which is improbable to promote optimal film ordering. So, attention has also been turned to octasubstituted purely symmetrical macrocycles, such as the octakis (dodecylo-xymethyl)-copper phthalocyanine **5a** (Fig. **8**) [288, 289]. Nevertheless, nonperipherally substituted octaalkoxy or octaalkyl phthalocyanines **6a,b** (Fig. **8**) form better monolayers, depending on the aliphatic chain length, with the octapentyloxy derivatives forming the best films [290].

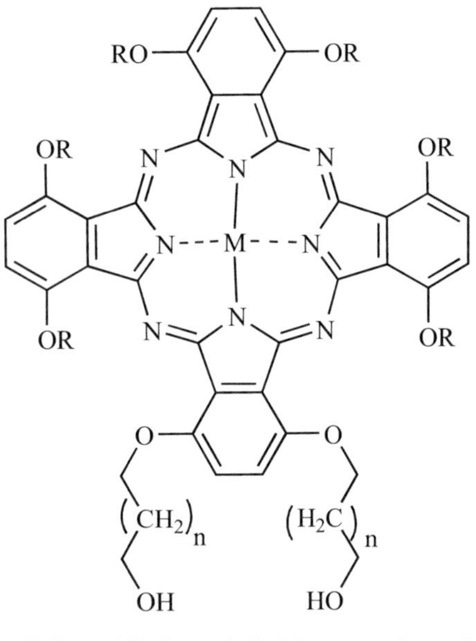

7: R = -Alkyl; n = 1, 3; M = several metals

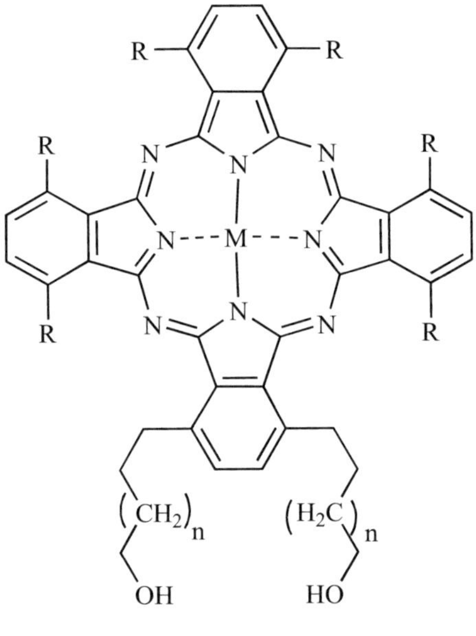

8: R = -Alkyl; n = 1, 3; M = several metals

Fig. (9). Non-peripherally substituted phthalocyanines for application in Langmuir-Blodgett films.

While the octasubstituted phthalocyanine derivatives described above possess an amphiphilic character, their properties are inferior to the phthalocyanines bearing long-chain alcohols **7** and **8** (Fig. **9**) [291 - 295].

As sensing and recognition of biologically and environmentally important ions has appeared, most of thin films prepared actually meet the requirements to act as powerful chemical sensors affording actual information on the localization and quantification of the targets of interest. For instance, films produced from phthalocyanines type **7** and **8** have been prepared and used as efficient NOx sensors [296, 297]. These materials possess hydrophilic or lipophilic chains, therefore assisting the preparation of an organized floating film, which revealed to be essential for an effective sensor. In subsequent years, Ding also prepared sensor films using phthalocyanine derivatives of type **9** (Fig. **10**), and successfully tested them in gas sensoring of NOx, ammonia and other amine containing gases [298 - 301].

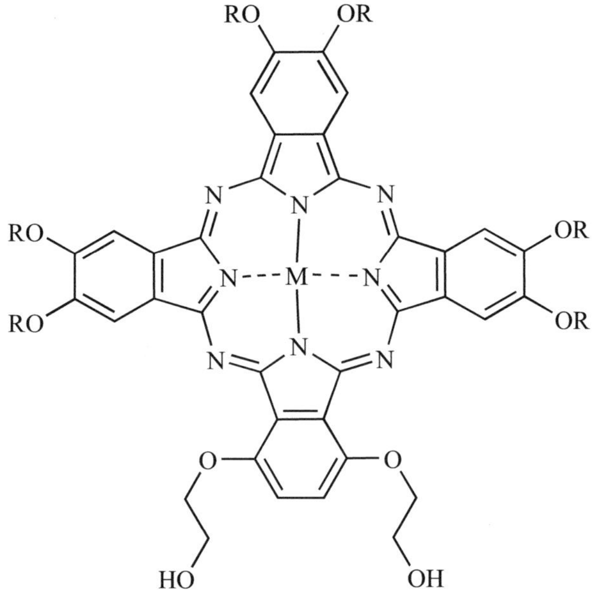

9: R = nC_5H_{11}; M = H_2 or several metals

Fig. (10). Peripherally substituted phthalocyanines for application in Langmuir-Blodgett films.

Applications as Fluorogenic Chemical Sensors

Besides from the intensive utilization of phthalocyanine Langmuir-Blodgett films as sensors, other analytical methods have been used, *e.g.* fluorescence, this is a strong tool due to its simplicity and high detection limit. Particularly, suitable probes in conjunction with fluorogenic methods are favorable to measure analytes, given that fluorimetry is highly sensitive, quickly performed, and nondestructive.

Several authors have prepared and used phthalocyanine derivatives as metal ion sensors [302 - 305]. For instance, Barret and Hoffman prepared polydentate based on porphyrazine derivatives **10** and **11** (Fig. **11**) [302] and demonstrated their ability in binding metal ions through the peripheral coordinating diamine or diazacrown groups, along with metal ion coordination.

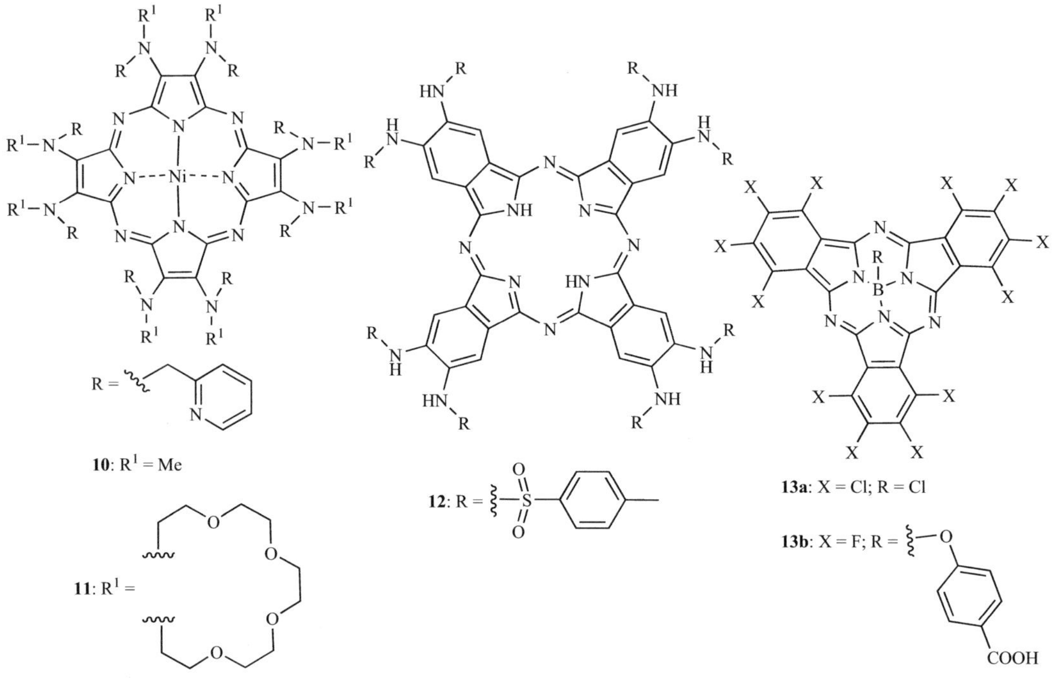

Fig. (11). Phthalocyanines for application as sensors.

Anions can also be detected using phthalocyanine derivatives by the same fluorometric means [306 - 309]. For instance, phthalocyanine **12** (Fig. **11**) [307], an octatosylaminophthalocyanine was used as anionic chromogenic chemosensor. The host:guest complexes formed could be restored back by acid treatment, without loss of sensoring capability, allowing the phthalocyanine based chemosensor to be reused.

Martínez-Mánez [308] and Torres [309] have also separately used subphthalocyanines **13a** [308] and **13b** [309] (Fig. **11**) for cyanide ion detection to the ppm level due to the selective reaction of this anion with the selected subphthalocyanines. The method was easy assessed, allowing selective visual detection of very low cyanide concentrations [309].

Applications as Semiconductors

Thin films of phthalocyanines are further important for application in conductor devices. Organic semiconductor devices are increasingly enjoying a widespread commercialization [310, 311]. Phthalocyanines can be used in organic electronics, as organic semiconductor single crystal assembly directly onto device structures. For instance, the mostly studied phthalocyanine for conducting purposes has been the unsubstituted copper phthalocyanine [140, 312 - 314], ever since it was the first phthalocyanine found with seconducting properties [140, 312]. Also zinc phthalocyanine [315, 316] and hexadecafluorinated copper phthalocyanine [317] are good examples of phthalocyanine derivatives used for this application.

Applications as Liquid Crystals

Nowadays, liquid-crystal displays are necessary to our everyday life. Usually these devices are based on self-organized stacked molecules. Vorländer described the rodlike (calamitic) counterparts, in 1907 [318], while Chandrasekhar discovered the disklike (columnar, discotic) liquid crystals in 1977 [319], providing a wider range of applications, turning into appropriate compounds in several electronic/optical devices, such as photocopiers, laser printers, photovoltaic cells, light-emitting diodes, holographic data storage centers and field-effect transistors. Mesomorphic phthalocyanines [179, 320, 321] provide effective materials, possessing strong visible and NIR absorption but nevertheless, an easy alignment and/or long-time mesophase stability are the most important features desirable for potential applications, [322].

For instance, comparison of side chain effect on the mesophases produced by phthalocyanines of the type **14a-e** (Fig. **12**) [323], revealed that phthalocyanines **14c,d** bearing ramified alkyloxy chains displayed higher phase transition than linear side chain bearing phthalocyanines **14a,b**, consequently reducing the melting transition. Very interesting was (*S*)-citronellol substituted phthalocyanine compound **14e** [324], which formed a room temperature chiral columnar mesophase and a high temperature achiral rectangular columnar mesophase [325].

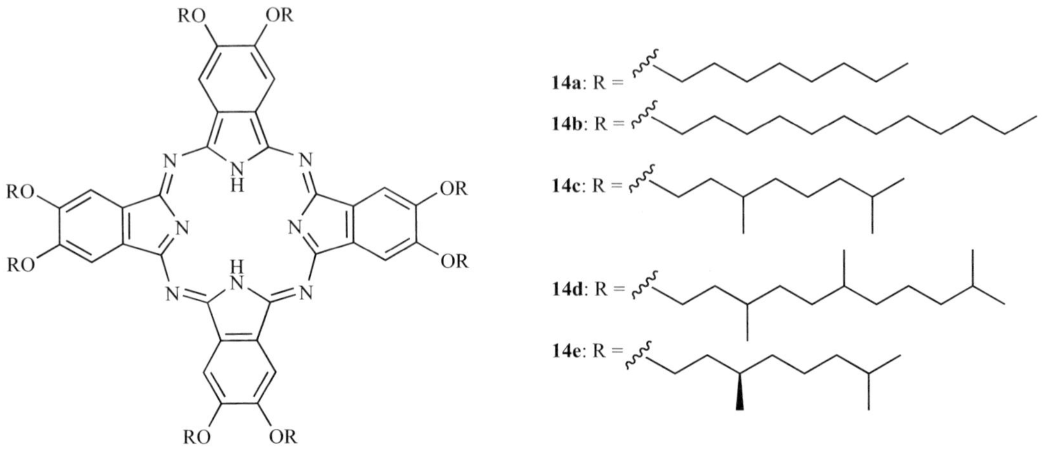

Fig. (12). Octa alkoxy-substituted phthalocyanines for application as liquid crystals.

Exploiting the capabilities of the chiral substituent, Nolte and coworkers have further investigated the columnar properties of phthalocyanine **15** (Fig. **13**) and observed a molecular arrangement comprising one "edge-on" lamellar phase and two "face-on" phases.

Applications as Light-harvesting Modules for Organic Photovoltaics and Dye-sensitized Solar Cells

Solar cells based on organic materials are enjoying rising attention. Usually, these materials are mostly constituted by a bulk heterojunction resultant from the interaction of donor and acceptor materials, combining the advantages of higher conversion efficiency and straightforward fabrication. Phthalocyanines conjugated system produces intense absorption spectra, possessing very high molar absorptivity (ε) values exceeding 10^5 $Lmol^{-1}cm^{-1}$, high chemical and thermal stabilities and appropriate redox properties.

Tang reported the first phthalocyanine based heterojunction solar cell, constituted of acceptor perylene derivative **16** and donor copper phthalocyanine (**17a**) (Fig. **14**) [326], achieving efficiencies of *ca* 1%. When zinc phthalocyanine (ZnPc, **17b**) was employed as a donor instead of **17a** conversion efficiency of 1.3% was obtained [327]. Furthermore, other different acceptors (**17c-e**) were used and the efficiencies obtained were usually lower than 1% [328 - 330].

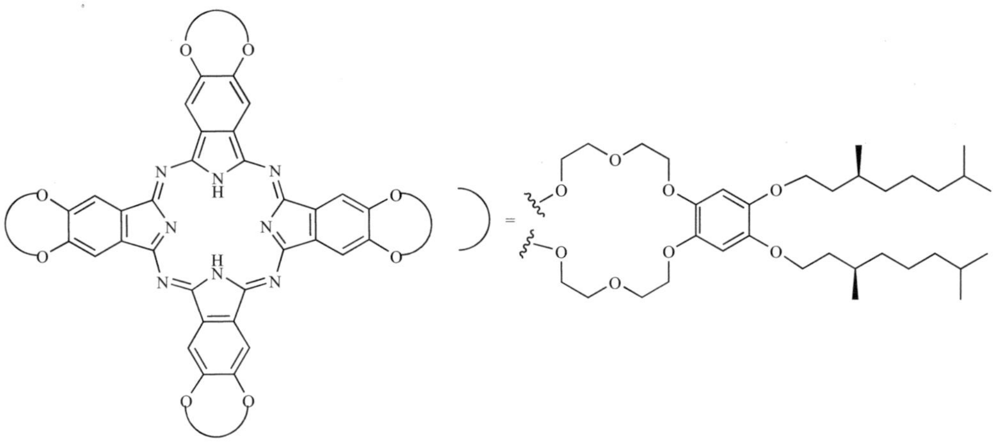

15

Fig. (13). 18-Crown-6 substituted phthalocyanines for application as liquid crystals.

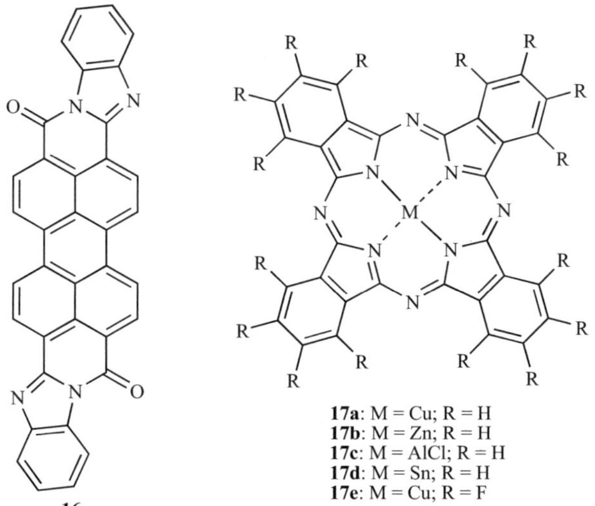

16

17a: M = Cu; R = H
17b: M = Zn; R = H
17c: M = AlCl; R = H
17d: M = Sn; R = H
17e: M = Cu; R = F

Fig. (14). Perylene-phthalocyanine system for photovoltaics.

To improve the process ability of a phthalocyanine: [6 , 6]-phenyl-C_{61}-butyric acid methyl ester blend (phthalocyanine:PCBM) and to expand the spectral absorbance window to 350-550 nm, Ru(II) phthalocyanine **18** complex was prepared and evaluated in a blended Pc:PCBM blend bulk heterojunction solar cell (Fig. **15**) [331]. The blend with a 2:1 ratio of acceptor:donor produced efficiencies up to 1.6%.

Phthalocyanines can be used in solar cells with mainly three different attachment methods: adsorption, metal-ligand interactions and anchoring using sulfoxy or carboxy substituents.

While the first method produces cells with very limited efficiency, the two last anchoring methods provide the requested ability for efficient light harvesting.

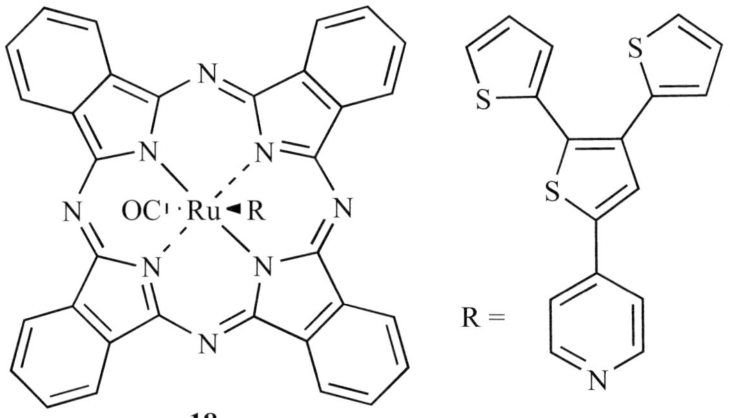

18

Fig. (15). Ruthenium(II) phthalocyanine complex.

Many photosensitizers of the phthalocyanine family have been tested in DSSCs, with different efficiencies, for instance the dye **19** gave 3.1% efficiency with a PEDOT:PSS-coated fluorine-doped tin oxide (FTO) counter electrode (Fig. **16**). Additional development of the system was obtained by pretreatment of TiO_2 surface with dilute HNO_3, displaying a 4.1% overall efficiency [332]. Using chenodeoxycholic acid as co-adsorbent, dye **20** (Fig. **16**) gave an efficiency of 3.6% [333] and 7.7% when tested as a blended DSSC. The use of Pc dyes with two carboxy ligands that had no macrocyle core conjugation was also reported

[334]. The dye **21,** in the presence of chenodeoxycholic acid (Fig. **16**) gave an energy conversion efficiency of 3.1%.

A high efficiency sole phthalocyanine DSSC was reported by Kimura [335]. Phthalocyanine **22** (Fig. **16**) displayed a conversion efficiency of 4.6%, most probably due to diphenylphenoxy moieties bulkiness, resultant from low dye aggregation.

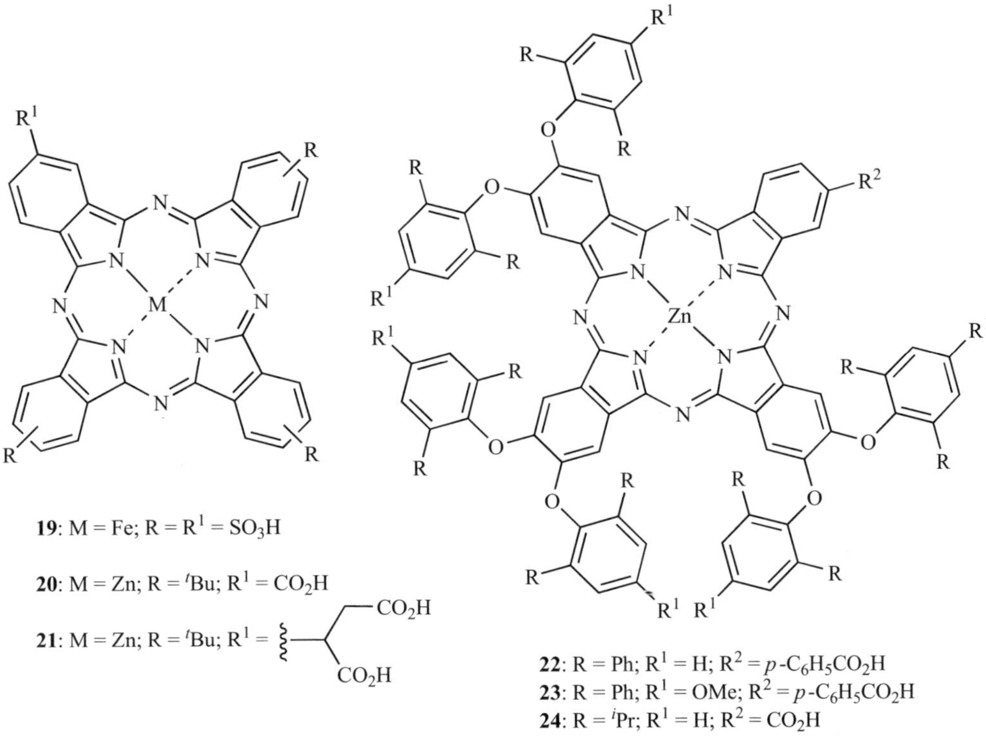

Fig. (16). Phthalocyanines for application in DSSCs.

The same group later prepared compound **23** (Fig. **16**), bearing electron-donating methoxy groups in the 2,6-diphenylphenoxy substituents [336], displaying a higher efficiency of 5.3%. In sequence, the same group reported an efficiency of 5.9%, using phthalocyanine **24**, (Fig. **16**) [337], with 2,6-diisopropylphenol groups at the periphery of the phthalocyanine and direct attachment of a carboxylic acid.

Besides zinc phthalocyanines, attention has also been given to ruthenium, titanium and silicium derivatives, profiting from their axial coordination ability, preventing molecular aggregation and thus alleviating the unfavorable recombination kinetics and poor injection between the dye and the TiO_2. In this context, phthalocyanine **25** (Fig. **17**), bearing peripheral carboxylic acid groups and two pyridine axial ligands, was compared with phthalocyanine **20** in DSSCs, just differing in the central metal plus its own coordinated axial ligands [338].

Further, Ru phthalocyanine derivatives **26** (Fig. **17**) were reported [339], failing to improve the photovoltaic performance. While no enthusiastic results have been reported for titanium [340, 341] and silicium phthalocyanines [342] of the same type, more recent examples describing the synthesis and application of silicon compounds **27** and **28** (Fig. **17**) were reported [343].

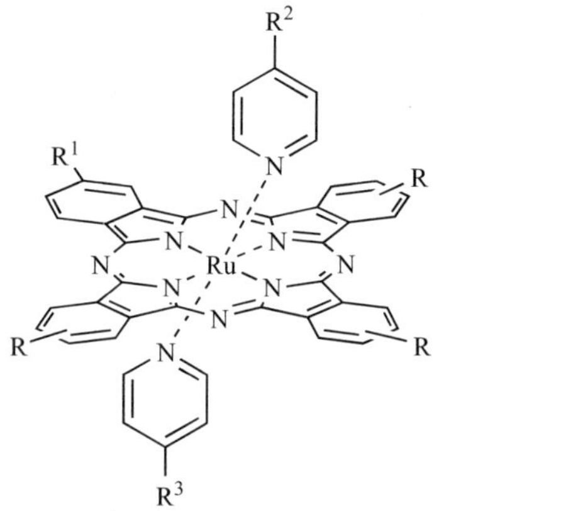

25: R = ^{t}Bu; R^1 = CO_2H; R^2 = R^3 = CH_3
26: R = R^1 = ^{t}Bu; R^2 = CO_2H; R^3 = OMe, Me; NPh_3, NPh_2

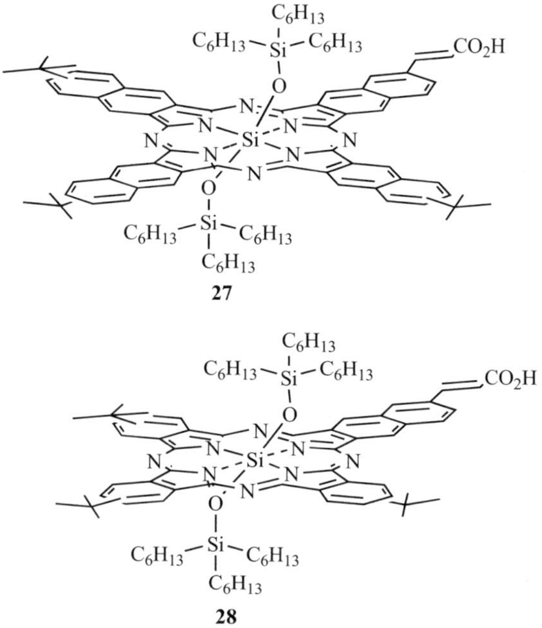

Fig. (17). Axially substituted phthalocyanines for application in DSSCs.

Characterized by its red absorption features, **28** demonstrated an efficiency of 4.5%, ascribed to the improvement of dye's light-harvesting capability by direct naphthalene unit fusion, along with reduced molecular aggregation by axial introduction of the alkylsilane chains. Nevertheless, naphthalocyanine **27** only presented a 0.9% efficiency, which could be attributed to the weak dye regeneration caused by the HOMO energy level of **27**, in opposition to **28**, as demonstrated by cyclic voltammetry.

Applications in Nonlinear Optics

Nonlinear optical (NLO) materials can control optical signal processing applications, *e.g.* all-optical switches, modulators and high speed electro-optical, as well as optical signals in optical communications.

On from the 80s, organic materials appeared as important subjects for NLO applications [344 - 346], since they exhibit fast and large nonlinearities and are usually easy to integrate and process into optical devices. Furthermore, organic materials offer the advantage of fine-tuning, modification of the chemical structure. Particularly, phthalocyanines can have large nonlinearities and picosecond response times, having the potential to be integrated into optical devices.

Nonlinear refraction and absorption are nonlinear optical features directly related. Generally, several mechanisms, both non-parametric and parametric, contribute to the nonlinearities and a consistent evaluation of the responsible mechanisms requires experiments involving different pulse lengths. Optical limiting consists of a sample transmittance decrease under fluence illumination or high intensity. Optical limiting is also regarded as the reverse of saturable absorption where increase in transmittance is seen at high illumination. Phthalocyanines are promising metallo-organic materials for optical power limiting in the visible and NIR spectral range, due to their suitable photophysical properties.

Phthalocyanine's first report on optical limiting appeared for chloroaluminium phthalocyanine **29** (Fig. **18**) [347], and later on, the heavy-atom effect has been explored to obtain remarkable improvements on the phthalocyanine's reverse saturable absorption. Thus, several phthalocyanine series containing In(III),

Ga(III), Al(III), Pb(II) and Sn(IV), Ge(IV), Si(IV) as central atoms were studied [348]. For example, a lead phthalocyanine bearing 4 β-cumylphenoxy groups **30** [349 - 351] and a chloro-indium phthalocyanine **31** [218] are still among the best benchmark materials (Fig. **19**) [352].

Several groups have been very proficient on the synthesis of several series of phthalocyanines in order to establish unambiguous structure property relationships. For instance, Hanack's group reported, in 1998, the highly soluble axially substituted aryl and alkyl indium phthalocyanine compounds (**32**) [353], extending their studies to several sets of phthalocyanines (**33**) (Fig. **18**) [221, 354]

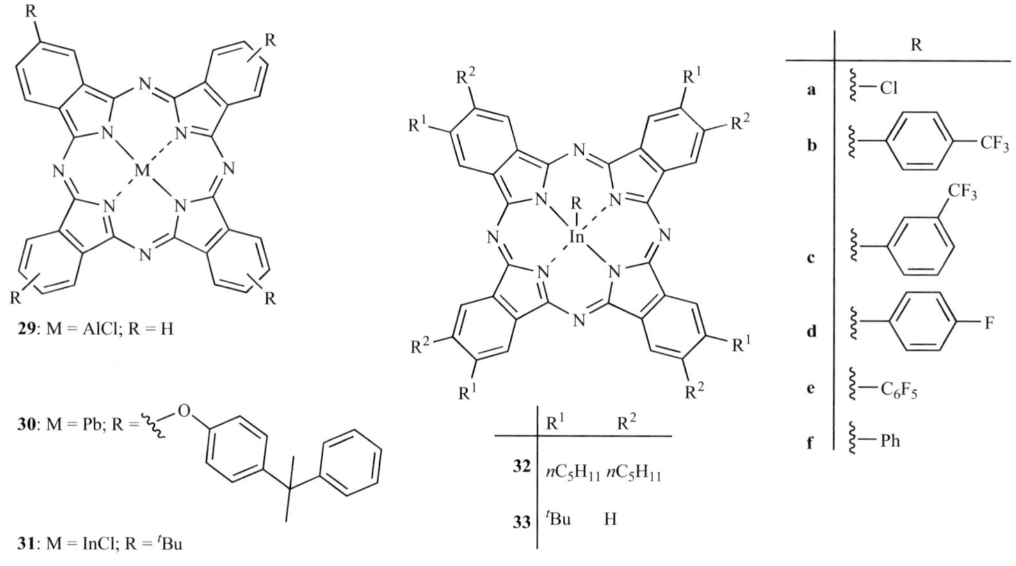

Fig. (18). Phthalocyanines for NLO applications.

Phthalocyanines used as nonlinear transmitters frequently suffer from solution aggregation, causing upper triplet state fast decays [355 - 357]. Changing axial ligands in phthalocyanines proved to be a valuable tool for suppressing the typical aggregation.

Thus, the axially substituted phthalocyanine complexes **32b** and **32e**, as well as their related **33b** and **33e** displayed higher nonlinear absorption coefficients, lower limiting thresholds and lower transmission at high fluences. Snow and Shirk [358] also prepared a substituted phthalocyanine **34** (Fig. **19**) for optical limiting thin film production, displaying self-healing properties after irradiation.

The presence of peripheral electron withdrawing groups effect in Pc's was also analyzed [359]. Their presence, as in Pcs **35** (Fig. **19**) or **36** (Fig. **20**) improved the OL effect, probably due to a transition dipole moment increase between the excited states accountable for the OL effect. Several other studies have been carried out, always with the purpose of establishing standard methodologies for measuring the optical power limiting effects of phthalocyanines and derivatives [164, 360 - 364].

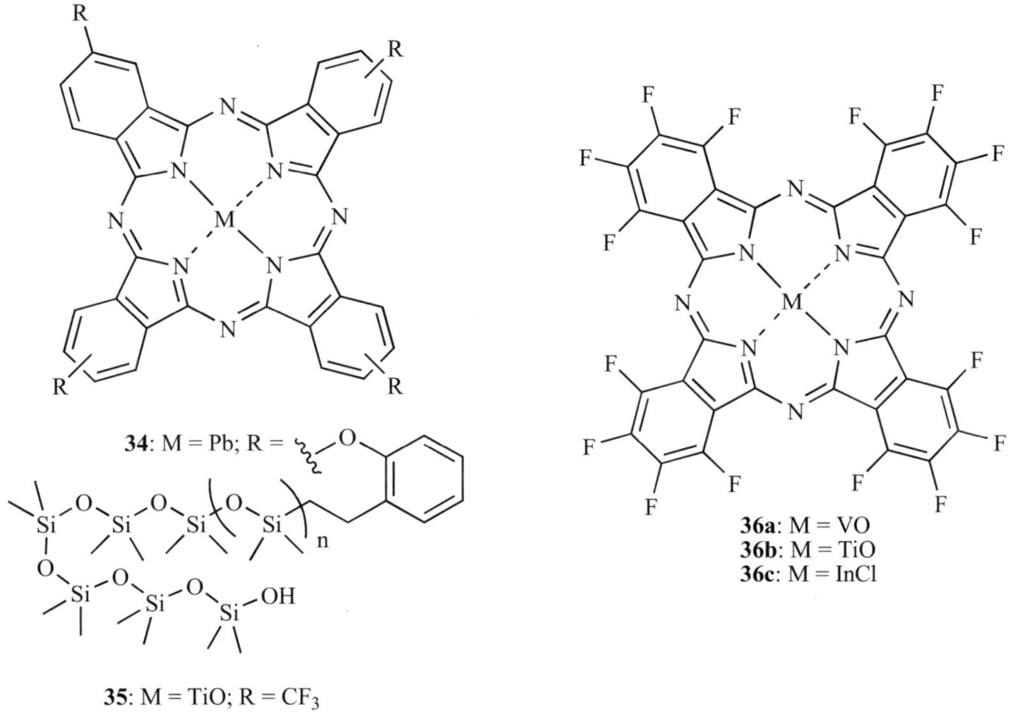

34: M = Pb; R =

35: M = TiO; R = CF_3

36a: M = VO
36b: M = TiO
36c: M = InCl

Fig. (19). Phthalocyanines for NLO applications.

Extension of conjugation in going from phthalocyanines to naphthalocyanines causes an increase in π-electron delocalization and electronic polarizability in both linear and nonlinear optical regimes [365, 366]. The structure property relationships that were carried out afterward were similar to the studies made for phthalocyanines, as the structural similarity induced quite related results. Thus, the optical limiting properties of naphthalocyanines were first reported in 1991 [367], and other authors continued their efforts in finding suitable naphthalocyanines for optical power limiting [368 - 371].

Through appropriate peripheral substitution, naphthalocyanines can give an increased transmitting window, shifted more to the red, which is desirable for the fabrication of optical limiters effective in the broad visible-light range including the red-light region. A strong naphthalocyanine's Q-band bathochromic shift upon alkoxy-substitution in the peripheral positions closest to the core (α- or 1,6-positions) can be observed, due to mixing the oxygen p_z-orbitals with π-HOMO of the macrocycle and distortion of the naphthalocyanine-core geometry from the planar, causing the destabilization of π -HOMO.

Thus, α-substituted alkoxy naphthalocyanines **37a** and **37b** (Fig. **20**) presented their maximum transmittance at 635 nm in contrast to 560nm for other naphthalocyanines [371], and the absorption maximum of **37c** was shifted to 910nm leaving open the 520-780nm window in which RSA is observed [372]. Nevertheless, no significant improvement was observed, for other substitution patterns were tested. For instance, compounds **38a,b** (Fig. **20**) were also prepared [208, 373], where compound **38b** has shown high solubility and no aggregation at $2x10^{-3}$M concentrations in chloroform and an optical limiting threshold of 10 nJ.

The presence of heavy atoms like bromine in peripherally of the naphthalocyanines was also focused [374]. It could be seen that the influence of bromine substituent on the excited state absorption was significant and, therefore, the optical limiting behavior of **39a** was better than the one of **39b** (Fig. **21**). Later, a comparison of the effect of the presence of halogens at the periphery was evaluated [375 - 377]. For instance, the comparison of nonlinear transmittance of naphthalocyanines **40a,b** (Fig. **21**) [377] demonstrated that bromine substituted naphthalocyanine rings display significantly OL effect enhancement when high-intensity nanosecond laser pulses at 532nm are utilized.

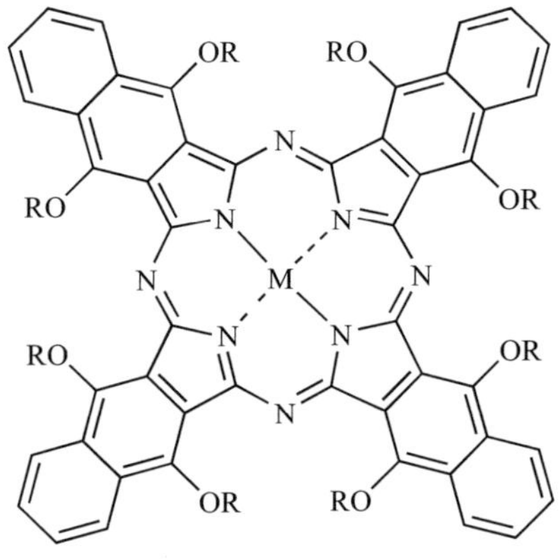

37a: M = InCl; R = nC_4H_9
37b: M = SnF_2; R = nC_4H_9
37c: M = Pb; R = nC_5H_{11}

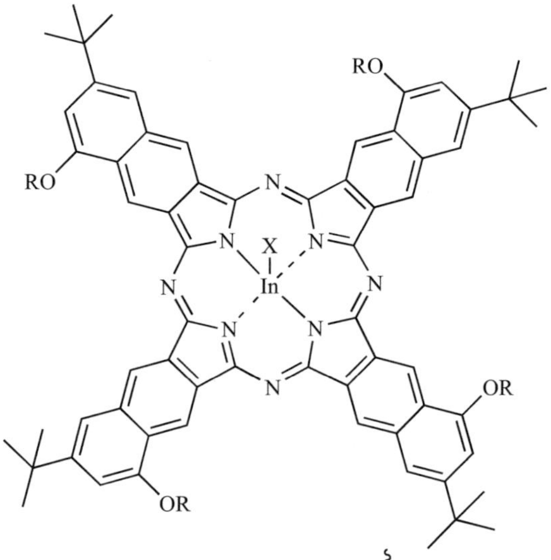

38a: R = 2-ethylhexyl; X = —Cl

38b: R = 2-ethylhexyl; X = 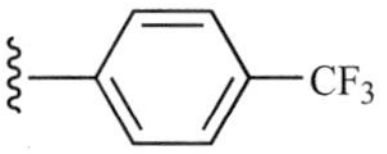

Fig. (20). Naphthalocyanines for NLO applications.

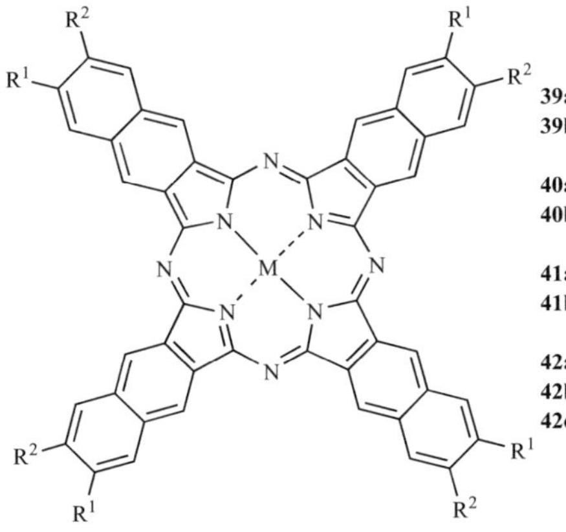

39a: M = Pb; R^1 = Br R^2 = H
39b: M = Pb; $R^1 = R^2$ = H

40a: M = InCl; $R^1 = R^2 = (^tBu)_2PhO$
40b: M = InCl; R^1 = Br; $R^2 = (^tBu)_2PhO$

41a: M = $Si[(OSi(n\text{-}C_6H_{13})_3]_2$; R^1 = Br; $R^2 = (^tBu)_2PhO$
41b: M = $Si[(OSi(n\text{-}C_6H_{13})_3]_2$; $R^1 = R^2$ = Br

42a: M = InCl; $R^1 = R^2 = (^tBu)_2PhO$
42b: M = InBr; $R^1 = R^2 = (^tBu)_2PhO$
42c: M = InI; $R^1 = R^2 = (^tBu)_2PhO$

$(^tBu)_2PhO$ = 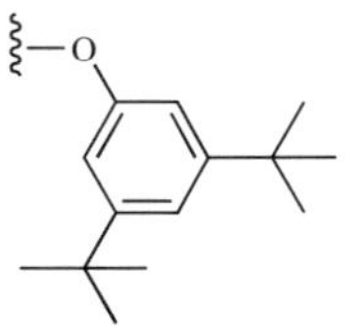

Fig. (21). Naphthalocyanines for NLO applications.

The NLO properties two Si naphthalocyanines **41a,b** (Fig. **21**) [378] were measured, showing that the increasing number of peripheral bromines from **41a** to **41b** resulted in a triplet excited-state lifetime decrease and consequent triplet excited-state quantum yield increase. The improved nonlinear transmittance (NLT) could be mainly attributed to a triplet excited state absorption to ground-state absorption cross-section increased ratio.

In order to acquire the differences in the OPL capability upon variation of the axial ligand on naphthalocyanines, the axial halogen effect on the OL of indium naphthalocyanines **42a-c** (Fig. **21**) was reported [379]. Good OL effect of 532nm picosecond and nanosecond laser pulses using these complexes was observed; however little influence of halogens at axial positions was found.

Applications as Photosensitizers for Photodynamic Cancer Therapy

As already well noted, phthalocyanines absorb strongly in the red and NIR part of the visible. Phthalocyanines are particularly interesting for photomedical applications, and have been used for photodynamic therapy of cancer since 1985 [232, 380]. In some cases photodynamic therapy treatment has some advantages over conventional cancer therapies such as chemotherapy, radiation and surgical treatments. It permits selective damage of diseased tissues, due to interaction of individually non-toxic components (photosensitizer, light and oxygen).

Phthalocyanines belong to the group of 2nd generation photosensitizers, exhibiting high extinction coefficients between 670 and 750 nm. Variation of the peripheral and axial substituents can modify the aggregation tendency, biodistribution, pharmacokinetics, solubility, as well as fine-tuning of NIR absorbance [381, 382]. For instance, the use of detergents can minimize the natural aggregation tendency of phthalocyanines in aqueous buffer (pH 7.4), allowing these compounds to keep their photochemical activity [383, 384].

One of the most known (and successful) cases of phthalocyanine use for photodynamic therapy regards the report of sulfonated aluminium phthalocyanine derivative **43** (also known as **Photosens**®, a registered trademark by Russian researchers) (Fig. **22**), which started clinical trials against breast, skin, gastrointestinal tract and lung cancers [385 - 388]. More recently, silicon phthalocyanine complex **44** (also known as **Pc4**) (Fig. **22**) was successfully tested in blood sterilization, and against colon, ovarian, and breast cancers [237, 385, 389 - 394]. Cationic phthalocyanine derivatives are believed to localize in the mitochondrion which are organelles crucial to cell survival and the site of oxidative phosphorylation [395 - 397].

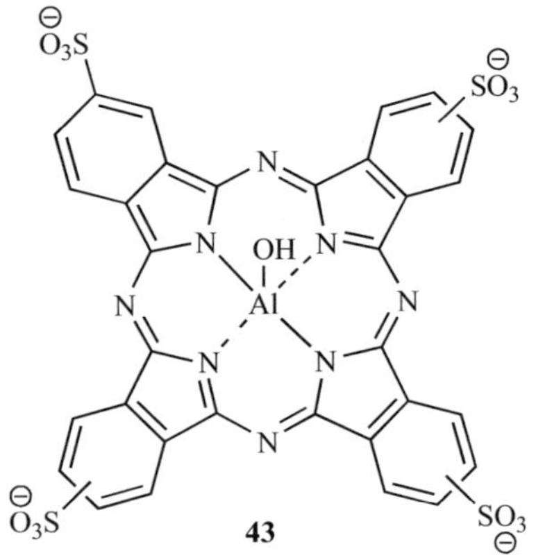

43

a.k.a. **Photosens®**

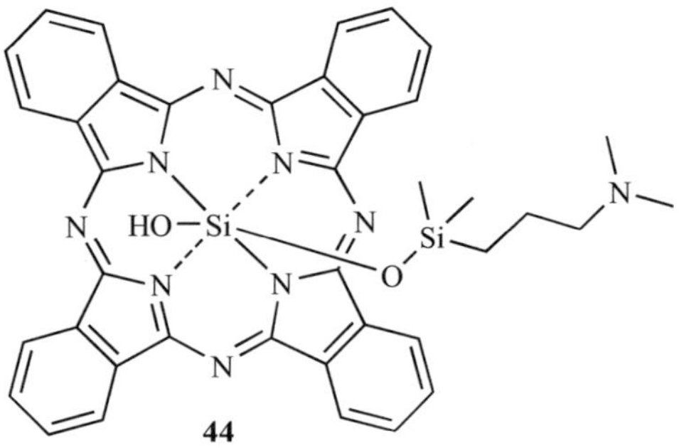

44

a.k.a. **Pc4**

Fig. (22). Water soluble phthalocyanines for PDT applications.

Applications as Antibacterial Composites

The application of photodynamic therapy to kill microbial pathogens is currently an emerging area of research. Nevertheless, the first recorded observation of photodynamic processes was the inactivation of a microorganism (paramecia) by Oscar Raab, more than 100 years ago [398]. The development of our current lifestyle has naturally induced bacteriological proliferation, and photodynamic therapy can be seen as a practical alternative, since photosensitizers can act on parasitic protozoa, bacteria and fungi, upon light activation [242 - 244, 399].

Gram bacteria can be classified as gram-negative and gram-positive bacteria. Gram-positive bacteria possess thick peptidoglycan layers in the cell wall that encases their cell membrane, while gram-negative bacteria posses thinner peptidoglycan layer sandwiched between the inner cell membrane and the bacterial outer membrane. Regardless of their thicker peptidoglycan layer, gram-positive bacteria are more susceptible to antibiotics than gram-negative ones, due to the latter's relatively impervious lipid bacterial outer membrane [400]. Usually, neutral and anionic photosensitizers are efficiently bound to Gram (+) bacteria, and they effectively inactivate these Gram (+) bacteria after illuminating with appropriate light, which is not the case for Gram (-) bacteria [400].

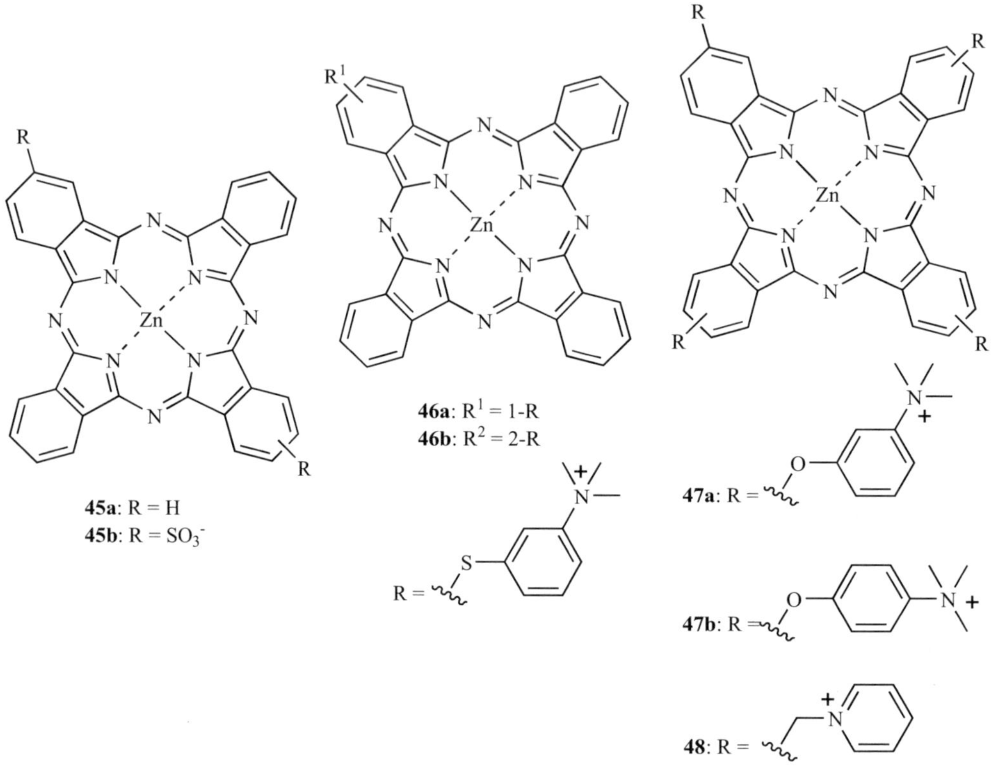

Fig. (23). Cationic phthalocyanines for application in antibactericidal applications.

Since photosensitizers bearing cationic charge increase the permeability of the outer membrane of Gram (-), these may effectively increase killing Gram (-) bacteria and many phthalocyanine type sensitizers have already been tested for photodynamic inactivation of bacteria. For instance Jori and van Lier studied the effect of pre-treatment of the bacteria with ethylenediaminetetraacetic-acid (EDTA) [401, 402]. Treatment with EDTA made Gram (-) cells lose up to 50% of their lipopolysaccharides, caused by removal of divalent cations upon EDTA treatment. The photosensitizers **45** (Fig. **24**) were then used to efficiently inactive the Gram (-) bacteria, through induced photosensitivity and the analysis on irradiated cells suggested that the cytoplasmic membrane was an important target of the photo process.

Several research groups, including Jori [251, 403] and Brown [247, 250], among others [404 - 406] are quite active in understanding the mechanisms of action of phthalocyanines in the use of phthalocyanines for bacteria inactivation. For instance, Jori prepared unsymetrically substituted Zn(II)-phthalocyanines **46ab** (Fig. **23**) [403] which were used efficiently against *Candida albicans*. They have also prepared the cationic Zn(II) phthalocyanines **47a,b** (Fig. **23**) [251] bearing peripheral cationic substituents which have shown to be efficient photoantimicrobial agents: against *Staphylococcus aureus* upon short irradiation times in the presence of low phthalocyanine concentrations (0.1 μM). Brown also prepared cationic phthalocyanine **48** and used it successfully for Gram-positive and Gram-negative bacteria photoinactivation [247, 250].

Applications in Molecular Imaging

Phthalocyanine molecules can be used to detect and image diseased tissues. The ideal fluorophore should own a large Stokes shift, minimum photobleaching, high quantum yield of fluorescence, besides high target affinity. The research and development of late generation photosensitizers spotted the potential in the detection and treatment of diseases that a unique therapeutic and imaging platform could have.

Switching between imaging and therapeutic modalities in a single platform is a particularly suitable use for photosensitizers in imaging modalities such as radio, nuclear, fluorescence and magnetic resonance imaging, combined with photodynamic, photoacoustic or photothermal therapies, besides hyperthermia, chemo and gene therapy [407].

Lier group has promoted the ^{64}Cu labeling of sulfo-phthalocyanine derivatives **49a-e** (Fig. **24**) [408, 409]. In the first case, a mixture of phthalocyanines with different degrees of sulfonation was labeled with ^{64}Cu source ($^{64}Cu(OAc)_2$) [408], by conventional heating, yielding a mixture of sulfonated ^{64}Cuphthalocyanine derivatives, appropriate for positron emission tomography (PET) studies.

Liu [410, 411] also synthesized several zinc phthalocyanines substituted with galactose moieties **50a-e** (Fig. **24**) and studied them as near infrared fluorescent probes, using liver tumor-bearing athymic nude mice as models.

The authors observed that in the tumor imaging of galactose substituted zinc phthalocyanines, the fluorescence images were best acquired at 24 h after intravenous injection, since due to probe distribution, the fluorescence was found in the whole body with no selectivity at initial hours. Luminescence signals were observed in the liver cancer, whereas no luminescence signal was observed in the control group.

Vicente also prepared zinc(II) phthalocyanine **51** substituted with a monoclonal antibody (MAb) [412] (Fig. **24**). The zinc phthalocyanine:MAb bioconjugate (3:1 ratio) was *in vitro* tested for its human colorectal HT-29 cells target capacity, by fluorescence.

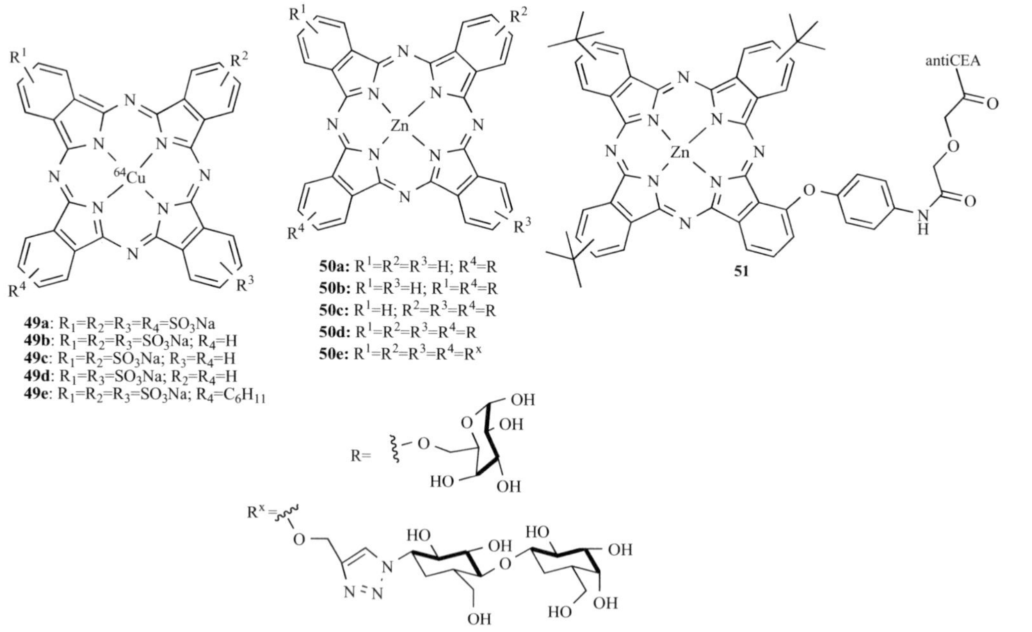

Fig. (24). Phthalocyanine conjugates for application in medical imaging.

EXPERIMENTAL SECTION

Synthesis of Water Soluble Tetrasulfonated Phthalocyanine (Scheme 4)

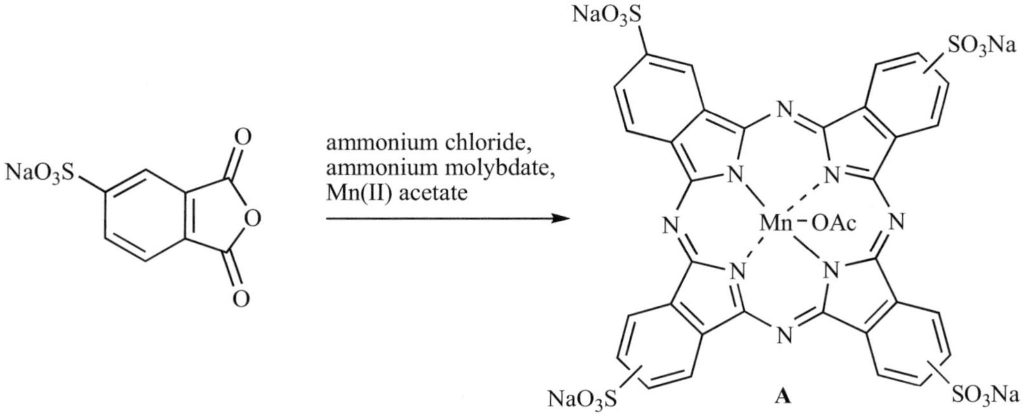

Scheme (4). Water soluble tetrasulfonated phthalocyanine. Procedure for total synthesis of phthalocyanine A can be directly found in Reference [413].

Synthesis of (2,3)-type Octa-alkoxy Substituted Phthalocyanines (Scheme 5)

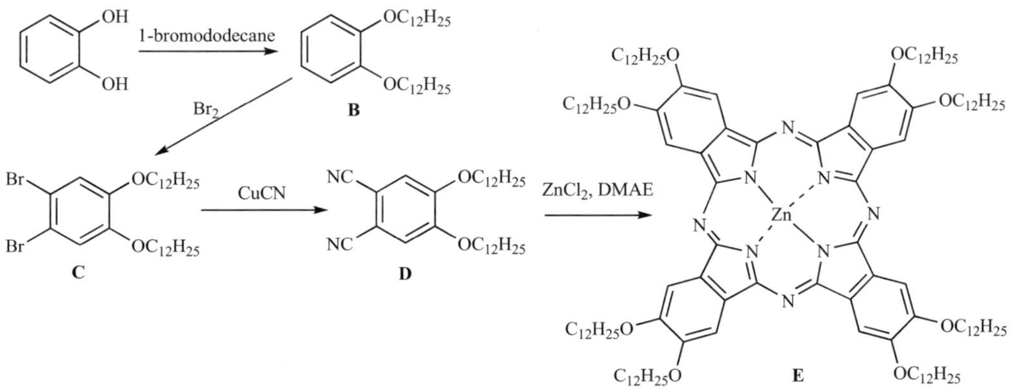

Scheme (5). 2,3-Octa-alkoxy substituted phthalocyanine. Procedure for total synthesis of phthalocyanine E can be directly found in Reference [414].

Synthesis of (1,4)-type Octa-alkoxy Substituted Phthalocyanines (Scheme 6)

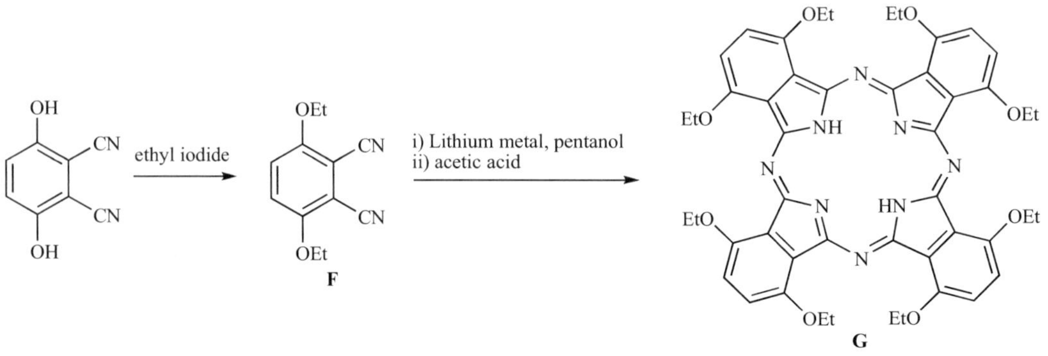

Scheme (6). 2,3-Octa-alkoxy substituted phthalocyanine. Procedure for total synthesis of phthalocyanine G can be directly found in Reference [415].

Synthesis of Sulfonamide Substituted Phthalocyanines (Scheme 7)

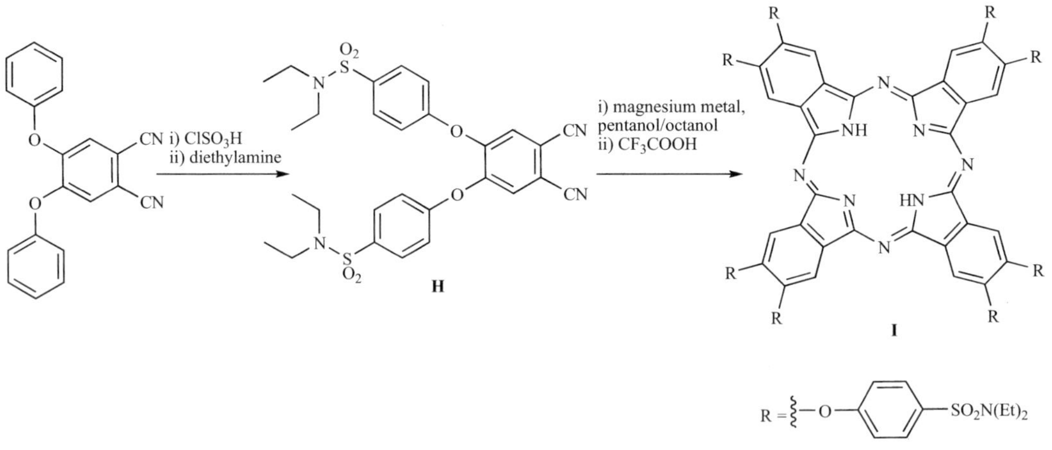

Scheme (7). Sulfonamide substituted phthalocyanine. Procedure for total synthesis of phthalocyanine I can be directly found in Reference [416].

Synthesis of Tetra Substituted Glucopyranosyl Phthalocyanines (Scheme 8)

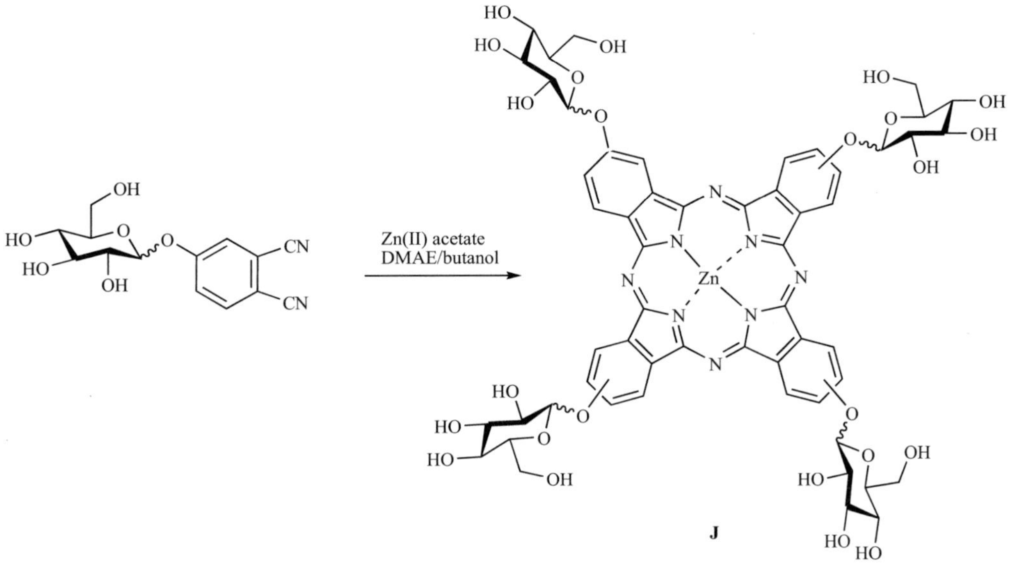

J

Scheme (8). Tetra glycosylated zinc phthalocyanine. Procedure for total synthesis of phthalocyanine I can be directly found in Reference [417].

Synthesis of Unsymmetrically Substituted (AAAB) Phthalocyanines (Scheme9)

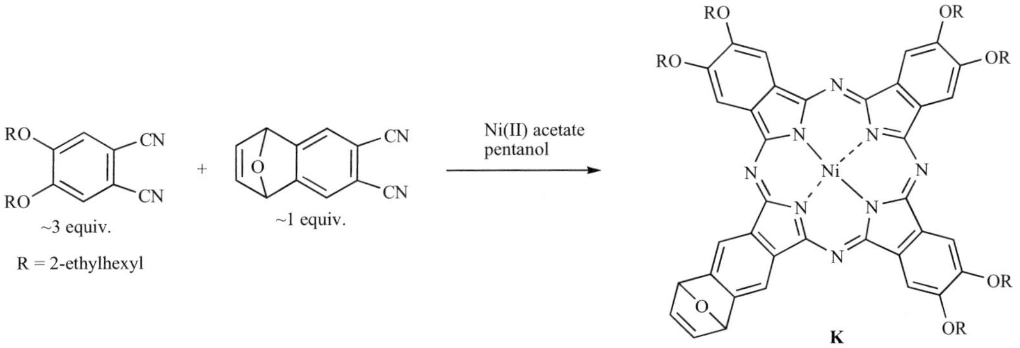

K

Scheme (9). Unsymmetrically substituted (AAAB) nickel phthalocyanines. Procedure for total synthesis of phthalocyanine K can be directly found in Reference [22].

Synthesis of Subphthalocyanines (Scheme 10)

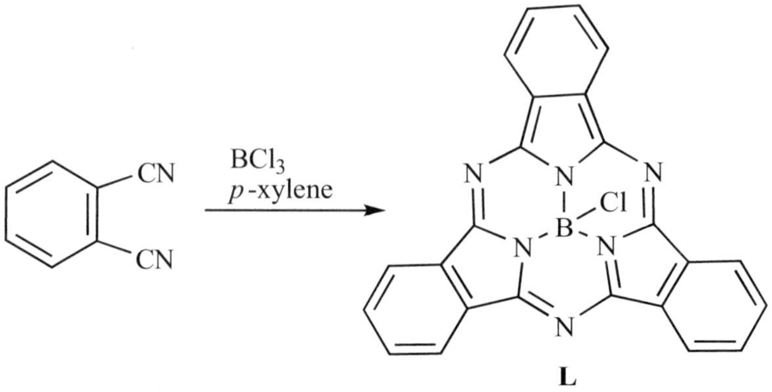

Scheme (10). Subphthalocyanine synthesis. Procedure for total synthesis of subphthalocyanine L can be directly found in Reference [418].

Synthesis of (AABB) Substituted Phthalocyanines (Scheme 11)

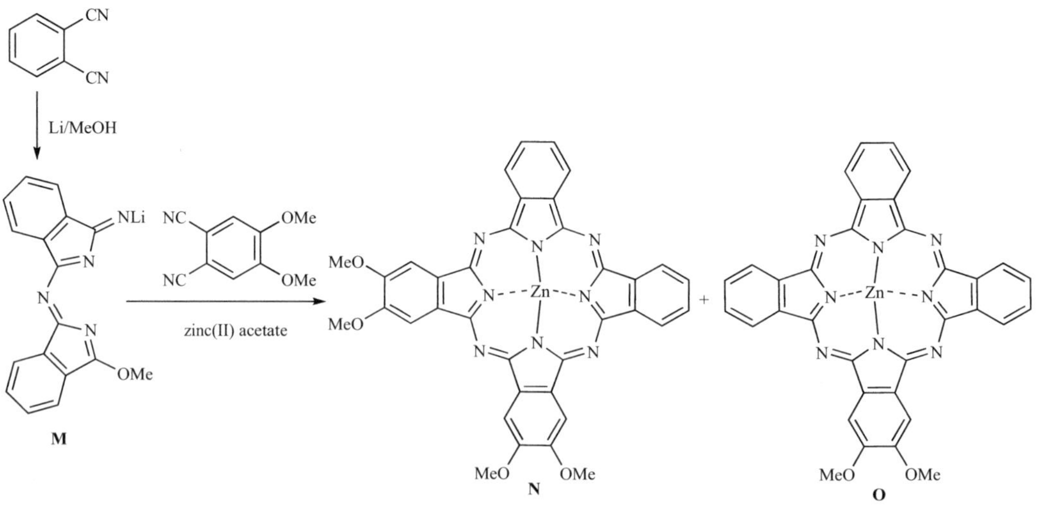

Scheme (11). Synthesis of (AABB) substituted phthalocyanine. Procedure for total synthesis of phthalocyanines N and O can be directly found in Reference [419].

Synthesis of (ABAB) Substituted Phthalocyanines (Scheme 12)

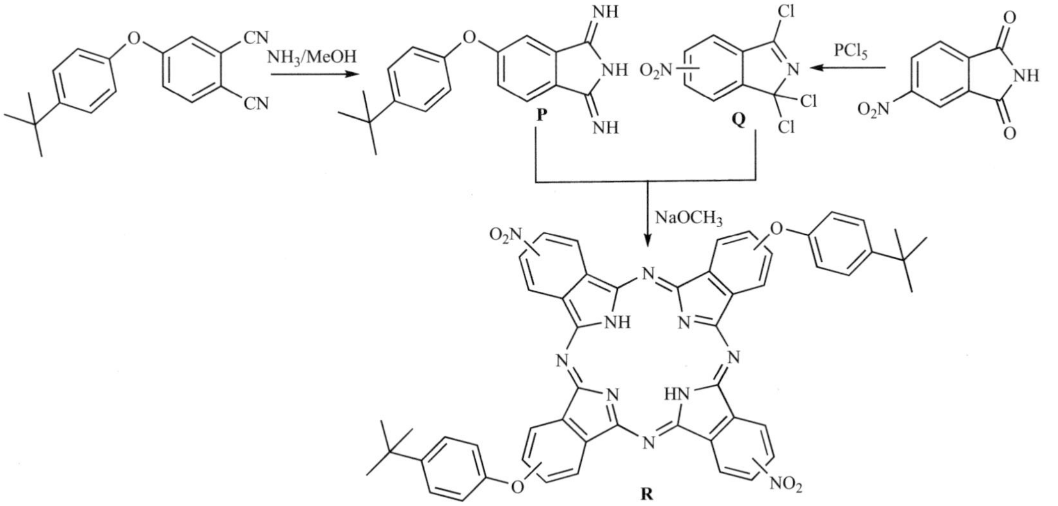

Scheme (12). Synthesis of (ABAB) substituted phthalocyanine. Procedure for total synthesis of phthalocyanine R can be directly found in Reference [48].

Synthesis of an Octa-substituted Naphthalocyanine (Scheme 13)

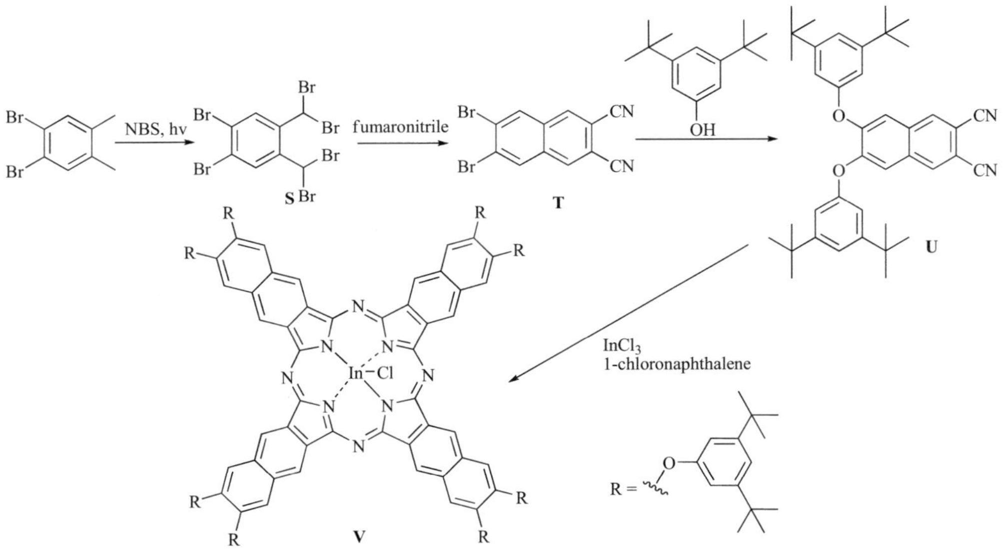

Scheme (13). Octa-substituted naphthalocyanine. Procedure for total synthesis of naphthalocyanine V can be directly found in Reference [377].

CONCLUDING REMARKS

Phthalocyanines, as NIR dyes, are appealing molecules for a variety of applications. Their synthetic versatility can simply allow almost any type of application known in fields ranging from catalysis, to materials science and biomedicinal applications.

As far as any scientist imagination can go, so can the possible applications for materials based on phthalocyanines and related compounds. The possibility of widely varying the substitution pattern on phthalocyanine structures is probably the main possible direction for successful applications; however not less important than the establishment of definite structure-property relationships for any desired purpose, something has been only partially achieved so far. Standardization of procedures will definitely boost the understanding and appliance of phthalocyanines in the main areas of science.

REFERENCES

[1] Braun, A.; Tcherniac, J. The products of the action of acetanhydride on phthalamide. *Ber. Dtsch. Chem. Ges,* **1907**, *40*, 2709-2714.
[http://dx.doi.org/10.1002/cber.190704002202]

[2] de Diesbach, H.; von der Weid, E. Some salt complexes of o-dinitriles with copper and pyridine. *Helv. Chim. Acta,* **1927**, *10*, 886-888.
[http://dx.doi.org/10.1002/hlca.19270100110]

[3] Dandridge, A.G.; Drescher, H.A.; Thomas, J. VAT dyes. Patent BP 322169,

[4] Linstead, R.P. Phthalocyanines part I: A new type of synthetic colouring matters. *J. Chem. Soc.,* **1934**, 1016-1017.
[http://dx.doi.org/10.1039/jr9340001016]

[5] Byrne, G.T.; Linstead, R.P.; Lowe, A.R. Phthalocyanines part II: The preparation of phthalocyanine and some metallic derivatives from o-cyanobenzamide and phthalimide. *J. Chem. Soc.,* **1934**, 1017-1022.
[http://dx.doi.org/10.1039/jr9340001017]

[6] Linstead, R.P.; Lowe, A.R. Phthalocyanines part III: Preliminary experiments on the preparation of phthalocyanines from phthalonitrile. *J. Chem. Soc.,* **1934**, 1022-1027.
[http://dx.doi.org/10.1039/jr9340001022]

[7] Dent, C.E.; Linstead, R.P. Phthalocyanines part IV: Copper phthalocyanines. *J. Chem. Soc.,* **1934**, 1027-1031.
[http://dx.doi.org/10.1039/jr9340001027]

[8] Dent, C.E.; Linstead, R.P.; Lowe, R. Phthalocyanines part VI: The structure of the phthalocyanines. *J.*

Chem. Soc., **1934**, 1033-1039.
[http://dx.doi.org/10.1039/jr9340001033]

[9] Linstead, R.P.; Dent, C.E. Molecular weights of the phthalocyanines. *Nature,* **1935**, *135*, 506-507. [http://dx.doi.org/10.1038/135506b0]

[10] Anderson, J.S.; Bradbrook, E.F.; Cook, A.H.; Linstead, R.P. Phthalocyanines and associated compounds. Part XIII. Absorption spectra. *J. Chem. Soc.,* **1938**, 1151-1156. [http://dx.doi.org/10.1039/jr9380001151]

[11] Bradbrook, E.F.; Linstead, R.P. Phthalocyanines. Part VIII 1,2-naphthalocyanines. *J. Chem. Soc.,* **1936**, 1744-1748. [http://dx.doi.org/10.1039/jr9360001744]

[12] Barret, P.A.; Dent, C.E.; Linstead, R.P. Phthalocyanines Part VII: Phthalocyanine as a coordinating group-A general investigation of the metallic derivatives. *J. Chem. Soc.,* **1936**, 1719-1736. [http://dx.doi.org/10.1039/jr9360001719]

[13] Meller, A.; Ossko, A. Phthalocyaninartige Bor-Komplexe. *Monatsh. Chem.,* **1972**, *103*, 150-155. [http://dx.doi.org/10.1007/BF00912939]

[14] Kietaibl, H. Die Kristall- und Molekülstruktur eines neuratigen phthalocyaninähnlichen Borkomplexes. *Monatsh. Chem.,* **1974**, *105*, 405-418. [http://dx.doi.org/10.1007/BF00907390]

[15] Hanack, M.; Heckmann, H.; Polley, R. *Methoden der Organischen Chemie (Houben–Weyl), vol. E9d,* 4th ed; Thieme Verlag: Stuttgart, **1998**, pp. 717-766.

[16] *The Phthalocyanines*; Leznoff , C.C.; Lever, A.B., Eds.; 1986-1993.

[17] McKeown, N.B. *In Phthalocyanine Materials-Synthesis, Structure and Function*; Cambridge University Press: Cambridge, **1998**.

[18] Moser, F.H.; Thomas, A.L. *The Phthalocyanines*; CRC Press: Boca Raton, FL, **1983**.

[19] Baumann, F.; Bienert, B.; Rösch, G.; Vollmann, H.; Wolf, W. Isoindolenine als Zwischenprodukte der Phthalocyanin-Synthese. *Angew. Chem.,* **1956**, *68*, 133-150. [http://dx.doi.org/10.1002/ange.19560680402]

[20] Owen, J.E.; Kenney, M.E. Aluminium and silicon hydroxyl and oxy systems of phthalocyanine type. *Inorg. Chem.,* **1962**, *1*, 334-336. [http://dx.doi.org/10.1021/ic50002a027]

[21] Cariati, F.; Galizzioli, D.; Morazzoni, F.; Busetto, C. New adducts of phthalocyaninatocobalt(II) with pyridine and 4-methylpyridine and their vibrational, magnetic, and electronic properties. Part I. Reactivity towards oxygen. *J. Chem. Soc., Dalton Trans.,* **1975**, *7*, 556-561. [http://dx.doi.org/10.1039/dt9750000556]

[22] Calvete, M.J.; Hanack, M. Symmetrically and unsymmetrically substituted phthalocyanines In: *Materials Syntheses-A practical guide*; Schubert, U.; Hüsing, N.; Laine, R., Eds.; , **2008**; pp. 217-225.

[23] McKeown, N.B. *The synthesis of symmetrical phthalocyanines, The Porphyrin Handbook*; Kadish, K.M.; Smith, K.M.; Guilard, R., Eds.; Academic Press: New York, **2003**, Vol. 15, pp. 61-124. [http://dx.doi.org/10.1016/B978-0-08-092389-5.50008-0]

[24] Sharman, W.M.; van Lier, J.E. *Synthesis of phthalocyanine precursors in The Porphyrin Handbook*; Kadish, K.M.; Smith, K.M.; Guilard, R., Eds.; Academic Press: New York, **2003**, Vol. 15, pp. 1-60.

[25] Hurley, T.J.; Robinson, M.A. Trotz, S. I. Complexes derived from 1,3-diiminoisoindoline-containing ligands. II. Stepwise formation of nickel phthalocyanine. *Inorg. Chem.,* **1967**, *6*, 389-392. [http://dx.doi.org/10.1021/ic50048a044]

[26] Oliver, S.W.; Smith, T.D. Oligomeric cyclization of dinitriles in the synthesis of phthalocyanines and related compounds: the role of the alkoxide anion. *J. Chem. Soc., Perkin Trans. 2,* **1987**, 1579-1582. [http://dx.doi.org/10.1039/p29870001579]

[27] Gaspard, S.; Maillard, P. Structure des phtalocyanines tetra tertio-butylees: mecanisme de la synthese. *Tetrahedron,* **1987**, *43*, 1083-1090. [http://dx.doi.org/10.1016/S0040-4020(01)90045-6]

[28] Molek, C.D.; Halfen, J.A.; Loe, J.C.; McGaff, E.W. Solventothermal synthesis and X-ray crystal structures of two nickel complexes with novel alkoxy-substituted phthalocyanine ligands. *Chem. Commun. (Camb.),* **2001**, 2644-2645. [http://dx.doi.org/10.1039/b109481p]

[29] van der Pol, J.F.; Neeleman, E.; Nolte, R.J.; Zwikker, J.W.; Drenth, W. Asymmetrically substituted liquid-crystalline phthalocyanines and side-chain polymers derived from them. *Makromol. Chem.,* **1989**, *190*, 2727-2745. [http://dx.doi.org/10.1002/macp.1989.021901107]

[30] Hanack, M.; Lang, M. Conducting stacked metallophthalocyanines and related compounds. *Adv. Mater.,* **1994**, *6*, 819-833. [http://dx.doi.org/10.1002/adma.19940061103]

[31] Sommerauer, M.; Rager, C.; Hanack, M. Separation of 2(3),9(10),16(17),23(24)-Tetrasubstituted phthalocyanines with newly developed HPLC phases. *J. Am. Chem. Soc.,* **1996**, *118*, 10085-10093. [http://dx.doi.org/10.1021/ja961009x]

[32] Hanack, M.; Schmid, G.; Sommerauer, M. Chromatographic separation of the four possible structural isomers of a tetrasubstituted phthalocyanine: tetrakis(2-ethylhexyloxy)phthalocyaninatonickel(II). *Angew. Chem.*, **1993**, *105*, 1540-1542. *Angew. Chem. Int. Ed. Engl.,* **1993**, *32*, 1422-1424. [http://dx.doi.org/10.1002/anie.199314221]

[33] Nemykin, V.N.; Lukyanets, E.A. Synthesis of substituted phthalocyanines. *ARKIVOC,* **2010**, *1*, 136-208. [http://dx.doi.org/10.3998/ark.5550190.0011.104]

[34] Lukyanets, E.A.; Nemykin, V.N. The key role of peripheral substituents in the chemistry of phthalocyanines and their analogs. *J. Porphyr. Phthalocyanines,* **2010**, *14*, 1-40. [http://dx.doi.org/10.1142/S1088424610001799]

[35] Elvidge, J.A.; Linstead, R.P. Conjugated macrocycles. Part XXVII. The formation of tetrazaporphins from imidines. Tribenzotetrazaporphin. *J. Chem. Soc.,* **1955**, 3536-3544. [http://dx.doi.org/10.1039/jr9550003536]

[36] Leznoff, C.C.; Hall, T.W. The synthesis of a soluble, unsymmetrical phthalocyanine on a polymer support. *Tetrahedron Lett.,* **1982**, *23*, 3023-3026.

[http://dx.doi.org/10.1016/S0040-4039(00)87523-1]

[37] de la Torre, G.; Claessens, C.G.; Torres, T. Phthalocyanines: the need for selective synthetic approaches. *Eur. J. Org. Chem.,* **2000**, 2821-2830. [http://dx.doi.org/10.1002/1099-0690(200008)2000:16<2821::AID-EJOC2821>3.0.CO;2-2]

[38] Leznoff, C.C. From early developments in multi-step organic synthesis on solid phases to multinuclear phthalocyanines. *Can. J. Chem.,* **2000**, *78*, 167-184. [http://dx.doi.org/10.1139/v00-007]

[39] Torres, T. Perspectives in the selective synthesis of phthalocyanines and related compounds. *J. Porphyr. Phthalocyanines,* **2000**, *4*, 325-331. [http://dx.doi.org/10.1002/(SICI)1099-1409(200006/07)4:4<325::AID-JPP225>3.0.CO;2-I]

[40] Tolbin, A.Y.; Tomilova, L.G.; Zefirov, N.S. Non-symmetrically substituted phthalocyanines: synthesis and structure modification. *Russ. Chem. Rev.,* **2007**, *76*, 681-692. [http://dx.doi.org/10.1070/RC2007v076n07ABEH003698]

[41] Mack, J.; Kobayashi, N. Low symmetry phthalocyanines and their analogues. *Chem. Rev.,* **2011**, *111*(2), 281-321. [http://dx.doi.org/10.1021/cr9003049] [PMID: 21175133]

[42] Makarov, S.G.; Suvorova, O.N.; Wöhrle, D. Conjugated di- and trinuclear phthalocyanines and their analogs. *J. Porphyr. Phthalocyanines,* **2011**, *15*, 791-808. [http://dx.doi.org/10.1142/S1088424611003835]

[43] Wang, A.; Long, L.; Zhang, C. Synthesis of unsymmetrical phthalocyanines: a brief overview. *Tetrahedron,* **2012**, *68*, 2433-2451. [http://dx.doi.org/10.1016/j.tet.2012.01.004]

[44] McKeown, N.; Chambrier, I.; Cook, M. Synthesis and characterisation of some 1,4,8,11,15,18,22,25-octa-alkyl- and 1,4,8,11,15,18-hexa-alkyl-22,25-bis(carboxypropyl)phthalocyanines. *J. Chem. Soc. Perkin Trans,* **1990**, *1*, 1169-1177. [http://dx.doi.org/10.1039/p19900001169]

[45] Kobayashi, N.; Kondo, R.; Nakajima, S-I.; Osa, T. New route to unsymmetrical phthalocyanine analogs by the use of structurally distorted subphthalocyanines. *J. Am. Chem. Soc.,* **1990**, *112*, 9640-9641. [http://dx.doi.org/10.1021/ja00182a034]

[46] Sastre, A.; Torres, T.; Hanack, M. Synthesis of novel unsymmetrical monoaminated phthalocyanines. *Tetrahedron Lett.,* **1995**, *36*, 8501-8504. [http://dx.doi.org/10.1016/0040-4039(95)01781-C]

[47] Kudrevich, S.V.; Gilbert, S.; van Lier, J.E. Syntheses of trisulfonated phthalocyanines and their derivatives using boron(III) subphthalocyanines as intermediates. *J. Org. Chem.,* **1996**, *61*, 5706-5707. [http://dx.doi.org/10.1021/jo9605354]

[48] Young, J.; Onyebuagu, W. Synthesis and characterization of di-disubstituted phthalocyanines. *J. Org. Chem.,* **1990**, *55*, 2155-2159. [http://dx.doi.org/10.1021/jo00294a032]

[49] Stihler, P.; Hauschel, B.; Hanack, M. Synthesis of a bisdienophilic phthalocyanine and of precursors for repetitive Diels-Alder reactions based on hemiporphyrazines and phthalocyanines. *Chem. Ber.*, **1997**, *130*, 801-806.
[http://dx.doi.org/10.1002/cber.19971300620]

[50] Nolan, K.J.; Hu, M.; Leznoff, C.C. "Adjacent" substituted phthalocyanines. *Synlett,* **1997**, 593-594.
[http://dx.doi.org/10.1055/s-1997-3226]

[51] Kobayashi, N. Optically active 'adjacent' type non-centrosymmetrically substituted phthalocyanines. *Chem. Commun. (Camb.),* **1998**, 487-488.
[http://dx.doi.org/10.1039/a708343b]

[52] Miwa, H.; Kobayashi, N. A novel naphthalocyanine isomer: A dinaphthophthalocyanine with C2v symmetry. *Chem. Lett.,* **1999**, *28*, 1303-1304.
[http://dx.doi.org/10.1246/cl.1999.1303]

[53] Kobayashi, N.; Miwa, H.; Isago, H.; Tomura, T. An adjacent dibenzotetraazaporphyrin: A structural intermediate between tetraazaporphyrin and phthalocyanine. *Inorg. Chem.,* **1999**, *38*(3), 479-485.
[http://dx.doi.org/10.1021/ic9806535] [PMID: 11673952]

[54] Fukuda, T.; Kobayashi, N. Efficient synthesis of a donor-acceptor phthalocyanine having adjacently-fused pyrazine rings. *Chem. Lett.,* **2002**, *31*, 866-867.
[http://dx.doi.org/10.1246/cl.2002.866]

[55] Gregory, P. Industrial applications of phthalocyanines. *J. Porphyr. Phthalocyanines,* **2000**, *4*, 432-437.
[http://dx.doi.org/10.1002/(SICI)1099-1409(200006/07)4:4<432::AID-JPP254>3.0.CO;2-N]

[56] Gregory, P. Steamrollers, sports cars and security: phthalocyanine progress through the ages. *J. Porphyr. Phthalocyanines,* **2000**, *3*, 468-476.
[http://dx.doi.org/10.1002/(SICI)1099-1409(199908/10)3:6/7<468::AID-JPP158>3.0.CO;2-K]

[57] *Phthalocyanines: Properties and Applications*; Leznoff, C. C.; Lever, A. B. P.; Publishers, V.C., Eds.; 1990-1996.

[58] *The Porphyrin Handbook*; Kadish, K.M.; Smith, K.M.; Guilard, R., Eds.; Academic Press: Boston, **1999**, Vols. 1-10, . Vols. 11-20, 2003

[59] *Handbook of Porphyrin Science*; Kadish, K.M.; Smith, K.M.; Guilard, R., Eds.; World Scientific Press: Singapore, **2010-2013**, Vol. 1-35, .

[60] Darwent, J.R.; Douglas, P.; Harriman, A.; Porter, G.; Richoux, M.C. Metal phthalocyanines and porphyrins as photosensitizers for reduction of water to hydrogen. *Coord. Chem. Rev.,* **1982**, *44*, 83-126.
[http://dx.doi.org/10.1016/S0010-8545(00)80518-4]

[61] Lever, A.B.; Hempstead, M.R.; Leznoff, C.C.; Liu, W.; Melnik, M.; Nevin, W.A.; Seymour, P. Recent studies in phthalocyanine chemistry. *Pure Appl. Chem.,* **1986**, *58*, 1467-1476.
[http://dx.doi.org/10.1351/pac198658111467]

[62] Zagal, J.H. Metallophthalocyanines as catalysts in electrochemical reactions. *Coord. Chem. Rev.,* **1992**, *119*, 89-136.
[http://dx.doi.org/10.1016/0010-8545(92)80031-L]

[63] Kaliya, O.L.; Lukyanets, E.A.; Vorozhtsov, G.N. Catalysis and photocatalysis by phthalocyanines for technology, ecology and medicine. *J. Porphyr. Phthalocyanines,* **1999**, *3*, 592-610. [http://dx.doi.org/10.1002/(SICI)1099-1409(199908/10)3:6/7<592::AID-JPP180>3.0.CO;2-G]

[64] Sorokin, A.B.; Tuel, A. Metallophthalocyanine functionalized silicas: catalysts for the selective oxidation of aromatic compounds. *Catal. Today,* **2000**, *57*, 45-59. [http://dx.doi.org/10.1016/S0920-5861(99)00312-0]

[65] Zagal, J.H.; Griveau, S.; Ozoemena, K.I.; Nyokong, T.; Bedioui, F. Carbon nanotubes, phthalocyanines and porphyrins: attractive hybrid materials for electrocatalysis and electroanalysis. *J. Nanosci. Nanotechnol.,* **2009**, *9*(4), 2201-2214. [http://dx.doi.org/10.1166/jnn.2009.SE15] [PMID: 19437959]

[66] Zagal, J.H.; Griveau, S.; Silva, J.F.; Nyokong, T.; Bedioui, F. Metallophthalocyanine-based molecular materials as catalysts for electrochemical reactions. *Coord. Chem. Rev.,* **2010**, *254*, 2755-2791. [http://dx.doi.org/10.1016/j.ccr.2010.05.001]

[67] Sorokin, A. Photocatalytic degradation of pollutants with emphasis on phthalocyanines and related complexes. In: *Photosensitizers in medicine, environment and security*; Nyokong, T.; Ahsen, V., Eds.; Springer: Dordrecht, **2012**; pp. 433-467.

[68] Sorokin, A.B. Phthalocyanine metal complexes in catalysis. *Chem. Rev.,* **2013**, *113*(10), 8152-8191. [http://dx.doi.org/10.1021/cr4000072] [PMID: 23782107]

[69] Calvete, M.J.; Silva, M.; Pereira, M.M.; Burrows, H.D. Inorganic helping organic: recent advances in catalytic heterogeneous oxidations by immobilised tetrapyrrolic macrocycles in micro and mesoporous supports. *RSC Advances,* **2013**, *3*, 22774-22789. [http://dx.doi.org/10.1039/c3ra44038a]

[70] Herron, N.; Stucky, G.D.; Tolman, C.A. Shape selectivity in hydrocarbon oxidations using zeolite encapsulated iron phthalocyanine catalysts. *J. Chem. Soc. Chem. Commun.,* **1986**, 1521-1522. [http://dx.doi.org/10.1039/c39860001521]

[71] Zagal, J.; Paez, M.; Tanaka, A.A.; dos Santos, J.R.; Linkous, C.A. Electrocatalytic activity of metal phthalocyanines for oxygen reduction. *J. Electroanal. Chem.,* **1992**, *339*, 13-30. [http://dx.doi.org/10.1016/0022-0728(92)80442-7]

[72] Parton, R.F.; Vankelecom, I.F.; Casselman, M.J.; Bezoukhanova, C.P.; Uytterhoeven, J.B.; Jacobs, P.A. An efficient mimic of cytochrome P-450 from a zeolite-encaged iron complex in a polymer membrane. *Nature,* **1994**, *370*(6490), 541-544. [http://dx.doi.org/10.1038/370541a0] [PMID: 8052309]

[73] Sorokin, A.; Meunier, B. Efficient H_2O_2 oxidation of chlorinated phenols catalyzed by supported iron phthalocyanines. *J. Chem. Soc. Chem. Commun.,* **1994**, *15*, 1799-1800. [http://dx.doi.org/10.1039/c39940001799]

[74] Chebotareva, T.; Nyokong, T. Metallophthalocyanine catalysed electroreduction of nitrate and nitrite ions in alkaline media. *J. Appl. Electrochem.,* **1997**, *27*, 975-981. [http://dx.doi.org/10.1023/A:1018466021838]

[75] Armengol, E.; Corma, A.; Fornes, V.; Garcia, H.; Primo, J. Cu^{2+}-phthalocyanine and Co2+-perfluorophthalocyanine incorporated inside Y faujasite and mesoporous MCM-41 as heterogeneous

catalysts for the oxidation of cyclohexane. *Appl. Catal. A Gen.,* **1999**, *181*, 305-312.
[http://dx.doi.org/10.1016/S0926-860X(98)00402-5]

[76] Sorokin, A.B.; Mangematin, S.; Pergrale, C. Selective oxidation of aromatic compounds with dioxygen and peroxides catalyzed by phthalocyanine supported catalysts. *J. Mol. Catal. Chem.,* **2002**, *182*, 267-281.
[http://dx.doi.org/10.1016/S1381-1169(01)00486-1]

[77] Sharma, V.B.; Jain, S.L.; Sain, B. Cobalt phthalocyanine catalyzed aerobic oxidation of secondary alcohols: an efficient and simple synthesis of ketones. *Tetrahedron Lett.,* **2003**, *44*, 383-386.
[http://dx.doi.org/10.1016/S0040-4039(02)02453-X]

[78] Chauhan, S.M.; Kumar, A.; Srinivas, K.A. Oxidation of thiols with molecular oxygen catalyzed by cobalt(II) phthalocyanines in ionic liquid. *Chem. Commun. (Camb.),* **2003**, (18), 2348-2349.
[http://dx.doi.org/10.1039/b305888c] [PMID: 14518908]

[79] Wohrle, D.; Suvorova, O.; Gerdes, R.; Bartels, O.; Lapok, L.; Baziakina, N.; Makarov, S.; Slodek, A. Efficient oxidations and photooxidations with molecular oxygen using metal phthalocyanines as catalysts and photocatalysts. *J. Porphyr. Phthalocyanines,* **2004**, *8*, 1020-1041.
[http://dx.doi.org/10.1142/S1088424604000398]

[80] Zhao, F.; Harnisch, F.; Schroder, U.; Scholz, F.; Bogdanoff, P.; Herrmann, I. Application of pyrolysed iron(II) phthalocyanine and CoTMPP based oxygen reduction catalysts as cathode materials in microbial fuel cells. *Electrochem. Commun.,* **2005**, *7*, 1405-1410.
[http://dx.doi.org/10.1016/j.elecom.2005.09.032]

[81] McKeown, N.B.; Budd, P.M.; Msayib, K.J.; Ghanem, B.S.; Kingston, H.J.; Tattershall, C.E.; Makhseed, S.; Reynolds, K.J.; Fritsch, D. Polymers of intrinsic microporosity (PIMs): bridging the void between microporous and polymeric materials. *Chemistry,* **2005**, *11*(9), 2610-2620.
[http://dx.doi.org/10.1002/chem.200400860] [PMID: 15651019]

[82] Iliev, V.; Tomova, D.; Bilyarska, L.; Eliyas, A.; Petrov, L. Photocatalytic properties of TiO2 modified with platinum and silver nanoparticles in the degradation of oxalic acid in aqueous solution. *Appl. Catal. B,* **2006**, *63*, 266-271.
[http://dx.doi.org/10.1016/j.apcatb.2005.10.014]

[83] Mackintosh, H.J.; Budd, P.M.; McKeown, N.B. Catalysis by microporous phthalocyanine and porphyrin network polymers. *J. Mater. Chem.,* **2008**, *18*, 573-578.
[http://dx.doi.org/10.1039/B715660J]

[84] Morozan, A.; Jousselme, B.; Palacin, S. Low-platinum and platinum-free catalysts for the oxygen reduction reaction at fuel cell cathodes. *Energy Environ. Sci.,* **2011**, *4*, 1238-1254.
[http://dx.doi.org/10.1039/c0ee00601g]

[85] Kockrick, E.; Lescouet, T.; Kudrik, E.V.; Sorokin, A.B.; Farrusseng, D. Synergistic effects of encapsulated phthalocyanine complexes in MIL-101 for the selective aerobic oxidation of tetralin. *Chem. Commun. (Camb.),* **2011**, *47*(5), 1562-1564.
[http://dx.doi.org/10.1039/C0CC04431H] [PMID: 21116537]

[86] Cao, R.; Thapa, R.; Kim, H.; Xu, X.; Gyu Kim, M.; Li, Q.; Park, N.; Liu, M.; Cho, J. Promotion of oxygen reduction by a bio-inspired tethered iron phthalocyanine carbon nanotube-based catalyst. *Nat. Commun.,* **2013**, *4*, 2076.

[http://dx.doi.org/10.1038/ncomms3076] [PMID: 23797710]

[87] Janata, J. Chemical sensors. *Anal. Chem.,* **1992**, *64*, R196-R219. [http://dx.doi.org/10.1021/ac00036a012]

[88] Simic-Glavaski, B. *Phthalocyanines: Properties and Applications*; Lever, A.B.; Leznoff, C.C., Eds.; Wiley-VCH: New York, **1993**, Vol. 3, .

[89] Simon, J.; Andre, J.J. Mol. In: *Semiconductors*; Springer: Berlin, **1995**.

[90] Simon, J.; Toupance, T. *Comprehensive Supramolecular Chemistry in Supramolecular Technology*; Reinhoudt, D.N., Ed.; Pergamon: London, **1996**.

[91] Zhou, R.; Josse, F.; Göpel, W.; Öztürk, Z.Z.; Bekaroglu, O. Phthalocyanines as sensitive materials for chemical sensors. *Appl. Organomet. Chem.,* **1996**, *10*, 557-577. [http://dx.doi.org/10.1002/(SICI)1099-0739(199610)10:8<557::AID-AOC521>3.0.CO;2-3]

[92] Guillaud, G.; Simon, J.; Germain, J.P. Metallophthalocyanines-gas sensors, resistors and field effect transistors. *Coord. Chem. Rev.,* **1998**, *178*, 1433-1484. [http://dx.doi.org/10.1016/S0010-8545(98)00177-5]

[93] Antonisse, M.M.; Reinhoudt, D.N. Potentiometric anion selective sensors. *Electroanal,* **1999**, *11*, 1035-1048. [http://dx.doi.org/10.1002/(SICI)1521-4109(199910)11:14<1035::AID-ELAN1035>3.0.CO;2-I]

[94] Valli, L. Phthalocyanine-based Langmuir-Blodgett films as chemical sensors. *Adv. Colloid Interface Sci.,* **2005**, *116*(1-3), 13-44. [http://dx.doi.org/10.1016/j.cis.2005.04.008] [PMID: 16112639]

[95] Bouvet, M. Phthalocyanine-based field-effect transistors as gas sensors. *Anal. Bioanal. Chem.,* **2006**, *384*(2), 366-373. [http://dx.doi.org/10.1007/s00216-005-3257-6] [PMID: 15933850]

[96] Ciosek, P.; Wróblewski, W. Sensor arrays for liquid sensing--electronic tongue systems. *Analyst (Lond.),* **2007**, *132*(10), 963-978. [http://dx.doi.org/10.1039/b705107g] [PMID: 17893798]

[97] Öztürk, Z.Z.; Kılınç, N.; Atilla, D.; Gürek, A.G.; Ahsen, V. Recent studies chemical sensors based on phthalocyanines. *J. Porphyr. Phthalocyanines,* **2009**, *13*, 1179-1187. [http://dx.doi.org/10.1142/S1088424609001522]

[98] Rodriguez-Mendez, M.L.; Gay, M.; de Saja, J.A. New insights into sensors based on radical bisphthalocyanines. *J. Porphyr. Phthalocyanines,* **2009**, *13*, 1159-1167. [http://dx.doi.org/10.1142/S1088424609001509]

[99] Paoletti, A.M.; Pennesi, G.; Rossi, G.; Generosi, A.; Paci, B.; Albertini, V.R. Titanium and ruthenium phthalocyanines for NO2 sensors: a mini-review. *Sensors (Basel),* **2009**, *9*(7), 5277-5297. [http://dx.doi.org/10.3390/s90705277] [PMID: 22346697]

[100] Bouvet, M.; Gaudillat, P.; Suisse, J.M. Phthalocyanine-based hybrid materials for chemosensing. *J. Porphyr. Phthalocyanines,* **2013**, *17*, 913-919. [http://dx.doi.org/10.1142/S1088424613300073]

[101] Trogler, W.C. Chemical sensing with semiconducting metal phthalocyanines. In *Molecular Electronic*

Structures of Transition Metal Complexes I, Structure and Bonding, Mingos, D.M.P.; Day, P.; Dahl. *J.P. Eds.,* **2012**, *142*, 91-117.

[102] Sauer, T.; Caseri, W.; Wegner, G.; Vogel, A.; Hoffmann, B. Developments of novel chemical sensor devices based on LB films from phthalocyaninato polysiloxane polymers. *J. Phys. D Appl. Phys.,* **1990**, *23*, 79-84. [http://dx.doi.org/10.1088/0022-3727/23/1/013]

[103] Passard, M.; Pauly, A.; Blanc, J.P.; Dogo, S.; Germain, J.P.; Maleysson, C. Doping mechanisms of phthalocyanines by oxidizing gases-application to gas sensors. *Thin Solid Films,* **1994**, *237*, 272-276. [http://dx.doi.org/10.1016/0040-6090(94)90272-0]

[104] Berger, O.; Fischer, W.J.; Adolphi, B.; Tierbach, S.; Melev, V.; Schreiber, J. Studies on phase transformations of Cu-phthalocyanine thin films. *J. Mater. Sci. Mater. Electron.,* **2000**, *11*, 331-346. [http://dx.doi.org/10.1023/A:1008933516940]

[105] Fleischer, M.; Simon, E.; Rumpel, E.; Ulmer, H.; Harbeck, M.; Wandel, M.; Fietzek, C.; Weimar, U.; Meixner, H. Detection of volatile compounds correlated to human diseases through breath analysis with chemical sensors. *Sens. Actuators B Chem.,* **2002**, *83*, 245-249. [http://dx.doi.org/10.1016/S0925-4005(01)01056-5]

[106] Basova, T.V.; Tasaltin, C.; Gurek, A.G.; Ebeoglu, M.A.; Ozturk, Z.Z.; Ahsen, V. Mesomorphic phthalocyanine as chemically sensitive coatings for chemical sensors. *Sens. Actuators B Chem.,* **2003**, *96*, 70-75. [http://dx.doi.org/10.1016/S0925-4005(03)00487-8]

[107] Umar, A.; Rahman, M.M.; Kim, S.H.; Hahn, Y.B. Zinc oxide nanonail based chemical sensor for hydrazine detection. *Chem. Commun. (Camb.),* **2008**, (2), 166-168. [http://dx.doi.org/10.1039/B711215G] [PMID: 18092075]

[108] Bohrer, F.I.; Colesniuc, C.N.; Park, J.; Ruidiaz, M.E.; Schuller, I.K.; Kummel, A.C.; Trogler, W.C. Comparative gas sensing in cobalt, nickel, copper, zinc, and metal-free phthalocyanine chemiresistors. *J. Am. Chem. Soc.,* **2009**, *131*(2), 478-485. [http://dx.doi.org/10.1021/ja803531r] [PMID: 19093852]

[109] Dossi, N.; Toniolo, R.; Pizzariello, A.; Carrilho, E.; Piccin, E.; Battiston, S.; Bontempelli, G. An electrochemical gas sensor based on paper supported room temperature ionic liquids. *Lab Chip,* **2012**, *12*(1), 153-158. [http://dx.doi.org/10.1039/C1LC20663J] [PMID: 22076475]

[110] Erbahar, D.D.; Gurol, I.; Gumus, G.; Musluoglu, E.; Ozturk, Z.Z.; Ahsen, V.; Harbeck, M. Pesticide sensing in water with phthalocyanine based QCM sensors. *Sens. Actuators B Chem.,* **2012**, *173*, 562-568. [http://dx.doi.org/10.1016/j.snb.2012.07.041]

[111] Pochekailov, S.; Nozar, J.; Nespurek, S.; Rakusan, J.; Karaskova, M. Interaction of nitrogen dioxide with sulfonamide-substituted phthalocyanines: towards NO2 gas sensor. *Sens. Actuators B Chem.,* **2012**, *169*, 1-9. [http://dx.doi.org/10.1016/j.snb.2011.12.087]

[112] Cook, M.J. Thin film formulations of substituted phthalocyanines. *J. Mater. Chem.,* **1996**, *6*, 677-689. [http://dx.doi.org/10.1039/jm9960600677]

[113] Dimitrakopoulos, C.D.; Malenfant, P.R. Organic thin film transistors for large area electronics. *Adv. Mater.,* **2002**, *14*, 99-117.
[http://dx.doi.org/10.1002/1521-4095(20020116)14:2<99::AID-ADMA99>3.0.CO;2-9]

[114] Li, L.; Tang, Q.; Li, H.; Hu, W.; Yang, X.; Shuai, Z.; Liu, Y.; Zhu, D. Organic thin-film transistors of phthalocyanines. *Pure Appl. Chem.,* **2008**, *80*, 2231-2240.
[http://dx.doi.org/10.1351/pac200880112231]

[115] Hussain, S.A.; Bhattacharjee, D. Langmuir-Blodgett films and molecular electronics. *Mod. Phys. Lett. B,* **2009**, *23*, 3437-3451.
[http://dx.doi.org/10.1142/S0217984909021508]

[116] Roberts, G.G.; Petty, M.C.; Barker, S.; Fowler, M.T.; Thomas, N.J. Electronic devices incorporating stable phthalocyanine Langmuir-Blodgett films. *Thin Solid Films,* **1985**, *132*, 113-123.
[http://dx.doi.org/10.1016/0040-6090(85)90463-8]

[117] Cook, M.J.; Dunn, A.J.; Daniel, F.M.; Hart, R.C.; Richardson, R.M.; Rose, S.J. New asymmetric substitution of phthalocyanines: derivatives designed for deposition as Langmuir-Blodgett films. *Thin Solid Films,* **1988**, *159*, 469-478.
[http://dx.doi.org/10.1016/0040-6090(88)90662-1]

[118] Palacin, S.; Lesieur, P.; Stefanelli, I.; Barraud, A. Structural studies of intermolecular interactions in pure and diluted films of a redox-active phthalocyanine. *Thin Solid Films,* **1988**, *159*, 83-90.
[http://dx.doi.org/10.1016/0040-6090(88)90619-0]

[119] Burghard, M.; Schmelzer, M.; Roth, S.; Haisch, P.; Hanack, M. Langmuir-Blodgett film formation of a series of peripherally octasubstituted metallophthalocyanines. *Langmuir,* **1994**, *10*, 4265-4269.
[http://dx.doi.org/10.1021/la00023a057]

[120] Baker, S.; Petty, M.C.; Roberts, G.G.; Twigg, M.V. The preparation and properties of stable metal-free phthalocyanine Langmuir-Blodgett films. *Thin Solid Films,* **1983**, *99*, 53-99.
[http://dx.doi.org/10.1016/0040-6090(83)90359-0]

[121] Baker, S.; Roberts, G.G.; Petty, M.C. Phthalocyanine Langmuir-Blodgett film gas detector. *IEE Proc., I, Commun. Speech Vis.,* **1983**, *130*, 260-263.

[122] Casstevens, M.K.; Samoc, M.; Pfleger, J.; Prasad, P.N. Dynamics of 3rd order nonlinear optical processes in Langmuir-Blodgett and evaporated films of phthalocyanines. *J. Chem. Phys.,* **1990**, *92*, 2019-2024.
[http://dx.doi.org/10.1063/1.458035]

[123] Kobayashi, N.; Lam, H.; Nevin, W.A.; Janda, P.; Leznoff, C.C.; Koyama, T.; Monden, A.; Shirai, H. Synthesis, spectroscopy, electrochemistry, Langmuir-Blodgett film formation, and molecular-orbital calculations of planar binuclear phthalocyanines. *J. Am. Chem. Soc.,* **1994**, *116*, 879-890.
[http://dx.doi.org/10.1021/ja00082a007]

[124] Apetrei, C.; Alessio, P.; Constantino, C.J.; de Saja, J.A.; Rodriguez-Mendez, M.L.; Pavinatto, F.J.; Ramos Fernandes, E.G.; Zucolotto, V.; Oliveira, O.N., Jr Biomimetic biosensor based on lipidic layers containing tyrosinase and lutetium bisphthalocyanine for the detection of antioxidants. *Biosens. Bioelectron.,* **2011**, *26*(5), 2513-2519.
[http://dx.doi.org/10.1016/j.bios.2010.10.047] [PMID: 21123042]

[125] Barszcz, B.; Bogucki, A.; Biadasz, A.; Bursa, B.; Wrobel, D.; Graja, A. Molecular orientation and spectral investigations of Langmuir-Blodgett films of selected copper phthalocyanines. *J. Photochem. Photobiol. Chem.,* **2011**, *218*, 48-57.
[http://dx.doi.org/10.1016/j.jphotochem.2010.12.002]

[126] Kan, J.; Chen, Y.; Qi, D.; Liu, Y.; Jiang, J. High-performance air-stable ambipolar organic field-effect transistor based on tris(phthalocyaninato) europium(III). *Adv. Mater.,* **2012**, *24*(13), 1755-1758.
[http://dx.doi.org/10.1002/adma.201200006] [PMID: 22389017]

[127] Law, K.Y. Organic photoconductive materials-recent trends and developments. *Chem. Rev.,* **1993**, *93*, 449-486.
[http://dx.doi.org/10.1021/cr00017a020]

[128] Burrows, P.E.; Shen, Z.; Bulovic, V.; McCarty, D.M.; Forrest, S.R.; Cronin, J.A.; Thompson, M.E. Relationship between electroluminescence and current transport in organic heterojunction light-emitting devices. *J. Appl. Phys.,* **1996**, *79*, 7991-8006.
[http://dx.doi.org/10.1063/1.362350]

[129] Laudise, R.A.; Kloc, C.; Simpkins, P.G.; Siegrist, T. Physical vapor growth of organic semiconductors. *J. Cryst. Growth,* **1998**, *187*, 449-454.
[http://dx.doi.org/10.1016/S0022-0248(98)00034-7]

[130] de la Torre, G.; Nicolau, M.; Torres, T. Phthalocyanines: synthesis, supramolecular organization, and physical properties. In: *Supramolecular Photosensitive and Electroactive materials*; Nalwa, H.S., Ed.; Academic Press: San Diego, **2001**; pp. 1-111.
[http://dx.doi.org/10.1016/B978-012513904-5/50003-X]

[131] Shen, Y.; Hosseini, A.R.; Wong, M.H.; Malliaras, G.G. How to make ohmic contacts to organic semiconductors. *Chem Phys Chem,* **2004**, *5*(1), 16-25.
[http://dx.doi.org/10.1002/cphc.200300942] [PMID: 14999839]

[132] Papageorgiou, N.; Salomon, E.; Angot, T.; Layet, J.M.; Giovanelli, L.; Le Lay, G. Physics of ultra-thin phthalocyanine films on semiconductors. *Prog. Surf. Sci.,* **2004**, *77*, 139-170.
[http://dx.doi.org/10.1016/j.progsurf.2005.01.001]

[133] Gregg, B.A.; Chen, S.G.; Cormier, R.A. Coulomb forces and doping in organic semiconductors. *Chem. Mater.,* **2004**, *16*, 4586-4599.
[http://dx.doi.org/10.1021/cm049625c]

[134] Liu, S.H.; Wang, W.C.; Briseno, A.L.; Mannsfeld, S.C.; Bao, Z.N. Controlled deposition of crystalline organic semiconductors for field-effect-transistor applications. *Adv. Mater.,* **2009**, *21*, 1217-1232.
[http://dx.doi.org/10.1002/adma.200802202]

[135] Hains, A.W.; Liang, Z.; Woodhouse, M.A.; Gregg, B.A. Molecular semiconductors in organic photovoltaic cells. *Chem. Rev.,* **2010**, *110*(11), 6689-6735.
[http://dx.doi.org/10.1021/cr9002984] [PMID: 20184362]

[136] Jiang, L.; Dong, H.L.; Hu, W.P. Organic single crystal field-effect transistors: advances and perspectives. *J. Mater. Chem.,* **2010**, *20*, 4994-5007.
[http://dx.doi.org/10.1039/b925875b]

[137] Zhang, Y.X.; Cai, X.; Bian, Y.Z.; Jiang, J.Z. Organic semiconductors of phthalocyanine compounds

for field effect transistors (FETs) in functional phthalocyanine molecular materials. *Structure and Bonding,* **2010**, *135*, 275-321.
[http://dx.doi.org/10.1007/978-3-642-04752-7_9]

[138] Kobayashi, H.; Kobayashi, A.; Tajima, H. Studies on molecular conductors: from organic semiconductors to molecular metals and superconductors. *Chem. Asian J.,* **2011**, *6*(7), 1688-1704.
[http://dx.doi.org/10.1002/asia.201100061] [PMID: 21612013]

[139] Lin, Y.; Li, Y.; Zhan, X. Small molecule semiconductors for high-efficiency organic photovoltaics. *Chem. Soc. Rev.,* **2012**, *41*(11), 4245-4272.
[http://dx.doi.org/10.1039/c2cs15313k] [PMID: 22453295]

[140] Eley, D.D. Phthalocyanines as semiconductors. *Nature,* **1948**, *162*(4125), 819-819.
[http://dx.doi.org/10.1038/162819a0] [PMID: 18103107]

[141] Fan, F.R.; Faulkner, L.R. Phthalocyanine thin films as semiconductor electrodes. *J. Am. Chem. Soc.,* **1979**, *101*, 4779-4787.
[http://dx.doi.org/10.1021/ja00511a001]

[142] Giraudeau, A.; Fan, F.R.; Bard, A.J. Semiconductor electrodes. 30. Spectral sensitization of the semiconductors normal-TiO2 and normal-WO3 with metal phthalocyanines. *J. Am. Chem. Soc.,* **1980**, *102*, 5137-5142.
[http://dx.doi.org/10.1021/ja00536a001]

[143] Turek, P.; Petit, P.; Andre, J.J.; Simon, J.; Even, R.; Boudjema, B.; Guillaud, G.; Maitrot, M. A new series of molecular semiconductors-phthalocyanine radicals. *J. Am. Chem. Soc.,* **1987**, *109*, 5119-5122.
[http://dx.doi.org/10.1021/ja00251a012]

[144] Bao, Z.; Lovinger, A.J.; Dodabalapur, A. Organic field-effect transistors with high mobility based on copper phthalocyanine. *Appl. Phys. Lett.,* **1996**, *69*, 3066-3068.
[http://dx.doi.org/10.1063/1.116841]

[145] Parthasarathy, G.; Burrows, P.E.; Khalfin, V.; Kozlov, V.G.; Forrest, S.R. A metal-free cathode for organic semiconductor devices. *Appl. Phys. Lett.,* **1998**, *72*, 2138-2140.
[http://dx.doi.org/10.1063/1.121301]

[146] Pfeiffer, M.; Leo, K.; Zhou, X.; Huang, J.S.; Hofmann, M.; Werner, A.; Blochwitz-Nimoth, J. Doped organic semiconductors: physics and application in light emitting diodes. *Org. Electron.,* **2003**, *4*, 89-103.
[http://dx.doi.org/10.1016/j.orgel.2003.08.004]

[147] Jones, B.A.; Facchetti, A.; Wasielewski, M.R.; Marks, T.J. Tuning orbital energetics in arylene diimide semiconductors. materials design for ambient stability of n-type charge transport. *J. Am. Chem. Soc.,* **2007**, *129*(49), 15259-15278.
[http://dx.doi.org/10.1021/ja075242e] [PMID: 17999505]

[148] Cinchetti, M.; Heimer, K.; Wüstenberg, J.P.; Andreyev, O.; Bauer, M.; Lach, S.; Ziegler, C.; Gao, Y.; Aeschlimann, M. Determination of spin injection and transport in a ferromagnet/organic semiconductor heterojunction by two-photon photoemission. *Nat. Mater.,* **2009**, *8*(2), 115-119.
[http://dx.doi.org/10.1038/nmat2334] [PMID: 19029891]

[149] Coffey, D.C.; Ferguson, A.J.; Kopidakis, N.; Rumbles, G. Photovoltaic charge generation in organic semiconductors based on long-range energy transfer. *ACS Nano,* **2010**, *4*(9), 5437-5445. [http://dx.doi.org/10.1021/nn101106b] [PMID: 20735062]

[150] Urdampilleta, M.; Klyatskaya, S.; Cleuziou, J.P.; Ruben, M.; Wernsdorfer, W. Supramolecular spin valves. *Nat. Mater.,* **2011**, *10*(7), 502-506. [http://dx.doi.org/10.1038/nmat3050] [PMID: 21685902]

[151] Kim, Y.; Kim, J.P.; Kwon, O.S.; Cho, I.H. The synthesis and application of thermally stable dyes for ink-jet printed LCD color filters. *Dyes Pigm.,* **2009**, *81*, 45-52. [http://dx.doi.org/10.1016/j.dyepig.2008.09.006]

[152] Takano, S.; Enokida, T.; Kambata, A. A new polymer of metal-free phthalocyanine. *Chem. Lett.,* **1984**, *13*, 2037-2040. [http://dx.doi.org/10.1246/cl.1984.2037]

[153] Loutfy, R.O.; Hor, A.M.; Di Paola-Baranyl, G.; Hsiao, C.K. Electrophotographic photoreceptors incorporating aggregated phthalocyanines. *J. Imaging Sci. Technol.,* **1985**, *29*, 116-121.

[154] Enokida, T.; Hirohashi, R. Electrophotographic dual-layered photoreceptors incorporating copperphthalocyanines. *J. Imaging Sci. Technol.,* **1992**, *36*, 135-141.

[155] Pai, D.M.; Springett, B.E. Physics of electrophotography. *Rev. Mod. Phys.,* **1993**, *65*, 163-212. [http://dx.doi.org/10.1103/RevModPhys.65.163]

[156] Zhou, S.; Jin, X.; Hu, W.; Liu, Y.; Liu, S. An interesting phenomenon of impurity sulfur in phthalocyanine photoreceptor device. *Solid State Commun.,* **1999**, *112*, 269-273. [http://dx.doi.org/10.1016/S0038-1098(99)00352-X]

[157] Murayama, T. Organic photoconductive materials used for electrophotographic organic photoreceptors. *J. Synth. Org. Chem. Jpn.,* **1999**, *57*, 541-551. [http://dx.doi.org/10.5059/yukigoseikyokaishi.57.541]

[158] Kobayashi, S.; Iwasaki, K.; Sasaki, H.; Oh-Hara, S.; Nishizawa, M.; Katayose, M. Organic optical storage medium with a read-out stability under 1 mW laser irradiation. *Jpn. J. Appl. Phys.,* **1991**, *30*, L114-L116. [http://dx.doi.org/10.1143/JJAP.30.L114]

[159] Ao, R.; Kümmert, L.; Haarer, D. Present limits of data storage using dye molecules in solid matrices. *Adv. Mater.,* **1995**, *5*, 495-499. [http://dx.doi.org/10.1002/adma.19950070522]

[160] Seto, J.; Tamura, S.; Asai, N.; Kishii, N.; Kijima, Y.; Matsuzawa, N. Macrocyclic functional dyes: applications to optical disk media, photochemical hole burning and nonlinear optics. *Pure Appl. Chem.,* **1996**, *68*, 1429-1434. [http://dx.doi.org/10.1351/pac199668071429]

[161] Gan, F.X. Some considerations of organic materials for high density optical disk data storage. *Chin. Sci. Bull.,* **2000**, *45*, 572-576. [http://dx.doi.org/10.1007/BF02887109]

[162] Majumdar, H.S.; Bandyopadhyay, A.; Pal, A.J. Data-storage devices based on layer-by-layer self-assembled films of a phthalocyanine derivative. *Org. Electron.,* **2003**, *4*, 39-44.

[http://dx.doi.org/10.1016/S1566-1199(03)00007-7]

[163] Makarov, N.S.; Rebane, A.; Drobizhev, M.; Wolleb, H.; Spahni, H. Optimizing two-photon absorption for volumetric optical data storage. *J. Opt. Soc. Am. B,* **2007**, *24*, 1874-1885. [http://dx.doi.org/10.1364/JOSAB.24.001874]

[164] Dini, D.; Calvete, M.J.; Hanack, M.; Amendola, V.; Meneghetti, M. Large two-photon absorption cross sections of hemiporphyrazines in the excited state: the multiphoton absorption process of hemiporphyrazines with different central metals. *J. Am. Chem. Soc.,* **2008**, *130*(37), 12290-12298. [http://dx.doi.org/10.1021/ja802678u] [PMID: 18722439]

[165] Gao, H.J.; Gao, L. Scanning tunneling microscopy of functional nanostructures on solid surfaces: Manipulation, self-assembly, and applications. *Prog. Surf. Sci.,* **2010**, *85*, 28-91. [http://dx.doi.org/10.1016/j.progsurf.2009.10.001]

[166] Lian, K.; Li, R.; Wang, H.; Zhang, J.; Gamota, D. Printed flexible memory devices using copper phthalocyanine. *Mater. Sci. Eng. B: Adv. Funct. Solid State Mater,* **2010**, *167*, 12-16. [http://dx.doi.org/10.1016/j.mseb.2010.01.020]

[167] Schultz, H.; Lehmann, H.; Rein, M.; Hanack, M. Phthalocyaninatometal and related complexes with special electrical and optical properties. In: *Structure and Bonding*; , **1991**; 74, pp. 41-154.

[168] Giroudgodquin, A.M.; Maitlis, P.M. Metallomesogens-metal-complexes in organized fluid phases. *Angew. Chem. Int. Ed. Engl.,* **1991**, *30*, 375-402. [http://dx.doi.org/10.1002/anie.199103751]

[169] Espinet, P.; Esteruelas, M.A.; Oro, L.A.; Serrano, J.L.; Sola, E. Transition-metal liquid-crystal--advanced materials within the reach of the coordination chemist. *Coord. Chem. Rev.,* **1992**, *117*, 215-274. [http://dx.doi.org/10.1016/0010-8545(92)80025-M]

[170] Simon, J.; Bassoul, P. Phthalocyanine based liquid crystals: towards submicronic devices. In: *Phthalocyanines: Properties and Applications*; Leznoff, C.C.; Lever, A.B., Eds.; VCH Publ. Inc.: New York, **1993**; Vol. 2, p. 223.

[171] Cammidge, A.N.; Bushby, R.J. Demus, D.; Goodby, J.; Gray, G.W.; Spiess, H.-W.; Vill, V. Synthesis and structural features. In: *Handbook of Liquid Crystals*; Wiley-VCH Verlag GmbH: Weinheim, **1998**; Vol. 2-Part III, pp. 693-748.

[172] Ohta, K.; Hatsusaka, K.; Sugibayashi, M.; Ariyoshi, M.; Ban, K.; Maeda, F.; Naito, R.; Nishizawa, K.; van de Craats, A.M.; Warman, J.M. Discotic liquid crystalline semiconductors. *Mol. Cryst. Liquid Cryst,* **2003**, *397*, 25-45. [http://dx.doi.org/10.1080/714965592]

[173] Sleven, J.; Cardinaels, T.; Görller-Walrand, C.; Binnemans, K. Liquid-crystalline metallophthalocyanines containing late first-row transition metals. *ARKIVOC,* **2003**, *4*, 68-82. [http://dx.doi.org/10.3998/ark.5550190.0004.406]

[174] Laschat, S.; Baro, A.; Steinke, N.; Giesselmann, F.; Hägele, C.; Scalia, G.; Judele, R.; Kapatsina, E.; Sauer, S.; Schreivogel, A.; Tosoni, M. Discotic liquid crystals: from tailor-made synthesis to plastic electronics. *Angew. Chem. Int. Ed. Engl.,* **2007**, *46*(26), 4832-4887. [http://dx.doi.org/10.1002/anie.200604203] [PMID: 17568461]

[175] Chen, Z.; Lohr, A.; Saha-Möller, C.R.; Würthner, F. Self-assembled pi-stacks of functional dyes in solution: structural and thermodynamic features. *Chem. Soc. Rev.,* **2009**, *38*(2), 564-584. [http://dx.doi.org/10.1039/B809359H] [PMID: 19169466]

[176] Chen, S.; Eichhorn, S.H. Ionic discotic liquid crystals. *Isr. J. Chem.,* **2012**, *52*, 830-843. [http://dx.doi.org/10.1002/ijch.201200046]

[177] Cook, M.J.; Daniel, M.F.; Harrison, K.J.; Mckeown, N.B.; Thomson, A.J. 1,4,8,11,15,18,22,25-Oct--alkyl phthalocyanines-new discotic liquid-crystal materials. *J. Chem. Soc. Chem. Commun.,* **1987**, 1086-1088. [http://dx.doi.org/10.1039/c39870001086]

[178] van der Pol, J.F.; Neelman, E.; Zwikker, J.W.; Nolte, R.J.; Drenth, W.; Aerts, J.; Visser, R.; Picken, S.J. Homologous series of liquid-crystalline metal free and copper octa-n-alkoxyphthalocyanines. *Liq. Cryst.,* **1989**, *6*, 577-592. [http://dx.doi.org/10.1080/02678298908034177]

[179] Simon, J.; Sirlin, C. Mesomorphic molecular materials for electronics, opto-electronics, iono-electronics: octaalkyl-phthalocyanine derivatives. *Pure Appl. Chem.,* **1989**, *61*, 1625-1630. [http://dx.doi.org/10.1351/pac198961091625]

[180] Lelievre, D.; Bosio, L.; Simon, J.; Andre, J.J.; Bensebaa, F. Dimeric substituted copper phthalocyanine liquid-crystals-synthesis, characterization, and magnetic properties. *J. Am. Chem. Soc.,* **1992**, *114*, 4475-4479. [http://dx.doi.org/10.1021/ja00038a005]

[181] Schouten, P.G.; Warman, J.M.; Dehaas, M.P.; Vannostrum, C.F.; Gelinck, G.H.; Nolte, R.J.; Copyn, M.J.; Zwikker, J.W.; Engel, M.K.; Hanack, M.; Chang, Y.H.; Ford, W.T. The effect of structural modifications on charge migration in mesomorphic phthalocyanines. *J. Am. Chem. Soc.,* **1994**, *116*, 6880-6894. [http://dx.doi.org/10.1021/ja00094a048]

[182] van Nostrum, C.F.; Picken, S.J.; Schouten, A.J.; Nolte, R.J. Synthesis and supramolecular chemistry of novel liquid-crystalline crown ether-substituted phthalocyanines-toward molecular wires and molecular ionoelectronics. *J. Am. Chem. Soc.,* **1995**, *117*, 9957-9965. [http://dx.doi.org/10.1021/ja00145a004]

[183] Van Nostrum, C.F.; Nolte, R.J. Functional supramolecular materials: self-assembly of phthalocyanines and porphyrazines. *Chem. Commun. (Camb.),* **1996**, 2385-2392. [http://dx.doi.org/10.1039/cc9960002385]

[184] Fox, J.M.; Katz, T.J.; Van Elshocht, S.; Verbiest, T.; Kauranen, M.; Persoons, A.; Thongpanchang, T.; Krauss, T.; Brus, L. Synthesis, self-assembly, and nonlinear optical properties of conjugated helical metal phthalocyanine derivatives. *J. Am. Chem. Soc.,* **1999**, *121*, 3453-3459. [http://dx.doi.org/10.1021/ja983633a]

[185] McNeill, J.D.; O'Connor, D.B.; Barbara, P.F. Imaging organic device function with near-field scanning optical microscopy. *J. Chem. Phys.,* **2000**, *112*, 7811-7821. [http://dx.doi.org/10.1063/1.481386]

[186] Swarts, J.C.; Langner, E.H.; Krokeide-Hove, N.; Cook, M.J. Synthesis and electrochemical

characterisation of some long chain 1,4,8,11,15,18,22,25-octa-alkylated metal-free and zinc phthalocyanines possessing discotic liquid crystalline properties. *J. Mater. Chem.,* **2001**, *11*, 434-443. [http://dx.doi.org/10.1039/b006123i]

[187] van de Craats, A.M.; Warman, J.M. The core-size effect on the mobility of charge in discotic liquid crystalline materials. *Adv. Mater.,* **2001**, *13*, 130-133. [http://dx.doi.org/10.1002/1521-4095(200101)13:2<130::AID-ADMA130>3.0.CO;2-L]

[188] Zhang, Q.M.; Li, H.; Poh, M.; Xia, F.; Cheng, Z.Y.; Xu, H.; Huang, C. An all-organic composite actuator material with a high dielectric constant. *Nature,* **2002**, *419*(6904), 284-287. [http://dx.doi.org/10.1038/nature01021] [PMID: 12239563]

[189] Elemans, J.A.; van Hameren, R.; Nolte, R.J.; Rowan, A.E. Molecular materials by self-assembly of porphyrins, phthalocyanines, and perylenes. *Adv. Mater.,* **2006**, *18*, 1251-1266. [http://dx.doi.org/10.1002/adma.200502498]

[190] Bottari, G.; de la Torre, G.; Guldi, D.M.; Torres, T. Covalent and noncovalent phthalocyanine-carbon nanostructure systems: synthesis, photoinduced electron transfer, and application to molecular photovoltaics. *Chem. Rev.,* **2010**, *110*(11), 6768-6816. [http://dx.doi.org/10.1021/cr900254z] [PMID: 20364812]

[191] Peumans, P.; Yakimov, A.; Forrest, S.R. Small molecular weight organic thin-film photodetectors and solar cells. *J. Appl. Phys.,* **2003**, *93*, 3693-3723. [http://dx.doi.org/10.1063/1.1534621]

[192] Rand, B.P.; Genoe, J.; Heremans, P.; Poortmans, J. Solar cells utilizing small molecular weight organic semiconductors. *Prog. Photovolt. Res. Appl.,* **2007**, *15*, 659-676. [http://dx.doi.org/10.1002/pip.788]

[193] Imahori, H.; Umeyama, T.; Ito, S. Large pi-aromatic molecules as potential sensitizers for highly efficient dye-sensitized solar cells. *Acc. Chem. Res.,* **2009**, *42*(11), 1809-1818. [http://dx.doi.org/10.1021/ar900034t] [PMID: 19408942]

[194] Martinez-Diaz, V.M.; Torres, T. *Handbook of Porphyrin Science*; Kadish, K.M.; Smith, K.M.; Guilard, R., Eds.; World Scientific Publ.: Singapore, **2010**, Vol. 10, pp. 141-182.

[195] Walter, M.G.; Rudine, A.B.; Wamser, C.C. Porphyrins and phthalocyanines in solar photovoltaic cells. *J. Porphyr. Phthalocyanines,* **2010**, *14*, 759-792. [http://dx.doi.org/10.1142/S1088424610002689]

[196] Li, X.Y.; Wang, H.X.; Wu, H.X. Phthalocyanines and their analogs applied in dye-sensitized solar cell. , **2010**. [http://dx.doi.org/10.1007/978-3-642-04752-7_8]

[197] Sonar, P.; Lim, J.P.; Chan, K.L. Organic non-fullerene acceptors for organic photovoltaics. *Energy Environ. Sci.,* **2011**, *4*, 1558-1574. [http://dx.doi.org/10.1039/c0ee00668h]

[198] Imahori, H.; Kurotobi, K.; Walter, M.G.; Rudine, A.B.; Wamser, C.C. *Handbook of Porphyrin Science*; Kadish, K.M.; Smith, K.M.; Guilard, R., Eds.; World Scientific Publ.: Singapore, **2012**, Vol. 18, pp. 57-123.

[199] Schlettwein, D.; Nyokong, T. *Handbook of Porphyrin Science*; Kadish, K.M.; Smith, K.M.; Guilard,

R., Eds.; World Scientific Publ.: Singapore, **2012**, Vol. 24, pp. 389-420.

[200] Giribabu, L.; Sudhakar, K.; Velkannan, V. Phthalocyanines: potential alternative sensitizers to Ru(II) polypyridyl complexes for dye-sensitized solar cells. *Curr. Sci.,* **2012**, *102*, 991-1000.

[201] Schlenker, C.W.; Thompson, M.E. Current challenges in organic photovoltaic solar energy conversion. *Top. Curr. Chem.,* **2012**, *312*, 175-212.
[http://dx.doi.org/10.1007/128_2011_219] [PMID: 21837556]

[202] Ragoussi, M.E.; Ince, M.; Torres, T. Recent advances in phthalocyanine-based sensitizers for dye-sensitized solar cells. *Eur. J. Org. Chem.,* **2013**, 6475-6489.
[http://dx.doi.org/10.1002/ejoc.201301009]

[203] Singh, V.K.; Kanaparthi, R.K.; Giribabu, L. Emerging molecular design strategies of unsymmetrical phthalocyanines for dye-sensitized solar cell applications. *RSC Advances,* **2014**, *4*, 6970-6984.
[http://dx.doi.org/10.1039/c3ra45170d]

[204] Zhang, L.; Wang, L. Recent research progress on optical limiting property of materials based on phthalocyanine, its derivatives, and carbon nanotubes. *J. Mater. Sci.,* **2008**, *43*, 5692-5701.
[http://dx.doi.org/10.1007/s10853-008-2826-4]

[205] Nalwa, H.S.; Shirk, J.A. *Phthalocyanines: Properties and Applications*; Leznoff, C.C.; Lever, A.B.P., Eds.; VCH Publ. Inc.: New York, **1996**, Vol.4, p. 79.

[206] de la Torre, G.; Vazquez, P.; Agullo-Lopez, F.; Torres, T. Phthalocyanines and related compounds: organic targets for nonlinear optical applications. *J. Mater. Chem.,* **1998**, *8*, 1671-1683.
[http://dx.doi.org/10.1039/a803533d]

[207] Hanack, M.; Schneider, T.; Barthel, M.; Shirk, J.S.; Flom, S.R.; Pong, R.G. Indium phthalocyanines and naphthalocyanines for optical limiting. *Coord. Chem. Rev.,* **2001**, *219-221*, 235-258.
[http://dx.doi.org/10.1016/S0010-8545(01)00327-7]

[208] Hanack, M.; Dini, D.; Barthel, M.; Vagin, S. Conjugated macrocycles as active materials in nonlinear optical processes: optical limiting effect with phthalocyanines and related compounds. *Chem. Rec.,* **2002**, *2*(3), 129-148.
[http://dx.doi.org/10.1002/tcr.10024] [PMID: 12112866]

[209] Dini, D.; Hanack, M. *The porphyrin handbook: Physical Properties of phthalocyanine-based materials*; Kadish, K. M.; Smith, K. M.; Guilard, R., Eds.; Academic Press: San Diego, **2003**, Vol. 17, pp. 1-36.
[http://dx.doi.org/10.1016/B978-0-08-092391-8.50007-8]

[210] Flom, S.R. Nonlinear optical properties of phthalocyanines. In: *Porphyrin Handbook*; Kadish, K.M.; Smith, K.M.; Guilard, R., Eds.; Academic Press: San Diego, **2003**; Vol. 19, pp. 179-190.
[http://dx.doi.org/10.1016/B978-0-08-092393-2.50011-1]

[211] O'Flaherty, S.M.; Hold, S.V.; Cook, M.J.; Torres, T.; Chen, Y.; Hanack, M.; Blau, W.J. Molecular engineering of peripherally and axially modified phthalocyanines for optical limiting and nonlinear optics. *Adv. Mater.,* **2003**, *15*, 19-32.
[http://dx.doi.org/10.1002/adma.200390002]

[212] de la Torre, G.; Vázquez, P.; Agulló-López, F.; Torres, T. Role of structural factors in the nonlinear optical properties of phthalocyanines and related compounds. *Chem. Rev.,* **2004**, *104*(9), 3723-3750.

[http://dx.doi.org/10.1021/cr030206t] [PMID: 15352778]

[213] Calvete, M.; Yang, G.Y.; Hanack, M. Porphyrins and phthalocyanines as materials for optical limiting. *Synth. Met.,* **2004**, *141*, 231-243.
[http://dx.doi.org/10.1016/S0379-6779(03)00407-7]

[214] Chen, Y.; Hanack, M.; Araki, Y.; Ito, O. Axially modified gallium phthalocyanines and naphthalocyanines for optical limiting. *Chem. Soc. Rev.,* **2005**, *34*(6), 517-529.
[http://dx.doi.org/10.1039/b416368k] [PMID: 16137164]

[215] Wei, T.H.; Hagan, D.J.; Sence, M.J.; van Stryland, E.W.; Perry, J.W.; Coulter, D.R. Direct measurements of nonlinear absorption and refraction in solutions of phthalocyanines. *Appl. Phys., B Photophys. Laser Chem.,* **1992**, *54*, 46-51.
[http://dx.doi.org/10.1007/BF00331733]

[216] Nalwa, H.S.; Saito, T.; Kakuta, A.; Iwayanagi, T., III -Order nonlinear-optical properties of polymorphs of oxotitanium phthalocyanine. *J. Phys. Chem.,* **1993**, *97*, 10515-10517.
[http://dx.doi.org/10.1021/j100143a002]

[217] Diaz-Garcia, M.A.; Ledoux, I.; Duro, J.A.; Torres, T.; Agullo-Lopez, F.; Zyss, J., III -Order nonlinear-optical properties of soluble octasubstituted phthalocyanines. *J. Phys. Chem.,* **1994**, *98*, 8761-8764.
[http://dx.doi.org/10.1021/j100086a028]

[218] Perry, J.W.; Mansour, K.; Lee, I.Y.; Wu, X.L.; Bedworth, P.V.; Chen, C.T.; Ng, D.; Marder, S.R.; Miles, P.; Wada, T.; Tian, M.; Sasabe, H. Organic optical limiter with a strong nonlinear absorptive response. *Science,* **1996**, *273*, 1533-1536.
[http://dx.doi.org/10.1126/science.273.5281.1533]

[219] delaTorre, G.; Torres, T.; Agullo-Lopez, F. The phthalocyanine approach to second harmonic generation. *Adv. Mater.,* **1997**, *9*, 265-269.
[http://dx.doi.org/10.1002/adma.19970090320]

[220] del Rey, B.; Keller, U.; Torres, T.; Rojo, G.; Agullo-Lopez, F.; Nonell, S.; Marti, C.; Brasselet, S.; Ledoux, I.; Zyss, J. Synthesis and nonlinear optical, photophysical, and electrochemical properties of subphthalocyanines. *J. Am. Chem. Soc.,* **1998**, *120*, 12808-12817.
[http://dx.doi.org/10.1021/ja980508q]

[221] Shirk, J.S.; Pong, R.G.; Flom, S.R.; Heckmann, H.; Hanack, M. Effect of axial substitution on the optical limiting properties of indium phthalocyanines. *J. Phys. Chem. A,* **2000**, *104*, 1438-1449.
[http://dx.doi.org/10.1021/jp993254j]

[222] Maya, E.M.; Snow, A.W.; Shirk, J.S.; Pong, R.G.; Flom, S.R.; Roberts, G.L. Synthesis, aggregation behavior and nonlinear absorption properties of lead phthalocyanines substituted with siloxane chains. *J. Mater. Chem.,* **2003**, *13*, 1603-1613.
[http://dx.doi.org/10.1039/b301566a]

[223] Rosenthal, I. Phthalocyanines as photodynamic sensitizers. *Photochem. Photobiol.,* **1991**, *53*(6), 859-870.
[http://dx.doi.org/10.1111/j.1751-1097.1991.tb09900.x] [PMID: 1886943]

[224] Henderson, B.W.; Dougherty, T.J. How does photodynamic therapy work? *Photochem. Photobiol.,* **1992**, *55*(1), 145-157.

[http://dx.doi.org/10.1111/j.1751-1097.1992.tb04222.x] [PMID: 1603846]

[225] Pass, H.I. Photodynamic therapy in oncology: mechanisms and clinical use. *J. Natl. Cancer Inst.,* **1993**, *85*(6), 443-456.
[http://dx.doi.org/10.1093/jnci/85.6.443] [PMID: 8445672]

[226] Bonnet, R. Photosensitizers of the porphyrin and phthalocyanine series for photodynamic therapy. *Chem. Soc. Rev.,* **1995**, *24*, 19-33.
[http://dx.doi.org/10.1039/cs9952400019]

[227] Ochsner, M. Photophysical and photobiological processes in the photodynamic therapy of tumours. *J. Photochem. Photobiol. B,* **1997**, *39*(1), 1-18.
[http://dx.doi.org/10.1016/S1011-1344(96)07428-3] [PMID: 9210318]

[228] Allen, C.M.; Sharman, W.M.; Van Lier, J.E. Current status of phthalocyanines in the photodynamic therapy of cancer. *J. Porphyr. Phthalocyanines,* **2001**, *5*, 161-169.
[http://dx.doi.org/10.1002/jpp.324]

[229] DeRosa, M.C.; Crutchley, R.J. Photosensitized singlet oxygen and its applications. *Coord. Chem. Rev.,* **2002**, *233*, 351-371.
[http://dx.doi.org/10.1016/S0010-8545(02)00034-6]

[230] Detty, M.R.; Gibson, S.L.; Wagner, S.J. Current clinical and preclinical photosensitizers for use in photodynamic therapy. *J. Med. Chem.,* **2004**, *47*(16), 3897-3915.
[http://dx.doi.org/10.1021/jm040074b] [PMID: 15267226]

[231] Bullous, A.J.; Alonso, C.M.; Boyle, R.W. Photosensitiser-antibody conjugates for photodynamic therapy. *Photochem. Photobiol. Sci.,* **2011**, *10*(5), 721-750.
[http://dx.doi.org/10.1039/c0pp00266f] [PMID: 21409207]

[232] Benhur, E.; Rosenthal, I. The phthalocyanines-a new class of mammalian cells photosensitizers with a potential for cancer phototherapy. *Int. J. Radiat. Biol.,* **1985**, *47*, 145-147.
[http://dx.doi.org/10.1080/09553008514550211]

[233] Chan, W.S.; Svensen, R.; Phillips, D.; Hart, I.R. Cell uptake, distribution and response to aluminium chloro sulphonated phthalocyanine, a potential anti-tumour photosensitizer. *Br. J. Cancer,* **1986**, *53*(2), 255-263.
[http://dx.doi.org/10.1038/bjc.1986.43] [PMID: 2937437]

[234] Reddi, E.; Zhou, C.; Biolo, R.; Menegaldo, E.; Jori, G. Lipossome-administered or LDL-administered Zn(II) pahthalocyanine as a photodynamic agent for tumors. 1. Pharmacokinetic properties and phototherapeutic efficiency. *Br. J. Cancer,* **1990**, *61*, 407-411.
[http://dx.doi.org/10.1038/bjc.1990.89] [PMID: 2328207]

[235] Agarwal, M.L.; Clay, M.E.; Harvey, E.J.; Evans, H.H.; Antunez, A.R.; Oleinick, N.L. Photodynamic therapy induces rapid cell death by apoptosis in L5178Y mouse lymphoma cells. *Cancer Res.,* **1991**, *51*(21), 5993-5996.
[PMID: 1933862]

[236] Gomer, C.J. Preclinical examination of first and second generation photosensitizers used in photodynamic therapy. *Photochem. Photobiol.,* **1991**, *54*(6), 1093-1107.
[http://dx.doi.org/10.1111/j.1751-1097.1991.tb02133.x] [PMID: 1775531]

[237] Oleinick, N.L.; Antunez, A.R.; Clay, M.E.; Rihter, B.D.; Kenney, M.E. New phthalocyanine photosensitizers for photodynamic therapy. *Photochem. Photobiol.,* **1993**, *57*(2), 242-247. [http://dx.doi.org/10.1111/j.1751-1097.1993.tb02282.x] [PMID: 8451285]

[238] Horowitz, B.; Williams, B.; Rywkin, S.; Prince, A.M.; Pascual, D.; Geacintov, N.; Valinsky, J. Inactivation of viruses in blood with aluminum phthalocyanine derivatives. *Transfusion,* **1991**, *31*(2), 102-108. [http://dx.doi.org/10.1046/j.1537-2995.1991.31291142938.x] [PMID: 1847558]

[239] Nyamekye, I. Vascular and other non-tumour applications of photodynamic therapy. *Lasers Med. Sci.,* **1996**, *11*, 3-10. [http://dx.doi.org/10.1007/BF02161287]

[240] Jori, G.; Brown, S.B. Photosensitized inactivation of microorganisms. *Photochem. Photobiol. Sci.,* **2004**, *3*(5), 403-405. [http://dx.doi.org/10.1039/b311904c] [PMID: 15122355]

[241] Maisch, T. Anti-infective agents in medicinal chemistry. *Antiinfect. Agents Med. Chem.,* **2007**, *6*, 145-150. [http://dx.doi.org/10.2174/187152107780361634]

[242] Jori, G. Photodynamic therapy of microbial infections: state of the art and perspectives. *J. Environ. Pathol. Toxicol. Oncol.,* **2006**, *25*(1-2), 505-519. [http://dx.doi.org/10.1615/JEnvironPatholToxicolOncol.v25.i1-2.320] [PMID: 16566738]

[243] Donnelly, R.F.; McCarron, P.A.; Tunney, M.M. Antifungal photodynamic therapy. *Microbiol. Res.,* **2008**, *163*(1), 1-12. [http://dx.doi.org/10.1016/j.micres.2007.08.001] [PMID: 18037279]

[244] Maisch, T. A new strategy to destroy antibiotic resistant microorganisms: antimicrobial photodynamic treatment. *Mini Rev. Med. Chem.,* **2009**, *9*(8), 974-983. [http://dx.doi.org/10.2174/138955709788681582] [PMID: 19601890]

[245] Jori, G.; Fabris, C.; Soncin, M.; Ferro, S.; Coppellotti, O.; Dei, D.; Fantetti, L.; Chiti, G.; Roncucci, G. Photodynamic therapy in the treatment of microbial infections: basic principles and perspective applications. *Lasers Surg. Med.,* **2006**, *38*(5), 468-481. [http://dx.doi.org/10.1002/lsm.20361] [PMID: 16788934]

[246] Calin, M.A.; Parasca, S.V. Light sources for photodynamic inactivation of bacteria. *Lasers Med. Sci.,* **2009**, *24*(3), 453-460. [http://dx.doi.org/10.1007/s10103-008-0588-5] [PMID: 18622686]

[247] Minnock, A.; Vernon, D.I.; Schofield, J.; Griffiths, J.; Parish, J.H.; Brown, S.B. Mechanism of uptake of a cationic water-soluble pyridinium zinc phthalocyanine across the outer membrane of Escherichia coli. *Antimicrob. Agents Chemother.,* **2000**, *44*(3), 522-527. [http://dx.doi.org/10.1128/AAC.44.3.522-527.2000] [PMID: 10681312]

[248] Berg, K.; Bommer, J.C.; Moan, J. Evaluation of sulfonated aluminum phthalocyanines for use in photochemotherapy. A study on the relative efficiencies of photoinactivation. *Photochem. Photobiol.,* **1989**, *49*(5), 587-594. [http://dx.doi.org/10.1111/j.1751-1097.1989.tb08428.x] [PMID: 2755994]

[249] Rywkin, S.; Ben-Hur, E.; Malik, Z.; Prince, A.M.; Li, Y.S.; Kenney, M.E.; Oleinick, N.L.; Horowitz, B. New phthalocyanines for photodynamic virus inactivation in red blood cell concentrates. *Photochem. Photobiol.,* **1994**, *60*(2), 165-170.
[http://dx.doi.org/10.1111/j.1751-1097.1994.tb05085.x] [PMID: 7938215]

[250] Minnock, A.; Vernon, D.I.; Schofield, J.; Griffiths, J.; Parish, J.H.; Brown, S.T. Photoinactivation of bacteria. Use of a cationic water-soluble zinc phthalocyanine to photoinactivate both gram-negative and gram-positive bacteria. *J. Photochem. Photobiol. B,* **1996**, *32*(3), 159-164.
[http://dx.doi.org/10.1016/1011-1344(95)07148-2] [PMID: 8622179]

[251] Soncin, M.; Fabris, C.; Busetti, A.; Dei, D.; Nistri, D.; Roncucci, G.; Jori, G. Approaches to selectivity in the Zn(II)-phthalocyanine-photosensitized inactivation of wild-type and antibiotic-resistant Staphylococcus aureus. *Photochem. Photobiol. Sci.,* **2002**, *1*(10), 815-819.
[http://dx.doi.org/10.1039/b206554a] [PMID: 12656484]

[252] Mantareva, V.; Kussovski, V.; Angelov, I.; Wöhrle, D.; Dimitrov, R.; Popova, E.; Dimitrov, S. Non-aggregated Ga(III)-phthalocyanines in the photodynamic inactivation of planktonic and biofilm cultures of pathogenic microorganisms. *Photochem. Photobiol.Sci.,* **2011**, *10*(1), 91-102.
[http://dx.doi.org/10.1039/B9PP00154A] [PMID: 21031201]

[253] Giuliani, F.; Martinelli, M.; Cocchi, A.; Arbia, D.; Fantetti, L.; Roncucci, G. *In vitro* resistance selection studies of RLP068/Cl, a new Zn(II) phthalocyanine suitable for antimicrobial photodynamic therapy. *Antimicrob. Agents Chemother.,* **2010**, *54*(2), 637-642.
[http://dx.doi.org/10.1128/AAC.00603-09] [PMID: 20008782]

[254] Ragàs, X.; Agut, M.; Nonell, S. Singlet oxygen in *Escherichia coli*: New insights for antimicrobial photodynamic therapy. *Free Radic. Biol. Med.,* **2010**, *49*(5), 770-776.
[http://dx.doi.org/10.1016/j.freeradbiomed.2010.05.027] [PMID: 20638940]

[255] Mantareva, V.; Angelov, I.; Kussovski, V.; Dimitrov, R.; Lapok, L.; Wöhrle, D. Photodynamic efficacy of water-soluble Si(IV) and Ge(IV) phthalocyanines towards *Candida albicans* planktonic and biofilm cultures. *Eur. J. Med. Chem.,* **2011**, *46*(9), 4430-4440.
[http://dx.doi.org/10.1016/j.ejmech.2011.07.015] [PMID: 21816518]

[256] Ke, M-R.; Eastel, J.M.; Ngai, K.L.; Cheung, Y-Y.; Chan, P.K.; Hui, M.; Ng, D.K.; Lo, P-C. Oligolysine-conjugated zinc(II) phthalocyanines as efficient photosensitizers for antimicrobial photodynamic therapy. *Chem. Asian J.,* **2014**, *9*(7), 1868-1875.
[http://dx.doi.org/10.1002/asia.201402025] [PMID: 24799418]

[257] Sekkat, N.; van den Bergh, H.; Nyokong, T.; Lange, N. Like a bolt from the blue: phthalocyanines in biomedical optics. *Molecules,* **2012**, *17*(1), 98-144.
[PMID: 22198535]

[258] Josefsen, L.B.; Boyle, R.W. Unique diagnostic and therapeutic roles of porphyrins and phthalocyanines in photodynamic therapy, imaging and theranostics. *Theranostics,* **2012**, *2*(9), 916-966.
[http://dx.doi.org/10.7150/thno.4571] [PMID: 23082103]

[259] Calvete, M.J.; Simões, A.V.; Henriques, C.A.; Pinto, S.M.; Pereira, M.M. Tetrapyrrolic macrocycles: potentialities in medical imaging technologies. *Curr. Org. Synth.,* **2014**, *11*, 127-140.
[http://dx.doi.org/10.2174/15701794113106660090]

[260] Trivedi, E.R.; Vesper, B.J.; Weitman, H.; Ehrenberg, B.; Barrett, A.G.; Radosevich, J.A.; Hoffman, B.M. Chiral bis-acetal porphyrazines as near-infrared optical agents for detection and treatment of cancer. *Photochem. Photobiol.,* **2010**, *86*(2), 410-417. [http://dx.doi.org/10.1111/j.1751-1097.2009.00681.x] [PMID: 20074089]

[261] Costa, S.M.; Andrade, S.M.; Togashi, D.M.; Paulo, P.M.; Laia, C.A.; Viseu, M.I.; Silva, A.M. Optical spectroscopy and photochemistry of porphyrins and phthalocyanines. *J. Porphyr. Phthalocyanines,* **2009**, *13*, 509-517. [http://dx.doi.org/10.1142/S1088424609000589]

[262] Basu, B.; Satapathy, S.; Bhatnagar, A.K. merox and related metal phthalocyanine catalyzed oxidation processes. *Catal. Rev.,* **1993**, *35*, 571-609. [http://dx.doi.org/10.1080/01614949308013917]

[263] Nyokong, T. Effects of substituents on the photochemical and photophysical properties of main group metal phthalocyanines. *Coord. Chem. Rev.,* **2007**, *251*, 1707-1722. [http://dx.doi.org/10.1016/j.ccr.2006.11.011]

[264] Loas, A.; Gerdes, R.; Zhang, Y.; Gorun, S.M. Broadening the reactivity spectrum of a phthalocyanine catalyst while suppressing its nucleophilic, electrophilic and radical degradation pathways. *Dalton Trans.,* **2011**, *40*(19), 5162-5165. [http://dx.doi.org/10.1039/c1dt10458f] [PMID: 21468396]

[265] Christendat, D.; David, M-A.; Morin, S.; Lever, A.B.; Kadish, K.M.; Shao, J. Synthesis and characterization of highly soluble hexadecachloro- and hexadecafluorophthalocyanine ruthenium(II) complexes. *J. Porphyr. Phthalocyanines,* **2005**, *9*, 626-636. [http://dx.doi.org/10.1142/S1088424605000733]

[266] Gabrielov, A.G.; Balkus, K.J. Jr.; Bell, S.L.; Bedioui, F.; Devynck, J. Faujasite-type zeolites modified with iron perfluorophthalocyanines: Synthesis and characterization. *Microporous Mater.,* **1994**, *2*, 119-126. [http://dx.doi.org/10.1016/0927-6513(93)E0044-H]

[267] Balkus, K.J. Jr.; Gabrielov, A.G.; Bell, S.L.; Bedioui, F.; Roué, L.; Devynck, J. Zeolite encapsulated cobalt(II) and copper(II) perfluorophthalocyanines. Synthesis and characterization. *Inorg. Chem.,* **1994**, *33*, 67-72. [http://dx.doi.org/10.1021/ic00079a013]

[268] Bench, B.A.; Beveridge, A.; Sharman, W.M.; Diebold, G.J.; van Lier, J.E.; Gorun, S.M. Introduction of bulky perfluoroalkyl groups at the periphery of zinc perfluorophthalocyanine: chemical, structural, electronic, and preliminary photophysical and biological effects. *Angew. Chem. Int. Ed. Engl.,* **2002**, *41*(5), 747-750. [http://dx.doi.org/10.1002/1521-3773(20020301)41:5<747::AID-ANIE747>3.0.CO;2-J] [PMID: 12491323]

[269] Bench, B.A.; Brennessel, W.W.; Lee, H-J.; Gorun, S.M. Synthesis and structure of a biconcave cobalt perfluorophthalocyanine and its catalysis of novel oxidative carbon-phosphorus bonds formation by using air. *Angew. Chem. Int. Ed. Engl.,* **2002**, *41*(5), 750-754. [http://dx.doi.org/10.1002/1521-3773(20020301)41:5<750::AID-ANIE750>3.0.CO;2-6] [PMID: 12491324]

[270] Alvaro, M.; Carbonell, E.; Esplá, M.; Garcia, H. Iron phthalocyanine supported on silica or encapsulated inside zeolite Y as solid photocatalysts for the degradation of phenols and sulfur heterocycles. *Appl. Catal. B,* **2005**, *57*, 37-42.
[http://dx.doi.org/10.1016/j.apcatb.2004.10.003]

[271] Silva, M.; Calvete, M.J.; Gonçalves, N.P.; Burrows, H.D.; Sarakha, M.; Fernandes, A.; Ribeiro, M.F.; Azenha, M.E.; Pereira, M.M. Zinc(II) phthalocyanines immobilized in mesoporous silica Al-MCM-41 and their applications in photocatalytic degradation of pesticides. *J. Hazard. Mater.,* **2012**, *233-234*, 79-88.
[http://dx.doi.org/10.1016/j.jhazmat.2012.06.059] [PMID: 22835771]

[272] Zalomaeva, O.V.; Sorokin, A.B.; Kholdeeva, O.A. Clean catalytic oxidation of 8-hydroxyquinoline to quinoline-5,8-dione with tBuOOH in the presence of covalently bound FePcS–SiO2 catalysts. *Green Chem.,* **2010**, *12*, 1076-1082.
[http://dx.doi.org/10.1039/b923674k]

[273] Sirotin, S.V.; Tolbin, A.Y.; Moskovskaya, I.F.; Abramchuk, S.S.; Tomilova, L.G.; Romanovsky, B.V. Heterogenized Fe(III) phthalocyanine: Synthesis, characterization and application in liquid-phase oxidation of phenol. *J. Mol. Catal. Chem.,* **2010**, *319*, 39-45.
[http://dx.doi.org/10.1016/j.molcata.2009.11.017]

[274] DeOliveira, E.; Neri, C.R.; Ribeiro, A.O.; Garcia, V.S.; Costa, L.L.; Moura, A.O.; Prado, A.G.; Serra, O.A.; Iamamoto, Y. Hexagonal mesoporous silica modified with copper phthalocyanine as a photocatalyst for pesticide 2,4-dichlorophenoxiacetic acid degradation. *J. Colloid Interface Sci.,* **2008**, *323*(1), 98-104.
[http://dx.doi.org/10.1016/j.jcis.2008.04.025] [PMID: 18455176]

[275] Forrest, S.R. Ultrathin organic films grown by organic molecular beam deposition and related techniques. *Chem. Rev.,* **1997**, *97*(6), 1793-1896.
[http://dx.doi.org/10.1021/cr941014o] [PMID: 11848893]

[276] Fabian, J.; Nakazumi, H.; Matsuoka, M. Near-infrared absorbing dyes. *Chem. Rev.,* **1992**, *92*, 1197-1226.
[http://dx.doi.org/10.1021/cr00014a003]

[277] Armstrong, N.R. Phthalocyanines and porphyrins as materials. *J. Porphyr. Phthalocyanines,* **2000**, *4*, 414-417.
[http://dx.doi.org/10.1002/(SICI)1099-1409(200006/07)4:4<414::AID-JPP247>3.0.CO;2-B]

[278] Alexander, A.E. 381. Monolayers of porphyrins and related compounds. *J. Chem. Soc.,* **1937**, 1813-1816.
[http://dx.doi.org/10.1039/jr9370001813]

[279] Yamamoto, H.; Sugyama, T.; Tanaka, M. Electrochromism of metal-free phthalocyanine Langmuir-Blodgett films. *Jpn. J. Appl. Phys.,* **1985**, *24*, L305-L307.
[http://dx.doi.org/10.1143/JJAP.24.L305]

[280] George, R.D.; McMillan, P.F.; Burrows, V.A.; Hervig, R. Structure and composition of thin films of peripherally unsubstituted phthalocyanine deposited using the Langmuir-Blodgett technique. *Thin Solid Films,* **1991**, *203*, 303-316.
[http://dx.doi.org/10.1016/0040-6090(91)90138-N]

[281] Kanezaki, E. Electronic states of zinc phthalocyanine in various films; its association profile. *J. Coord. Chem.,* **1988**, *18*, 113-120.
[http://dx.doi.org/10.1080/00958978808080695]

[282] Yan, W.; Zhou, Y.; Wang, X.; Chen, W.; Xi, S. Determination of the molecular orientation of a phthalocyanine derivative in a Langmuir–Blodgett film by polarized UV–VIS spectra. *J. Chem. Soc. Chem. Commun.,* **1992**, 873-875.
[http://dx.doi.org/10.1039/c39920000873]

[283] Nichog, K.; Waragai, K.; Taomoto, A.; Saito, Y.; Asakawa, S. Lead phthalocyanine Langmuir-Blodgett films. *Thin Solid Films,* **1989**, *179*, 297-301.
[http://dx.doi.org/10.1016/0040-6090(89)90198-3]

[284] Ogawa, K.; Yonehara, H.; Shoji, T.; Kinashita, S-I.; Maekawa, E.; Nakahara, H.; Fukuda, K. In-plane anisotropy in Langmuir-Blodgett films of a copper phthalocyanine derivative. *Thin Solid Films,* **1989**, *178*, 439-443.
[http://dx.doi.org/10.1016/0040-6090(89)90335-0]

[285] Itoh, H.; Koyama, T.; Hanabusa, K.; Masuda, E.; Shirai, H.; Hayakawa, T. Functional metallomacrocycle derivatives and their polymers. Part 23. Biaxially orientated Langmuir–Blodgett films of substituted metallophthalocyanines. *J. Chem. Soc., Dalton Trans.,* **1989**, 1543-1546.
[http://dx.doi.org/10.1039/dt9890001543]

[286] Fujiki, M.; Tabei, H.; Imamura, S. Self-assembling features of soluble nickel phthalocyanines. *J. Phys. Chem.,* **1988**, *92*, 1281-1285.
[http://dx.doi.org/10.1021/j100316a053]

[287] Fujiki, M.; Tabei, H.; Kunhara, T. In-plane dichroisms of phthalocyanine Langmuir-Blodgett films. *Langmuir,* **1988**, *4*, 1123-1128.
[http://dx.doi.org/10.1021/la00083a011]

[288] Kalina, D.W.; Crane, S.W. Langmuir-Blodgett films of soluble copper octa(dodecoxymethyl) phthalocyanine. *Thin Solid Films,* **1985**, *134*, 109-119.
[http://dx.doi.org/10.1016/0040-6090(85)90123-3]

[289] Nakahara, H.; Fukuda, K.; Kitahara, K.; Nishi, H. Langmuir-Blodgett films of octa-alkyl phthalocyanines. *Thin Solid Films,* **1989**, *178*, 361.
[http://dx.doi.org/10.1016/0040-6090(89)90323-4]

[290] Cook, M.J.; Dunn, A.J.; Daniel, M.F.; Hart, R.C.; Richardson, R.M.; Roser, S.J. Fabrication of ordered Langmuir-Blodgett multilayers of octa-n-alkoxy phthalocyanines. *Thin Solid Films,* **1988**, *159*, 395-404.
[http://dx.doi.org/10.1016/0040-6090(88)90653-0]

[291] Cook, M.J.; McMurdo, J.; Miles, D.A.; Poynter, R.H.; Simmons, J.M.; Haslam, S.D.; Richardson, R.M.; Welford, K. Monolayer behaviour and Langmuir–Blodgett film properties of some amphiphilic phthalocyanines: factors influencing molecular organisation within the film assembly. *J. Mater. Chem.,* **1994**, *4*, 1205-1213.
[http://dx.doi.org/10.1039/jm9940401205]

[292] McKeown, N.B.; Cook, M.J.; Chambrier, I. Synthesis and characterisation of some

1,4,8,11,15,18,22,25-octa-alkyl- and 1,4,8,11,15,18-hexa-alkyl-22,25-bis(carboxypropyl) phthalocyanines. *J. Chem.* Soc. *Perkin Trans,* **1990**, *1*, 1169-1177. [http://dx.doi.org/10.1039/p19900001169]

[293] McKeown, N.B.; Cook, M.J.; Thomson, A.J.; Harrison, K.J.; Daniel, M.F.; Richardson, R.M.; Roser, S.J. New asymmetric substitution of phthalocyanines: Derivatives designed for deposition as Langmuir-Blodgett films. *Thin Solid Films,* **1988**, *159*, 469-478. [http://dx.doi.org/10.1016/0040-6090(88)90662-1]

[294] Cook, M.J.; McKeown, N.B.; Simmons, J.M.; Thomson, A.J.; Daniel, M.F.; Harmon, K.J.; Richardson, R.M.; Roser, S.J. Spectroscopic and X-ray diffraction study of Langmuir–Blodgett films of some 1,4,8,11,15,18-hexaalkyl-22,25-bis(carboxypropyl)phthalocyanines. *J. Mater. Chem.,* **1991**, *1*, 121-127. [http://dx.doi.org/10.1039/jm9910100121]

[295] Cook, M.J. 1,4,8,11,15,18,22,25-octasubstituted phthalocyanines: the contrasting effects of alkyl and alkoxy substituents on molecular self-assembly. *J. Mater. Sci. Mater. Electron.,* **1994**, *5*, 117-128.

[296] Cole, A.; McIlroy, R.J.; Thorpe, S.C.; Cook, M.J.; McMurdo, J.; Ray, A.K. Substituted phthalocyanine gas sensors. *Sens. Actuators B Chem.,* **1993**, *13-14*, 416-419. [http://dx.doi.org/10.1016/0925-4005(93)85415-7]

[297] Crouch, D.; Thorpe, S.C.; Cook, M.J.; Chambrier, I.; Ray, A.K. Langmuir-Blodgett films of an asymmetrically substitute phthalocyanine: Improved gas-sensing properties. *Sens. Actuators B Chem.,* **1994**, *19*, 411-414. [http://dx.doi.org/10.1016/0925-4005(93)01017-X]

[298] Ding, X.; Shen, S.; Zhou, Q.; Xu, H. The synthesis of asymmetrically substituted amphiphilic phthalocyanines and their gas-sensing properties. *Dyes Pigm.,* **1999**, *40*, 187-191. [http://dx.doi.org/10.1016/S0143-7208(98)00049-7]

[299] Ding, X.; Xu, H. Gas-sensing properties of asymmetrically substituted amphiphilic phthalocyanines. *Sens. Actuators B Chem.,* **2000**, *65*, 108-110. [http://dx.doi.org/10.1016/S0925-4005(99)00409-8]

[300] Ding, X.; Xu, H.; Zhang, L.; Jiang, D.; Lu, A. The anisotropic gas response behavior of LB films of an amphiphilic asymmetrically substituted phthalocyanine. *Thin Solid Films,* **1999**, *340*, 271-273. [http://dx.doi.org/10.1016/S0040-6090(98)01371-6]

[301] Ding, X.; Xu, H.; Zhang, L.; Jiang, D.; Lu, A. A novel amphiphilic zinc phthalocyanine LB films as gas sensor material and its interaction with NH3. *Mol. Cryst. Liq. Cryst. (Phila. Pa.),* **1999**, *337*, 481-484. [http://dx.doi.org/10.1080/10587259908023482]

[302] Beall, L.S.; Mani, N.S.; White, A.J.; Williams, D.J.; Barrett, A.G.; Hoffman, B.M. Porphyrazines and norphthalocyanines bearing nitrogen donor pockets: metal sensor properties. *J. Org. Chem.,* **1998**, *63*(17), 5806-5817. [http://dx.doi.org/10.1021/jo9802574] [PMID: 11672181]

[303] Ehrlich, L.A.; Skrdla, P.J.; Jarrell, W.K.; Sibert, J.W.; Armstrong, N.R.; Saavedra, S.S.; Barrett, A.G.; Hoffman, B.M. Preparation of polyetherol-appended sulfur porphyrazines and investigations of peripheral metal ion binding in polar solvents. *Inorg. Chem.,* **2000**, *39*(18), 3963-3969.

[http://dx.doi.org/10.1021/ic991033m] [PMID: 11198848]

[304] Yarasir, M.N.; Kandaz, M.; Koca, A.; Salih, B. Polytopic cation receptor functional phthalocyanines: Synthesis, characterization, electrochemistry and metal ion binding. *Polyhedron,* **2007**, *26*, 1139-1147.
[http://dx.doi.org/10.1016/j.poly.2006.10.002]

[305] Chaure, S.; Paul, D.; Vadagma, P.; Ray, A.K. Spectroscopic investigation of sulfonate phthalocyanine to probe enzyme reactions for heavy metals detection. *J. Hazard. Mater.,* **2010**, *173*(1-3), 253-257.
[http://dx.doi.org/10.1016/j.jhazmat.2009.08.077] [PMID: 19740602]

[306] Hill, J.P.; Subbaiyan, N.K.; D'Souza, F.; Xie, Y.; Sahu, S.; Sanchez-Ballester, N.M.; Richards, G.J.; Mori, T.; Ariga, K. Antioxidant-substituted tetrapyrazinoporphyrazine as a fluorescent sensor for basic anions. *Chem. Commun. (Camb.),* **2012**, *48*(33), 3951-3953.
[http://dx.doi.org/10.1039/c2cc30712j] [PMID: 22422058]

[307] Farinha, A.S.; Calvete, M.J.; Paz, F.A.; Tomé, A.C.; Cavaleiro, J.A.; Sessler, J.L.; Tomé, J.P. Octatosylaminophthalocyanine: a reusable chromogenic anion chemosensor. *Sens. Actuators B Chem.,* **2014**, *201*, 387-394.
[http://dx.doi.org/10.1016/j.snb.2014.03.115]

[308] Ros-Lis, J.V.; Martínez-Máñez, R.; Soto, J. Subphthalocyanines as fluoro-chromogenic probes for anions and their application to the highly selective and sensitive cyanide detection. *Chem. Commun. (Camb.),* **2005**, (42), 5260-5262.
[http://dx.doi.org/10.1039/b510710e] [PMID: 16244721]

[309] Palomares, E.; Martínez-Díaz, M.V.; Torres, T.; Coronado, E. A highly sensitive hybrid colorimetric and fluorometric molecular probe for cyanide sensing based on a subphthalocyanine dye. *Adv. Funct. Mater.,* **2006**, *16*, 1166-1170.
[http://dx.doi.org/10.1002/adfm.200500517]

[310] Malliaras, G.G.; Scott, J.C. The roles of injection and mobility in organic light emitting diodes. *J. Appl. Phys.,* **1998**, *83*, 5399-5403.
[http://dx.doi.org/10.1063/1.367369]

[311] Carter, S.A.; Angelopoulos, M.; Karg, S.; Brock, P.J.; Scott, J.C. Polymeric anodes for improved polymer light-emitting diode performance. *Appl. Phys. Lett.,* **1997**, *70*, 2067-2069.
[http://dx.doi.org/10.1063/1.118953]

[312] Karan, S.; Mallik, B. Templating effects and optical characterization of copper (II) phthalocyanine nanocrystallites thin film: nanoparticles, nanoflowers, nanocabbages, and nanoribbons. *J. Phys. Chem. C,* **2007**, *111*, 7352-7365.
[http://dx.doi.org/10.1021/jp070302o]

[313] Forsythe, E.W.; Abkowitz, M.A.; Gao, Y.L. Tuning the carrier injection efficiency for organic light-emitting diodes. *J. Phys. Chem. B,* **2000**, *104*, 3948-3952.
[http://dx.doi.org/10.1021/jp993793o]

[314] Tang, Q.X.; Li, H.X.; Song, Y.B.; Xu, W.; Hu, W.P.; Jiang, L.; Liu, Y.Q.; Wang, X.K.; Zhu, D.B. In situ patterning of organic single-crystalline nanoribbons on a SiO2 surface for the fabrication of various architectures and high-quality transistors. *Adv. Mater.,* **2006**, *18*, 3010-3014.
[http://dx.doi.org/10.1002/adma.200600542]

[315] Zeis, R.; Siegrist, T.; Kloc, C. Single-crystal field-effect transistors based on copper phthalocyanine. *Appl. Phys. Lett.,* **2005**, *86*, 022103.
[http://dx.doi.org/10.1063/1.1849438]

[316] de Boer, R.W.; Stassen, A.F.; Craciun, M.F.; Mulder, C.L.; Molinari, A.; Rogge, S.; Morpurgo, A.F. Ambipolar Cu- and Fe-phthalocyanine single-crystal field-effect transistors. *Appl. Phys. Lett.,* **2005**, *86*, 262109.
[http://dx.doi.org/10.1063/1.1984093]

[317] Tang, Q.; Li, H.; Liu, Y.; Hu, W. High-performance air-stable n-type transistors with an asymmetrical device configuration based on organic single-crystalline submicrometer/nanometer ribbons. *J. Am. Chem. Soc.,* **2006**, *128*(45), 14634-14639.
[http://dx.doi.org/10.1021/ja064476f] [PMID: 17090049]

[318] Vorländer, D. Einfluß der molekularen Gestalt auf den krystallinisch-flüssigen Zustand. *Ber. Dtsch. Chem. Ges.,* **1907**, *40*, 1970-1972.
[http://dx.doi.org/10.1002/cber.190704002101]

[319] Chandrasekhar, S.; Sadashiva, B.K.; Suresh, K.A. Liquid crystals of disc-like molecules. *Pramana,* **1977**, *9*, 471-480.
[http://dx.doi.org/10.1007/BF02846252]

[320] Engel, M.K.; Bassoul, P.; Bosio, L.; Lehmann, H.; Hanack, M.; Simon, J. Mesomorphic molecular materials. Influence of chain length on the structural properties of octa-alkyl substituted phthalocyanines. *Liq. Cryst.,* **1993**, *15*, 709-722.
[http://dx.doi.org/10.1080/02678299308036489]

[321] Cook, M.J. Properties of some alkyl substituted phthalocyanines and related macrocycles. *Chem. Rec.,* **2002**, *2*(4), 225-236.
[http://dx.doi.org/10.1002/tcr.10028] [PMID: 12203905]

[322] Eichhorn, H.; Bruce, D.W.; Wöhrle, D. Amphitropic mesomorphic phthalocyanines-a new approach to highly ordered layers. *Adv. Mater.,* **1998**, *10*, 419-422.
[http://dx.doi.org/10.1002/(SICI)1521-4095(199803)10:5<419::AID-ADMA419>3.0.CO;2-U]

[323] Schouten, P.G.; van der Pol, J.F.; Zwikker, J.W.; Drenth, W.; Picken, S.J. Peripherally octasubstituted phthalocyanines with branched alkoxy chains. *Mol. Cryst. Liq. Cryst. (Phila. Pa.),* **1991**, *195*, 291-305.
[http://dx.doi.org/10.1080/00268949108030908]

[324] Engelkamp, H.; van Nostrum, C.F.; Picken, S.J.; Nolte, R.J. Shish kebab-like chirality. *Chem. Commun. (Camb.),* **1998**, 979-980.
[http://dx.doi.org/10.1039/a801246f]

[325] Saettel, N.; Katsonis, N.; Marchenko, A.; Teulade-Fichou, M-P.; Fichou, D. Triazatrinaphthylene, a three-fold symmetry planar conjugated system with two-dimensional self-assembly properties. *J. Mater. Chem.,* **2005**, *15*, 3175-3180.
[http://dx.doi.org/10.1039/b504704h]

[326] Tang, C.W. Two-layer organic photovoltaic cell. *Appl. Phys. Lett.,* **1986**, *48*, 183-185.
[http://dx.doi.org/10.1063/1.96937]

[327] Erten, S.; Meghdadi, F.; Gunes, S.; Koeppe, R.; Sariciftci, N.S.; Icli, S. Donor-acceptor heterojunction solar cells based on perylene dimide and perylene bisbenzimidazole. *Eur. Phys. J. Appl. Phys.,* **2006**, *36*, 225-229.
[http://dx.doi.org/10.1051/epjap:2006146]

[328] Susarova, D.K.; Troshin, P.A.; Hoglinger, D.; Koeppe, R.; Babenko, S.D.; Lyubovskaya, R.N.; Razumov, V.F.; Sariciftci, N.S. Donor–acceptor complex formation in evaporated small molecular organic photovoltaic cells. *Sol. Energy Mater. Sol. Cells,* **2010**, *94*, 803-811.
[http://dx.doi.org/10.1016/j.solmat.2009.12.027]

[329] Shin, W.S.; Jeong, H.H.; Kim, M.K.; Jin, S.H.; Kim, M.R.; Lee, J.K.; Lee, J.W.; Gal, Y.S. Effects of functional groups at perylene diimide derivatives on organic photovoltaic device application. *J. Mater. Chem.,* **2006**, *16*, 384-390.
[http://dx.doi.org/10.1039/B512983D]

[330] Lu, Y.; Chen, H.; Qiao, F.; Hu, X.; Ng, S-C. Synthesis, characterization and photovoltaic application of N,N′-bisfluorenyl substituted perylene bisimide. *Sol. Energy Mater. Sol. Cells,* **2010**, *94*, 2036-2041.
[http://dx.doi.org/10.1016/j.solmat.2010.06.009]

[331] Fischer, M.K.; López-Duarte, I.; Wienk, M.M.; Martínez-Díaz, M.V.; Janssen, R.A.; Bäuerle, P.; Torres, T. Functionalized dendritic oligothiophenes: ruthenium phthalocyanine complexes and their application in bulk heterojunction solar cells. *J. Am. Chem. Soc.,* **2009**, *131*(24), 8669-8676.
[http://dx.doi.org/10.1021/ja901537d] [PMID: 19530734]

[332] Balraju, P.; Kumar, M.; Roy, M.S.; Sharma, G.D. Dye sensitized solar cells (DSSCs) based on modified iron phthalocyanine nanostructured TiO2 electrode and PEDOT:PSS counter electrode. *Synth. Met.,* **2009**, *159*, 1325-1331.
[http://dx.doi.org/10.1016/j.synthmet.2009.03.001]

[333] Yum, J.H.; Jang, S.R.; Humphry-Baker, R.; Grätzel, M.; Cid, J.J.; Torres, T.; Nazeeruddin, M.K. Effect of coadsorbent on the photovoltaic performance of zinc pthalocyanine-sensitized solar cells. *Langmuir,* **2008**, *24*(10), 5636-5640.
[http://dx.doi.org/10.1021/la800087q] [PMID: 18435553]

[334] Reddy, P.Y.; Giribabu, L.; Lyness, C.; Snaith, H.J.; Vijaykumar, C.; Chandrasekharam, M.; Lakshmikantam, M.; Yum, J.H.; Kalyanasundaram, K.; Grätzel, M.; Nazeeruddin, M.K. Efficient sensitization of nanocrystalline TiO2 films by a near-IR-absorbing unsymmetrical zinc phthalocyanine. *Angew. Chem. Int. Ed. Engl.,* **2007**, *46*(3), 373-376.
[http://dx.doi.org/10.1002/anie.200603098] [PMID: 17146815]

[335] Mori, S.; Nagata, M.; Nakahata, Y.; Yasuta, K.; Goto, R.; Kimura, M.; Taya, M. Enhancement of incident photon-to-current conversion efficiency for phthalocyanine-sensitized solar cells by 3D molecular structuralization. *J. Am. Chem. Soc.,* **2010**, *132*(12), 4054-4055.
[http://dx.doi.org/10.1021/ja9109677] [PMID: 20205478]

[336] Kimura, M.; Nomoto, H.; Masaki, N.; Mori, S. Carboxyethynyl anchoring ligands: a means to improving the efficiency of phthalocyanine-sensitized solar cells. *Angew. Chem. Int. Ed.,* **2012**, *51*, 4371-4374.
[http://dx.doi.org/10.1002/anie.201108610]

[337] Kimura, M.; Nomoto, H.; Suzuki, H.; Ikeuchi, T.; Matsuzaki, H.; Murakami, T.N.; Furube, A.; Masaki, N.; Griffith, M.J.; Mori, S. Molecular design rule of phthalocyanine dyes for highly efficient near-IR performance in dye-sensitized solar cells. *Chemistry,* **2013**, *19*(23), 7496-7502. [http://dx.doi.org/10.1002/chem.201300716] [PMID: 23576330]

[338] Listorti, A.; López-Duarte, I.; Martínez-Díaz, M.V.; Torres, T.; DosSantos, T.; Barnes, P.R.; Durrant, J.R. Zn(II) *versus* Ru(II) phthalocyanine-sensitised solar cells. A comparison between singlet and triplet electron injectors. *Energy Environ. Sci.,* **2010**, *3*, 1573-1579. [http://dx.doi.org/10.1039/c0ee00083c]

[339] Morandeira, A.; López-Duarte, I.; O'Regan, B.; Martínez-Díaz, M.V.; Forneli, A.; Palomares, E.; Torres, T.; Durrant, J.R. Ru(II)-phthalocyanine sensitized solar cells: the influence of co-adsorbents upon interfacial electron transfer kinetics. *J. Mater. Chem.,* **2009**, *19*, 5016-5026. [http://dx.doi.org/10.1039/b904179f]

[340] Palomares, E.; Martínez-Diaz, M.V.; Haque, S.A.; Torres, T.; Durrant, J.R. State selective electron injection in non-aggregated titanium phthalocyanine sensitised nanocrystalline TiO2 films. *Chem. Commun. (Camb.),* **2004**, (18), 2112-2113. [http://dx.doi.org/10.1039/b407860h] [PMID: 15368000]

[341] Rodríguez-Morgade, M.S.; Pellejà, L.; Torres, T.; Palomares, E. Ti(IV) phthalocyanines for dye sensitized solar cells. *J. Porphyr. Phthalocyanines,* **2013**, *17*, 814-820. [http://dx.doi.org/10.1142/S1088424613500454]

[342] Martín-Gomis, L.; Barea, E.M.; Fernández-Lázaro, F.; Bisquert, J.; Sastre-Santos, A. Dye sensitized solar cells using non-aggregated silicon phthalocyanines. *J. Porphyr. Phthalocyanines,* **2011**, *15*, 1004-1010. [http://dx.doi.org/10.1142/S1088424611003914]

[343] Lim, B.; Margulis, G.Y.; Yum, J-H.; Unger, E.L.; Hardin, B.E.; Grätzel, M.; McGehee, M.D.; Sellinger, A. Silicon-naphthalo/phthalocyanine-hybrid sensitizer for efficient red response in dye-sensitized solar cells. *Org. Lett.,* **2013**, *15*(4), 784-787. [http://dx.doi.org/10.1021/ol303436q] [PMID: 23384416]

[344] *Nonlinear Optical Properties of Organic Molecules and Crystals*; Chemla, D.S.; Zyss, J., Eds.; Academic Press: Orlando, FL, **1987**, Vol. 1 and 2, .

[345] *Introduction to Nonlinear Effects in Molecules and Polymers*; Prasad, P.N.; Williams, D.; Wiley, J., Eds.; John Wiley: New York, **1991**.

[346] *Nonlinear Optics of Organic Molecules and Polymers*; Nalwa, H.S.; Miyata, S.; Press, C.R., Eds.; CRC Press: Boca Raton, FL, **1997**.

[347] Shirk, J.S.; Lindle, J.R.; Bartoli, F.J.; Hoffman, C.A.; Kafafi, Z.H.; Snow, A. Off,resonant third,order optical nonlinearities of metal,substituted phthalocyanines. *Appl. Phys. Lett.,* **1989**, *55*, 1287-1288. [http://dx.doi.org/10.1063/1.101634]

[348] Sanghadasa, M.; Wu, B.; Clark, R.D.; Guo, H.; Penn, B.G. Third-order nonlinear optical properties of metal phthalocyanines with butoxy and decyloxy substituents. *Proc. SPIE, Int. Soc. Opt. Eng.,* **1997**. [http://dx.doi.org/10.1117/12.284248]

[349] Shirk, J.S.; Pong, R.G.; Bartoli, F.J.; Snow, A.W. Optical limiter using a lead phthalocyanine. *Appl.*

Phys. Lett., **1993**, *63*, 1880-1882.
[http://dx.doi.org/10.1063/1.110635]

[350] Shirk, J.S.; Pong, R.G.; Flom, S.R.; Bartoli, F.J.; Boyle, M.E.; Snow, A.W. Lead phthalocyanine reverse saturable absorption optical limiters. *Pure Appl. Opt. J. Eur. Opt. Soc. Part A,* **1996**, *5*, 701-709.
[http://dx.doi.org/10.1088/0963-9659/5/5/023]

[351] Nalwa, H.S.; Kobayashi, S. Large third-order optical non-linearities of spin-cast thin films of novel metallo-naphthalocyanines. *J. Porphyr. Phthalocyanines,* **1998**, *2*, 21-30.
[http://dx.doi.org/10.1002/(SICI)1099-1409(199801/02)2:1<21::AID-JPP45>3.0.CO;2-M]

[352] Shirk, J.S.; Lindle, J.R.; Bartoli, F.J.; Boyle, M.E. Third-order optical nonlinearities of bis(phthalocyanines). *J. Phys. Chem.,* **1992**, *96*, 5847-5852.
[http://dx.doi.org/10.1021/j100193a038]

[353] Hanack, M.; Heckmann, H. Soluble chloro- and aryl(phthalocyaninato)indium(III) complexes: synthesis and characterization. *Eur. J. Inorg. Chem.,* **1998**, 367-373.
[http://dx.doi.org/10.1002/(SICI)1099-0682(199803)1998:3<367::AID-EJIC367>3.0.CO;2-5]

[354] Dini, D.; Barthel, M.; Hanack, M. Phthalocyanines as active materials for optical limiting. *Eur. J. Org. Chem.,* **2001**, 3759-3769.
[http://dx.doi.org/10.1002/1099-0690(200110)2001:20<3759::AID-EJOC3759>3.0.CO;2-U]

[355] Hosoda, M.; Wada, T.; Yamada, A.; Garito, A.F.; Sasabe, H. Phases and third-order optical nonlinearities in tetravalent metallophthalocyanine thin films. *Jpn. J. Appl. Phys.,* **1991**, *30*, L1486-L1488.
[http://dx.doi.org/10.1143/JJAP.30.L1486]

[356] Ho, Z.Z.; Ju, C.Y.; Hetherington, W.M., Jr. Third harmonic generation in phthalocyanines. *J. Appl. Phys.,* **1987**, *62*, 716-718.
[http://dx.doi.org/10.1063/1.339749]

[357] Snow, A.W.; Jarvis, N.L. Molecular association and monolayer formation of soluble phthalocyanine compounds. *J. Am. Chem. Soc.,* **1984**, *106*, 4706-4711.
[http://dx.doi.org/10.1021/ja00329a012]

[358] Maya, E.M.; Snow, A.W.; Shirk, J.S.; Flom, S.R.; Pong, R.G.; Roberts, G.L. Silicon substituted phthalocyanines for optical limiting applications. *Polymer Prepr.,* **2001**, *42*, 253-257.

[359] Nalwa, H.S.; Kakuta, A. Third-order non-linear optical properties of donor- and acceptor-substituted metallophthalocyanines. *Thin Solid Films,* **1995**, *254*, 218-223.
[http://dx.doi.org/10.1016/0040-6090(94)06260-R]

[360] Dini, D.; Calvete, M.J.; Hanack, M.; Chen, W.; Ji, W. Synthesis of axially substituted gallium, indium and thallium phthalocyanines with nonlinear optical properties. *ARKIVOC,* **2006**, *3*, 77-96.

[361] Dini, D.; Calvete, M.J.; Hanack, M.; Amendola, V.; Meneghetti, M. Demonstration of the optical limiting effect for an hemiporphyrazine. *Chem. Commun. (Camb.),* **2006**, (22), 2394-2396.
[http://dx.doi.org/10.1039/b601591c] [PMID: 16733591]

[362] Dini, D.; Calvete, M.J.; Hanack, M.; Meneghetti, M. Indium phthalocyanines with different axial ligands: a study of the influence of the structure on the photophysics and optical limiting properties. *J.*

Phys. Chem. A, **2008**, *112*(37), 8515-8522.
[http://dx.doi.org/10.1021/jp802960s] [PMID: 18714959]

[363] Amendola, V.; Dini, D.; Polizzi, S.; Shen, J.; Kadish, K.M.; Calvete, M.J.; Hanack, M.; Meneghetti, M. Self-healing of gold nanoparticles in the presence of zinc phthalocyanines and their very efficient nonlinear absorption performances. *J. Phys. Chem. C,* **2009**, *113*, 8688-8695.
[http://dx.doi.org/10.1021/jp810921w]

[364] Carvalho, E.F.; Calvete, M.J.; Cavaleiro, J.A.; Dini, D.; Meneghetti, M.; Tomé, A.C. Synthesis and high ranked NLT properties of new sulfonamide-substituted indium phthalocyanines. *Inorg. Chim. Acta,* **2010**, *363*, 3945-3950.
[http://dx.doi.org/10.1016/j.ica.2010.07.062]

[365] Hanack, M.; Datz, R.; Fay, K.; Fischer, K.; Keppeler, U.; Koch, J.; Metz, J.; Mezger, M.; Schneider, O.; Schulze, H-J. *Handbook of Conducting Polymers*; Skotheim, T.A., Ed.; Marcel Dekker: New York, **1986**, Vol. 1, p. 133.

[366] Nalwa, H.S.; Hanack, M.; Pawlowski, G.; Engel, M.K. Third-order nonlinear optical properties of porphyrazine, phthalocyanine and naphthalocyanine germanium derivatives: Demonstrating the effect of π-conjugation length on third-order optical nonlinearity of two-dimensional molecules. *Chem. Phys.,* **1999**, *245*, 17-26.
[http://dx.doi.org/10.1016/S0301-0104(99)00145-7]

[367] Perry, J.W.; Khundar, L.R.; Coulter, D.R.; Alvarez, D.; Marder, S.R.; Wei, T.; Sense, M.J.; van Stryland, E.W.; Hagan, D.J. *Organic Materials for Nonlinear Optics and Photonics*; Messier, J., Ed.; Kluwer Academic Publishers: Dordrecht, **1991**, Vol. 194, p. NATO ASI Series E.369.
[http://dx.doi.org/10.1007/978-94-011-3370-8_27]

[368] Robertson, J.; Milsom, P.; Duignan, J.; Bourhill, G. Spatial redistribution of energy in a nanosecond laser pulse by an organic optical limiter. *Opt. Lett.,* **2000**, *25*(17), 1258-1260.
[http://dx.doi.org/10.1364/OL.25.001258] [PMID: 18066185]

[369] Kulchin, Y.N.; Vitrik, O.B.; Lantsov, A.D. Correlation method of processing single fiber multimode interferometer (SFMI) signals using a charge coupled device. *Proc. SPIE,* **2007**, *6595*, 659508.

[370] Dasari, R.R.; Sartin, M.M.; Cozzuol, M.; Barlow, S.; Perry, J.W.; Marder, S.R. Synthesis and linear and nonlinear absorption properties of dendronised ruthenium(II) phthalocyanine and naphthalocyanine. *Chem. Commun. (Camb.),* **2011**, *47*(15), 4547-4549.
[http://dx.doi.org/10.1039/c0cc05688j] [PMID: 21399800]

[371] Miles, P.A. Material figures of merit for saturated excited-state absorptive limiters. *Proc. SPIE,* **1994**, *2143*, 251.
[http://dx.doi.org/10.1117/12.173825]

[372] Shirk, J.S.; Pong, R.G.; Flom, S.R.; Boyle, M.E.; Snow, A.W. Materials for reverse saturable absorption optical limiters. *Proc. MRS,* **1995**, *374*, 201-205.
[http://dx.doi.org/10.1557/PROC-374-201]

[373] Schneider, T.; Heckmann, H.; Barthel, M.; Hanack, M. Synthesis and characterization of soluble chloro- and Aryl(naphthalocyaninato)indium(III) complexes and their precursors. *Eur. J. Org. Chem.,* **2001**, *16*, 3055-3065.
[http://dx.doi.org/10.1002/1099-0690(200108)2001:16<3055::AID-EJOC3055>3.0.CO;2-E]

[374] Hongan, Y.; Qing, C.; Yiqun, W.; Chunying, H.; Xia, Z.; Jianhong, Z.; Yuxiao, W.; Yinglin, S. Optical limiting of metallic naphthalocyanine compound. *Mater. Lett.,* **2003**, *57*, 3302-3304. [http://dx.doi.org/10.1016/S0167-577X(03)00052-1]

[375] Dini, D.; Calvete, M.; Vagin, S.; Hanack, M.; Eriksson, A.; Lopes, C. Analysis of the nonlinear transmission properties of some naphthalocyanines. *J. Porphyr. Phthalocyanines,* **2006**, *10*, 1165-1171. [http://dx.doi.org/10.1142/S1088424606000521]

[376] Dini, D.; Meneghetti, M.; Calvete, M.J.; Arndt, T.; Liddiard, C.; Hanack, M. Tetrabrominated lead naphthalocyanine for optical power limiting. *Chemistry,* **2010**, *16*(4), 1212-1220. [http://dx.doi.org/10.1002/chem.200901499] [PMID: 19937866]

[377] Dini, D.; Calvete, M.J.; Hanack, M.; Pong, R.G.; Flom, S.R.; Shirk, J.S. Nonlinear transmission of a tetrabrominated naphthalocyaninato indium chloride. *J. Phys. Chem. B,* **2006**, *110*(25), 12230-12239. [http://dx.doi.org/10.1021/jp0571776] [PMID: 16800543]

[378] Li, Y.; Dini, D.; Calvete, M.J.; Hanack, M.; Sun, W. Photophysics and nonlinear optical properties of tetra- and octabrominated silicon naphthalocyanines. *J. Phys. Chem. A,* **2008**, *112*(3), 472-480. [http://dx.doi.org/10.1021/jp0771116] [PMID: 18163601]

[379] Sun, W.; Wang, G.; Li, Y.; Calvete, M.J.; Dini, D.; Hanack, M. Axial halogen ligand effect on photophysics and optical power limiting of some indium naphthalocyanines. *J. Phys. Chem. A,* **2007**, *111*(17), 3263-3270. [http://dx.doi.org/10.1021/jp071152k] [PMID: 17419601]

[380] Brasseur, N.; Ali, H.; Autenrieth, D.; Langlois, R.; van Lier, J.E. Biological activities of phthalocyanines--III. Photoinactivation of V-79 Chinese hamster cells by tetrasulfophthalocyanines. *Photochem. Photobiol.,* **1985**, *42*(5), 515-521. [http://dx.doi.org/10.1111/j.1751-1097.1985.tb01603.x] [PMID: 4089034]

[381] Nyokong, T.; Antunes, E. Photochemical and photophysical properties of metallophthalocyanines. In: *Handbook of Porphyrin Science with Applications to Chemistry, Physics, Materials Science, Engineering, Biology and Medecine*; Kadish, K.M.; Smith, K.M.; Guilard, R., Eds.; World Scientific: Singapore, **2010**; pp. 247-349.

[382] Rio, Y.; Salomé Rodríguez-Morgade, M.; Torres, T. Modulating the electronic properties of porphyrinoids: a voyage from the violet to the infrared regions of the electromagnetic spectrum. *Org. Biomol. Chem.,* **2008**, *6*(11), 1877-1894. [http://dx.doi.org/10.1039/b800617b] [PMID: 18480898]

[383] Josefsen, L.B.; Boyle, R.W. Photodynamic therapy and the development of metal-based photosensitisers. *Metal-Based Drugs,* **2008**, *2008*, 276109. [http://dx.doi.org/10.1155/2008/276109]

[384] Ali, H.; van Lier, J.E. Metal complexes as photo- and radiosensitizers. *Chem. Rev.,* **1999**, *99*(9), 2379-2450. [http://dx.doi.org/10.1021/cr980439y] [PMID: 11749485]

[385] Sharman, W.M.; Allen, C.M.; van Lier, J.E. Photodynamic therapeutics: basic principles and clinical applications. *Drug Discov. Today,* **1999**, *4*(11), 507-517.

[http://dx.doi.org/10.1016/S1359-6446(99)01412-9] [PMID: 10529768]

[386] Sobolev, A.S.; Stranadko, E.F. Photodynamic therapy in Russia. Clinical and fundamental aspects. *International Photodynamics,* **1997**, *1*, pp. 2-3.Eurocommunication PublicationsWest Sussex, UK

[387] Sokolov, V.V.; Chissov, V.I.; Filonenko, E.V.; Yakubovskaya, R.I.; Sukhin, G.M.; Galpern, M.G.; Vorozhtsov, G.N.; Gulin, A.V.; Zhitkova, M.B.; Zharkova, N.N.; Kozlov, D.N.; Smirnov, V.V. First clinical results with a new drug for PDT. *Photodynamic Therapy of Cancer II. Proceedings of the photo-optical instrumentation engineers.,* **1995**, *2325*, pp. 364-366.Washington, BellinghamSPIE [http://dx.doi.org/10.1117/12.199168]

[388] Zharkova, N.N.; Kozlov, D.N.; Smirnov, V.V.; Sokolov, V.V.; Chissov, V.I.; Filonenko, E.V.; Sukhin, G.M.; Galpern, M.G.; Vorozhtsov, G.N. Jori, G. ; Moan, J. ; Star, W.M. Fluorescence observations of patients in the course of photodynamic therapy of cancer with the photosensitizer PHOTOSENS. *Photodynamic Therapy of Cancer II. Proceedings of the photo-optical in-strumentation engineers.,* **1995**, *2325*, pp. 400-403.Washington, BellinghamSPIE [http://dx.doi.org/10.1117/12.199176]

[389] Whitacre, C.M.; Feyes, D.K.; Satoh, T.; Grossmann, J.; Mulvihill, J.W.; Mukhtar, H.; Oleinick, N.L. Photodynamic therapy with the phthalocyanine photosensitizer Pc 4 of SW480 human colon cancer xenografts in athymic mice. *Clin. Cancer Res.,* **2000**, *6*(5), 2021-2027. [PMID: 10815928]

[390] Lo, P.C.; Leung, S.C.; Chan, E.Y.; Fong, W-P.; Ko, W-H.; Ng, D.K. Photodynamic effects of a novel series of silicon(IV) phthalocyanines against human colon adenocarcinoma cells. *Photodiagn. Photodyn. Ther.,* **2007**, *4*(2), 117-123. [http://dx.doi.org/10.1016/j.pdpdt.2007.03.001] [PMID: 25047345]

[391] Zaidi, S.I.; Agarwal, R.; Eichler, G.; Rihter, B.D.; Kenney, M.E.; Mukhtar, H. Photodynamic effects of new silicon phthalocyanines: in vitro studies utilizing rat hepatic microsomes and human erythrocyte ghosts as model membrane sources. *Photochem. Photobiol.,* **1993**, *58*(2), 204-210. [http://dx.doi.org/10.1111/j.1751-1097.1993.tb09550.x] [PMID: 8415911]

[392] Colussi, V.C.; Feyes, D.K.; Mulvihill, J.W.; Li, Y-S.; Kenney, M.E.; Elmets, C.A.; Oleinick, N.L.; Mukhtar, H. Phthalocyanine 4 (Pc 4) photodynamic therapy of human OVCAR-3 tumor xenografts. *Photochem. Photobiol.,* **1999**, *69*(2), 236-241. [http://dx.doi.org/10.1111/j.1751-1097.1999.tb03280.x] [PMID: 10048316]

[393] Whitacre, C.M.; Satoh, T.H.; Xue, Ly.; Gordon, N.H.; Oleinick, N.L. Photodynamic therapy of human breast cancer xenografts lacking caspase-3. *Cancer Lett.,* **2002**, *179*(1), 43-49. [http://dx.doi.org/10.1016/S0304-3835(01)00853-9] [PMID: 11880181]

[394] George, J.E., III; Ahmad, Y.; Varghai, D.; Li, X.; Berlin, J.; Jackowe, D.; Jungermann, M.; Wolfe, M.S.; Lilge, L.; Totonchi, A.; Morris, R.L.; Peterson, A.; Lust, W.D.; Kenney, M.E.; Hoppel, C.L.; Sun, J.; Oleinick, N.L.; Dean, D. Pc 4 photodynamic therapy of U87-derived human glioma in the nude rat. *Lasers Surg. Med.,* **2005**, *36*(5), 383-389. [http://dx.doi.org/10.1002/lsm.20185] [PMID: 15965990]

[395] Ball, D.J.; Wood, S.R.; Vernon, D.I.; Griffiths, J.; Dubbelman, T.M.; Brown, S.B. The characterisation of three substituted zinc phthalocyanines of differing charge for use in photodynamic therapy. A comparative study of their aggregation and photosensitising ability in relation to mTHPC and

polyhaematoporphyrin. *J. Photochem. Photobiol. B,* **1998**, *45*(1), 28-35.
[http://dx.doi.org/10.1016/S1011-1344(98)00156-0] [PMID: 9819897]

[396] Fabris, C.; Soncin, M.; Miotto, G.; Fantetti, L.; Chiti, G.; Dei, D.; Roncucci, G.; Jori, G. Zn(II)-phthalocyanines as phototherapeutic agents for cutaneous diseases. Photosensitization of fibroblasts and keratinocytes. *J. Photochem. Photobiol. B,* **2006**, *83*(1), 48-54.
[http://dx.doi.org/10.1016/j.jphotobiol.2005.12.006] [PMID: 16427302]

[397] Mantareva, V.; Kussovski, V.; Angelov, I.; Borisova, E.; Avramov, L.; Schnurpfeil, G.; Wöhrle, D. Photodynamic activity of water-soluble phthalocyanine zinc(II) complexes against pathogenic microorganisms. *Bioorg. Med. Chem.,* **2007**, *15*(14), 4829-4835.
[http://dx.doi.org/10.1016/j.bmc.2007.04.069] [PMID: 17517508]

[398] Raab, O. On the effects of fluorescent substances on infusoria. *Z. Biol.,* **1890**, *39*, 524-546.

[399] Wainwright, M. Photodynamic antimicrobial chemotherapy (PACT). *J. Antimicrob. Chemother.,* **1998**, *42*(1), 13-28.
[http://dx.doi.org/10.1093/jac/42.1.13] [PMID: 9700525]

[400] Malik, Z.; Ladan, H.; Nitzan, Y. Photodynamic inactivation of Gram-negative bacteria: problems and possible solutions. *J. Photochem. Photobiol. B,* **1992**, *14*(3), 262-266.
[http://dx.doi.org/10.1016/1011-1344(92)85104-3] [PMID: 1432395]

[401] Bertoloni, G.; Rossi, F.; Valduga, G.; Jori, G.; van Lier, J. Photosensitizing activity of water- and lipid-soluble phthalocyanines on *Escherichia coli. FEMS Microbiol. Lett.,* **1990**, *59*(1-2), 149-155.
[http://dx.doi.org/10.1111/j.1574-6968.1990.tb03814.x] [PMID: 2125956]

[402] Bertoloni, G.; Rossi, F.; Valduga, G.; Jori, G.; Ali, H.; van Lier, J.E. Photosensitizing activity of water- and lipid-soluble phthalocyanines on prokaryotic and eukaryotic microbial cells. *Microbios,* **1992**, *71*(286), 33-46.
[PMID: 1406343]

[403] Giuntini, F.; Nistri, D.; Chiti, G.; Fantetti, L.; Jori, G.; Roncucci, G. Synthesis of trimethylammonium-phenylthio-substituted phthalocyanines with different pattern of substitution. *Tetrahedron Lett.,* **2003**, *44*, 515-517.
[http://dx.doi.org/10.1016/S0040-4039(02)02597-2]

[404] Paardekopper, M.; van Gompel, A.E.; van Steveninck, J.; van de Broek, J. The effect of photodynamic treatment of yeast with the sensitiser chloroalluminium phthalocyanine on various cellular parameters. *J. Photochem. Photobiol. B,* **1995**, *62*, 561-567.
[http://dx.doi.org/10.1111/j.1751-1097.1995.tb02385.x]

[405] Schuster, F.L.; Visvesvara, G.S. Amebae and ciliated protozoa as causal agents of waterborne zoonotic disease. *Vet. Parasitol.,* **2004**, *126*(1-2), 91-120.
[http://dx.doi.org/10.1016/j.vetpar.2004.09.019] [PMID: 15567581]

[406] Dutta, S.; Ray, D.; Kolli, B.K.; Chang, K.P. Photodynamic sensitization of Leishmania amazonensis in both extracellular and intracellular stages with aluminum phthalocyanine chloride for photolysis in vitro. *Antimicrob. Agents Chemother.,* **2005**, *49*(11), 4474-4484.
[http://dx.doi.org/10.1128/AAC.49.11.4474-4484.2005] [PMID: 16251285]

[407] Kelkar, S.S.; Reineke, T.M. Theranostics: combining imaging and therapy. *Bioconjug. Chem.,* **2011**,

22(10), 1879-1903.
[http://dx.doi.org/10.1021/bc200151q] [PMID: 21830812]

[408] Faulkner, A.S.; Rousseau, J.A.; Langlois, R.; Berard, V.; Lecomte, R.; Bérnard, F. van Lier. J.E. Copper-64 labeled sulfophthalocyanines for positron emission tomography (PET) imaging in tumor bearing rats. *J. Porphyr. Phthalocyanines,* **2008**, *12*, 49-53.
[http://dx.doi.org/10.1142/S1088424608000078]

[409] Ranyuk, E.R.; Cauchon, N.; Ali, H.; Lecomte, R.; Guérin, B.; van Lier, J.E. PET imaging using 64Cu-labeled sulfophthalocyanines: synthesis and biodistribution. *Bioorg. Med. Chem. Lett.,* **2011**, *21*(24), 7470-7473.
[http://dx.doi.org/10.1016/j.bmcl.2011.09.121] [PMID: 22050887]

[410] Lv, F.; He, X.; Wu, L.; Liu, T. Lactose substituted zinc phthalocyanine: a near infrared fluorescence imaging probe for liver cancer targeting. *Bioorg. Med. Chem. Lett.,* **2013**, *23*(6), 1878-1882.
[http://dx.doi.org/10.1016/j.bmcl.2012.12.103] [PMID: 23395663]

[411] Lv, F.; Li, Y.; Cao, B.; Liu, T. Galactose substituted zinc phthalocyanines as near infrared fluorescence probes for liver cancer imaging. *J. Mater. Sci. Mater. Med.,* **2013**, *24*(3), 811-819.
[http://dx.doi.org/10.1007/s10856-012-4820-2] [PMID: 23183962]

[412] Sehgal, I.; Li, H.; Ongarora, B.; Devillier, D.; Vicente, M.G. Synthesis and biological investigations of a ZnPc-antiCEA bioconjugate for imaging of colorectal cancer. *J. Porphyr. Phthalocyanines,* **2013**, *17*(1-2), 150-156.
[http://dx.doi.org/10.1142/S108842461250143X] [PMID: 25328375]

[413] Saini, S.K.; Jena, A.; Dey, J.; Sharma, A.K.; Singh, R. MnPcS4: a new MRI contrast enhancing agent for tumor localisation in mice. *Magn. Reson. Imaging,* **1995**, *13*(7), 985-990.
[http://dx.doi.org/10.1016/0730-725X(95)00046-J] [PMID: 8583877]

[414] Sleven, J.; Cardinaels, T.; Görller-Walrand, C.; Binnemans, K. Liquid-crystalline metallophthalocyanines containing late first-row transition metals. *ARKIVOC,* **2003**, *4*, 68-82.
[http://dx.doi.org/10.3998/ark.5550190.0004.406]

[415] Cook, M.J.; Dunn, A.J.; Howe, S.D.; Thomson, A.J. Octa-alkoxy phthalocyanine and naphthalocyanine derivatives: dyes with Q-band absorption in the far red or near infrared. *J. Chem. Soc., Perkin Trans. 1,* **1988**, 2453-2458.
[http://dx.doi.org/10.1039/p19880002453]

[416] Carvalho, E.F.; Calvete, M.J.; Tome, A.C.; Cavaleiro, J.A. Synthesis of sulfonamide-substituted phthalocyanines. *Tetrahedron Lett.,* **2009**, *50*, 6882-6885.
[http://dx.doi.org/10.1016/j.tetlet.2009.09.128]

[417] Alvarez-Mico, X.; Calvete, M.J.; Hanack, M.; Ziegler, T. The first example of anomeric glycoconjugation to phthalocyanines. *Tetrahedron Lett.,* **2006**, *47*, 3283-3286.
[http://dx.doi.org/10.1016/j.tetlet.2006.03.029]

[418] Claessens, C.G.; González-Rodríguez, D.; del Rey, B.; Torres, T.; Mark, G.; Schuchmann, H-P.; von Sonntag, C.; MacDonald, J.G.; Nohr, R.S. Highly efficient synthesis of chloro- and phenoxy-substituted subphthalocyanines. *Eur. J. Org. Chem.,* **2003**, 2547-2551.
[http://dx.doi.org/10.1002/ejoc.200300169]

[419] Nolan, K.J.; Hu, M.; Leznoff, C.C. "Adjacent" substituted phthalocyanines. *Synlett,* **1997**, 593-594. [http://dx.doi.org/10.1055/s-1997-3226]

CHAPTER 2

Polymethines as Top Materials

Abstract: Polymethines are planar, conjugated compounds made up from an odd number of methine groups (CH) linked together by alternating single and double bonds, also known as cyanines. These highly conjugated compounds can absorb light at wide range of wavelengths, ranging from ultraviolet to infra red, just depending on the number of alternating double bonds present in their structure. Herein is presented a conspectus on the field of cyanines, where emphasis is set on their preparation and functionalization methods, along with more detailed examples on several promising applications with 165 references provided. Some synthetic details on the preparation of state-of-the-art compounds are given as well.

Keywords: Biolabeling, Cyanines, Infrared dyes, Near infrared dyes, Polymethines, Ultra violet dyes.

INTRODUCTION

Polymethines are planar, conjugated compounds made up from an odd number of methine groups (CH) linked together by alternating single and double bonds [1, 2]. These compounds differ from related polyenes, which are compounds made up from an even number of methine groups. This family of compounds is also known as cyanines, a non-systematic name of this synthetic dye family.

The word cyanine derives from the English word "cyan", meaning blue-green, being derived from the Greek "kyanos" which means a rather different color: "dark blue".

Cyanines are cationic compounds possessing a +1 charge, being classified in three types: i) hemicyanines, ii) open chain cyanines and iii) closed chain cyanines (Fig. **1**). Inside that classification they can also be symmetric or unsymmetric, depending on the nature of each of the open or closed chain groups at the extremities of the compounds. Each nitrogen is an independent part of a heteroaromatic moiety, such as imidazole, pyrrole, pyridine, thiazole, indole,

Mário J. F. Calvete

benzothiazole, quinoline, *etc* [3 - 5] and, depending on their structure, their absorption spectra can cover the spectrum from UV to IR.

In 1856, Williams reported the first carbocyanine [6]. Not much importance was given to this type of compounds, as it was an extremely light sensitive type of molecules. The great value of carbocyanine dyes was not discovered until almost 20 years later when Vogel used them to increase sensitivity of a photographic plate [7]. Cyanine dyes were also recognized as natural products dyes [8 - 11], as observed by Wyler [8, 9] and Musso [10] (Fig. **2**).

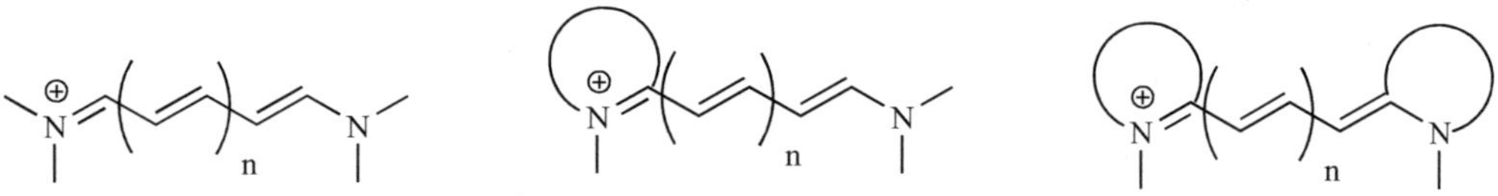

Fig. (1). General cyanine structures.

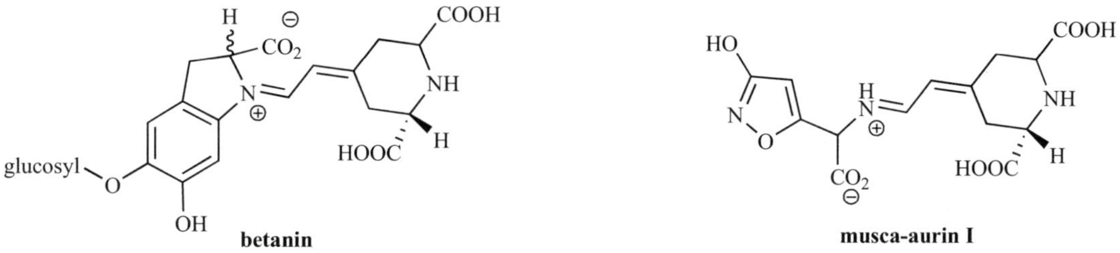

Fig. (2). Natural carbocyanine dyes.

Since then, several synthetic method for the preparation of polymethine dyes with varied molecular structures have been reported [11 - 20], including monomethines [21 - 27], dimethines [28 - 32], trimethines [33 - 36], tetramethines [37, 38], pentamethines [39 - 43], heptamethines [44 - 48] and various other cyanine dyes [49 - 51].

SYNTHESIS OF CYANINE DERIVATIVES

The synthesis of polymethine cyanine derivatives is typically accomplished through a condensation reaction between a polyene-chain precursor and two nucleophilic *aza*-heterocycles bearing an activated methyl group.

In general, in the synthesis of most methine dyes, a counterpart contains a methylene group activated by a quaternized nitrogen atom and another component has a reactive group with an auxochrome (Fig. **3**).

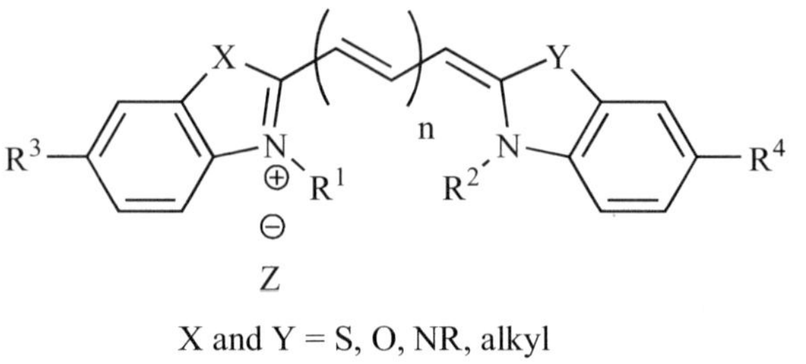

Fig. (3). General cyanine structure.

For instance, monomethine cyanine dyes can be synthesized by heating a quaternary salt of heterocyclic compounds and sulfobetaines under basic conditions (Scheme **1**) [52, 53].

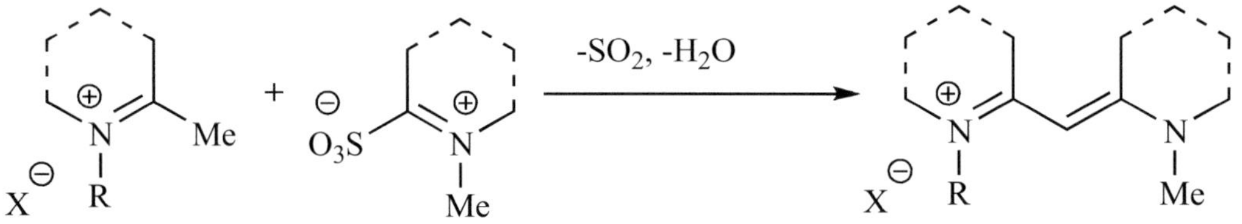

Scheme (1). Methine dyes.

Other approach regards the reaction of coumarins and 2- or 4-methyl quaternary salts, in the presence of piperidine (Scheme **2**) [54].

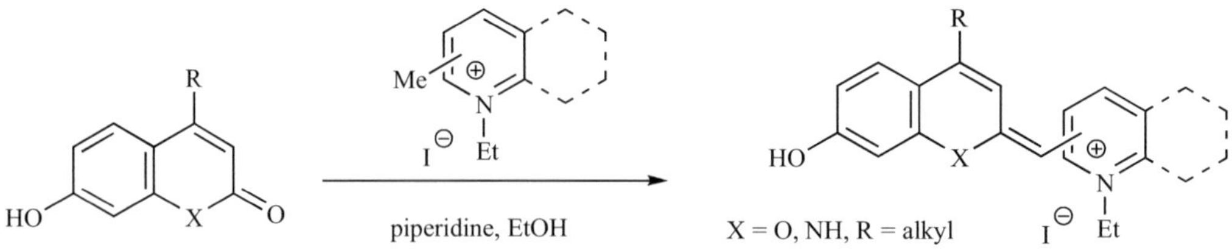

Scheme (2). Monomethine cyanine dyes.

Koenig [55] developed a method for the synthesis of symmetrical trimethine cyanine dyes (Scheme **3**), which was later extended [56, 57].

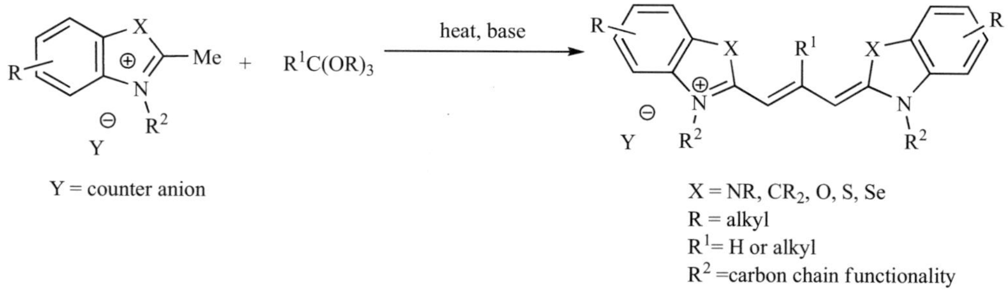

Scheme (3). Trimethine cyanine dyes.

A more versatile method was also developed, which can be applied for the synthesis of either symmetrical or unsymmetrical trimethines (Scheme **4**) [57 - 60].

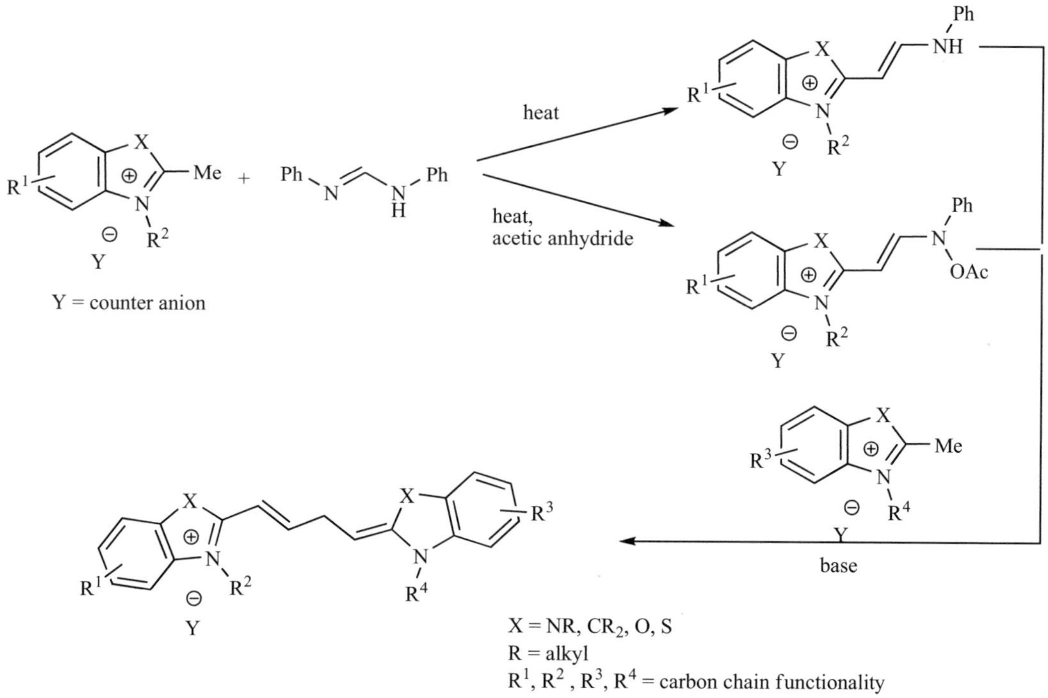

Scheme (4). N,N' -Diphenylformamidine method for cyanine synthesis.

The major synthetic route to the subclass of pentamethine cyanine compounds deals with the condensation of malonodialdehyde derivatives with cationic heterocyclic compounds possessing an activated methyl group [61 - 64] (Scheme **5**).

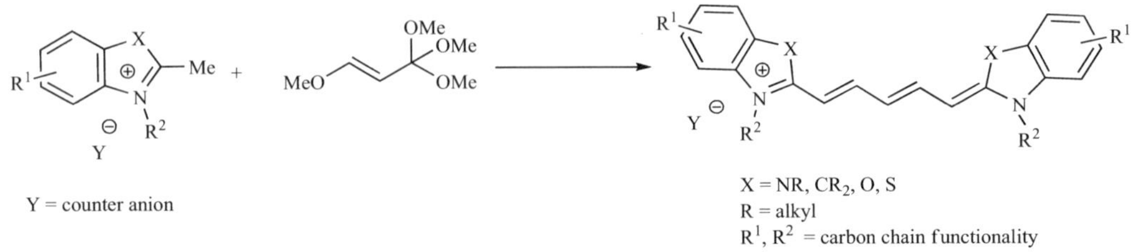

Scheme (5). Pentamethine cyanines.

Heptamethine cyanine dyes are also very interesting dyes. The first time this type of dyes was prepared was in 1977 [65], by reaction of a heterocyclic salt (iodide) containing an activated methyl group and a cyclohexanone derived Vielsmeier–Haack reagent. More recently, Strekowski introduced recently a stabilized diphenyl iminium salt reagent for an easier reaction to form the corresponding dyes (R^1 = COOEt) (Scheme **6**) [66, 67].

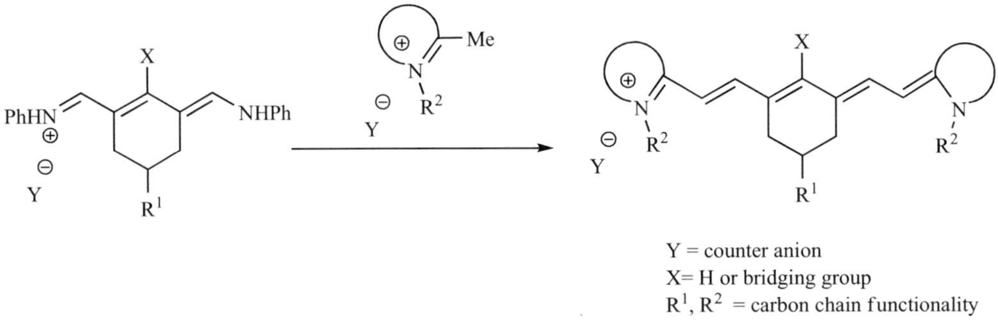

Scheme (6). Heptamethine cyanine dyes synthesized by the "imine" method.

This alternative synthetic approach started to be a "favorite" tool, since allows directed functionalization for any desired application, giving the fluorophore bearing the desired reactive groups [4, 5, 68 - 70]. By substitution of the bridging atom (*e.g.* Cl or Br) with different nucleophiles, a variety of heptamethine cyanine dyes can be obtained and used for many applications [67, 70 - 79].

Cy3 and Cy5 are the mainly studied cyanine dyes. While Cy3 dyes posses an orange fluorescence, Cy5 fluoresces in the red region. They can be substituted with a variety of reactive groups, normally for further chemical linkage to protein molecules or nucleic acids, which is the main use for these molecules. The Cy3 and Cy5 specific nomenclature was proposed for the first time in 1989 by Waggoner [80] and is considered non-standard, since it does not gives any information of their chemical structures. Thus, various structures can be designated Cy3 and Cy5 in the literature.

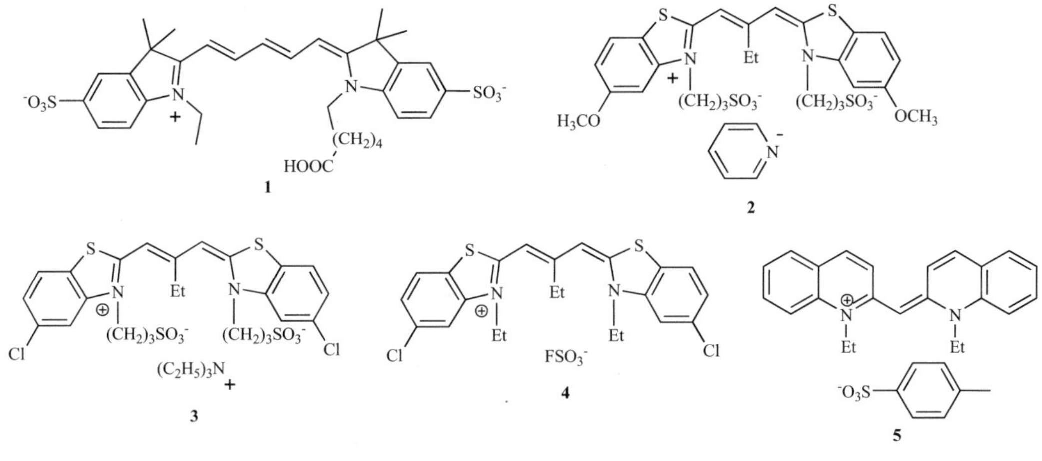

Fig. (4). Carbocyanines for application as photoactive materials.

APPLICATIONS OF POLYMETHINE DYES

Polymethine dyes posses a wide range of spectral, photophysical and photochemical properties, with the possibility of being effortlessly modified. The structural modulations that can be applied to theses dyes are, in most cases, intended either at increasing the emission efficiency or shifting the absorption to a longer wavelength. Thus they mostly find applications as photoactive materials, [74, 81 - 94], including in light harvesting systems and photovoltaics [95 - 101] and in biomedical applications, including as probes for biological systems [67, 80, 102 - 118], as antitumor agents, [119 - 123], in imaging/diagnostics [112, 124 - 136], among other applications [137 - 143].

Applications as Photoactive Materials

Materials that allow tuning in a predictable and accurate fashion by using an external stimulus offer a wide number of applications, in several areas in science and technology ranging from data storage to environmental cleanup.

For instance, Heileman [85] demonstrated the use of a commercially available unmodified Cy5 carbocyanine dye **1** (Fig. **4**) derivative as a reversible optical switch, for which the fluorescence state could be, upon irradiation, re-established at room temperature, after perceptible photobleaching.

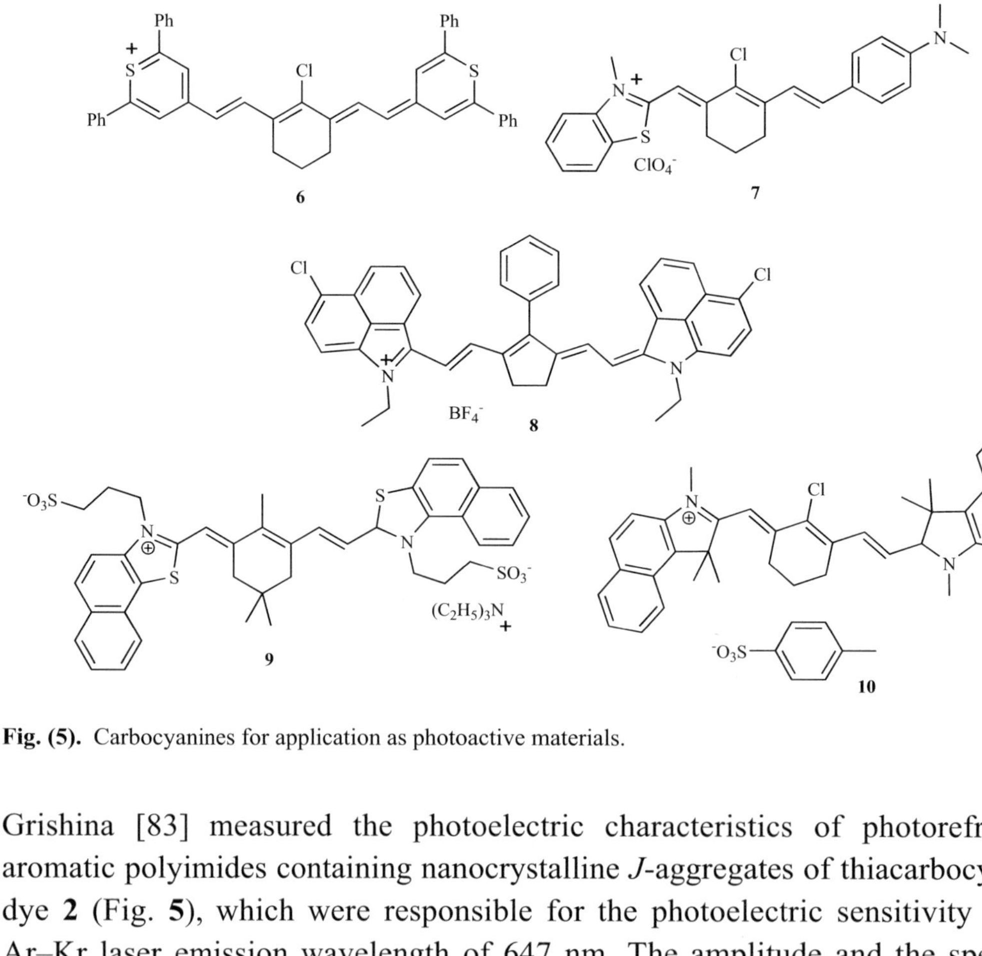

Fig. (5). Carbocyanines for application as photoactive materials.

Grishina [83] measured the photoelectric characteristics of photorefractive aromatic polyimides containing nanocrystalline *J*-aggregates of thiacarbocyanine dye **2** (Fig. **5**), which were responsible for the photoelectric sensitivity at the Ar–Kr laser emission wavelength of 647 nm. The amplitude and the speed of writing of the photorefractive diffraction were determined and observed that the dependence of the quantum yield of free charge carriers on the applied electric field was linear.

Lanzafame [89] also studied the *J*-aggregates of cyanine dyes **3-5** (Fig. **5**) adsorbed onto silver halide microcrystals, for their excited state dynamics, namely fluorescence rates, non-radiative relaxation and charge transfer, observing that the rate of charge transfer increased with the aggregate sizes.

Carbocyanine **6** (Fig. **6**) was also studied as a lasing dye [92], due to its high quantum efficiency and broad spectral range tuning, displaying a 43% conversion efficiency and $\Delta\lambda = 63$ nm tunable range, using **6** in a polyurethane matrix.

Polymethines, when properly modulated, show fairly good NIR fluorescence properties, and have been used in light emitting diodes (LED). Suzuki reported a device based on ionic dye **7** (Fig. **6**), doped in poly(N-vinylcarbazole) and 2- (4-biphenylyl)-5-(4-tert-butylphenyl)-1,3,4-oxidiazole [144] for its NIR electro-luminescence properties. Since the energy level of **7** was between that of poly(N-vinylcarbazole), the dye acted as an efficient radiative-recombination center, originating LED NIR emission at 800 nm with 0.015% efficiency. Nevertheless, the electroluminescence quantum efficiency reached 1%, with repeated application of reverse bias, due to the improved hole and electron injections.

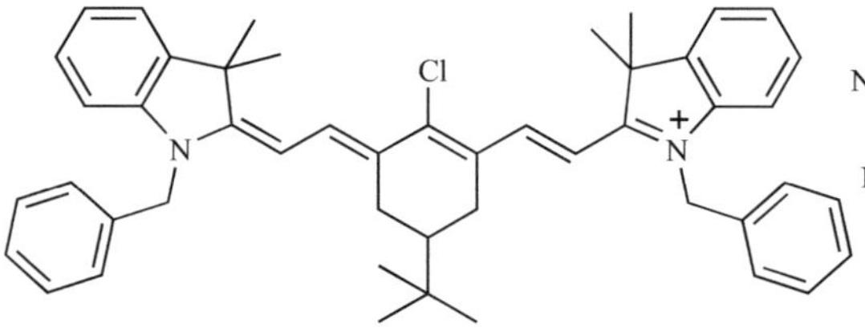

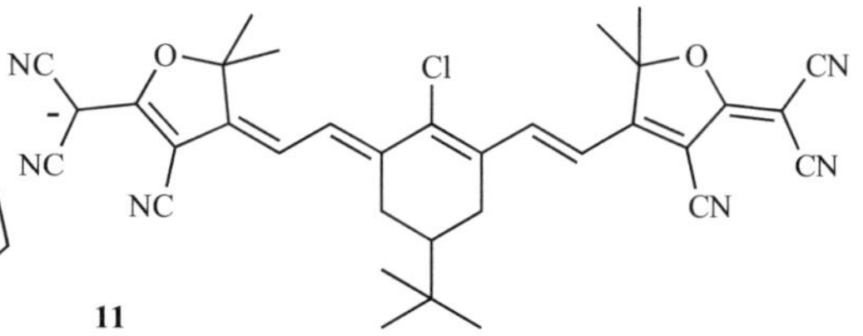

11

Fig. (6). Cyanine dyes for application as photoactive materials.

Similarly, the doping the polymethine dye (1% weight) and 2-(4-biphenylyl)- 5-(4-tert-butylphenyl)-1,3,4-oxidiazole (30% weight) in poly(N-vinylcarbazole), the electroluminescence properties of compound **8** were evaluated [145]. The electroluminescence spectrum showed 2 peaks at 1100 and 990 nm, attributed to monomeric dye and aggregated dyes, respectively.

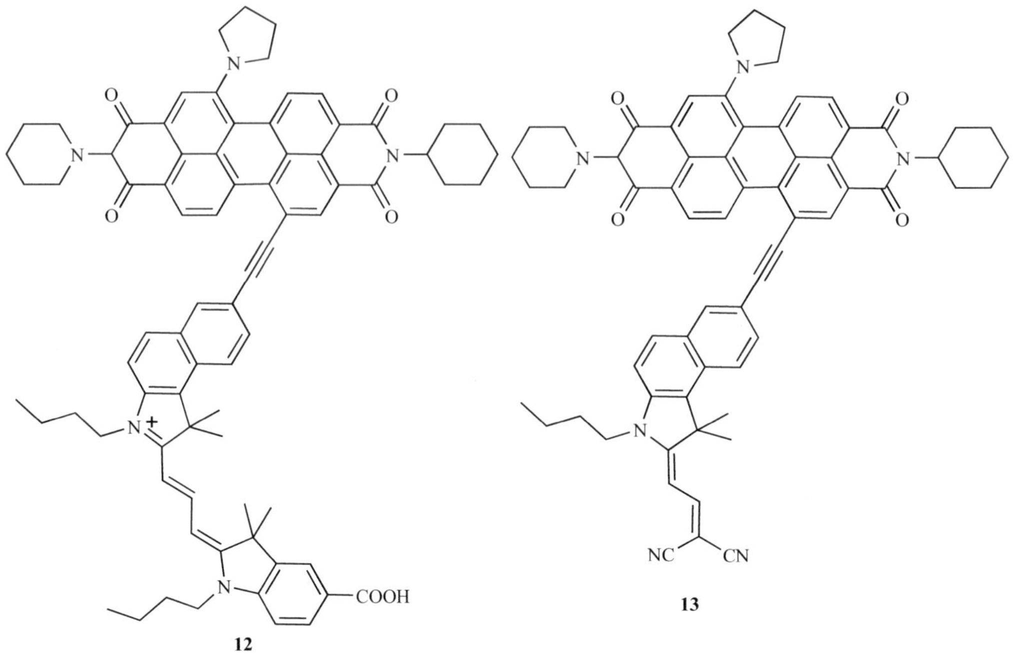

Fig. (7). Cationic NIR absorbing cyanine dye dyad.

J-aggregates of dye **9** were also studied for their fluorescence in polyimide films [146], while the absorption peak of **10** polyimide film J-aggregate was showed at 767 nm, together with a sharp electroluminescence band at 815 nm, displaying a low 0.01% quantum efficiency [147].

To acquire the light harvesting properties of a polymethine dye, Bouit [100] prepared a NIR absorbing ionic dye **11** by association of two cyanine dyes (Fig. 7), exhibiting fine light-harvesting properties in the NIR region. Cyclic voltammetry measurements of **11** revealed amphoteric redox character. Additionally, the authors measured the photovoltaic activity for chromophore **11** on bulky solar cell heterojunction using [6, 6]-phenyl-C_{61}-butyric acid methyl ester as fullerenic acceptor, displaying a 0.4% power conversion efficiency.

Jin reported cyanines **12** and **13**, which were prepared by linking a perylene imide derivative with benzo[e]indole and cyanine derivatives [148]. The power conversion efficiency of the dye sensitized solar cell based on **12** and **13** was

0.34% and 1.38%, respectively.

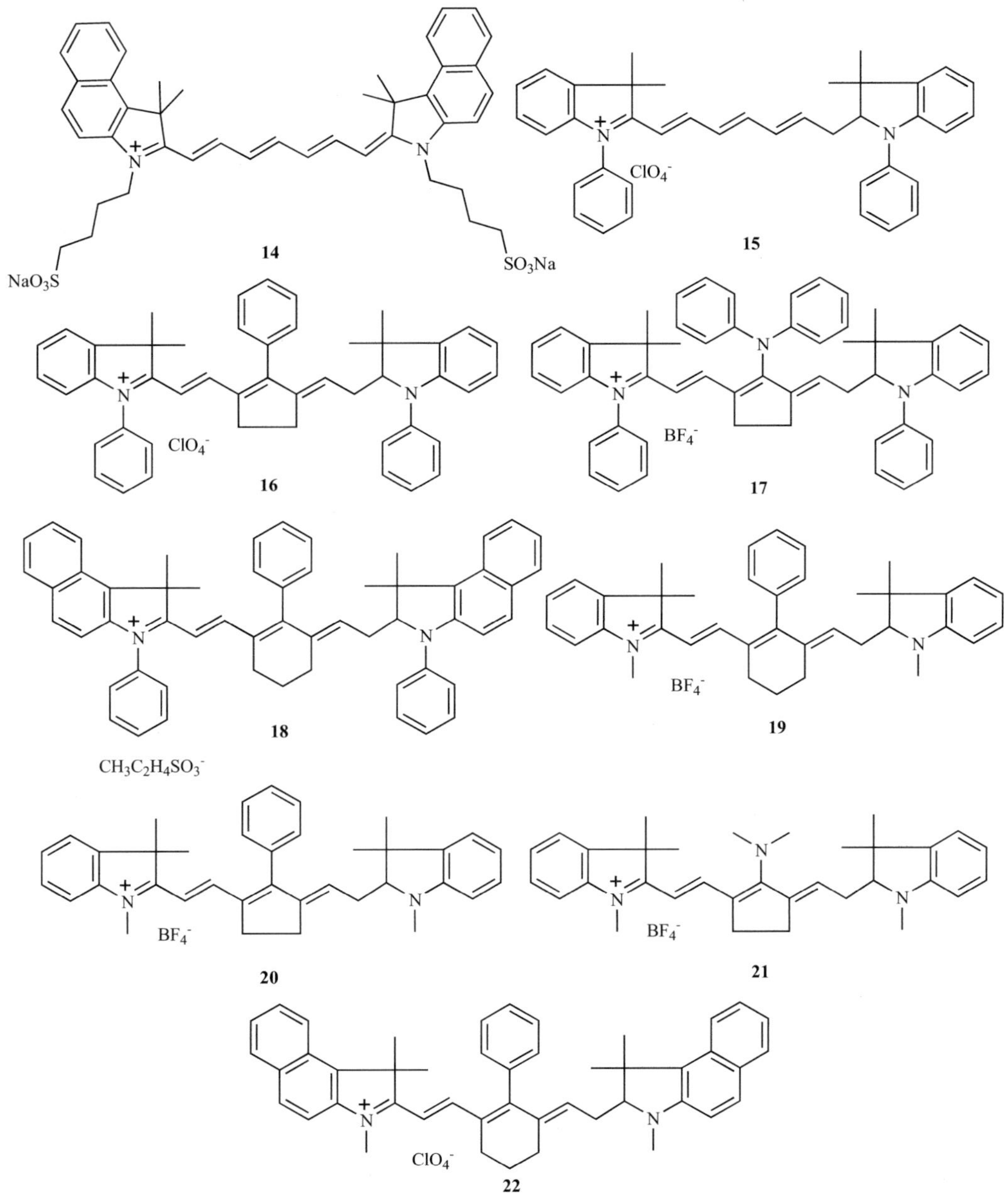

Fig. (8). NIR absorbing cyanine dyes for solar cell application.

Applications in Nonlinear Optics

Nonlinear refraction and absorption are nonlinear optical features directly related. Generally, several mechanisms, both non-parametric and parametric, contribute to the nonlinearities and a consistent evaluation of the responsible mechanisms requires experiments involving different pulse lengths. Optical limiting consists of a sample transmittance decrease under fluence illumination or high intensity. Optical limiting is also regarded as the reverse of saturable absorption where increase in transmittance is seen at high illumination. Phthalocyanines are promising metallo-organic materials for optical power limiting in the visible and NIR spectral range, due to their suitable photophysical properties.

The first reported optical limiting properties of polymethine dyes was given by Penzkofer (Fig. **8**) [149]. Nonlinear energy transmission measurements using femtosecond pulses for concentrated aqueous indocyanine green (**14**) solutions were performed to determine its NLT properties. Formation of J-aggregates in aqueous solution at high concentration was carried out. The nonlinear transmittance behavior could be observed from the plot of energy transmission *versus* cell position, where the transmission was measured over a moving cell sample at a z-axis.

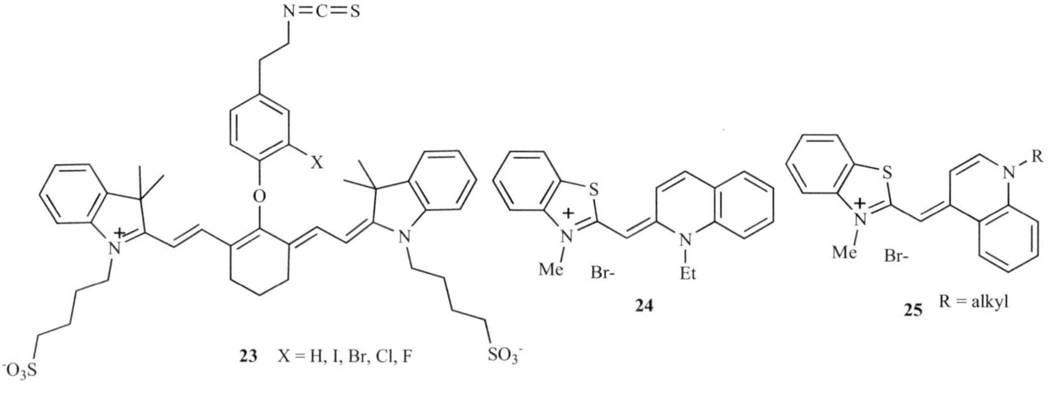

Fig. (9). Carbocyanine dyes for NLO application.

The nonlinear properties of a set of different polymethine dyes **15–22** (Fig. **9**) were studied by van Stryland [150 - 152]. Most of the studied compounds were NIR absorbing, showing optical limiting performance in ethanol solution and in an elastopolymeric material (polyurethane acrylate) [150, 151]. However, some inconsistency on the results caused by apparent degradation of the compounds was observed. Nevertheless, dominant nonlinearity in all the studied dyes was found, showing reverse saturable absorption behavior, leading to very suitable low optical limiting thresholds [151].

The value of the excited state lifetimes in polyurethane acrylate was higher than in ethanol, accounted for the importance of polymeric media on excited-state relaxation processes of the dye molecules [153].

Applications in Biomedicine

When polymethine dyes form complexes with biomacromolecules, usually an increase in fluorescence is observed. Additionally, they are able to form ordered aggregates (dimers, H- and J-aggregates), both conjugated with biomacromolecules such as proteins and nucleic acids.

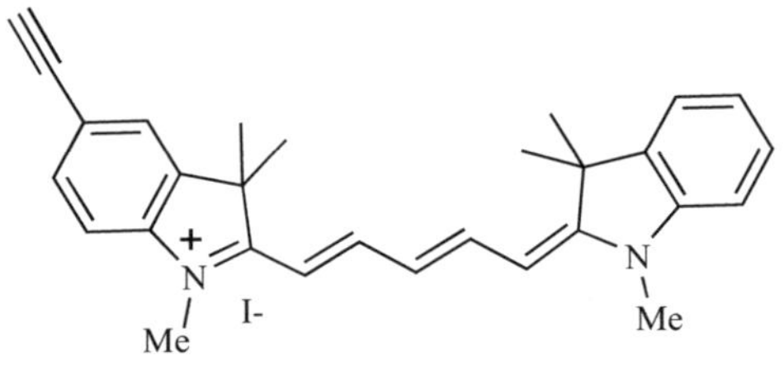

26

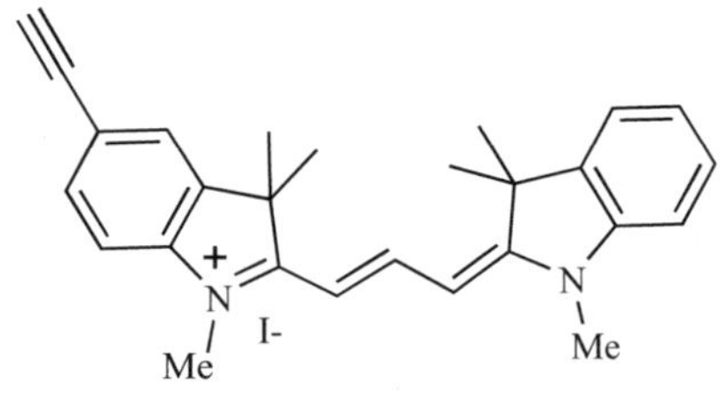

27

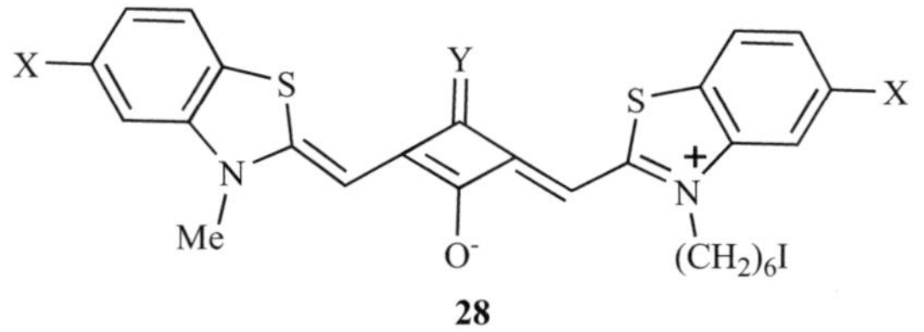

28

a: X = H; Y = O
b: X = F; Y = O
c: X = H; Y = $C(CN)_2$

Fig. (10). Near-IR fluorescent polymethine dyes.

For instance Hammer and Soper [110] prepared several near-IR fluorescent polymethine dyes **23** (Fig. **10**) bearing a heavy atom, appropriate for DNA sequencing. Deligeorgiev [111] also prepared dyes **24** and **25** (Fig. **10**), which did not reveal fluorescent properties in aqueous solution. However, in the presence of DNA, a large fluorescence enhancement was observed.

Balasubramanian [114] reported the attachment of cyanine dyes **26** and **27** (Fig. **11**) to DNA *via* a rigid linker, which showed useful fluorescence and Förster resonance energy transfer properties without altering the stability of duplex DNA. Asseline [115] also reported the synthesis of a squaraine dye with fluorinated benzothiazole rings **28** (Fig. **11**).

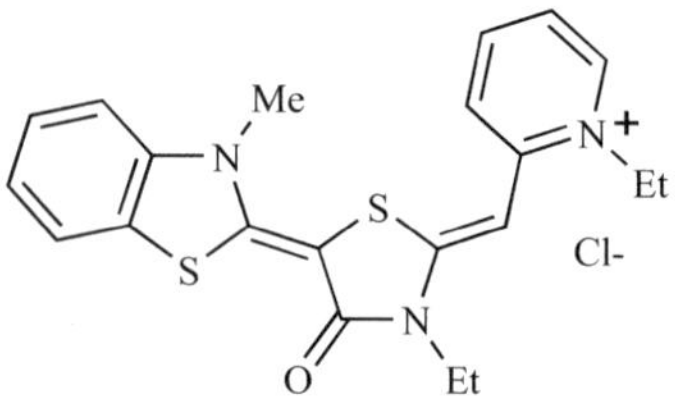

29

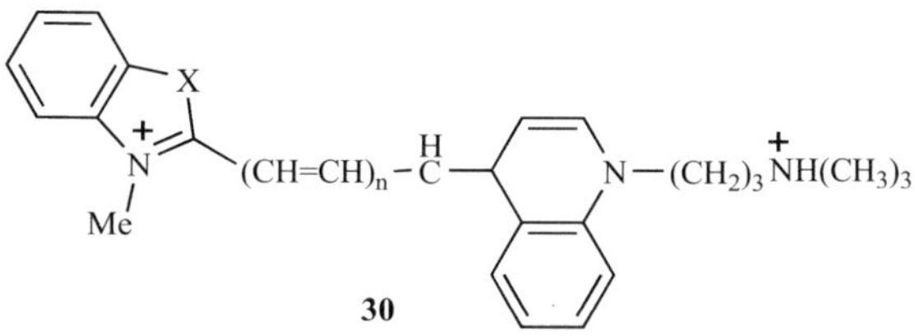

30

Fig. (11). Cyanine dyes for application in medicine.

This new labeled conjugate possessed enhanced chemical stability and photophysical properties when compared with the corresponding non-fluorinated and dicyanosquaraines.

These compounds were used to label a series of oligonucleotides with several lengths and sequences, being the fluorinated squaraine conjugates the best compounds, showing high quantum yields (0.27–0.39) and emission wavelengths >670 nm.

Kawakami [119] reported compound **29** (Fig. **12**) possessing a benzothiazole moiety, which exhibited very favorable biological and chemical properties, being even currently studied as an antitumor agent in clinical trials for solid tumor treatment.

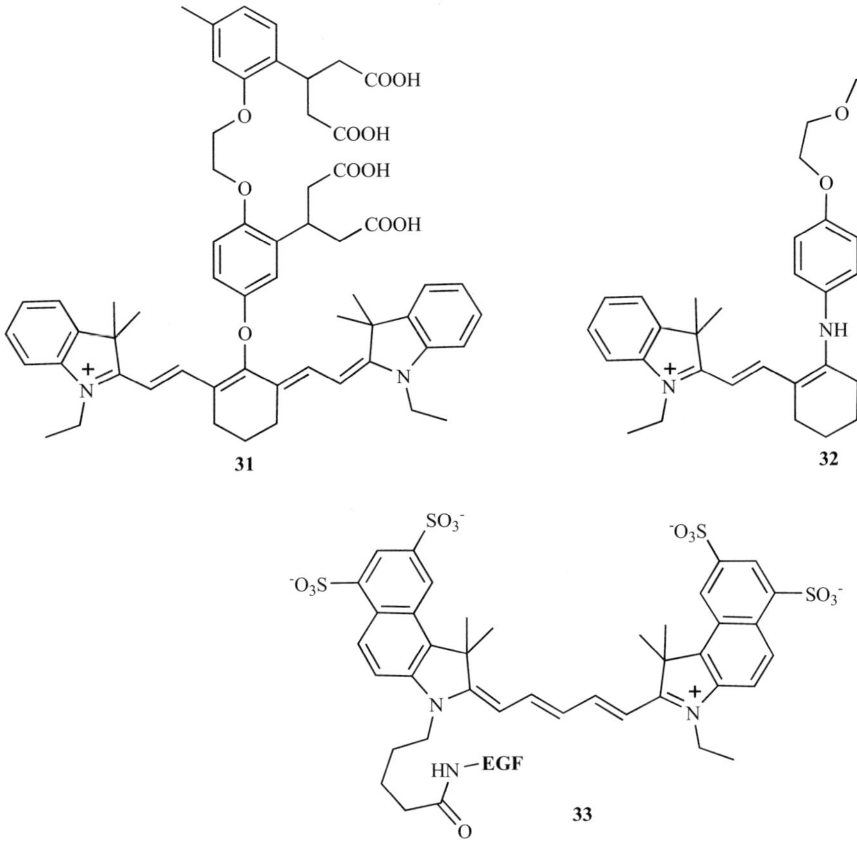

31 **32** **33**

Fig. (12). Carbocyanine dyes for biomedical applications.

Jones and Singer [120] developed cyanine **30**, named SYTOX (Fig. **12**), an unsymmetrical cationic cyanine that binds to the cellular nucleic acids, by means of fluorescence. Akkaya and Ozmen reported dye **31** (Fig. **13**), a fluorescent tricarbocyanine, in combination with bis(2-aminophenoxy)-ethane-*N,N,N',N'*-tetraacetic acid (BAPTA) unit, which acts as fluorescence quencher and calcium receptor [154]. Also Patonay's group described the use of probe **32**, composed of a benzo-15-crown-5 binded with a heptamethine cyanine dye (Fig. **13**) [155], for selective binding to Li^+ ions.

Li reported the synthesis of probe **33** (Fig. **13**) [126], along with biological tests, to acquire the specificity of **33** in the detection of epidermal growth factor (EGF) receptor through fluorescence, by continuous-wave fluorescence imaging, obtained through an intensified charge-coupled device camera. Measurement of time-fluorescence intensity in mice established that **33** accumulated only in MDA-MB-468 tumors. Moreover, little fluorescence was noticed in MDA-M--435 tumors 24 h after injection, which meant that excretion occurred efficiently.

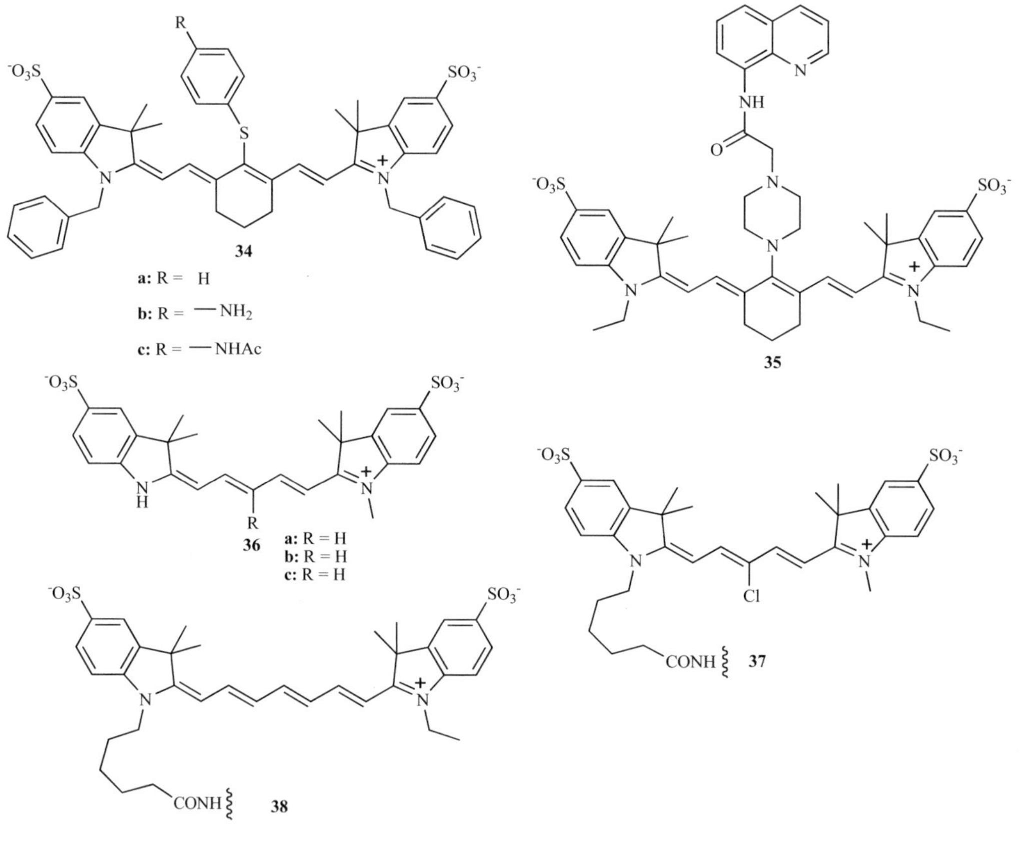

Fig. (13). Cyanine dyes for application in medicine.

Later, Peng reported the photoinduced electron transfer in heptamethine cyanine dyes, adjusted by altering electron donating ability at the polymethine chain's central position [156]. Compounds **34a-c** (Fig. **14**), bearing a 4-aminophenyl-thio group, displayed low fluorescence quantum yield due to efficient photoinduced electron transfer to the heptamethine dye. Nonetheless, the intensity of fluorescence from **34** was considerably improved by acid addition or transition metal ion (Zn^{2+}, Fe^{3+}, Al^{3+} or Ce^{3+}) coordination, as it diminished substituent's HOMO energy level, thus suppressing photoinduced electron transfer [156]. Lin also prepared **35**-Cu^{2+}, a fluorescent probe for sulfide anions (Fig. **14**) [157]. While free compound **35** was fluorescent, the introduction of one equivalent of Cu^{2+} caused a nearly total fluorescence quenching.

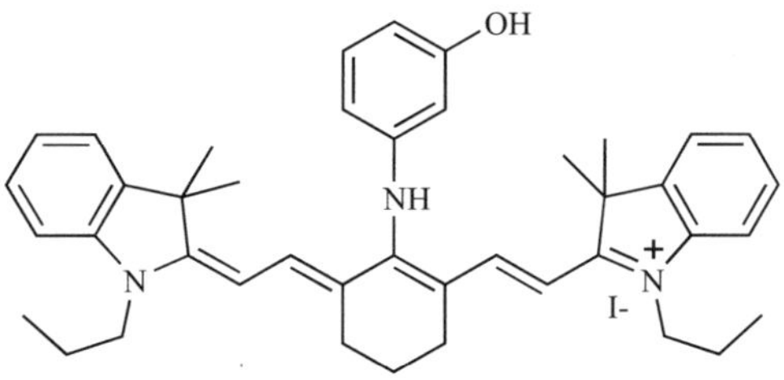

39

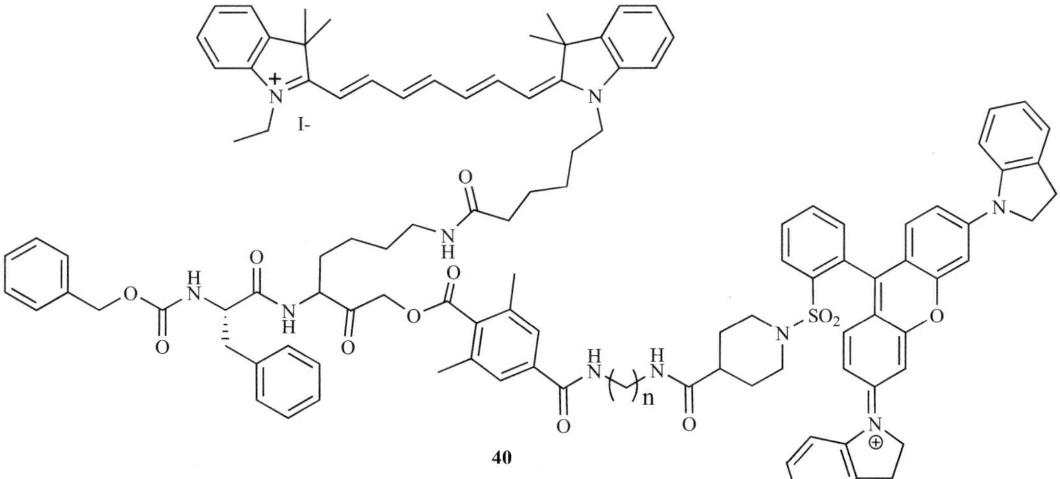

40

Fig. (14). Carbocyanine dyes for biomedical applications.

The pH, measured inside the cells, can be of high importance in several cellular actions, namely cell metastasis, apoptosis, proliferation and drug resistance. Taking that in consideration, Weissleder and Hilderbrand reported pH-responsive NIR dyes **36a–c** (Fig. **14**) [158]. Dye **36b** was employed to distinguish samples of mouse urine and blood, by fluorescence reflectance imaging. Additionally, a ratiometric NIR pH fluorescence probe was also developed, by linking the pH-insensitive compound **38** (Fig. **14**) and pH-sensitive cyanine dye **37** (Fig. **14**) (pKa = 6.2) to bacteriophage surfaces, registering the pH variations in RAW cells [159, 160].

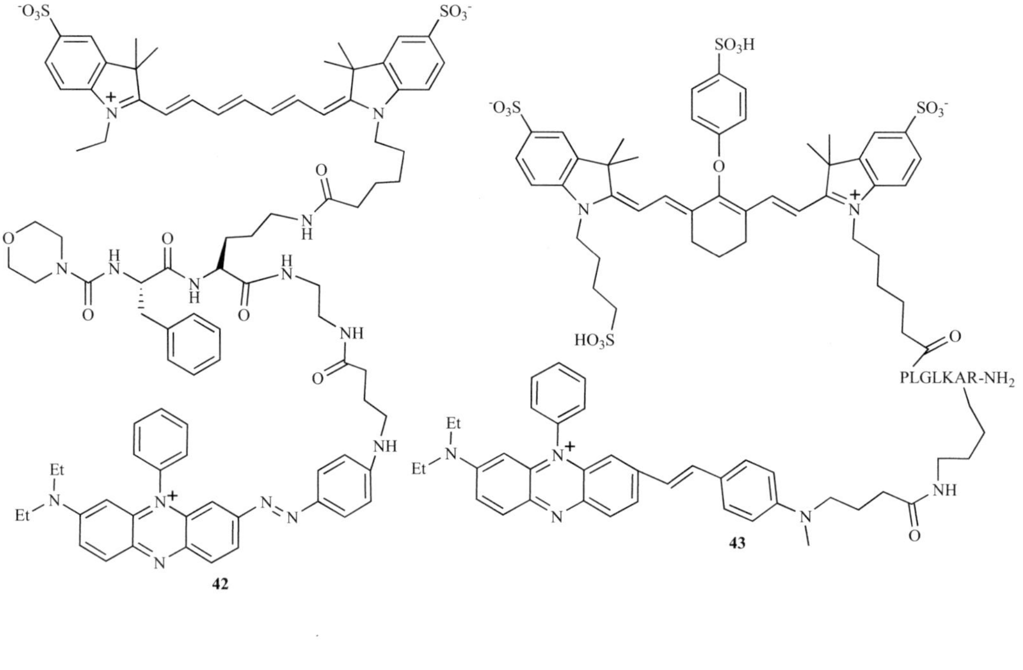

42 **43**

Fig. (15). Cyanine dyes for application in medicine.

Tang also described dye **39** (Fig. **15**), a dual fluorescent NIR pH probe, consisting of an aminophenol-based pH modulator and a NIR tricarbocyanine dye backbone [161]. Upon protonation, **39** showed an absorbing band at 558 nm and an emitting peak at 615 nm, with a 10-fold increase in fluorescence intensity.

In the same year, Bogyo's group described the fluorogenic probe **40** (Fig. **15**), for cysteine protease activity imaging, by connecting a carbocyanine through a cysteine protease-sensitive linker to a fluorescence quenching group [162].

In 2011, Wendt reported probe **41** (Fig. **16**) for cathepsin S activity detection [163], where black hole quencher-3 (BHQ-3) effectively quenched NIR fluorescence of probe **41**. After cathepsin S addition, a fluorescence turn-on signal could be observed, caused by enzymatic peptide bond cleavage. In 2011, Linder demonstrated that the polymethine component's fluorescence of probe **42** (Fig. **16**) could also be effectively quenched by BHQ-3, however, with fluorescence recovery after protease cleavage [164].

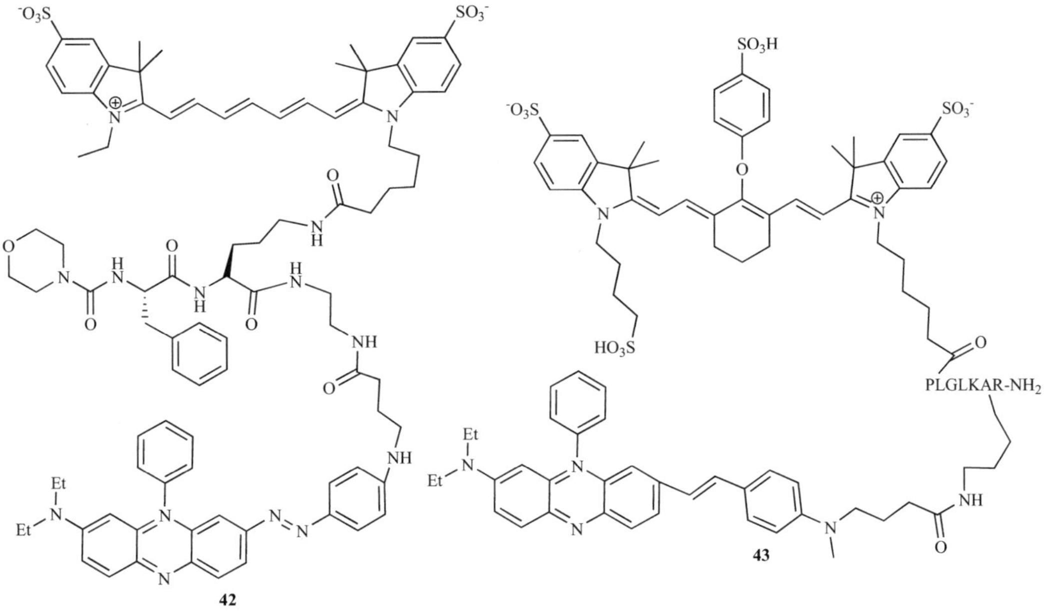

Fig. (16). Polymethine probes.

EXPERIMENTAL SECTION

Synthesis of Monomethine Dyes (Scheme 7)

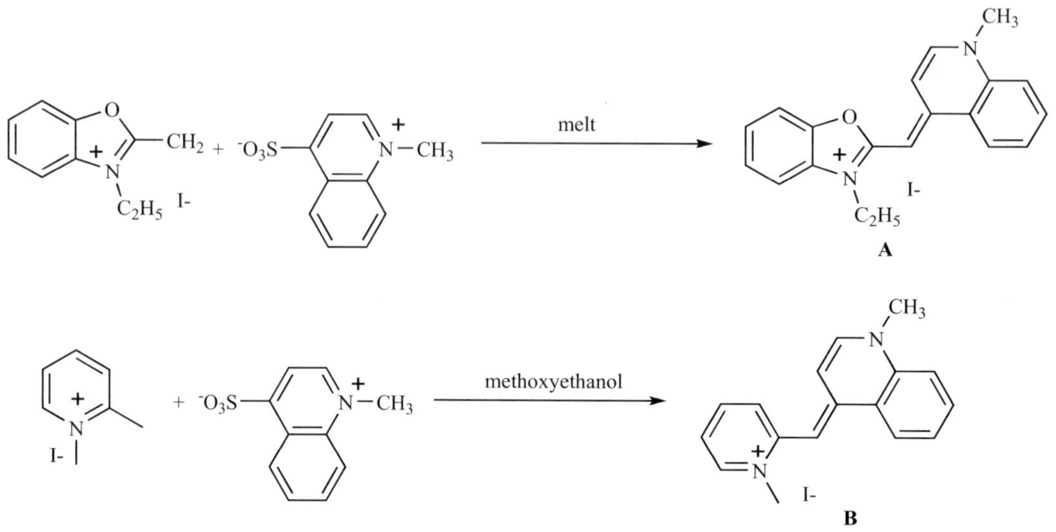

Scheme (7). Monomethine dyes. Procedure for total synthesis of polymethines A and B can be directly found in Reference [53].

Synthesis of Trimethine and Pentamethine Dyes (Scheme 8)

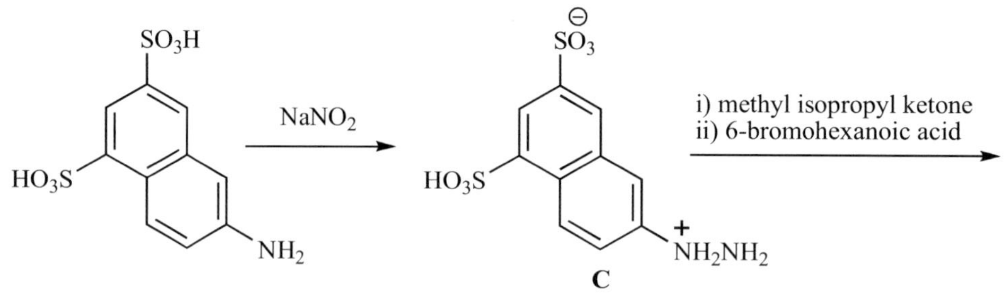

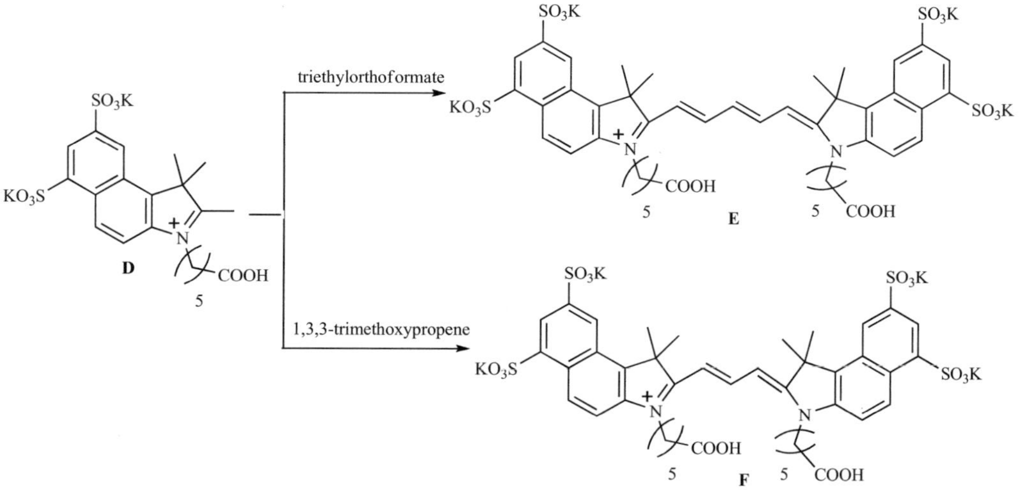

Scheme (8). Trimethine and pentamethine dyes. Procedure for total synthesis of polymethines E and F can be directly found in Reference [61].

Synthesis of Heptamethine Cyanine Dyes by the N,N'-diphenylformamidine Method (Scheme 9)

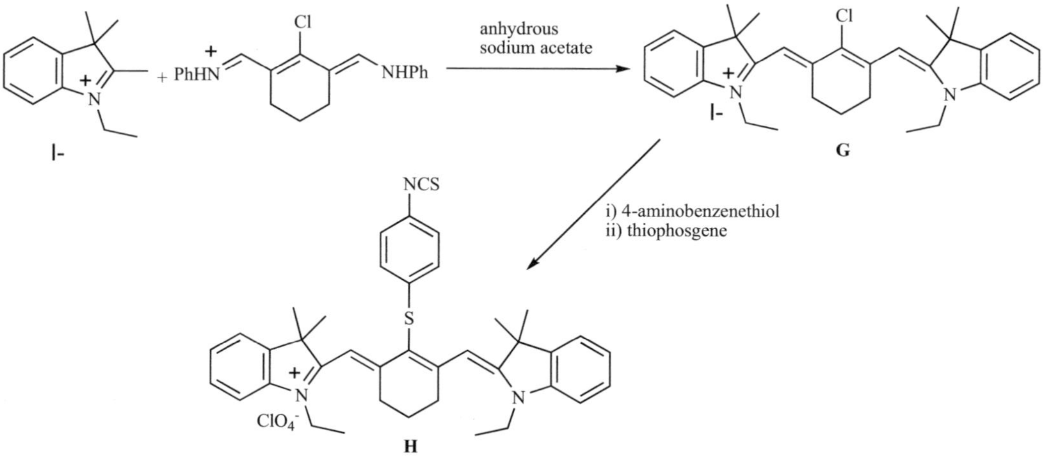

Scheme (9). Heptamethine cyanine dyes. Procedure for total synthesis of polymethine H can be directly found in Reference [67].

Functionalization of Heptamethine Dyes with Amines (Scheme 10)

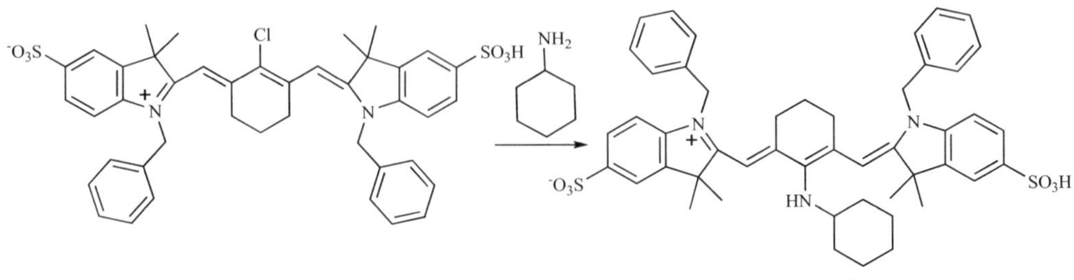

I

Scheme (10). Functionalization of heptamethine dyes with amines. Procedure for total synthesis of polymethine I can be directly found in Reference [165].

CONCLUDING REMARKS

Near infrared (NIR) absorbing/emitting dyes are of utmost importance for a variety of applications, rendering the consideration of scientists, particularly those from multidisciplinary areas. To date, a myriad of types of NIR organic compounds are known at date, but polymethine dyes, particularly cyanines have achieved credit as being one of the most resourceful dyes over the past two decades.

The cyanines offer great chemical and photochemical stability, both of which are two advantageous qualities for a dye, especially for aiming commercial applications. Additionally, modified cyanine dyes containing anchoring supports for attachment to a variety of functionalities may offer this class of dyes an exceptional prospect in many applications, from materials science to biomedical applications.

The field of polymethine dyes can be greatly improved by the intensification of new architectures, naturally matched by the potential for new applications, at the frontier between chemistry, physics, biology and engineering science.

REFERENCES

[1] Kachkovski, A.D.; Dekhtyar, M.L. Electronic properties of polymethine compounds: 2. Electron donor ability and relative stability. *Dyes Pigm.,* **1983**, *22*, 83-97. [http://dx.doi.org/10.1016/0143-7208(93)85008-N]

[2] *Heterocyclic Polymethine Dyes: Synthesis, Properties and Applications in Topics in Heterocyclic Chemistry*; Strekowski, L., Ed.; Springer: Berlin, **2008**.
[http://dx.doi.org/10.1007/978-3-540-79064-8]

[3] Patonay, G.; Salon, J.; Sowell, J.; Strekowski, L. Noncovalent labeling of biomolecules with red and near- infrared dyes. *Molecules,* **2004**, *9*(3), 40-49.
[http://dx.doi.org/10.3390/90300040] [PMID: 18007410]

[4] Narayanan, N.; Patonay, G. A new method for the synthesis of heptamethine cyanine dyes: synthesis of new near-infrared fluorescent labels. *J. Org. Chem.,* **1995**, *60*, 2391-2395.
[http://dx.doi.org/10.1021/jo00113a018]

[5] Flanagan, J.H., Jr; Khan, S.H.; Menchen, S.; Soper, S.A.; Hammer, R.P. Functionalized tricarbocyanine dyes as near-infrared fluorescent probes for biomolecules. *Bioconjug. Chem.,* **1997**, *8*(5), 751-756.
[http://dx.doi.org/10.1021/bc970113g] [PMID: 9327141]

[6] Williams, C.H. Researches on chinoline and its homologues. *Trans. R. Soc. Edinb.,* **1856**, *21*, 377-402.
[http://dx.doi.org/10.1017/S0080456800032208]

[7] Vogel, H. Ueber die Lichtempfindlichkeit des Bromsilbers für die sogenannten chemisch unwirksamen Farben. *Annal. Phys.,* **1874**, *226*, 453-459.
[http://dx.doi.org/10.1002/andp.18742261112]

[8] Wyler, H. Das Experiment: Papierelektrophorese. *Chem. Unserer Zeit,* **1969**, *3*, 111-115.
[http://dx.doi.org/10.1002/ciuz.19690030404]

[9] Wyler, H. Die Betalaine. *Chem. Unserer Zeit,* **1969**, *3*, 146-151.
[http://dx.doi.org/10.1002/ciuz.19690030504]

[10] Musso, H. The pigments of fly agaric, *Amanita muscaria. Tetrahedron,* **1979**, *35*, 2843-2853.
[http://dx.doi.org/10.1016/S0040-4020(01)99498-0]

[11] Reichardt, C. Chiral polymethine dyes: A remarkable but forgotten conjugated π system. *J. Phys. Org. Chem.,* **1995**, *8*, 761-773.
[http://dx.doi.org/10.1002/poc.610081202]

[12] Hamer, F.M. *The Chemistry of Heterocyclic Compounds*; Weissberger, A., Ed.; Interscience: New York, NY, **1964**, p. 18.
[http://dx.doi.org/10.1002/9780470186794]

[13] Mishra, A.; Behera, R.K.; Behera, P.K.; Mishra, B.K.; Behera, G.B. Cyanines during the 1990s: A review *Chem. Rev.,* **2000**, *100*, 1973-2012.

[14] Venkatraman, K. *The chemistry of synthetic dyes*; Academic Press: New York, **1952**, Vol. 2, p. 1143.

[15] Ficken, G.E. *The chemistry of synthetic dyes*; Venkatraman, K., Ed.; Academic Press: New York, **1971**, Vol. 4, p. 21.

[16] Gamon, N.; Reicherdt, C. Darstellung eines β-Phospha-trimethincyanin-Farbstoffes. *Angew. Chem.,* **1977**, *89*, 418-419.
[http://dx.doi.org/10.1002/ange.19770890618]

[17] Gieseking, R.L.; Mukhopadhyay, S.; Risko, C.; Marder, S.R.; Brédas, J.L. 25th anniversary article: Design of polymethine dyes for all-optical switching applications: guidance from theoretical and computational studies. *Adv. Mater.,* **2014**, *26*(1), 68-83.
[http://dx.doi.org/10.1002/adma.201302676] [PMID: 24302357]

[18] Henary, M.; Levitz, A. Synthesis and applications of unsymmetrical carbocyanine dyes. *Dyes Pigm.,* **2013**, *99*, 1107-1116.
[http://dx.doi.org/10.1016/j.dyepig.2013.08.001]

[19] Panigrahi, M.; Dash, S.; Patel, S.; Mishra, B.K. Syntheses of cyanines: a review. *Tetrahedron,* **2012**, *68*, 781-805.
[http://dx.doi.org/10.1016/j.tet.2011.10.069]

[20] Kulinich, A.V.; Ishchenko, A.A. Merocyanine dyes: synthesis, structure, properties and applications. *Russ. Chem. Rev.,* **2009**, *78*, 141-164.
[http://dx.doi.org/10.1070/RC2009v078n02ABEH003900]

[21] Gadjev, N.I.; Deligeorgiev, T.G.; Kim, S.H. Preparation of monomethine cyanine dyes as noncovalent labels for nucleic acids. *Dyes Pigm.,* **1999**, *40*, 181-186.
[http://dx.doi.org/10.1016/S0143-7208(98)00046-1]

[22] Gadjev, N.I.; Deligeorgiev, T.G.; Timtcheva, I.; Maximova, V.; Vassilev, A.; Jacobson, J.P.; Drexhage, K.H. Homodimeric monomethine cyanine dyes SOSO-1 and TOTO-1-6C-synthesis and fluorescence properties in the presence of nucleic acids. *Dyes Pigm.,* **2004**, *61*, 79-84.
[http://dx.doi.org/10.1016/j.dyepig.2003.10.006]

[23] Gadjev, N.I.; Deligeorgiev, T.G.; Vassilev, A.; Drexhage, K.H.; Yarmoluk, S.M. Synthesis of novel monomeric cyanine dyes containing mercapto and thioacetyl substituents for nucleic acid detection. *Dyes Pigm.,* **2006**, *70*, 185-191.
[http://dx.doi.org/10.1016/j.dyepig.2005.05.010]

[24] Gadjev, N.I.; Deligeorgiev, T.G.; Timtcheva, I.; Maximova, V.; Vassilev, A.; Katerinopoulos, H.E.; Foukaraki, E. Synthesis of homodimeric monomethine cyanine dyes as noncovalent nucleic acid labels and their absorption and fluorescence spectral characteristics. *Dyes Pigm.,* **2000**, *44*, 131-136.
[http://dx.doi.org/10.1016/S0143-7208(99)00084-4]

[25] Abd el Aal, R.M.; Younis, M. Synthesis and antimicrobial activity of certain novel monomethine cyanine dyes. *Dyes Pigm.,* **2004**, *60*, 205-214.
[http://dx.doi.org/10.1016/S0143-7208(03)00131-1]

[26] Yarmouluk, S.M.; Kovalska, V.B.; Kocheshev, I.A. Interaction of cyanine dyes with nucleic acids-XXVII: synthesis and spectral properties of novel homodi- and homotrimeric monomethine cyanine dyes. *Dyes Pigm.,* **2001**, *50*, 21-28.
[http://dx.doi.org/10.1016/S0143-7208(01)00023-7]

[27] Vasilev, A.; Deligeorgiev, T.G.; Gadjev, N.; Drexhage, K.H. Synthesis of novel monomeric and homodimeric cyanine dyes based on oxazolo[4,5-b]pyridinium and quinolinium end groups for nucleic acid detection. *Dyes Pigm.,* **2005**, *66*, 135-142.
[http://dx.doi.org/10.1016/j.dyepig.2004.09.019]

[28] Koraiem, A.I.; Khalafalla, A.K.; Soliman, H.A. Synthesis and spectra of pyrazolo(4,5-b)pyrido-(2-

3-c)pyrazine dimethine cyanine dyes. *Dyes Pigm.,* **1990**, *12*, 27-34.
[http://dx.doi.org/10.1016/0143-7208(90)85003-7]

[29] Zhang, X.H.; Wang, L.Y.; Zhai, G.H.; Wen, Z.Y.; Zhang, Z.X. Microwave-assisted solvent-free synthesis of some dimethine cyanine dyes, spectral properties and TD-DFT/PCM calculations. *Bull. Korean Chem. Soc.,* **2007**, *28*, 2382-2388.
[http://dx.doi.org/10.5012/bkcs.2007.28.12.2382]

[30] Koraiem, A.I.; Abd el Aal, R.M.; Khalil, Z.H.; El-Kodey, A.M. The use of oxonium salts in the synthesis of mono-, β-substituted dimethine and styryl cyanine dyes. *Dyes Pigm.,* **2005**, *66*, 201-209.
[http://dx.doi.org/10.1016/j.dyepig.2004.09.015]

[31] Koraiem, A.I.; Abd el Aal, R.M.; El-Deen, N.M. The use of N-bridgehead heterocyclic indolizinium ylide in the synthesis of aza-cyanine dyes. *Dyes Pigm.,* **2006**, *68*, 235-242.
[http://dx.doi.org/10.1016/j.dyepig.2005.01.007]

[32] Shindy, H.A.; El-Maghraby, M.A.; Eissa, F.M. Novel quinone cyanine dyes: synthesis and spectral studies. *Dyes Pigm.,* **2002**, *52*, 79-87.
[http://dx.doi.org/10.1016/S0143-7208(01)00077-8]

[33] Koraiem, A.I.; Khalil, Z.H.; El-Hamed, R.M. Synthesis of some β- and ψ-substituted azotrimethine cyanine dyes. *Dyes Pigm.,* **1990**, *13*, 197-204.
[http://dx.doi.org/10.1016/0143-7208(90)80019-L]

[34] Koraiem, A.I.; Khalil, Z.H.; El-Hamed, R.M. Synthesis and spectra of some asymmetric trimethine cyanine dyes. *Dyes Pigm.,* **1990**, *13*, 289-299.
[http://dx.doi.org/10.1016/0143-7208(90)85027-L]

[35] Meguellati, K.; Spichty, M.; Ladame, S. Reversible synthesis and characterization of dynamic imino analogues of trimethine and pentamethine cyanine dyes. *Org. Lett.,* **2009**, *11*(5), 1123-1126.
[http://dx.doi.org/10.1021/ol802913b] [PMID: 19193121]

[36] Koraiem, A.I.; El-Maghraby, M.A.; Khalafalla, A.K.; Shindey, H.A. Studies on the synthesis and solvatochromic behaviour of mono- and tri-methine cyanines: methine cyanine dyes, synthesis and solvatochromism. *Dyes Pigm.,* **1989**, *11*, 47-63.
[http://dx.doi.org/10.1016/0143-7208(89)85025-9]

[37] Koraiem, A.I.; El-Maghraby, M.A.; El-Hamed, R.M. Synthesis and electronic spectra of some mono-, di- and tri-cationic di- and tetra-methine cyanine dyes. *Dyes Pigm.,* **1990**, *14*, 59-72.
[http://dx.doi.org/10.1016/0143-7208(90)87006-O]

[38] Wolinska, E.; Strekowski, L. Visible bis(indolium tetramethine hemicyanine) dyes with a spacer derived from oligo(ethylene glycol). *Heterocycl. Commun.,* **2009**, *15*, 407-409.
[http://dx.doi.org/10.1515/HC.2009.15.6.407]

[39] Reidlinger, C.; Dworczak, R.; Fabian, W.M.; Junek, H. Structure-color correlations of penta- and heptamethines: Syntheses with nitriles XCIV. *Dyes Pigm.,* **1994**, *24*, 185-204.
[http://dx.doi.org/10.1016/0143-7208(94)80009-X]

[40] Voitenko, Z.; Mazieres, M.R.; Sanchez, M.; Wolf, J.G. Newly substituted pentamethine merocyanines. Part 1: Synthesis, physical properties, and synthetic applications. *Tetrahedron,* **2001**, *57*, 1059-1066.
[http://dx.doi.org/10.1016/S0040-4020(00)01095-4]

[41] Dai, Z.; Peng, B. Synthesis and characterization of indodicarbocyanines. *Dyes Pigm.,* **1998**, *36*, 169-175.
[http://dx.doi.org/10.1016/S0143-7208(97)00002-8]

[42] El-Aal, R.M.; Koraiem, A.I. Synthesis and photophysics in organic solvents of meso-substituted pentamethine and related metal complexes cyanine dyes. *Dyes Pigm.,* **2002**, *54*, 121-129.
[http://dx.doi.org/10.1016/S0143-7208(02)00034-7]

[43] Mazieres, M.R.; Duprat, C.; Bellan, J.; Wolf, J.G. Synthesis and characterisation of new phosphonate labelled cyanines. *Dyes Pigm.,* **2007**, *74*, 404-409.
[http://dx.doi.org/10.1016/j.dyepig.2006.03.003]

[44] Otsuka, A.; Funabiki, K.; Sugiyama, N.; Mase, H.; Yoshida, T.; Minoura, H.; Matsui, M. Design and synthesis of near-infrared-active heptamethine-cyanine dyes to suppress aggregation in a dye-sensitized porous zinc oxide solar cell. *Chem. Lett.,* **2008**, *37*, 176-177.
[http://dx.doi.org/10.1246/cl.2008.176]

[45] Lee, H.; Mason, J.C.; Achilefu, S. Synthesis and spectral properties of near-infrared aminophenyl-, hydroxyphenyl-, and phenyl-substituted heptamethine cyanines. *J. Org. Chem.,* **2008**, *73*(2), 723-725.
[http://dx.doi.org/10.1021/jo701793h] [PMID: 18095702]

[46] El-Aal, R.M. Thiazoline and thiazoloxazole in synthesis of novel meso-substituted mono-, tri-, and hepta-methine cyanine dyes. *Dyes Pigm.,* **2004**, *61*, 251-261.
[http://dx.doi.org/10.1016/j.dyepig.2003.10.014]

[47] Nagao, Y.; Sakai, T.; Kozawa, K.; Urano, T. Synthesis and properties of barbiturate indolenine heptamethinecyanine dyes. *Dyes Pigm.,* **2007**, *73*, 344-352.
[http://dx.doi.org/10.1016/j.dyepig.2006.01.039]

[48] Lee, H.; Mason, J.C.; Achilefu, S. Heptamethine cyanine dyes with a robust C-C bond at the central position of the chromophore. *J. Org. Chem.,* **2006**, *71*(20), 7862-7865.
[http://dx.doi.org/10.1021/jo061284u] [PMID: 16995699]

[49] Shandura, M.P.; Poronik, Y.M.; Kovtun, Y.P. Substituted xanthylocyanines. II. Pyroninocyanines. *Dyes Pigm.,* **2005**, *66*, 171-177.
[http://dx.doi.org/10.1016/j.dyepig.2004.09.011]

[50] Poronik, Y.M.; Shandura, M.P.; Kovtun, Y.P. Substituted xanthylocyanines. III. Dyes containing non-symmetrically substituted xanthylium core. *Dyes Pigm.,* **2007**, *72*, 199-207.
[http://dx.doi.org/10.1016/j.dyepig.2005.08.017]

[51] Kim, J.S.; Kodagahally, R.; Strekowski, L.; Patonay, G. A study of intramolecular H-complexes of novel bis(heptamethine cyanine) dyes. *Talanta,* **2005**, *67*(5), 947-954.
[http://dx.doi.org/10.1016/j.talanta.2005.04.025] [PMID: 18970263]

[52] Clark, L.M. XXXIV.-The quaternary salts of benzoxazoles. *J. Chem. Soc.,* **1926**, 232-235.
[http://dx.doi.org/10.1039/jr9262900232]

[53] Deligeorgiev, T.G.; Zaneva, D.A.; Katerinopoulos, H.E.; Kolev, V.N. A novel method for the preparation of monomethine cyanine dyes. *Dyes Pigm.,* **1999**, *41*, 49-54.
[http://dx.doi.org/10.1016/S0143-7208(98)00061-8]

[54] Abd El-Aal, R.M.; Koraierm, A.I. Synthesis, absorption spectra studies and biological activity of some novel conjugated dyes. *J. Chil. Chem. Soc.,* **2000**, *47*, 389-395. [http://dx.doi.org/10.1002/jccs.200000052]

[55] Koenig, W. Über Indolenino-cyanine (Indocyanine). *Ber. Dtsch. Chem. Ges.,* **1924**, *57*, 685-692. [http://dx.doi.org/10.1002/cber.19240570420]

[56] Hamer, F.M. CCCLXXIV.-A general method for the preparation of carbocyanine dyes. *J. Chem. Soc.,* **1927**, 2796-2804. [http://dx.doi.org/10.1039/jr9270002796]

[57] Peng, Z-H.; Qun, L.; Zhou, X-F.; Carroll, S.; Geise, H.J.; Peng, B.X.; Dommisse, R.; Carleer, R. Synthesis and physical performance of indole and benzimidazole cyanine dyes. *J. Mater. Chem.,* **1996**, *6*, 559-565. [http://dx.doi.org/10.1039/jm9960600559]

[58] Rowe, F.M.; Twitchett, H.J. 380. A reaction of certain diazosulphonates derived from β-naphthol-1-sulphonic acid. Part XVI. The constitution and reactions of 1-methoxy-3-(4′- and 3′-nitroaryl)-4-methylene-3: 4-dihydrophthalazines. *J. Chem. Soc.,* **1936**, 1704-1713. [http://dx.doi.org/10.1039/jr9360001704]

[59] Peng, Z-H.; Geise, H.J.; Zhou, X-F.; Peng, B-X.; Carleer, R.; Dommisse, R. Structure of benzimidazole cyanine dyes, their spectroscopy and their performance in photographic emulsions. *Liebigs Ann. Chem.,* **1997**, 27-33. [http://dx.doi.org/10.1002/jlac.199719970107]

[60] De Rossi, U.; Moll, J.; Spieles, M.; Bach, G.; Daehne, S. Control of the J-aggregation phenomenon by variation of the N-alkyl-substituents. *J. Prakt. Chem.,* **1995**, *337*, 203-208. [http://dx.doi.org/10.1002/prac.19953370144]

[61] Mujumdar, S.R.; Mujumdar, R.B.; Grant, C.M.; Waggoner, A.S. Cyanine-labeling reagents: sulfobenzindocyanine succinimidyl esters. *Bioconjug. Chem.,* **1996**, *7*(3), 356-362. [http://dx.doi.org/10.1021/bc960021b] [PMID: 8816960]

[62] Jung, M.E.; Kim, W-J. Practical syntheses of dyes for difference gel electrophoresis. *Bioorg. Med. Chem.,* **2006**, *14*(1), 92-97. [http://dx.doi.org/10.1016/j.bmc.2005.07.068] [PMID: 16183291]

[63] Chipon, B.; Clavé, G.; Bouteiller, C.; Massonneau, M.; Renard, P-Y.; Romieu, A. Synthesis and post-synthetic derivatization of a cyanine-based amino acid. Application to the preparation of a novel water-soluble NIR dye. *Tetrahedron Lett.,* **2006**, *47*, 8279-8284. [http://dx.doi.org/10.1016/j.tetlet.2006.09.100]

[64] Mahmoud, R.; El-Aal, A. Synthesis and absorption spectra of new polymethine cyanine dyes. *Dyes Pigm.,* **2002**, *52*, 129-136. [http://dx.doi.org/10.1016/S0143-7208(01)00088-2]

[65] Makin, S.M.; Boiko, I.I.; Shavrygina, O.A. Synthesis and study of tricarbocyanines containing five- and six-membered cycles in the chromophore. *Zh. Org. Khim.,* **1977**, *13*, 1189-1192.

[66] Lipowska, M.; Patonay, G.; Strekowski, L. A novel near-infrared cyanine dye for bio analytical applications. *Heterocycl. Commun.,* **1995**, *1*, 427-430.

[http://dx.doi.org/10.1515/HC.1995.1.5-6.427]

[67] Strekowski, L.; Lipowska, M.; Patonay, G. Substitution reactions of a nucleofugal group in heptamethine cyanine dyes. Synthesis of an isothiocyanato derivative for labeling of proteins with a near-infrared chromophore. *J. Org. Chem.,* **1992**, *57*, 4578-4580.
[http://dx.doi.org/10.1021/jo00043a009]

[68] Narayanan, N.; Strekowski, L.; Lipowska, M.; Patonay, G. A new method for the synthesis of heptamethine cyanine dyes: synthesis of new near-Infrared fluorescent labels. *J. Org. Chem.,* **1997**, *62*, 9387-9387.
[http://dx.doi.org/10.1021/jo974022o]

[69] Song, F.; Peng, X.; Lu, E.; Zhang, R.; Chen, X.; Song, B. Syntheses, spectral properties and photostabilities of novel water-soluble near-infrared cyanine dyes. *J. Photochem. Photobiol. Chem.,* **2004**, *168*, 53-57.
[http://dx.doi.org/10.1016/j.jphotochem.2004.05.012]

[70] Peng, X.; Song, F.; Lu, E.; Wang, Y.; Zhou, W.; Fan, J.; Gao, Y. Heptamethine cyanine dyes with a large stokes shift and strong fluorescence: a paradigm for excited-state intramolecular charge transfer. *J. Am. Chem. Soc.,* **2005**, *127*(12), 4170-4171.
[http://dx.doi.org/10.1021/ja043413z] [PMID: 15783189]

[71] Gorecki, T.; Patonay, G.; Strekowski, L. Synthesis of novel near-infrared cyanine dyes for metal ion determination. *J. Heterocycl. Chem.,* **1996**, *33*, 1871-1876.
[http://dx.doi.org/10.1002/jhet.5570330652]

[72] Narayanan, M.; Little, G.; Raghavachari, R.; Gibson, J.; Lugade, A.; Prescott, C.; Reiman, K.; Draney, D. *Near-infrared dyes for high technology applications*; Kluwer: Dordrecht, The Netherlands, **1998**, p. 141.
[http://dx.doi.org/10.1007/978-94-011-5102-3_7]

[73] Yang, C.; Shimelis, O.; Zhou, X.; Li, G.; Bayle, C.; Nertz, M.; Lee, H.; Strekowski, L.; Patonay, G.; Couderc, F.; Giese, R.W. Handling and detection of 0.8 amol of a near-infrared cyanine dye by capillary electrophoresis with laser-induced fluorescence detection. *J. Chromatogr. A,* **2002**, *979*(1-2), 307-314.
[http://dx.doi.org/10.1016/S0021-9673(02)01441-3] [PMID: 12498262]

[74] Gallaher, D.L.; Johnson, M.E. Development of near-infrared fluorophoric labels for the determination of fatty acids separated by capillary electrophoresis with diode laser induced fluorescence detection. *Analyst (Lond.),* **1999**, *124*, 1541-1546.
[http://dx.doi.org/10.1039/a904541d]

[75] Zaheer, A.; Lenkinski, R.E.; Mahmood, A.; Jones, A.G.; Cantley, L.C.; Frangioni, J.V. *In vivo* near-infrared fluorescence imaging of osteoblastic activity. *Nat. Biotechnol.,* **2001**, *19*(12), 1148-1154.
[http://dx.doi.org/10.1038/nbt1201-1148] [PMID: 11731784]

[76] Kiyose, K.; Kojima, H.; Urano, Y.; Nagano, T. Development of a ratiometric fluorescent zinc ion probe in near-infrared region, based on tricarbocyanine chromophore. *J. Am. Chem. Soc.,* **2006**, *128*(20), 6548-6549.
[http://dx.doi.org/10.1021/ja060399c] [PMID: 16704241]

[77] Zhou, L-C.; Zhao, G-J.; Liu, J-F.; Han, K-L.; Wu, Y-K.; Peng, X-J.; Sun, M-T. The charge transfer

mechanism and spectral properties of a near-infrared heptamethine cyanine dye in alcoholic and aprotic solvents. *J. Photochem. Photobiol. Chem.,* **2007**, *187*, 305-310. [http://dx.doi.org/10.1016/j.jphotochem.2006.10.027]

[78] Chen, Y.; Gryshuk, A.; Achilefu, S.; Ohulchansky, T.; Potter, W.; Zhong, T.; Morgan, J.; Chance, B.; Prasad, P.N.; Henderson, B.W.; Oseroff, A.; Pandey, R.K. A novel approach to a bifunctional photosensitizer for tumor imaging and phototherapy. *Bioconjug. Chem.,* **2005**, *16*(5), 1264-1274. [http://dx.doi.org/10.1021/bc050177o] [PMID: 16173807]

[79] Tarazi, L.; George, A.; Patonay, G.; Strekowski, L. Spectral characterization of a novel near-infrared cyanine dye: a study of its complexation with metal ions. *Talanta,* **1998**, *46*(6), 1413-1424. [http://dx.doi.org/10.1016/S0039-9140(98)00012-5] [PMID: 18967271]

[80] Ernst, L.A.; Gupta, R.K.; Mujumdar, R.B.; Waggoner, A.S. Cyanine dye labeling reagents for sulfhydryl groups. *Cytometry,* **1989**, *10*(1), 3-10. [http://dx.doi.org/10.1002/cyto.990100103] [PMID: 2917472]

[81] Mendes, P.M. Stimuli-responsive surfaces for bio-applications. *Chem. Soc. Rev.,* **2008**, *37*(11), 2512-2529. [http://dx.doi.org/10.1039/b714635n] [PMID: 18949123]

[82] Yagai, S.; Kitamura, A. Recent advances in photoresponsive supramolecular self-assemblies. *Chem. Soc. Rev.,* **2008**, *37*(8), 1520-1529. [http://dx.doi.org/10.1039/b703092b] [PMID: 18648678]

[83] Grishina, A.D.; Pereshivko, L.Y.; Krivenko, T.V.; Shapiro, B.I.; Vannikov, A.V. Photoelectric properties of polyimide containing J-aggregates of cyanine dyes. *High Energy Chem.,* **2003**, *37*, 169-176. [http://dx.doi.org/10.1023/A:1024199219759]

[84] Raymo, F.M.; Tomasulo, M. Fluorescence modulation with photochromic switches. *J. Phys. Chem. A,* **2005**, *109*(33), 7343-7352. [http://dx.doi.org/10.1021/jp052440o] [PMID: 16834100]

[85] Heilemann, M.; Margeat, E.; Kasper, R.; Sauer, M.; Tinnefeld, P. Carbocyanine dyes as efficient reversible single-molecule optical switch. *J. Am. Chem. Soc.,* **2005**, *127*(11), 3801-3806. [http://dx.doi.org/10.1021/ja044686x] [PMID: 15771514]

[86] Tyutyulkov, N.; Fabian, J.; Mehlhorn, A.; Dietz, F.; Tadjer, A. *Polymethine Dyes*; Kliment Ohridski University Press: Sofia, **1991**.

[87] Yang, X.B.; Ding, S.T.; Yang, Y.S.; Zhou, X.H. A review of near-infrared laser protective dyes. *Chin. J. Org. Chem.,* **2002**, *22*, 33-41.

[88] Ishchenko, A.A.; Grabchuk, G.P. Physical and chemical problems of the creation of photostable converters of light energy on the basis of dyed polymers. *Theor. Exp. Chem.,* **2009**, *45*, 143-167. [http://dx.doi.org/10.1007/s11237-009-9078-5]

[89] Lanzafame, J.M.; Min, L.; Miller, R.J.; Muenter, A.A.; Parkinson, B.A. Electron injection from adsorbed oxazine into SnS_2. *Mol. Cryst. Liq. Cryst. (Phila. Pa.),* **1991**, *194*, 287-292. [http://dx.doi.org/10.1080/00268949108041178]

[90] Möbius, D. Scheibe aggregates. *Adv. Mater.,* **1995**, *7*, 437-444.

[http://dx.doi.org/10.1002/adma.19950070503]

[91] Lanzafame, J.M.; Muenter, A.A.; Brumbaugh, D.V. The effect of J-aggregate size on photoinduced charge transfer processes for dye-sensitized silver halides. *Chem. Phys.,* **1996**, *210*, 79-89. [http://dx.doi.org/10.1016/0301-0104(96)00121-8]

[92] Bezrodnyi, V.I.; Ishchenko, A.A. High efficiency lasing of a dye-doped polymer laser with 1.06 μm pumping. *Appl. Phys. B,* **2001**, *73*, 283-285. [http://dx.doi.org/10.1007/s003400100646]

[93] Kameyama, A.; Kaneda, Y.; Yamanaka, H.; Yoshimine, H.; Narimatsu, H.; Shinohara, Y. Detection of oligosaccharides labeled with cyanine dyes using matrix-assisted laser desorption/ionization mass spectrometry. *Anal. Chem.,* **2004**, *76*(15), 4537-4542. [http://dx.doi.org/10.1021/ac049897z] [PMID: 15283599]

[94] Padilha, L.A.; Webster, S.; Hu, H.H.; Przhonska, O.V.; Hagan, D.J.; Van Stryland, E.W.; Bondar, M.V.; Davydenko, I.G.; Slominsky, Y.L.; Kachkovski, A.D. Excited state absorption and decay kinetics of near IR polymethine dyes. *Chem. Phys.,* **2008**, *352*, 97-105. [http://dx.doi.org/10.1016/j.chemphys.2008.05.007]

[95] Tian, H.; Meng, F.S. Solar cells based on cyanine and polymethine dyes In: *Organic photovoltaics: mechanisms, materials and devices*; Sun, S.S; Sariciftci, N.S, Eds.; Boca Raton: CRC Press, **2005**; 99, pp. 313-329/.Optical engineering

[96] Qian, G.; Wang, Z.Y. Near-infrared organic compounds and emerging applications. *Chem. Asian J.,* **2010**, *5*(5), 1006-1029. [http://dx.doi.org/10.1002/asia.200900596] [PMID: 20352644]

[97] Ishchenko, A.A. Photonics and molecular design of dye-doped polymers for modern light-sensitive materials. *Pure Appl. Chem.,* **2008**, *80*, 1525-1538. [http://dx.doi.org/10.1351/pac200880071525]

[98] Emmelius, M.; Pawlowski, G.; Vollman, H.W. Materials for optical data storage. *Angew. Chem. Int. Ed. Engl.,* **1989**, *28*, 1445-1471. [http://dx.doi.org/10.1002/anie.198914453]

[99] Fenenko, L.I.; Ishchenko, A.A.; Verbitsky, A.B.; Vertsimakha, Y.I. Photovoltaic effects in composites of V2O5 xerogel with polymethine dye. *Mol. Cryst. Liq. Cryst. (Phila. Pa.),* **2005**, *426*, 157-169. [http://dx.doi.org/10.1080/15421400590891056]

[100] Bouit, P.A.; Rauh, D.; Neugebauer, S.; Delgado, J.L.; Di Piazza, E.; Rigaut, S.; Maury, O.; Andraud, C.; Dyakonov, V.; Martin, N. A "cyanine-cyanine" salt exhibiting photovoltaic properties. *Org. Lett.,* **2009**, *11*(21), 4806-4809. [http://dx.doi.org/10.1021/ol901764t] [PMID: 19795845]

[101] Gusiak, N.B.; Kobasa, I.M.; Kurek, S.S. New dyes for dye-sensitized solar cells and photocatalysis: Verifying thermodynamic requirements for electron transfer. *Funct. Mater. Lett.,* **2014**, *7*, 1450030. [http://dx.doi.org/10.1142/S1793604714500301]

[102] Haugland, R.P.; Spence, M.T.; Johnson, I.D. *Handbook of fluorescent probes and research chemicals,* 6th Ed.; Eugene, OR, USA, **1996**.

[103] Yarmoluk, S.; Kovalska, V.; Losytskyy, M. Symmetric cyanine dyes for detecting nucleic acids.

Biotech. Histochem., **2008**, *83*(3-4), 131-145.
[http://dx.doi.org/10.1080/10520290802383684] [PMID: 18802811]

[104] Levitus, M.; Ranjit, S. Cyanine dyes in biophysical research: the photophysics of polymethine fluorescent dyes in biomolecular environments. *Q. Rev. Biophys.,* **2011**, *44*(1), 123-151.
[http://dx.doi.org/10.1017/S0033583510000247] [PMID: 21108866]

[105] Tatikolov, A.S. Polymethine dyes as spectral-fluorescent probes for biomacromolecules. *J. Photochem. Photobiol. Photochem. Rev.,* **2012**, *13*, 55-90.
[http://dx.doi.org/10.1016/j.jphotochemrev.2011.11.001]

[106] Soper, S.A.; Mattingly, Q.L. Steady-state and picosecond laser fluorescence studies of nonradiative pathways in tricarbocyanine dyes: implications to the design of near-IR fluorochromes with high fluorescence efficiencies. *J. Am. Chem. Soc.,* **1994**, *116*, 3744-3742.
[http://dx.doi.org/10.1021/ja00088a010]

[107] Das, S.; Thomas, K.G.; Thomas, K.J.; Kamat, P.V.; George, M.V. Photochemistry of squaraine dyes. 8. Photophysical properties of crown ether squaraine fluoroionophores and their metal ion complexes. *J. Phys. Chem.,* **1994**, *98*, 9291-9296.
[http://dx.doi.org/10.1021/j100088a033]

[108] Terpetschnig, E.; Szmacinski, H.; Ozinskas, A.; Lakowicz, J.R. Synthesis of squaraine-- -hydroxysuccinimide esters and their biological application as long-wavelength fluorescent labels. *Anal. Biochem.,* **1994**, *217*(2), 197-204.
[http://dx.doi.org/10.1006/abio.1994.1109] [PMID: 8203747]

[109] Hossain, M.Z.; Ernst, L.A.; Nagy, J.I. Utility of intensely fluorescent cyanine dyes (Cy3) for assay of gap junctional communication by dye-transfer. *Neurosci. Lett.,* **1995**, *184*(1), 71-74.
[http://dx.doi.org/10.1016/0304-3940(94)11171-E] [PMID: 7739811]

[110] Flanagan, J.H., Jr; Owens, C.V.; Romero, S.E.; Waddell, E.; Kahn, S.H.; Hammer, R.P.; Soper, S.A. Near-infrared heavy-atom-modified fluorescent dyes for base-calling in DNA-sequencing applications using temporal discrimination. *Anal. Chem.,* **1998**, *70*(13), 2676-2684.
[http://dx.doi.org/10.1021/ac980018g] [PMID: 9666731]

[111] Deligeorgiev, T.G.; Zaneva, D.A.; Kim, S.H.; Sabnis, R.W. Preparation of monomethine cyanine dyes for nucleic acid detection. *Dyes Pigm.,* **1998**, *37*, 205-211.
[http://dx.doi.org/10.1016/S0143-7208(97)80088-5]

[112] Berezin, M.Y.; Lee, H.; Akers, W.; Achilefu, S. Near infrared dyes as lifetime solvatochromic probes for micropolarity measurements of biological systems. *Biophys. J.,* **2007**, *93*(8), 2892-2899.
[http://dx.doi.org/10.1529/biophysj.107.111609] [PMID: 17573433]

[113] Blower, M.D.; Feric, E.; Weis, K.; Heald, R. Genome-wide analysis demonstrates conserved localization of messenger RNAs to mitotic microtubules. *J. Cell Biol.,* **2007**, *179*(7), 1365-1373.
[http://dx.doi.org/10.1083/jcb.200705163] [PMID: 18166649]

[114] Fegan, A.; Shirude, P.S.; Balasubramanian, S. Rigid cyanine dye nucleic acid labels. *Chem. Commun. (Camb.),* **2008**, (17), 2004-2006.
[http://dx.doi.org/10.1039/b801629a] [PMID: 18536802]

[115] Renard, B.L.; Aubert, Y.; Asseline, U. Fluorinated squaraine as near-IR label with improved properties

for the labeling of oligonucleotides. *Tetrahedron Lett.,* **2009**, *50*, 1897-1901. [http://dx.doi.org/10.1016/j.tetlet.2009.02.029]

[116] Guo, Z.Q.; Nam, S.; Park, S.; Yoon, J. A highly selective ratiometric near-infrared fluorescent cyanine sensor for cysteine with remarkable shift and its application in bioimaging. *Chem. Sci. (Camb.),* **2012**, *3*, 2760-2765. [http://dx.doi.org/10.1039/c2sc20540h]

[117] Viseu, M.I.; Tatikolov, A.S.; Correia, R.F.; Costa, S.M. Time evolution of monomers and aggregates of a polymethine dye probe the dynamics of model vesicles and micelles. *J. Photochem. Photobiol. Chem.,* **2014**, *280*, 54-62. [http://dx.doi.org/10.1016/j.jphotochem.2014.02.007]

[118] Grabolle, M.; Pauli, J.; Brehm, R.; Resch-Genger, U. Structural control of dye-protein binding, aggregation and hydrophilicity in a series of asymmetric cyanines. *Dyes Pigm.,* **2014**, *103*, 118-126. [http://dx.doi.org/10.1016/j.dyepig.2013.11.027]

[119] Kawakami, M.; Koya, K.; Ukai, T.; Tatsuta, N.; Ikegawa, A.; Ogawa, K.; Shishido, T.; Chen, L.B. Structure-activity of novel rhodacyanine dyes as antitumor agents. *J. Med. Chem.,* **1998**, *41*(1), 130-142. [http://dx.doi.org/10.1021/jm970590k] [PMID: 9438030]

[120] Jones, L.J.; Singer, V.L. Synthesis of squaraine-*N*-hydroxysuccinimide esters and their biological application as long-wavelength fluorescent labels. *Anal. Biochem.,* **2001**, *293*, 8-15. [http://dx.doi.org/10.1006/abio.2001.5116] [PMID: 11373072]

[121] Krieg, M.; Bilitz, J.M.; Srichai, M.B.; Redmond, R.W. Effects of structural modifications on the photosensitizing properties of dialkylcarbocyanine dyes in homogeneous and heterogeneous solutions. *Biochim. Biophys. Acta,* **1994**, *1199*(2), 149-156. [http://dx.doi.org/10.1016/0304-4165(94)90110-4] [PMID: 8123664]

[122] Webster, S.; Peceli, D.; Hu, H.; Padilha, L.A.; Przhonska, O.V.; Masunov, A.E.; Gerasov, A.O.; Kachkovski, A.D.; Slominsky, Y.L.; Tolmachev, A.I.; Kurdyukov, V.V.; Viniychuk, O.O.; Barrasso, E.; Lepkowicz, R.; Hagan, D.J.; van Stryland, E.W. Near-unity quantum yields for intersystem crossing and singlet oxygen generation in polymethine-like molecules: design and experimental realization. *J. Phys. Chem. Lett.,* **2010**, *1*, 2354-2360. [http://dx.doi.org/10.1021/jz100381v]

[123] Harriman, A.; Luengo, G.; Gulliya, K.S. In vitro photodynamic activity of kryptocyanine. *Photochem. Photobiol.,* **1990**, *52*(4), 735-740. [http://dx.doi.org/10.1111/j.1751-1097.1990.tb08674.x] [PMID: 2089420]

[124] Licha, K.; Riefke, B.; Ntziachristos, V.; Becker, A.; Chance, B.; Semmler, W. Hydrophilic cyanine dyes as contrast agents for near-infrared tumor imaging: synthesis, photophysical properties and spectroscopic *in vivo* characterization. *Photochem. Photobiol.,* **2000**, *72*(3), 392-398. [http://dx.doi.org/10.1562/0031-8655(2000)072<0392:HCDACA>2.0.CO;2] [PMID: 10989611]

[125] Choi, H.S.; Nasr, K.; Alyabyev, S.; Feith, D.; Lee, J.H.; Kim, S.H.; Ashitate, Y.; Hyun, H.; Patonay, G.; Strekowski, L.; Henary, M.; Frangioni, J.V. Synthesis and *in vivo* fate of zwitterionic near-infrared fluorophores. *Angew. Chem. Int. Ed. Engl.,* **2011**, *50*(28), 6258-6263. [http://dx.doi.org/10.1002/anie.201102459] [PMID: 21656624]

[126] Ke, S.; Wen, X.; Gurfinkel, M.; Charnsangavej, C.; Wallace, S.; Sevick-Muraca, E.M.; Li, C. Near-infrared optical imaging of epidermal growth factor receptor in breast cancer xenografts. *Cancer Res.,* **2003**, *63*(22), 7870-7875.
[PMID: 14633715]

[127] Recknagel, P.; Claus, R.A.; Neugebauer, U.; Bauer, M.; Gonnert, F.A. *In vivo* imaging of hepatic excretory function in the rat by fluorescence microscopy. *J. Biophotonics,* **2012**, *5*(7), 571-581.
[http://dx.doi.org/10.1002/jbio.201100118] [PMID: 22271709]

[128] Chen, X.; Conti, P.S.; Moats, R.A. In vivo near-infrared fluorescence imaging of integrin alphavbeta3 in brain tumor xenografts. *Cancer Res.,* **2004**, *64*(21), 8009-8014.
[http://dx.doi.org/10.1158/0008-5472.CAN-04-1956] [PMID: 15520209]

[129] Hirche, C.; Murawa, D.; Mohr, Z.; Kneif, S.; Hünerbein, M. ICG fluorescence-guided sentinel node biopsy for axillary nodal staging in breast cancer. *Breast Cancer Res. Treat.,* **2010**, *121*(2), 373-378.
[http://dx.doi.org/10.1007/s10549-010-0760-z] [PMID: 20140704]

[130] Pavlik, C.; Biswal, N.C.; Gaenzler, F.C.; Morton, M.D.; Kuhn, L.T.; Claffey, K.P.; Zhu, Q.; Smith, M.B. Synthesis and fluorescent characteristics of imidazoleeindocyanine green conjugates. *Dyes Pigm.,* **2011**, *89*, 9-15.
[http://dx.doi.org/10.1016/j.dyepig.2010.08.008]

[131] Heo, J.; Lim, C.K.; Whang, D.R.; Shin, J.; Jeong, S.Y.; Park, S.Y.; Kwon, I.C.; Kim, S. Self-deprotonation and colorization of 1,3-bis(dicyanomethylidene)indan in polar media: a facile route to a minimal polymethine dye for NIR fluorescence imaging. *Chemistry,* **2012**, *18*(28), 8699-8704.
[http://dx.doi.org/10.1002/chem.201103042] [PMID: 22689413]

[132] Wang, X.; Sun, J.; Zhang, W.H.; Ma, X.X.; Lv, J.Z.; Tang, B. A near-infrared ratiometric fluorescent probe for rapid and highly sensitive imaging of endogenous hydrogen sulfide in living cells. *Chem. Sci. (Camb.),* **2013**, *4*, 2551-2556.
[http://dx.doi.org/10.1039/c3sc50369k]

[133] James, N.S.; Chen, Y.; Joshi, P.; Ohulchanskyy, T.Y.; Ethirajan, M.; Henary, M.; Strekowsk, L.; Pandey, R.K. Evaluation of polymethine dyes as potential probes for near infrared fluorescence imaging of tumors: part - 1. *Theranostics,* **2013**, *3*(9), 692-702.
[http://dx.doi.org/10.7150/thno.5922] [PMID: 24019854]

[134] Toutchkine, A.; Nguyen, D.V.; Hahn, K.M. Simple one-pot preparation of water-soluble, cysteine-reactive cyanine and merocyanine dyes for biological imaging. *Bioconjug. Chem.,* **2007**, *18*(4), 1344-1348.
[http://dx.doi.org/10.1021/bc060376n] [PMID: 17542551]

[135] Fei, X.; Gu, Y.; Ban, Y.; Liu, Z.; Zhang, B. Thiazole Orange derivatives: synthesis, fluorescence properties, and labeling cancer cells. *Bioorg. Med. Chem.,* **2009**, *17*(2), 585-591.
[http://dx.doi.org/10.1016/j.bmc.2008.11.083] [PMID: 19103495]

[136] James, N.S.; Ohulchanskyy, T.Y.; Chen, Y.; Joshi, P.; Zheng, X.; Goswami, L.N.; Pandey, R.K. Comparative tumor imaging and PDT Efficacy of HPPH conjugated in the mono- and di-forms to various polymethine cyanine dyes: part - 2. *Theranostics,* **2013**, *3*(9), 703-718.
[http://dx.doi.org/10.7150/thno.5923] [PMID: 24019855]

[137] Saito, K.; Yokoyama, H. Transient photocurrent response in merocyanine Langmuir-Blodgett films. *Thin Solid Films,* **1994**, *234*, 526-529.
[http://dx.doi.org/10.1016/0040-6090(93)04042-Q]

[138] Kabatc, J.; Bajorek, A.; Dobosz, R. Bichromophoric hemicyanine dyes as fluorescence probes applied for monitoring of the photochemically initiated polymerization. *J. Mol. Struct.,* **2011**, *985*, 95-104.
[http://dx.doi.org/10.1016/j.molstruc.2010.10.027]

[139] Anaya, C.; Church, N.; Lewis, J.P. Detection and identification of bacterial cell surface proteins by fluorescent labeling. *Proteomics,* **2007**, *7*(2), 215-219.
[http://dx.doi.org/10.1002/pmic.200600175] [PMID: 17205604]

[140] Karp, N.A.; Lilley, K.S. Maximising sensitivity for detecting changes in protein expression: experimental design using minimal CyDyes. *Proteomics,* **2005**, *5*(12), 3105-3115.
[http://dx.doi.org/10.1002/pmic.200500083] [PMID: 16035117]

[141] Xu, Y.F.; Liu, Y.; Qian, X.H. Novel cyanine dyes as fluorescent pH sensors: PET, ICT mechanism or resonance effect? *J. Photochem. Photobiol. Chem.,* **2007**, *190*, 1-8.
[http://dx.doi.org/10.1016/j.jphotochem.2007.03.004]

[142] Maether, M.P.; Desoubzdanne, D.; Izquierdo, A.; Guieu, V.; Maturano, M.; André-Barrès, C.; Valentin, A.; Jullian, V.; Chevalley, S.; Maynadier, M.; Vial, H.; Payrastre, C. Synthesis and antimalarial properties of streptocyanine dyes. *Chem.Med.Chem,* **2009**, *4*(8), 1327-1332.
[http://dx.doi.org/10.1002/cmdc.200900051] [PMID: 19551799]

[143] Calvete, M.J. Near-infrared absorbing organic materials with nonlinear transmission properties. *Int. Rev. Phys. Chem.,* **2012**, *31*, 319-366.
[http://dx.doi.org/10.1080/0144235X.2012.713210]

[144] Suzuki, H. Self-enhancement in the electroluminescence of a near-infrared ionic dye. *Appl. Phys. Lett.,* **2000**, *76*, 1543-1545.
[http://dx.doi.org/10.1063/1.126090]

[145] Suzuki, H. Infrared electroluminescence from an organic ionic dye containing no rare-earth ions. *Appl. Phys. Lett.,* **2002**, *80*, 3256-3258.
[http://dx.doi.org/10.1063/1.1476711]

[146] Maltsev, E.I.; Lypenko, D.A.; Bobinkin, V.V.; Tameev, A.R.; Kirillov, S.V.; Shapiro, B.I.; Schoo, H.F.; Vannikov, A.V. Near-infrared electroluminescence in polymer composites based on organic nanocrystals. *Appl. Phys. Lett.,* **2002**, *81*, 3088-3090.
[http://dx.doi.org/10.1063/1.1515880]

[147] Xuan, Y.; Qian, G.; Wang, Z.; Ma, D. Near-infrared polymer light-emitting diodes based on infrared dye doped poly(N-vinylcarbazole) film. *Thin Solid Films,* **2008**, *516*, 7891-7893.
[http://dx.doi.org/10.1016/j.tsf.2007.09.046]

[148] Jin, Y.; Hua, J.; Wu, W.; Ma, X.; Meng, F. Synthesis, characterization and photovoltaic properties of two novel near-infrared absorbing perylene dyes containing benzo[e]indole for dye-sensitized solar cells. *Synth. Met.,* **2008**, *158*, 64-71.
[http://dx.doi.org/10.1016/j.synthmet.2007.12.005]

[149] Wittmann, M.; Rotermund, F.; Weigand, R.; Penzkofer, A. Saturable absorption and absorption

recovery of indocyanine green J-aggregates in water. *Appl. Phys. B,* **1998**, *66*, 453-459. [http://dx.doi.org/10.1007/s00340005041 8]

[150] Przhonska, O.V.; Lim, J.H.; Hagan, D.J.; van Stryland, E.W.; Bondar, M.V.; Slominsky, Y.L. Nonlinear light absorption of polymethine dyes in liquid and solid media. *J. Opt. Soc. Am. B,* **1998**, *15*, 802-809. [http://dx.doi.org/10.1364/JOSAB.15.000802]

[151] Lim, J.H.; Przhonska, O.V.; Khodja, S.; Yang, S.; Ross, T.S.; Hagan, D.J.; van Stryland, E.W.; Bondar, M.V.; Slominsky, Y.L. Polymethine and squarylium molecules with large excited-state absorption. *Chem. Phys.,* **1999**, *245*, 79-97. [http://dx.doi.org/10.1016/S0301-0104(99)00086-5]

[152] Lepkowicz, R.; Kobyakov, A.; Hagan, D.J.; van Stryland, E.W. Picosecond optical limiting in reverse saturable absorbers: a theoretical and experimental study. *J. Opt. Soc. Am. B,* **2002**, *19*, 94-101. [http://dx.doi.org/10.1364/JOSAB.19.000094]

[153] Przhonska, O.; Slominsky, Y.; Stahl, U.; Daehne, S. Excited-state relaxation processes in polymethine dye molecules in polymeric media. *J. Lumin.,* **1996**, *69*, 105-113. [http://dx.doi.org/10.1016/0022-2313(96)00080-4]

[154] Ozmen, B.; Akkaya, E.U. Infrared fluorescence sensing of submicromolar calcium: pushing the limits of photoinduced electron transfer. *Tetrahedron Lett.,* **2000**, *41*, 9185-9188. [http://dx.doi.org/10.1016/S0040-4039(00)01662-2]

[155] Tarazi, L.; Choi, H.; Mason, J.C.; Sowell, J.; Strekowski, L.; Patonay, G. Characterization of a novel crown ether-bearing near-infrared heptamethine cyanine dye. A study of fluorescence quenching by lithium. *Microchem. J.,* **2002**, *72*, 55-62. [http://dx.doi.org/10.1016/S0026-265X(01)00150-3]

[156] Song, F.; Peng, X.; Lu, E.; Wang, Y.; Zhou, W.; Fan, J. Tuning the photoinduced electron transfer in near-infrared heptamethine cyanine dyes. *Tetrahedron Lett.,* **2005**, *46*, 4817-4820. [http://dx.doi.org/10.1016/j.tetlet.2005.04.089]

[157] Cao, X.; Lin, W.; He, L. A near-infrared fluorescence turn-on sensor for sulfide anions. *Org. Lett.,* **2011**, *13*(17), 4716-4719. [http://dx.doi.org/10.1021/ol201932c] [PMID: 21809838]

[158] Hilderbrand, S.A.; Weissleder, R. Optimized pH-responsive cyanine fluorochromes for detection of acidic environments. *Chem. Commun. (Camb.),* **2007**, (26), 2747-2749. [http://dx.doi.org/10.1039/b703764c] [PMID: 17594041]

[159] Kim, Y.; Hilderbrand, S.A.; Weissleder, R.; Tung, C-H. Macroscale assembly of peptide nanotubes. *Chem. Commun. (Camb.),* **2007**, 2299-2731. [http://dx.doi.org/10.1039/b700741h] [PMID: 17534523]

[160] Hilderbrand, S.A.; Kelly, K.A.; Niedre, M.; Weissleder, R. Near infrared fluorescence-based bacteriophage particles for ratiometric pH imaging. *Bioconjug. Chem.,* **2008**, *19*(8), 1635-1639. [http://dx.doi.org/10.1021/bc800188p] [PMID: 18666791]

[161] Tang, B.; Liu, X.; Xu, K.; Huang, H.; Yang, G.; An, L. A dual near-infrared pH fluorescent probe and its application in imaging of HepG2 cells. *Chem. Commun. (Camb.),* **2007**, (36), 3726-3728.

[http://dx.doi.org/10.1039/b707173f] [PMID: 17851608]

[162] Blum, G.; von Degenfeld, G.; Merchant, M.J.; Blau, H.M.; Bogyo, M. Noninvasive optical imaging of cysteine protease activity using fluorescently quenched activity-based probes. *Nat. Chem. Biol.,* **2007**, *3*(10), 668-677.
[http://dx.doi.org/10.1038/nchembio.2007.26] [PMID: 17828252]

[163] Caglič, D.; Globisch, A.; Kindermann, M.; Lim, N-H.; Jeske, V.; Juretschke, H-P.; Bartnik, E.; Weithmann, K.U.; Nagase, H.; Turk, B.; Wendt, K.U. Functional *in vivo* imaging of cysteine cathepsin activity in murine model of inflammation. *Bioorg. Med. Chem.,* **2011**, *19*(3), 1055-1061.
[http://dx.doi.org/10.1016/j.bmc.2010.10.028] [PMID: 21130662]

[164] Linder, K.E.; Metcalfe, E.; Nanjappan, P.; Arunachalam, T.; Ramos, K.; Skedzielewski, T.M.; Marinelli, E.R.; Tweedle, M.F.; Nunn, A.D.; Swenson, R.E. Synthesis, *in vitro* evaluation, and *in vivo* metabolism of fluor/quencher compounds containing IRDye 800CW and Black Hole Quencher-3 (BHQ-3). *Bioconjug. Chem.,* **2011**, *22*(7), 1287-1297.
[http://dx.doi.org/10.1021/bc100457s] [PMID: 21639144]

[165] Peng, X.; Song, F.; Lu, E.; Wang, Y.; Zhou, W.; Fan, J.; Gao, Y. Heptamethine cyanine dyes with a large stokes shift and strong fluorescence: a paradigm for excited-state intramolecular charge transfer. *J. Am. Chem. Soc.,* **2005**, *127*(12), 4170-4171.
[http://dx.doi.org/10.1021/ja043413z] [PMID: 15783189]

CHAPTER 3

Porphyrins as Top Materials

Abstract: Porphyrins are colored aromatic macrocycles, composed of four pyrrole units, linked by four methine bridges, giving origin to a highly conjugated macrocycle, known for its remarkable stability. Their remarkable features have given them a wide fame, and continue to pave the way for multiple purposes nowadays. Herein is presented a conspectus on the field of porphyrins, where emphasis is set on their preparation and functionalization methods, along with more detailed examples on several promising applications with 508 references provided. Some synthetic details on the preparation of state-of-the-art compounds are given as well.

Keywords: Energy conversion, Photocatalysis, Photodynamic therapy, Photosynthesis, Porphyrins, Porphyrinoids, Photodynamic therapy.

INTRODUCTION

Porphyrins and their related compounds are colored aromatic macrocycles, composed of four pyrrole units, linked by four methylenic bridges, giving a highly conjugated macrocycle known for its remarkable stability. Porphyrins contain a total of 22 π-electrons, with 18 π-electrons participating in the delocalization pathway each time, which can create up to six different 18 π-electron delocalization pathways (Fig. **1**).

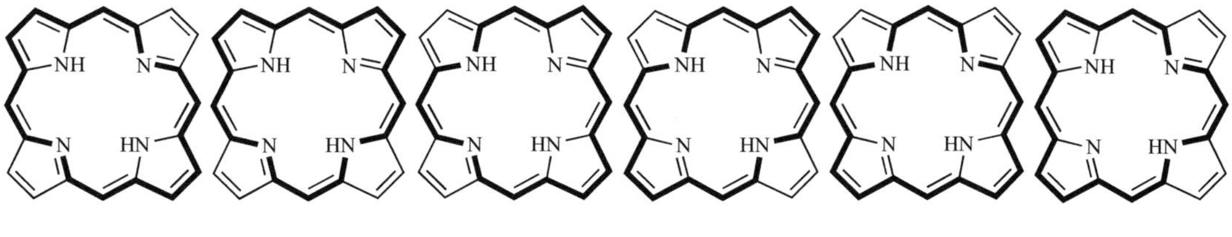

Fig. (1). 18 π-electron delocalization pathways.

Porphyrins follow Huckel's [4n + 2] rule for aromaticity (n = 4), meaning that their two peripheral double bonds can suffer additional reactions without delocalization pathway interfering. Other porphyrin related macrocycles include

Mário J. F. Calvete

benzoporphyrins (similar to porphyrins, but with β-fused benzene rings) [1 - 5], corroles (constituted by four pyrrole units intercalated with 3 methine bridges, having two of the pyrrole units directly connected) [6 - 10] and subporphyrins (the lower porphyrin homologues, constituted by three pyrrole units interconnected by three methine bridges) [11 - 14].

Nomenclature of porphyrins is an evolving issue. The first system of nomenclature was developed by Fischer in 1934 [15]. Here, substitution can occur at two positions along the periphery of the macrocycle in the β-carbons of the pyrrole fragments or at the four *meso* -positions which are the methine bridges between the α-carbons of the pyrrole fragments (Fig. **2**).

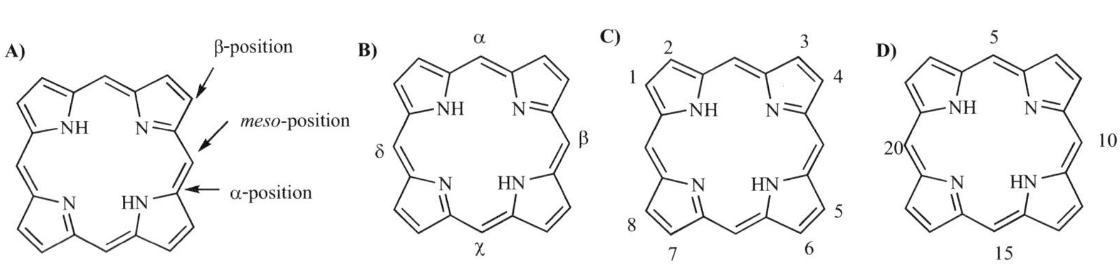

Fig. (2). Porphyrin nomenclature developed by Fischer. **A)** porphyrin positions; **B)** notation for meso-positions; **C)** numbering for β-positions; **D)** numbering for meso-positions.

This nomenclature is still used today; however, a more appropriate system had to be developed, to name more complicated structures, where Fischer's nomenclature failed. The International Union of Pure and Applied Chemistry (IUPAC) nomenclature was adopted in early 1970's [16]. This nomenclature differs from Fischer's in that all the carbons and nitrogens in the porphyrin macrocycle are given a number. Looking at the basic porphyrin system (porphine) the β-positions then become 2,3,7,8,12,13,17,18 instead of 1-8. The *meso* -positions are numbered 5,10,15,20, in place of the Greek letters Fischer previously assigned (Fig. **2**).

Porphyrins are naturally occurring compounds (Fig. **3**). Bottom-line, one can even say that these pigments are so important that life would simply not be possible without them. Photosynthesis and respiration are at the heart of what makes life possible and in both phenomena, porphyrinoids are omnipresent. In remote times, when the atmosphere in our planet was void of free oxygen, the evolution of oxygenic photosynthesis by chlorophyll (the process involves the oxidation of water forming molecular oxygen releasing it into the atmosphere) caused the change of Earth atmosphere from anaerobic to aerobic leading to a dramatic increase in biodiversity.

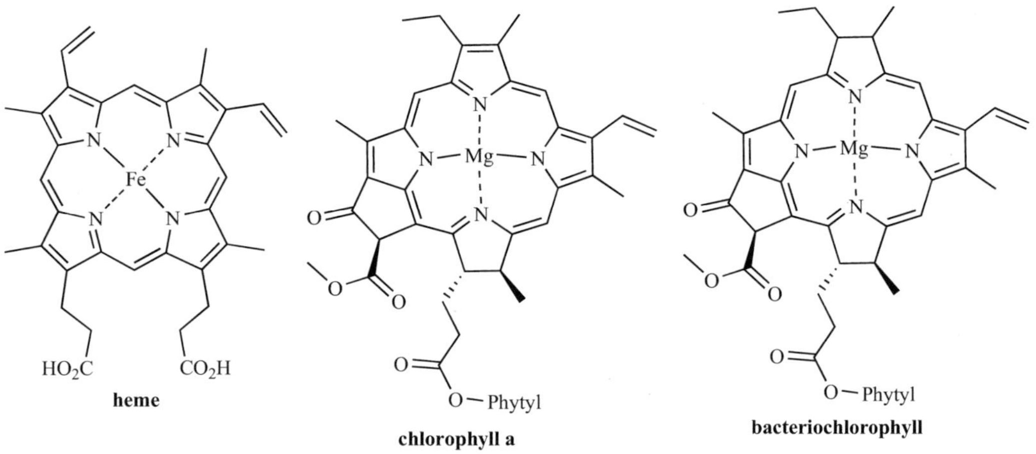

Fig. (3). Naturally occurring porphyrin derivatives.

Life has evolved using oxygen leading to organisms and plants containing chloroplasts for photosynthesis and eventually hemoglobin in the red blood cells for the respiration of vertebrates.

Understanding events related to these compounds became the life goal of several scientists, whose findings are well documented [17]. For instance, iron in human blood was first reported in 1747, when Menghini burned human blood to ash separating particles of iron with a magnet. Later, the experiments by Joseph Priestly in 1774 proved that the oxygen consumed by fire and breathing animals could be restored to the air by plants. This ignited interest in understanding the role of oxygen in respiration. In 1864, Hoppe-Seyler was able to isolate the red blood cell iron-containing metalloprotein responsible for oxygen-transport, naming it hemoglobin. Later, in 1871, Hoppe-Seyler isolated porphyrins from blood and demonstrated they were pyrrole derivatives, being later credited to reporting similarities between chlorophyll and heme, in 1879. Only in 1912 the correct structure for porphyrins was proposed by Kuester, though not accepted at the time. Later, Milroy reported the first general porphyrin synthesis in 1918 [18], while Hans Fischer published in 1926 a *de novo* synthesis of chlorohemin, proving the structure proposed by Kuester almost twenty years earlier [19], being later in 1930 awarded with the Nobel Prize for his work on porphyrins. Between 1934 and 1940 Hans Fischer and his porphyrin group, established in Munich, Germany, published three volumes that became a compendium of laboratory methods for synthesis of porphyrin compounds [15].

The development in subsequent years was intense, with many groups dealing with the understanding of such compounds, either isolating them or pursuing laboratory methods for their synthesis. Many books have been published throughout the years, namely the book "Porphyrins and Metalloporphyrins" by Falk, in 1964 [20], which possessed a reasonably detailed laboratory methods. This book was later followed in 1975 by a "Porphyrins and Metalloporphyrins" new version (edited by Kevin M. Smith) [21], and Dolphin's more traditional series named "The Porphyrins" (1978-1980) [22]. Very much appealing is also the book by Milgrom, published in 1997, a great introduction to the porphyrins and their role in life [23]. More recently, a comprehensive compendium of porphyrin chemistry synthetic methodologies and reactivity profiles is presented in "The Porphyrin Handbook", consisting of two ten-volume sets comprising 122 chapters, 98 of them dedicated to porphyrins [24]. Since then, the follow-on series (35 volumes to date) by the same editors (Kevin M. Smith, Karl M. Kadish and Roger Guilard) entitled "The Handbook of Porphyrin Science" has featured so far 180 chapters, including 159 regarding porphyrins [25].

SYNTHESIS OF PORPHYRIN DERIVATIVES

In general, there are several routes to afford porphyrins and have been widely covered along the years [26 - 33]. The first route involves the functionalization of a naturally obtained porphyrin system, such as chlorophylls *a* or *b* , bacteriochlorophylls and heme (Fig. **3**), which are then functionalized to give the teraget porphyrin. The second route deals with total synthesis, starting with pyrrolic subunits. While the first route is severely limited by the type of functional groups that may be inserted, the second route, on the other hand, offers vast possibilities.

The simplest porphyrins are symmetric and are easily prepared through tetramerization of monopyrrolic units (Scheme **1**). Regardless of the preparation method, symmetric porphyrins can be categorized as octa-ß substituted (Scheme **1a**) or tetra-*meso* substituted porphyrins (Scheme **1b**).

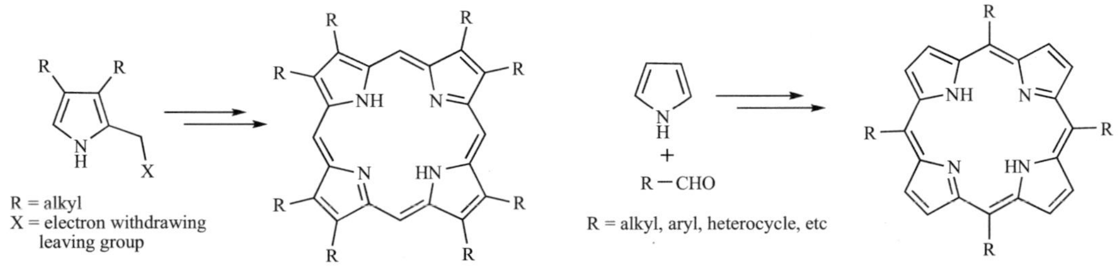

Scheme (1). Synthesis of symmetric porphyrins *via* the tetramerization of monopyrrolic subunits.

If, for instance, the aldehyde used is benzaldehyde, the obtained porphyrin is the *meso*-tetraphenylporphyrin (or 5,10,15,20-tetraphenylporphyrin), considered by all researchers in the porphyrin field the easiest porphyrin to synthesize (Scheme **2**). The route was first reported by Rothemund [34, 35], and later modified/improved by Adler and Longo [36]. While the former preferred to use sealed glass tubes to obtain the porphyrin in low yields, the second used refluxing propionic acid to obtain the corresponding *meso*-tetraarylporphyrin in 20-30% yields, however, always with some amounts of impurities, arising from deficient air oxidation. This issue was later solved, when Pereira and Gonsalves introduced nitrobenzene as both co-solvent (with acetic acid) and oxidizer, at refluxing temperature [37 - 39], obtaining pure porphyrins in 20-35% yield. Lindsey's group also optimized the synthesis [31, 40], obtaining excellent yields for a extensive variety of tetraarylporphyrins, using a highly diluted two-step reaction between pyrrole and arylaldehydes, using a Lewis acid catalyst (typically BF_3-etherate), followed by oxidation of porphyrinogen to porphyrin using quinones, such as 2,3-dichloro-5,6-dicyanobenzoquinone (DDQ).

Since *meso* substituted porphyrin are not natural, during many decades scientists have looked to ß-substituted porphyrins such as octa-ß-alkylporphyrins (OAP) as natural mimics [29]. However, methodologies to prepare porphyrins such OAP are quite intricate, due to the bigger difficulties in 3,4-disubstituted pyrrole preparations. Octa- ß substituted porphyrins can be prepared by self-condensation of pyrroles bearing electron-withdrawing leaving groups at 2-methinic positions, in the presence of bases, followed by heating in acid to yield the target porphyrins. Also monopyrrole tetramerization reactions are favored for pyrroles bearing the same substituents, because of acid catalyzed pyrrole scrambling reactions that can occur in presence of dissimilar pyrrole substituents [41].

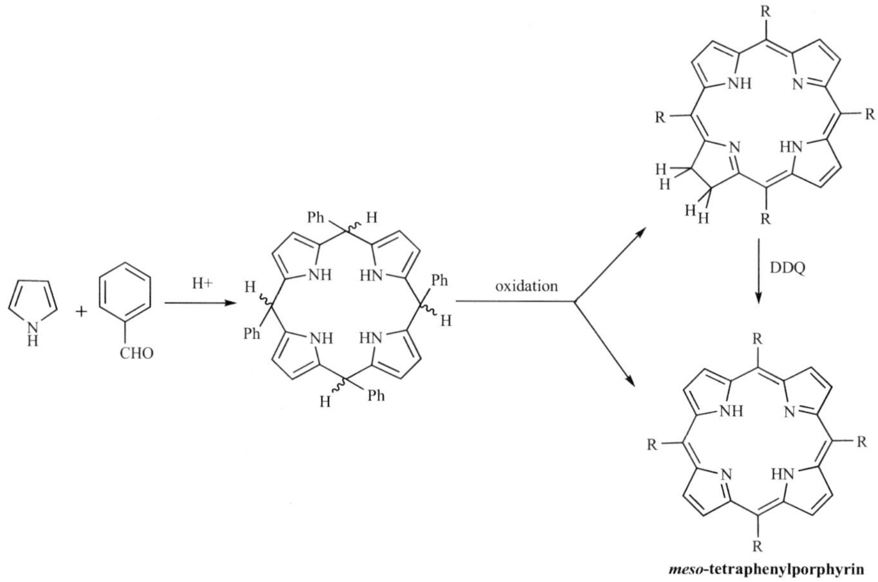

Scheme (2). *meso* -tetraphenylporphyrin synthesis.

Another major synthetic approach to prepare porphyrins is the so-called MacDonald [2+2] porphyrin synthesis [42], which was similarly developed by Woodward, on the synthesis of chlorophyll-a [43, 44]. In this approach, there is the need to prepare the precursory dipyrromethanes, which are key intermediates in the synthesis of the desired porphyrins. For the preparation of ß-substituted porphyrins, it is necessary to properly functionalize the pyrrole sub-units, which then are transformed into dipyrromethanes. These intermediates can be classified as either unsymmetrical or symmetrical, based on the interpyrrolic 5-carbon. Symmetric dipyrromethanes are usually synthesized by bromomethylpyrrole self-condensation [15], or by 2-acetoxymethylpyrrole heating [45], affording the corresponding 1,9-dicarboxylic acids through catalytic hydrogenation and hydrolysis. The corresponding 1,9-diformyl dipyrromethanes (Scheme **3**) are then formed by Vilsmeier formylation. On the other hand, unsymmetrically substituted dipyrro-methanes are obtained by condensation of 2-unsubstituted pyrroles with 2-acetoxymethylpyrroles in acetic acid using *p*-toluenesulfonic acid [46] or montmorillonite K-10 clay as catalysts (Scheme **4**) [47, 48].

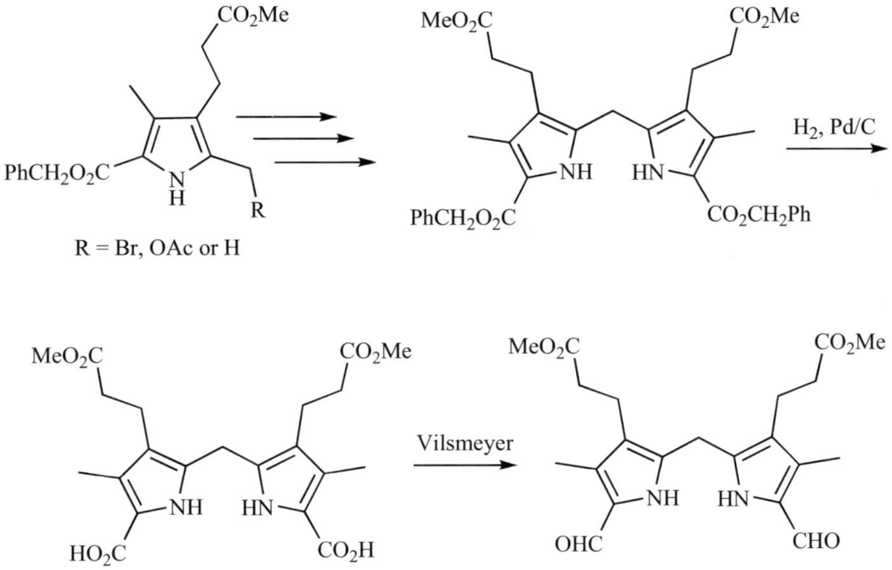

Scheme (3). Synthesis of dialdehyde substituted dipyrromethanes.

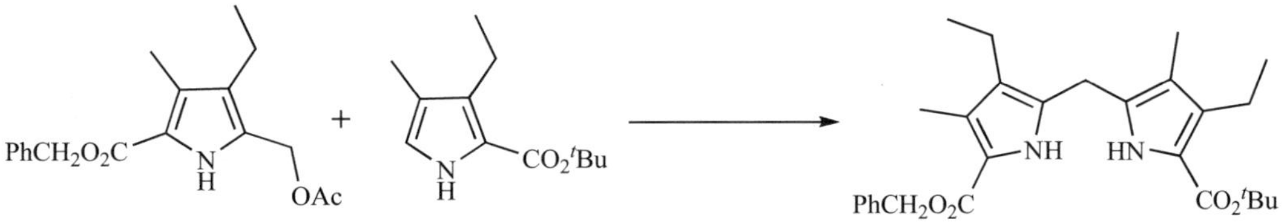

Scheme (4). Synthesis of ester substituted dipyrromethanes.

As expected, condensation of two unsymmetrical dipyrromethanes will produce two porphyrin products, due to dipyrromethane reaction in both two orientations related by a 180° rotation of one of the dipyrromethanes (Scheme **5**).

Meso-substituted porphyrins can also be prepared using a modified MacDonald [2+2] methodology [49 - 52]. Until the early 90's, almost all reported dipyrromethanes possessed *β*-pyrrole substituents, concomitantly lacking *meso* substituents, which could be attributed to a possible instability of *meso*-substituted dipyrromethanes. In this regard, the challenge in the isolation of measurable quantities of dipyrromethanes from the reaction of an aldehyde and unsubstituted pyrrole lied at the dipyrromethane stage, particularly in preventing oligomerization.

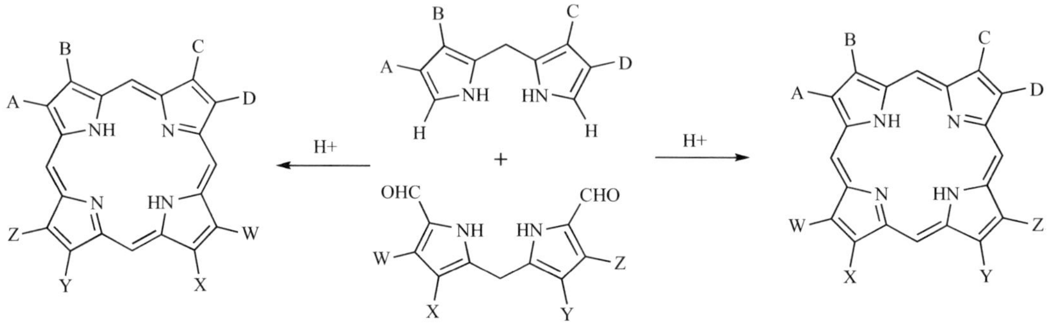

Scheme (5). Synthesis of porphyrins *via* dipyrromethane condensation.

Lindsey developed a fairly simple method to prepare *meso* substituted dipyrromethanes from small scale to 100 g scale [53 - 55], by using pyrrole as solvent and a Lewis acid (TFA, $BF_3{\cdot}O(Et)_2$ or $InCl_3$) in pyrrole/aldehyde ratio (25:1 to 100:1), therefore, blocking oligomerization further than dipyrromethane stage (Scheme **6**).

Other authors later have improved the synthesis of *meso*-substituted dipyrromethanes (mostly aryl) [56 - 60] in order in increase the dipyrromethane offer. Another way to prepare porphyrins is based on the [2+2] route developed by MacDonald.

This [3+1] route to porphyrins employs the MacDonald method; nevertheless, in this methodology a tripyrrane reacts with a pyrrole bearing the two “bridging” carbon atoms, hence the designation [3+1] (Scheme **7**).

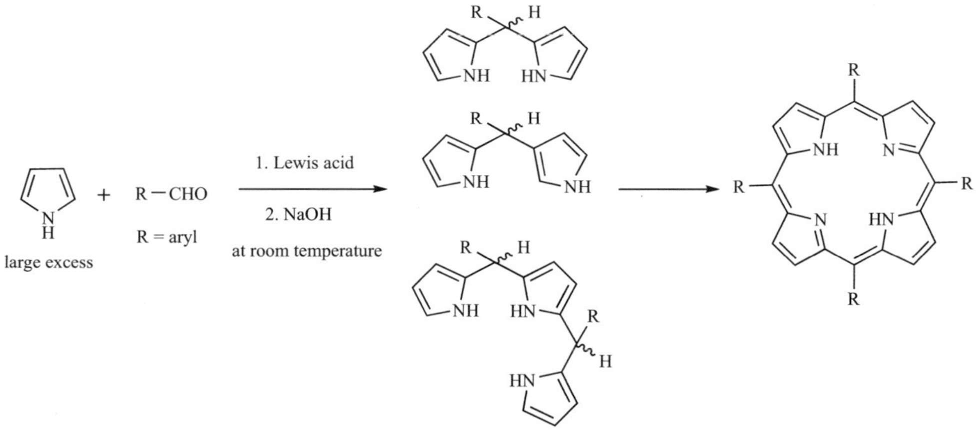

Scheme (6). Lindsey method for meso-substituted porphyrin synthesis.

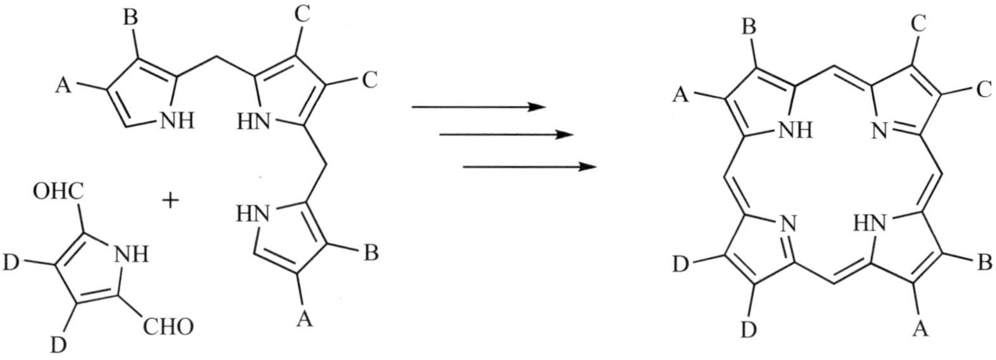

Scheme (7). MacDonald's based [3+1] route for porphyrin synthesis.

Using this protocol allows the synthesis of porphyrins in which a required functionality is located on a monopyrrole sub-unit to be introduced at the last step. This chemistry was then simplified by Sessler's advances on tripyrrane syntheses (Scheme **8**) [61, 62]. For instance, this route yields only symmetrically substituted tripyrranes; however, the straight forwardness on the synthesis compensated this limitation.

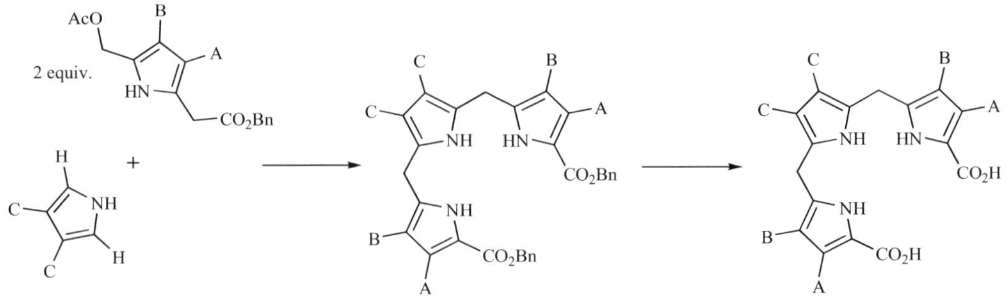

Scheme (8). Tripyrrane synthesis.

Thus, the reaction depicted in Scheme (**8**) represents a synthetic pathway to obtain tripyrrane-1,14-dicarboxylic acid derivatives. Boudif and Momenteau [63, 64] reported the first example of the [3+1] approach to porphyrins. Lash's group extended this general approach on their porphyrin syntheses [65 - 69], while Sessler reported the synthesis of many other porphyrins [70]. Smith's group also developed a [3+1] procedure which avoids the use of acid catalysis, assuring the stability of the tripyrrane reactant against pyrrole ring-scrambling reactions [71].

The methodologies referred above allow the synthesis of most of all porphyrin types, including symmetrically and unsymmetrically substituted porphyrins. For the later, besides the above approaches, other appealing possibilities arise. For instance, the synthesis of A_3B *meso*-substituted porphyrins, usually relies on the use of statistical cross condensation reactions between mixtures of aldehydes with equal amounts of pyrrole, varying the ratios between them, favoring a specified statistical product, according to (Table **1**).

Table 1. Expected product ratio from the statistical condensation of two aldehyde with pyrrole.

A:B	**AAAA**	**AAAB**	**ABAB**	**AABB**	**ABBB**	**BBBB**
1:1	6.25	25	12.5	25	25	6.25
3:1	31.6	42.2	7.0	14.1	4.7	0.4
9:1	65.6	29.2	1.6	3.2	0.4	0.01
Permutations	1	4	2	4	4	1

Nevertheless, for the preparation of symmetric ABAB *meso*-substituted porphyrins, one can use the accessible dipyrromethane [2+2] condensations (Scheme **9**) [50, 72 - 74].

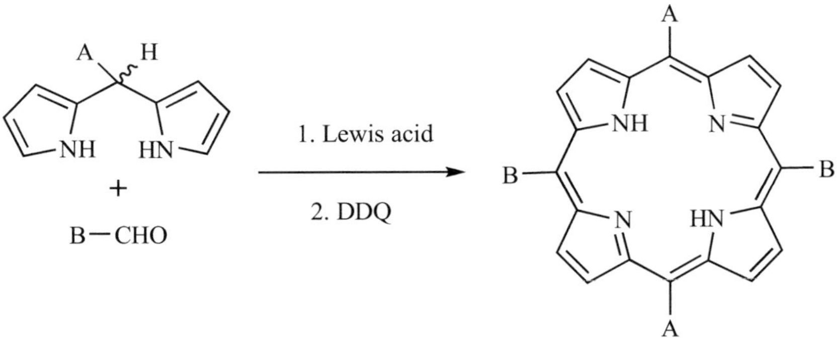

Scheme (9). Synthesis of ABAB *meso* substituted porphyrins.

In fact, this is the easiest route for the preparation of *trans-meso*-substituted porphyrins. The acid-catalysis conditions necessary for the cyclization introduce the possibility of forming mixtures of porphyrins, which mainly depends on the nature of the *meso* substituent in the dipyrromethane [75, 76].

Porphyrin Functionalization

Using the reactivity of porphyrins represents a very important tool to obtain any corresponding porphyrin derivative with the desired functionalities. Several important reviews describe methodologies for porphyrin functionalization [33, 77 - 83].

The most electronically reactive positions at the porphyrin periphery are the *meso* -positions the preferential sites for additions and substitution reactions without loss of their aromaticity, as well as cross-coupling, oxidation and reduction reactions, to increase the variability of functionalization.

Halogenations are characteristic reactions in order to introduce functionality on porphyrins. For instance, fluorination reactions are commonly carried away with this purpose. Typical fluorination reagents include cesium fluoroxysulfate [84], *N*-fluoropyridinium triflates [85] and fluorides of cobalt or silver [86], while for chlorination the used reagents are chlorine [87, 88] or *N*-chlorosuccinimide (NCS)

[89 - 94]. Similarly, bromine [87, 90, 95, 96] and *N*-bromosuccinimide (NBS) [87, 97 - 102] can be used for bromination, besides HBr/H_2O_2 [86, 103], copper(II) bromide [95] and $PhSeBr_3$ [85]. A particular aspect arises from bromination reactions, since polybromination can lead to mixtures of regioisomers when unsubstituted β- and *meso*-positions are accessible at the periphery. Nonetheless, in the case of *meso*-tetraarylporphyrins, such as *meso*-tetraphenylporphyrin, tetra-bromination can be achieved regioselectively using a slight excess of NBS using chloroform as solvent, yielding the corresponding 2,3,12,13-tetr--bromoporphyrins, (Scheme **10**) [104].

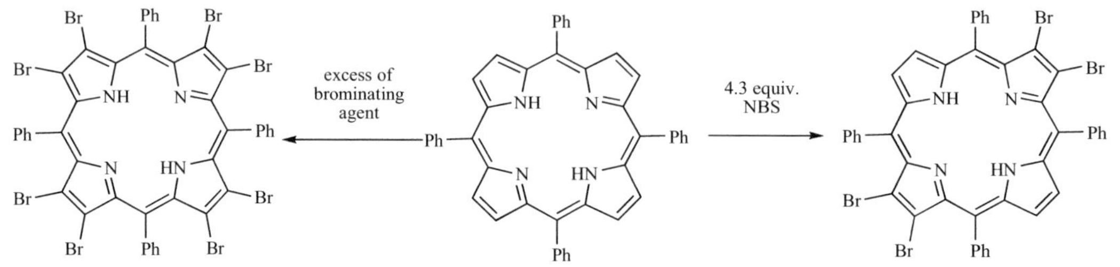

Scheme (10). Bromination of *meso* -substituted porphyrins.

Also mono- and di-iodination can be achieved using bis(trifluoroacetoxy)-iodobenzene-iodine in chloroform/pyridine [105, 106]. The *meso*-dibrominated porphyrin can be obtained in high yield and regioselectively, by reacting 5,15-diphenylporphyrn with NBS in chloroform/pyridine mixture (Scheme **11**) [107].

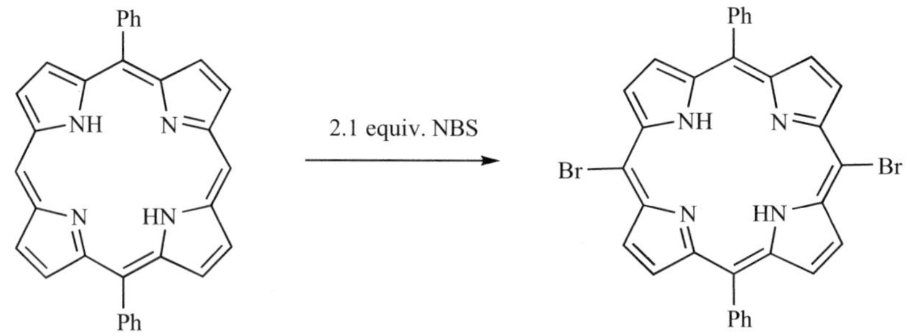

Scheme (11). Synthesis of *meso* -dibrominated porphyrin.

Normally, halogenated porphyrins serve as starting materials for a several reactions. For instance, the nucleophilic substitution of halogen substituents by cyanide [93, 108, 109], thiolate ions [104] and nitrite [110], has also been reported. The main use of bromo- and iodo-porphyrins is as starting materials for cross-coupling reactions, through palladium(0)-catalyzed Sonogashira, Suzuki, Heck and Stille type reactions, originating a range of aryl, ethynyl and alkenyl-functionalized porphyrin systems, as well as dimeric and oligomeric porphyrins, also synthesized through these methodologies [111 - 122].

In palladium(0)-catalyzed coupling reactions metalloporphyrins are often used, to avoid macrocycle metallation by palladium and/or copper, occasionally used as co-catalyst. Both *meso*- and β-bromoporphyrins have been functionalized by palladium(0)-catalyzed couplings with amines, amides, alcohols, sulfides, carbazates, hydrazones, phosphines [126 - 134] and boronates [135]. As example, dodecaarylporphyrins can be prepared using a Suzuki coupling reaction of a metal-free porphyrin with arylboronic acids (Scheme **12**) [123 - 125].

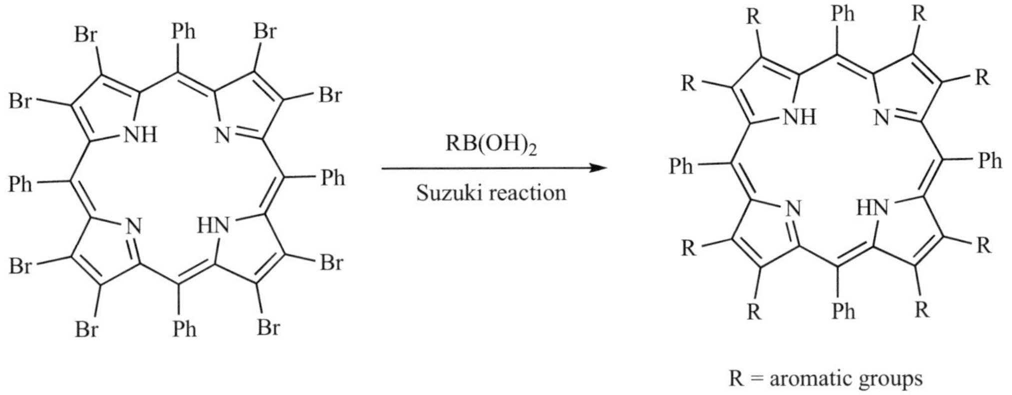

Scheme (12). Suzuki coupling reaction of brominated porphyrins.

Brominated porphyrins have also been coupled with iodo-alkyl fluorides in the presence of tris(dibenzylideneacetone)dipalladium(0), triphenylarsine and copper(0) in DMSO to produce *meso* - and β-perfluoroalkylated porphyrins [136], and with potassium organotrifluoroborates in the presence of $Pd(dppf)Cl_2$, producing an array of functionalized porphyrins [137].

Suzuki-type couplings have also been performed on boronated porphyrins, prepared from palladium-catalyzed reactions of bromoporphyrins with pinacol borane or bispinacol borane (Scheme **13**). This type of porphyrins can go through a diversity of palladium-catalyzed coupling reactions with aryl halides [138 - 145].

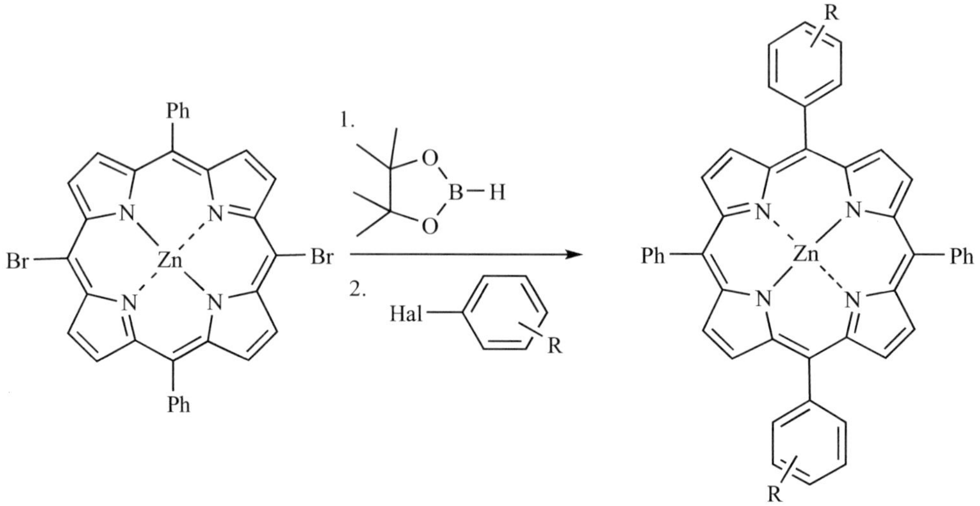

Scheme (13). Palladium-catalyzed reactions of bromoporphyrins with boranes.

Sonogashira coupling reactions can be used for synthesizing a variety of ethynyl-substituted porphyrins, usually by reaction of a *meso*- or β-halogenated porphyrin with acetylene compounds (Scheme **14**) [100, 127, 129, 146 - 151].

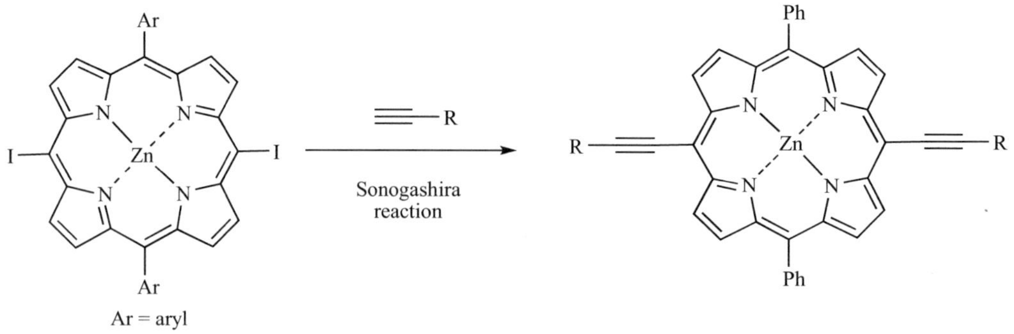

Scheme (14). Sonogashira coupling reaction of porphyrins.

Alkenyl and aryl-substituted porphyrins can also be prepared by Heck and Stille coupling reactions [152 - 156]. For instance, Stille-type reactions were first used in the preparation of protoporphyrin IX dimethyl ester from dibromo-deuteroporphyrin IX dimethyl ester (Scheme **15**) [157].

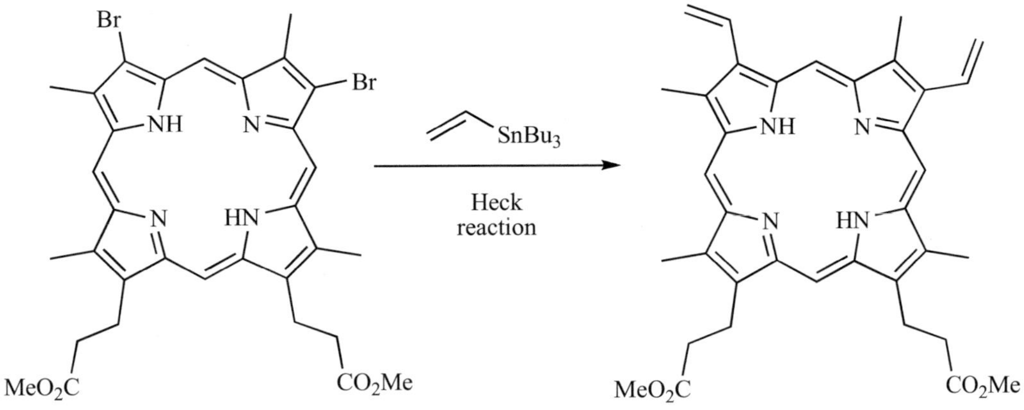

Scheme (15). Stille reactions of porphyrins.

Mono-substituted alkenes under Heck conditions likely produce the corresponding *trans*-alkenylporphyrins, particularly for large, sterically demanding substituents. For example, bromoporphyrins give the corresponding *trans*-alkenylporphyrins [134] and *meso*-tetraaryl zinc(II)-porphyrin can be prepared using Heck conditions (Scheme **16**) [137]. Other porphyrin were also synthesized using the same strategy [143, 158 - 160].

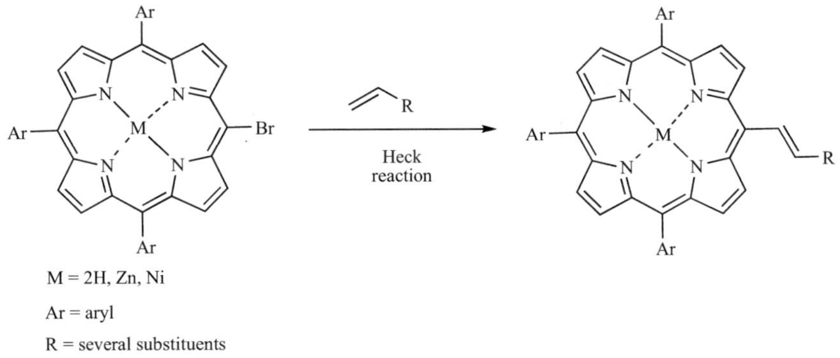

Scheme (16). Heck reactions of porphyrins.

Porphyrins can be nitrated with nitric acid [161 - 165], with N_2O_4 [166 - 169], nitrite ion [170 - 172] or nitrate salts (usually of copper, zinc or silver) [173], *via* the corresponding porphyrin β-cation radicals. In most cases, nitro-substituted porphyrins are used as starting materials for several reactions, including reduction to the corresponding amines using $SnCl_2$/HCl [174], or with $NaBH_4$ and 10% Pd/C in methanol [175], and β-aminoporphyrins have shown to be able to experience acylation, diazotization and diverse reactions with carbonyl compounds [176 - 179]. Besides reduction, *meso*- and β-nitroporphyrins can undergo *ipso* substitution of the nitro group, giving an array of functionalized porphyrin macrocycles [120, 180 - 184], nucleophilic Michael-type additions at the adjacent β-position [107, 119, 128, 167 - 169, 185, 186] to produce functionalized porphyrins (Scheme **17**).

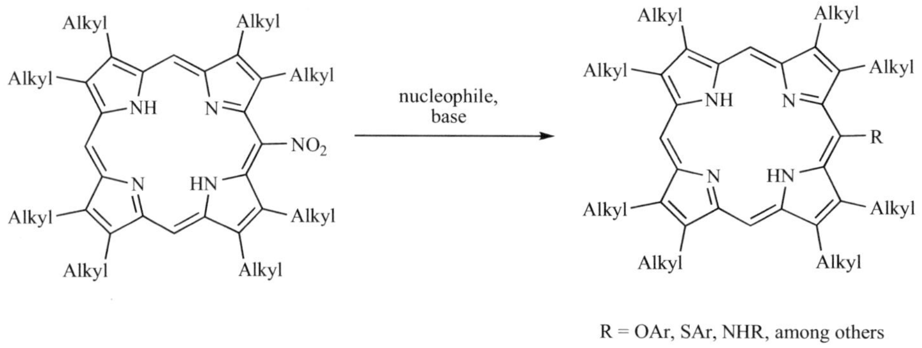

Scheme (17). Nitro group ipso substitution on porphyrins.

The reaction of porphyrins using DMF/$POCl_3$, followed by treatment with aqueous base introduces a formyl group at the *meso*- or the β-position of porphyrin macrocycles (Vilsmeier formylation) [187 - 190]. Altering the formylating reagent, can logically introduce different functionalization (for instance, the sterically hindered *N,N*-diisobutylformamide/$POCl_3$ Vilsmeier reagent gives 2-formylvinylporphyrins [191, 192]. Other regioselective formylation methods use trimethyl orthoformate in TFA [80, 193, 194], or through oxidation of (2-pyridyldimethylsilyl)methylporphyrins [195] or 1,3-dithianylporphyrins [80, 196]. Formyl groups at the periphery of porphyrins can undergo a diversity of reactions, including reductions, oxidations and nucleophilic additions, leading to a variety of functionalized macrocycles [197 - 209].

Using porphyrins to obtain their reduced derivatives is also possible. Chlorins, isobacteriochlorins and bacteriochlorins (Fig. **4**) can be prepared from porphyrins, through several reducing reactions, being appropriately called hydroporphyrins.

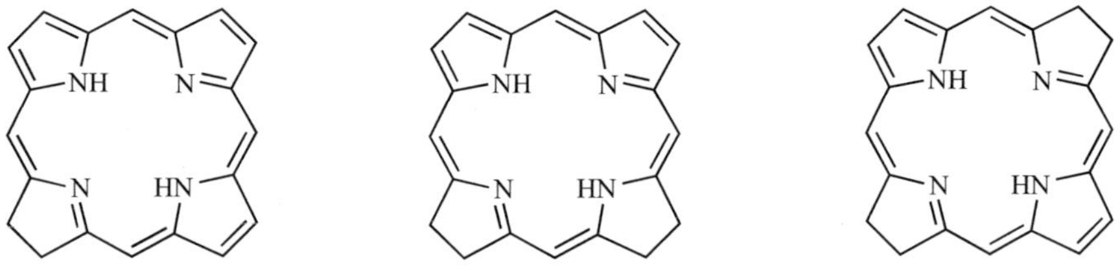

chlorin **isobacteriochlorin** **bacteriochlorin**

Fig. (4). Reduced porphyrin derivatives.

Hydroporphyrins are a class of tetrapyrrolic substances, normally resulting from porphyrins, by saturation of double bonds at the periphery of the macrocycle, consequently having fewer π bonds.

There are two main methods for preparing hydroporphyrins, either by ring construction or by reactions performed on the corresponding porphyrins. Here only the latter methodology will be discussed, as this method involves porphyrin transformation.

Reduction of porphyrins is one of the most known available methods for obtaining hydroporphyrins, which can be achieved under catalytic hydrogenation conditions with H_2/Pd or *p* -toluenesulfonylhydrazide, in the presence of a base.

The latter reagent is definitely the most used in hydroporphyrin chemistry, especially when dealing with *meso*-substituted porphyrins, used in a reaction developed Whitlock [210]. A benchmark example involves the synthesis of tetra(*m*-hydroxyphenyl)chlorin (*m*THPC, Foscan®), a clinically accepted photosensitizer for photodynamic therapy, through reduction of the corresponding porphyrin to bacteriochlorin, followed by oxidation with a mild oxidizer, like *o*-chloranil (Scheme **18**) [211 - 215]. It turns necessary to use *o*-chloranil to reduce the formed bacteriochlorin, since it is rather difficult to selectively reduce the corresponding porphyrin, giving origin to a mixture of bacteriochlorin and chlorin (mostly bacteriochlorin).

On the other hand, *o*-chloranil reveals to be a mild oxidizer, transforming only the present bacteriochlorin to chlorin, without danger of over-oxidation back to porphyrin.

Tetrahydroporphyrins can be obtained by further reduction of chlorins similarly to the above mentioned. Metalloporphyrins often give isobacteriochlorins, while metal-free porphyrins preferentially give bacteriochlorins. Again, typical preparations make use of the Whitlock method; however, in 2010, a solvent-free methodology for *meso*-aryl halogenated bacteriochlorin synthesis was reported, managing to reduce utilization of toxic solvents, through adjustment of Whitlock's diimide reduction method, where the solid porphyrins can be grounded together with solid p-toluenesulfonylhydrazide (large excess) and heated up to 140 °C (beyond its melting point), to promote reaction in liquid phase [216, 217].

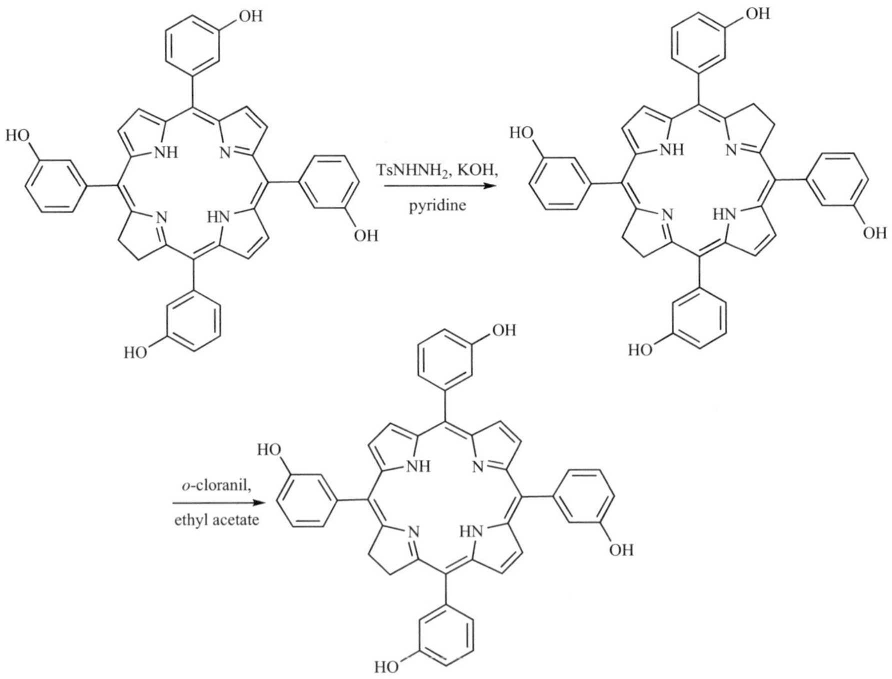

Scheme (18). Synthesis of tetra(*m*-hydroxyphenyl)chlorin.

It might be strange to use oxidation reactions to obtain the reduced forms of porphyrins, but for instance osmium tetroxide can selectively react at the β,β'-double bonds of porphyrins producing the corresponding *cis*-dihydroxychlorins; for example reaction of tetraphenylporphyrin with stoichiometric amount of OsO_4 followed by hydrogen sulfide treatment produces 2,3-*vic*-di-hydroxy-*meso*-tetraphenylchlorin (scheme **19**) [218 - 220].

Another important synthetic methodology for the preparation of hydroporphyrins refers to the cycloaddition reactions (Scheme **20**). Porphyrins bearing suitable substituent groups can take part in pericyclic reactions as dienophiles, dienes,1,3-dipoles and as dipolarophiles.

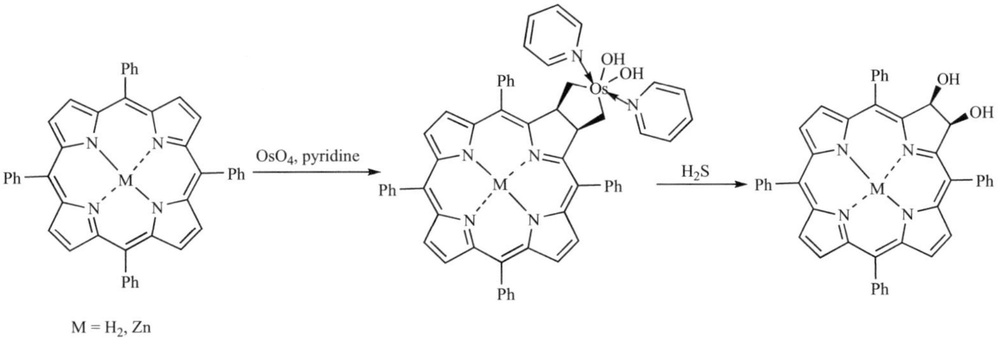

Scheme (19). Synthesis of 2,3-*vic* -di-hydroxy-meso-tetraphenylchlorin via OsO_4 oxidation.

For instance, porphyrins substituted with vinyl groups at β-pyrrole positions can act as dienes in Diels-Alder reactions. The first reports using a vinyl porphyrin were concerning the use of protoporphyrin-IX with tetracyanoethylene or dimethylace-tylenedicarboxylate [221 - 225], and later the same type of porphyrins was used as diene in reaction with maleic anhydride [226].

Meso substituted symmetrical porphyrins can also react as dienophiles, giving chlorins as main products, through Diels-Alder reactions with orthoquinobenzodimethanes, and polycyclic aromatic compounds. The main drawback resides in the necessity of sole use of porphyrins bearing electron withdrawing groups, to avoid oxidation of the corresponding chlorins to porphyrins [212, 227].

In the case of participation as dipolarophiles, porphyrins can react with several 1,3-dipoles, such as azomethine ylides [228 - 230], nitrones [231, 232] and diazoalkanes [233 - 236] to produce chlorins (Scheme **20**).

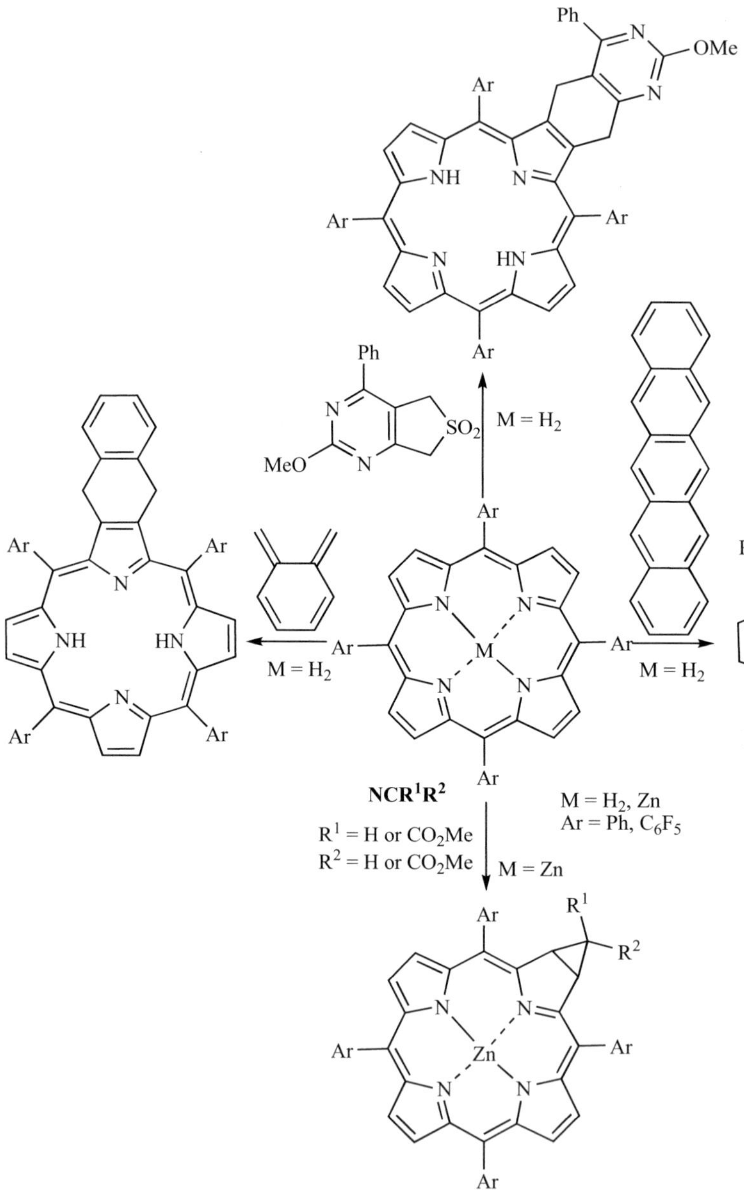

Scheme (20). Dipolar additions for the synthesis of chlorins.

For instance, the first study of the reactivity of *meso*-substituted porphyrins with azomethine ylides, in situ generated through paraformaldehyde and sarcosine was in 1999 [228] and several other studies followed these findings [229, 230].

APPLICATIONS OF PORPHYRINS

The remarkable features possessed by porphyrin and their related compounds have rendered them high applicability in catalysis [237 - 264], as chemical sensors [265 - 279] in supramolecular chemistry [280 - 293], as nonlinear optical materials [144, 294 - 303], in optoelectronics [304 - 313], in photovoltaics, including solar cells [147, 314 - 328], in polymer based materials [264, 329 - 350], in photodynamic inactivation of bacteria [351 - 365], in photodynamic therapy [211, 366 - 380] and disease diagnostics [381 - 391].

Applications in Catalysis

One of the most important applications of porphyrins is catalysis and they have been thoroughly studied as ligands to stabilize metals, tuning their reactivity. For instance, iron and manganese porphyrins are very important catalysts for epoxidation reactions. Mansuy and co-workers were the first to report the use of chlorinated Mn-porphyrin complex **1** (Fig. **5**) in epoxidation of alkenes in the presence of H_2O_2 [392]. These halogenated porphyrins were necessary, in order to avoid catalyst oxidation, while additives such as imidazole [392 - 395] or combinations of carboxylic acids and imidazoles [396 - 398] were used to assure high reactivity.

For instance, Quici reported porphyrins modified with pendant ligands, where porphyrin **2**, (Fig. **5**) tethered with imidazole groups and carboxylic acid arms gave enhanced epoxidation rates, when compared to non-modified porphyrins [397, 398].

Iron porphyrins are also active catalysts for epoxidation, however, selectivities and conversions are usually lower than when using their Mn(III) counterparts as catalysts. Traylor made important discoveries in this area, including a survey on suitable reaction conditions [399 - 401]. Electron-deficient porphyrins **3** (Fig. **5**) privileged alkene epoxidation with H_2O_2, using also other electron-withdrawing substituted porphyrins as catalysts [162, 402, 403].

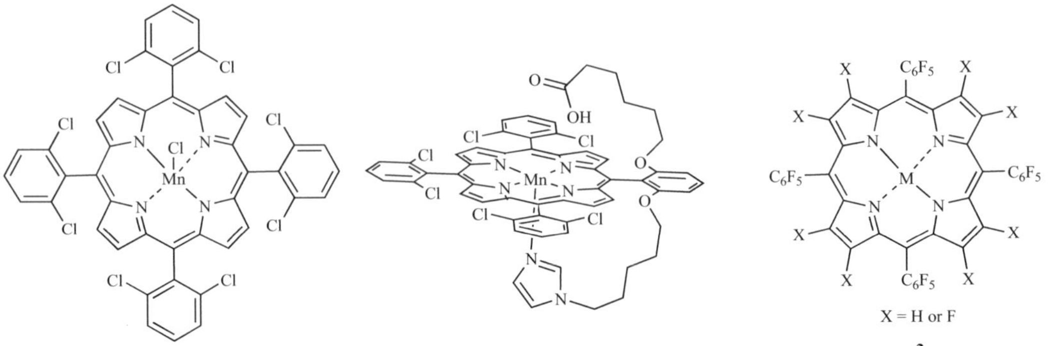

1 2 3

Fig. (5). Halogenated porphyrins for catalytic applications.

Asymmetric epoxidation is of upmost importance and chiral metalloporphyrins have also been widely studied. The modifiable periphery of porphyrins and their rigid macrocyclic core confer them attractive templates for preparing asymmetric catalysts. To attach optically active groups onto metalloporphyrin ring several strategies have been utilized [404, 405]. Additionally, Berkessel [406] and Che [407 - 409] reported high catalytic system's efficiency using other chiral porphyrins for non-functionalized olefin asymmetric epoxidation.

In 1993, Hevesi described the use of the chiral Mn(III) complex **4** (Fig. **6**) in asymmetric epoxidation of olefins [410], while in 1996, Gross reported for the first time the chiral ruthenium porphyrin **5a** (Fig. **6**) mediated alkene epoxidation [411], studying the solvent effect on chiral epoxy styrene's enantioselectivity. Later, new catalysts **5b,c** of the ruthenium porphyrins type were reported (Fig. **6**) [412], proving to be the most selective catalysts for trans-disubstituted and terminal olefin asymmetric epoxidation. Compound **5b** displayed higher enantioselectivity than **5a**; however, catalyst **5c** revealed to be the best catalyst for chiral induction for trans- and terminal olefins.

In 1997, Berkessel also synthesized an enantiomerically pure ruthenium carbonyl porphyrin **6a** (Fig. **6**) [406], for olefin catalytic asymmetric epoxidation, giving epoxides in high yields, also with high enantiomeric excess. Later, the same group described again the use of catalysts **6b-d** for asymmetric epoxidation of several olefin substrates [413].

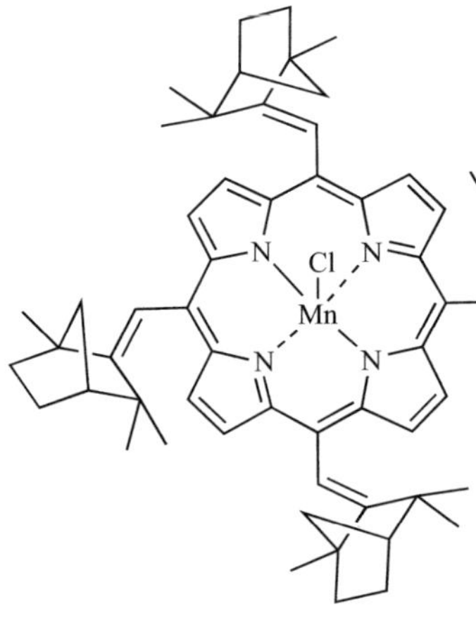

4

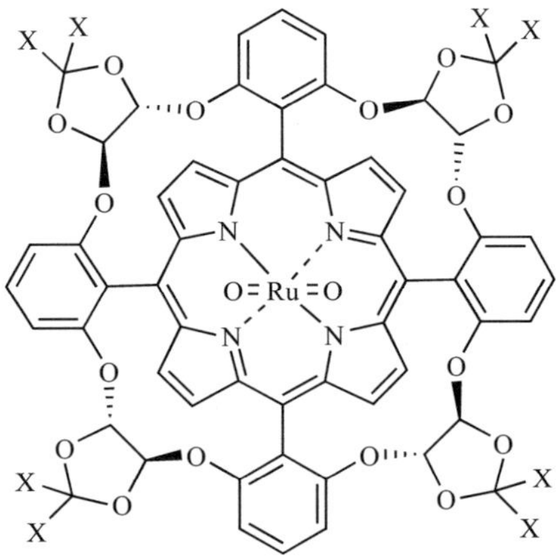

5a: X = Me
5b: X = Ph
5c: X = *m*-Cl-Ph
5d: X = Et
5e: X = -$(CH_2)_2$-

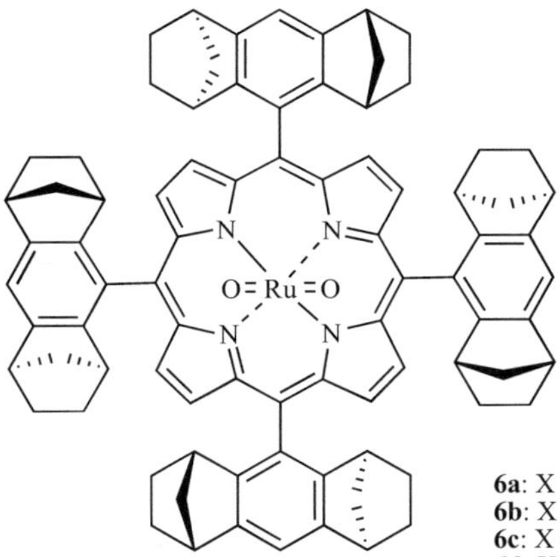

6a: X = H
6b: X = OMe
6c: X = Me
6d: X = CF_3

Fig. (6). Chiral porphyrins for catalytic applications.

In 1998, Quici group reported the synthesis of atropisomerically pure Mn(III) porphyrins **7a-b**, (Fig. 7) bearing a [2, 2]-*para* -cyclophane group, and used them as catalysts in terminal olefin epoxidation, using aqueous NaClO, PhIO 30% and H_2O_2 as oxygen donors [414]; however, no noteworthy enantioselectivity was achieved with any of the oxygen donors.

In 2003, Lindsay Smith also studied the chiral iron porphyrins **8a-d** (Fig. 7) and the structural dependence of the substrates and catalysts, the solvent and the temperature on the enantiomeric excess [415].

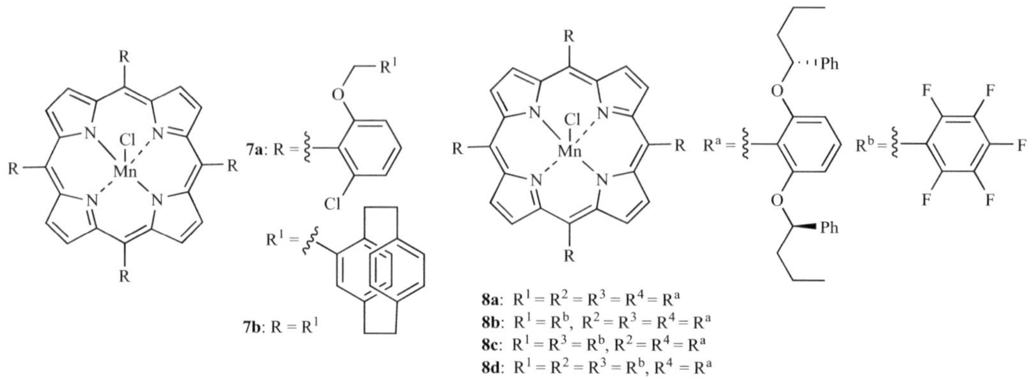

Fig. (7). *meso*-substituted porphyrins for catalytic applications.

Heterogeneous catalysis has been also a very important issue when using porphyrins as catalysts. Several examples exist of well succeeded research on this subject, including the use of immobilized porphyrins on several inorganic supports on the oxidation of several pollutants, along with epoxidation reactions. Azenha [416] studied the use of mesoporous molecular sieves as supports for immobilization of 5,10,15,20-tetra (2-fluorophenyl)porphyrin (**9a**) by ship-in-a-bottle and adsorption routes, respectively. Photodegradation of the pesticides mecoprop and 2,3,5-trimethylphenol was performed using **9a** immobilized inside MCM-41 material as catalyst (Fig. **8**), displaying good degradation results. Afterward, 5,10,15,20-tetra (2,6-dichlorophenyl)-porphyrin, **10a** was also immobilized by ship-in-a-bottle method inside MCM-41, as well as their corresponding copper metal complexes **9b** and **10b** [417].

Maldotti [418] described the method of chlorosulfonated (*meso*-tetra(2,--dichlorophenyl)porphyrin) **11** linkage to amorphous SiO_2 and MCM-41 and studied it on oxidation reactions. The 1,4-pentanediol oxidation rate using **11** immobilized onto SiO_2 was ca. four times lower than that obtained with **11** immobilized onto MCM-41, which was attributed to the mesoporous MCM-41 high surface area.

Moghadam [419] also prepared a iron porphyrin encapsulated in a zeolite, using the zeolite post-synthesis method. In this case, NaY or NaX zeolite was synthesized around the cationic Mn(III) 5,10,15,20-tetra(4-N-methylpyridyl)porphyrin **12** yielding the corresponding catalysts. These materials affectively catalyzed the epoxidation of a set of alkenes using $NaIO_4$ as oxidant, yielding the epoxides highly selectively.

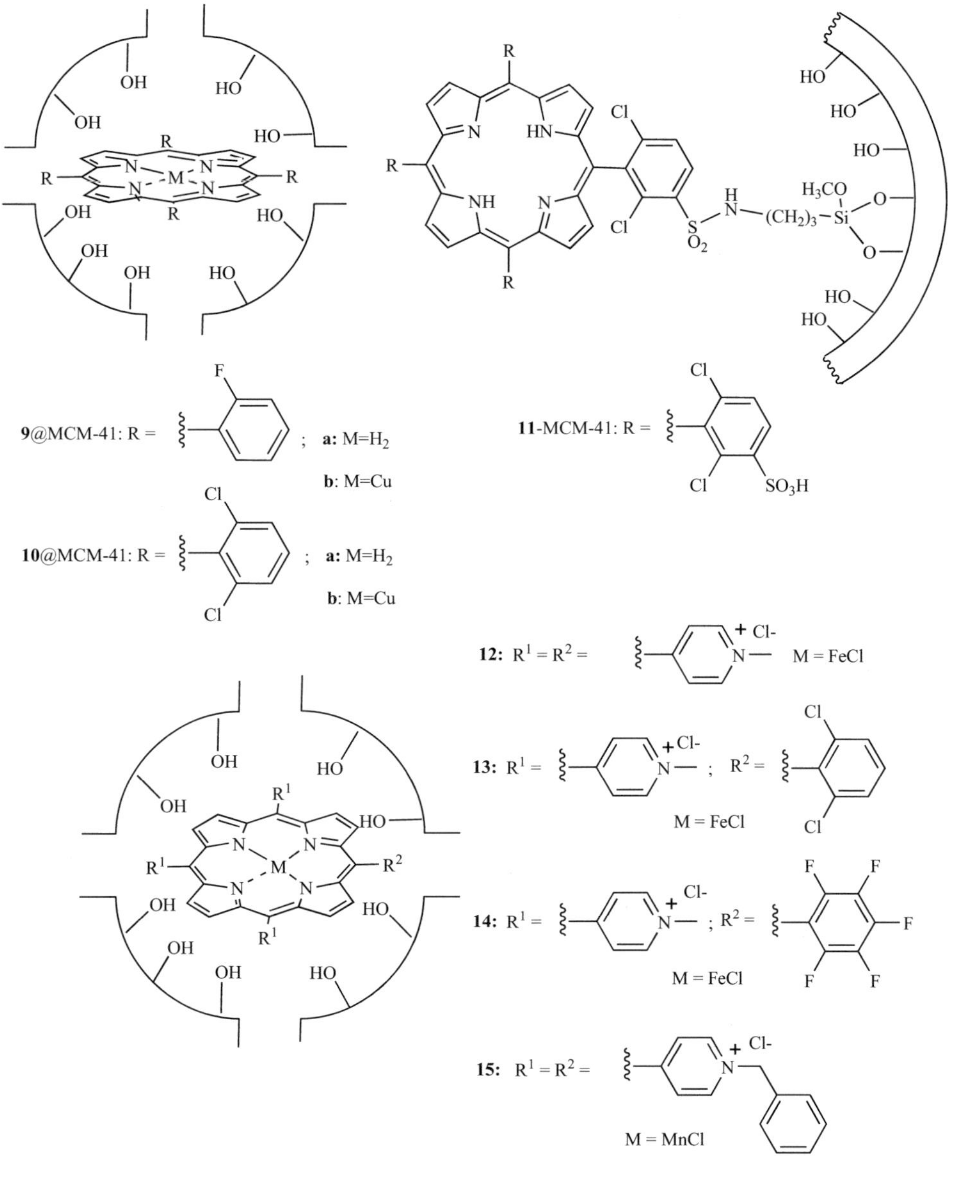

Fig. (8). Immobilized porphyrins for catalytic applications.

Viana Rosa also reported the preparation of chloro Fe(III) 5-mono(2,6-dichlo-o-phenyl)-10,15,20-tris(4-N-methylpyridyl)porphyrin **13** [420] and chloro Fe(III) mono(pentafluo-rophenyl)-tris-(4-N-methylpyridyl) porphyrin **14** [421], respectively encapsulated in NaX or NaY by post-synthesis method. The catalysts were then used in (*Z*)-cyclooctene, cyclohexane and adamantane oxidation, with enantioselectivities up to 99% and very high conversions. Using a similar approach, compound **15** [422] was also tested the catalyst in the oxidation of several terpenes using H_2O_2/ammonium acetate as oxidant, performing comparisons between homogeneous and heterogeneous catalysis, obtaining better results when heterogeneous systems were applied.

Applications as Chemical Sensors

The sensing and recognition of environmentally and biologically important ions has appeared as a noteworthy goal in the chemical sensors field, where fluorescence has turned to be a powerful analytical detection method, due to its high detection limit and straightforwardness. Particularly, suitable probes in conjunction with fluorogenic methods are favorable to measure analytes, given that fluorimetry is highly sensitive, nondestructive, quickly performed, and can proportionate real information on the quantification and localization of the targets of interest.

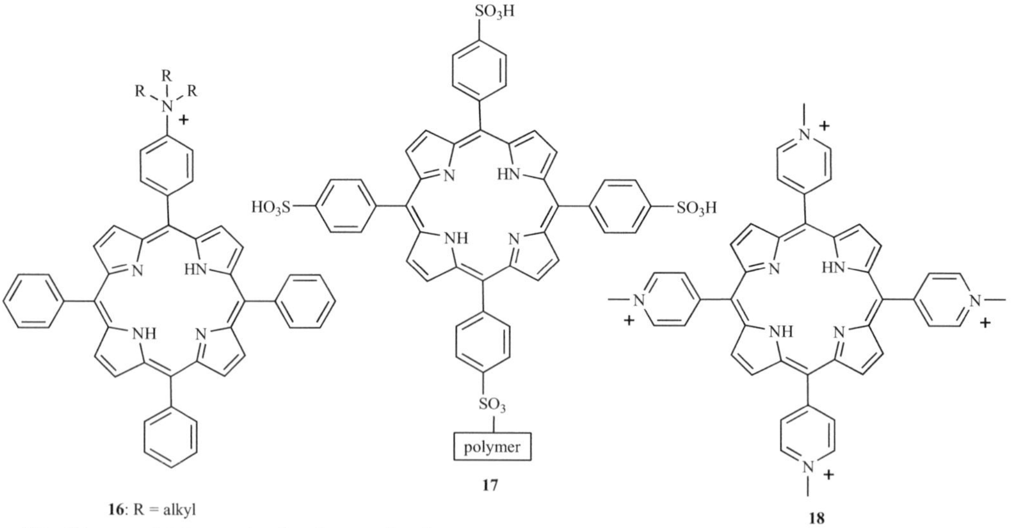

16: R = alkyl

17

18

Fig. (9). Water soluble porphyrins for application as sensors.

The chemical sensors based on porphyrins can be modified by synthetic modifications on the π-aromatic system, the peripheral substituents and coordinated metal. For instance, Paolesse reported the deposition of amphiphilic porphyrin derivative **16** (Fig. **9**) on glass surfaces [423, 424]. The tetraphenylporphyrin's periphery was functionalized with a cationic tetraalkylammonium group, promoting molecular self-aggregation in ethanol-water mixtures, which was transferred onto glass. The device revealed to be a good sensor for Cd^{2+}, Cu^{2+}, Pb^{2+} measured by fluorescence, which was quenched by cationionic interaction. Moreover, presented a strong sensitivity to Hg^{2+} ions whose quenching exceeded of orders of magnitude those observed for the previous ions.

Other authors [425] designed an optic heavy metal ions detector. The sensing layer consisted of 5,10,15,20-tetra(*p*-sulfonatophenyl)porphyrin **17** (Fig. **9**), covalently immobilized on a polymeric matrix. Differentiated absorbance spectra were observed upon ion complexation, such as Pb(II) ,Hg(II) or Cd(II), enabling the ions to be simultaneously detected. A similar study was later performed using cationic porphyrin **18** [426], either grafted on glass surfaces or adsorbed in sol-gel matrix, displaying quite low detection limits.

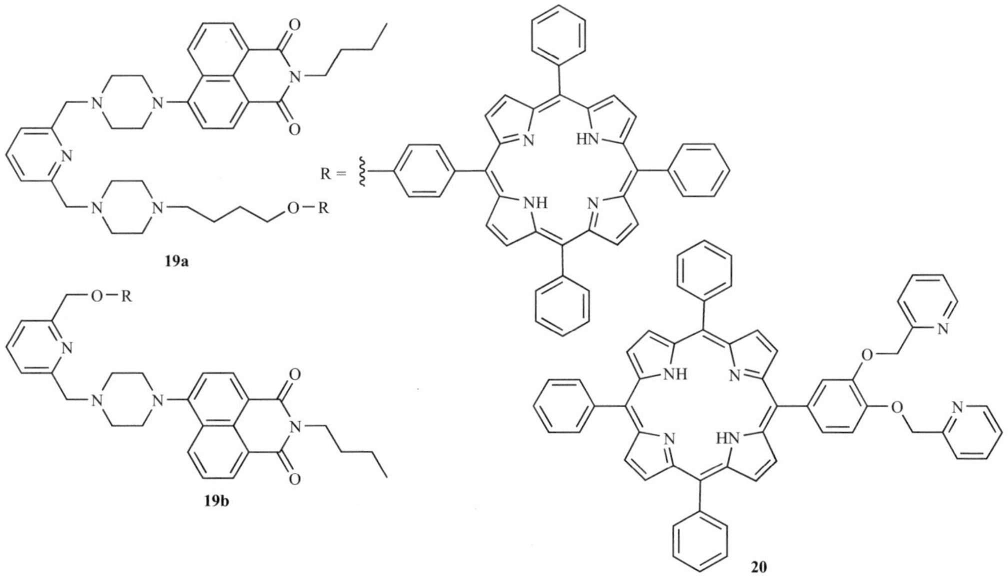

Fig. (10). Fluorescent probes for sensing applications.

Later, the fluorescent probe **19** (Fig. **10**), a naphthalimide-porphyrin hybrid was used for ratiometric detection of Hg^{2+} in living cells and aqueous solutions [427]. The probe's signal change was due to its structure, as it contained two independent Hg^{2+}-sensitive fluorophores, having their maximal excitation wavelengths at the same range. The same group developed also porphyrin derivative **20** (Fig. **10**), containing two 2-(oxymethyl)pyridine units for use as chemosensor for metal ions recognition [428]. Compound **20** exhibited highly selectivity ratiometric fluorescence response to Zn^{2+} and a 5.5×10^{-8} M detection limit. The authors also reported pH independent response behavior of **20** to Zn^{2} in medium condition (pH 4.0-8.0).

Starnes [429] prepared porphyrin **21** Fig. (**11**) and evaluated it as sensor for hydrogen sulfate, dihydrogen phosphate, acetate and several halides, leading to a high binding constant for fluoride anions.

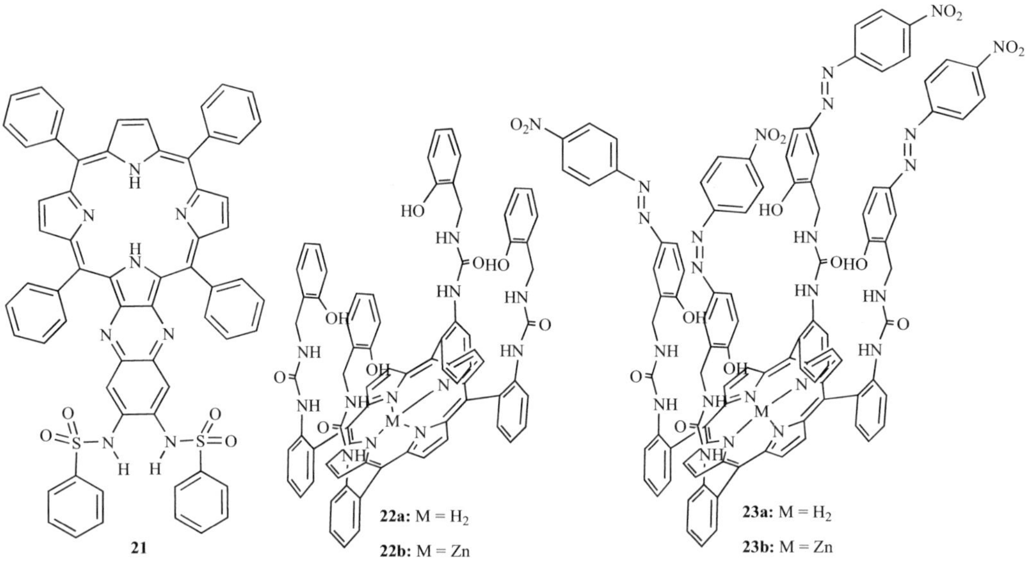

Fig. (11). Porphyrin derivatives for sensing.

Porphyrin derivatives **22a-b** and **23a-b** Fig. (**11**) with color-reporting and/or H-bond donors chromophoric units were reported, being evaluated as anion-binding receptors [430]. Compounds **22a-b** showed to be selective in the of AcO^{-} >

$H_2PO_4^-$ >> Cl^-, Br^-, I^- order in DMSO-d_6. Additionally, F^-, $H_2PO_4^-$ and AcO^- were also detected by color change tests using **23a-b** in organic solution.

Palladium and platinum porphyrins are most commonly used for oxygen sensing, due to their high quantum yields and strong phosphorescence [431].

Due to the presence of the electron-withdrawing perfluorophenyl substituents present in **25** Fig. (**12**), this platinum porphyrin [432, 433] was reported as a photostable sensing material, with reduced electron density and high redox potentials.

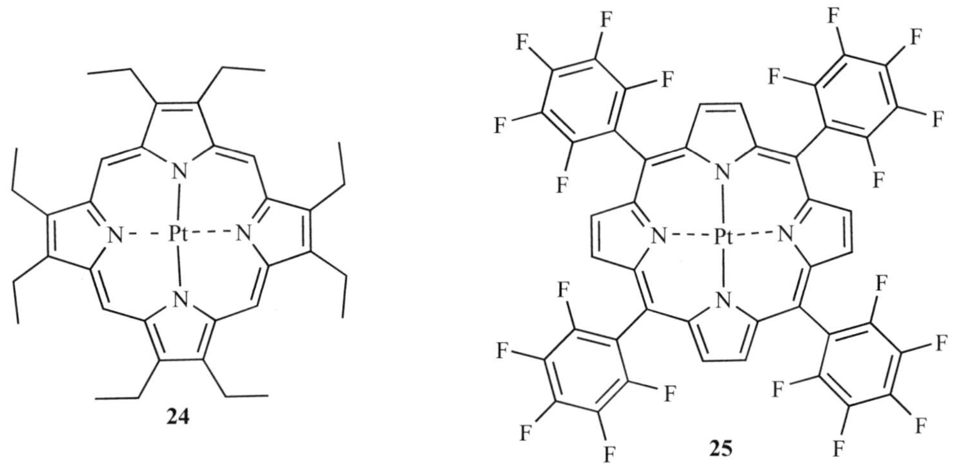

Fig. (12). Porphyrin for oxygen sensing.

Detection of amine is of particular interest because amines are found in many wastewater effluents from agriculture, pharmaceuticals, industry and food processing, inducing toxicological responses at low concentrations [434]. For instance, an optical sensor for determination of aliphatic amines was reported in 2000 [435], based on the reversible increase of fluorescence of heptakis(2,6-di-O-isobutyl)-(-cyclodextrin hosting tetraphenylporphyrin complex **26** Fig. (**13**), immobilized in poly(vinyl chloride), demonstrating high selectivity for long chain primary amines, with a 3.0 x10^{-7} mol/L detection limit.

Also Meyerhoff reported an optical sensor based on a polymer film for gas-phase amine detection at sub-ppm levels [278]. The sensor was constituted by indium(III) octaethylporphyrin hydroxide ion-bridged dimer species **27** Fig. (**13**) within a thin poly(vinyl chloride) film.

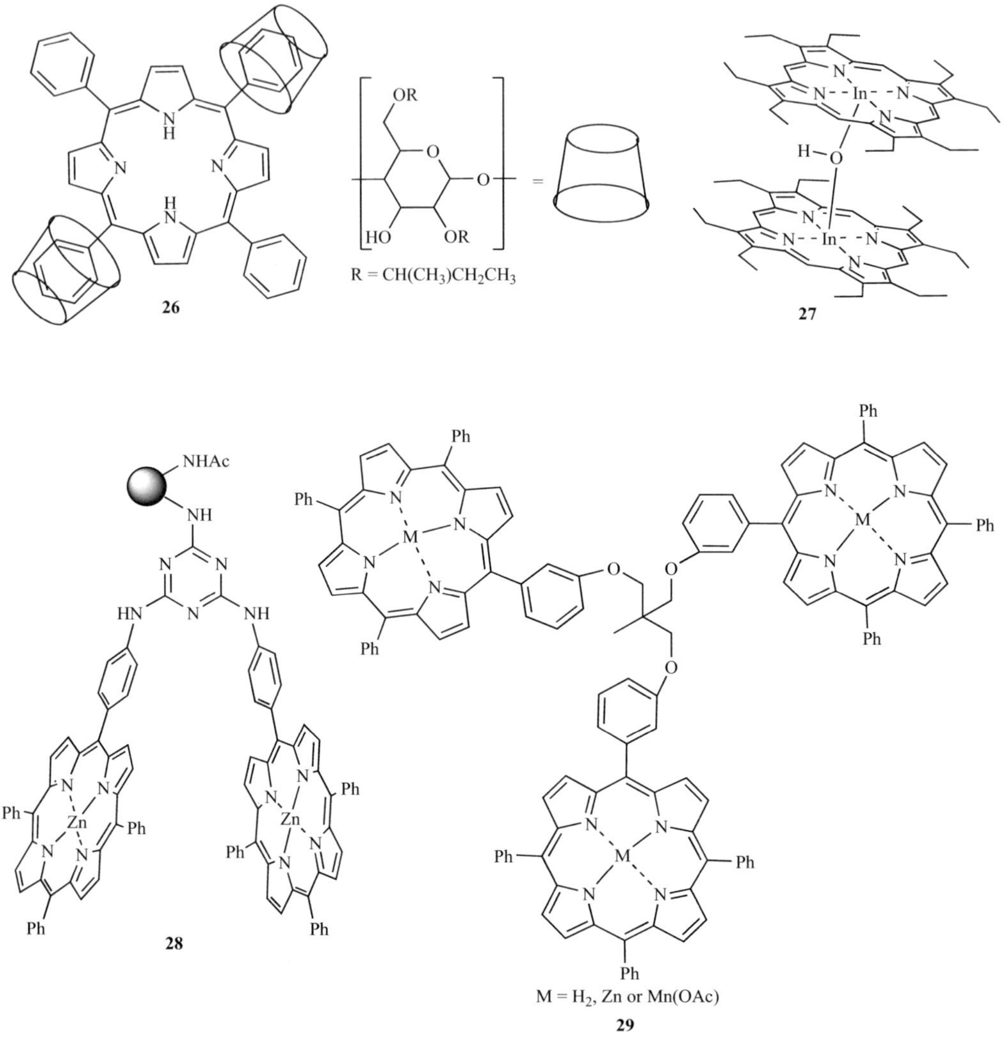

Fig. (13). Porphyrin based amine sensors.

Carofiglio reported the synthesis of a melamine-bridged bis-porphyrin tweezers **28** Fig. (**13**) [436]. These materials were appropriate for sensing diamines, for instance, cadaverine, a biologically active diamine important in food freshness monitoring. Dendronic sensors of type **29** Fig. (**13**) [279] were also synthesized by Burrows and Pereira, where this architecture led to a synergic effect, with very low detection limits, in the range of 5.0 x 10^{-7} M for a variety of amines.

Applications as Nonlinear Optical Materials

There is a substantial interest in nonlinear optical compounds, in order to control optical signals in optical signal processing applications, *e.g.* all-optical switches, high speed electro-optical and modulators. Organic materials, like porphyrins, can have large nonlinearities, being able to be integrated in optical devices more economically and easily than inorganic materials.

Nonlinear refraction and absorption are nonlinear optical features directly related. Generally, several mechanisms, both non-parametric and parametric, contribute to the nonlinearities and a consistent evaluation of the responsible mechanisms requires experiments involving different pulse lengths. Optical limiting consists of a sample transmittance decrease under fluence illumination or high intensity. Optical limiting is also regarded as the reverse of saturable absorption (RSA) where increase in transmittance is seen at high illumination. Phthalocyanines are promising metallo-organic materials for optical power limiting in the visible and NIR spectral range, due to their suitable photophysical properties.

Blau reported for the first time the effect of RSA in porphyrins [437], using 80 ps laser pulses at 532 nm. They found that ground-state absorption cross-sections of porphyrins **30a-c** were smaller than the corresponding excited-state absorption cross-sections (Fig. **14**).

Since then, substantial research to improve the RSA properties of porphyrin based optical limiters has been carried out. For instance, the optical limiting data in terms of excited-state absorption cross-section and limiting threshold, for octabromote-traphenylporphyrins **31a-f** Fig. (**14**) have shown the trend: **31f>31a>30c>C_{60}>30a** [438, 439], with complex **31f** displaying the best performance [440].

The optical limiting effect of several tetratolylporphyrin metal complexes **32** Fig. (**14**) was also described [441]. All complexes were found to display a strong NLO absorption at both 532 and 600 nm.

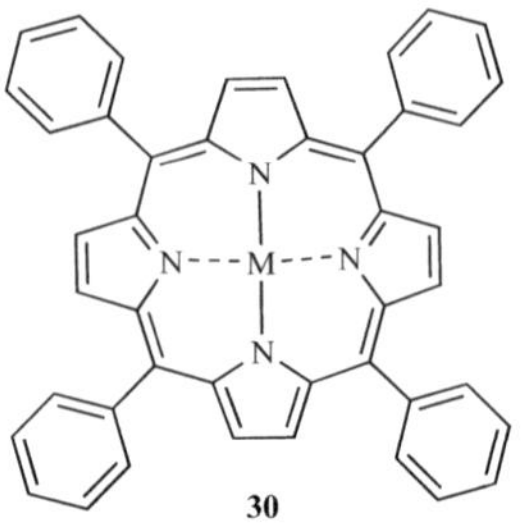

30

a: M = Zn
b: M = Co
c: M = 2H

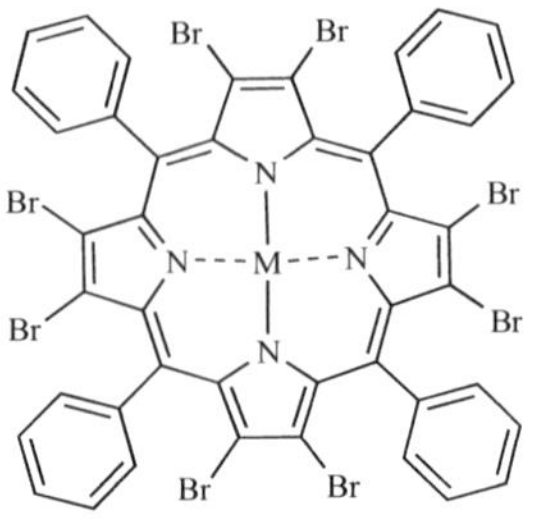

31

a: M = 2H
b: M = Co
c: M = Cu
d: M = Pb
e: M = Cd
f: M = Zn

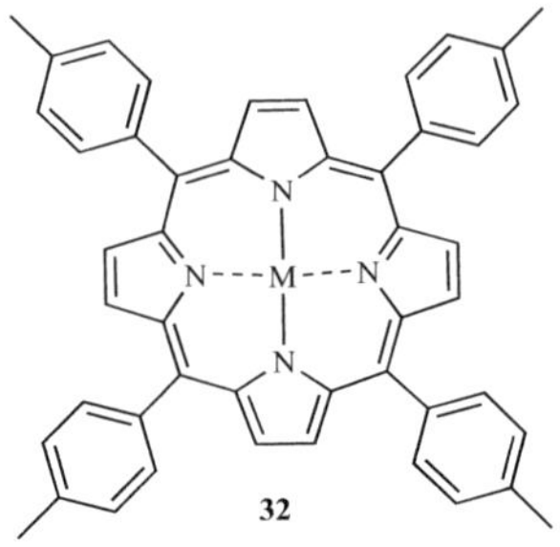

32

M = 2H, Co, Cu, Ni, Zn, Ag, Cd, Hg
CrCl, MnCl, FeCl, AuCl
InOH, POH, V(O)
$Ge(OH)_2$, $Sn(OH)_2$

Fig. (14). Porphyrin derivatives for NLO applications.

Another reported system was based on tetrabenzoporphyrin **33** Fig. (**15**) [442], where NLT effect was dependent on laser fluence [443, 444]. Later, the effect of *meso*-substituents with electron donating and electron withdrawing groups was also investigated for tetrabenzoporphyrins of type **34** (Fig. **15**) [445].

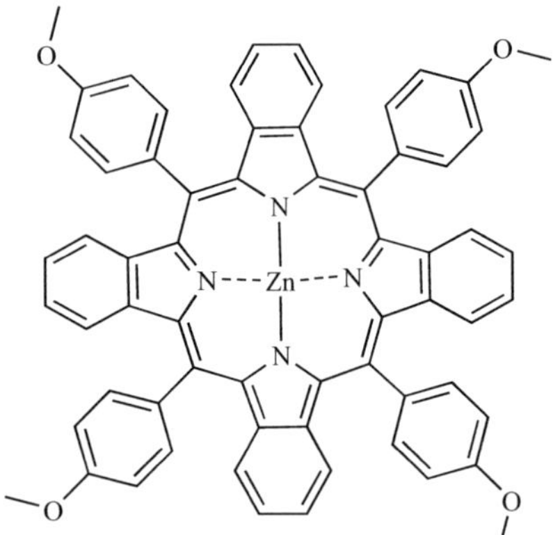

33

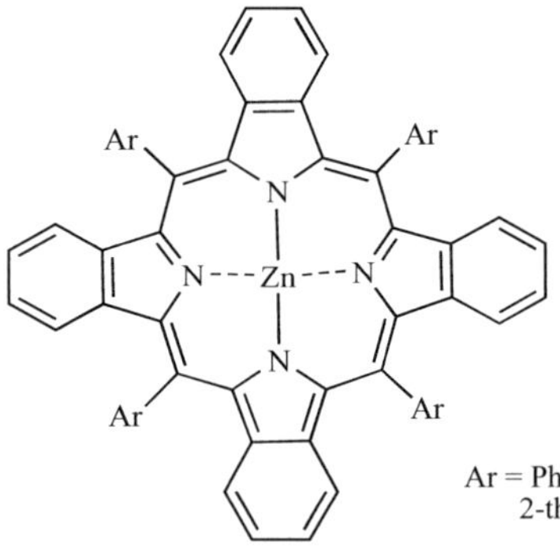

Ar = Phenyl,
2-thienyl

34

Fig. (15). Benzoporphyrins for NLO.

Other authors have shown that the optical limiting performance of **35a** (Fig. **16**) was better than that of the corresponding metal free porphyrin at 532 nm [446].

Related to this study, other comprehensive studies using pump/probe procedures established that introducing heavy metal atoms, such Pb(II) (**35b**) induced a strong influence on most of NLO parameters [447].

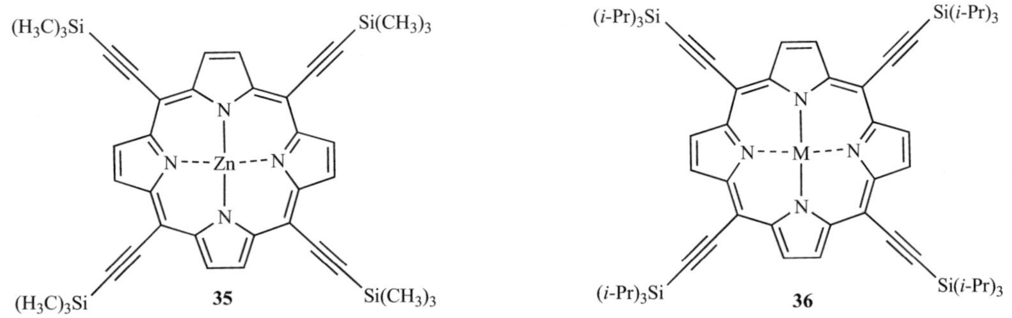

35

a: M = Zn
b: M = Pb

36

M = 2H, Pb, GaCl, InCl, TlCl, $SnCl_2$
$AlOSiPh_3$, $Ge(OSiPh_3)_2$

Fig. (16). Acetylene-type substituted porphyrins for nonlinear optics.

The metal complexes with Ge, Sn, Pb and Al, Ga, In, Tl, of tetrakis(4-triisopropy-lsilylethynyl) porphyrin **36** were also studied (Fig. **16**) [294], with complexes of Pb(II), In(III) and Tl(III) offering the best response at 532 nm.

Polymeric porphyrins were synthesized with attached groups at the *meso* positions (**37**) (Fig. **17**) [448]. In compound **37**, the delocalized π-system was extended, maintaining a symmetric and planar structure. Their optical limiting properties were evaluated at 532 nm and the relative transmittance decreased to 40, 60 and 65% at input fluences of 0.01 J/cm^2 for monomer, dimer and polymer thin films, respectively.

Fused porphyrinic structures were also synthesized (**38**) (Fig. **17**), in order to study the enhanced extended conjugation effect on the excited-state lifetime on porphyrin type materials [449]. Photophysical excited-state dynamics studies at picosecond regime were carried out and it was demonstrated that fused porphyrins show very short excited-state lifetimes. The authors suggested that this type of fused structures have a ~10 ns fluorescence lifetime, but the singlet excited state was quenched by fast nonradiative decay. Nevertheless, when bromine or iodine was present (**39a-c**) (Fig. **17**) [450], the ISC became fast enough to compete with nonradiative decay.

Porphyrinic related materials such as expanded pentaazadentate metal complex **40** has also been studied for their properties (Fig. **17**) [451]. This complex exhibited high yields of triplet excited-states and high intersystem crossing rates at 532nm, showing strong fluence dependence.

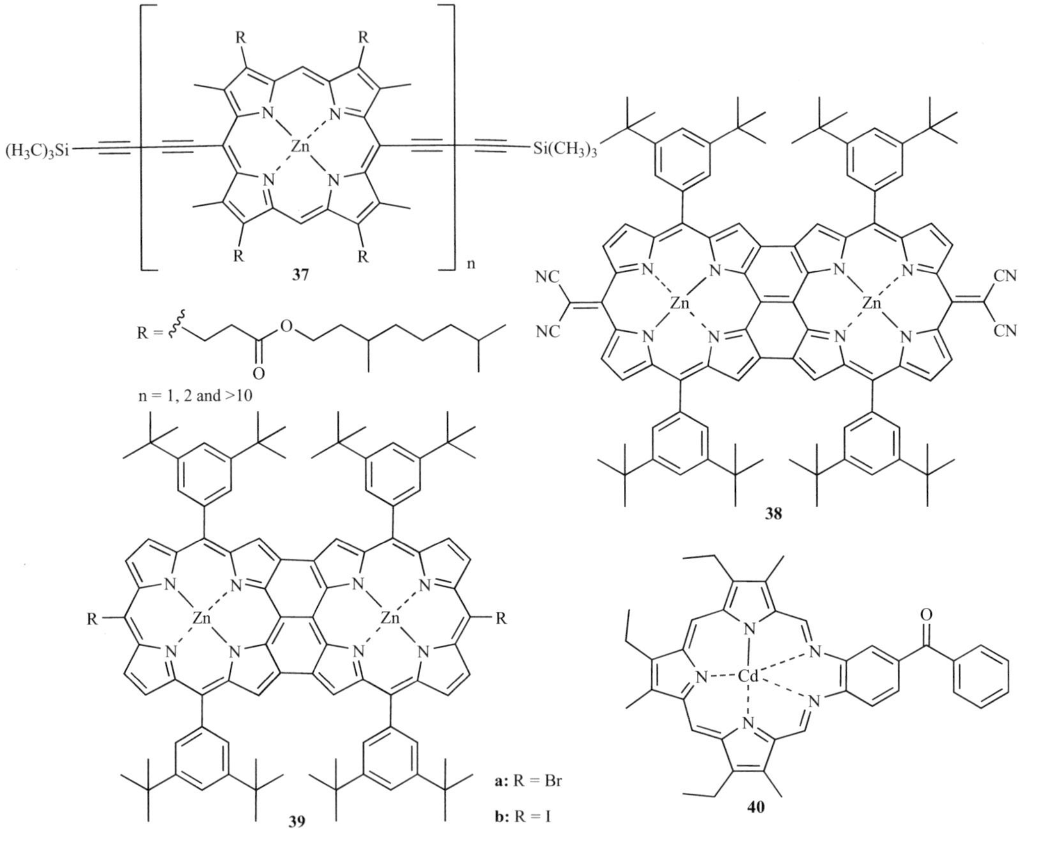

Fig. (17). π-Extended porphyrins for NLO.

Applications in Solar Cells

Solar cells based on organic materials are enjoying rising attention. Usually, these materials are mostly constituted by a bulk heterojunction resultant from the interaction of donor and acceptor materials, combining the advantages of higher conversion efficiency and straightforward fabrication. Hence, porphyrin dyes utilization as the basis of dye-sensitized solar cells (DSSC) has demonstrated solar conversion efficiencies reaching silicon based photovoltaic devices. The use of porphyrins in DSSCs was first demonstrated by Kay and Grätzel in 1993 [323, 452], where they used a copper chlorophyllin sensitizer with efficiency $\eta = 2.6\%$. At the time, the finding was not even considered extraordinary, as the same authors had achieved 10% efficiency with a ruthenium based dye months before [453]. In the next decade there was nearly no developments on the use (-linked porphyrins for sensitized solar cells, until 2004, when Officer and Grätzel [454] reported several Zn(II) porphyrins, where compound **1** (Fig. **18**) reached $\eta = 4.8\%$ [454]. They subsequently reported several other zinc porphyrins, where the best one (**42**, Fig. **18**), obtained $\eta = 5.6\%$, using chenodeoxycholic acid as co-adsorbent [325]. Later, Grätzel reported other sensitizers, with the best one (**43**, Fig. **18**) attaining $\eta = 7.1\%$ [455], paving the way for the growth of porphyrin-type sensitizers designed to increase DSSC efficiency.

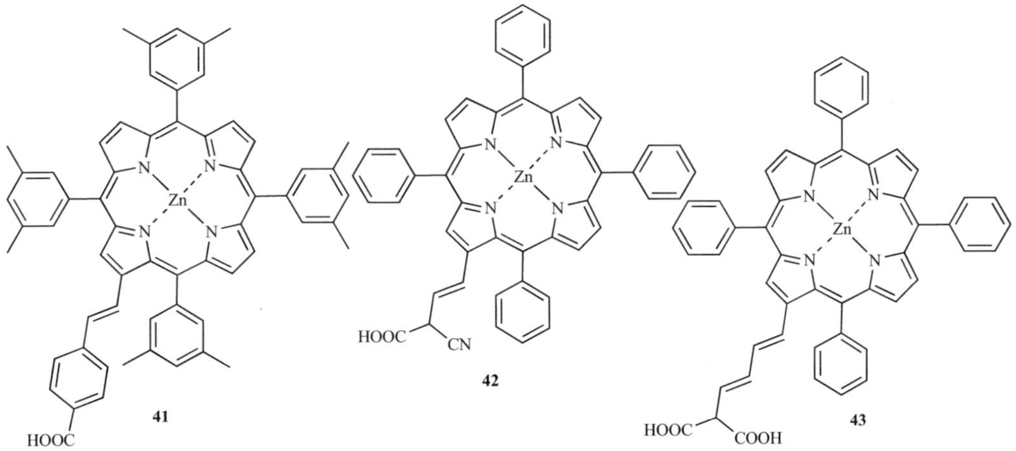

41 42 43

Fig. (18). Porphyrin derivatives for solar cell applications.

The concept of using carboxylic acid anchoring groups was continued by Kim [321], who reported the β-functionalized zinc porphyrin **44** (Fig. **19**), bearing a diarylamino group, exhibiting the high performance of $\eta = 7.5\%$.

Though the first *meso*-substituted porphyrin sensitized solar cell was reported in 2000 by Wamser and Cherian [327], with η ~ 3.5%, no major advances were reported for this type of porphyrin sensitizers, until 2007, when Galoppini [456] reported the zinc porphyrins of type **45** (Fig. **19**), attempting to diminish dye aggregation.

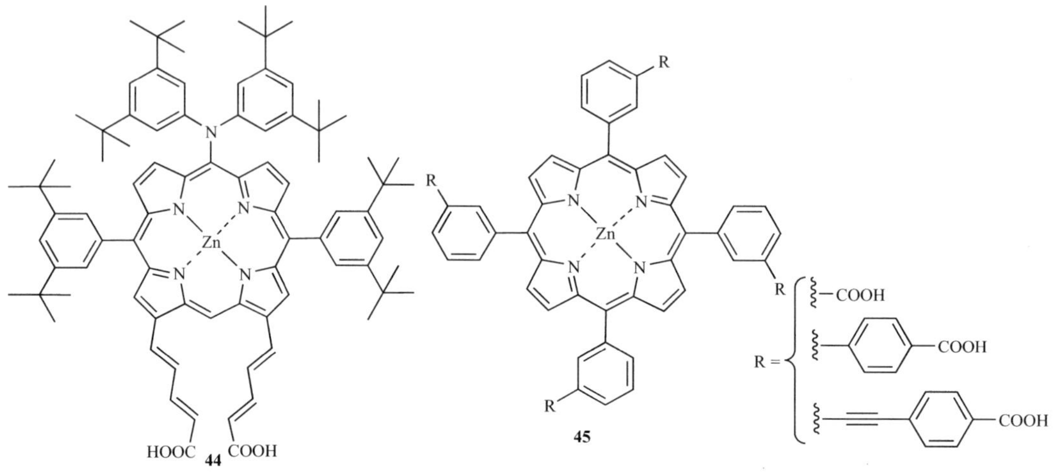

Fig. (19). Porphyrins for DSSC applications.

Yeh, Diau and co-workers introduced 3,5-Di-tertbutylphenyl groups at the porphyrin ring *meso*-positions, trying to solve the porphyrin aggregation issue [147, 457, 458]. The authors reported the *meso*-substituted zinc porphyrin derivatives **46-47** (Fig. **20**), obtaining η = 6.0% for **46** [147]. Later, they also prepared dye **47**, reaching η = 6.8% [457, 458]. Lin and Diau [459, 460] extended the π-conjugation on porphyrins by *meso*-position functionalization. They reported fluorene-functionalized porphyrin **48** and pyrene-functionalized porphyrin **49** obtaining impressive performances of η = 8.1% [459] and η = 10.1% (Fig. **20**) [460]. Since it was demonstrated that one of the best ways to prevent dye aggregation is through substitution with long alkyl chains, in 2011 Grätzel [322, 461, 462] reported several studies of an optimized performance of porphyrin **50** (Fig. **21**), in presence of an organic dye (**51**) achieving a very impressive efficiency of power conversion of η = 13.1%, the highest reported to date, for porphyrin sensitized solar cells.

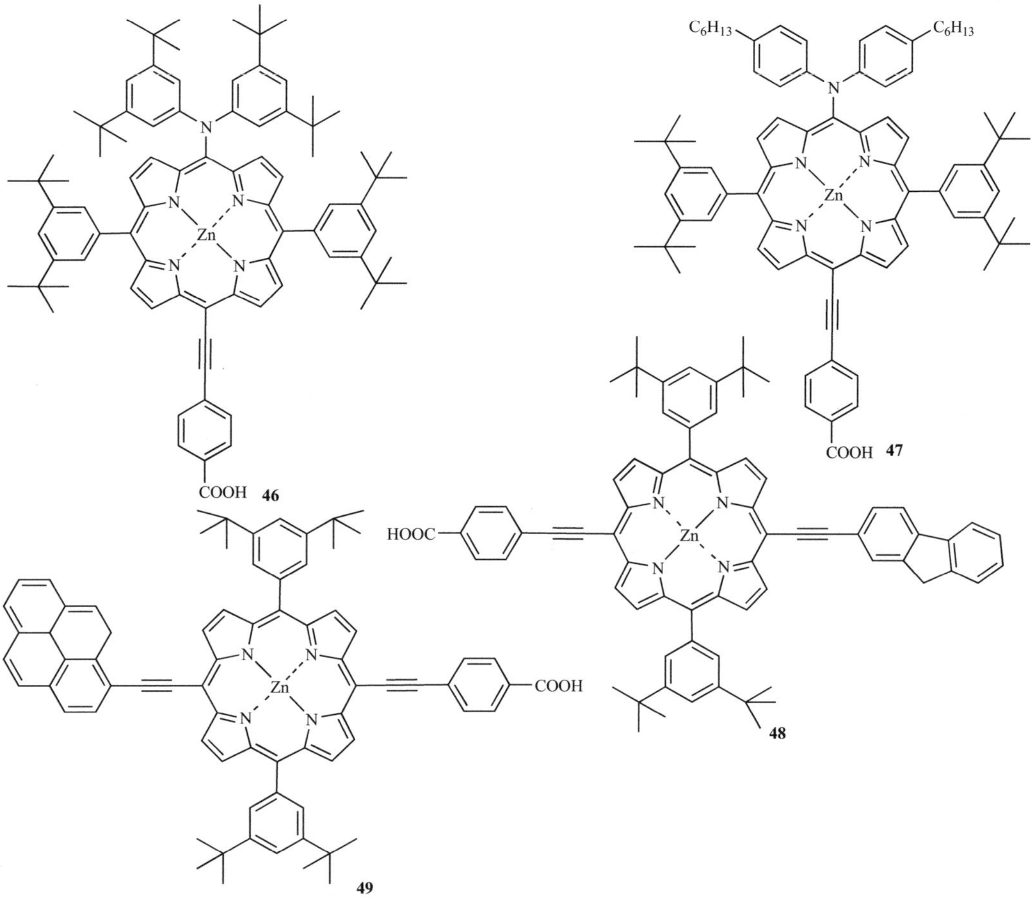

Fig. (20). Porphyrin derivatives for solar cell applications.

Applications as Polymer Based Porphyrinic Materials

The development of polymer based porphyrin materials has been mainly inspired by several interesting porphyrin derivatives' functions in biological systems, since porphyrin mimic several important biological functions, *e.g.* with protein scaffold interaction. Hence, three-dimensional architectures like dendrimers may be considered appealing synthons, due to their morphological resemblance to proteins.

Porphyrins based dendrimers were first reported in 1993 by Aida and Inoue, where the porphyrin unit was positioned as the core of a poly(benzyl ether) dendrimer (**52**) (Fig. **22**) [463]. Closely later, Diederich reported the synthesis of poly(ether-amide) dendrimer porphyrins **53** (Fig. **23**) [339]. Later, in 1996, Suslick reported the poly(aryl ester) Mn porphyrin-cored dendrimers **54** (Fig. **24**) [340].

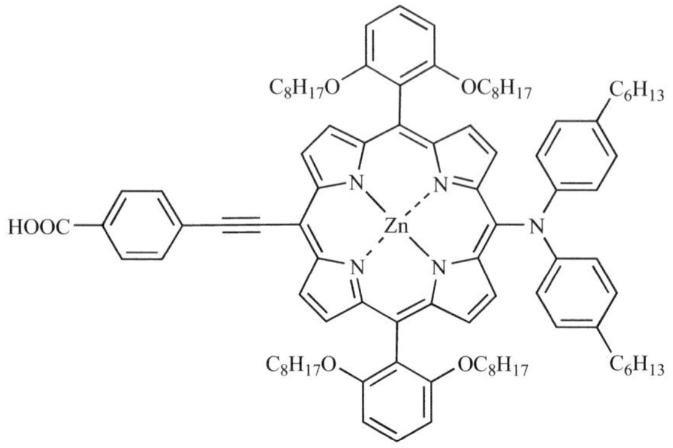

50

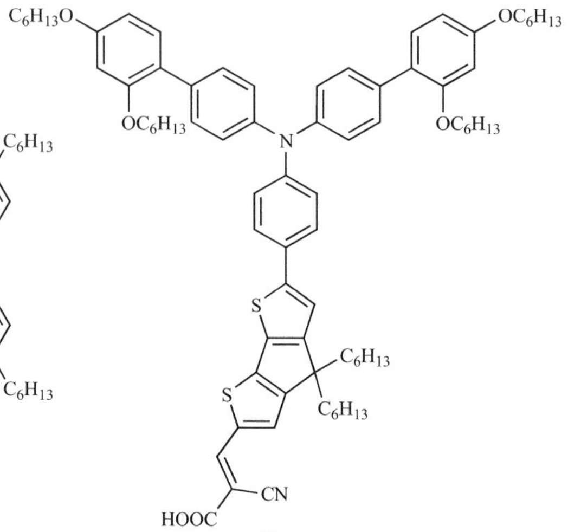

51

Fig. (21). Porphyrin based system for solar cells.

As said above, the first porphyrin dendrimer (**52**) was planned to mimic hemoproteins [463]. Suslick also developed manganese porphyrins protected by poly(aryl ester) dendrimers, in order to mimic the cytochrome P450 [340, 464].

Furthermore, dendrimer **54** was used as catalysts in alkene epoxidation. When used in the epoxidation of 1:1 mixtures of cyclic and linear alkenes and dienes, the authors also observed that compound **54** showed much higher selectivity for the less hindered double bonds than the nondendritic analogue **55**.

The use of dendrimers based on porphyrins as light-harvesters has also attracted particular attention. Hawker, Fréchet, and co-workers reported intramolecular energy transfer properties of dendrimers **56a-c** (Fig. **25**) [330]. Interestingly, they found that dendrimer **56c** exhibited a large energy transfer efficiency value of 83.9%. Moreover, these dendrimers showed just a slight diminishing in energy transfer efficiency from 89.7% to 83.9%, when the generation number increased from 3 (**56a**) to 5 (**56c**).

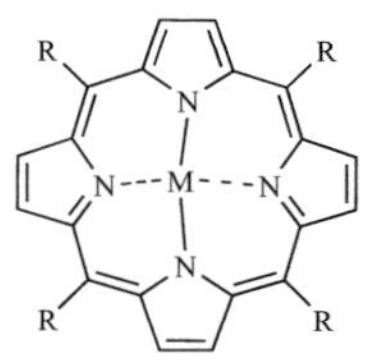

52

a: M = 2H
b: M = Zn
c: M = Fe
d: M = FeCl

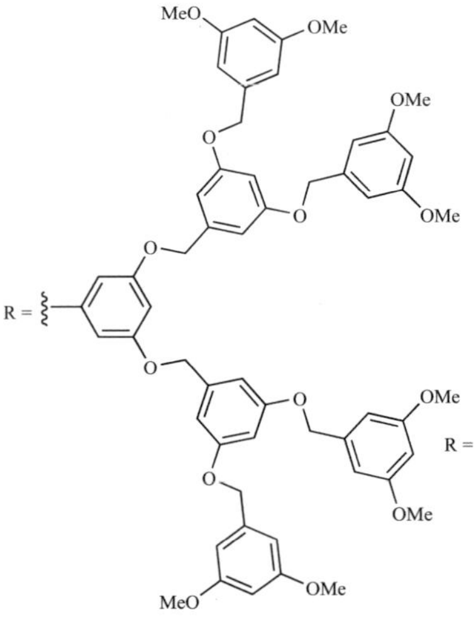

G1

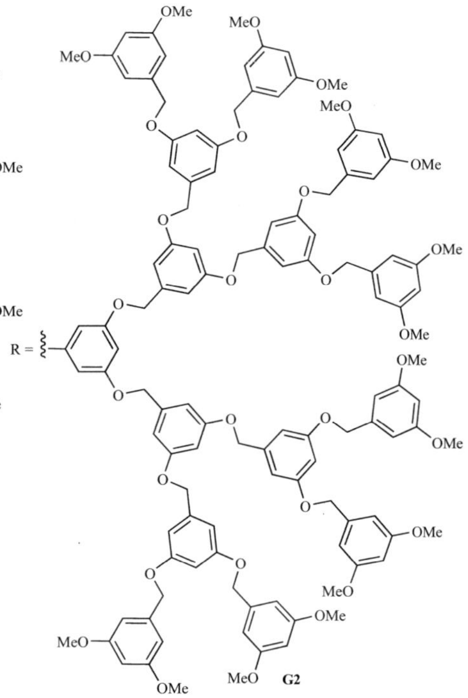

G2

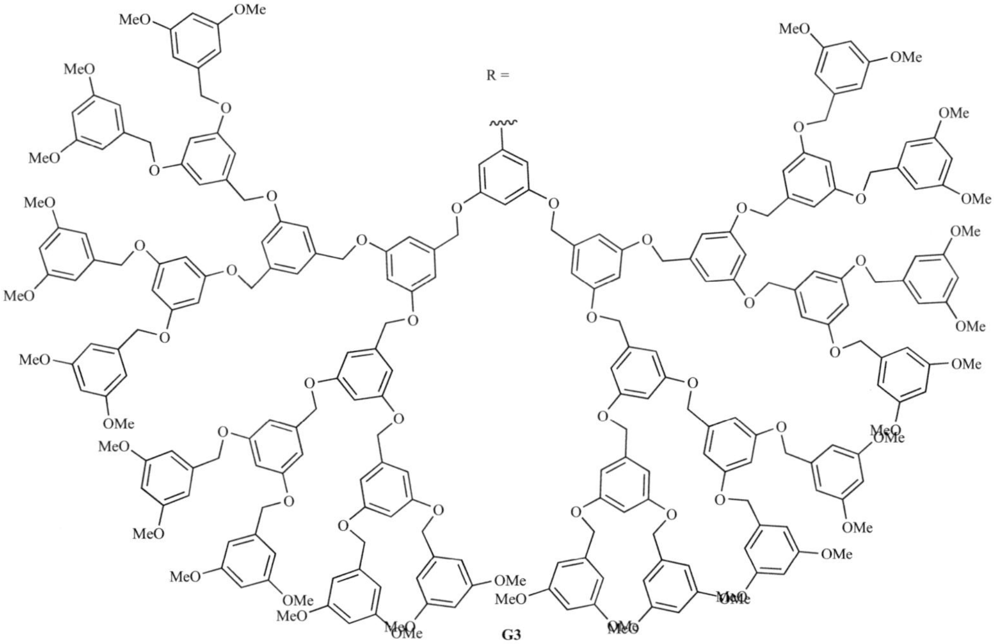

G3

Fig. (22). Porphyrin based dendrimer.

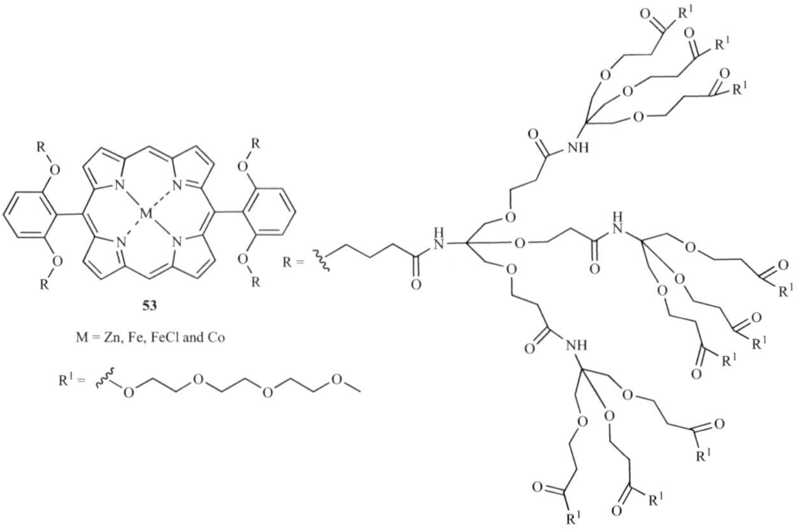

Fig. (23). Porphyrin centered dendrimer.

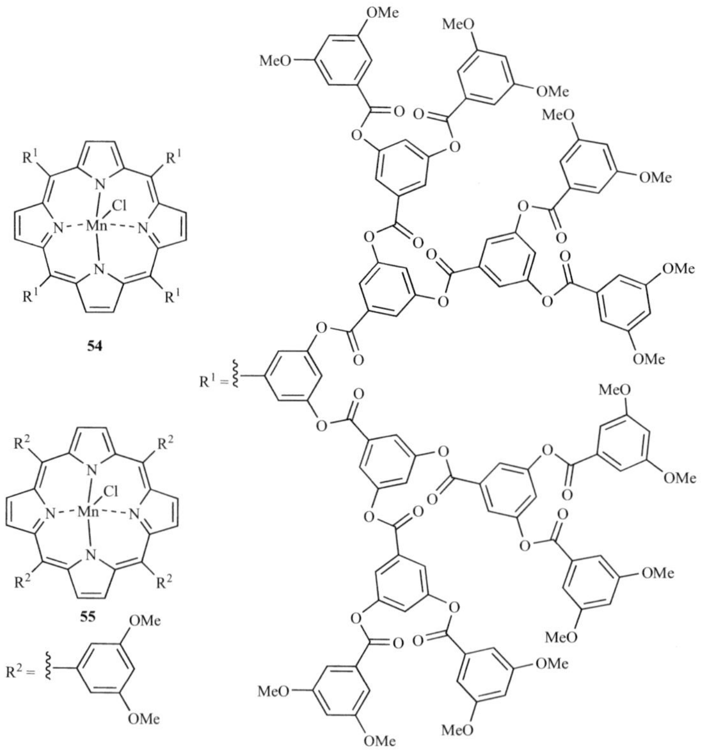

Fig. (24). Manganese porphyrin based dendrimer.

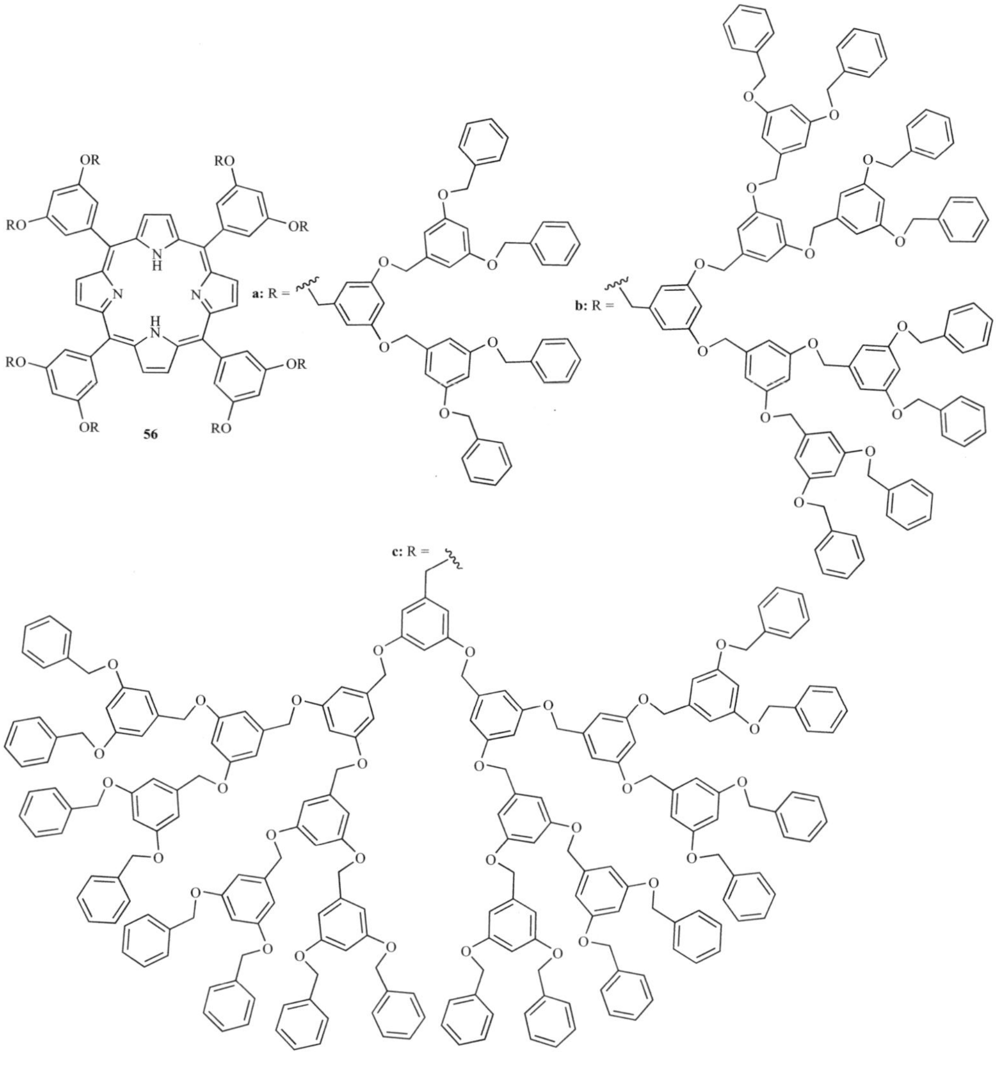

Fig. (25). Porphyrin based dendrimer.

Amino acid-based dendrimers offer unique properties including stereoselective sites for non-covalent interactions with photosensitizing molecules. Porphyrins have been conjugated with branched dendritic polyaspartic dendrimers (**57-59**) (Fig. **26**) and tested *in vitro* to evaluate their potential as photosensitizers [465]. With increasing dendron generation it was observed an increase in singlet oxygen and fluorescence quantum yields. *In vitro* studies performed in *HeLa* cells demonstrated that **57-59** conjugates were non-toxic in darkness and phototoxic after photoexcitation.

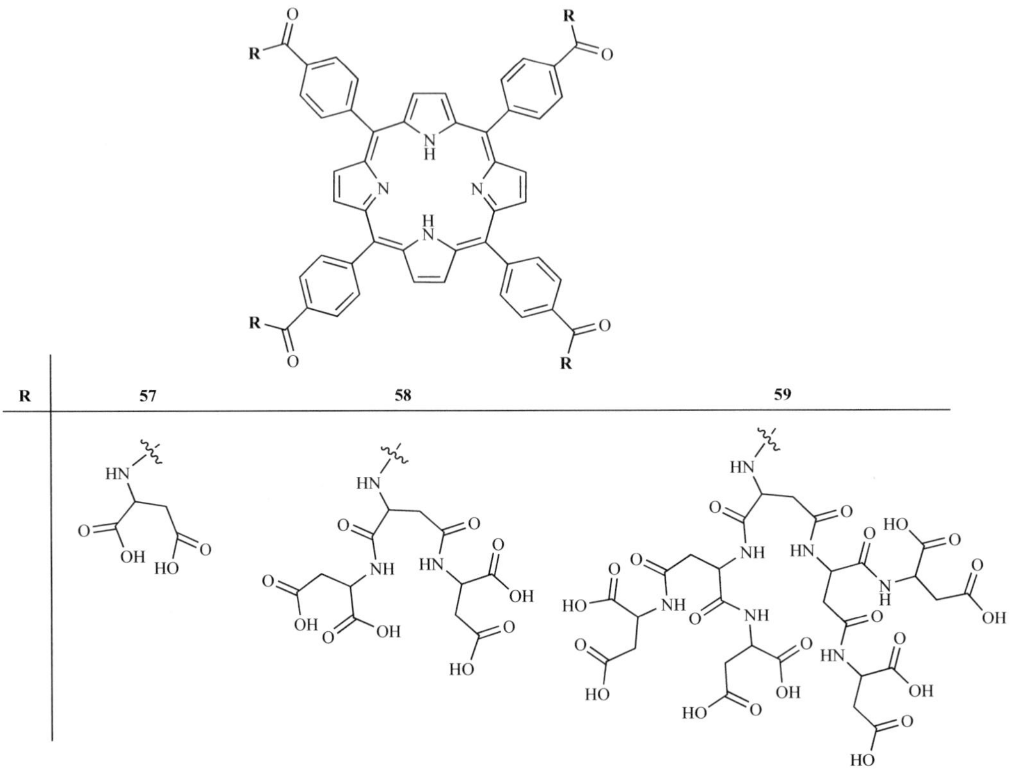

Fig. (26). Amino acid based porphyrin dendrimer.

Applications in Antimicrobial Photodynamic Inactivation

The application of photodynamic therapy as a means to kill microbial pathogens is currently an emerging area of research. Nevertheless, the first recorded observation of photodynamic processes was the inactivation of a microorganism (paramecia) by Oscar Raab, more than 100 years ago [466]. The development of our current lifestyle has naturally induced bacteriological proliferation; thus, nowadays, it is imperative to develop efficient and innovative strategies against microbial diseases.

Photodynamic inactivation represents a practical alternative, since most antibiotic drugs usually operate differently from photodynamic sensitizers on microbial cells. Positively charged porphyrin photosensitizers, at physiological pH values [360], can directly photoinactivate Gram(+) and Gram(-)bacteria. This type of porphyrins are commonly characterized by bearing quaternized nitrogen atoms, albeit substitution with amino groups having adequately strong basic properties also allows neutral pH protonation. The nature of the substituents generally plays a significant role on the binding extent with microbial cells [353, 467]. Structure-activity studies [468] suggested that hydrophilic compounds, bind readily with Gram(-) bacteria, while amphiphilic derivatives (di- or tetracationic porphyrins) display higher affinity for Gram(+) bacteria.

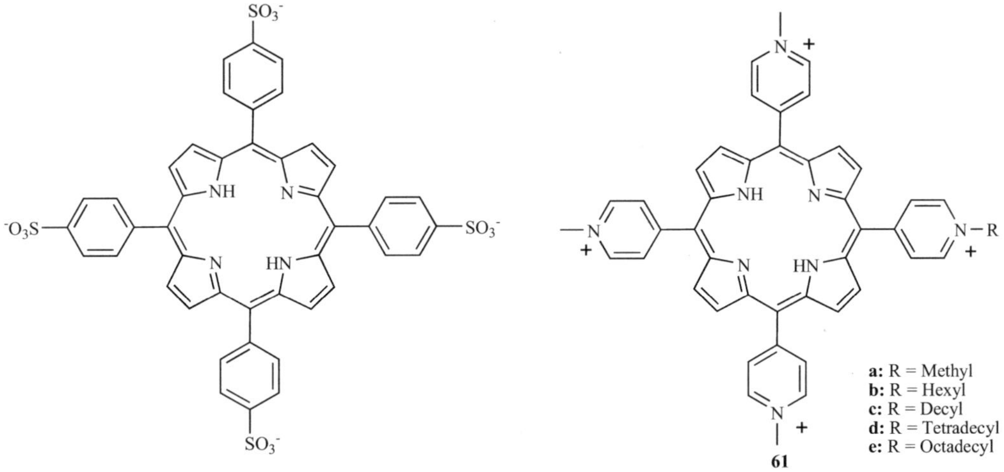

60 **61**

Fig. (27). Water soluble porphyrin derivatives for antimicrobial applications.

For instance, Jori reported the comparison of photosensitizing effect of anionic *meso* (tetra-4-sulfonatophenyl)porphyrin **(60)** (Fig. **27**) and cationic (tetra-*N*-methyl-pyridyl)porphyrin (**61a**) (Fig. **27**) on the survival of Gram(-) *E. coli* and Gram(+) *Enteroccus seriolicida* . The cationic porphyrin showed to be quite effective against Gram(-), while the anionic showed activity only against Gram(+) [360, 469]. Introduction of longer alkyl chains tetra-*meso* (*N*-alky--pyridyl)-substituted porphyrins (**61b-e**) (Fig. **27**), resulted in progressive hydrophobicity increase, as reported again by Jori's group [468]. Maximum efficiency was obtained using porphyrin substituted with one *N*-tetradecyl moiety. However, spectroscopic studies with porphyrins bearing longer chains revealed that heavy aggregation resulted in lower efficiency, as reported by Wilder-Smith. [470].

Also, Hamblin [471, 472] found that attachment of a neutral chlorin with a polylysine chain (**62**) (Fig. **29**) induced both Gram(+) and Gram(-) bacteria photoinactivation. Similar results were obtained by Cavaleiro and Jori, using porphyrin-polylysine conjugates **63** (Fig. **28**) [361].

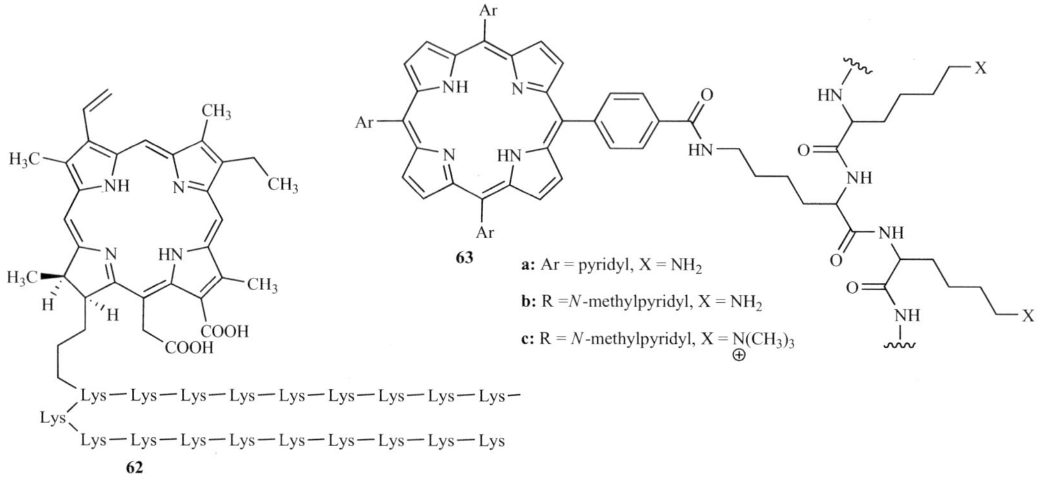

Fig. (28). Water soluble porphyrin derivatives for microbial inactivation.

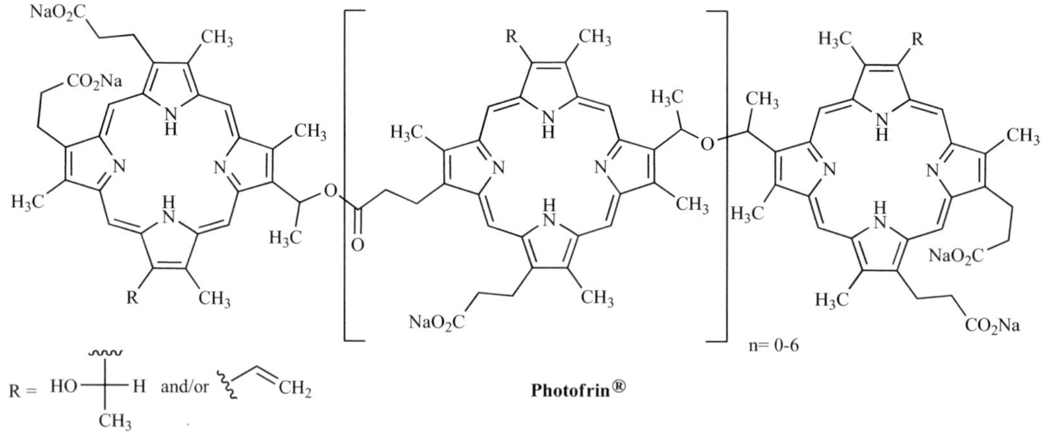

Fig. (29). Photofrin® PDT drug.

Applications in Photodynamic Therapy

Photodynamic therapy (PDT) is considered an interesting treatment modality for several diseases, particularly cancer. In photodynamic therapy, light, oxygen, and a photosensitizer are coupled, producing a selective therapeutic action. In fact, a photosensitizer of insignificant toxicity (in the dark) can accumulate preferentially in fast dividing cells. Then, a synchronized laser light beam is directed to unhealthy tissues. The drug is activated by light, inducing cytotoxicity. Soon after treatment, damaged cells become necrotic or suitably modified.

The ideal PDT photosensitizer should have powerful absorption in the visible region of the spectrum, high extinction coefficient high singlet oxygen quantum yield, low dark toxicity, be soluble and stable in the body, efficient accumulation in tumor tissues and easy excretion after treatment. Porphyrin-type compounds, including chlorins and bacteriochlorins are quite suitable dye sensitizers for PDT [473]. The combination of light and a photosensitizer in PDT was reported for the first time in the beginning of the twentieth century [474]; however, the approval of a drug by Food and Drug Administration (FDA) occured only in the 1990s for **Photofrin**® Fig. (**29**), a mixture of ether-linked porphyrin-type dimers [372, 475], being known as the first generation photosensitizer. The use of this first approved sensitizer possesses several drawbacks, such as low red region of visible spectrum absorption, skin phototoxicity, being a complex mixture arising from the synthetic method. Hence improvement of the photosensitizers became an intense target of research.

Belonging to the porphyrin family, *meta* -tetra(hydroxyphenyl)porphyrin (**64**) Fig. (**30**) and 5,10,15,20-tetrakis(4-sulfonatophenyl)-21H,23H-porphyrin (**65**) Fig. (**30**) are second generation PDT sensitizers [211, 476, 477]. Porphyrin **64**, revealed to be 25-30 times more potent than **Photofrin**® in tumor photo-necrosis [211], possessing, however, some skin phototoxicity. On the other hand, compound **65**, revealed it to be a good candidate for treatment of basal cell carcinoma [476, 477].

There are some photosensitizers tested for PDT that are from chlorin type, including a benzoporphyrin derivative (**Visudyne**®) (Fig. **30**), *meta*-

tetra(hydroxyphenyl)chlorin (**Foscan®**) (Fig. **30**), tin ethyl etiopurpurin (**Purlytin™**) (Fig. **30**) and *N* -aspartyl chlorin e6 (**Aptocine®**) (Fig. **30**).

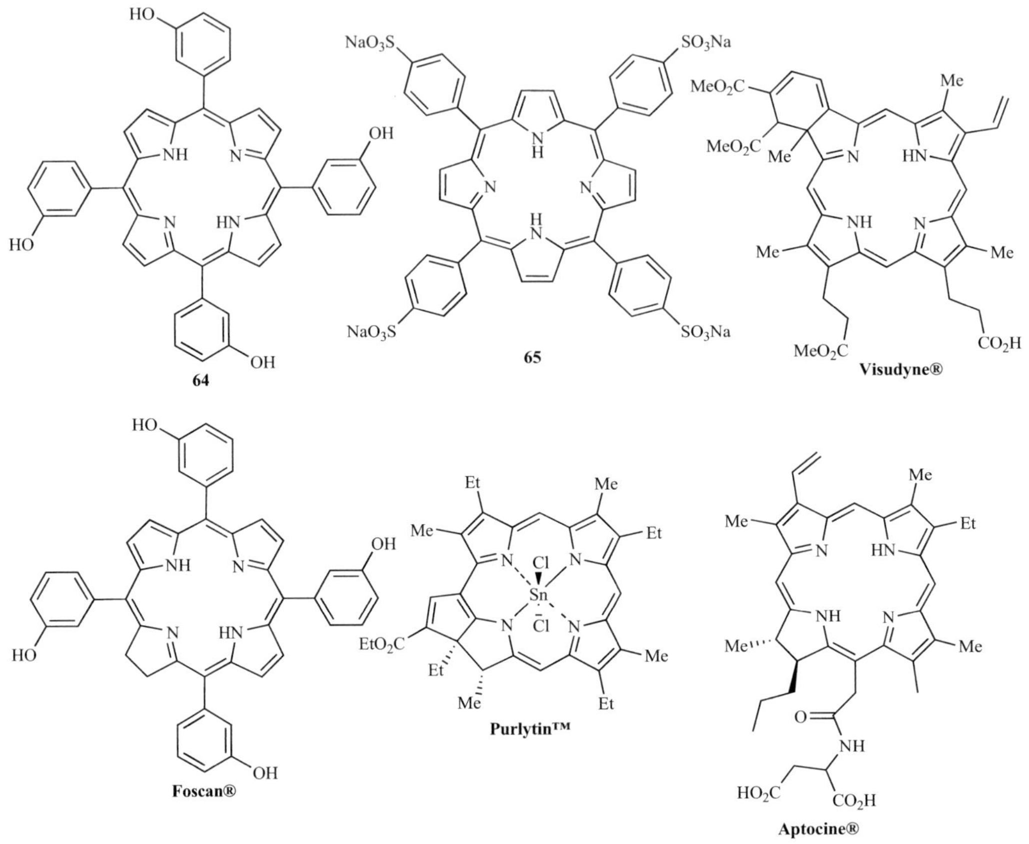

Fig. (30). Several PDT photosensitizers.

Visudyne® is light-activated at 689 nm, having lower skin phototoxicity time interval than **Photofrin®**, due to fast tissue and plasma pharmacokinetics, permitting faster body drug excretion [478]. The US FDA licenced in 1999 the use of **Visudyne®** in ophthalmology (age-related macular degeneration) [479]. Furthermore, photosensitizer Verteporfin was also studied and displayed an health improvement in patients suffering from skin cancer [480]. Another important photosensitizer, the chlorin **Foscan®**, prepared by reducing one of the pyrrole rings at the corresponding porphyrin was also tested in scalp and neck cancer [327]. PDT treatment with **Foscan®** was also successfully used in the treatment of prostate, breast and pancreatic cancers [479, 481, 482]. However, a major **Foscan®** disadvantage concerns high skin photosensitivity in quite a number of patients. Additionally, **Purlytin™** is being tested in Phase I/II trials for basal cell carcinoma, metastatic breast adenocarcinoma and Kaposi's sarcoma [375], displaying deep tissue penetration. The drug's disadvantage is the likelihood of skin photosensitivity and dark toxicity. **Aptocine®** is another photosensitizer that is light activated at 664 nm, having potentiality in liver, brain, fibrosarcoma and oral cancer PDT treatment, being also approved in Japan in 2003 for lung cancer treatment [483, 484]. Similarly to **Visudyne®**, **Aptocine®** provokes negligible skin photosensitivity, unlike **Photofrin®**.

Pheophorbides are derived from chlorophyll. In this context, Dougherty reported the synthesis and evaluation of 2-(1-hexyloxyethyl)-2-devinyl pyropheophorbide (**Photochlor®**) (Fig. **31**) [485]. **Photochlor®** is currently being tested in several clinical trials, *e.g.* for esophageal cancer, basal cell skin cancer, lung cancer, neck and head cancer, carcinoma of the oropharynx and carcinoma of the oral cavity.

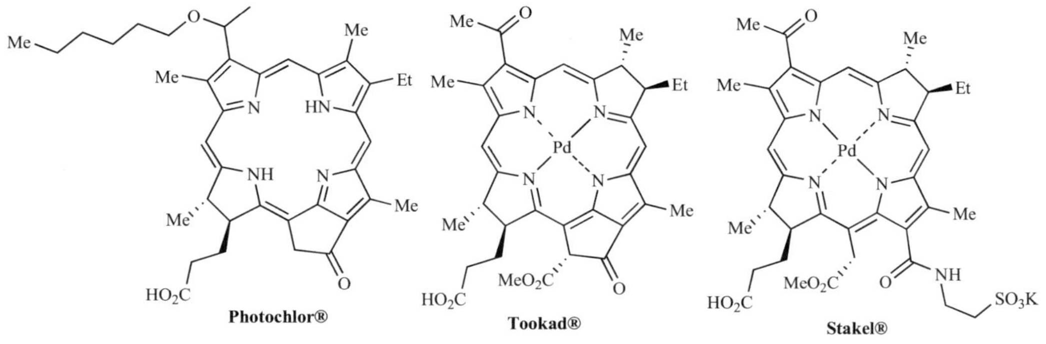

Fig. (31). Bacteriochlorin based photosensitizers.

Moreover, bacteriopheophorbides are similar to bacteriochlorins, possessing a λ_{max} around 740-800 nm. For instance, a derivative of bacteriochlorphyll-a, isolated from bacteria (**Tookad®**) (Fig. **31**) possesses fast body drug excretion with low skin accumulation [486], for which has been tested in some clinical trials for prostate cancer treatment. The advantages of minimum skin photosensitivity and deeper tissue penetration than **Photofrin®** have rendered this drug superior efficiency when compared with other clinically used photosensitizers [375]. Another water soluble derivative of bacteriochlorphyll-a (**Stakel®**) Fig. (**31**) is currently involved in clinical trials for prostate cancer treatment [487], along with bacteriochlorin **Luzitin®** (Fig. **32**) for head and neck cancers [488, 489].

A texaphyrin (**Lutrin®**) (Fig. **32**), a pentaaza cored porphyrinoid analog was reported by Sessler [490], being soluble in water, absorbing light at 732 nm, being currently under trial scrutiny for prostate cancer treatment [491].

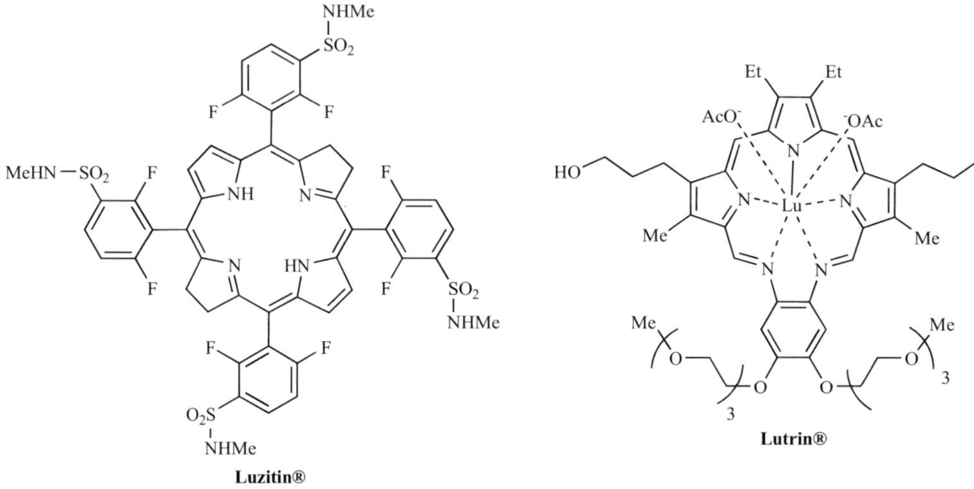

Fig. (32). Third generation photosensitizers.

Applications in Disease Diagnostics

Porphyrin molecules might behave uniquely in disease therapy, since they can be used in imaging, detecting and treating different forms of diseased tissue and various different cancers. A significant current center of attention is on the tumor tissue clinical image, through photosensitizer targeted delivery in biomedical multimodal theranostic platforms. The ideal fluorophore should also possess high fluorescence quantum yield, minimum photobleaching, large Stokes shift and targeted high affinity.

The research and development of late generation photosensitizers spotted the potential that a multimodal (imaging-therapeutic) platform can have in the disease detection/treatment, avoiding the use of multiple individually administered agents, surpassing variability issues in the uptake, localization, pharmacodynamic and pharmacokinetic behavior of image/therapy agents. Switching between imaging modality and therapeutic/cytotoxic modality in a single platform is a particularly suitable use for photosensitizers in imaging modalities such as magnetic resonance-, fluorescence-, radio- and nuclear imaging, along with as photodynamic, photoacoustic and photothermal treatment therapies, as well as hyperthermia, nuclear- gene- and chemo-therapy [492].

Hence, tetrapyrrolic macrocycles are logic and amenable targets for contrast agents for molecular imaging, namely in MRI (magnetic resonance imaging, PET (positron emission tomography) and FI (fluorescence imaging) modalities.

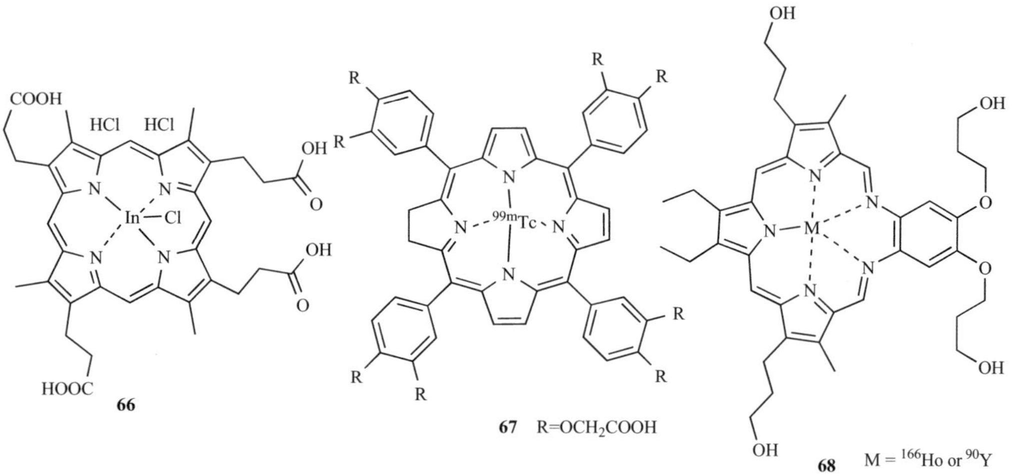

66

67 R=OCH$_2$COOH

68 M = ^{166}Ho or ^{90}Y

Fig. (33). Porphyrin derivatives for imaging disease diagnostics.

PET is based in the direct detection and quantification of the distribution of molecules, marked with a positron emitter, that when administered to the living body, can perform specific functions providing information of the metabolic functions involved. This technique is used in oncology to identify and localize tumors and also to follow their evolution [493]. Kulkarni [494] reported the first radiolabeled porphyrin derivative, the naturally occurring coproporphyrin III with ^{111}In (**66**) (Fig. **33**), concluding that the lipid pool in atherometous lesions was the predominant site of radiotracer uptake, since unstable atherosclerotic plaques are pathologically characterized by attenuated fibrous cap, large lipid pool and intense macrophage infiltration.

Later, Murugesan [495] reported the water soluble photosensitizer **67**, (Fig. **33**), a chlorin and bearing ^{99m}Tc as central labeled moiety, while Konirová [496] reported the radioactive labeling of a texaphyrin with radionuclides ^{166}Ho and ^{90}Y (**68** - Fig. **33**), obtaining high labeling.

Due to a suitable half-life decay ($T_{1/2}$=6.73 days), the radionuclide ^{177}Lu was also used in targeted radiotherapeutics, through the synthesis of conjugate **69** Fig. (**34**) [497]. Several parameters, such as incubation temperature, concentration and reaction time were optimized, in order to achieve maximum labeling.

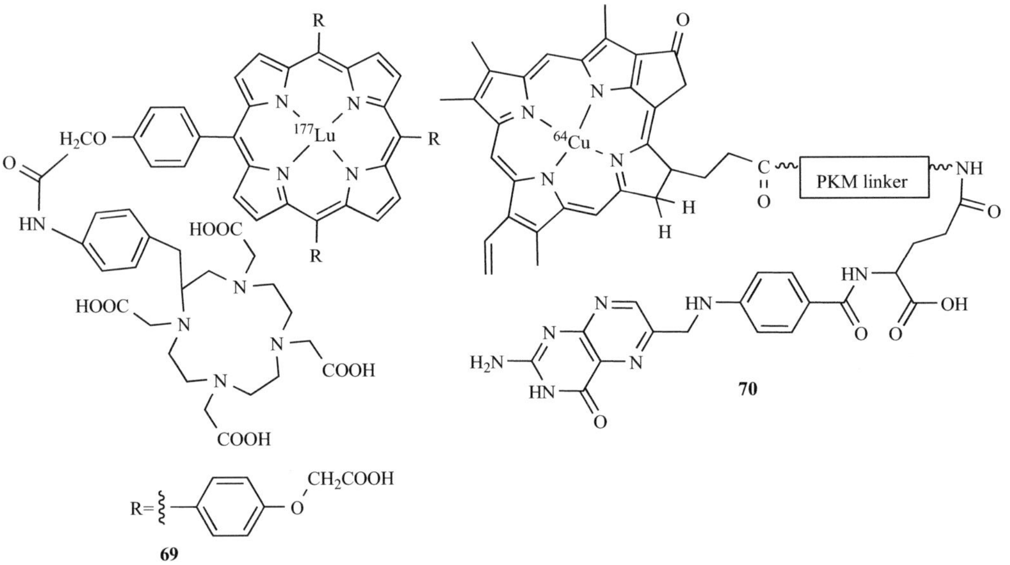

69

70

Fig. (34). Porphyrin conjugates for imaging disease diagnostics.

Biological studies revealed promising internalization and times of retention in Swiss mice fibrosarcoma tumors having noteworthy tumor/muscle and tumor/blood ratios.

In 2011, Zheng [388, 498] reported the application of porphyrin-polypeptid--folate **70** (Fig. **34**) in PET imaging, a conjugate composed by a unit of pyropheophorbide, a folate unit for targeting delivery and a short peptide as modifying linker.

The authors managed to incorporate ^{64}Cu in the porphyrin corc, showing that the 12.7 h half-life of labeled copper was compatible with porphyrin conjugate pharmokinetics, providing adequate time for both the radiolabeling chemistry and accumulation of ^{64}Cu-radiotracer at tumor sites.

MRI is based on the detection of molecules that contain nuclei that possess an odd number of protons and/or of neutrons, having very high spatial resolution for morphological and functional imaging. Several studies were carried out using porphyrin derivatives as MRI contrast agents. For instance, Lippard and Jasanoff [499] prepared manganese (III) porphyrin derivatives **71a,b** for MRI analysis (Fig. **35**). The compounds were tested for membrane permeability, optical activity and relaxivity of intracranially injected rats, with promising results.

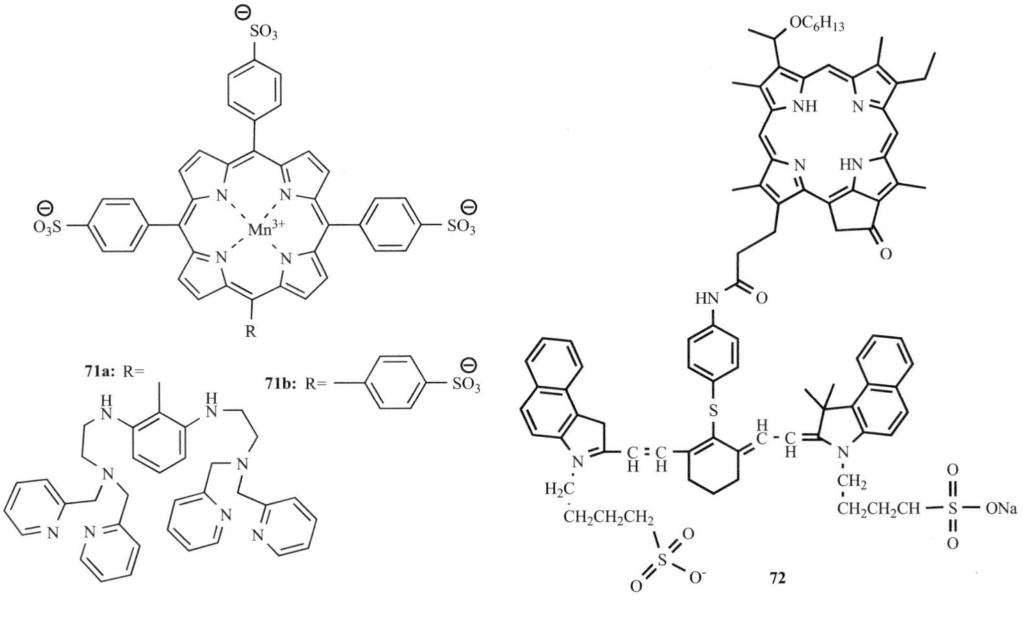

Fig. (35). Porphyrin conjugated derivatives for imaging disease diagnostics.

Although fluorescence imaging technology is a method widely used in physics, chemistry, and medicine for quite some time, its *in vivo* application became more popular only recently. This technique is commonly based on fluorescence marking, using exogenous fluorophores, which bind selectively to specific tissues targets.

As an example, in 2005, Pandey [500] prepared porphyrin conjugate **72** (Fig. **36**), in which a chlorophyll derivative was conjugated with a cyanine dye. The authors found the conjugate to be both an efficient photosensitizer as well as tumor imaging agent.

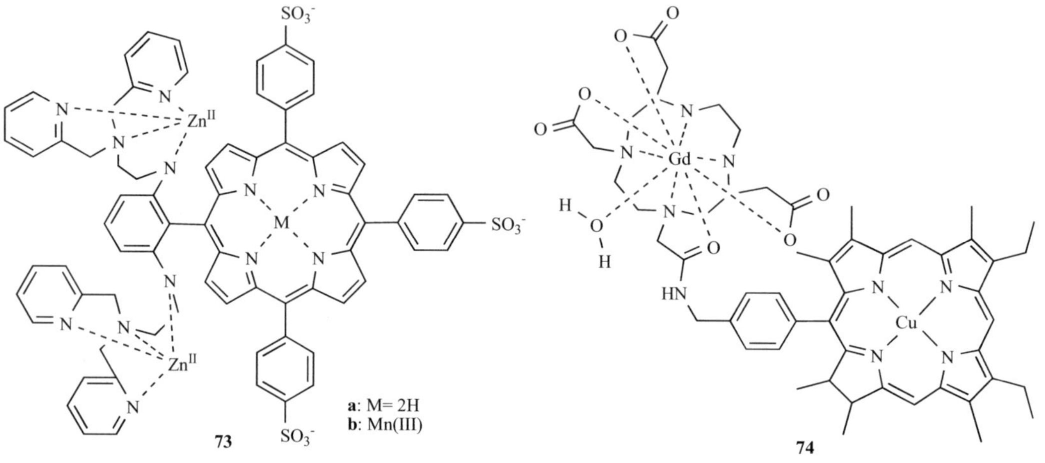

Fig. (36). Bifunctional based imaging probes.

Using the synergic effect of two modalities in the same molecule, Lippard [501] described the synthesis of a water-soluble porphyrin substituted with zinc binding units (dipicolylamine) in free-base (**73a**) and Mn(III) forms (**73b**) (Fig. **36**). While the first was tested as a fluorescent zinc sensor, the latter acted as MRI zinc sensor.

More recently, a bifunctional chelator **74**, featuring a DO3A and a porphyrin unit was reported (Fig. **36**) [502]. Selective metallation of the DO3A counterpart with gadolinium (III) salt, followed by porphyrin metallation with copper (II) salt, afforded the bimodal complex. The relaxivity of the Gd(III) complex was comparable to commercially available contrast agents, showing that the first coordination sphere remained identical to the related Gd(DO3A)(H_2O), with one coordinated water molecule.

EXPERIMENTAL SECTION

Synthesis of 2,3,7,8,12,13,17,18-octaethylporphyrin (Scheme 21)

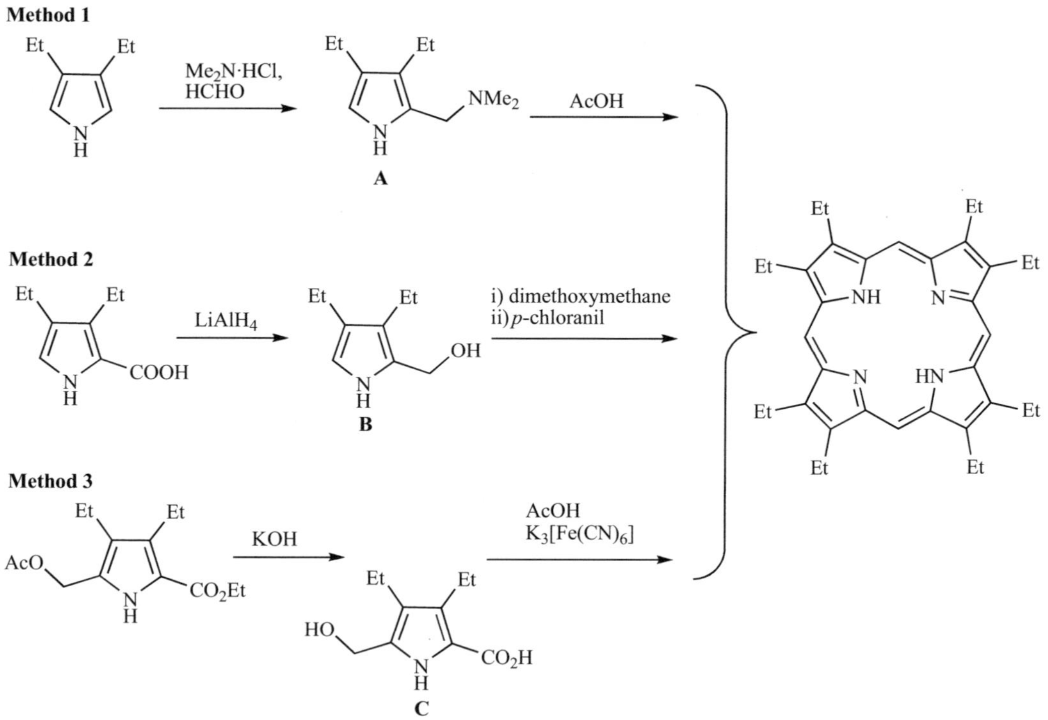

Scheme (21). Synthesis of 2,3,7,8,12,13,17,18-octaethylporphyrin. Procedure for total synthesis of porphyrin can be directly found in References [503] (method 1), [504] (method 2) and [505] (method 3).

Synthesis of 5-phenyldipyrromethane (Scheme 22)

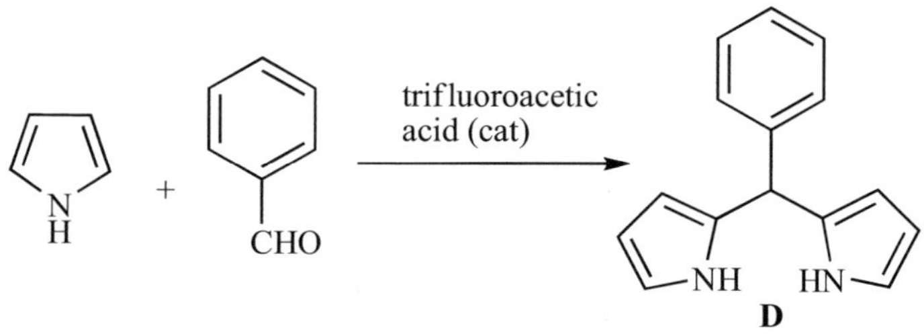

Scheme (22). Synthesis of 5-phenyldipyrromethane. Procedure for total synthesis of dipyrromethane D can be directly found in Reference [53].

Synthesis of 5,15-diphenylporphyrin (Scheme 23)

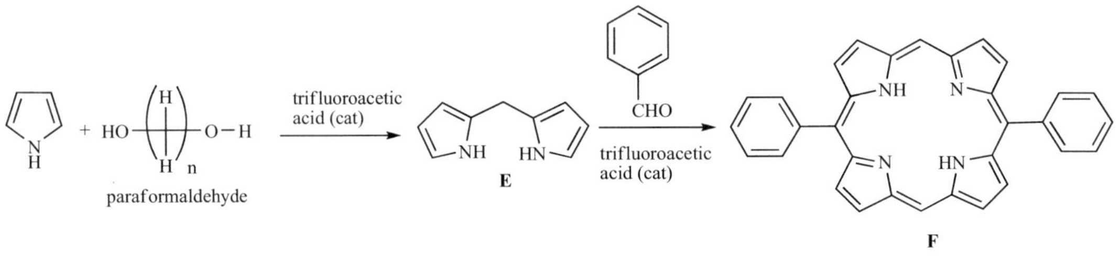

Scheme (23). Synthesis of 5,15-diphenylporphyrin. Procedure for total synthesis of porphyrin F can be directly found in References [55, 73].

Synthesis of Unsymmetric Meso-substituted Porphyrins *via* Suzuki Coupling Reaction (Scheme 24)

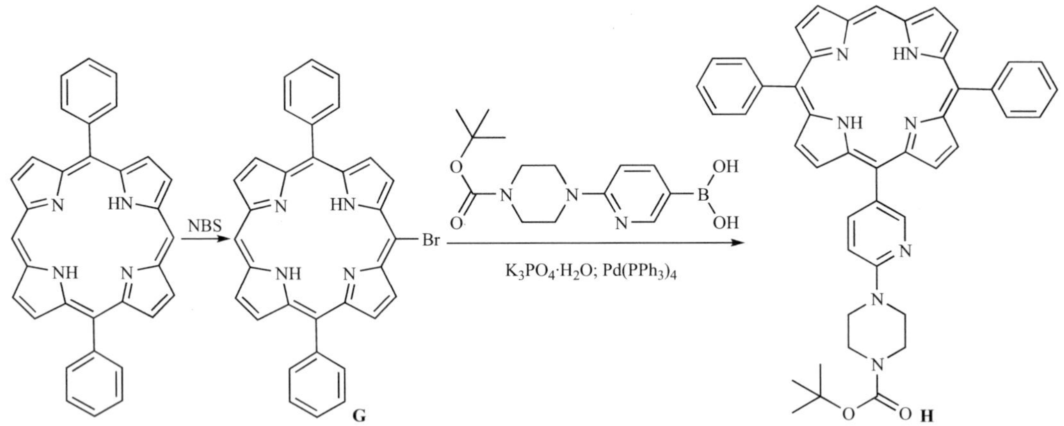

Scheme (24). Suzuki coupling reaction of brominated porphyrin. Procedure for total synthesis of porphyrin H can be directly found in References [107, 507].

Synthesis of [5-(4-carboxyphenyl)-10,15,20-tris(4-pyridyl)porphyrin] (Scheme 25)

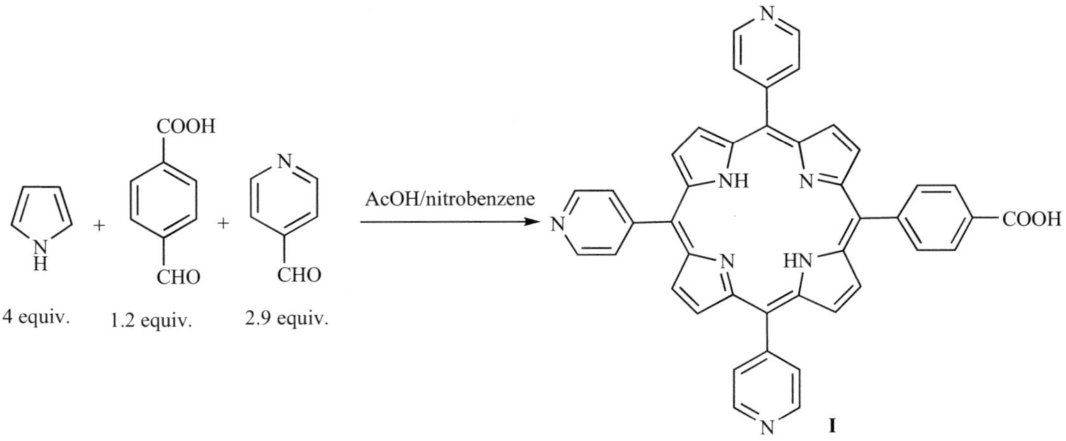

I

Scheme (25). Synthesis of AAAB meso-substituted porphyrin. Procedure for total synthesis of porphyrin I can be directly found in Reference [361].

Synthesis of Meso-substituted Halogenated Bacteriochlorins (Scheme 26)

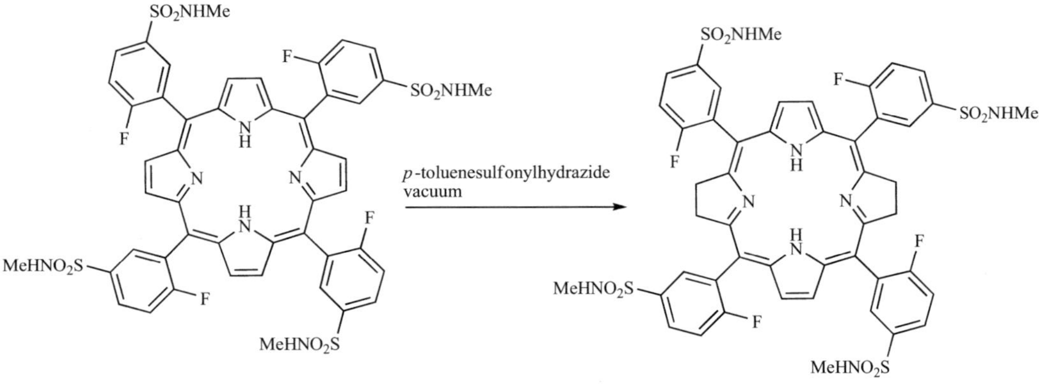

J

Scheme (26). Synthesis of meso-substituted halogenated bacteriochlorin. Procedure for total synthesis of bacteriochlorin J can be directly found in Reference [216].

Synthesis of beta-substituted Bacteriochlorins (Scheme 27)

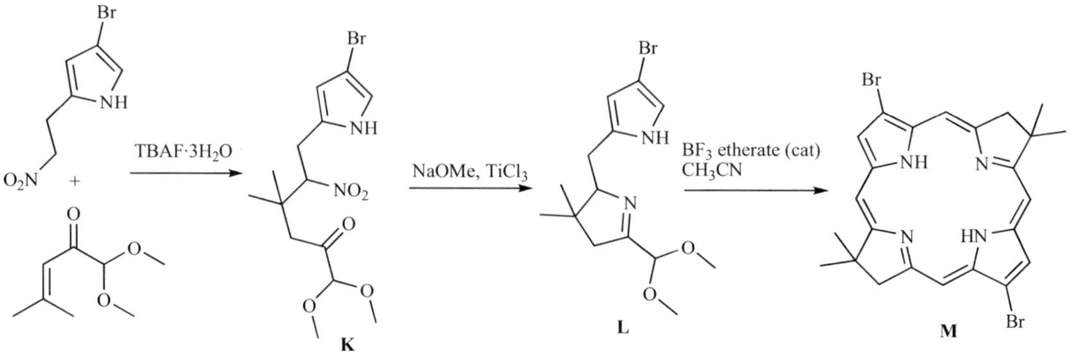

Scheme (27). Synthesis of beta-substituted bacteriochlorins. Procedure for total synthesis of bacteriochlorin M can be directly found in Reference [508].

CONCLUDING REMARKS

Porphyrins are naturally appealing molecules, holding an impressive synthetic versatility, and simply allowing almost any type of application known in fields ranging from catalysis, to materials science and biomedicine applications.

Porphyrins offer great chemical and photochemical stability, both of which are two advantageous qualities for a dye, especially for aiming commercial applications. The field of porphyrin dyes can be highly improved by the intensification of new architectures, naturally matched by the potential for new applications, at the frontier between chemistry, physics, biology and engineering science.

The possibility of widely varying the substitution pattern on porphyrin structures is probably the main possible direction for successful applications; however not less important than the establishment of definite structure-property relationships for any desired purpose, something has been only partially achieved so far. Standardization of procedures will definitely boost the understanding and appliance of porphyrins in the main areas of science.

REFERENCES

[1] Carvalho, C.M.; Brocksom, T.J.; de Oliveira, K.T. Tetrabenzoporphyrins: synthetic developments and applications. *Chem. Soc. Rev.,* **2013**, *42*(8), 3302-3317. [http://dx.doi.org/10.1039/c3cs35500d] [PMID: 23364186]

[2] Galanin, N.E.; Shaposhnikov, G.P.; Koifman, O.I. Methods for synthesis of meso-substituted tetrabenzoporphyrins. *Russ. Chem. Rev.,* **2013**, *82*, 412-428. [http://dx.doi.org/10.1070/RC2013v082n05ABEH004317]

[3] Lash, T.D. Modification of the porphyrin chromophore by ring fusion: identifying trends due to annelation of the porphyrin nucleus. *J. Porphyr. Phthalocyanines,* **2001**, *5*, 267-288. [http://dx.doi.org/10.1002/jpp.313]

[4] Finikova, O.S.; Cheprakov, A.V.; Beletskaya, I.P.; Carroll, P.J.; Vinogradov, S.A. Novel versatile synthesis of substituted tetrabenzoporphyrins. *J. Org. Chem.,* **2004**, *69*(2), 522-535. [http://dx.doi.org/10.1021/jo0350054] [PMID: 14725469]

[5] Helberger, J.H.; von Rebay, A.; Hever, D.B. The effect of metals on o-cyanacetophenone as well as on 3-methyl phtalimidine, synthesis of tetrabenzoporphin. *Justus Liebigs Ann. Chem.,* **1938**, *533*, 197-215. [http://dx.doi.org/10.1002/jlac.19385330110]

[6] Santos, C.I.; Barata, J.F.; Calvete, M.J.; Vale, L.S.; Dini, D.; Meneghetti, M.; Neves, M.G.; Faustino, M.A.; Tome, A.C.; Cavaleiro, J.A. Synthesis and functionalization of corroles. An insight on their nonlinear optical absorption properties. *Curr. Org. Synth.,* **2014**, *11*, 29-41. [http://dx.doi.org/10.2174/15701794113106660084]

[7] Aviv, I.; Gross, Z. Corrole-based applications. *Chem. Commun. (Camb.),* **2007**, *20*(20), 1987-1999. [http://dx.doi.org/10.1039/b618482k] [PMID: 17713061]

[8] Gryko, D.T. Recent advances in the synthesis of corroles and core-modified corroles. *Eur. J. Org. Chem.,* **2002**, 1735-1743. [http://dx.doi.org/10.1002/1099-0690(200206)2002:11<1735::AID-EJOC1735>3.0.CO;2-K]

[9] Johnson, A.W.; Kay, I.T.; Corroles, I. Synthesis. *J. Chem. Soc.,* **1965**, 1620-1629. [http://dx.doi.org/10.1039/jr9650001620] [PMID: 20282466]

[10] Johnson, A.W.; Price, R. The synthesis of derivatives of corrole (pentadehydrocorrin). *J. Am. Chem. Soc.,* **1960**, 1649-1653. [http://dx.doi.org/10.1039/jr9600001649]

[11] Claessens, C.G.; González-Rodríguez, D.; Rodríguez-Morgade, M.S.; Medina, A.; Torres, T. Subphthalocyanines, subporphyrazines, and subporphyrins: singular nonplanar aromatic systems. *Chem. Rev.,* **2014**, *114*(4), 2192-2277. [http://dx.doi.org/10.1021/cr400088w] [PMID: 24568182]

[12] Osuka, A.; Tsurumaki, E.; Tanaka, T. Subporphyrins: A legitimate ring-contracted porphyrin with versatile electronic and optical properties. *Bull. Chem. Soc. Jpn.,* **2011**, *84*, 679-697. [http://dx.doi.org/10.1246/bcsj.20110118]

[13] Inokuma, Y.; Osuka, A. Subporphyrins: emerging contracted porphyrins with aromatic 14pi-electronic systems and bowl-shaped structures: rational and unexpected synthetic routes. *Dalton Trans.,* **2008**, (19), 2517-2526.
[http://dx.doi.org/10.1039/b719808f] [PMID: 18443691]

[14] Inokuma, Y.; Kwon, J.H.; Ahn, T.K.; Yoo, M-C.; Kim, D.; Osuka, A. Tribenzosubporphines: synthesis and characterization. *Angew. Chem. Int. Ed. Engl.,* **2006**, *45*(6), 961-964.
[http://dx.doi.org/10.1002/anie.200503426] [PMID: 16381040]

[15] Fischer, H.; Orth, H. *Die Chemie des Pyrrols*; Akademische Verlagsgesellschaft: Leipzig, **1934-1940**, Vol. I-III, .

[16] Dixon, H.B.; Cornish-Bowden, A.; Liebecq, C.; Loening, K.L.; Moss, G.P.; Reedijk, J.; Velick, S.F.; Venetianer, P.; Vliegenthart, J.F. Nomenclature of tetrapyrroles. *Pure Appl. Chem.,* **1987**, *59*, 779-832.

[17] Sheldon, R.A. *Metalloporphyrins in Catalytic Oxidations*; Marcel Dekker, Inc.: New York, **1994**.

[18] Milroy, J.A. Observations on some metallic compounds of hematoporphyrin. *Biochem. J.,* **1918**, *12*(4), 318-338.
[PMID: 16742768]

[19] Fischer, H.; Zeile, K. Porphyrin syntheses. XXII. Syntheses of hematoporphyrin, protoporphyrin and hemin. *Justus Liebigs Ann. Chem.,* **1929**, *468*, 98-116.
[http://dx.doi.org/10.1002/jlac.19294680104]

[20] Falk, J.E. *Porphyrins and metalloporphyrins: their general, physical and coordination chemistry, and laboratory methods*; Elsevier: Amsterdam, **1964**.

[21] *Porphyrins and Metalloporphyrins*; Smith, K.M., Ed.; Elsevier: Amsterdam, **1975**.

[22] *The porphyrins*; Dolphin, D., Ed.; 1978-1980.

[23] Milgrom, L.R. *The colours of life: an introduction to the chemistry of porphyrins and related compounds*; Oxford University Press: New York, **1997**.

[24] *The Porphyrin Handbook*; Kadish, K.M; Smith, K.M; Guilard, R, Eds.; Academic Press: Boston, **1999**, Vols. 1-10, . , **2003**, Vols. 11-20, .

[25] *Handbook of Porphyrin Science, Kadish, K.M.; Smith, K.M*; Guilard, R., Ed.; 2010-2013.

[26] Jackson, A.H.; Smith, K.M. Total synthesis of pyrrole pigments. In: *Total synthesis of natural products*; ApSimon, J.W., Ed.; Wiley: New York, **1973**; Vol. 1, pp. 143-278. 1984; Vol. 6, pp 237-280.

[27] Baltazzi, E.; Krimen, L.I. Recent advances in chemistry of pyrrole. *Chem. Rev.,* **1963**, *63*, 511-556.
[http://dx.doi.org/10.1021/cr60225a004]

[28] Gossauer, A. *Die Chemie der Pyrrole*; Springer Verlag: Berlin, **1974**.
[http://dx.doi.org/10.1007/978-3-642-51118-9]

[29] Smith, K.M.; Vicente, M.G.H. Product Class 8: Porphyrins and Related Compounds, Section 17.8 In: *Science of Synthesis*; Georg Thieme Verlag: Stuttgart, **2003**.

[30] Smith, K.M. Strategies for the synthesis of octaalkylporphyrin systems. In: *The Porphyrin Handbook*;

Kadish, K.M.; Smith, K.M.; Guilard, R., Eds.; Academic Press: Boston, **1999**; Vol. 1, pp. 1-44.

[31] Lindsey, J.S. Synthesis of meso-substituted porphyrins. In: *The Porphyrin Handbook*; Kadish, K.M.; Smith, K.M.; Guilard, R., Eds.; Academic Press: Boston, **1999**; Vol. 1, pp. 45-118.

[32] Smith, K.M. Cyclizations of a,c-biladiene salts to give porphyrins and their derivatives. In: *The Porphyrin Handbook*; Kadish, K.M.; Smith, K.M.; Guilard, R., Eds.; Academic Press: Boston, **1999**; Vol. 1, pp. 119-144.

[33] Vicente, M.D.; Smith, K.M. Syntheses and functionalizations of porphyrin macrocycles. *Curr. Org. Synth.,* **2014**, *11*(1), 3-28.
[http://dx.doi.org/10.2174/15701794113106660083] [PMID: 25484638]

[34] Rothemund, P. Formation of porphyrins from pyrrole and aldehydes. *J. Am. Chem. Soc.,* **1935**, *57*, 2010-2011.
[http://dx.doi.org/10.1021/ja01313a510]

[35] Rothemund, P.; Menotti, A.R. The synthesis of 5,10,15,20-tetraphenylporphine. *J. Am. Chem. Soc.,* **1941**, *63*, 267-270.
[http://dx.doi.org/10.1021/ja01846a065]

[36] Adler, A.D.; Longo, F.R.; Finarelli, J.D.; Goldmacher, J.; Assour, J.; Korsakoff, L. A simplified synthesis for meso-tetraphenylporphine. *J. Org. Chem.,* **1967**, *32*, 476-477.
[http://dx.doi.org/10.1021/jo01288a053]

[37] Gonsalves, A.M.; Pereira, M.M. A new look into the Rothemund meso-tetraalkyl and tetraarylporphyrin synthesis. *J. Heterocycl. Chem.,* **1985**, *22*, 931-933.
[http://dx.doi.org/10.1002/jhet.5570220363]

[38] Gonsalves, A.M.; Varejao, J.M.; Pereira, M.M. Some new aspects related to the synthesis of mesosubstituted porphyrins. *J. Heterocycl. Chem.,* **1991**, *28*, 635-640.
[http://dx.doi.org/10.1002/jhet.5570280317]

[39] Johnstone, R.A.; Nunes, M.L.; Pereira, M.M.; Gonsalves, A.M.; Serra, A.C. Improved syntheses of 5,10,15,20-tetrakisaryl- and tetrakisalkylporphyrins. *Heterocycles,* **1996**, *43*, 1423-1437.
[http://dx.doi.org/10.3987/COM-96-7436]

[40] Lindsey, J.S.; Schreiman, I.C.; Hsu, H.C.; Kearney, P.C.; Marguerettaz, A.M. Rothemund and Adler-Longo reactions revisited: synthesis of tetraphenylporphyrins under equilibrium conditions. *J. Org. Chem.,* **1987**, *52*, 827-836.
[http://dx.doi.org/10.1021/jo00381a022]

[41] Mauzerall, D. The thermodynamic stability of porphyrinogens. *J. Am. Chem. Soc.,* **1960**, *82*, 2601-2603.
[http://dx.doi.org/10.1021/ja01495a052]

[42] Arsenault, G.P.; Bullock, E.; MacDonald, S.F. Pyrromethanes and porphyrins therefrom. *J. Am. Chem. Soc.,* **1960**, *82*, 4384-4389.
[http://dx.doi.org/10.1021/ja01501a066]

[43] Woodward, R.B. Total synthese des Chlorophylls. *Angew. Chem.,* **1960**, *72*, 651-662.
[http://dx.doi.org/10.1002/ange.19600721803]

[44] Woodward, R.B.; Ayer, W.A.; Beaton, J.M.; Bickelhaupt, F.; Bonnett, R.; Buchschacher, P.; Closs, G.L.; Dutler, H.; Hannah, J.; Hauck, F.P.; Ito, S.; Langemann, A.; Le Goff, E.; Leimgruber, W.; Lwowski, W.; Sauer, J.; Valenta, Z.; Volz, H. The total synthesis of chlorophyll a. *Tetrahedron,* **1990**, *46*, 7599-7659.
[http://dx.doi.org/10.1016/0040-4020(90)80003-Z]

[45] Mironov, A.F.; Ovsepyan, T.R.; Evstigneeva, R.P.; Preobrazenskii, N.A. Synthetic studies in the dipyrrylmethane series. Synthesis of 4,4-dimethyl-3,3-bis(carbomethoxyethyl)dipyrrylmethane. *Zh. Obshch. Khim,* **1965**, *35*, 324-328.

[46] Cavaleiro, J.A.; Gonsalves, A.M.d'A.R.; Kenner, G.W.; Smith, K.M. Synthesis of pyrromethanes and a tripyrrane. *J. Chem. Soc., Perkin Trans. 1,* **1973**, 2471-2478.
[http://dx.doi.org/10.1039/p19730002471]

[47] Jackson, A.H.; Pandey, R.K.; Rao, K.R.; Roberts, E. Reactions on solid supports, Part II: a convenient method for synthesis of pyrromethanes using a montmorillonite clay as catalyst. *Tetrahedron Lett.,* **1985**, *26*, 793-796.
[http://dx.doi.org/10.1016/S0040-4039(00)89139-X]

[48] Freeman, B.A.; Smith, K.M. Alternative syntheses of pyrromethanes and porphyrins using acid-modified Montmorillonite K-10 Clay. *Synth. Commun.,* **1999**, *29*, 1843-1855.
[http://dx.doi.org/10.1080/00397919908086174]

[49] Wallace, D.M.; Leung, S.H.; Senge, M.O.; Smith, K.M. Rational tetraarylporphyrin syntheses-tetraarylporphyrins form MacDonald route. *J. Org. Chem.,* **1993**, *58*, 7245-7257.
[http://dx.doi.org/10.1021/jo00077a056]

[50] Lindsey, J.S. Synthetic routes to meso-patterned porphyrins. *Acc. Chem. Res.,* **2010**, *43*(2), 300-311.
[http://dx.doi.org/10.1021/ar900212t] [PMID: 19863076]

[51] Geier, G.R.; Lindsey, J.S. Investigation of porphyrin-forming reactions. Part 2. Examination of the reaction course in two-step, one-flask syntheses of meso-substituted porphyrins. *J. Chem. Soc., Perkin Trans.,* **2001**, *2*, 687-700.
[http://dx.doi.org/10.1039/b009092l]

[52] Geier, G.R.; Lindsey, J.S. Effects of aldehyde or dipyrromethane substituents on the reaction course leading to meso-substituted porphyrins. *Tetrahedron,* **2004**, *60*, 11435-11444.
[http://dx.doi.org/10.1016/j.tet.2004.09.081]

[53] Lee, C-H.; Lindsey, J.S. One-flask synthesis of *meso* -substituted dipyrromethanes and their application in the synthesis of *trans* -substituted porphyrin building blocks. *Tetrahedron,* **1994**, *50*, 11427-11440.
[http://dx.doi.org/10.1016/S0040-4020(01)89282-6]

[54] Laha, J.K.; Dhanalekshmi, S.; Taniguchi, M.; Ambroise, A.; Lindsey, J.S. A scalable synthesis of meso-substituted dipyrromethanes. *Org. Process Res. Dev.,* **2003**, *7*, 799-812.
[http://dx.doi.org/10.1021/op034083q]

[55] Littler, B.J.; Miller, M.A.; Hung, C-H.; Wagner, R.W.; O'Shea, D.F.; Boyle, P.D.; Lindsey, J.S. Refined synthesis of 5-substituted dipyrromethanes. *J. Org. Chem.,* **1999**, *64*, 1391-1396.
[http://dx.doi.org/10.1021/jo982015+]

[56] Naik, R.; Joshi, P.; Kaiwar, S.P.; Deshpande, R.K. Facile synthesis of meso-substituted dipyrromethanes and porphyrins using cation exchange resins. *Tetrahedron,* **2003**, *59*, 2207-2213. [http://dx.doi.org/10.1016/S0040-4020(03)00245-X]

[57] Sobral, A.J.; Rebanda, N.G.; da Silva, M.; Lampreia, S.H.; Silva, M.R.; Beja, A.M.; Paixao, J.A.; Gonsalves, A.M. One-step synthesis of dipyrromethanes in water. *Tetrahedron Lett.,* **2003**, *44*, 3971-3973. [http://dx.doi.org/10.1016/S0040-4039(03)00785-8]

[58] Singh, K.; Behal, S.; Deb, P.K. Efficient and versatile single-pot alternative approach to dipyrromethanes. *Synth. Commun.,* **2005**, *35*, 929-934. [http://dx.doi.org/10.1081/SCC-200051691]

[59] Zoli, L.; Cozzi, P.G. Electrophilic activation of aldehydes "on water": a facile route to dipyrromethanes. *Chem Sus Chem,* **2009**, *2*(3), 218-220. [http://dx.doi.org/10.1002/cssc.200900023] [PMID: 19229895]

[60] Faugeras, P.A.; Boens, B.; Elchinger, P.H.; Vergnaud, J.; Teste, K.; Zerrouki, R. Synthesis of meso-substituted dipyrromethanes using iodine-catalysis. *Tetrahedron Lett.,* **2010**, *35*, 4630-4632. [http://dx.doi.org/10.1016/j.tetlet.2010.06.122]

[61] Sessler, J.L.; Lynch, V.; Johnson, M.R. Synthesis and crystal structure of a novel tripyrrane-containing porphyrinogen-like macrocycle. *J. Org. Chem.,* **1987**, *52*, 4394-4397. [http://dx.doi.org/10.1021/jo00228a048]

[62] Sessler, J.L.; Cyr, M.J.; Lynch, V. Synthetic and structural studies of sapphyrin, a 22-p-electron pentapyrrolic "expanded porphyrin". *J. Am. Chem. Soc.,* **1990**, *112*, 2810-2813. [http://dx.doi.org/10.1021/ja00163a059]

[63] Boudif, A.; Momenteau, M. Synthesis of a porphyrin-2,3-diacrylic acid using a new '3 + 1' type procedure. *J. Chem. Soc. Chem. Commun.,* **1994**, 2069-2070. [http://dx.doi.org/10.1039/c39940002069]

[64] Boudif, A.; Momenteau, M. A new convergent method for porphyrin synthesis based on a '3+1' condensation. *J. Chem. Soc., Perkin Trans. 1,* **1996**, 1235-1242. [http://dx.doi.org/10.1039/p19960001235]

[65] Lash, T.D.; Novak, B.H. Tetraphenanthro[9,10-*b* :9,10-*g* :9,10-*l* :9,10-*q*]-porphyrin, a new highly conjugated porphyrin system. *Angew. Chem. Int. Ed. Engl.,* **1995**, *34*, 683-685. [http://dx.doi.org/10.1002/anie.199506831]

[66] Lash, T.D.; Novak, B.H. New highly conjugated porphyrin chromophores: Synthesis of mono- and diphenanthroporphyrins. *Tetrahedron Lett.,* **1995**, *36*, 4381-4284. [http://dx.doi.org/10.1016/0040-4039(95)00818-W]

[67] Lin, Y.; Lash, T.D. Porphyrin synthesis by the "3 + 1" methodology: A superior approach for the preparation of porphyrins with fused 9,10-phenanthroline subunits. *Tetrahedron Lett.,* **1995**, *36*, 9441-9444. [http://dx.doi.org/10.1016/0040-4039(95)02100-0]

[68] Lash, T.D. Porphyrin synthesis by the "3+1" approach: new applications for an old methodology. *Chemistry,* **1996**, *2*, 1197-1200.

[http://dx.doi.org/10.1002/chem.19960021004]

[69] Lash, T.D. Porphyrins with exocyclic rings. Part 9. Synthesis of porphyrins by the '3+1' approach. *J. Porphyr. Phthalocyanines,* **1997**, *1*, 29-144.
[http://dx.doi.org/10.1002/(SICI)1099-1409(199701)1:1<29::AID-JPP7>3.0.CO;2-S]

[70] Sessler, J.L.; Genge, J.W.; Urbach, A.; Sanson, P.A. '3+1' approach to monofunctionalized alkyl porphyrins. *Synlett,* **1996**, *2*, 187-188.
[http://dx.doi.org/10.1055/s-1996-5350]

[71] Nguyen, L.T.; Senge, M.O.; Smith, K.M. Simple methodology for syntheses of porphyrins possessing multiple peripheral substituents with an element of symmetry. *J. Org. Chem.,* **1996**, *61*, 998-1003.
[http://dx.doi.org/10.1021/jo951870f]

[72] Brueckner, C.; Posakony, J.J.; Johnson, C.K.; Boyle, R.W.; James, B.R.; Dolphin, D. Novel and improved syntheses of 5,15-diphenylporphyrin and its dipyrrolic precursors. *J. Porphyr. Phthalocyanines,* **1998**, *2*, 455-466.
[http://dx.doi.org/10.1002/(SICI)1099-1409(199811/12)2:6<455::AID-JPP67>3.0.CO;2-C]

[73] Manka, J.S.; Lawrence, D.S. High yield synthesis of 5,15-diarylporphyrins. *Tetrahedron Lett.,* **1989**, *30*, 6989-6992.
[http://dx.doi.org/10.1016/S0040-4039(01)93405-7]

[74] Ryppa, C.; Senge, M.O.; Hatscher, S.S.; Kleinpeter, E.; Wacker, P.; Schilde, U.; Wiehe, A. Synthesis of mono- and disubstituted porphyrins: A- and 5,10-A2-type systems. *Chemistry,* **2005**, *11*(11), 3427-3442.
[http://dx.doi.org/10.1002/chem.200500001] [PMID: 15798971]

[75] Littler, B.J.; Ciringh, Y.; Lindsey, J.S. Investigation of conditions giving minimal scrambling in the synthesis of *trans* -porphyrins from dipyrromethanes and aldehydes. *J. Org. Chem.,* **1999**, *64*(8), 2864-2872.
[http://dx.doi.org/10.1021/jo982452o] [PMID: 11674358]

[76] Geier, G.R., III; Littler, B.J.; Lindsey, J.S. Investigation of porphyrin-forming reactions. Part 3. The origin of scrambling in dipyrromethane + aldehyde condensations yielding *trans* -A2B--tetraarylporphyrins. *J. Chem. Soc., Perkin Trans. 2,* **2001**, 701-711.
[http://dx.doi.org/10.1039/b009098k]

[77] Senge, M.O.; Kalisch, W.W.; Bischoff, I. The reaction of porphyrins with organolithium reagents. *Chemistry,* **2000**, *6*(15), 2721-2738.
[http://dx.doi.org/10.1002/1521-3765(20000804)6:15<2721::AID-CHEM2721>3.0.CO;2-Z] [PMID: 10985721]

[78] Jaquinod, L. Functionalization of 5,10,15,20-tetra-substituted porphyrins. In: *The Porphyrin Handbook*; Kadish, K.M.; Smith, K.M.; Guilard, R., Eds.; Academic Press: Boston, MA, **2000**; Vol. 1, pp. 201-237.

[79] Senge, M.O. Nucleophilic substitution as a tool for the synthesis of unsymmetrical porphyrins. *Acc. Chem. Res.,* **2005**, *38*(9), 733-743.
[http://dx.doi.org/10.1021/ar0500012] [PMID: 16171316]

[80] Dahms, K.; Senge, M.O.; Bakar, M.B. Exploration of meso-substituted formylporphyrins and their

Grignard and Wittig reactions. *Eur. J. Org. Chem.,* **2007**, *23*, 3833-3848.
[http://dx.doi.org/10.1002/ejoc.200700380]

[81] Senge, M.O. Stirring the porphyrin alphabet soup--functionalization reactions for porphyrins. *Chem. Commun. (Camb.),* **2011**, *47*(7), 1943-1960.
[http://dx.doi.org/10.1039/c0cc03984e] [PMID: 21218237]

[82] Takanami, T. Functionalization of porphyrins through C-C bond formation reactions with functional group-bearing organometallic reagents. *Heterocycl.,* **2013**, *87*, 1659-1689.
[http://dx.doi.org/10.3987/REV-13-775]

[83] Yorimitsu, H.; Osuka, A. Organometallic approaches for direct modification of peripheral C-H bonds in porphyrin cores. *Asian J. Org. Chem.,* **2013**, *2*, 356-373.
[http://dx.doi.org/10.1002/ajoc.201200183]

[84] Andrews, L.E.; Bonnett, R.; Kozyrev, A.N.; Appelman, E.H. *meso* -Reactivity of porphyrins and related compounds. Part 10. Direct fluorination of octaethylporphyrin with caesium fluoroxysulphate. *J. Chem. Soc., Perkin Trans. 1,* **1988**, 1735-1738.
[http://dx.doi.org/10.1039/p19880001735]

[85] Naruta, Y.; Tani, F.; Maruyama, K. *Meso* -perfluorination of porphyrins with *N* -fluoropyridinium triflate. *Tetrahedron Lett.,* **1992**, *33*, 1069-1072.
[http://dx.doi.org/10.1016/S0040-4039(00)91863-X]

[86] Tsuchiya, S.; Seno, M. Novel synthetic method of phenol from benzene catalysed by perfluorinated hemin. *Chem. Lett.,* **1989**, 263-266.
[http://dx.doi.org/10.1246/cl.1989.263]

[87] Lin, J.J.; Gerzevske, K.R.; Liddell, P.A.; Senge, M.O.; Olmstead, M.M.; Khoury, R.G.; Weeth, B.E.; Tsao, S.A.; Smith, K.M. Metal-catalyzed oxidative cyclizations of a,c-biladiene salts bearing 1- and/or 19-arylmethyl substituents: Macrocyclic products and their chemistry. *J. Org. Chem.,* **1997**, *62*(13), 4266-4276.
[http://dx.doi.org/10.1021/jo9619138] [PMID: 11671746]

[88] Lin, J.J.; Holmes, R.T.; Smith, K.M. Oxidative cyclization of a,c-biladienes: The role of the metal ion. *J. Porphyr. Phthalocyanines,* **1998**, *2*, 363-368.
[http://dx.doi.org/10.1002/(SICI)1099-1409(199807/10)2:4/5<363::AID-JPP91>3.0.CO;2-C]

[89] Wijesekera, T.; Matsumono, A.; Dolphin, D.; Lexa, D. Perchlorinated and highly chlorinated *meso* -tetraphenylporphyrins. *Angew. Chem. Int. Ed. Engl.,* **1990**, *29*, 1028-1030.
[http://dx.doi.org/10.1002/anie.199010281]

[90] Gonsalves, A.M.d'A.R.; Johnstone, R.A.W.; Pereira, M.M.; Shaw, J.; Sobral, J.F.N. Metal-assisted reactions. Part 22. Synthesis of perhalogenated porphyrins and their use as oxidation catalysts. *Tetrahedron Lett.,* **1991**, *32*, 1355-1358.
[http://dx.doi.org/10.1016/S0040-4039(00)79666-3]

[91] Chorghade, M.S.; Dolphin, D.; Dupré, D.; Hill, D.R.; Lee, E.C.; Wijesekera, T.P. Improved protocols for the synthesis and halogenation of sterically hindered metalloporphyrins. *Synthesis,* **1996**, 1320-1324.
[http://dx.doi.org/10.1055/s-1996-4401]

[92] Wijesekera, T.; Dupré, D.; Cader, M.R.; Dolphin, D. Porphyrins bearing halogens at the meso-phenyl and beta-pyrrolic positions: Synthesis and spectral properties. *Bull. Soc. Chim. Fr.,* **1996**, *133*, 765-765.

[93] Vicente, M.G.; Smith, K.M. Functionalizations of the alkyl substituents in octa-alkylporphyrins. *Tetrahedron,* **1991**, *47*, 6887-6894.
[http://dx.doi.org/10.1016/S0040-4020(01)96144-7]

[94] Hoffmann, P.; Robert, A.; Meunier, B. Preparation and catalytic activities of the manganese and iron derivatives of Br8TMP and Cl12TMP, two robust porphyrin ligands obtained by halogenation of tetramesitylporphyrin. *Bull. Soc. Chim. Fr.,* **1992**, *129*, 85-97.

[95] Bhyrappa, P.; Krishnan, V. Octabromotetraphenylporphyrin and its metal derivatives: electronic structure and electrochemical properties. *Inorg. Chem.,* **1991**, *30*, 239-245.
[http://dx.doi.org/10.1021/ic00002a018]

[96] Hariprasad, G.; Dahal, S.; Maiya, B.G. *meso* -Substituted octabromoporphyrins: synthesis, spectroscopy, electrochemistry and electronic structure. *J. Chem. Soc., Dalton Trans.,* **1996**, 3429-3436.
[http://dx.doi.org/10.1039/dt9960003429]

[97] Gerzevske, K.R.; Lin, J.J.; Smith, K.M. Kinetic *vs.* thermodynamic product distributions of macrocyclic tetrapyrrole cyclization products from 1,19-disubstituted a,c-biladiene salts. *Heterocycles,* **1994**, *37*, 207-212.
[http://dx.doi.org/10.3987/COM-93-S29]

[98] Almeida, J.A.; Kenner, G.W.; Rimmer, J.; Smith, K.M. A general stepwise synthesis of unsymmetrically substituted porphyrins. *Tetrahedron,* **1976**, *32*, 1793-1799.

[99] Hoffmann, P.; Labat, G.; Robert, A.; Meunier, B. Highly selective bromination of tetramesitylporphyrin: an easy access to robust metalloporphyrins, MBr8TMP and M-Br8TMPS. Examples of application in catalytic oxygenation and oxidation reactions. *Tetrahedron Lett.,* **1990**, *31*, 1991-1994.
[http://dx.doi.org/10.1016/S0040-4039(00)88897-8]

[100] DiMagno, S.G.; Lin, V.S Therien. M.J. Facile elaboration of porphyrins *via* metal-mediated cross-coupling. *J. Org. Chem.,* **1993**, *58*, 5983-5993.
[http://dx.doi.org/10.1021/jo00074a027]

[101] Tagliatesta, P.; Li, J.; Autret, M.; Van Caemelbecke, E.; Villard, A.; D'Souza, F.; Kadish, K.M. Electrochemistry and spectral characterization of oxidized and reduced (TPPBrx)FeCl where TPPBrx is the dianion of β-pyrrole brominated tetraphenylporphyrin and x varies from 0 to 8. *Inorg. Chem.,* **1996**, *35*(19), 5570-5576.
[http://dx.doi.org/10.1021/ic960148c] [PMID: 11666748]

[102] Shea, K.M.; Jaquinod, L.; Khoury, R.G.; Smith, K.M. Dodecasubstituted chlorins (dihydroporphyrins). *Chem. Commun. (Camb.),* **1998**, 759-760.
[http://dx.doi.org/10.1039/a708766g]

[103] Boschi, T.; Paolesse, R.; Tagliatesta, P. Transition-metal-catalyzed cyclization of [*a,c*] biladiene salts as an efficient route to the synthesis of alkyl porphyrins. *Inorg. Chim. Acta,* **1990**, *168*, 83-87.

[http://dx.doi.org/10.1016/S0020-1693(00)88020-2]

[104] Crossley, M.J.; Burn, P.L.; Chew, S.S.; Cuttance, F.B.; Newsom, I.A. Regiospecific introduction of four substituents to porphyrin systems at antipodal pyrrolenic positions. *J. Chem. Soc. Chem. Commun.,* **1991**, 1564-1566.
[http://dx.doi.org/10.1039/c39910001564]

[105] Boyle, R.W.; Johnson, C.K.; Dolphin, D. Iodination and Heck alkynylation of 5,15-diphenylporphyrin. A convenient entry to asymmetrically *meso* substituted porphyrins. *J. Chem. Soc. Chem. Commun.,* **1995**, 527-528.
[http://dx.doi.org/10.1039/c39950000527]

[106] Shanmugathasan, S.; Johnson, C.K.; Edwards, C.; Matthews, E.K.; Dolphin, D.; Boyle, R.W. Regioselective halogenation and palladium-catalysed couplings on 5,15-diphenylporphyrin. *J. Porphyr. Phthalocyanines,* **2000**, *4*, 228-232.
[http://dx.doi.org/10.1002/(SICI)1099-1409(200004/05)4:3<228::AID-JPP199>3.0.CO;2-7]

[107] Monteiro, C.J.; Pereira, M.M.; Vicente, M.G.; Arnaut, L.G. Photophysical properties of unsymmetric meso-substituted porphyrins synthesized *via* the Suzuki coupling reaction. *Tetrahedron,* **2012**, *68*, 8783-8788.
[http://dx.doi.org/10.1016/j.tet.2012.08.002]

[108] Callot, H.J. Bromination of meso-tetraphenylporphine. Structure and reactivity of products. *Tetrahedron Lett.,* **1973**, *50*, 4987-4990.
[http://dx.doi.org/10.1016/S0040-4039(01)87629-2]

[109] Callot, H.J. Bomination of m-tetraphenylporphine. Preparation of alkyl derivatives and polycyanoporphines. *Bull. Soc. Chim. Fr.,* **1974**, 1492-1496.

[110] Arnold, D.P.; Bott, R.C.; Eldridge, H.; Elms, F.M.; Smith, G.; Zojaji, M. Functionalization of 5,15-diphenylporphyrin: preparation and x-ray crystal structures of *meso* nitro, bromo, and trimethylsilylethynyl derivatives. *Aust. J. Chem.,* **1997**, *50*, 495-503.
[http://dx.doi.org/10.1071/C96193]

[111] Gauler, R.; Risch, N. Synthesis of carbon-carbon linked porphyrinoligomers by utilization of Heck-type coupling of natural tetrapyrroles. *Tetrahedron Lett.,* **1997**, *38*, 223-224.
[http://dx.doi.org/10.1016/S0040-4039(96)02280-0]

[112] Shultz, D.A.; Gwaltney, K.P.; Lee, H. A modified procedure for Sonogashira couplings: synthesis and characterization of a bisporphyrin, 1,1-bis[zinc(II) 5‘-ethynyl-10‘,15‘, 20‘-trimesitylporphyrinyl]methylenecyclohexane. *J. Org. Chem.,* **1998**, *63*, 4034-4038.
[http://dx.doi.org/10.1021/jo980098o]

[113] Vicente, M.G.; Jaquinod, L.; Smith, K.M. Oligomeric porphyrin arrays. *Chem. Commun. (Camb.),* **1999**, 1771-1782.
[http://dx.doi.org/10.1039/a902334h]

[114] Setsune, J-I. Palladium chemistry in recent porphyrin research. *J. Porphyr. Phthalocyanines,* **2004**, *8*, 93-102.
[http://dx.doi.org/10.1142/S1088424604000088]

[115] Fazekas, M.; Pintea, M.; Senge, M.O.; Zawadzka, M. Synthetic strategies and porphyrin building

blocks for unsymmetrical multichromophores. *Tetrahedron Lett.,* **2008**, *49*, 2236-2239. [http://dx.doi.org/10.1016/j.tetlet.2008.02.029]

[116] Esdaile, L.J.; Senge, M.O.; Arnold, D.P. New palladium catalysed reactions of bromoporphyrins: synthesis and crystal structures of nickel(ii) complexes of primary 5-aminoporphyrin, 5,5′-bis(porphyrinyl) secondary amine, and 5-hydroxyporphyrin. *Chem. Commun. (Camb.),* **2006**, (40), 4192-4194.
[http://dx.doi.org/10.1039/b608365j] [PMID: 17031428]

[117] Gao, G-Y.; Ruppel, J.V.; Fields, K.B.; Xu, X.; Chen, Y.; Zhang, X.P. Synthesis of diporphyrins *via* palladium-catalyzed C-O bond formation: effective access to chiral diporphyrins. *J. Org. Chem.,* **2008**, *73*(13), 4855-4858.
[http://dx.doi.org/10.1021/jo800443n] [PMID: 18540646]

[118] Bringmann, G.; Götz, D.C.; Gulder, T.A.; Gehrke, T.H.; Bruhn, T.; Kupfer, T.; Radacki, K.; Braunschweig, H.; Heckmann, A.; Lambert, C. Axially chiral β,β'-bisporphyrins: synthesis and configurational stability tuned by the central metals. *J. Am. Chem. Soc.,* **2008**, *130*(52), 17812-17815.
[http://dx.doi.org/10.1021/ja8055886] [PMID: 19055342]

[119] Aratani, N.; Osuka, A. Synthesis of *meso-meso* linked hybrid porphyrin arrays by Pd-catalyzed cross-coupling reaction. *Org. Lett.,* **2001**, *3*(26), 4213-4216.
[http://dx.doi.org/10.1021/ol0168770] [PMID: 11784180]

[120] Peng, X.; Nakamura, Y.; Aratani, N.; Kim, D.; Osuka, A. 1,4-Phenylene bridged *meso-meso* linked diporphyrin array. *Tetrahedron Lett.,* **2004**, *45*, 4981-4984.
[http://dx.doi.org/10.1016/j.tetlet.2004.04.026]

[121] Frampton, M.J.; Akdas, H.; Cowley, A.R.; Rogers, J.E.; Slagle, J.E.; Fleitz, P.A.; Drobizhev, M.; Rebane, A.; Anderson, H.L. Synthesis, crystal structure, and nonlinear optical behavior of beta-unsubstituted meso-meso E-vinylene-linked porphyrin dimers. *Org. Lett.,* **2005**, *7*(24), 5365-5368.
[http://dx.doi.org/10.1021/ol0520525] [PMID: 16288507]

[122] Tokuji, S.; Yurino, T.; Aratani, N.; Shinokubo, H.; Osuka, A. Palladium-catalyzed dimerization of meso-bromoporphyrins: highly regioselective meso-beta coupling through unprecedented remote C-H bond cleavage. *Chemistry,* **2009**, *15*(45), 12208-12211.
[http://dx.doi.org/10.1002/chem.200902143] [PMID: 19834950]

[123] Chan, K.S.; Zhou, X.; Au, M.T.; Tam, C.Y. Synthesis of *beta* -aryl substituted porphyrins by palladium catalyzed Suzuki cross-coupling reactions. *Tetrahedron,* **1995**, *51*, 3129-3136. [http://dx.doi.org/10.1016/0040-4020(95)00069-K]

[124] Zhou, X.; Tse, M.K.; Wan, T.S.; Chan, K.S. Synthesis of β-mono-, tetra-, and octasubstituted sterically bulky porphyrins *via* Suzuki cross coupling. *J. Org. Chem.,* **1996**, *61*(11), 3590-3593.
[http://dx.doi.org/10.1021/jo952205+] [PMID: 11667202]

[125] Muzzi, C.M.; Medforth, C.J.; Voss, L.; Cancilla, M.; Lebrilla, C.; Ma, J-G.; Shelnutt, J.A.; Smith, K.M. Novel dodecaarylporphyrin: synthesis and dynamic properties. *Tetrahedron Lett.,* **1999**, *40*, 6159-6162.
[http://dx.doi.org/10.1016/S0040-4039(99)01105-3]

[126] Chen, Y.; Zhang, X.P. Facile and efficient synthesis of *meso* -arylamino- and alkylamino-substituted porphyrins *via* palladium-catalyzed amination. *J. Org. Chem.,* **2003**, *68*(11), 4432-4438.

[http://dx.doi.org/10.1021/jo034063m] [PMID: 12762748]

[127] Gao, G-Y.; Colvin, A.J.; Chen, Y.; Zhang, X.P. Versatile synthesis of *meso*- aryloxy- and alkoxy-substituted porphyrins *via* palladium-catalyzed C-O cross-coupling reactions. *Org. Lett.,* **2003**, *5*(18), 3261-3264.
[http://dx.doi.org/10.1021/ol035081t] [PMID: 12943402]

[128] Takanami, T.; Hayashi, M.; Hino, F.; Suda, K. Palladium-catalyzed *meso* amination and amidation of porphyrins: marked acceleration with the Ni(II) central metal ion. *Tetrahedron Lett.,* **2003**, *44*, 7353-7357.
[http://dx.doi.org/10.1016/S0040-4039(03)01837-9]

[129] Esdaile, L.J.; McMurtrie, J.C.; Turner, P.; Arnold, D.P. Modified porphyrinoids from carbazates and hydrazones and the first crystal structure of a diiminoporphodimethene. *Tetrahedron Lett.,* **2005**, *46*, 6931-6935.
[http://dx.doi.org/10.1016/j.tetlet.2005.08.004]

[130] Gao, G-Y.; Chen, Y.; Zhang, X.P. General synthesis of *meso*- amidoporphyrins *via* palladium-catalyzed amidation. *Org. Lett.,* **2004**, *6*(11), 1837-1840.
[http://dx.doi.org/10.1021/ol049440b] [PMID: 15151427]

[131] Chen, Y.; Gao, G.Y.; Zhang, X.P. Palladium-mediated synthesis of novel *meso* -chiral porphyrins: cobalt-catalyzed cyclopropanation. *Tetrahedron Lett.,* **2005**, *46*, 4965-4969.
[http://dx.doi.org/10.1016/j.tetlet.2005.05.089]

[132] Gao, G-Y.; Colvin, A.J.; Chen, Y.; Zhang, X.P. Synthesis of *meso*- arylsulfanyl- and alkylsulfanyl-substituted porphyrins *via* palladium-mediated C-S bond formation. *J. Org. Chem.,* **2004**, *69*(25), 8886-8892.
[http://dx.doi.org/10.1021/jo048552d] [PMID: 15575770]

[133] Atefi, F.; McMurtrie, J.C.; Turner, P.; Duriska, M.; Arnold, D.P. *Meso* -porphyrinylphosphine oxides: mono- and bidentate ligands for supramolecular chemistry and the crystal structures of monomeric [10,20-diphenylporphyrinatonickel(II)-5,15-diyl]-bis-[P(O)Ph(2)] and polymeric self-coordinated {[10,20-diphenylporphyrinatozinc(II)-5,15-diyl]-bis-[P(O)Ph(2)]}. *Inorg. Chem.,* **2006**, *45*(16), 6479-6489.
[http://dx.doi.org/10.1021/ic060372u] [PMID: 16878962]

[134] Matano, Y.; Matsumoto, K.; Nakao, Y.; Uno, H.; Sakaki, S.; Imahori, H. Regioselective β-metalation of *meso* -phosphanylporphyrins. Structure and optical properties of porphyrin dimers linked by peripherally fused phosphametallacycles. *J. Am. Chem. Soc.,* **2008**, *130*(14), 4588-4589.
[http://dx.doi.org/10.1021/ja710542e] [PMID: 18345675]

[135] Locos, O.B.; Dahms, K.; Senge, M.O. Allenylporphyrins: a new motif on the porphyrin periphery. *Tetrahedron Lett.,* **2009**, *50*, 2566-2569.
[http://dx.doi.org/10.1016/j.tetlet.2009.03.097]

[136] Liu, C.; Chen, Q.Y. General and efficient synthesis of *meso* - and β-perfluoroalkylated porphyrins *via* Pd-catalyzed cross-coupling reaction. *Synlett,* **2005**, 1306-1310.

[137] Horn, S.; Cundell, B.; Senge, M.O. Exploration of the reaction of potassium organotrifluoroborates with porphyrins. *Tetrahedron Lett.,* **2009**, *50*, 2562-2565.
[http://dx.doi.org/10.1016/j.tetlet.2009.03.098]

[138] Hao, E.; Fronczek, F.R.; Vicente, M.G. Carborane functionalized pyrroles and porphyrins *via* the Suzuki cross-coupling reaction. *Chem. Commun. (Camb.),* **2006**, (47), 4900-4902.
[http://dx.doi.org/10.1039/b611571c] [PMID: 17136240]

[139] Hyslop, A.G.; Kellett, M.A.; Iovine, P.M.; Therien, M.J. Suzuki porphyrins: New synthons for the fabrication of porphyrin-containing supramolecular assemblies. *J. Am. Chem. Soc.,* **1998**, *120*, 12676-12677.
[http://dx.doi.org/10.1021/ja982410h]

[140] Iovine, P.M.; Kellett, M.A.; Redmore, N.P.; Therien, M.J. Syntheses and 1H NMR spectroscopy of rigid, cofacially aligned, porphyrin-bridge-quinone systems in which the interplanar separations between the porphyrin, aromatic bridge, and quinone are less than the sum of their respective van der Waals radii. *J. Am. Chem. Soc.,* **2000**, *122*, 8717-8727.
[http://dx.doi.org/10.1021/ja000759a]

[141] Chng, L.L.; Chang, C.J.; Nocera, D.G. *Meso* -tetraaryl cofacial bisporphyrins delivered by Suzuki cross-coupling. *J. Org. Chem.,* **2003**, *68*(10), 4075-4078.
[http://dx.doi.org/10.1021/jo026610u] [PMID: 12737594]

[142] Chng, L.L.; Chang, C.J.; Nocera, D.G. Catalytic O[bond]O activation chemistry mediated by iron hangman porphyrins with a wide range of proton-donating abilities. *Org. Lett.,* **2003**, *5*(14), 2421-2424.
[http://dx.doi.org/10.1021/ol034581j] [PMID: 12841745]

[143] Chang, C.J.; Chng, L.L.; Nocera, D.G. Proton-coupled O-O activation on a redox platform bearing a hydrogen-bonding scaffold. *J. Am. Chem. Soc.,* **2003**, *125*(7), 1866-1876.
[http://dx.doi.org/10.1021/ja028548o] [PMID: 12580614]

[144] Zhang, T-G.; Zhao, Y.; Asselberghs, I.; Persoons, A.; Clays, K.; Therien, M.J. Design, synthesis, linear, and nonlinear optical properties of conjugated (porphinato)zinc(II)-based donor-acceptor chromophores featuring nitrothiophenyl and nitrooligothiophenyl electron-accepting moieties. *J. Am. Chem. Soc.,* **2005**, *127*(27), 9710-9720.
[http://dx.doi.org/10.1021/ja0402553] [PMID: 15998075]

[145] Sazanovich, I.V.; Balakumar, A.; Muthukumaran, K.; Hindin, E.; Kirmaier, C.; Diers, J.R.; Lindsey, J.S.; Bocian, D.F.; Holten, D. Excited-state energy-transfer dynamics of self-assembled imine-linked porphyrin dyads. *Inorg. Chem.,* **2003**, *42*(21), 6616-6628.
[http://dx.doi.org/10.1021/ic034558u] [PMID: 14552614]

[146] Ali, H.; van Lier, J.E. Synthesis of β-substituted porphyrins using palladium catalysed reactions. *Tetrahedron,* **1994**, *50*, 11933-11944.
[http://dx.doi.org/10.1016/S0040-4020(01)89306-6]

[147] Lee, C-W.; Lu, H-P.; Lan, C-M.; Huang, Y-L.; Liang, Y-R.; Yen, W-N.; Liu, Y-C.; Lin, Y-S.; Diau, E.W.; Yeh, C-Y. Novel zinc porphyrin sensitizers for dye-sensitized solar cells: synthesis and spectral, electrochemical, and photovoltaic properties. *Chemistry,* **2009**, *15*(6), 1403-1412.
[http://dx.doi.org/10.1002/chem.200801572] [PMID: 19097125]

[148] Poon, K.W.; Liu, W.; Chan, P-K.; Yang, Q.; Chan, T-W.; Mak, T.C.; Ng, D.K. Tetrapyrrole derivatives substituted with ferrocenylethynyl moieties. synthesis and electrochemical studies. *J. Org. Chem.,* **2001**, *66*(5), 1553-1559.

[http://dx.doi.org/10.1021/jo0004308] [PMID: 11262097]

[149] Huang, X.; Zhu, C.; Zhang, S.; Li, W.; Guo, Y.; Zhan, X.; Liu, Y.; Bo, Z. Porphyrin-dithienothiophene pi-conjugated copolymers: synthesis and their applications in field-effect transistors and solar cells. *Macromolecules,* **2008**, *41*, 6895-6902.
[http://dx.doi.org/10.1021/ma801407u]

[150] Jiblaoui, A.; Baudequin, C.; Chaleix, V.; Ducourthial, G.; Louradour, F.; Ramondenc, Y.; Sol, V.; Leroy-Lhez, S. An easy one-pot desilylation/copper-free Sonogashira cross-coupling reaction assisted by tetrabutylammonium fluoride (TBAF): synthesis of highly π-conjugated porphyrins. *Tetrahedron,* **2013**, *69*, 5098-5103.
[http://dx.doi.org/10.1016/j.tet.2013.04.092]

[151] Yu, Z.; Ptaszek, M. Multifunctional bacteriochlorins from selective palladium-coupling reactions. *Org. Lett.,* **2012**, *14*(14), 3708-3711.
[http://dx.doi.org/10.1021/ol3015545] [PMID: 22783931]

[152] Ishizuka, T.; Yamasaki, H.; Osuka, A.; Furuta, H. Syntheses of aryl- and arylethynyl-substituted N-confused porphyrins. *Tetrahedron,* **2007**, *63*, 5137-5147.
[http://dx.doi.org/10.1016/j.tet.2007.04.007]

[153] Locos, O.B.; Arnold, D.P. The Heck reaction for porphyrin functionalisation: synthesis of meso-alkenyl monoporphyrins and palladium-catalysed formation of unprecedented meso-beta ethene-linked diporphyrins. *Org. Biomol. Chem.,* **2006**, *4*(5), 902-916.
[http://dx.doi.org/10.1039/b516989e] [PMID: 16493475]

[154] Odobel, F.; Suzenet, F.; Blart, E.; Quintard, J-P. An efficient synthetic approach to highly conjugated porphyrin-based assemblies containing a bipyridine moiety. *Org. Lett.,* **2000**, *2*(2), 131-133.
[http://dx.doi.org/10.1021/ol990356j] [PMID: 10814264]

[155] Bhyrappa, P.; Velkannan, V. Tetrakis(2-thienyl)-mesotetraphenylporphyrins: Synthesis, structural and electrochemical redox properties. *Inorg. Chim. Acta,* **2012**, *387*, 64-73.
[http://dx.doi.org/10.1016/j.ica.2011.12.053]

[156] Smith, M.J.; Clegg, W.; Nguyen, K.A.; Rogers, J.E.; Pachter, R.; Fleitz, P.A.; Anderson, H.L. Synthesis and crystal structure of a push-pull quinoidal porphyrin: a nanoporous framework assembled from cyclic trimer aggregates. *Chem. Commun. (Camb.),* **2005**, (19), 2433-2435.
[http://dx.doi.org/10.1039/b502776d] [PMID: 15886761]

[157] Minnetian, O.M.; Morris, I.K.; Snow, K.M.; Smith, K.M. New syntheses and reactions of some halogenated porphyrins. *J. Org. Chem.,* **1989**, *54*, 5567-5574.
[http://dx.doi.org/10.1021/jo00284a033]

[158] Deshpande, R.; Jiang, L.; Schmidt, G.; Rakovan, J.; Wang, X.; Wheeler, K.; Wang, H. A concise approach to the synthesis of *opp* -dibenzoporphyrins through the Heck reaction. *Org. Lett.,* **2009**, *11*(19), 4251-4253.
[http://dx.doi.org/10.1021/ol901615f] [PMID: 19775180]

[159] Jiang, L.; Engle, J.T.; Sirk, L.; Hartley, C.S.; Ziegler, C.J.; Wang, H. Triphenylene-fused porphyrins. *Org. Lett.,* **2011**, *13*(12), 3020-3023.
[http://dx.doi.org/10.1021/ol200853g] [PMID: 21604789]

[160] Jiang, L.; Zaenglein, R.A.; Engle, J.T.; Mittal, C.; Hartley, C.S.; Ziegler, C.J.; Wang, H. Water-soluble ionic benzoporphyrins. *Chem. Commun. (Camb.),* **2012**, *48*(55), 6927-6929. [http://dx.doi.org/10.1039/c2cc31057k] [PMID: 22673796]

[161] Dahal, S.; Krishnan, V. Charge transfer excited states of zinc(II) derivatives of β-substituted dinitrotetraphenylporphyrin. *Photochem. Photobiol. A: Chem,* **1995**, *89*, 105-112. [http://dx.doi.org/10.1016/1010-6030(94)04017-V]

[162] Bartoli, J.F.; Battioni, P.; De Foor, W.R.; Mansuy, D. Synthesis and remarkable properties of iron β-polynitroporphyrins as catalysts for monooxygenation reactions. *J. Chem. Soc. Chem. Commun.,* **1994**, 23-24. [http://dx.doi.org/10.1039/c39940000023]

[163] Johnson, A.W.; Oldfield, D. The nitration and hydroxylation of ætioporphyrim I. *J. Chem. Soc.,* **1965**, 4303-4312. [http://dx.doi.org/10.1039/jr9650004303]

[164] Drach, J.E.; Longo, F.R. Electrophilic substitution on porphin. I. Nitration. *J. Org. Chem.,* **1974**, *39*(22), 3282-3284. [http://dx.doi.org/10.1021/jo00936a027] [PMID: 4424935]

[165] Ozette, K.; Battioni, P.; Leduc, P.; Bartoli, J-F.; Mansuy, D. A new manganese-β-heptanitro-porphyrin with extreme redox potentials: spectral, electrochemical and catalytic properties. *Inorg. Chim. Acta,* **1998**, *272*, 4-6. [http://dx.doi.org/10.1016/S0020-1693(97)05896-9]

[166] Fanning, J.C.; Mandel, F.S.; Gray, T.L.; Datta-Gupta, N. The reaction of metalloporphyrins with nitrogen dioxide. *Tetrahedron,* **1979**, *35*, 1251-1255. [http://dx.doi.org/10.1016/0040-4020(79)80049-6]

[167] Gong, L-C.; Dolphin, D. Nitrooctaethylporphyrins: synthesis, optical and redox properties. *Can. J. Chem.,* **1985**, *63*, 401-405. [http://dx.doi.org/10.1139/v85-066]

[168] Catalano, M.M.; Crossley, M.J.; Harding, M.M.; King, L.G. Control of reactivity at the porphyrin periphery by metal ion co-ordination: a general method for specific nitration at the β-pyrrolic position of 5,10,15,20-tetraarylporphyrins. *J. Chem. Soc. Chem. Commun.,* **1984**, 1535-1536. [http://dx.doi.org/10.1039/c39840001535]

[169] Shea, K.M.; Jaquinod, L.; Smith, K.M. Dihydroporphyrin synthesis: new methodology. *J. Org. Chem.,* **1998**, *63*(20), 7013-7021. [http://dx.doi.org/10.1021/jo980965p] [PMID: 11672326]

[170] Baldwin, J.E.; Crossley, M.J.; DeBernardis, J.F. Efficient peripheral functionalization of porphyrins. *Tetrahedron,* **1982**, *38*, 685-693. [http://dx.doi.org/10.1016/0040-4020(82)80211-1]

[171] Barnett, G.H.; Smith, K.M. Reactions of some metalloporphyrin and metallochlorin β-cation radicals with nitrite. *J. Chem. Soc. Chem. Commun.,* **1974**, 772-773. [http://dx.doi.org/10.1039/c39740000772]

[172] Osuka, A.; Shimidzu, H. *meso,meso* -Linked porphyrin arrays. *Angew. Chem. Int. Ed. Engl.,* **1997**, *36*,

135-137.
[http://dx.doi.org/10.1002/anie.199701351]

[173] Watanabe, E.; Nishimura, S.; Ogoshi, H.; Yoshida, Z. Orientation of electrophilic meso-substitution in metallooctaethylporphyrins. *Tetrahedron,* **1975**, *31*, 1385-1390.
[http://dx.doi.org/10.1016/0040-4020(75)87068-2]

[174] Crossley, M.J.; Govenlock, L.J.; Prashar, J.K. Synthesis of porphyrin-2,3,12,13- and -2,3,7,- -tetraones: building blocks for the synthesis of extended porphyrin arrays. *J. Chem. Soc. Chem. Commun.,* **1995**, 2379-2380.
[http://dx.doi.org/10.1039/c39950002379]

[175] Baldwin, J.E.; DeBernardis, J.F. Efficient peripheral functionalization of capped porphyrins. *J. Org. Chem.,* **1977**, *42*, 3986-3987.
[http://dx.doi.org/10.1021/jo00444a054]

[176] Hombrecher, H.K.; Gherdan, V.M.; Ohm, S.; Cavaleiro, J.A.; Neves, M.G.; Condesso, M.F. Synthesis and electrochemical investigation of β-alkyloxy substituted *meso* -tetraphenylporphyrins. *Tetrahedron,* **1993**, *49*, 8569-8578.
[http://dx.doi.org/10.1016/S0040-4020(01)96263-5]

[177] Crossley, M.J.; Hambley, T.W.; Mackay, L.G.; Try, A.C.; Walton, R.J. Porphyrin analogues of Tr ger's base: large chiral cavities with a bimetallic binding site *Chem. Soc. Chem. Commun.,* **1995**, 1077-1079.

[178] Crossley, M.J.; King, G.K.; Newsom, I.A.; Sheehan, C.S. Investigation of a 'reverse' approach to extended porphyrin systems, Synthesis of a 2,3-diaminoporphyrin and its reactions with α-diones. *J. Chem. Soc., Perkin Trans. 1,* **1996**, 2675-2684.
[http://dx.doi.org/10.1039/p19960002675]

[179] Johnson, C.K.; Dolphin, D. Syntheses of chlorins possessing fused nitrogen containing rings. *Tetrahedron Lett.,* **1998**, *39*, 4619-4622.
[http://dx.doi.org/10.1016/S0040-4039(98)00853-3]

[180] Crossley, M.J.; King, L.G.; Pyke, S.M. A new and highly efficient synthesis of hydroxyporphyrins. *Tetrahedron,* **1987**, *43*, 4569-4577.
[http://dx.doi.org/10.1016/S0040-4020(01)86899-X]

[181] Crossley, M.J.; Burn, P.L.; Langford, S.J.; Pyke, S.M.; Stark, A.G. A new method for the synthesis of porphyrin-α-diones that is applicable to the synthesis of trans-annular extended porphyrin systems. *J. Chem. Soc. Chem. Commun.,* **1991**, 1567-1568.
[http://dx.doi.org/10.1039/c39910001567]

[182] Crossley, M.J.; Gosper, J.J.; Wilson, M.G. A general reductive denitration method for regiospecific deuteriation of the porphyrin nucleus: synthesis of [20-2H1]mesoporphyrin IX dimethyl Ester. *J. Chem. Soc. Chem. Commun.,* **1985**, 1798-1799.
[http://dx.doi.org/10.1039/c39850001798]

[183] Beavington, R.; Rees, P.A.; Burn, P.L. A study on the oxidation of 2-hydroxyporphyrins to porphyrin-α-diones. *J. Chem. Soc., Perkin Trans. 1,* **1998**, 2847-2852.
[http://dx.doi.org/10.1039/a803633k]

[184] Crossley, M.J.; King, L.G.; Pyke, S.M.; Tansey, C.W. Reaction of 5-nitrooctaethylporphyrins with nucleophiles. *J. Porphyr. Phthalocyanines,* **2002**, *6*, 685-694. [http://dx.doi.org/10.1142/S1088424602000816]

[185] Jaquinod, L.; Gros, C.P.; Khoury, R.G.; Smith, K.M. A convenient synthesis of functionalized tetraphenylchlorins. *Chem. Commun. (Camb.),* **1996**, 2581-2582. [http://dx.doi.org/10.1039/cc9960002581]

[186] Luguya, R.; Fronczek, F.R.; Smith, K.M.; Vicente, M.G. Synthesis of novel carboranylchlorins with dual application in boron neutron capture therapy (BNCT) and photodynamic therapy (PDT). *Appl. Radiat. Isot.,* **2004**, *61*(5), 1117-1123. [http://dx.doi.org/10.1016/j.apradiso.2004.05.068] [PMID: 15308202]

[187] Smith, K.M.; Bisset, G.M.; Case, J.J.; Tabba, H.D. Polyformylation of copper(II) porphyrins. *Tetrahedron Lett.,* **1980**, *21*, 3747-3750. [http://dx.doi.org/10.1016/0040-4039(80)80168-7]

[188] Smith, K.M.; Bisset, G.M.; Tabba, H.D. Polyformylation of copper(II) complexes of octa-alkyl porphyrins. *J. Chem. Soc., Perkin Trans. 1,* **1982**, 581-585. [http://dx.doi.org/10.1039/p19820000581]

[189] Momenteau, M.; Loock, B.; Bisagni, E. Five-coordinate iron(II) porphyrins derived from *meso* -α,β,γ,δtetraphenylporphin: synthesis, characterization, and coordinating properties. *Can. J. Chem.,* **1979**, *57*, 1804-1813. [http://dx.doi.org/10.1139/v79-288]

[190] Buchler, J.W.; Dreher, C.; Herget, G. Metallkomplexe mit tetrapyrrolliganden, XLVIII. Vilsmeier-formylierung von metallporphyrinen mit CoII, NiII, PdII, PtII, CuII, ZnII, CoIII, CrIII, MnIII, FeIII, AlIII, SiIV und PtIV in abhängigkeit vom zentralmetall. *Liebigs Ann. Chem.,* **1988**, 43-54. [http://dx.doi.org/10.1002/jlac.198819880110]

[191] Vicente, M.G.; Rezzano, I.N.; Smith, K.M. Efficient new syntheses of benzochlorins, benzoisobacteriochlorins and benzobacteriochlorins. *Tetrahedron Lett.,* **1990**, *31*, 1365-1368. [http://dx.doi.org/10.1016/S0040-4039(00)88807-3]

[192] Vicente, M.G.; Smith, K.M. Vilsmeier reactions of porphyrins and chlorins with 3-(dimethylamino)acrolein. New syntheses of benzochlorins, benzoisobacteriochlorins and dibenzobacteriochlorins and reductive coupling of porphyrins and chlorins using low-valent titanium complexes. *J. Org. Chem.,* **1991**, *56*, 4407-4418. [http://dx.doi.org/10.1021/jo00014a016]

[193] Montforts, F-P.; Scheurich, G.; Meier, A.; Haake, G.; Hoper, F. An improved method for the preparation of formyldeuteroporphyrins. Synthesis of biologically relevant porphyrins. *Tetrahedron Lett.,* **1991**, *32*, 3477-3480. [http://dx.doi.org/10.1016/0040-4039(91)80810-S]

[194] Haake, G.; Meier, A.; Montforts, F-P.; Scheurich, G.; Zimmermann, G. Synthese von chlorinen aus dem roten blutfarbstoff hämin. *Liebigs Ann. Chem.,* **1992**, 325-336. [http://dx.doi.org/10.1002/jlac.199219920159]

[195] Takanami, T.; Wakita, A.; Sawaizumi, A.; Iso, K.; Onodera, H.; Suda, K. One-pot synthesis of *meso* -

formylporphyrins by S(N)Ar reaction of 5,15-disubstituted porphyrins with (2-pyridyldimethylsilyl)methyllithium. *Org. Lett.,* **2008**, *10*(4), 685-687. [http://dx.doi.org/10.1021/ol703107t] [PMID: 18198882]

[196] Senge, M.O.; Hatscher, S.S.; Wiehe, A.; Dahms, K.; Kelling, A. The dithianyl group as a synthon in porphyrin chemistry: condensation reactions and preparation of formylporphyrins under basic conditions. *J. Am. Chem. Soc.,* **2004**, *126*(42), 13634-13635. [http://dx.doi.org/10.1021/ja045223u] [PMID: 15493918]

[197] Clezy, P.S.; Lim, C.L.; Shannon, J.S. Chemistry of pyrrolic compounds. XXVII. Some aspects of the mass spectra and chemistry of meso-substituted porphyrins. *Aust. J. Chem.,* **1974**, *27*, 1103-1120. [http://dx.doi.org/10.1071/CH9741103]

[198] Arnold, D.P.; Johnson, A.W.; Mahendran, M. Some reactions of *meso* formyloctaethylporhyrin. *J. Chem. Soc., Perkin Trans. 1,* **1978**, 366-370. [http://dx.doi.org/10.1039/p19780000366]

[199] Runge, S.; Senge, M.O. Reaction of β-formylporphyrins with organometallic reagents. A facile method for the preparation of porphyrins with exocyclic double bonds. *Tetrahedron,* **1999**, *55*, 10375-10390. [http://dx.doi.org/10.1016/S0040-4020(99)00579-7]

[200] Hoper, F.; Montforts, F-P. A simple synthesis of functionalized *rac* -chlorins *via* allylboration of porphyrin aldehydes. *Liebigs Ann. Chem.,* **1995**, 1033-1038. [http://dx.doi.org/10.1002/jlac.1995199506146]

[201] Smith, K.M.; Bisset, G.M.; Bushell, M.J. *meso* -Methylporphyrins and chlorins. *Bioorg. Chem.,* **1980**, *9*, 1-26. [http://dx.doi.org/10.1016/0045-2068(80)90030-9]

[202] Smith, N.W.; Smith, K.M. Preparation of bacteriopetroporphyrins by partial synthesis from the *Chlorobium* chlorophylls. *Energy Fuels,* **1990**, *4*, 675-688. [http://dx.doi.org/10.1021/ef00024a011]

[203] Smith, K.M.; Bisset, G.M. General synthesis of hydrocarbon-soluble porphyrins. *J. Org. Chem.,* **1979**, *44*, 2077-2081. [http://dx.doi.org/10.1021/jo01327a006]

[204] Smith, K.M.; Bisset, G.M.; Bushell, M.J. Recent studies on The *chlorobium* chlorophylls (bacteriochlorophylls-C). *Int. J. Biochem.,* **1980**, *12*(5-6), 695-700. [http://dx.doi.org/10.1016/0020-711X(80)90146-9] [PMID: 7450123]

[205] Smith, K.M.; Bisset, G.M. *Meso* (methine) functionalization of octaalkylporphyrins. *J. Chem. Soc., Perkin Trans. 1,* **1981**, 2625-2630. [http://dx.doi.org/10.1039/p19810002625]

[206] Bonfantini, E.E.; Officer, D.L. The synthesis of butadiene-bridged porphyrin dimers and styryl porphyrins using a porphyrin-derived Wittig reagent. *Tetrahedron Lett.,* **1993**, *34*, 8531-8534. [http://dx.doi.org/10.1016/S0040-4039(00)61377-1]

[207] Burrell, A.K.; Officer, D.L. Functionalizing porphyrins *via* Wittig reactions: a building block approach. *Synlett,* **1998**, 1297-1307.

[http://dx.doi.org/10.1055/s-1998-1938]

[208] Yeh, C-Y.; Miller, S.E.; Carpenter, S.D.; Nocera, D.G. Structurally homologous beta- and meso-amidinium porphyrins. *Inorg. Chem.,* **2001**, *40*(14), 3643-3646. [http://dx.doi.org/10.1021/ic001387+] [PMID: 11421722]

[209] Ol'shevskaya, V.A.; Zaitsev, A.V.; Luzgina, V.N.; Kondratieva, T.T.; Ivanov, O.G.; Kononova, E.G.; Petrovskii, P.V.; Mironov, A.F.; Kalinin, V.N.; Hofmann, J.; Shtil, A.A. Novel boronated derivatives of 5,10,15,20-tetraphenylporphyrin: synthesis and toxicity for drug-resistant tumor cells. *Bioorg. Med. Chem.,* **2006**, *14*(1), 109-120. [http://dx.doi.org/10.1016/j.bmc.2005.07.067] [PMID: 16185886]

[210] Whitlock, H.W., Jr; Hanauer, R.; Oester, M.Y.; Bower, B.K. Diimide reduction of porphyrins. *J. Am. Chem. Soc.,* **1969**, *91*, 7485-7489. [http://dx.doi.org/10.1021/ja01054a044]

[211] Bonnett, R. Photosensitizers of the porphyrin and phthalocyanine series for photodynamic therapy. *Chem. Soc. Rev.,* **1995**, *24*, 19-33. [http://dx.doi.org/10.1039/cs9952400019]

[212] Galezowski, M.; Gryko, D.T. Recent advances in the synthesis of hydroporphyrins. *Curr. Org. Chem.,* **2007**, *11*, 1310-1338. [http://dx.doi.org/10.2174/138527207782023157]

[213] Varamo, M.; Loock, B.; Maillard, P.; Grierson, D.S. Development of strategies for the regiocontrolled synthesis of meso-5,10,20-triaryl-2,3-chlorins. *Org. Lett.,* **2007**, *9*(23), 4689-4692. [http://dx.doi.org/10.1021/ol0713382] [PMID: 17939670]

[214] Genady, A.R.; Gabel, D. Synthesis of triazole- and pyridine-bridged porphyrin–chlorin and porphyrin dimers. *Tetrahedron Lett.,* **2003**, *44*, 2915-2917. [http://dx.doi.org/10.1016/S0040-4039(03)00468-4]

[215] Gravier, J.; Schneider, R.; Frochot, C.; Bastogne, T.; Schmitt, F.; Didelon, J.; Guillemin, F.; Barberi-Heyob, M. Improvement of meta-tetra(hydroxyphenyl)chlorin-like photosensitizer selectivity with folate-based targeted delivery. synthesis and *in vivo* delivery studies. *J. Med. Chem.,* **2008**, *51*(13), 3867-3877. [http://dx.doi.org/10.1021/jm800125a] [PMID: 18553957]

[216] Pereira, M.M.; Monteiro, C.J.; Simões, A.V.; Pinto, S.M.; Abreu, A.R.; Sá, G.F.; Silva, E.F.; Rocha, L.B.; Dabrowski, J.M.; Formosinho, S.J.; Simões, S.; Arnaut, L.G. Synthesis and photophysical characterization of a library of photostable halogenated bacteriochlorins: an access to near infrared chemistry. *Tetrahedron,* **2010**, *66*, 9545-9551. [http://dx.doi.org/10.1016/j.tet.2010.09.106]

[217] Pereira, M.M.; Abreu, A.R.; Goncalves, N.P.; Calvete, M.J.; Simões, A.V.; Monteiro, C.J.; Arnaut, L.G.; Eusébio, M.E.; Canotilho, J. An insight into solvent-free diimide porphyrin reduction: a versatile approach for meso-aryl hydroporphyrin synthesis. *Green Chem.,* **2012**, *14*, 1666-1672. [http://dx.doi.org/10.1039/c2gc35126a]

[218] Sutton, J.M.; Fernandez, N.; Boyle, R.W. *J.* Functionalized diphenylchlorins and bacteriochlorins: their synthesis and bioconjugation for targeted photodynamic therapy and tumour cell imaging. *Porphyrins Phthalocyanines,* **2000**, *4*, 655-658.

[http://dx.doi.org/10.1002/1099-1409(200011)4:7<655::AID-JPP245>3.0.CO;2-8]

[219] MacAlpine, J.K.; Boch, R.; Dolphin, D. Evaluation of tetraphenyl-2,3-dihydroxychlorins as potential photosensitizers. *J. Porphyr. Phthalocyanines,* **2002**, *6*, 146. [http://dx.doi.org/10.1142/S1088424602000191]

[220] Brückner, C.; Dolphin, D. β,β'-dihydroxylation of meso-tetraphenylchlorins and metallochlorins. *Tetrahedron Lett.,* **1995**, *36*, 9425. [http://dx.doi.org/10.1016/0040-4039(95)02052-7]

[221] Inhoffen, H.H.; Brockmann, H.; Bliesener, K-M. Zur weiteren Kenntnis des Chlorophylls und des Hämins, XXX1) Photoprotoporphyrine und ihre Umwandlung in Spirographis- sowie Isospirographis-porphyrin. *Justus Liebigs Ann. Chem.,* **1969**, *730*, 173-185. [http://dx.doi.org/10.1002/jlac.19697300118]

[222] DiNello, R.K.; Dolphin, D. Reactions of protoporphyrin with tetracyanoethylene. *J. Org. Chem.,* **1980**, *45*, 5196-5204. [http://dx.doi.org/10.1021/jo01313a035]

[223] Morgan, A.R.; Pangka, V.S.; Dolphin, D. Ready syntheses of benzoporphyrins *via* Diels–Alder reactions with protoporphyrin IX. *J. Chem. Soc. Chem. Commun.,* **1984**, 1047-1048. [http://dx.doi.org/10.1039/c39840001047]

[224] Pangka, V.S.; Morgan, A.R.; Dolphin, A. The Diels-Alder reactions of protoporphyrin IX dimethyl ester with electron-deficient alkynes. *J. Org. Chem.,* **1986**, *51*, 1094-1100. [http://dx.doi.org/10.1021/jo00357a029]

[225] Faustino, M.A.; Neves, M.G.; Vicente, M.G.; Silva, A.M.; Cavaleiro, J.A. Diels-Alder reactions of Ni(II) β-vinyl-meso-tetraarylporphyrins. *Tetrahedron Lett.,* **1996**, *37*, 3569-3570. [http://dx.doi.org/10.1016/0040-4039(96)00623-5]

[226] Silva, A.M.; Tomé, A.C.; Neves, M.G.; Cavaleiro, J.A.; Kappe, C.V. Porphyrins in Diels–Alder reactions. Improvements on the synthesis of barrelene-fused chlorins using microwave irradiation. *Tetrahedron Lett.,* **2005**, *46*, 4723-4726. [http://dx.doi.org/10.1016/j.tetlet.2005.05.047]

[227] Cavaleiro, J.A.; Neves, M.G.; Tomé, A.C. Cycloaddition reactions of porphyrins. *ARKIVOC,* **2003**, *14*, 107-130. [http://dx.doi.org/10.3998/ark.5550190.0004.e11]

[228] Silva, A.M.; Tomé, A.C.; Neves, M.G.; Silva, A.M.; Cavaleiro, J.A. 1,3-dipolar cycloaddition reactions of porphyrins with azomethine ylides. *J. Org. Chem.,* **2005**, *70*(6), 2306-2314. [http://dx.doi.org/10.1021/jo048349i] [PMID: 15760219]

[229] Silva, A.M.; Tomé, A.C.; Neves, M.G.; Cavaleiro, J.A.; Perrone, D.; Dondoni, A. Porphyrins in 1,3-dipolar cycloadditions with sugar azomethine ylides. Synthesis of pyrrolidinoporphyrin glycoconjugates. *Synlett,* **2005**, 857-859.

[230] Zheng, X.; Pandey, R.K. Porphyrin-carbohydrate conjugates: impact of carbohydrate moieties in photodynamic therapy (PDT). *Anticancer. Agents Med. Chem.,* **2008**, *8*(3), 241-268. [http://dx.doi.org/10.2174/187152008783961897] [PMID: 18393785]

[231] Desjardins, A.; Flemming, J.; Sternberg, E.D.; Dolphin, D. Nitrogen extrusion from pyrazoline-

substituted porphyrins and chlorins using long wavelength visible light. *Chem. Commun. (Camb.),* **2002**, (22), 2622-2623.
[http://dx.doi.org/10.1039/b208368j] [PMID: 12510267]

[232] Silva, A.M.; Tomé, A.C.; Neves, M.G.; Cavaleiro, J.A. Porphyrins in 1,3-dipolar cycloaddition reactions: Synthesis of a novel pyrazoline-fused chlorin and a pyrazole-fused porphyrin. *Synlett,* **2002**, 1155-1157.
[http://dx.doi.org/10.1055/s-2002-32581]

[233] Liu, X.G.; Feng, Y.Q.; Chen, X.; Li, F.; Li, X.G. Porphyrins as dipolarophiles in 1,3-dipolar cycloaddition reactions with nitrile oxide. *Synlett,* **2005**, 1030-1032.

[234] Li, X.F.; Zhuang, J.P.; Li, Y.L.; Liu, H.B.; Wang, S.; Zhu, D.B. Synthesis of isoxazoline-fused chlorins and bacteriochlorins by 1,3-dipolar cycloaddition reaction of porphyrin with nitrile oxide. *Tetrahedron Lett.,* **2005**, *46*, 1555-1559.
[http://dx.doi.org/10.1016/j.tetlet.2004.12.138]

[235] Callot, H.J. La dicyclohexylcarbodiimide, agent de deshydratation intramoleculaire des cetols. *Tetrahedron Lett.,* **1972**, *11*, 1011-1012.
[http://dx.doi.org/10.1016/S0040-4039(01)84496-8]

[236] Gomes, A.T.; Leão, R.A.; Alonso, C.M.; Neves, M.G.; Faustino, M.A.; Tomé, A.C.; Silva, A.M.; Pinheiro, S.; Souza, M.C.; Ferreira, V.F.; Cavaleiro, J.A. A new insight into the catalytic decomposition of ethyl diazoacetate in the presence of meso-tetraarylporphyrin (=5,10,15,20-tetraary--21H,23H-porphine) complexes. *Helv. Chim. Acta,* **2008**, *91*, 2270-2283.
[http://dx.doi.org/10.1002/hlca.200890247]

[237] Meunier, B. Metalloporphyrins as versatile catalysts for oxidation reactions and oxidative cleavage. *Chem. Rev.,* **1992**, *92*, 1411-1456.
[http://dx.doi.org/10.1021/cr00014a008]

[238] Collman, J.P.; Wagenknecht, P.S.; Hutchinson, J.E. Molecular catalysts for multielectron redox reactions of small molecules-the cofacial metalloporphyrin approach. *Angew. Chem. Int. Ed. Engl.,* **1994**, *33*, 1537-1554.
[http://dx.doi.org/10.1002/anie.199415371]

[239] Bedioui, F. Zeolite-encapsulated and clay-intercalated metal porphyrin, phthalocyanine and Schiff-base complexes as models for biomimetic oxidation cataluysts-an overview. *Coord. Chem. Rev.,* **1995**, *144*, 39-68.
[http://dx.doi.org/10.1016/0010-8545(94)08000-H]

[240] Dolphin, D.; Traylor, T.G.; Xie, L.Y. Polyhaloporphyrins: Unusual ligands for metals and metal-catalyzed oxidations. *Acc. Chem. Res.,* **1997**, *30*, 251-259.
[http://dx.doi.org/10.1021/ar960126u]

[241] Rives, V.; Ulibarri, M.A. Layered double hydroxides (LDH) intercalated with metal coordination compounds and oxometalates. *Coord. Chem. Rev.,* **1999**, *181*, 61-120.
[http://dx.doi.org/10.1016/S0010-8545(98)00216-1]

[242] Simonneaux, G.; Le Maux, P. Optically active ruthenium porphyrins: chiral recognition and asymmetric catalysis. *Coord. Chem. Rev.,* **2002**, *228*, 43-60.
[http://dx.doi.org/10.1016/S0010-8545(02)00009-7]

[243] Lane, B.S.; Burgess, K. Metal-catalyzed epoxidations of alkenes with hydrogen peroxide. *Chem. Rev.,* **2003**, *103*(7), 2457-2473.
[http://dx.doi.org/10.1021/cr020471z] [PMID: 12848577]

[244] Müller, P.; Fruit, C. Enantioselective catalytic aziridinations and asymmetric nitrene insertions into CH bonds. *Chem. Rev.,* **2003**, *103*(8), 2905-2920.
[http://dx.doi.org/10.1021/cr020043t] [PMID: 12914485]

[245] Grigoropoulou, G.; Clark, J.H.; Elings, J.A. Recent developments on the epoxidation of alkenes using hydrogen peroxide as an oxidant. *Green Chem.,* **2003**, *5*, 1-7.
[http://dx.doi.org/10.1039/b208925b]

[246] Xia, Q.H.; Ge, H.Q.; Ye, C.P.; Liu, Z.M.; Su, K.X. Advances in homogeneous and heterogeneous catalytic asymmetric epoxidation. *Chem. Rev.,* **2005**, *105*(5), 1603-1662.
[http://dx.doi.org/10.1021/cr0406458] [PMID: 15884785]

[247] Meunier, B.; de Visser, S.P.; Shaik, S. Mechanism of oxidation reactions catalyzed by cytochrome p450 enzymes. *Chem. Rev.,* **2004**, *104*(9), 3947-3980.
[http://dx.doi.org/10.1021/cr020443g] [PMID: 15352783]

[248] Rose, E.; Andrioletti, B.; Zrig, S.; Quelquejeu-Ethève, M. Enantioselective epoxidation of olefins with chiral metalloporphyrin catalysts. *Chem. Soc. Rev.,* **2005**, *34*(7), 573-583.
[http://dx.doi.org/10.1039/b405679p] [PMID: 15965539]

[249] Nam, W. High-valent iron(IV)-oxo complexes of heme and non-heme ligands in oxygenation reactions. *Acc. Chem. Res.,* **2007**, *40*(7), 522-531.
[http://dx.doi.org/10.1021/ar700027f] [PMID: 17469792]

[250] Bauer, E.B. Recent advances in iron catalysis in organic synthesis. *Curr. Org. Chem.,* **2008**, *12*, 1341-1369.
[http://dx.doi.org/10.2174/138527208786241556]

[251] Lu, H.; Zhang, X.P. Catalytic C-H functionalization by metalloporphyrins: recent developments and future directions. *Chem. Soc. Rev.,* **2011**, *40*(4), 1899-1909.
[http://dx.doi.org/10.1039/C0CS00070A] [PMID: 21088785]

[252] Calvete, M.J.; Silva, M.; Pereira, M.M.; Burrows, H.D. Inorganic helping organic: recent advances in catalytic heterogeneous oxidations by immobilised tetrapyrrolic macrocycles in micro and mesoporous supports. *RSC Advances,* **2013**, *3*, 22774-22789.
[http://dx.doi.org/10.1039/c3ra44038a]

[253] Nakagaki, S.; Ferreira, G.K.; Marcal, A.L.; Ciuffi, K.J. *Curr. Org. Synth.,* **2014**, *11*, 67-88.
[http://dx.doi.org/10.2174/15701794113106660087]

[254] Collman, J.P.; Denisevich, P.; Konai, Y.; Marrocco, M.; Koval, C.; Anson, F.C. Electrode catalysis of the 4-electron reduction of oxygen to water by dicobalt face-to-face porphyrin. *J. Am. Chem. Soc.,* **1980**, *102*, 6027-6036.
[http://dx.doi.org/10.1021/ja00539a009]

[255] Groves, J.T.; Nemo, T.E. Epoxidation reactions catalyzed by iron porphyrins. Oxygen-tranfers from iodosylbenzene. *J. Am. Chem. Soc.,* **1983**, *105*, 5786-5791.
[http://dx.doi.org/10.1021/ja00356a015]

[256] Groves, J.T.; Myers, R.S. Catalytic asymmetric epoxidations with chiral iron porphyrins. *J. Am. Chem. Soc.,* **1983**, *105*, 5791-5796.
[http://dx.doi.org/10.1021/ja00356a016]

[257] Groves, J.T.; Nemo, T.E. Aliphatic hydroxylation catalyzed by iron porphyrin complexes. *J. Am. Chem. Soc.,* **1983**, *105*, 6243-6248.
[http://dx.doi.org/10.1021/ja00358a009]

[258] Nappa, M.J.; Tolman, C.A. Steric and electronic control of iron porphyrin catalyzed hydrocarbon oxidations. *Inorg. Chem.,* **1985**, *24*, 4711-4719.
[http://dx.doi.org/10.1021/ic00220a056]

[259] Groves, J.T.; Quinn, R. Aerobic epoxidation of olefins with ruthenium porphyrin catalysts. *J. Am. Chem. Soc.,* **1985**, *107*, 5790-5792.
[http://dx.doi.org/10.1021/ja00306a029]

[260] Dawson, J.H. Probing structure-function relations in heme-containing oxygenases and peroxidases. *Science,* **1988**, *240*(4851), 433-439.
[http://dx.doi.org/10.1126/science.3358128] [PMID: 3358128]

[261] Collman, J.P.; Zhang, X.; Lee, V.J.; Uffelman, E.S.; Brauman, J.I. Regioselective and enantioselective epoxidation catalyzed by metalloporphyrins. *Science,* **1993**, *261*(5127), 1404-1411.
[http://dx.doi.org/10.1126/science.8367724] [PMID: 8367724]

[262] Grinstaff, M.W.; Hill, M.G.; Labinger, J.A.; Gray, H.B. Mechanism of catalytic oxygenation of alkanes by halogenated iron porphyrins. *Science,* **1994**, *264*(5163), 1311-1313.
[http://dx.doi.org/10.1126/science.8191283] [PMID: 8191283]

[263] Abrahams, B.F.; Hoskins, B.F.; Michail, D.M.; Robson, R. Assembly of porphyrin building-blocks into network structures with large channels. *Nature,* **1994**, *369*, 727-729.
[http://dx.doi.org/10.1038/369727a0]

[264] Kosal, M.E.; Chou, J.H.; Wilson, S.R.; Suslick, K.S. A functional zeolite analogue assembled from metalloporphyrins. *Nat. Mater.,* **2002**, *1*(2), 118-121.
[http://dx.doi.org/10.1038/nmat730] [PMID: 12618827]

[265] Anderson, H.L. Building molecular wires from the colours of life: conjugated porphyrin oligomers. *Chem. Commun. (Camb.),* **1999**, 2323-2330.
[http://dx.doi.org/10.1039/a904209a]

[266] Biesaga, M.; Pyrzyńska, K.; Trojanowicz, M. Porphyrins in analytical chemistry. A review. *Talanta,* **2000**, *51*(2), 209-224.
[http://dx.doi.org/10.1016/S0039-9140(99)00291-X] [PMID: 18967853]

[267] Sessler, J.L.; Camiolo, S.; Gale, P.A. Pyrrolic and polypyrrolic anion binding agents. *Coord. Chem. Rev.,* **2003**, *240*, 17-55.
[http://dx.doi.org/10.1016/S0010-8545(03)00023-7]

[268] Amao, Y. Probes and polymers for optical sensing of oxygen. *Mikrochim. Acta,* **2003**, *143*, 1-12.
[http://dx.doi.org/10.1007/s00604-003-0037-x]

[269] Borovkov, V.V.; Hembury, G.A.; Inoue, Y. Origin, control, and application of supramolecular chirogenesis in bisporphyrin-based systems. *Acc. Chem. Res.,* **2004**, *37*(7), 449-459.

[http://dx.doi.org/10.1021/ar0302437] [PMID: 15260507]

[270] Drain, C.M.; Goldberg, I.; Sylvain, I.; Falber, A. Synthesis and applications of supramolecular porphyrinic materials, in *Functional Molecular Materials* , Schluter, A.D. Ed. *Top. Curr. Chem.,* **2005**, *245*, 55-88.

[271] Stich, M.I.; Fischer, L.H.; Wolfbeis, O.S. Multiple fluorescent chemical sensing and imaging. *Chem. Soc. Rev.,* **2010**, *39*(8), 3102-3114.
[http://dx.doi.org/10.1039/b909635n] [PMID: 20571676]

[272] Di Natale, C.; Monti, D.; Paolesse, R. Chemical sensitivity of porphyrin assemblies. *Mater. Today,* **2010**, *13*, 37-43.
[http://dx.doi.org/10.1016/S1369-7021(10)70127-9]

[273] Giancane, G.; Valli, L. State of art in porphyrin Langmuir-Blodgett films as chemical sensors. *Adv. Colloid Interface Sci.,* **2012**, *171-172*, 17-35.
[http://dx.doi.org/10.1016/j.cis.2012.01.001] [PMID: 22309985]

[274] Lv, Y.Y.; Wu, S.W.; Li, J. Porphyrinic polymers for gas sensing: An overview. *Curr. Org. Chem.,* **2014**, *18*, 475-488.
[http://dx.doi.org/10.2174/138527281804140303130547]

[275] Malinski, T.; Taha, Z.; Grunfeld, S.; Patton, S.; Kapturczak, M.; Tomboulian, P. Diffusion of nitric oxide in the aorta wall monitored in situ by porphyrinic microsensors. *Biochem. Biophys. Res. Commun.,* **1993**, *193*(3), 1076-1082.
[http://dx.doi.org/10.1006/bbrc.1993.1735] [PMID: 8323533]

[276] Papkovsky, D.B.; Ponomarev, G.V.; Trettnak, W.; Oleary, P. Phosphorescent complexes of porphyrin ketones-optical properties and application to oxygen sensing. *Anal. Chem.,* **1995**, *67*, 4112-4117.
[http://dx.doi.org/10.1021/ac00118a013]

[277] Mills, A.; Lepre, A. Controlling the response characteristics of luminescent porphyrin plastic film sensors for oxygen. *Anal. Chem.,* **1997**, *69*, 4653-4659.
[http://dx.doi.org/10.1021/ac970430g]

[278] Qin, W.; Parzuchowski, P.; Zhang, W.; Meyerhoff, M.E. Optical sensor for amine vapors based on dimer-monomer equilibrium of indium(III) octaethylporphyrin in a polymeric film. *Anal. Chem.,* **2003**, *75*(2), 332-340.
[http://dx.doi.org/10.1021/ac0205356] [PMID: 12553770]

[279] Pinto, S.M.; Lourenço, M.A.; Calvete, M.J.; Abreu, A.R.; Rosado, M.T.; Burrows, H.D.; Pereira, M.M. Synthesis of new metalloporphyrin triads: efficient and versatile tripod optical sensor for the detection of amines. *Inorg. Chem.,* **2011**, *50*(17), 7916-7918.
[http://dx.doi.org/10.1021/ic200727f] [PMID: 21819057]

[280] Philp, D.; Stoddart, J.F. Self-assembly in natural and unnatural systems. *Angew. Chem. Int. Ed. Engl.,* **1996**, *35*, 1155-1196.
[http://dx.doi.org/10.1002/anie.199611541]

[281] Toma, H.E.; Araki, K. Supramolecular assemblies of ruthenium complexes and porphyrins. *Coord. Chem. Rev.,* **2000**, *196*, 307-329.
[http://dx.doi.org/10.1016/S0010-8545(99)00041-7]

[282] Beletskaya, I.; Tyurin, V.S.; Tsivadze, A.Y.; Guilard, R.; Stern, C. Supramolecular chemistry of metalloporphyrins. *Chem. Rev.,* **2009**, *109*(5), 1659-1713.
[http://dx.doi.org/10.1021/cr800247a] [PMID: 19301872]

[283] Yokoyama, T.; Yokoyama, S.; Kamikado, T.; Okuno, Y.; Mashiko, S. Selective assembly on a surface of supramolecular aggregates with controlled size and shape. *Nature,* **2001**, *413*(6856), 619-621.
[http://dx.doi.org/10.1038/35098059] [PMID: 11675782]

[284] De Feyter, S.; De Schryver, F.C. Two-dimensional supramolecular self-assembly probed by scanning tunneling microscopy. *Chem. Soc. Rev.,* **2003**, *32*(3), 139-150.
[http://dx.doi.org/10.1039/b206566p] [PMID: 12792937]

[285] Vriezema, D.M.; Comellas Aragonès, M.; Elemans, J.A.; Cornelissen, J.J.; Rowan, A.E.; Nolte, R.J. Self-assembled nanoreactors. *Chem. Rev.,* **2005**, *105*(4), 1445-1489.
[http://dx.doi.org/10.1021/cr0300688] [PMID: 15826017]

[286] Elemans, J.A.; Van Hameren, R.; Nolte, R.J.; Rowan, A.E. Molecular materials by self-assembly of porphyrins, phthalocyanines, and perylenes. *Adv. Mater.,* **2006**, *18*, 1251-1266.
[http://dx.doi.org/10.1002/adma.200502498]

[287] Kay, E.R.; Leigh, D.A.; Zerbetto, F. Synthetic molecular motors and mechanical machines. *Angew. Chem. Int. Ed. Engl.,* **2007**, *46*(1-2), 72-191.
[http://dx.doi.org/10.1002/anie.200504313] [PMID: 17133632]

[288] Drain, C.M.; Varotto, A.; Radivojevic, I. Self-organized porphyrinic materials. *Chem. Rev.,* **2009**, *109*(5), 1630-1658.
[http://dx.doi.org/10.1021/cr8002483] [PMID: 19253946]

[289] Anderson, S.; Anderson, H.L.; Bashall, A.; McPartlin, M.; Sanders, J.K. Assembly and crystal structure of a photoactive array of give porphyrins. *Angew. Chem. Int. Ed. Engl.,* **1995**, *34*, 1096-1099.
[http://dx.doi.org/10.1002/anie.199510961]

[290] Drain, C.M.; Russell, K.C.; Lehn, J.M. Self-assembly of a multi-porphyrin supramolecular macrocycle by hydrogen bond molecular recognition. *Chem. Commun. (Camb.),* **1996**, 337-338.
[http://dx.doi.org/10.1039/cc9960000337]

[291] Drain, C.M.; Nifiatis, F.; Vasenko, A.; Batteas, J.D. Porphyrin tessellation by design: Metal-mediated self-assembly of large arrays and tapes. *Angew. Chem. Int. Ed.,* **1998**, *37*, 2344-2347.
[http://dx.doi.org/10.1002/(SICI)1521-3773(19980918)37:17<2344::AID-ANIE2344>3.0.CO;2-B]

[292] Michelsen, U.; Hunter, C.A. Self-assembled porphyrin polymers. *Angew. Chem. Int. Ed. Engl.,* **2000**, *39*(4), 764-767.
[http://dx.doi.org/10.1002/(SICI)1521-3773(20000218)39:4<764::AID-ANIE764>3.0.CO;2-6] [PMID: 10760862]

[293] Roberts, D.A.; Schmidt, T.W.; Crossley, M.J.; Perrier, S. Tunable self-assembly of triazole-linked porphyrin-polymer conjugates. *Chemistry,* **2013**, *19*(38), 12759-12770.
[http://dx.doi.org/10.1002/chem.201301133] [PMID: 23939811]

[294] Qureshi, F.M.; Martin, S.J.; Long, X.; Bradley, D.D.; Henari, F.Z.; Blau, W.J.; Smith, E.C.; Wang, C.H.; Kar, A.K.; Anderson, H.L. Optical limiting properties of a zinc porphyrin polymer and its dimer

and monomer model compounds. *Chem. Phys.,* **1998**, *231*, 87-94.
[http://dx.doi.org/10.1016/S0301-0104(98)00081-0]

[295] Calvete, M.; Yang, G.Y.; Hanack, M. Porphyrins and phthalocyanines as materials for optical limiting. *Synth. Met.,* **2004**, *141*, 231-243.
[http://dx.doi.org/10.1016/S0379-6779(03)00407-7]

[296] Senge, M.O.; Fazekas, M. Notaras, *E.G.* A.; Blau, W.J.; Zawadzka, M.; Locos, O. B.; Mhuircheartaigh, E. M. N. Nonlinear optical properties of porphyrins. *Adv. Mater.,* **2007**, *19*, 2737-2774.
[http://dx.doi.org/10.1002/adma.200601850]

[297] Saito, S.; Osuka, A. Expanded porphyrins: intriguing structures, electronic properties, and reactivities. *Angew. Chem. Int. Ed. Engl.,* **2011**, *50*(19), 4342-4373.
[http://dx.doi.org/10.1002/anie.201003909] [PMID: 21491551]

[298] Calvete, M.J. Near-infrared absorbing organic materials with nonlinear transmission properties. *Int. Rev. Phys. Chem.,* **2012**, *31*, 319-366.
[http://dx.doi.org/10.1080/0144235X.2012.713210]

[299] Suslick, K.S.; Chen, C.T.; Meredith, G.R.; Cheng, L.T. Push-pull porphyrins and nolinear optical materials. *J. Am. Chem. Soc.,* **1992**, *114*, 6928-6930.
[http://dx.doi.org/10.1021/ja00043a055]

[300] Anderson, H.L.; Martin, S.J.; Bradley, D.D. Synthesis and 3rd-order nonlinear optical properties of a conjugated porphyrin polymer. *Angew. Chem. Int. Ed. Engl.,* **1994**, *33*, 655-657.
[http://dx.doi.org/10.1002/anie.199406551]

[301] LeCours, S.M.; Guan, H.W.; DiMagno, S.G.; Wang, C.H.; Therien, M.J. Push-pull arylethynyl porphyrins: New chromophores that exhibit large molecular first-order hyperpolarizabilities. *J. Am. Chem. Soc.,* **1996**, *118*, 1497-1503.
[http://dx.doi.org/10.1021/ja953610l]

[302] Ogawa, K.; Zhang, T.; Yoshihara, K.; Kobuke, Y. Large third-order optical nonlinearity of self-assembled porphyrin oligomers. *J. Am. Chem. Soc.,* **2002**, *124*(1), 22-23.
[http://dx.doi.org/10.1021/ja0169015] [PMID: 11772054]

[303] Zawadzka, M.; Wang, J.; Blau, W.J.; Senge, M.O. Solvent effect on the nonlinear absorption of 5,10-A(2)B(2) meso substituted porphyrins. *Photochem. Photobiol. Sci.,* **2013**, *12*(10), 1811-1823.
[http://dx.doi.org/10.1039/c3pp50156f] [PMID: 23857545]

[304] Liu, C.Y.; Bard, A.J. Optoelectronic properties and memories based on organic single-crystal thin films. *Acc. Chem. Res.,* **1999**, *32*, 235-245.
[http://dx.doi.org/10.1021/ar980031n]

[305] Gust, D.; Moore, T.A.; Moore, A.L. Mimicking photosynthetic solar energy transduction. *Acc. Chem. Res.,* **2001**, *34*(1), 40-48.
[http://dx.doi.org/10.1021/ar9801301] [PMID: 11170355]

[306] Holten, D.; Bocian, D.F.; Lindsey, J.S. Probing electronic communication in covalently linked multiporphyrin arrays. A guide to the rational design of molecular photonic devices. *Acc. Chem. Res.,* **2002**, *35*(1), 57-69.

[http://dx.doi.org/10.1021/ar970264z] [PMID: 11790089]

[307] Choi, M.S.; Yamazaki, T.; Yamazaki, I.; Aida, T. Bioinspired molecular design of light-harvesting multiporphyrin arrays. *Angew. Chem. Int. Ed. Engl.,* **2004**, *43*(2), 150-158. [http://dx.doi.org/10.1002/anie.200301665] [PMID: 14695602]

[308] Langa, F.; Gomez-Escalonilla, M.J.; de la Cruz, P. Carbon nanotubes and porphyrins: an exciting combination for optoelectronic devices. *J. Porphyr. Phthalocyanines,* **2007**, *11*, 348-358. [http://dx.doi.org/10.1142/S1088424607000394]

[309] Liu, C.Y.; Tang, H.J.; Bard, A.J. Effect of orientation of porphyrin single-crystal slices on optoelectronic properties. *J. Phys. Chem.,* **1996**, *100*, 3587-3591. [http://dx.doi.org/10.1021/jp9516093]

[310] Yang, S.I.; Seth, J.; Strachan, J.P.; Gentemann, S.; Kim, D.; Holten, D.; Lindsey, J.S.; Bocian, D.F. Ground and excited state electronic properties of halogenated tetraarylporphyrins. Tuning the building blocks for porphyrin-based photonic devices. *J. Porphyr. Phthalocyanines,* **1999**, *3*, 117-147. [http://dx.doi.org/10.1002/(SICI)1099-1409(199902)3:2<117::AID-JPP110>3.0.CO;2-X]

[311] Bahr, J.L.; Kodis, G.; de la Garza, L.; Lin, S.; Moore, A.L.; Moore, T.A.; Gust, D. Photoswitched singlet energy transfer in a porphyrin-spiropyran dyad. *J. Am. Chem. Soc.,* **2001**, *123*(29), 7124-7133. [http://dx.doi.org/10.1021/ja010058t] [PMID: 11459493]

[312] Liddell, P.A.; Kodis, G.; Moore, A.L.; Moore, T.A.; Gust, D. Photonic switching of photoinduced electron transfer in a dithienylethene-porphyrin-fullerene triad molecule. *J. Am. Chem. Soc.,* **2002**, *124*(26), 7668-7669. [http://dx.doi.org/10.1021/ja026327c] [PMID: 12083915]

[313] Hu, J.S.; Guo, Y.G.; Liang, H.P.; Wan, L.J.; Jiang, L. Three-dimensional self-organization of supramolecular self-assembled porphyrin hollow hexagonal nanoprisms. *J. Am. Chem. Soc.,* **2005**, *127*(48), 17090-17095. [http://dx.doi.org/10.1021/ja0553912] [PMID: 16316256]

[314] Li, L.L.; Diau, E.W. Porphyrin-sensitized solar cells. *Chem. Soc. Rev.,* **2013**, *42*(1), 291-304. [http://dx.doi.org/10.1039/C2CS35257E] [PMID: 23023240]

[315] Wasielewski, M.R. Self-assembly strategies for integrating light harvesting and charge separation in artificial photosynthetic systems. *Acc. Chem. Res.,* **2009**, *42*(12), 1910-1921. [http://dx.doi.org/10.1021/ar9001735] [PMID: 19803479]

[316] Hardin, B.E.; Snaith, H.J.; McGehee, M.D. The renaissance of dye-sensitized solar cells. *Nat. Photonics,* **2012**, *6*, 162-169. [http://dx.doi.org/10.1038/nphoton.2012.22]

[317] Imahori, H.; Umeyama, T.; Ito, S. Large pi-aromatic molecules as potential sensitizers for highly efficient dye-sensitized solar cells. *Acc. Chem. Res.,* **2009**, *42*(11), 1809-1818. [http://dx.doi.org/10.1021/ar900034t] [PMID: 19408942]

[318] Campbell, W.M.; Burrell, A.K.; Officer, D.L.; Jolley, K.W. Porphyrins as light harvesters in the dye-sensitised TiO2 solar cell. *Coord. Chem. Rev.,* **2004**, *248*, 1363-1379. [http://dx.doi.org/10.1016/j.ccr.2004.01.007]

[319] Wöhrle, D.; Meissner, D. Organic solar cells. *Adv. Mater.,* **1991**, *3*, 129-138.

[http://dx.doi.org/10.1002/adma.19910030303]

[320] Walter, M.G.; Rudine, A.B.; Wamser, C.C. Porphyrins and phthalocyanines in solar photovoltaic cells. *J. Porphyr. Phthalocyanines,* **2010**, *14*, 759-792.
[http://dx.doi.org/10.1142/S1088424610002689]

[321] Ishida, M.; Park, S.W.; Hwang, D.; Koo, Y.B.; Sessler, J.L.; Kim, D.Y.; Kim, D. Donor-substituted β-functionalized porphyrin dyes on hierarchically structured mesoporous TiO_2 spheres. Highly efficient dye-sensitized solar cells. *J. Phys. Chem. C,* **2011**, *115*, 19343-19354.
[http://dx.doi.org/10.1021/jp202307b]

[322] Yella, A.; Lee, H.W.; Tsao, H.N.; Yi, C.; Chandiran, A.K.; Nazeeruddin, M.K.; Diau, E.W.; Yeh, C.Y.; Zakeeruddin, S.M.; Grätzel, M. Porphyrin-sensitized solar cells with cobalt (II/III)-based redox electrolyte exceed 12 percent efficiency. *Science,* **2011**, *334*(6056), 629-634.
[http://dx.doi.org/10.1126/science.1209688] [PMID: 22053043]

[323] Kay, A.; Gratzel, M. Artificial photosynthesis 1. Photosensitization of TiO2 solar cells with chlorophyll derivatives and related natural porphyrins. *J. Phys. Chem.,* **1993**, *97*, 6272-6277.
[http://dx.doi.org/10.1021/j100125a029]

[324] Bessho, T.; Zakeeruddin, S.M.; Yeh, C.Y.; Diau, E.W.; Grätzel, M. Highly efficient mesoscopic dye-sensitized solar cells based on donor-acceptor-substituted porphyrins. *Angew. Chem. Int. Ed. Engl.,* **2010**, *49*(37), 6646-6649.
[http://dx.doi.org/10.1002/anie.201002118] [PMID: 20687058]

[325] Wang, Q.; Campbell, W.M.; Bonfantani, E.E.; Jolley, K.W.; Officer, D.L.; Walsh, P.J.; Gordon, K.; Humphry-Baker, R.; Nazeeruddin, M.K.; Grätzel, M. Efficient light harvesting by using green Zn-porphyrin-sensitized nanocrystalline TiO2 films. *J. Phys. Chem. B,* **2005**, *109*(32), 15397-15409.
[http://dx.doi.org/10.1021/jp052877w] [PMID: 16852953]

[326] Martínez-Díaz, M.V.; de la Torre, G.; Torres, T. Lighting porphyrins and phthalocyanines for molecular photovoltaics. *Chem. Commun. (Camb.),* **2010**, *46*(38), 7090-7108.
[http://dx.doi.org/10.1039/c0cc02213f] [PMID: 20835465]

[327] Cherian, S.; Wamser, C.C. Adsorption and photoactivity of tetra(4-carboxyphenyl)porphyrin (TCPP) on nanoparticulate TiO2. *J. Phys. Chem. B,* **2000**, *104*, 3624-3629.
[http://dx.doi.org/10.1021/jp994459v]

[328] Odobel, F.; Blart, E.; Lagree, M.; Villieras, M.; Boujtita, H.; El Murr, N.; Caramori, S.; Bignozzi, C.A. Porphyrin dyes for TiO2 sensitization. *J. Mater. Chem.,* **2003**, *13*, 502-510.
[http://dx.doi.org/10.1039/b210674d]

[329] Dandliker, P.J.; Diederich, F.; Zingg, A.; Gisselbrecht, J.P.; Gross, M.; Louati, A.; Sanford, E. Dendrimers with porphyrin cores: Synthetic models for globular heme proteins. *Helv. Chim. Acta,* **1997**, *80*, 1773-1801.
[http://dx.doi.org/10.1002/hlca.19970800603]

[330] Harth, E.M.; Hecht, S.; Helms, B.; Malmstrom, E.E.; Fréchet, J.M.; Hawker, C.J. The effect of macromolecular architecture in nanomaterials: a comparison of site isolation in porphyrin core dendrimers and their isomeric linear analogues. *J. Am. Chem. Soc.,* **2002**, *124*(15), 3926-3938.
[http://dx.doi.org/10.1021/ja025536u] [PMID: 11942830]

[331] Suslick, K.S.; Bhyrappa, P.; Chou, J.H.; Kosal, M.E.; Nakagaki, S.; Smithenry, D.W.; Wilson, S.R. Microporous porphyrin solids. *Acc. Chem. Res.,* **2005**, *38*(4), 283-291. [http://dx.doi.org/10.1021/ar040173j] [PMID: 15835875]

[332] Li, W.S.; Aida, T. Dendrimer porphyrins and phthalocyanines. *Chem. Rev.,* **2009**, *109*(11), 6047-6076. [http://dx.doi.org/10.1021/cr900186c] [PMID: 19769361]

[333] Makiura, R.; Kitagawa, H. Porous porphyrin nanoarchitectures on surfaces. *Eur. J. Inorg. Chem.,* **2010**, *24*, 3715-3724. [http://dx.doi.org/10.1002/ejic.201000730]

[334] Nakagaki, S.; Ferreira, G.K.; Ucoski, G.M.; Dias de Freitas Castro, K.A. Chemical reactions catalyzed by metalloporphyrin-based metal-organic frameworks. *Molecules,* **2013**, *18*(6), 7279-7308. [http://dx.doi.org/10.3390/molecules18067279] [PMID: 23792922]

[335] Gu, Z.Y.; Park, J.; Raiff, A.; Wei, Z.W.; Zhou, H.C. Metal-organic frameworks as biomimetic catalysts. *ChemCatChem,* **2014**, *6*, 67-75. [http://dx.doi.org/10.1002/cctc.201300493]

[336] Zhao, M.; Ou, S.; Wu, C.D. Porous metal-organic frameworks for heterogeneous biomimetic catalysis. *Acc. Chem. Res.,* **2014**, *47*(4), 1199-1207. [http://dx.doi.org/10.1021/ar400265x] [PMID: 24499017]

[337] Figueira, F.; Pereira, P.M.; Silva, S.; Cavaleiro, J.A.; Tome, J.P. Porphyrins and phthalocyanines decorated with dendrimers: synthesis and biomedical applications. *Curr. Org. Synth.,* **2014**, *11*, 110-126. [http://dx.doi.org/10.2174/15701794113106660089]

[338] Zhuang, C.F.; Liu, J.L.; Dai, W.; Wu, Z.P.; Wang, Y.; Gao, S. Synthesis and applications in catalysis of porphyrinic metal-organic frameworks. *Prog. Chem.,* **2014**, *26*, 277-292.

[339] Dandliker, P.J.; Diederich, F.; Gross, M.; Knobler, C.B.; Louati, A.; Sanford, E.M. Dendritic porphyrins-modulating redox potentials of electroactive chromophores with pendant multifunctionality. *Angew. Chem. Int. Ed. Engl.,* **1994**, *33*, 1739-1742. [http://dx.doi.org/10.1002/anie.199417391]

[340] Bhyrappa, P.; Young, J.K.; Moore, J.S.; Suslick, K.S. Dendrimer-metalloporphyrins: Synthesis and catalysis. *J. Am. Chem. Soc.,* **1996**, *118*, 5708-5711. [http://dx.doi.org/10.1021/ja953474k]

[341] Jiang, D.L.; Aida, T. Morphology-dependent photochemical events in aryl ether dendrimer porphyrins: Cooperation of dendron subunits for singlet energy transduction. *J. Am. Chem. Soc.,* **1998**, *120*, 10895-10901. [http://dx.doi.org/10.1021/ja9823520]

[342] Krishna Kumar, R.; Balasubramanian, S.; Goldberg, I. Supramolecular multiporphyrin architecture. Coordination polymers and open networks in crystals of tetrakis(4-cyanophenyl) and tetrakis(4-nitrophenyl)metalloporphyrin. *Inorg. Chem.,* **1998**, *37*(3), 541-552. [http://dx.doi.org/10.1021/ic971259u] [PMID: 11670306]

[343] Hagrman, D.; Hagrman, P.J.; Zubieta, J. Solid-state coordination chemistry: The self-assembly of

microporous organic-inorganic hybrid frameworks constructed from tetrapyridylporphyrin and bimetallic oxide chains or oxide clusters. *Angew. Chem. Int. Ed. Engl.,* **1999**, *38*(21), 3165-3168. [http://dx.doi.org/10.1002/(SICI)1521-3773(19991102)38:21<3165::AID-ANIE3165>3.0.CO;2-O] [PMID: 10556890]

[344] Finikova, O.; Galkin, A.; Rozhkov, V.; Cordero, M.; Hägerhäll, C.; Vinogradov, S. Porphyrin and tetrabenzoporphyrin dendrimers: tunable membrane-impermeable fluorescent pH nanosensors. *J. Am. Chem. Soc.,* **2003**, *125*(16), 4882-4893. [http://dx.doi.org/10.1021/ja0341687] [PMID: 12696908]

[345] Kitagawa, S.; Noro, S.; Nakamura, T. Pore surface engineering of microporous coordination polymers. *Chem. Commun. (Camb.),* **2006**, (7), 701-707. [http://dx.doi.org/10.1039/B511728C] [PMID: 16465313]

[346] Alkordi, M.H.; Liu, Y.; Larsen, R.W.; Eubank, J.F.; Eddaoudi, M. Zeolite-like metal-organic frameworks as platforms for applications: on metalloporphyrin-based catalysts. *J. Am. Chem. Soc.,* **2008**, *130*(38), 12639-12641. [http://dx.doi.org/10.1021/ja804703w] [PMID: 18759392]

[347] Shultz, A.M.; Farha, O.K.; Hupp, J.T.; Nguyen, S.T. A Catalytically active, permanently microporous MOF with metalloporphyrin struts. *J. Am. Chem. Soc.,* **2009**, *131*, 4204-4205. [http://dx.doi.org/10.1021/ja900203f]

[348] Farha, O.K.; Shultz, A.M.; Sarjeant, A.A.; Nguyen, S.T.; Hupp, J.T. Active-site-accessible, porphyrinic metal-organic framework materials. *J. Am. Chem. Soc.,* **2011**, *133*(15), 5652-5655. [http://dx.doi.org/10.1021/ja111042f] [PMID: 21446651]

[349] Fateeva, A.; Chater, P.A.; Ireland, C.P.; Tahir, A.A.; Khimyak, Y.Z.; Wiper, P.V.; Darwent, J.R.; Rosseinsky, M.J. A water-stable porphyrin-based metal-organic framework active for visible-light photocatalysis. *Angew. Chem. Int. Ed. Engl.,* **2012**, *51*(30), 7440-7444. [http://dx.doi.org/10.1002/anie.201202471] [PMID: 22696508]

[350] Zou, C.; Zhang, Z.; Xu, X.; Gong, Q.; Li, J.; Wu, C.D. A multifunctional organic-inorganic hybrid structure based on Mn(III)-porphyrin and polyoxometalate as a highly effective dye scavenger and heterogenous catalyst. *J. Am. Chem. Soc.,* **2012**, *134*(1), 87-90. [http://dx.doi.org/10.1021/ja209196t] [PMID: 22148189]

[351] Jori, G.; Brown, S.B. Photosensitized inactivation of microorganisms. *Photochem. Photobiol. Sci.,* **2004**, *3*(5), 403-405. [http://dx.doi.org/10.1039/b311904c] [PMID: 15122355]

[352] Calzavara-Pinton, P.G.; Venturini, M.; Sala, R. A comprehensive overview of photodynamic therapy in the treatment of superficial fungal infections of the skin. *J. Photochem. Photobiol. B,* **2005**, *78*(1), 1-6. [http://dx.doi.org/10.1016/j.jphotobiol.2004.06.006] [PMID: 15629243]

[353] Jori, G. Photodynamic therapy of microbial infections: state of the art and perspectives. *J. Environ. Pathol. Toxicol. Oncol.,* **2006**, *25*(1-2), 505-519. [http://dx.doi.org/10.1615/JEnvironPatholToxicolOncol.v25.i1-2.320] [PMID: 16566738]

[354] Costa, L.; Faustino, M.A.; Neves, M.G.; Cunha, A.; Almeida, A. Photodynamic inactivation of mammalian viruses and bacteriophages. *Viruses,* **2012**, *4*(7), 1034-1074.

[http://dx.doi.org/10.3390/v4071034] [PMID: 22852040]

[355] Maisch, T. Anti-microbial photodynamic therapy: useful in the future? *Lasers Med. Sci.,* **2007**, *22*(2), 83-91.
[http://dx.doi.org/10.1007/s10103-006-0409-7] [PMID: 17120167]

[356] Donnelly, R.F.; McCarron, P.A.; Tunney, M.M. Antifungal photodynamic therapy. *Microbiol. Res.,* **2008**, *163*(1), 1-12.
[http://dx.doi.org/10.1016/j.micres.2007.08.001] [PMID: 18037279]

[357] Javed, F.; Samaranayake, L.P.; Romanos, G.E. Treatment of oral fungal infections using antimicrobial photodynamic therapy: a systematic review of currently available evidence. *Photochem. Photobiol. Sci.,* **2014**, *13*(5), 726-734.
[http://dx.doi.org/10.1039/c3pp50426c] [PMID: 24686309]

[358] Nitzan, Y.; Dror, R.; Ladan, H.; Malik, Z.; Kimel, S.; Gottfried, V. Structure-activity relationship of porphines for photoinactivation of bacteria. *Photochem. Photobiol.,* **1995**, *62*(2), 342-347.
[http://dx.doi.org/10.1111/j.1751-1097.1995.tb05279.x] [PMID: 7480142]

[359] Minnock, A.; Vernon, D.I.; Schofield, J.; Griffiths, J.; Parish, J.H.; Brown, S.T. Photoinactivation of bacteria. Use of a cationic water-soluble zinc phthalocyanine to photoinactivate both gram-negative and gram-positive bacteria. *J. Photochem. Photobiol. B,* **1996**, *32*(3), 159-164.
[http://dx.doi.org/10.1016/1011-1344(95)07148-2] [PMID: 8622179]

[360] Merchat, M.; Bertolini, G.; Giacomini, P.; Villanueva, A.; Jori, G. Meso-substituted cationic porphyrins as efficient photosensitizers of gram-positive and gram-negative bacteria. *J. Photochem. Photobiol. B,* **1996**, *32*(3), 153-157.
[http://dx.doi.org/10.1016/1011-1344(95)07147-4] [PMID: 8622178]

[361] Tomé, J.P.; Neves, M.G.; Tomé, A.C.; Cavaleiro, J.A.; Soncin, M.; Magaraggia, M.; Ferro, S.; Jori, G. Synthesis and antibacterial activity of new poly-S-lysine-porphyrin conjugates. *J. Med. Chem.,* **2004**, *47*(26), 6649-6652.
[http://dx.doi.org/10.1021/jm040802v] [PMID: 15588101]

[362] Demidova, T.N.; Hamblin, M.R. Effect of cell-photosensitizer binding and cell density on microbial photoinactivation. *Antimicrob. Agents Chemother.,* **2005**, *49*(6), 2329-2335.
[http://dx.doi.org/10.1128/AAC.49.6.2329-2335.2005] [PMID: 15917529]

[363] Tomé, J.P.; Neves, M.G.; Tomé, A.C.; Cavaleiro, J.A.; Mendonça, A.F.; Pegado, I.N.; Duarte, R.; Valdeira, M.L. Synthesis of glycoporphyrin derivatives and their antiviral activity against herpes simplex virus types 1 and 2. *Bioorg. Med. Chem.,* **2005**, *13*(12), 3878-3888.
[http://dx.doi.org/10.1016/j.bmc.2005.04.015] [PMID: 15911304]

[364] Prasanth, C.S.; Karunakaran, S.C.; Paul, A.K.; Kussovski, V.; Mantareva, V.; Ramaiah, D.; Selvaraj, L.; Angelov, I.; Avramov, L.; Nandakumar, K.; Subhash, N. Antimicrobial photodynamic efficiency of novel cationic porphyrins towards periodontal Gram-positive and Gram-negative pathogenic bacteria. *Photochem. Photobiol.,* **2014**, *90*(3), 628-640.
[http://dx.doi.org/10.1111/php.12198] [PMID: 24164211]

[365] Mesquita, M.Q.; Menezes, J.C.; Neves, M.G.; Tomé, A.C.; Cavaleiro, J.A.; Cunha, Â.; Almeida, A.; Hackbarth, S.; Röder, B.; Faustino, M.A. Photodynamic inactivation of bioluminescent Escherichia coli by neutral and cationic pyrrolidine-fused chlorins and isobacteriochlorins. *Bioorg. Med. Chem.*

Lett., **2014**, *24*(3), 808-812.
[http://dx.doi.org/10.1016/j.bmcl.2013.12.097] [PMID: 24424133]

[366] Henderson, B.W.; Dougherty, T.J. How does photodynamic therapy work? *Photochem. Photobiol.,* **1992**, *55*(1), 145-157.
[http://dx.doi.org/10.1111/j.1751-1097.1992.tb04222.x] [PMID: 1603846]

[367] Pass, H.I. Photodynamic therapy in oncology: mechanisms and clinical use. *J. Natl. Cancer Inst.,* **1993**, *85*(6), 443-456.
[http://dx.doi.org/10.1093/jnci/85.6.443] [PMID: 8445672]

[368] Dougherty, T.J.; Gomer, C.J.; Henderson, B.W.; Jori, G.; Kessel, D.; Korbelik, M.; Moan, J.; Peng, Q. Photodynamic therapy. *J. Natl. Cancer Inst.,* **1998**, *90*(12), 889-905.
[http://dx.doi.org/10.1093/jnci/90.12.889] [PMID: 9637138]

[369] Oleinick, N.L.; Evans, H.H. The photobiology of photodynamic therapy: cellular targets and mechanisms. *Radiat. Res.,* **1998**, *150*(5) Suppl., S146-S156.
[http://dx.doi.org/10.2307/3579816] [PMID: 9806617]

[370] Ackroyd, R.; Kelty, C.; Brown, N.; Reed, M. The history of photodetection and photodynamic therapy. *Photochem. Photobiol.,* **2001**, *74*(5), 656-669.
[http://dx.doi.org/10.1562/0031-8655(2001)074<0656:THOPAP>2.0.CO;2] [PMID: 11723793]

[371] Vicente, M.G. Porphyrin-based sensitizers in the detection and treatment of cancer: recent progress. *Curr. Med. Chem. Anticancer Agents,* **2001**, *1*(2), 175-194.
[http://dx.doi.org/10.2174/1568011013354769] [PMID: 12678766]

[372] Moan, J.; Peng, Q. An outline of the hundred-year history of PDT. *Anticancer Res.,* **2003**, *23*(5A), 3591-3600.
[PMID: 14666654]

[373] Detty, M.R.; Gibson, S.L.; Wagner, S.J. Current clinical and preclinical photosensitizers for use in photodynamic therapy. *J. Med. Chem.,* **2004**, *47*(16), 3897-3915.
[http://dx.doi.org/10.1021/jm040074b] [PMID: 15267226]

[374] Nyman, E.S.; Hynninen, P.H. Research advances in the use of tetrapyrrolic photosensitizers for photodynamic therapy. *J. Photochem. Photobiol. B,* **2004**, *73*(1-2), 1-28.
[http://dx.doi.org/10.1016/j.jphotobiol.2003.10.002] [PMID: 14732247]

[375] O'Connor, A.E.; Gallagher, W.M.; Byrne, A.T. Porphyrin and nonporphyrin photosensitizers in oncology: preclinical and clinical advances in photodynamic therapy. *Photochem. Photobiol.,* **2009**, *85*(5), 1053-1074.
[http://dx.doi.org/10.1111/j.1751-1097.2009.00585.x] [PMID: 19682322]

[376] Ormond, A.B.; Freeman, H.S. Dye sensitizers for photodynamic therapy. *Materials (Basel),* **2013**, *6*, 817-840.
[http://dx.doi.org/10.3390/ma6030817]

[377] Moan, J.; Berg, K. The photodegradation of porphyrins in cells can be used to estimate the lifetime of singlet oxygen. *Photochem. Photobiol.,* **1991**, *53*(4), 549-553.
[http://dx.doi.org/10.1111/j.1751-1097.1991.tb03669.x] [PMID: 1830395]

[378] Kachynski, A.V.; Pliss, A.; Kuzmin, A.N.; Ohulchanskyy, T.Y.; Baev, A.; Qu, J.; Prasad, P.N.

Photodynamic therapy by in situ nonlinear photon conversion. *Nat. Photonics,* **2014**, *8*, 455-461. [http://dx.doi.org/10.1038/nphoton.2014.90]

[379] Spiller, W.; Kliesch, H.; Wohrle, D.; Hackbarth, S.; Röder, B.; Schnurpfeil, G. Singlet oxygen quantum yields of different photosensitizers in polar solvents and micellar solutions. *J. Porphyr. Phthalocyanines,* **1998**, *2*, 145-158.
[http://dx.doi.org/10.1002/(SICI)1099-1409(199803/04)2:2<145::AID-JPP60>3.0.CO;2-2]

[380] Rapozzi, V.; Zorzet, S.; Zacchigna, M.; Della Pietra, E.; Cogoi, S.; Xodo, L.E. Anticancer activity of cationic porphyrins in melanoma tumour-bearing mice and mechanistic *in vitro* studies. *Mol. Cancer,* **2014**, *13*, 75.
[http://dx.doi.org/10.1186/1476-4598-13-75] [PMID: 24684778]

[381] Berg, K.; Selbo, P.K.; Weyergang, A.; Dietze, A.; Prasmickaite, L.; Bonsted, A.; Engesaeter, B.Ø.; Angell-Petersen, E.; Warloe, T.; Frandsen, N.; Høgset, A. Porphyrin-related photosensitizers for cancer imaging and therapeutic applications. *J. Microsc.,* **2005**, *218*(Pt 2), 133-147.
[http://dx.doi.org/10.1111/j.1365-2818.2005.01471.x] [PMID: 15857375]

[382] Celli, J.P.; Spring, B.Q.; Rizvi, I.; Evans, C.L.; Samkoe, K.S.; Verma, S.; Pogue, B.W.; Hasan, T. Imaging and photodynamic therapy: mechanisms, monitoring, and optimization. *Chem. Rev.,* **2010**, *110*(5), 2795-2838.
[http://dx.doi.org/10.1021/cr900300p] [PMID: 20353192]

[383] Ethirajan, M.; Chen, Y.; Joshi, P.; Pandey, R.K. The role of porphyrin chemistry in tumor imaging and photodynamic therapy. *Chem. Soc. Rev.,* **2011**, *40*(1), 340-362.
[http://dx.doi.org/10.1039/B915149B] [PMID: 20694259]

[384] Zhang, Y.; Lovell, J.F. Porphyrins as theranostic agents from prehistoric to modern times. *Theranostics,* **2012**, *2*(9), 905-915.
[http://dx.doi.org/10.7150/thno.4908] [PMID: 23082102]

[385] Josefsen, L.B.; Boyle, R.W. Unique diagnostic and therapeutic roles of porphyrins and phthalocyanines in photodynamic therapy, imaging and theranostics. *Theranostics,* **2012**, *2*(9), 916-966.
[http://dx.doi.org/10.7150/thno.4571] [PMID: 23082103]

[386] Brown, D.M.; Kaiser, P.K.; Michels, M.; Soubrane, G.; Heier, J.S.; Kim, R.Y.; Sy, J.P.; Schneider, S. Ranibizumab *versus* verteporfin for neovascular age-related macular degeneration. *N. Engl. J. Med.,* **2006**, *355*(14), 1432-1444.
[http://dx.doi.org/10.1056/NEJMoa062655] [PMID: 17021319]

[387] Liu, T.W.; Chen, J.; Burgess, L.; Cao, W.; Shi, J.; Wilson, B.C.; Zheng, G. Multimodal bacteriochlorophyll theranostic agent. *Theranostics,* **2011**, *1*, 354-362.
[http://dx.doi.org/10.7150/thno/v01p0354] [PMID: 21938263]

[388] Shi, J.; Liu, T.W.; Chen, J.; Green, D.; Jaffray, D.; Wilson, B.C.; Wang, F.; Zheng, G. Transforming a targeted porphyrin theranostic agent into a PET imaging probe for cancer. *Theranostics,* **2011**, *1*, 363-370.
[http://dx.doi.org/10.7150/thno/v01p0363] [PMID: 21938264]

[389] Jing, L.; Liang, X.; Li, X.; Yang, Y.; Dai, Z. Covalent attachment of Mn-porphyrin onto doxorubicin-loaded poly(lactic acid) nanoparticles for potential magnetic resonance imaging and pH-sensitive drug

delivery. *Acta Biomater.,* **2013**, *9*(12), 9434-9441.
[http://dx.doi.org/10.1016/j.actbio.2013.08.018] [PMID: 23962645]

[390] Liang, X.; Li, X.; Jing, L.; Yue, X.; Dai, Z. Theranostic porphyrin dyad nanoparticles for magnetic resonance imaging guided photodynamic therapy. *Biomaterials,* **2014**, *35*(24), 6379-6388.
[http://dx.doi.org/10.1016/j.biomaterials.2014.04.094] [PMID: 24818886]

[391] Waghorn, P.A. Radiolabelled porphyrins in nuclear medicine. *J. Labelled Comp. Radiopharm.,* **2014**, *57*(4), 304-309.
[http://dx.doi.org/10.1002/jlcr.3166] [PMID: 24353174]

[392] Renaud, J-P.; Battioni, P.; Bartoli, J.F.; Mansuy, D. A very efficient system for alkene epoxidation by hydrogen peroxide: catalysis by manganese-porphyrins in the presence of imidazole. *J. Chem. Soc. Chem. Commun.,* **1985**, 888-889.
[http://dx.doi.org/10.1039/c39850000888]

[393] Battioni, P.; Renaud, J-P.; Bartoli, J.F.; Reina-Artiles, M.; Fort, M.; Mansuy, D. Monooxygenase-like oxidation of hydrocarbons by hydrogen peroxide catalyzed by manganese porphyrins and imidazole: selection of the best catalytic system and nature of the active oxygen species. *J. Am. Chem. Soc.,* **1988**, *110*, 8462-8470.
[http://dx.doi.org/10.1021/ja00233a023]

[394] Baciocchi, E.; Boschi, T.; Cassioli, L.; Galli, C.; Lapi, A.; Tagliatesta, P. Epoxidation and hydroxylation reactions catalysed by β-tetrahalogeno and β-octahalogeno manganese porphyrins. *Tetrahedron Lett.,* **1997**, *38*, 7283-7286.
[http://dx.doi.org/10.1016/S0040-4039(97)01692-4]

[395] Battioni, P.; Renaud, J.P.; Bartoli, J.F.; Mansay, D.J. Hydroxylation of alkanes by hydrogen peroxide: and efficient system using manganese porphyrins and imidazole as catalysts. *J. Chem. Soc. Chem. Commun.,* **1986**, 341-343.
[http://dx.doi.org/10.1039/c39860000341]

[396] Anelli, P.L.; Banfi, S.; Montanari, F.; Quici, S. Synergistic effect of lipophilic carboxylic acids and heterocyclic axial ligands in alkene epoxidation by hydrogen peroxide catalysed by manganese(III) tetra-aryl porphyrins. *J. Chem. Soc. Chem. Commun.,* **1989**, 779-780.
[http://dx.doi.org/10.1039/c39890000779]

[397] Banfi, S.; Legramandi, F.; Montanari, F.; Pozzi, G.; Quici, S. Biomimetic models of cytochrome P-450. A doubly tailed manganese(III)–tetraaryl porphyrin; an extremely efficient catalyst for hydrocarbon oxygenations promoted by 30% H2O2. *J. Chem. Soc. Chem. Commun.,* **1991**, 1285-1287.
[http://dx.doi.org/10.1039/c39910001285]

[398] Anelli, P.L.; Banfi, L.; Legramandi, F.; Montanari, F.; Pozzi, G.; Quici, S. Tailed MnIII-tetraarylporphyrins bearing an axial ligand and/or a carboxylic group: self-consistent catalysts for H2O2 or NaOCl alkene epoxidation. *J. Chem. Soc., Perkin Trans. 1,* **1993**, 1345-1357.
[http://dx.doi.org/10.1039/p19930001345]

[399] Traylor, T.G.; Fann, W-P.; Bandyopadhyay, D. A common heterolytic mechanism for reactions of iodosobenzenes, peracids, hydroperoxides, and hydrogen peroxide with iron(III) porphyrins. *J. Am. Chem. Soc.,* **1989**, *111*, 8009-8010.

[http://dx.doi.org/10.1021/ja00202a049]

[400] Traylor, T.G.; Xu, F. A biomimetic model for catalase: the mechanisms of reaction of hydrogen peroxide and hydroperoxides with iron(III) porphyrins. *J. Am. Chem. Soc.,* **1987**, *109*, 6201-6202. [http://dx.doi.org/10.1021/ja00254a059]

[401] Traylor, T.G.; Tsuchiya, S.; Byun, Y-S.; Kim, C. High-yield epoxidations with hydrogen peroxide and tert-butyl hydroperoxide catalyzed by iron(III) porphyrins: heterolytic cleavage of hydroperoxides. *J. Am. Chem. Soc.,* **1993**, 2775-2781. [http://dx.doi.org/10.1021/ja00060a027]

[402] Traylor, T.G.; Kim, C.; Richards, J.L.; Xu, F.; Perrin, C.L. Reactions of iron(III) porphyrins with oxidants. Structure-reactivity studies. *J. Am. Chem. Soc.,* **1995**, *117*, 3468-3474. [http://dx.doi.org/10.1021/ja00117a015]

[403] Nam, W.; Jin, S.W.; Lim, M.H.; Ryu, J.Y.; Kim, C. Anionic ligand effect on the nature of epoxidizing intermediates in iron porphyrin complex-catalyzed epoxidation reactions. *Inorg. Chem.,* **2002**, *41*(14), 3647-3652. [http://dx.doi.org/10.1021/ic011145p] [PMID: 12099867]

[404] Halterman, R.L.; Jan, S.T.; Nimmons, H.L.; Standlee, D.J.; Khan, M.A. Synthesis and catalytic reactivity of D4-symmetric dinorbornabenzene-derived metallotetraarylporphyrins. *Tetrahedron,* **1997**, *53*, 11257-11276. [http://dx.doi.org/10.1016/S0040-4020(97)00729-1]

[405] Halterman, R.L.; Jan, S.T.; Abdulwali, A.H.; Standlee, D.J. Synthesis and catalytic reactivity of sterically and electronically modified D4-symmetric metallotetraarylporphyrins. *Tetrahedron,* **1997**, *53*, 11277-11296. [http://dx.doi.org/10.1016/S0040-4020(97)00730-8]

[406] Berkessel, A.; Frauenkron, M. Catalytic asymmetric epoxidation with a chiral ruthenium porphyrin and N-oxides. *J. Chem. Soc., Perkin Trans. 1,* **1997**, 2265-2266. [http://dx.doi.org/10.1039/a704275b]

[407] Lai, T.S.; Zhang, R.; Cheung, K.K.; Kwong, H.L.; Che, C.M. Aerobic enantioselective alkene epoxidation by a chiral trans-dioxo(D4-porphyrinato)ruthenium(VI) complex. *Chem. Commun. (Camb.),* **1998**, 1583-1584. [http://dx.doi.org/10.1039/a802009d]

[408] Zhang, R.; Yu, W-Y.; Lai, T-S.; Che, C-M. Enantioselective epoxidation of trans-disubstituted alkenes by D2-symmetric chiral dioxoruthenium(VI) porphyrins. *Chem. Commun. (Camb.),* **1999**, 409-410. [http://dx.doi.org/10.1039/a808776h]

[409] Zhang, R.; Yu, W.Y.; Wong, K.Y.; Che, C.M. Highly efficient asymmetric epoxidation of alkenes with a D(4)-symmetric chiral dichlororuthenium(IV) porphyrin catalyst. *J. Org. Chem.,* **2001**, *66*(24), 8145-8153. [http://dx.doi.org/10.1021/jo010329i] [PMID: 11722218]

[410] Proess, G.; Hevesi, L. Synthesis of a new chiral porphyrin ligand and its Mn (III) complex as catalyst in asymmetric alkene epoxidation. *J. Mol. Catal.,* **1993**, *80*, 395-401. [http://dx.doi.org/10.1016/0304-5102(93)85012-I]

[411] Gross, Z.; Ini, S.; Kapon, M.; Cohen, S. First utilization of a homochiral ruthenium porphyrin as enantioselective epoxidation catalyst. *Tetrahedron Lett.,* **1996**, *37*, 7325-7328. [http://dx.doi.org/10.1016/0040-4039(96)01599-7]

[412] Gross, Z.; Ini, S. Asymmetric catalysis by a chiral ruthenium porphyrin: epoxidation, hydroxylation, and partial kinetic resolution of hydrocarbons. *Org. Lett.,* **1999**, *1*, 2077-2080. [http://dx.doi.org/10.1021/ol991131b]

[413] Berkessel, A.; Kaiser, P.; Lex, J. Electronically tuned chiral ruthenium porphyrins: extremely stable and selective catalysts for asymmetric epoxidation and cyclopropanation. *Chemistry,* **2003**, *9*(19), 4746-4756. [http://dx.doi.org/10.1002/chem.200305045] [PMID: 14566882]

[414] Rispens, M.T.; Manfredi, A.; Pozzi, G.; Banfi, S.; Quici, S. [2.2]-para-Cyclophane-4-carbaldehyde as building-block for chiral ligands: Part II: Epoxidation of alkenes catalyzed by the Mn(III)-complex of an atropoisomerically pure (α,β,α,β)-tetraarylporphyrin. *J. Mol. Catal. Chem.,* **1998**, *136*, 13-22. [http://dx.doi.org/10.1016/S1381-1169(98)00011-9]

[415] Smith, J.R.; Reginato, G. Asymmetric alkene epoxidation catalysed by a novel family of chiral metalloporphyrins: effect of structure on catalyst activity, stability and enantioselectivity. *Org. Biomol. Chem.,* **2003**, *1*(14), 2543-2549. [http://dx.doi.org/10.1039/b304212j] [PMID: 12956074]

[416] Silva, M.; Azenha, M.E.; Pereira, M.M.; Burrows, H.D.; Sarakha, M.; Ribeiro, M.F.; Fernandes, A.; Monsanto, P.; Castanheira, F. Immobilization of 5,10,15,20-tetrakis-(2-fluorophenyl)porphyrin into MCM-41 and NaY: Routes toward photodegradation of pesticides. *Pure Appl. Chem.,* **2009**, *81*, 2025-2033. [http://dx.doi.org/10.1351/PAC-CON-08-12-04]

[417] Silva, M.; Azenha, M.E.; Pereira, M.M.; Burrows, H.D.; Sarakha, M.; Forano, C.; Ribeiro, M.F. Fernandes, Immobilization of halogenated porphyrins and their copper complexes in MCM-41: Environmentally friendly photocatalysts for the degradation of pesticides. *Appl. Catal. B,* **2010**, *100*, 1-9. [http://dx.doi.org/10.1016/j.apcatb.2010.07.033]

[418] Molinari, A.; Maldotti, A.; Bratovcic, A.; Magnacca, G. Fe(III)-porphyrin heterogenized on MCM-41: Matrix effects on the oxidation of 1,4-pentanediol. *Catal. Today,* **2011**, *161*, 64-69. [http://dx.doi.org/10.1016/j.cattod.2010.09.004]

[419] Moghadam, M.; Tangestaninejad, S.; Mirkhani, V.; Mohammadpoor-Baltork, I.; Moosavifar, M. Host (nanocavity of zeolite-Y or X)–guest (manganese (III) tetrakis[4-N-methylpyridinum]porphyrin) nanocomposite materials as efficient catalysts for biomimetic alkene epoxidation with sodium periodate: Shape-selective epoxidation of linear alkenes. *J. Mol. Catal. Chem.,* **2009**, *302*, 68-75. [http://dx.doi.org/10.1016/j.molcata.2008.11.036]

[420] Rosa, I.L.; Manso, C.M.; Serra, O.A.; Iamamoto, Y. Biomimetical catalytic activity of iron(III) porphyrins encapsulated in the zeolite X. *J. Mol. Catal. Chem.,* **2000**, *160*, 199-208. [http://dx.doi.org/10.1016/S1381-1169(00)00214-4]

[421] Skrobot, F.C.; Rosa, I.L.; Marques, A.P.; Martins, P.R.; Rocha, J.; Valente, A.A.; Iamamoto, Y. Asymmetric cationic methyl pyridyl and pentafluorophenyl porphyrin encapsulated in zeolites: A

cytochrome P-450 model. *J. Mol. Catal. Chem.,* **2005**, *237*, 86-92.
[http://dx.doi.org/10.1016/j.molcata.2005.05.001]

[422] Skrobot, F.C.; Valente, A.A.; Neves, G.; Rosa, I.; Rocha, J.; Cavaleiro, J.A. Monoterpenes oxidation in the presence of Y zeolite-entrapped manganese(III) tetra(4-N-benzylpyridyl)porphyrin. *J. Mol. Catal. Chem.,* **2003**, *201*, 211-222.
[http://dx.doi.org/10.1016/S1381-1169(03)00181-X]

[423] Monti, D.; Venanzi, M.; Russo, M.; Bussetti, G.; Goletti, C.; Montalti, M.; Zaccheroni, N.; Prodi, L.; Rella, R.; Manera, M.G.; Mancini, G.; Di Natale, C.; Paolesse, R. Spontaneous deposition of amphiphilic porphyrin films on glass. *New J. Chem.,* **2004**, *28*, 1123-1128.
[http://dx.doi.org/10.1039/b403591g]

[424] Dolci, L.S.; Marzocchi, E.; Montalti, M.; Prodi, L.; Monti, D.; Di Natale, C.; D'Amico, A.; Paolesse, R. Amphiphilic porphyrin film on glass as a simple and selective solid-state chemosensor for aqueous Hg2+. *Biosens. Bioelectron.,* **2006**, *22*(3), 399-404.
[http://dx.doi.org/10.1016/j.bios.2006.05.013] [PMID: 16793254]

[425] Czolk, R.; Reichert, J.; Ache, H.J. An optical sensor for the detection of heavy metal ions. *Sens. Actuators B Chem.,* **1992**, *7*, 540-543.
[http://dx.doi.org/10.1016/0925-4005(92)80360-A]

[426] Delmarre, D.; Meallet, R.; Bied-Charreton, C.; Pansu, R.B. Heavy metal ions detection in solution, in sol-gel and with grafted porphyrin monolayers. *J. Photochem. Photobiol. Chem.,* **1999**, *124*, 23-28.
[http://dx.doi.org/10.1016/S1010-6030(99)00046-5]

[427] Li, C-Y.; Zhang, X-B.; Qiao, L.; Zhao, Y.; He, C-M.; Huan, S-Y.; Lu, L-M.; Jian, L-X.; Shen, G-L.; Yu, R-Q. Naphthalimide-porphyrin hybrid based ratiometric bioimaging probe for Hg2+: well-resolved emission spectra and unique specificity. *Anal. Chem.,* **2009**, *81*(24), 9993-10001.
[http://dx.doi.org/10.1021/ac9018445] [PMID: 19904913]

[428] Li, C-Y.; Zhang, X-B.; Dong, Y-Y.; Ma, Q-J.; Han, Z-X.; Zhao, Y.; Shen, G-L.; Yu, R-Q. A porphyrin derivative containing 2-(oxymethyl)pyridine units showing unexpected ratiometric fluorescent recognition of Zn2+ with high selectivity. *Anal. Chim. Acta,* **2008**, *616*(2), 214-221.
[http://dx.doi.org/10.1016/j.aca.2008.04.024] [PMID: 18482606]

[429] Starnes, S.D.; Arungundram, S.; Saunders, C.H. Anion sensors based on β,β'-disubstituted porphyrin derivatives. *Tetrahedron Lett.,* **2002**, *43*, 7785-7788.
[http://dx.doi.org/10.1016/S0040-4039(02)01828-2]

[430] Lee, C.; Lee, D.H.; Hong, J-I. Colorimetric anion sensing by porphyrin-based anion receptors. *Tetrahedron Lett.,* **2001**, *42*, 8665-8668.
[http://dx.doi.org/10.1016/S0040-4039(01)01876-7]

[431] Kalyanasundaram, K. *Photochemistry of polypyridine and porphyrin complexes*; Academic Press: New York, **1992**.

[432] Lee, S-K.; Okura, I. Optical sensor for oxygen using a porphyrin-doped sol–gel glass. *Analyst (Lond.),* **1997**, *122*, 81-84.
[http://dx.doi.org/10.1039/a604885d]

[433] Lee, S-K.; Okura, I. Photostable optical oxygen sensing material: platinum

tetrakis(pentafluorophenyl)porphyrin immobilized in polystyrene. *Anal. Commun.,* **1997**, *34*, 185-188. [http://dx.doi.org/10.1039/a701130j]

[434] Gong, W-L.; Sears, K.J.; Alleman, J.E.; Blatchley, E.R., III Toxicity of model aliphatic amines and their chlorinated forms. *Environ. Toxicol. Chem.,* **2004**, *23*(2), 239-244. [http://dx.doi.org/10.1897/03-109] [PMID: 14982367]

[435] Yang, R.H.; Wang, K.M.; Xiao, D.; Yang, X.H. A host-guest optical sensor for aliphatic amines based on lipophilic cyclodextrin. *Fresenius J. Anal. Chem.,* **2000**, *367*, 429-435. [http://dx.doi.org/10.1007/s002160000353]

[436] Elisa, L.; Baldini, F.; Giannetti, A.; Trono, C.; Carofiglio, T. Solid-supported Zn(II) porphyrin tweezers as optical sensors for diamines. *Chem. Commun. (Camb.),* **2010**, *46*(21), 3678-3680. [http://dx.doi.org/10.1039/c000954g] [PMID: 20396810]

[437] Blau, W.; Byrne, H.; Dennis, W.M.; Kelly, J.M. Reverse saturable absorption in tetraphenylporphyrins. *Opt. Commun.,* **1985**, *56*, 25-29. [http://dx.doi.org/10.1016/0030-4018(85)90059-8]

[438] Bhyrappa, P.; Nethaji, M.; Krishnan, V. Structure of nonplanar octabromotetraphenyl porphyrin and kinetics of rapid metalation reactions. *Chem. Lett.,* **1993**, *22*, 869-872. [http://dx.doi.org/10.1246/cl.1993.869]

[439] Birnbaum, E.R.; Hodge, J.A.; Grinstaff, M.W.; Schaefer, W.P.; Henling, L.; Labinger, J.A.; Bercaw, J.E.; Gray, H.B. 19F NMR spectra and structures of halogenated porphyrins. *Inorg. Chem.,* **1995**, *34*, 3625-3632. [http://dx.doi.org/10.1021/ic00118a010]

[440] Perry, J.W.; Mansour, K.; Lee, S.; Wu, X-L.; Bedworth, P.V.; Chen, C-T.; Ng, D.; Marder, S.R.; Miles, P.; Wada, T.; Tian, M.; Sasabe, H. Organic optical limiter with a strong nonlinear absorptive response. *Science,* **1996**, *273*, 1533-1536. [http://dx.doi.org/10.1126/science.273.5281.1533]

[441] Mishra, S.R.; Rawat, H.S.; Laghate, M. Nonlinear absorption and optical limiting IN metalloporphyrins. *Opt. Commun.,* **1998**, *147*, 328-332. [http://dx.doi.org/10.1016/S0030-4018(97)00614-7]

[442] Guha, S.; Kang, K.; Porter, P.; Roach, J.F.; Remy, D.E.; Aranda, F.J.; Rao, D.V. Third-order optical nonlinearities of metallotetrabenzoporphyrins and a platinum poly-yne. *Opt. Lett.,* **1992**, *17*(4), 264-266. [http://dx.doi.org/10.1364/OL.17.000264] [PMID: 19784296]

[443] Wood, G.L.; Mott, A.G.; Miller, M.J.; Nakashima, M.; Roach, J.F. Nonlinear optical properties of tetrabenzporphyrins. *Proc. MRS,* **1995**, *374*, 267-274. [http://dx.doi.org/10.1557/PROC-374-267]

[444] Tang, N.; Su, W.; Krein, D.M.; Mclean, D.G.; Brant, M.C.; Fleitz, P.A.; Brandelik, D.M.; Sutherland, R.L.; Cooper, T.M. Optical limiting in modified meso-alkynyl porphyrins. *Proc. MRS,* **1997**, *479*, 47-52. [http://dx.doi.org/10.1557/PROC-479-47]

[445] Wang, D.; Sun, W.; Dong, S.; Si, J.; Li, C. Nonlinear optical properties of pentaazadentate expanded

porphyrins and application in optical limiting. *Proc. MRS,* **1997**, *479*, 41-46. [http://dx.doi.org/10.1557/PROC-479-41]

[446] McEwan, K.J.; Bourhill, G.; Robertson, J.M.; Anderson, H.L. The nonlinear optical characterization of meso-substituted porphyrin dyes. *J. Nonlinear Opt. Phys. Mater.,* **2000**, *9*, 451-468. [http://dx.doi.org/10.1142/S0218863500000364]

[447] Krivokapic, A.; Anderson, H.L.; Bourhill, G.; Ives, R.; Clark, S.; McEwan, K.J. Meso-tetra-alkynyl porphyrins for optical limiting-a survey of group III and IV metal complexes. *Adv. Mater.,* **2001**, *13*, 652-656. [http://dx.doi.org/10.1002/1521-4095(200105)13:9<652::AID-ADMA652>3.0.CO;2-3]

[448] Su, W.J.; Cooper, T.M.; Tang, N.; Krein, D.; Jiang, H.; Brandelik, D.; Fleitz, P.; Brant, M.C.; Mclean, D.G. Structure-optical property relationships of porphyrins. *Proc. MRS,* **1997**, *479*, 313-318. [http://dx.doi.org/10.1557/PROC-479-313]

[449] Gratz, H.; Penzkofer, A. Singlet–singlet excited-state absorption and triplet–triplet absorption of meso-tetraphenylporphine. *Chem. Phys.,* **2000**, *254*, 363-374. [http://dx.doi.org/10.1016/S0301-0104(00)00049-5]

[450] McEwan, K.J.; Fleitz, P.A.; Rogers, J.E.; Slagle, J.E.; McLean, D.G.; Akdas, H.; Katterle, M.; Blake, I.M.; Anderson, H.L. Reverse saturable absorption in the near-Infrared by fused porphyrin dimers. *Adv. Mater.,* **2004**, *16*, 1933-1935. [http://dx.doi.org/10.1002/adma.200400492]

[451] Sun, W.; Byeon, C.C.; Lawson, C.M.; Gray, G.M.; Wang, D. Third-order nonlinear optical properties of an expanded porphyrin cadmium complex. *Appl. Phys. Lett.,* **2000**, *77*, 1759-1761. [http://dx.doi.org/10.1063/1.1311321]

[452] Kay, A.; Humphry-Baker, R.; Grätzel, M. Artificial Photosynthesis. 2. Investigations on the mechanism of photosensitization of nanocrystalline TiO2 solar cells by chlorophyll derivatives. *J. Phys. Chem.,* **1994**, *98*, 952-959. [http://dx.doi.org/10.1021/j100054a035]

[453] Nazeeruddin, M.K.; Kay, A.; Rodicio, I.; Humphry-Baker, R.; Muller, E.; Liska, P.; Vlachopoulos, N.; Grätzel, M. Conversion of light to electricity by cis-X2bis(2,2'-bipyridyl-4-4'-dicarboxylate)ruthenium(II) charge-transfer sensitizers (X = Cl-, Br-, I-, CN-, and SCN-) on nanocrystalline titanium dioxide electrodes. *J. Am. Chem. Soc.,* **1993**, *115*, 6382-6390. [http://dx.doi.org/10.1021/ja00067a063]

[454] Nazeeruddin, M.K.; Humphry-Baker, R.; Officer, D.L.; Campbell, W.M.; Burrell, A.K.; Grätzel, M. Application of metalloporphyrins in nanocrystalline dye-sensitized solar cells for conversion of sunlight into electricity. *Langmuir,* **2004**, *20*(15), 6514-6517. [http://dx.doi.org/10.1021/la0496082] [PMID: 15248744]

[455] Campbell, W.M.; Jolley, K.W.; Wagner, P.; Wagner, K.; Walsh, P.J.; Gordon, K.C.; Schmidt-Mende, L.; Nazeeruddin, M.K.; Wang, Q.; Grätzel, M.; Officer, D.L. Highly efficient porphyrin sensitizers for dye-sensitized solar cells. *J. Phys. Chem. C,* **2007**, *111*, 11760-11762. [http://dx.doi.org/10.1021/jp0750598]

[456] Rochford, J.; Chu, D.; Hagfeldt, A.; Galoppini, E. Tetrachelate porphyrin chromophores for metal oxide semiconductor sensitization: effect of the spacer length and anchoring group position. *J. Am.*

Chem. Soc., **2007**, *129*(15), 4655-4665.
[http://dx.doi.org/10.1021/ja068218u] [PMID: 17385856]

[457] Lu, H.P.; Tsai, C.Y.; Yen, W.N.; Hsieh, C.P.; Lee, C.W.; Yeh, C.Y.; Diau, E.W. Control of dye aggregation and electron injection for highly efficient porphyrin sensitizers adsorbed on semiconductor films with varying ratios of coadsorbate. *J. Phys. Chem. C,* **2009**, *113*, 20990-20997.
[http://dx.doi.org/10.1021/jp908100v]

[458] Hsieh, C-P.; Lu, H-P.; Chiu, C-L.; Lee, C-W.; Chuang, S-H.; Mai, C-L.; Yen, W-N.; Hsu, S-J.; Diau, E.W; Yeh, C-Y. Synthesis and characterization of porphyrin sensitizers with various electron-donating substituents for highly efficient dye-sensitized solar cells. *J. Mater. Chem.,* **2010**, *20*, 1127-1134.
[http://dx.doi.org/10.1039/B919645E]

[459] Wu, C.H.; Pan, T.Y.; Hong, S.H.; Wang, C.L.; Kuo, H.H.; Chu, Y.Y.; Diau, E.W.; Lin, C.Y. A fluorene-modified porphyrin for efficient dye-sensitized solar cells. *Chem. Commun. (Camb.),* **2012**, *48*(36), 4329-4331.
[http://dx.doi.org/10.1039/c2cc30892d] [PMID: 22446840]

[460] Wang, C-L.; Chang, Y-C.; Lan, C-M.; Lo, C-F.; Diau, E.W.; Lin, C-Y. Enhanced light harvesting with π-conjugated cyclic aromatic hydrocarbons for porphyrin-sensitized solar cells. *Energy Environ. Sci.,* **2011**, *4*, 1788-1795.
[http://dx.doi.org/10.1039/c0ee00767f]

[461] Tsao, H.N.; Yi, C.; Moehl, T.; Yum, J-H.; Zakeeruddin, S.M.; Nazeeruddin, M.K.; Grätzel, M. Cyclopentadithiophene bridged donor-acceptor dyes achieve high power conversion efficiencies in dye-sensitized solar cells based on the tris-cobalt bipyridine redox couple. *ChemSusChem,* **2011**, *4*(5), 591-594.
[http://dx.doi.org/10.1002/cssc.201100120] [PMID: 21557495]

[462] Dualeh, A.; De Angelis, F.; Fantacci, S.; Moehl, T.; Yi, C.; Kessler, F.; Baranoff, E.; Nazeeruddin, M.K.; Grätzel, M. Influence of donor groups of organic D–π–A dyes on open-circuit voltage in solid-state dye-sensitized Solar cells. *J. Phys. Chem. C,* **2012**, *116*, 1572-1578.
[http://dx.doi.org/10.1021/jp209691e]

[463] Jin, R-H.; Aida, T.; Inoue, S. 'Caged' porphyrin: the first dendritic molecule having a core photochemical functionality. *Chem. Commun.,* **1993**, 1260-1262.
[http://dx.doi.org/10.1039/c39930001260]

[464] Bhyrappa, P.; Young, J.K.; Moore, J.S.; Suslick, K.S. Shape selective epoxidation of alkenes by metalloporphyrin-dendrimers. *J. Mol. Catal. Chem.,* **1996**, *113*, 109-116.
[http://dx.doi.org/10.1016/S1381-1169(96)00161-6]

[465] Gao, B.X.; Liu, Y.L.; Yin, H.J.; Li, Y.X.; Bai, Q.Q.; Zhang, L.C. Watersoluble dendritic polyaspartic porphyrins: potential photosensitizers for photodynamic therapy. *New J. Chem.,* **2012**, *36*(1), 28-31.
[http://dx.doi.org/10.1039/C1NJ20733D]

[466] Raab, O. On the effects of fluorescent substances on infusoria. *Z. Biol.,* **1890**, *39*, 524-546.

[467] Wainwright, M. Photodynamic antimicrobial chemotherapy (PACT). *J. Antimicrob. Chemother.,* **1998**, *42*(1), 13-28.
[http://dx.doi.org/10.1093/jac/42.1.13] [PMID: 9700525]

[468] Reddi, E.; Ceccon, M.; Valduga, G.; Jori, G.; Bommer, J.C.; Elisei, F.; Latterini, L.; Mazzucato, U. Photophysical properties and antibacterial activity of meso-substituted cationic porphyrins. *Photochem. Photobiol.,* **2002**, *75*(5), 462-470. [http://dx.doi.org/10.1562/0031-8655(2002)075<0462:PPAAAO>2.0.CO;2] [PMID: 12017471]

[469] Jori, G. Photodynamische therapien in der mikrobiologie. In: *Klinische fluoreszenzdiagnostik und photodynamische therapie*; Szeimies , RM; Jocham , D; Landthaler , M, Eds.; Blackwell verlag: Berlin, **2003**.

[470] Wilder-Smith, C.H.; Wilder-Smith, P.; Grosjean, P.; van den Bergh, H.; Woodtli, A.; Monnier, P.; Dorta, G.; Meister, F.; Wagnières, G. Photoeradication of Helicobacter pylori using 5-aminolevulinic acid: preliminary human studies. *Lasers Surg. Med.,* **2002**, *31*(1), 18-22. [http://dx.doi.org/10.1002/lsm.10066] [PMID: 12124710]

[471] Soukos, N.S.; Ximenez-Fyvie, L.A.; Hamblin, M.R.; Socransky, S.S.; Hasan, T.; Hasan, T. Targeted antimicrobial photochemotherapy. *Antimicrob. Agents Chemother.,* **1998**, *42*(10), 2595-2601. [PMID: 9756761]

[472] Hamblin, M.R.; O'Donnell, D.A.; Murthy, N.; Rajagopalan, K.; Michaud, N.; Sherwood, M.E.; Hasan, T. Polycationic photosensitizer conjugates: effects of chain length and Gram classification on the photodynamic inactivation of bacteria. *J. Antimicrob. Chemother.,* **2002**, *49*(6), 941-951. [http://dx.doi.org/10.1093/jac/dkf053] [PMID: 12039886]

[473] Zimcik, P.; Miletin, M. Photodynamic Therapy. In: *Dyes and Pigments: New Research*; Lang, A.R., Ed.; Nova Science Publishers: New York, NY, USA, **2008**.

[474] MacDonald, I.J.; Dougherty, T.J. Basic principles of photodynamic therapy. *J. Porphyr. Phthalocyanines,* **2001**, *5*, 105-129. [http://dx.doi.org/10.1002/jpp.328]

[475] Kessel, D.; Thompson, P.; Musselman, B.; Chang, C.K. Probing the structure and stability of the tumor-localizing derivative of hematoporphyrin by reductive cleavage with LiAlH4. *Cancer Res.,* **1987**, *47*(17), 4642-4645. [PMID: 3621158]

[476] Stapleton, M.; Rhodes, L.E. Photosensitizers for photodynamic therapy of cutaneous disease. *J. Dermatolog. Treat.,* **2003**, *14*(2), 107-112. [http://dx.doi.org/10.1080/09546630310012127] [PMID: 12775318]

[477] Santoro, O.; Bandieramonte, G.; Melloni, E.; Marchesini, R.; Zunino, F.; Lepera, P.; De Palo, G. Photodynamic therapy by topical meso-tetraphenylporphinesulfonate tetrasodium salt administration in superficial basal cell carcinomas. *Cancer Res.,* **1990**, *50*(15), 4501-4503. [PMID: 2369729]

[478] Fingar, V.H.; Kik, P.K.; Haydon, P.S.; Cerrito, P.B.; Tseng, M.; Abang, E.; Wieman, T.J. Analysis of acute vascular damage after photodynamic therapy using benzoporphyrin derivative (BPD). *Br. J. Cancer,* **1999**, *79*(11-12), 1702-1708. [http://dx.doi.org/10.1038/sj.bjc.6690271] [PMID: 10206280]

[479] Rousset, N.; Bourre, L.; Thibaud, S. Sensitizers in photodynamic therapy. In: *Photodynamic Therapy*; Patrice, T., Ed.; Royal Society of Chemistry: Cambridge, UK, **2003**.

[http://dx.doi.org/10.1039/9781847551658-00059]

[480] Lui, H.; Hobbs, L.; Tope, W.D.; Lee, P.K.; Elmets, C.; Provost, N.; Chan, A.; Neyndorff, H.; Su, X.Y.; Jain, H.; Hamzavi, I.; McLean, D.; Bissonnette, R. Photodynamic therapy of multiple nonmelanoma skin cancers with verteporfin and red light-emitting diodes: two-year results evaluating tumor response and cosmetic outcomes. *Arch. Dermatol.,* **2004**, *140*(1), 26-32. [http://dx.doi.org/10.1001/archderm.140.1.26] [PMID: 14732656]

[481] Triesscheijn, M.; Ruevekamp, M.; Aalders, M.; Baas, P.; Stewart, F.A. Outcome of mTHPC mediated photodynamic therapy is primarily determined by the vascular response. *Photochem. Photobiol.,* **2005**, *81*(5), 1161-1167. [http://dx.doi.org/10.1562/2005-04-04-RA-474] [PMID: 15934792]

[482] Dolmans, D.E.; Fukumura, D.; Jain, R.K. Photodynamic therapy for cancer. *Nat. Rev. Cancer,* **2003**, *3*(5), 380-387. [http://dx.doi.org/10.1038/nrc1071] [PMID: 12724736]

[483] Kobayashi, W.; Liu, Q.; Nakagawa, H.; Sakaki, H.; Teh, B.; Matsumiya, T.; Yoshida, H.; Imaizumi, T.; Satoh, K.; Kimura, H. Photodynamic therapy with mono-L-aspartyl chlorin e6 can cause necrosis of squamous cell carcinoma of tongue: experimental study on an animal model of nude mouse. *Oral Oncol.,* **2006**, *42*(1), 46-50. [http://dx.doi.org/10.1016/j.oraloncology.2005.05.009] [PMID: 16266818]

[484] Usuda, J.; Kato, H.; Okunaka, T.; Furukawa, K.; Tsutsui, H.; Yamada, K.; Suga, Y.; Honda, H.; Nagatsuka, Y.; Ohira, T.; Tsuboi, M.; Hirano, T. Photodynamic therapy (PDT) for lung cancers. *J. Thorac. Oncol.,* **2006**, *1*(5), 489-493. [http://dx.doi.org/10.1097/01243894-200606000-00018] [PMID: 17409904]

[485] Lobel, J.; MacDonald, I.J.; Ciesielski, M.J.; Barone, T.; Potter, W.R.; Pollina, J.; Plunkett, R.J.; Fenstermaker, R.A.; Dougherty, T.J. 2-[1-hexyloxyethyl]-2-devinyl pyropheophorbide-a (HPPH) in a nude rat glioma model: implications for photodynamic therapy. *Lasers Surg. Med.,* **2001**, *29*(5), 397-405. [http://dx.doi.org/10.1002/lsm.10001] [PMID: 11891727]

[486] Brun, P.H.; DeGroot, J.L.; Dickson, E.F.; Farahani, M.; Pottier, R.H. Determination of the in vivo pharmacokinetics of palladium-bacteriopheophorbide (WST09) in EMT6 tumour-bearing Balb/c mice using graphite furnace atomic absorption spectroscopy. *Photochem. Photobiol. Sci.,* **2004**, *3*(11-12), 1006-1010. [http://dx.doi.org/10.1039/b403534h] [PMID: 15570387]

[487] Chevalier, S.; Anidjar, M.; Scarlata, E.; Hamel, L.; Scherz, A.; Ficheux, H.; Borenstein, N.; Fiette, L.; Elhilali, M. Preclinical study of the novel vascular occluding agent, WST11, for photodynamic therapy of the canine prostate. *J. Urol.,* **2011**, *186*(1), 302-309. [http://dx.doi.org/10.1016/j.juro.2011.03.039] [PMID: 21600602]

[488] Pereira, M.M.; Arnaut, L.G.; Formosinho, S.J.; Monteiro, C.J.P. Novel derivatives of porphyrin, particularly chlorins and/or bacteriochlorins and uses thereof in photodynamic therapy WO Patent 053707, 2006.

[489] Pereira, M.M.; Monteiro, C.J.; Simões, A.V.; Pinto, S.M.; Abreu, A.R.; Sá, G.F.; Silva, E.F.; Rocha, L.B.; Dabrowski, J.M.; Formosinho, S.J.; Simões, S.; Arnaut, L.G. Synthesis and photophysical

characterization of a library of photostable halogenated bacteriochlorins: an access to near infrared chemistry. *Tetrahedron,* **2010**, *66*, 9545-9551.
[http://dx.doi.org/10.1016/j.tet.2010.09.106]

[490] Mody, T.D.; Sessler, J.L. Texaphyrins: A new approach to drug development. *J. Porphyr. Phthalocyanines,* **2001**, *5*, 134-142.
[http://dx.doi.org/10.1002/jpp.326]

[491] Patel, H.; Mick, R.; Finlay, J.; Zhu, T.C.; Rickter, E.; Cengel, K.A.; Malkowicz, S.B.; Hahn, S.M.; Busch, T.M. Motexafin lutetium-photodynamic therapy of prostate cancer: short- and long-term effects on prostate-specific antigen. *Clin. Cancer Res.,* **2008**, *14*(15), 4869-4876.
[http://dx.doi.org/10.1158/1078-0432.CCR-08-0317] [PMID: 18676760]

[492] Kelkar, S.S.; Reineke, T.M. Theranostics: combining imaging and therapy. *Bioconjug. Chem.,* **2011**, *22*(10), 1879-1903.
[http://dx.doi.org/10.1021/bc200151q] [PMID: 21830812]

[493] Saha, G.B. *Basics of PET Imaging: Physics, Chemistry and Regulations,* 2nd ed; Springer: New York, **2010**.
[http://dx.doi.org/10.1007/978-1-4419-0805-6]

[494] Kulkarni, P.V.; Jain, D.; Narula, J. Radio indium and gallium labeled porphyrins for medical imaging. *AIP Conf. Proc.,* **2001**, *576*, 837.
[http://dx.doi.org/10.1063/1.1395433]

[495] Murugesan, S.; Shetty, S.J.; Srivastava, T.S.; Samuel, A.M.; Noronha, O.P. Preparation and biological evaluation of the new chlorin photosensitizer T3,4BCPC for detection and treatment of tumors. *J. Photochem. Photobiol. B,* **2002**, *68*(1), 33-38.
[http://dx.doi.org/10.1016/S1011-1344(02)00329-9] [PMID: 12208034]

[496] Konirová, R.; Ernestová, M.; Jedináková-Krísová, V.; Kral, V. Radioactively labelled porphyrin derivatives. *Czech. J. Phys.,* **2003**, *53*, A755-A761.
[http://dx.doi.org/10.1007/s10582-003-0098-4]

[497] Chakraborty, S.; Das, T.; Sarma, H.D.; Venkatesh, M.; Banerjee, S. Preparation and preliminary studies on 177Lu-labeled hydroxyapatite particles for possible use in the therapy of liver cancer. *Nucl. Med. Biol.,* **2008**, *35*(5), 589-597.
[http://dx.doi.org/10.1016/j.nucmedbio.2008.03.003] [PMID: 18589303]

[498] Stefflova, K.; Li, H.; Chen, J.; Zheng, G. Peptide-based pharmacomodulation of a cancer-targeted optical imaging and photodynamic therapy agent. *Bioconjug. Chem.,* **2007**, *18*(2), 379-388.
[http://dx.doi.org/10.1021/bc0602578] [PMID: 17298029]

[499] Lee, T.; Zhang, X.A.; Dhar, S.; Faas, H.; Lippard, S.J.; Jasanoff, A. *In vivo* imaging with a cell-permeable porphyrin-based MRI contrast agent. *Chem. Biol.,* **2010**, *17*(6), 665-673.
[http://dx.doi.org/10.1016/j.chembiol.2010.05.009] [PMID: 20609416]

[500] Chen, Y.; Gryshuk, A.; Achilefu, S.; Ohulchansky, T.; Potter, W.; Zhong, T.; Morgan, J.; Chance, B.; Prasad, P.N.; Henderson, B.W.; Oseroff, A.; Pandey, R.K. A novel approach to a bifunctional photosensitizer for tumor imaging and phototherapy. *Bioconjug. Chem.,* **2005**, *16*(5), 1264-1274.
[http://dx.doi.org/10.1021/bc050177o] [PMID: 16173807]

[501] Zhang, X.A.; Lovejoy, K.S.; Jasanoff, A.; Lippard, S.J. Water-soluble porphyrins as a dual-function molecular imaging platform for MRI and fluorescence zinc sensing. *Proc. Natl. Acad. Sci. USA,* **2007**, *104*(26), 10780-10785.
[http://dx.doi.org/10.1073/pnas.0702393104] [PMID: 17578918]

[502] Gros, C.P.; Eggenspiller, A.; Nonat, A.; Barbe, J-M.; Denat, F. New potential bimodal imaging contrast agents based on DOTA-like and porphyrin macrocycles. *MedChemComm,* **2011**, *2*, 119-125.
[http://dx.doi.org/10.1039/C0MD00205D]

[503] Whitlock, H.W.; Hanauer, R. Octaethylporphyrin. *J. Org. Chem.,* **1968**, *33*, 2169-2171.
[http://dx.doi.org/10.1021/jo01269a122]

[504] Ono, N.; Kawamura, H.; Bougauchi, M.; Maruyama, K. Porphyrin synthesis from nitrocompounds. *Tetrahedron,* **1990**, *46*, 7483-7496.
[http://dx.doi.org/10.1016/S0040-4020(01)89062-1]

[505] Inhoffen, H.H.; Fuhrhop, J-H.; Voigt, H.; Brockmann, H., Jr Zur weiteren Kenntnis des Chlorophylls und des Hämins, VI. Formylierung der meso-Kohlenstoffatome von Alkyl-substituierten Porphyrinen. *Justus Liebigs Ann. Chem.,* **1966**, *695*, 133-143.
[http://dx.doi.org/10.1002/jlac.19666950114]

[506] Barnett, G.H.; Hudson, M.F.; Smith, K.M. Concerning meso- tetraphenylporphyrin purification. *J. Chem. Soc. Perkin Trans. I,* **1975**, (14), 1401-1403.
[http://dx.doi.org/10.1039/p19750001401] [PMID: 1172508]

[507] Yeung, M.; Ng, A.C.; Drew, M.G.; Vorpagel, E.; Breitung, E.M.; McMahon, R.J.; Ng, D.K. Facile synthesis and nonlinear optical properties of push-pull 5,15-diphenylporphyrins. *J. Org. Chem.,* **1998**, *63*(21), 7143-7150.
[http://dx.doi.org/10.1021/jo971597c] [PMID: 11672352]

[508] Taniguchi, M.; Cramer, D.L.; Bhise, A.D.; Kee, H.L.; Bocian, D.F.; Holten, D.; Lindsey, J.S. Accessing the near-infrared spectral region with stable, synthetic, wavelength-tunable bacteriochlorins. *New J. Chem.,* **2008**, *32*, 947-958.
[http://dx.doi.org/10.1039/b717803d]

CHAPTER 4

BODIPY Dyes as Top Materials

Abstract: BODIPY dyes are structures based on two pyrrolic units linked through a methine bridge, complexated with boron(III) fluoride. These dyes posses usually a strong fluorescence, absorbing light in the visible range, and are known for their remarkable photophysical properties, gaining recognition as one of the highly versatile fluorophores with increasing popularity. Herein is presented a conspectus on the field of BODIPY dyes, where emphasis is set on their preparation and functionalization methods, along with more detailed examples on several promising applications with 165 references provided. Some synthetic details on the preparation of state-of-the-art compounds are given as well.

Keywords: Biolabeling, Bodipy, Boron dipyrromethene, Dyes, Fluorescence, Methine bridge, Pyrrole, Sensors, Solar cells.

INTRODUCTION

The 4,4'-difluoro-4-bora-3a,4a-diaza-*s*-indacene dyes, or BOron DIPYrromethenes, *aka* BODIPY dyes, are structures based on two pyrrolic units linked through a methine bridge, complexated with boron(III) fluoride, possessing π-electron conjugation. This structure can be illustrated as a boradiazaindacene, being considered as an indacene family member, with similar number assignments for any substituents, according to IUPAC rules.

The numbering of the related dipyrromethane and dipyrromethene systems is depicted in Fig. (**1**). While for a dipyrromethane and dipyrromethene these are identical, the boron complex is different, in analogy with indacene. In all three cases, the central carbon is considered the *meso*-position, arising from porphyrin nomenclature. Also, the positions adjacent to the nitrogen atoms are called α-positions, while the others are β-positions.

These compounds are dyes that generally absorb light in the visible range, and are often highly fluorescent [1, 2], having no charge. Their complexes are very stable,

Mário J. F. Calvete

only decomposing in strong acidic and basic conditions [3], possessing also low toxicity [4].

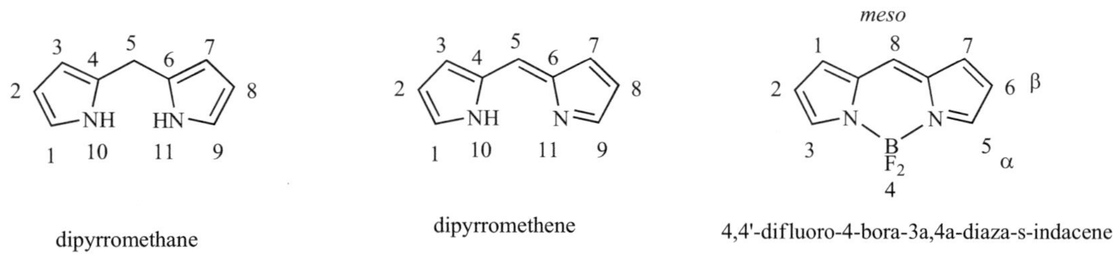

Fig. (1). Dipyrromethene system numbering.

SYNTHESIS OF BODIPY DERIVATIVES

In 1968, Treibs and Kreuzer [5] tried to perform the acylation of 2,4-dimethylpyrrole with acetic anhydride, using boron trifluoride as Lewis acid, instead they obtained a highly fluorescent compound (Scheme **1**). These findings however did not cause much interest at that time, and it was only in the end of the 70's-80's that this chemistry flourished [6 - 8]. When the biological labeling potential of these dyes was acknowledged [9, 10], various new BODIPY-type dyes were then designed and even commercialized [11].

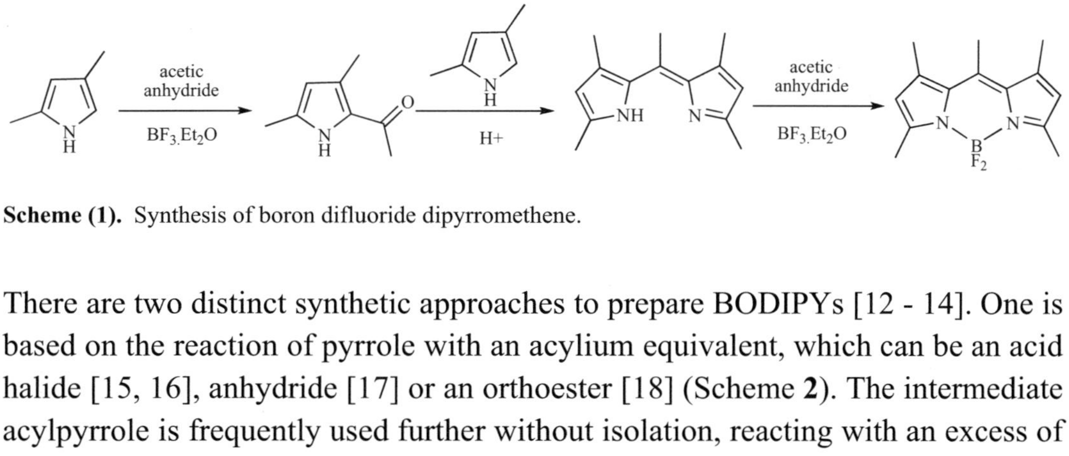

Scheme (1). Synthesis of boron difluoride dipyrromethene.

There are two distinct synthetic approaches to prepare BODIPYs [12 - 14]. One is based on the reaction of pyrrole with an acylium equivalent, which can be an acid halide [15, 16], anhydride [17] or an orthoester [18] (Scheme **2**). The intermediate acylpyrrole is frequently used further without isolation, reacting with an excess of pyrrole under acidic conditions to form the dipyrromethene. This approach allows the synthesis of asymmetric dipyrromethenes and, as an isolated acylpyrrole can be conjugated with a structurally different pyrrole unit under acidic condensation, which leads to the formation of the corresponding BODIPY dye, using an excess of base and boron trifluoride etherate.

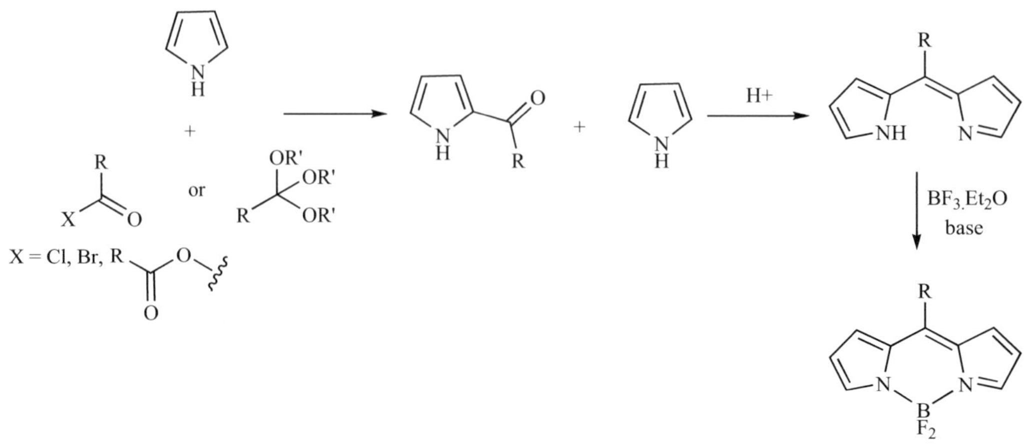

Scheme (2). Synthetic approaches to prepare BODIPYs.

Burgess [19] discovered, in 2008, that the second pyrrole equivalent is sometimes unnecessary for the dipyrromethene formation. He postulated a mechanism for the reaction, in which phosphorus oxychloride could substitute the aldehyde's oxygen, yielding a chlorinated azafulvene, which could be further reacted with the second pyrrolic aldehyde. Dipyrromethene could be obtained after chloride anion nucleophilic attack. This mechanism is only plausible in the case of 2-substituted pyrrole aldehydes, but the yields are exceptionally high, due to the fact the reaction is a one pot procedure and requires less purification procedures.

The other route, mostly followed by researchers, lies in the synthesis of the starting dipyrromethene unit by pyrrole condensation reaction, originated from the well-known porphyrin chemistry. In this case a carbonyl compound possessing high electrophilicity (aldehyde, acyl chloride or anhydride) leads to the formation of a methine bridge between two commonly substituted pyrrole units at 2- (or 5-) positions, to prevent polymerization. If a naked dipyrromethene is intended, an excess of pyrrole is needed to obtain satisfactory yields [20, 21]. Complexation of $BF_3 \cdot Et_2O$ using a base, *e.g.* amine [15, 22], originates the symmetric F-BODIPY dyes (Scheme **3**). 8-Substituted BODIPY's (substituted at the *meso*-position) are relatively easy to synthesize *via* this type of condensation of pyrroles with acyl chlorides [15, 16].

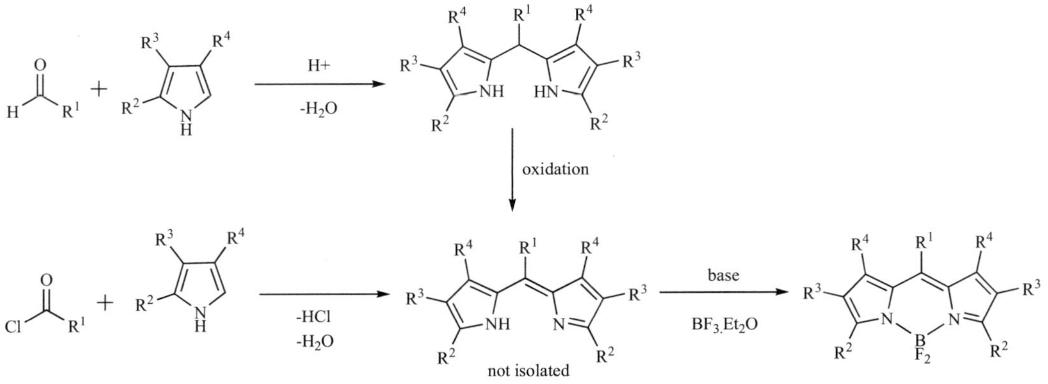

Scheme (3). Synthesis of symmetric BODIPYs.

Also acid anhydrides can be used to prepare BODIPYs instead of acid chlorides. Scheme **4** shows how a BODIPY derivative can be prepared from glutaric anhydride [17], by heating a substituted pyrrole with the anhydride at reflux in presence of $BF_3 \cdot Et_2O$, in presence of base.

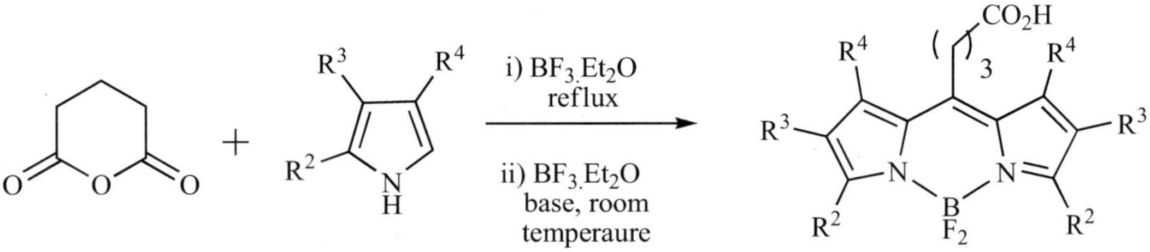

Scheme (4). Synthesis of carboxylic acid substituted BODIPYs.

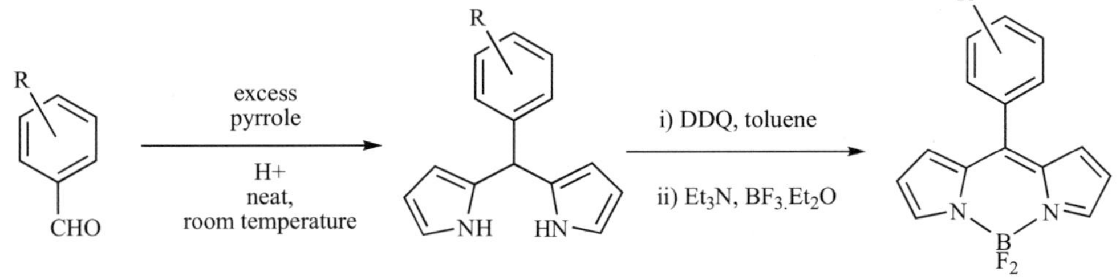

Scheme (5). Synthesis of aryl-substituted BODIPYs.

Preparations using aromatic aldehydes require oxidation steps, as depicted in Scheme (**5**). α,β-Unsubstituted BODIPYs can also be prepared by dissolving the aldehyde in pyrrole excess, further obtaining a dipyrromethane intermediary. The BODIPY dye is then obtained after oxidation with 2,3-dichloro-5,6-dicyano-1,4-benzoquinone (DDQ) and complexation with boron [23].

Extending π-electronic conjugation is an important tool for diverse applications. The main method, extending BODIPY conjugation consists in preparing vinyl, phenyl, or thiophene substituted at the 3-position of pyrrole, followed by condensation in the usual manner [24 - 26]. Other strategies include retro Diels-Alder reactions [27 - 29] or through reaction of substituted 2-acylacetophenones with ammonia to give the dibenzopyrromethene after reaction with BF_3•Et_2O [30 - 32]. Conjugation can also be enhanced by substitution of methinic bridge by a nitrogen atom, giving origin to azadipyrromethene chromophores, firstly reported in the 1940's [33 - 35], which further reacting with the boron source can give rise to aza-BODIPYs [36 - 41], whose π-electronic conjugation can be extended using similar procedures as for BODIPY's [42]. The general method to prepare these compounds deals with the conversion of 2,4-diarylpyrroles into their corresponding 5-nitroso derivatives, followed by reaction with another pyrrole unit (Scheme **6**).

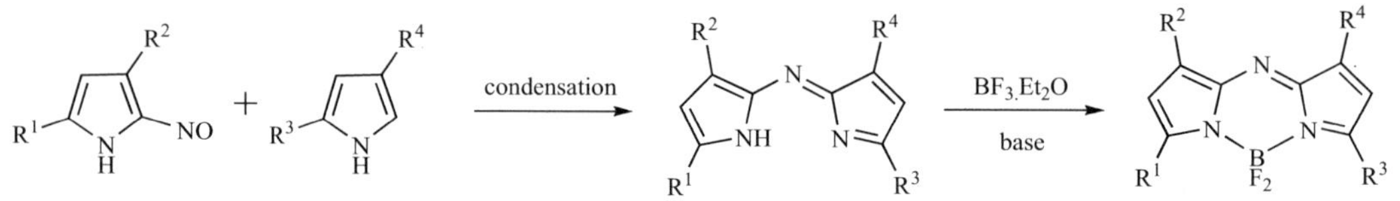

Scheme (6). Synthesis of aza-BODIPY dyes.

Functionalization of BODIPYs

A diversity of groups can be introduced at the BODIPY core, in order to obtain appropriately functionalized compounds. For instance, the presence of good leaving groups at the adjacent positions of the N atoms, the 3- and/or 5-positions, allows the straightforward introduction of groups through nucleophilic substitution reactions (Scheme **7**) [43 - 46].

Other strategies include substitution at those positions through couplings like Knoevenagel reactions, since 3- and 5-methyl bearing BODIPYs are acidic enough at those positions to be part in these reactions. Hence, styryl-BODIPYs can be prepared by reaction of aromatic aldehydes with 3,5-dimethyl- BODIPYs (Scheme **8**) [47 - 50].

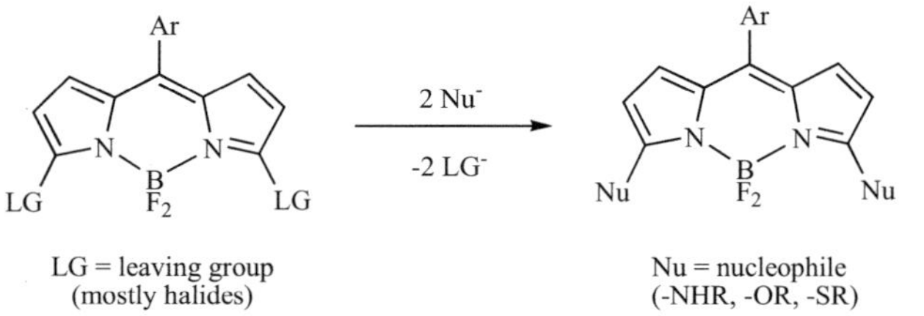

Scheme (7). Nucleophilic substitution reactions on BODIPYs.

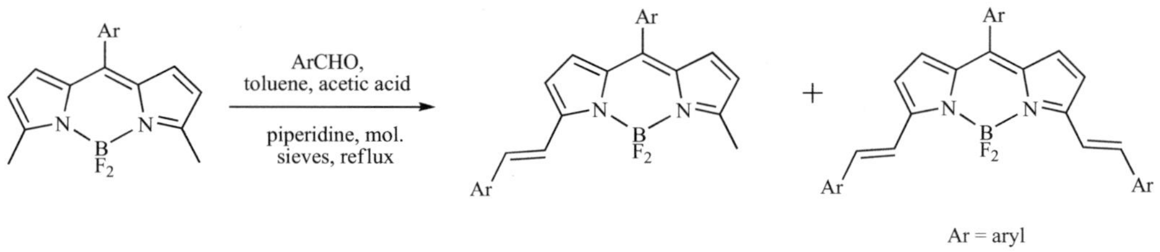

Scheme (8). Synthesis of styryl-BODIPY derivatives.

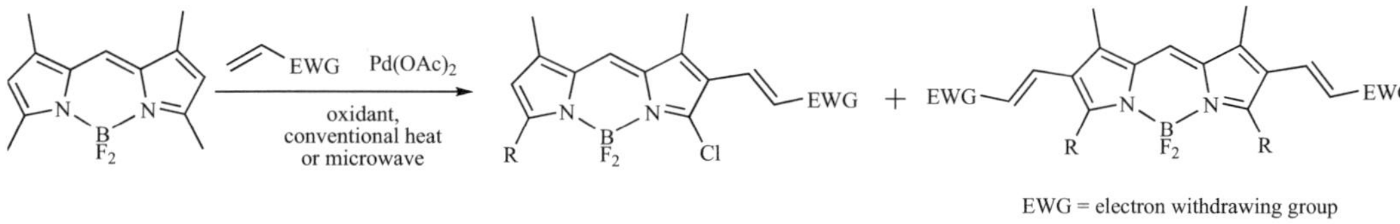

Scheme (9). Palladium catalyzed cross-coupling reactions of BODIPY dyes.

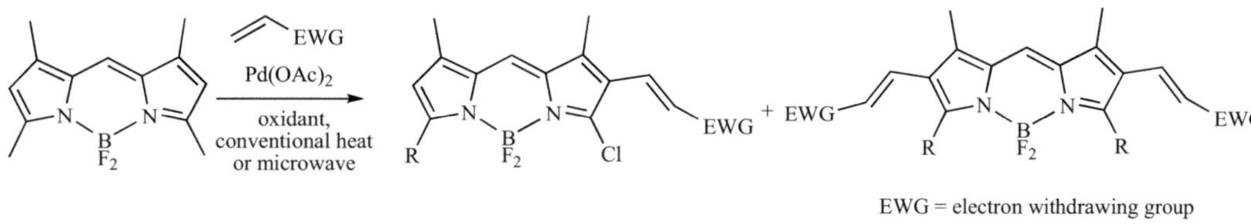

Scheme (10). Palladium mediated C-H functionalization of BODIPY dyes.

Palladium catalyzed is also a possibility to introduce additional functionalization the BODIPY core. So, new ethynylaryl, ethenylaryl and 3-, 5-aryl compounds can be obtained *via* Heck, Sonogashira, Suzuki and Stille couplings (Scheme **9**) [51] or through palladium mediated C-H functionalization (Scheme **10**) [52].

Several groups can also be introduced at the BODIPY dye core, through electrophilic substitutions. For instance, sulfonation [8, 15, 53 - 57] and nitration [58, 59] reactions have been reported, but the most researched electrophilic reactions deal with halogenation reactions [44, 45, 60 - 62] (Scheme **11**).

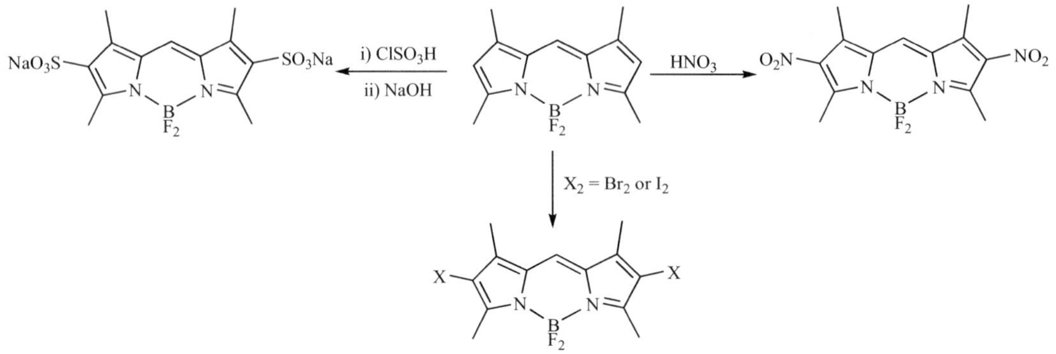

Scheme (11). Electrophilic substitutions on BODIPYs.

APPLICATIONS OF BODIPY DYES

BODIPY derivatives acquired noteworthy interest over the past decades as fluorescent dyes due to their photochemical stability and intense emission profiles. BODIPY dyes typically have sharp absorption bands with emission peaks that can be tuned from the green to the near IR [63, 64]. These dyes usually have high fluorescence quantum yields, high molar absorptivities, and relatively long excited-state lifetimes [65]. These properties of BODIPY dyes rendered them fame in a large range of applications, such as suitability for biological experiments [66, 67], including biolabeling [3, 23, 68 - 80], cell imaging [81 - 93]. Their properties have also been used as sensors [45, 50, 94 - 106], in nonlinear optics [107 - 115] and in dye sensitized solar cells [116 - 135].

Applications in Biolabeling

Labeling is a very common methodology, used for bioanalytical functions. The near-IR biological tissue absorbing window appears in the 600–900 nm range, where absorption and scattering of light, along with cellular autofluorescence

(300–550 nm) are strongly diminished [136]. Therefore, red- emitting and absorbing fluorophores present the best resolution and contrast in the detection of various bioanalytes and biological labeling tests [137 - 139]. Several requirements need to be met, including high absorption coefficients, high photostability, fairly long lifetimes (1-5 ns), presence of additional functional groups for bioconjugation, high fluorescence quantum yields under physiological conditions and good water-solubility [68, 140]. The synthesis of water-soluble bioconjugatable red-emitting BODIPYs remains less studied, regardless of their photophysical properties [66, 67]. There are, however, several already commercially available BODIPY dyes that can be linked to proteins, peptides, dextran, oligonucleotides, polystyrene microspheres and lipids [3, 74, 141 - 149].

Applications in Cell Imaging

Molecular tools to clarify the complex relationship between signaling molecules, biologically important ions inside the cells and macromolecules, play a central role in cell and molecular biology. Fluorescence spectroscopy is an extremely powerful tool in these fields and is used to visualize biological processes and biological relevant species both in tissues and living animals with high spatial and temporal resolution. Fluorescent probes, such as BODIPY type ones, with emission and absorption in the NIR region are favorable for *in vivo* bioimaging, because of minimum interference from background auto-fluorescence by biomolecules in the living systems, minimum photo-damage to biological samples and deep tissue penetration.

In 2001, NIR BODIPY dye **1**, bearing an aminophenyl-based pH receptor in the *meso*-position was reported (Fig. **2**) [47]. The fluorescence of probe **1** was considerably quenched in polar solvents, as the electron donating aminophenyl group caused an excited state intramolecular charge transfer process. Still, probe **1** protonation induced a 2000-fold fluorescence improvement due to excited state charge transfer suppression.

Later, two pH receptors linked at the 3 and 5-positions to the BODIPY molecule were reported (**2** and **3**) (Fig. **2**) [150, 151]. Upon protonation of the probes, it was observed a blue shift in emission and absorption spectra, since the intramolecular charge transfer was switched off by protonation.

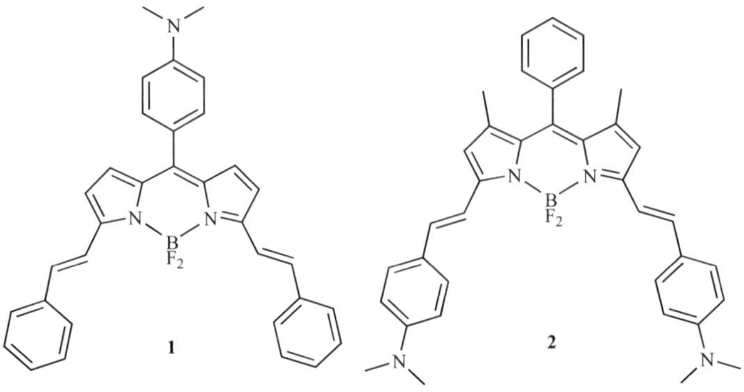

Fig. (2). Styryl-substituted BODIPYs for pH sensing.

In 2012, Nagano designed and reported NIR fluorescent of probe **4** for tumor, atherosclerosis, and inflammation markers denoted as matrix metalloproteinases, when compared to previously designed probes (Fig. **3**) [152]. Compound **4** was intracellularly well retained and rapidly internalized by free diffusion, after activation by extracellular matrix metalloproteinases. Under physiological conditions, due to fluorescence quenching, the fluorescence intensity of **4** was quite shallow; nevertheless, the fluorescence was regained after addition of metalloproteinase. The probe was then used to detect metalloproteinase activity in HT-1080 cells, as well as in the detection of metalloproteinase activity in xenograft tumor models from mice.

Also in 2012, Hill developed a BODIPY fluorophore as fluorescent antagonist, for which BODIPY-based compound **5** Fig. (**3**) demonstrated a slow off-rate and high affinity [153]. In 2013, other authors [154] reported fluorescent probes **6** and **7** (Fig. **3**) for protein antagonism studies, by conjugation of a far-red BODIPY fluorophores with the dipeptide, showing human CHO cell membrane binding.

Hill and Kellam reported far-red emitting BODIPY **8** (Fig. **3**) [155], a new ligand for human β-adrenoceptors (β-AR). The fluorescent β-blocker showed high affinity for the β2-AR, being also used for single living cell fluorescent labeling of the human β2-AR. Also Ono reported the red BODIPY probe **9**, Fig. (**4**), for β-amyloid (Aβ) brain plaques imaging [156]. The probe displayed absorption around 604 nm, along with high fluorescence at 648 nm.

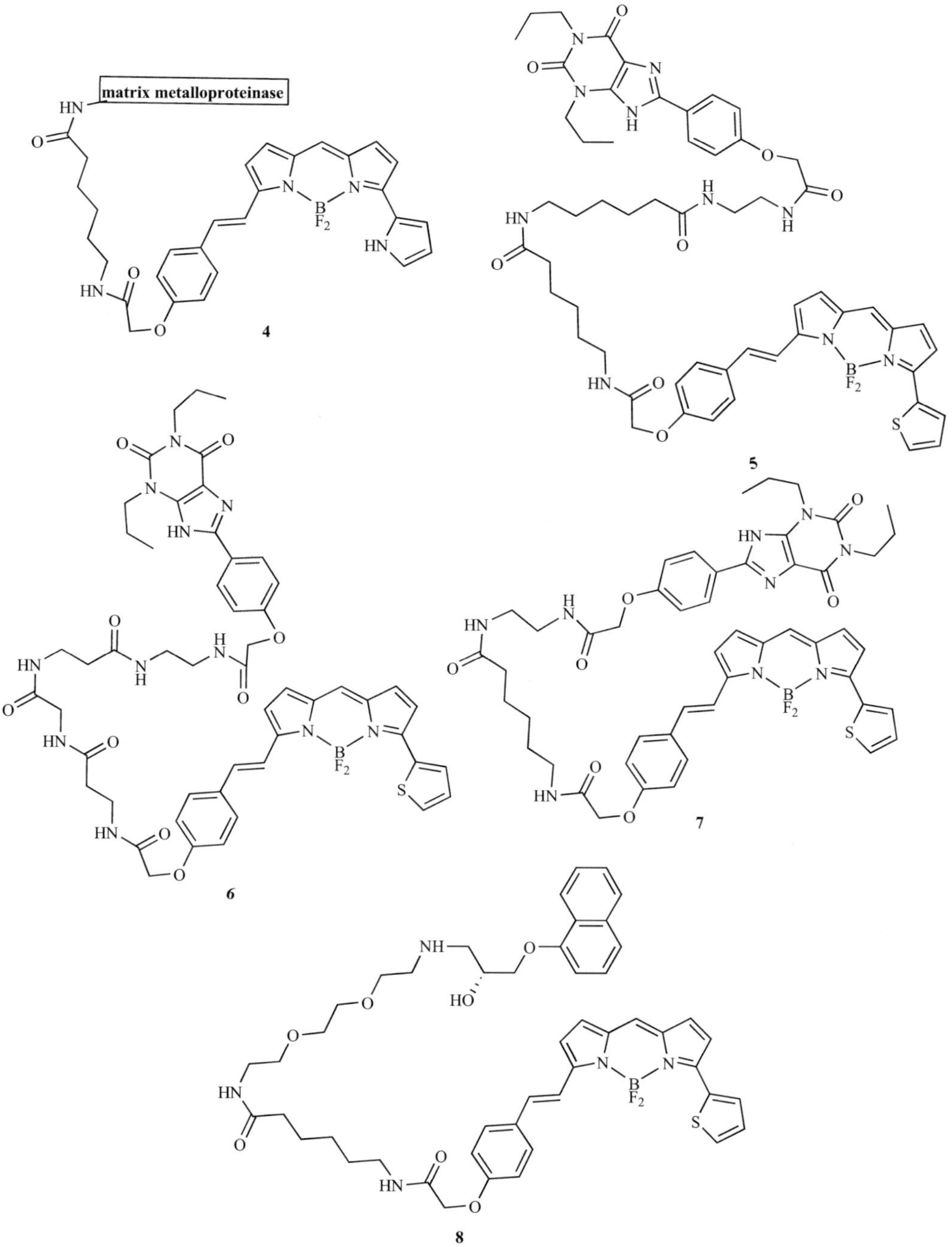

Fig. (3). NIR BODIPY based fluorescent probes.

Keyes [93] compared two polyethylene glycol substituted water-soluble BODIPY compounds, the visible emitting derivative **10** and the NIR emitting fluorophore **11** (Fig. **4**), having a 165 nm Stokes shift, as contrast reagents for murine brain tissue and live mammalian cells. The compounds exhibited good cell uptake and confocal fluorescence microscopy revealed that all dyes were localizing predominantly within the lipophilic organelles.

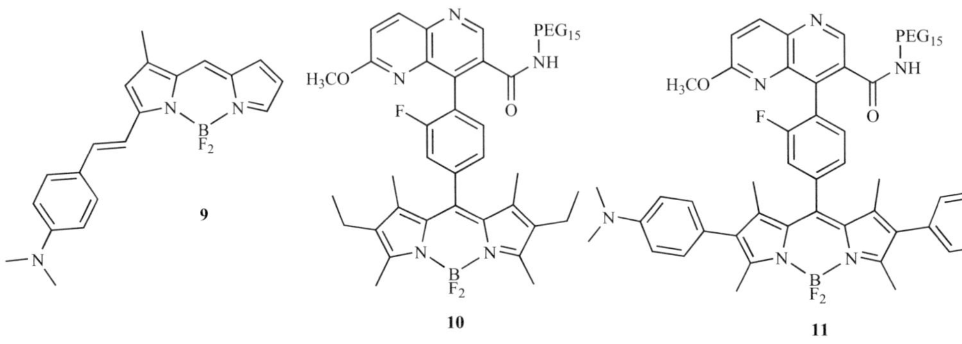

Fig. (4). BODIPY derivatives for cellular imaging.

Applications as Sensors

The sensing and recognition of environmentally and biologically important ions has appeared as a noteworthy goal in the chemical sensors field, where fluorescence has turned to be a powerful analytical detection method, due to its high detection limit and straightforwardness. Particularly, suitable probes in conjunction with fluorogenic methods are favorable to measure analytes, given that fluorimetry is highly sensitive, nondestructive, quickly performed, and can proportionate real information on the quantification and localization of the targets of interest.

A new BODIPY derivative **12** (Fig. **5**) was synthesized as ''off-on'' fluorescent chemodosimeter and fluorescent chemosensor for Cu^{2+} [157], displaying increased fluorescence upon Cu^{2+} chelation, reaching its maximum when 5 equivalents of Cu^{2+} were added, along with a 9 nm red-shift. The same authors reported also set of fluorophore-triazine tripodal based fluorescent systems **13-15** (Fig. **5**) [158].

Their binding abilities were evaluated using Ag^+, Cs^+, K^+, Na^+, Ca^{2+}, Cd^{2+}, Cu^{2+}, Co^{2+}, Mg^{2+}, Pb^{2+}, Hg^{2+}, Ni^{2+}, Mn^{2+} and Zn^{2+} ions (2 equivalents).

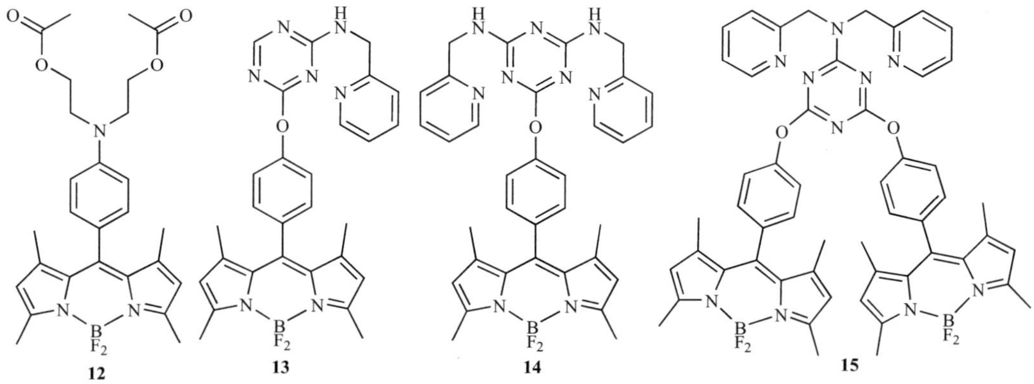

Fig. (5). BODIPY derivatives for sensing applications.

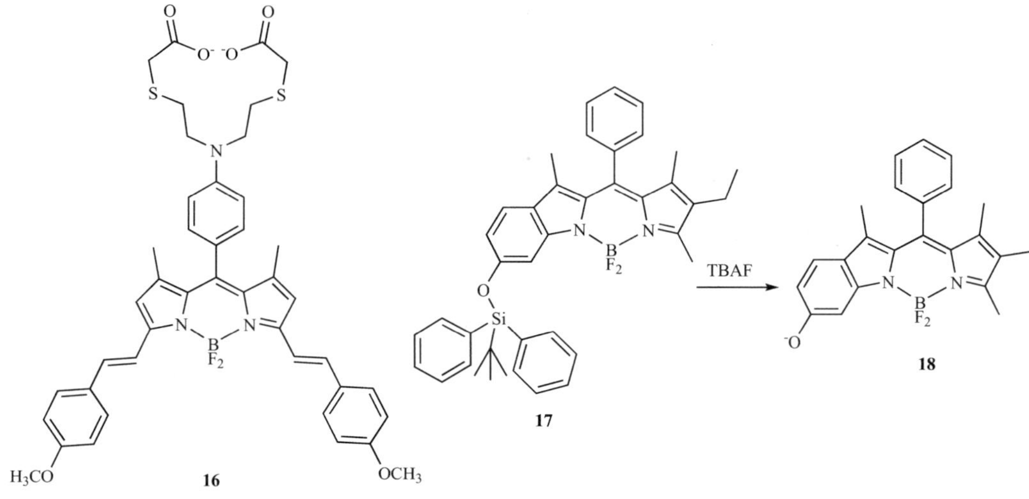

Fig. (6). BODIPY dyes for sensing applications.

In 2012, Guo synthesized probe **16** (Fig. **6**), designed for Hg^{2+} dosing [159]. Probe **16** displayed quite weak fluorescence at 655 nm but, upon Hg^{2+} addition, a considerable fluorescence turn-on signal was observed, caused by blocking of photoelectronic transfer. In 2012, Zhao developed 6-hydroxyindole based BODIPY dye **17** (Fig. **6**), for F^- detection, [160]. Probe **17** displayed strong fluorescence at 573 nm while, upon F^- addition, a considerable emission amplification at 676 nm was noted, ascribed to an F^- transformation of probe **17** into dye **18**.

Application in Nonlinear Optics

Nonlinear refraction and absorption are nonlinear optical features directly related. Generally, several mechanisms, both non-parametric and parametric, contribute to the nonlinearities and a consistent evaluation of the responsible mechanisms requires experiments involving different pulse lengths. Optical limiting consists of a sample transmittance decrease under fluence illumination or high intensity. Optical limiting is also regarded as the reverse of saturable absorption where increase in transmittance is seen at high illumination. Phthalocyanines are promising metallo-organic materials for optical power limiting in the visible and NIR spectral range, due to their suitable photophysical properties.

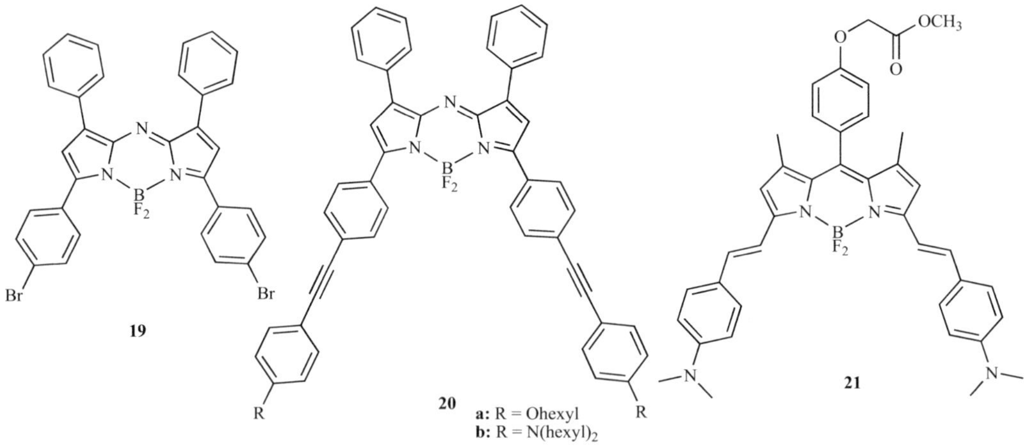

Fig. (7). Aza-BODIPY and BODIPY derivatives for NLO.

NIR absorbing aza-bodipy (aza-borondipyrromethene) dyes **19-20** (Fig. **7**) have been tested for its optical limiting properties [161, 162]. Chromophore **19**, possessing extended aromatic conjugation displayed quite interesting OL properties. Attempting further the absorption into the NIR and increase the OL properties, electron donating groups were then introduced at the periphery of the dye **20a-b** (Fig. **7**) [161].

These molecules showed a very intense NIR linear absorption around 745nm in dichloromethane. At telecommunication wavelengths (above 1200 nm), the compound **19** did not present any two-photon absorption response, in opposition to the electron-donor substituted compounds **20a-b**. While molecule **19** and **20a** showed a small nonlinear optic activity, chromophore **20b** presented good nonlinear optic properties between 1200 and 1450nm [162].

BODIPY chromophore **21** (Fig. **7**) was synthesized and its two-photon absorption properties were evaluated by direct nonlinear transmission (NLT) method in femto-second regime. This dye displayed a fairly large two-photon absorption cross-section at around 1310 nm and broad two-photon absorption spectral distribution in the near-IR range. Both good optical stabilization and optical limiting performances were achieved at the wavelength of 1310 nm [111].

Application in Dye Sensitized Solar Cells

There has been significant attention to fluorescent organic dyes in several electronic applications, such as photoinduced electron transfer materials, light-harvesting antennas and applications as solar cells. The foreseen collapse of energy derived from fossil resources and environmental restraints are motivating a research increase on renewable energies, namely solar energy.

Based on these premises, BODIPY dyes have emerged as attractive starting platforms for the advance in the design of active materials for (opto)electronic devices and solar cells.

For instance, the panchromatic trichromophoric system **22** (Fig. **8**) using antenna effect was also reported by Odobel for incorporation as a sensitizer in dye sensitized solar cells (DSSC) [163]. Compound **22** was prepared by Heck alkynylation and presented effective intramolecular electron transfer through all units. The antenna effect was evaluated by the photoaction spectrum, which featured all absorption bands of the chromophores.

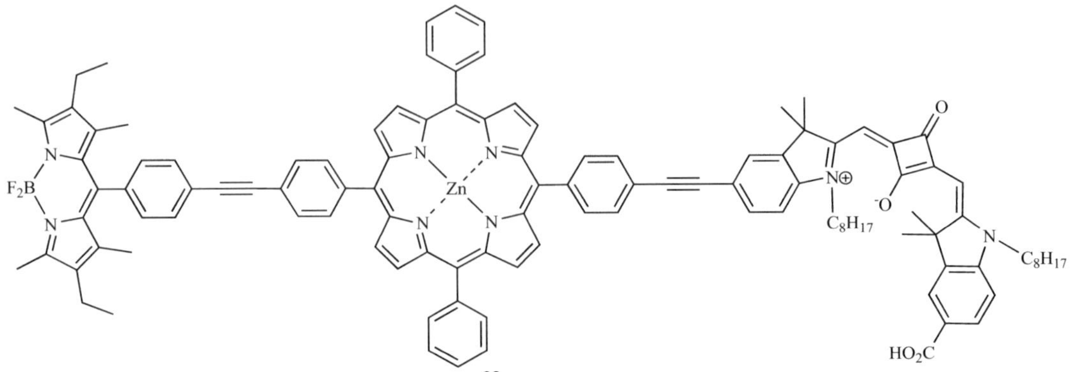

Fig. (8). BODIPY based triad for solar cell applications.

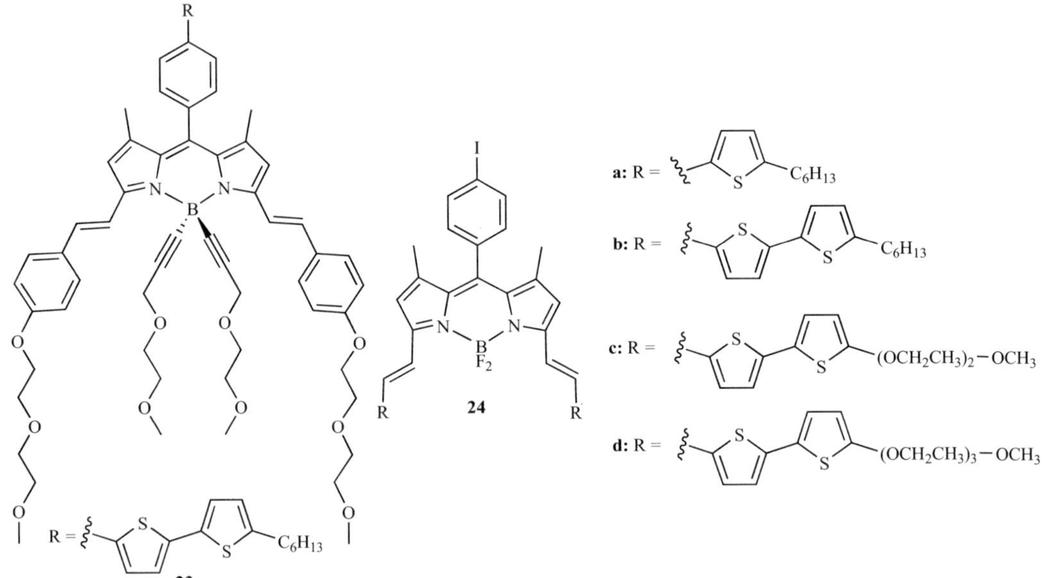

Fig. (9). Oligothiophene substituted BODIPY dyes for photovoltaics.

In 2010, Ziessel published the hybrid BODIPY-oligothiophene donor **23** (Fig. **9**) for bulky heterojunction solar cells with enhanced performance [164]. The authors associated the presence of a 5-hexyl-2,20-bithyenyl (BT) unit on the BODIPY with the increment of the power-conversion efficiency of the resulting solar cell from 1.3 to 2.2%, mostly caused by the large current density increase [125].

Later, Ziessel reached a 4.7% power-conversion efficiency in a bulky heterojunction solar cell, using bisvinyl-thienyl dyes **24a-d** (Fig. **9**) [165]. Akkaya [132] also report a series of panchromatic dyes **25a-d** and **26a-b** Fig. (**10**) obtaining an overall efficiency of 2.46% in liquid electrolyte solar cells, at the time the highest reported efficiency using BODIPYs as photosensitizers.

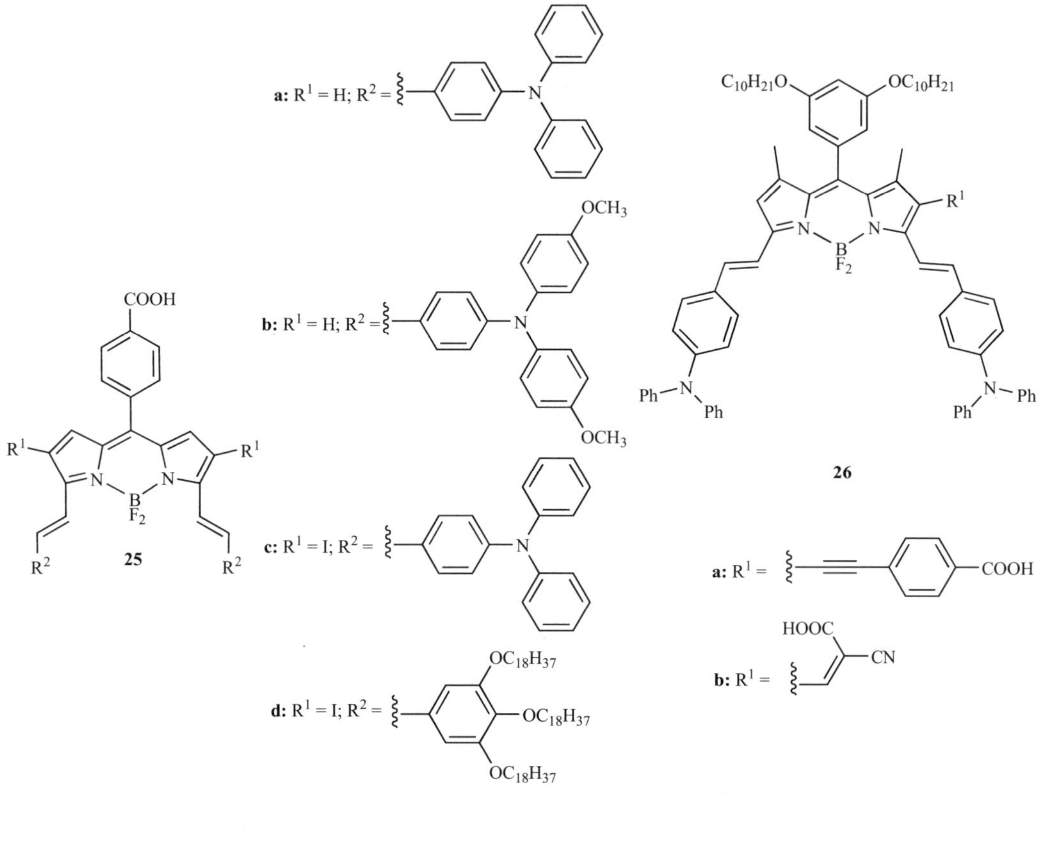

Fig. (10). Panchromatic BODIPY derivatives for solar cells.

EXPERIMENTAL SECTION

Synthesis of a BODIPY-triazine-based Tripod Fluorescent System (Scheme 12)

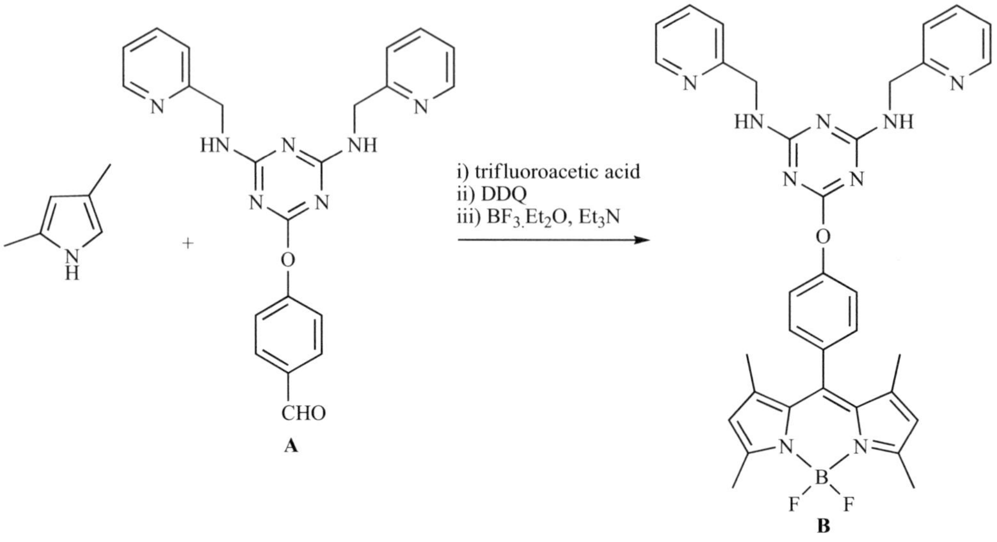

Scheme (12). Synthesis of a BODIPY-triazine-based tripod. Procedure for total synthesis of BODIPY B can be directly found in Reference [158].

Synthesis of an Extended BODIPY Sensor (Scheme 13)

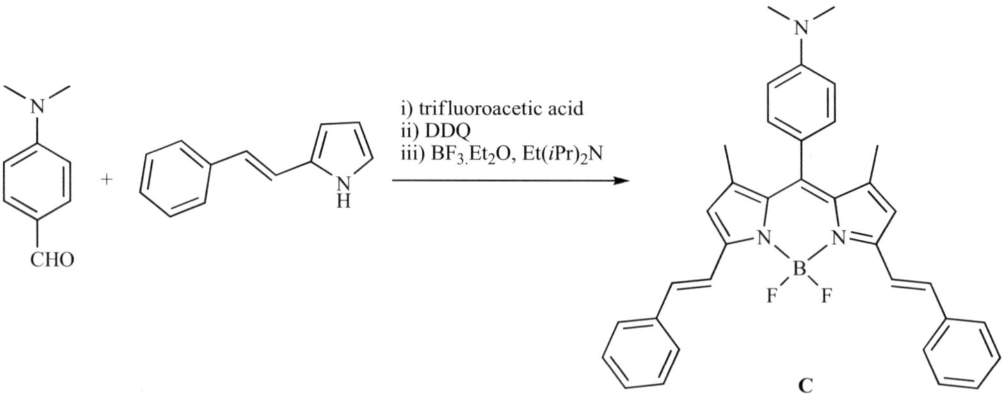

Scheme (13). Synthesis of styryl-substituted BODIPY. Procedure for total synthesis of BODIPY B can be directly found in Reference [47].

Synthesis of Extended BODIPY Dyes with Restricted Bond Rotation (Schemes 14 and 15)

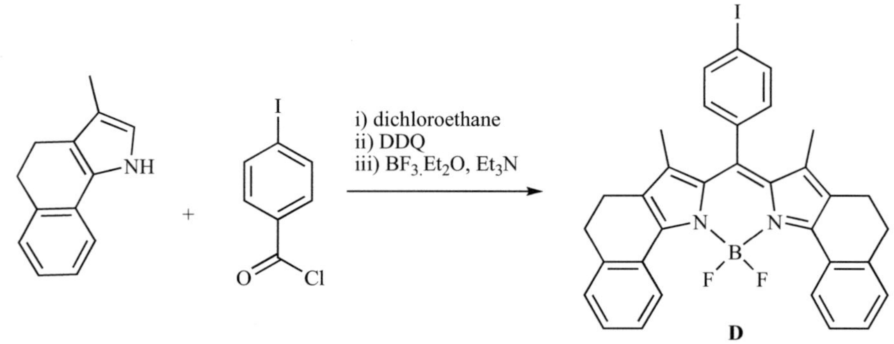

D

Scheme (14). Synthesis of rigid BODIPY derivative. Procedure for total synthesis of BODIPY D can be directly found in Reference [26].

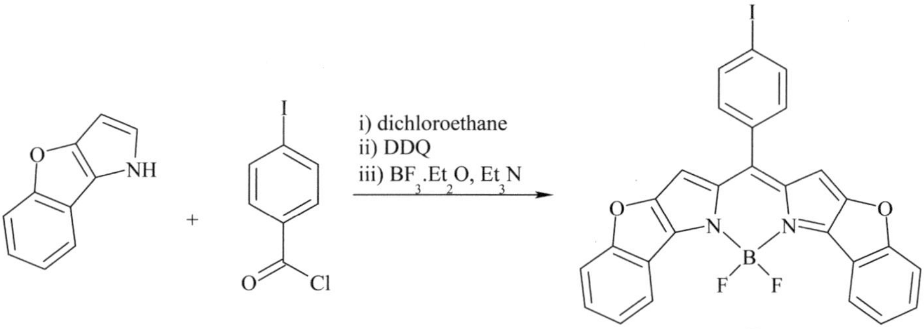

E

Scheme (15). Synthesis of rigid BODIPY dye. Procedure for total synthesis of BODIPY E can be directly found in Reference [6].

Synthesis of BF2-azadipyrromethenes (Scheme 16)

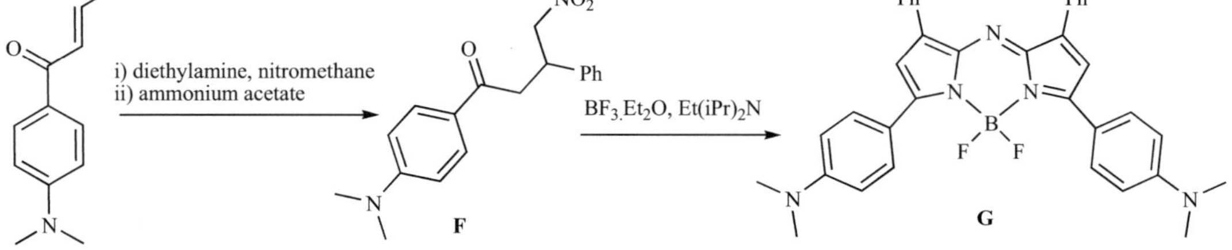

Scheme (16). Synthesis of BF2-azadipyrromethene. Procedure for total synthesis of BODIPY G can be directly found in Reference [40].

CONCLUDING REMARKS

The difluoro-boraindacene (BODIPY) compound family has achieved credit as being one of the most resourceful fluorophores over the past two decades.

The BODIPY nucleus offers superior photochemical and chemical stability, two advantageous qualities for a dye. Additionally, tailored BODIPY dyes bearing anchoring possibilities to a variety of engineered functionalities may offer this class of dyes an unmatchable prospect in many applications, from materials science to biomedical applications.

The unrestrained field of BODIPY dyes can be greatly improved by the intensification of new architectures, naturally matched by the potential for new applications, at the edge between chemistry, physics, biology, engineering and supramolecular science. Thoroughly enhanced properties can be achieved by straightforward synthetic approaches. Such stimulating features are, for sure, a source of encouragement for chemists to design, synthesize and develop new architectures.

REFERENCES

[1] Loudet, A.; Burgess, K. BODIPY dyes and their derivatives: syntheses and spectroscopic properties. *Chem. Rev.,* **2007**, *107*(11), 4891-4932.
[http://dx.doi.org/10.1021/cr078381n] [PMID: 17924696]

[2] Lu, H.; Mack, J.; Yang, Y.; Shen, Z. Structural modification strategies for the rational design of red/NIR region BODIPYs. *Chem. Soc. Rev.,* **2014**, *43*(13), 4778-4823.
[http://dx.doi.org/10.1039/c4cs00030g] [PMID: 24733589]

[3] Haugland, R. *Handbook of Fluorescent Probes and Research Chemicals,* 10th ed; Molecular Probes: Eugene, Oregon, **2005**.

[4] Alford, R.; Simpson, H.M.; Duberman, J.; Hill, G.C.; Ogawa, M.; Regino, C.; Kobayashi, H.; Choyke, P.L. Toxicity of organic fluorophores used in molecular imaging: literature review. *Mol. Imaging,* **2009**, *8*(6), 341-354.
[PMID: 20003892]

[5] Treibs, A.; Kreuzer, F. Difluorboryl-Komplexe von Di- und Tripyrrylmethenen. *Justus Liebigs Ann. Chem.,* **1968**, *718*, 208-223.
[http://dx.doi.org/10.1002/jlac.19687180119]

[6] Falk, H.; Hofer, O.; Lehner, H. Beiträge zur Chemie der Pyrrolpigmente, 1. Mitt.: Der induzierte Circulardichroismus einiger Pyrromethenderivate in cholesterischer Mesophase. *Monatsh. Chem.,* **1974**, *105*, 169-178.

[http://dx.doi.org/10.1007/BF00911302]

[7] E.; Pardoen, J.A.; Vankoeveringe, J.A.; Lugtenburg, J. Pyrromethene-BF2 complexes (4,4′-difluoro-4-bora-3a,4a-diaza-s-indacenes). Synthesis and luminescence properties. *Recl. Trav. Chim. Pays Bas,* **1977**, *96*, 306-309.

[8] Wories, H.J.; Koek, J.H.; Lodder, G.; Lugtenburg, J.; Fokkens, R.; Driessen, O.; Mohn, G.R. A novel water-soluble fluorescent probe: Synthesis, luminescence and biological properties of the sodium salt of the 4-sulfonato-3,3′,5,5′-tetramethyl-2,2′-pyrromethen-1,1′-BF2 complex. *Recl. Trav. Chim. Pays Bas,* **1985**, *104*, 288-291.
[http://dx.doi.org/10.1002/recl.19851041104]

[9] Haughland, R.P.; Kang, H.C. **1988**.Chemically reactive dipyrrometheneboron difluoride dyes *US Patent ,*

[10] Monsma, F.J., Jr; Barton, A.C.; Kang, H.C.; Brassard, D.L.; Haugland, R.P.; Sibley, D.R. Characterization of novel fluorescent ligands with high affinity for D1 and D2 dopaminergic receptors. *J. Neurochem.,* **1989**, *52*(5), 1641-1644.
[http://dx.doi.org/10.1111/j.1471-4159.1989.tb09220.x] [PMID: 2523470]

[11] Registered trademark of Molecular Probes *http://probes.invitrogen.com,*

[12] Wood, T.E.; Thompson, A. Advances in the chemistry of dipyrrins and their complexes. *Chem. Rev.,* **2007**, *107*(5), 1831-1861.
[http://dx.doi.org/10.1021/cr050052c] [PMID: 17430001]

[13] Pereira, N.A.; Melo, T.M. Recent developments in the synthesis of dipyrromethanes. A review. *Org. Prep. Proced. Int.,* **2014**, *46*, 183-213.
[http://dx.doi.org/10.1080/00304948.2014.903140]

[14] Ziessel, R.; Ulrich, G.; Harriman, A. The chemistry of Bodipy: a new El Dorado for fluorescence tools. *New J. Chem.,* **2007**, *31*, 496-501.
[http://dx.doi.org/10.1039/b617972j]

[15] Shah, M.; Thangaraj, K.; Soong, M.; Wolford, M.; Boyer, J.; Politzer, I.; Pavlopoulos, T. Pyrromethene–BF2 complexes as laser dyes:1. *Heteroatom Chem.,* **1990**, *1*, 389-399.
[http://dx.doi.org/10.1002/hc.520010507]

[16] Boyer, J.; Haag, A.; Sathyamoorthi, G.; Soong, M.; Thangaraj, K.; Pavlopoulos, T. Pyrromethene–BF2 complexes as laser dyes: 2. *Heteroatom Chem.,* **1993**, *4*, 39-49.
[http://dx.doi.org/10.1002/hc.520040107]

[17] Li, Z.; Mintzer, E.; Bittman, R. First synthesis of free cholesterol-BODIPY conjugates. *J. Org. Chem.,* **2006**, *71*(4), 1718-1721.
[http://dx.doi.org/10.1021/jo052029x] [PMID: 16468832]

[18] Yakubovskyi, V.; Shandura, M.; Mykola, P.; Kovtun, Y. Boradipyrromethenecyanines. *Eur. J. Org. Chem.,* **2009**, *19*, 3237-3243.
[http://dx.doi.org/10.1002/ejoc.200900192]

[19] Wu, L.; Burgess, K. A new synthesis of symmetric boraindacene (BODIPY) dyes. *Chem. Commun. (Camb.),* **2008**, *40*(40), 4933-4935.
[http://dx.doi.org/10.1039/b810503k] [PMID: 18931743]

[20] Littler, B.J.; Miller, M.A.; Hung, C-H.; Wagner, R.W.; O'Shea, D.F.; Boyle, P.D.; Lindsey, J.S. Refined synthesis of 5-substituted dipyrromethanes. *J. Org. Chem.,* **1999**, *64*, 1391-1396. [http://dx.doi.org/10.1021/jo982015+]

[21] Baruah, M.; Qin, W.; Basarić, N.; De Borggraeve, W.M.; Boens, N. BODIPY-based hydroxyaryl derivatives as fluorescent pH probes. *J. Org. Chem.,* **2005**, *70*(10), 4152-4157. [http://dx.doi.org/10.1021/jo0503714] [PMID: 15876108]

[22] Burghart, A. Kim, H. Welch, M.B. Thorensen, L.H. Reibenspies, J. Burgess, K. 3,5-Diaryl-4-4-difluoro-4-bora-3a,4a-diaza-s-indacene (BODIPY) dyes: synthesis, spectroscopic, electrochemical, and structural properties. *J. Org. Chem.,* **1999**, *64*, 7813-7819. [http://dx.doi.org/10.1021/jo990796o]

[23] Wagner, R.W.; Lindsey, J.S. Boron-dipyrromethene dyes for incorporation in synthetic multi-pigment light-harvesting arrays. *Pure Appl. Chem.,* **1996**, *68*, 1373-1380. [http://dx.doi.org/10.1351/pac199668071373]

[24] Kang, H.C.; Haughland, R.P. Long wavelength chemically reactive dipyrrometheneboron difluoride dyes and conjugates *US Patent 5451663,* **1993**.

[25] Thoresen, L.H.; Kim, H.; Welch, M.B.; Burghart, A.; Burgess, K. Synthesis of 3,5-diaryl-4,4-difluro-4-bora-3a,4a-diaza-s-indacene (BODIPY®) Dyes. *Synlett,* **1998**, 1276-1278. [http://dx.doi.org/10.1055/s-1998-1924]

[26] Chen, J.; Burghart, A.; Derecskei-Kovacs, A.; Burgess, K. 4,4-Difluoro-4-bora-3a,4a-diaza-s-indacene (BODIPY) dyes modified for extended conjugation and restricted bond rotations. *J. Org. Chem.,* **2000**, *65*(10), 2900-2906. [http://dx.doi.org/10.1021/jo991927o] [PMID: 10814176]

[27] Wada, M.; Ito, S.; Uno, H.; Murashima, T.; Ono, N.; Urano, T.; Urano, Y. Synthesis and optical properties of a new class of pyrromethene–BF2 complexes fused with rigid bicyclo rings and benzo derivatives. *Tetrahedron Lett.,* **2001**, *42*, 6711-6713. [http://dx.doi.org/10.1016/S0040-4039(01)01299-0]

[28] Shen, Z.; Röhr, H.; Rurack, K.; Uno, H.; Spieles, M.; Schulz, B.; Reck, G.; Ono, N. Boron-diindomethene (BDI) dyes and their tetrahydrobicyclo precursors--en route to a new class of highly emissive fluorophores for the red spectral range. *Chemistry,* **2004**, *10*(19), 4853-4871. [http://dx.doi.org/10.1002/chem.200400173] [PMID: 15372665]

[29] Ito, S.; Murashima, T.; Ono, N.; Uno, H. A new synthesis of benzoporphyrins using 4,7-dihydro-4-7-ethano-2H-isoindole as a synthon of isoindole. *Chem. Commun. (Camb.),* **1998**, 1661-1662. [http://dx.doi.org/10.1039/a803656j]

[30] Kang, H.C.; Haugland, R.P. Dibenzopyrrometheneboron difluoride dyes *U. S. Patent 5433896,* **1995**.

[31] Wu, Y.; Klaubert, D.H.; Kang, H.C.; Zhang, Y-Z. Long-wavelength dyes for infrared tracing and their use *U. S. Patent 6005113,* **1999**.

[32] Kotali, A.; Tsoungas, P.G. Oxidation of N-Aroylhydrazones of o-hydroxyaryl ketones withlead(IV)acetate: A facile route to aromatic o-diketones. *Tetrahedron Lett.,* **1987**, *28*, 4321-4322. [http://dx.doi.org/10.1016/S0040-4039(00)96497-9]

[33] Rogers, M.A. 156. 2: 4-Diarylpyrroles. Part I. Synthesis of 2: 4-diarylpyrroles and 2: 2′: 4: 4′-tetr--arylazadipyrromethines. *J. Chem. Soc.,* **1943**, 590-596. [http://dx.doi.org/10.1039/jr9430000590]

[34] Davies, W.H.; Rogers, M.A. 46. 2: 4-Diarylpyrroles. Part IV. The formation of acylated 5-amino-2: 4-diphenylpyrroles from β-benzoly-α-phenylpropionitrile and some notes on the Leuckart reaction. *J. Chem. Soc.,* **1944**, 126-131. [http://dx.doi.org/10.1039/jr9440000126]

[35] Knott, E.B. 225. β-Cycloylpropionitriles. Part II. Conversion into bis-2-(5-cyclylpyrrole)azamethin salts. *J. Chem. Soc.,* **1947**, 1196. [http://dx.doi.org/10.1039/jr9470001196]

[36] Sathyamoorthi, G.; Soong, M.L.; Ross, T.W.; Boyer, J.H. Fluorescent tricyclic β-azavinamidine–BF2 complexes. *Heteroatom Chem.,* **1993**, *4*, 603-608. [http://dx.doi.org/10.1002/hc.520040613]

[37] Killoran, J.; Allen, L.; Gallagher, J.F.; Gallagher, W.M.; O'Shea, D.F. Synthesis of BF2 chelates of tetraarylazadipyrromethenes and evidence for their photodynamic therapeutic behaviour. *Chem. Commun. (Camb.),* **2002**, (17), 1862-1863. [http://dx.doi.org/10.1039/b204317c] [PMID: 12271646]

[38] McDonnell, S.O.; Hall, M.J.; Allen, L.T.; Byrne, A.; Gallagher, W.M.; O'Shea, D.F. Supramolecular photonic therapeutic agents. *J. Am. Chem. Soc.,* **2005**, *127*(47), 16360-16361. [http://dx.doi.org/10.1021/ja0553497] [PMID: 16305199]

[39] Gallagher, W.M.; Allen, L.T.; O'Shea, C.; Kenna, T.; Hall, M.; Gorman, A.; Killoran, J.; O'Shea, D.F. A potent nonporphyrin class of photodynamic therapeutic agent: cellular localisation, cytotoxic potential and influence of hypoxia. *Br. J. Cancer,* **2005**, *92*(9), 1702-1710. [http://dx.doi.org/10.1038/sj.bjc.6602527] [PMID: 15841085]

[40] McDonnell, S.O.; O'Shea, D.F. Near-infrared sensing properties of dimethlyamino-substituted BF2-azadipyrromethenes. *Org. Lett.,* **2006**, *8*(16), 3493-3496. [http://dx.doi.org/10.1021/ol061171x] [PMID: 16869643]

[41] Gorman, A.; Killoran, J.; O'Shea, C.; Kenna, T.; Gallagher, W.M.; O'Shea, D.F. *In vitro* demonstration of the heavy-atom effect for photodynamic therapy. *J. Am. Chem. Soc.,* **2004**, *126*(34), 10619-10631. [http://dx.doi.org/10.1021/ja047649e] [PMID: 15327320]

[42] Zhao, W.; Carreira, E.M. Conformationally restricted aza-bodipy: a highly fluorescent, stable, near-infrared-absorbing dye. *Angew. Chem. Int. Ed. Engl.,* **2005**, *44*(11), 1677-1679. [http://dx.doi.org/10.1002/anie.200461868] [PMID: 15693049]

[43] Qin, W.; Rohand, T.; Baruah, M.; Stefan, A.; van der Auweraer, M.; Dehaen, W.; Boens, N. Solvent-dependent photophysical properties of borondipyrromethene dyes in solution. *Chem. Phys. Lett.,* **2006**, *420*, 562-568. [http://dx.doi.org/10.1016/j.cplett.2005.12.098]

[44] Rohand, T.; Baruah, M.; Qin, W.; Boens, N.; Dehaen, W. Functionalisation of fluorescent BODIPY dyes by nucleophilic substitution. *Chem. Commun. (Camb.),* **2006**, (3), 266-268.

[http://dx.doi.org/10.1039/B512756D] [PMID: 16391729]

[45] Baruah, M.; Qin, W.; Vallée, R.A.; Beljonne, D.; Rohand, T.; Dehaen, W.; Boens, N. A highly potassium-selective ratiometric fluorescent indicator based on BODIPY azacrown ether excitable with visible light. *Org. Lett.,* **2005**, *7*(20), 4377-4380. [http://dx.doi.org/10.1021/ol051603o] [PMID: 16178537]

[46] Goud, T.V.; Tutar, A.; Biellmann, J-F. Synthesis of 8-heteroatom-substituted 4,4-difluoro-4-b-ra-3a,4a-diaza-s-indacene dyes (BODIPY). *Tetrahedron,* **2006**, *62*, 5084-5091. [http://dx.doi.org/10.1016/j.tet.2006.03.036]

[47] Rurack, K.; Kollmannsberger, M.; Daub, J. A highly efficient sensor molecule emitting in the near infrared (NIR): 3,5-distyryl-8-(p-dimethylaminophenyl)difluoroboradiaza-s-indacene. *New J. Chem.,* **2001**, *25*, 289-292. [http://dx.doi.org/10.1039/b007379m]

[48] Yu, Y-H.; Descalzo, A.B.; Shen, Z.; Röhr, H.; Liu, Q.; Wang, Y-W.; Spieles, M.; Li, Y-Z.; Rurack, K.; You, X-Z. Mono- and di(dimethylamino)styryl-substituted borondipyrromethene and borondiindomethene dyes with intense near-infrared fluorescence. *Chem. Asian J.,* **2006**, *1*(1-2), 176-187. [http://dx.doi.org/10.1002/asia.200600042] [PMID: 17441053]

[49] Dost, Z.; Atilgan, S.; Akkaya, E.U. Distyryl-boradiazaindacenes: facile synthesis of novel near IR emitting fluorophores. *Tetrahedron,* **2006**, *62*, 8484-8488. [http://dx.doi.org/10.1016/j.tet.2006.06.082]

[50] Coskun, A.; Akkaya, E.U. Difluorobora-s-diazaindacene dyes as highly selective dosimetric reagents for fluoride anions. *Tetrahedron Lett.,* **2004**, *45*, 4947-4949. [http://dx.doi.org/10.1016/j.tetlet.2004.04.130]

[51] Rohand, T.; Qin, W.; Boens, N.; Dehaen, W. Palladium-catalyzed coupling reactions for the functionalization of BODIPY dyes with fluorescence spanning the visible spectrum. *Eur. J. Org. Chem.,* **2006**, 4658-4663. [http://dx.doi.org/10.1002/ejoc.200600531]

[52] Thivierge, C.; Bandichhor, R.; Burgess, K. Spectral dispersion and water solubilization of BODIPY dyes *via* palladium-catalyzed C-H functionalization. *Org. Lett.,* **2007**, *9*(11), 2135-2138. [http://dx.doi.org/10.1021/ol0706197] [PMID: 17455941]

[53] Boyer, J.H.; Morgan, L.R. Preparation of difluoroboradiaza-sindacene compounds and methods for using them *U. S. Patent 5189029,* **1993**.

[54] Boyer, J.H.; Morgan, L.R. Fluorescent chemical compositions useful as laser dyes and photodynamic therapy agents, and methods for their use *U.S. Patent 361936,* **1990**.

[55] Urano, T.; Nagasaka, H.; Tsuchiyama, M.; Ide, H. Photopolymerizable composition *U.S. Patent 5498641,* **1994**.

[56] Morgan, L.R.; Boyer, J.H. Heterocyclic compounds and their use in photodynamic therapy *WO Patent 9419355,* **1994**.

[57] Morgan, L.R.; Boyer, J.H. Boron difluoride compounds useful in photodynamic therapy and production of laser light *U.S. Patent 5189029,* **1995**.

[58] Pavlopoulos, T.G.; Boyer, J.H.; Shah, M.; Thangaraj, K.; Soong, M.L. Laser action from 2,6,8-position trisubstituted 1,3,5,7-tetramethylpyrromethene-BF(2) complexes: part 1. *Appl. Opt.,* **1990**, *29*(27), 3885-3886.
[http://dx.doi.org/10.1364/AO.29.003885] [PMID: 20577308]

[59] Pavlopoulos, T.G. Photostability of some pyrromethene laser dyes. *Proc. SPIE-Int. Soc. Opt. Eng.,* **1999**, pp. 1-7.

[60] Haugland, R.P.; Kang, H.C. . Reactive fluorescent dipyrrometheneboron difluoride dyes as molecular probes for biopolymers *U.S. Patent 4774339,* **1988**.

[61] Yogo, T.; Urano, Y.; Ishitsuka, Y.; Maniwa, F.; Nagano, T. Highly efficient and photostable photosensitizer based on BODIPY chromophore. *J. Am. Chem. Soc.,* **2005**, *127*(35), 12162-12163.
[http://dx.doi.org/10.1021/ja0528533] [PMID: 16131160]

[62] Strachan, J.P.; O'Shea, D.F.; Balasubramanian, T.; Lindsey, J.S. Rational synthesis of *meso*-substituted chlorin building blocks. *J. Org. Chem.,* **2000**, *65*(10), 3160-3172.
[http://dx.doi.org/10.1021/jo991942t] [PMID: 10814212]

[63] Umezawa, K.; Nakamura, Y.; Makino, H.; Citterio, D.; Suzuki, K. Bright, color-tunable fluorescent dyes in the visible-near-infrared region. *J. Am. Chem. Soc.,* **2008**, *130*(5), 1550-1551.
[http://dx.doi.org/10.1021/ja077756j] [PMID: 18193873]

[64] Umezawa, K.; Matsui, A.; Nakamura, Y.; Citterio, D.; Suzuki, K. Bright, color-tunable fluorescent dyes in the Vis/NIR region: establishment of new "tailor-made" multicolor fluorophores based on borondipyrromethene. *Chemistry,* **2009**, *15*(5), 1096-1106.
[http://dx.doi.org/10.1002/chem.200801906] [PMID: 19117043]

[65] Ulrich, G.; Ziessel, R.; Harriman, A. The chemistry of fluorescent bodipy dyes: versatility unsurpassed. *Angew. Chem. Int. Ed. Engl.,* **2008**, *47*(7), 1184-1201.
[http://dx.doi.org/10.1002/anie.200702070] [PMID: 18092309]

[66] Ptaszek, M. *Rational design of fluorophores for in vivo* applications in fluorescence-based biosensors: from concepts to applications; Morris, M.C, Ed.; , **2013**, 113, pp. 59-108.Progress in Molecular Biology and Translational Science

[67] Awuah, S.G.; You, Y. Boron dipyrromethene (BODIPY)-based photosensitizers for photodynamic therapy. *RSC Advances,* **2012**, *2*, 11169-11183.
[http://dx.doi.org/10.1039/c2ra21404k]

[68] Gonçalves, M.S. Fluorescent labeling of biomolecules with organic probes. *Chem. Rev.,* **2009**, *109*(1), 190-212.
[http://dx.doi.org/10.1021/cr0783840] [PMID: 19105748]

[69] Wang, D.; Fan, J.; Gao, X.; Wang, B.; Sun, S.; Peng, X. Carboxyl BODIPY dyes from bicarboxylic anhydrides: one-pot preparation, spectral properties, photostability, and biolabeling. *J. Org. Chem.,* **2009**, *74*(20), 7675-7683.
[http://dx.doi.org/10.1021/jo901149y] [PMID: 19772337]

[70] Niu, S.L.; Massif, C.; Ulrich, G.; Renard, P.Y.; Romieu, A.; Ziessel, R. Water-soluble red-emitting distyryl-borondipyrromethene (BODIPY) dyes for biolabeling. *Chemistry,* **2012**, *18*(23), 7229-7242.
[http://dx.doi.org/10.1002/chem.201103613] [PMID: 22544430]

[71] Romieu, A.; Massif, C.; Rihn, S.; Ulrich, G.; Ziessel, R.; Renard, P.Y. The first comparative study of the ability of different hydrophilic groups to water-solubilise fluorescent BODIPY dyes. *New J. Chem.,* **2013**, *37*, 1016-1027.
[http://dx.doi.org/10.1039/c3nj41093e]

[72] Ni, Y.; Zeng, L.; Kang, N.Y.; Huang, K.W.; Wang, L.; Zeng, Z.; Chang, Y.T.; Wu, J. *meso*-Ester and carboxylic acid substituted BODIPYs with far-red and near-infrared emission for bioimaging applications. *Chemistry,* **2014**, *20*(8), 2301-2310.
[http://dx.doi.org/10.1002/chem.201303868] [PMID: 24515608]

[73] Pagano, R.E.; Martin, O.C.; Kang, H.C.; Haugland, R.P. A novel fluorescent ceramide analogue for studying membrane traffic in animal cells: accumulation at the Golgi apparatus results in altered spectral properties of the sphingolipid precursor. *J. Cell Biol.,* **1991**, *113*(6), 1267-1279.
[http://dx.doi.org/10.1083/jcb.113.6.1267] [PMID: 2045412]

[74] Karolin, J.; Johansson, L.B; Strandberg, L.; Ny, T. Fluorescence and absorption spectroscopic properties of dipyrrometheneboron difluoride (BODIPY) derivatives in liquids, lipid membranes, and proteins. *J. Am. Chem. Soc.,* **1994**, *116*, 7801-7806.
[http://dx.doi.org/10.1021/ja00096a042]

[75] Li, X.; Traganos, F.; Melamed, M.R.; Darzynkiewicz, Z. Single-step procedure for labeling DNA strand breaks with fluorescein- or BODIPY-conjugated deoxynucleotides: detection of apoptosis and bromodeoxyuridine incorporation. *Cytometry,* **1995**, *20*(2), 172-180.
[http://dx.doi.org/10.1002/cyto.990200210] [PMID: 7664627]

[76] Tan, K.; Jaquinod, L.; Paolesse, R.; Nardis, S.; Di Natale, C.; Di Carlo, A.; Prodi, L.; Montalti, M.; Zaccheroni, N.; Smith, K.M. Synthesis and characterization of β-fused porphyrin-BODIPY® dyads. *Tetrahedron,* **2004**, *60*, 1099-1106.
[http://dx.doi.org/10.1016/j.tet.2003.11.072]

[77] Teske, C.A.; Schroeder, M.; Simon, R.; Hubbuch, J. Protein-labeling effects in confocal laser scanning microscopy. *J. Phys. Chem. B,* **2005**, *109*(28), 13811-13817.
[http://dx.doi.org/10.1021/jp050713+] [PMID: 16852729]

[78] Yee, M-C.; Fas, S.C.; Stohlmeyer, M.M.; Wandless, T.J.; Cimprich, K.A. A cell-permeable, activity-based probe for protein and lipid kinases. *J. Biol. Chem.,* **2005**, *280*(32), 29053-29059.
[http://dx.doi.org/10.1074/jbc.M504730200] [PMID: 15975929]

[79] Duheron, V.; Moreau, M.; Collin, B.; Sali, W.; Bernhard, C.; Goze, C.; Gautier, T.; Pais de Barros, J.P.; Deckert, V.; Brunotte, F.; Lagrost, L.; Denat, F. Dual labeling of lipopolysaccharides for SPECT-CT imaging and fluorescence microscopy. *ACS Chem. Biol.,* **2014**, *9*(3), 656-662.
[http://dx.doi.org/10.1021/cb400779j] [PMID: 24328371]

[80] Poirel, A.; Retailleau, P.; De Nicola, A.; Ziessel, R. Synthesis of water-soluble red-emitting thienyl-BODIPYs and bovine serum albumin labeling. *Chemistry,* **2014**, *20*(5), 1252-1257.
[http://dx.doi.org/10.1002/chem.201303988] [PMID: 24402814]

[81] Marks, D.L.; Bittman, R.; Pagano, R.E. Use of Bodipy-labeled sphingolipid and cholesterol analogs to examine membrane microdomains in cells. *Histochem. Cell Biol.,* **2008**, *130*(5), 819-832.
[http://dx.doi.org/10.1007/s00418-008-0509-5] [PMID: 18820942]

[82] Jameson, D.M.; Ross, J.A. Fluorescence polarization/anisotropy in diagnostics and imaging. *Chem. Rev.,* **2010**, *110*(5), 2685-2708.
[http://dx.doi.org/10.1021/cr900267p] [PMID: 20232898]

[83] Qian, X.; Xiao, Y.; Xu, Y.; Guo, X.; Qian, J.; Zhu, W. "Alive" dyes as fluorescent sensors: fluorophore, mechanism, receptor and images in living cells. *Chem. Commun. (Camb.),* **2010**, *46*(35), 6418-6436.
[http://dx.doi.org/10.1039/c0cc00686f] [PMID: 20589288]

[84] Yuan, L.; Lin, W.; Zheng, K.; He, L.; Huang, W. Far-red to near infrared analyte-responsive fluorescent probes based on organic fluorophore platforms for fluorescence imaging. *Chem. Soc. Rev.,* **2013**, *42*(2), 622-661.
[http://dx.doi.org/10.1039/C2CS35313J] [PMID: 23093107]

[85] Ni, Y.; Wu, J. Far-red and near infrared BODIPY dyes: synthesis and applications for fluorescent pH probes and bio-imaging. *Org. Biomol. Chem.,* **2014**, *12*(23), 3774-3791.
[http://dx.doi.org/10.1039/c3ob42554a] [PMID: 24781214]

[86] Drummen, G.P.; van Liebergen, L.C.; Op den Kamp, J.A.; Post, J.A. C11-BODIPY(581/591), an oxidation-sensitive fluorescent lipid peroxidation probe: (micro)spectroscopic characterization and validation of methodology. *Free Radic. Biol. Med.,* **2002**, *33*(4), 473-490.
[http://dx.doi.org/10.1016/S0891-5849(02)00848-1] [PMID: 12160930]

[87] Peng, X.; Du, J.; Fan, J.; Wang, J.; Wu, Y.; Zhao, J.; Sun, S.; Xu, T. A selective fluorescent sensor for imaging Cd^{2+} in living cells. *J. Am. Chem. Soc.,* **2007**, *129*(6), 1500-1501.
[http://dx.doi.org/10.1021/ja0643319] [PMID: 17283986]

[88] Lu, H.; Xiong, L.; Liu, H.; Yu, M.; Shen, Z.; Li, F.; You, X. A highly selective and sensitive fluorescent turn-on sensor for Hg^{2+} and its application in live cell imaging. *Org. Biomol. Chem.,* **2009**, *7*(12), 2554-2558.
[http://dx.doi.org/10.1039/b902912e] [PMID: 19503929]

[89] Sun, Z-N.; Wang, H-L.; Liu, F-Q.; Chen, Y.; Tam, P.K.; Yang, D.; Yang, D. BODIPY-based fluorescent probe for peroxynitrite detection and imaging in living cells. *Org. Lett.,* **2009**, *11*(9), 1887-1890.
[http://dx.doi.org/10.1021/ol900279z] [PMID: 19331349]

[90] Sekiya, M.; Umezawa, K.; Sato, A.; Citterio, D.; Suzuki, K. A novel luciferin-based bright chemiluminescent probe for the detection of reactive oxygen species. *Chem. Commun. (Camb.),* **2009**, (21), 3047-3049.
[http://dx.doi.org/10.1039/b903751a] [PMID: 19462082]

[91] Domaille, D.W.; Zeng, L.; Chang, C.J. Visualizing ascorbate-triggered release of labile copper within living cells using a ratiometric fluorescent sensor. *J. Am. Chem. Soc.,* **2010**, *132*(4), 1194-1195.
[http://dx.doi.org/10.1021/ja907778b] [PMID: 20052977]

[92] Zhu, H.; Fan, J.; Li, M.; Cao, J.; Wang, J.; Peng, X. A "distorted-BODIPY"-based fluorescent probe for imaging of cellular viscosity in live cells. *Chemistry,* **2014**, *20*(16), 4691-4696.
[http://dx.doi.org/10.1002/chem.201304296] [PMID: 24595961]

[93] Moriarty, R.D.; Martin, A.; Adamson, K.; O'Reilly, E.; Mollard, P.; Forster, R.J.; Keyes, T.E. The

application of water soluble, mega-Stokes-shifted BODIPY fluorophores to cell and tissue imaging. *J. Microsc.,* **2014**, *253*(3), 204-218.
[http://dx.doi.org/10.1111/jmi.12111] [PMID: 24467513]

[94] Coskun, A.; Baytekin, B.T.; Akkaya, E.U. Novel fluorescent chemosensor for anions *via* modulation of oxidative PET: A remarkable 25-fold enhancement of emission. *Tetrahedron Lett.,* **2003**, *44*, 5649-5651.
[http://dx.doi.org/10.1016/S0040-4039(03)01365-0]

[95] Rurack, K.; Kollmannsberger, M.; Resch-Genger, U.; Daub, J. A selective and sensitive fluoroionophore for Hg(II), Ag(I), and Cu(II) with virtually decoupled fluorophore and receptor units. *J. Am. Chem. Soc.,* **2000**, *122*, 968-969.
[http://dx.doi.org/10.1021/ja992630a]

[96] Yamada, Y.; Tomiyama, Y.; Morita, A.; Ikekita, M.; Aoki, S. BODIPY-based fluorescent redox potential sensors that utilize reversible redox properties of flavin. *ChemBioChem,* **2008**, *9*(6), 853-856.
[http://dx.doi.org/10.1002/cbic.200700718] [PMID: 18338353]

[97] Jeong, Y.; Yoon, J. Recent progress on fluorescent chemosensors for metal ions. *Inorg. Chim. Acta,* **2012**, *381*, 2-14.
[http://dx.doi.org/10.1016/j.ica.2011.09.011]

[98] Guliyev, R.; Buyukcakir, O.; Sozmen, F.; Bozdemir, O.A. Cyanide sensing *via* metal ion removal from a fluorogenic BODIPY complex. *Tetrahedron Lett.,* **2009**, *50*, 5139-5141.
[http://dx.doi.org/10.1016/j.tetlet.2009.06.117]

[99] Wang, J.G.; Hou, Y.J.; Li, C.; Zhang, B.W.; Wang, X.S. Selectivity tune of fluoride ion sensing for phenolic OH-containing BODIPY dyes. *Sens. Actuators B Chem.,* **2011**, *157*, 586-593.
[http://dx.doi.org/10.1016/j.snb.2011.05.027]

[100] Madhu, S.; Ravikanth, M. Synthesis, spectral, electrochemical, and anion binding properties of 3,5-bis(dipyrromethanyl) boron-dipyrromethenes. *Inorg. Chem.,* **2012**, *51*(7), 4285-4292.
[http://dx.doi.org/10.1021/ic202758r] [PMID: 22432414]

[101] Feng, L.; Li, H.; Niu, L.Y.; Guan, Y.S.; Duan, C.F.; Guan, Y.F.; Tung, C.H.; Yang, Q.Z. A fluorometric paper-based sensor array for the discrimination of heavy-metal ions. *Talanta,* **2013**, *108*, 103-108.
[http://dx.doi.org/10.1016/j.talanta.2013.02.073] [PMID: 23601876]

[102] Er, J.C.; Tang, M.K.; Chia, C.G.; Liew, H.; Vendrell, M.; Chang, Y.T. MegaStokes BODIPY-triazoles as environmentally sensitive turn-on fluorescent dyes. *Chem. Sci. (Camb.),* **2013**, *4*, 2168-2176.
[http://dx.doi.org/10.1039/c3sc22166k]

[103] Tikhomirova, K.V.; Khoroshutin, A.V.; Anisimov, A.V. Diethyl (biscyclohexano) BODIPY dicarboxylates. Chelation of alkaline-earth metal ions and sensor properties. *Russ. J. Org. Chem.,* **2014**, *50*, 229-239.
[http://dx.doi.org/10.1134/S1070428014020158]

[104] Wu, Y.; Peng, X.; Guo, B.; Fan, J.; Zhang, Z.; Wang, J.; Cui, A.; Gao, Y. Boron dipyrromethene fluorophore based fluorescence sensor for the selective imaging of Zn(II) in living cells. *Org. Biomol. Chem.,* **2005**, *3*(8), 1387-1392.

[http://dx.doi.org/10.1039/b501795e] [PMID: 15827633]

[105] Zeng, L.; Miller, E.W.; Pralle, A.; Isacoff, E.Y.; Chang, C.J. A selective turn-on fluorescent sensor for imaging copper in living cells. *J. Am. Chem. Soc.,* **2006**, *128*(1), 10-11. [http://dx.doi.org/10.1021/ja055064u] [PMID: 16390096]

[106] Pan, Z.H.; Luo, G.G.; Zhou, J.W.; Xia, J.X.; Fang, K.; Wu, R.B. A simple BODIPY-aniline-based fluorescent chemosensor as multiple logic operations for the detection of pH and CO2 gas. *Dalton Trans.,* **2014**, *43*(22), 8499-8507. [http://dx.doi.org/10.1039/c4dt00395k] [PMID: 24756338]

[107] Calvete, M.J. Near-infrared absorbing organic materials with nonlinear transmission properties. *Int. Rev. Phys. Chem.,* **2012**, *31*, 319-366. [http://dx.doi.org/10.1080/0144235X.2012.713210]

[108] Li, Y.; Liu, T.; Liu, H.; Tian, M.Z.; Li, Y. Self-assembly of intramolecular charge-transfer compounds into functional molecular systems. *Acc. Chem. Res.,* **2014**, *47*(4), 1186-1198. [http://dx.doi.org/10.1021/ar400264e] [PMID: 24666347]

[109] Zhu, M.; Jiang, L.; Yuan, M.J.; Liu, X.F.; Ouyang, C.B.; Zheng, H.Y.; Yin, X.D.; Zuo, Z.C.; Liu, H.B.; Li, Y.L. Efficient tuning nonlinear optical properties: synthesis and characterization of a series of novel poly(aryleneethynylene)s co-containing BODIPY, *J. Polym. Sci. part A. Polym. Chem.,* **2008**, *46*, 7401-7410. [http://dx.doi.org/10.1002/pola.23045]

[110] Harriman, A.; Mallon, L.J.; Elliot, K.J.; Haefele, A.; Ulrich, G.; Ziessel, R. Length dependence for intramolecular energy transfer in three- and four-color donor-spacer-acceptor arrays. *J. Am. Chem. Soc.,* **2009**, *131*(37), 13375-13386. [http://dx.doi.org/10.1021/ja9038856] [PMID: 19754187]

[111] Zheng, Q.D.; He, G.S.; Prasad, P.N. A novel near IR two-photon absorbing chromophore: Optical limiting and stabilization performances at an optical communication wavelength. *Chem. Phys. Lett.,* **2009**, *475*, 250-255. [http://dx.doi.org/10.1016/j.cplett.2009.05.040]

[112] Schmidt, E.Y.; Trofimov, B.A.; Mikhaleva, A.I.; Zorina, N.V.; Protzuk, N.I.; Petrushenko, K.B.; Ushakov, I.A.; Dvorko, M.Y.; Méallet-Renault, R.; Clavier, G.; Vu, T.T.; Tran, H.T.; Pansu, R.B. Synthesis and optical properties of 2-(benzo[b]thiophene-3-yl)pyrroles and a new BODIPY fluorophore (BODIPY = 4,4-difluoro-4-bora-3a,4a-diaza-s-indacene). *Chemistry,* **2009**, *15*(23), 5823-5830. [http://dx.doi.org/10.1002/chem.200802467] [PMID: 19396897]

[113] Wang, Y.C.; Zhang, D.K.; Zhou, H.; Ding, J.L.; Chen, Q.A.; Xiao, Y.; Qian, S.X. Nonlinear optical properties and ultrafast dynamics of three novel boradiazaindacene derivatives. *J. Appl. Phys.,* **2010**, *108*, 033520. [http://dx.doi.org/10.1063/1.3457859]

[114] Shi, W.J.; Lo, P.C.; Singh, A.; Ledoux-Rak, I.; Ng, D.K. Synthesis and second-order nonlinear optical properties of push-pull BODIPY derivatives. *Tetrahedron,* **2012**, *68*, 8712-8718. [http://dx.doi.org/10.1016/j.tet.2012.08.033]

[115] Jadhav, T.; Maragani, R.; Misra, R.; Sreeramulu, V.; Rao, D.N.; Mobin, S.M. Design and synthesis of

donor-acceptor pyrazabole derivatives for multiphoton absorption. *Dalton Trans.,* **2013**, *42*(13), 4340-4342.
[http://dx.doi.org/10.1039/c3dt33065f] [PMID: 23396416]

[116] Benniston, A.C.; Copley, G. Lighting the way ahead with boron dipyrromethene (Bodipy) dyes. *Phys. Chem. Chem. Phys.,* **2009**, *11*(21), 4124-4131.
[http://dx.doi.org/10.1039/b901383k] [PMID: 19458813]

[117] Roncali, J. Molecular bulk heterojunctions: an emerging approach to organic solar cells. *Acc. Chem. Res.,* **2009**, *42*(11), 1719-1730.
[http://dx.doi.org/10.1021/ar900041b] [PMID: 19580313]

[118] Qian, G.; Wang, Z.Y. Near-infrared organic compounds and emerging applications. *Chem. Asian J.,* **2010**, *5*(5), 1006-1029.
[http://dx.doi.org/10.1002/asia.200900596] [PMID: 20352644]

[119] Suzuki, S.; Kozaki, M.; Nozaki, K.; Okada, K. Recent progress in controlling photophysical processes of donor-acceptor arrays involving perylene diimides and boron-dipyrromethenes. *J. Photochem. Photobiol. Photochem. Rev.,* **2011**, *12*, 269-292.
[http://dx.doi.org/10.1016/j.jphotochemrev.2011.10.001]

[120] Ziessel, R.; Harriman, A. Artificial light-harvesting antennae: electronic energy transfer by way of molecular funnels. *Chem. Commun. (Camb.),* **2011**, *47*(2), 611-631.
[http://dx.doi.org/10.1039/C0CC02687E] [PMID: 20957235]

[121] El-Khouly, M.E.; Fukuzumi, S.; D'Souza, F. Photosynthetic antenna-reaction center mimicry by using boron dipyrromethene sensitizers. *ChemPhysChem,* **2014**, *15*(1), 30-47.
[http://dx.doi.org/10.1002/cphc.201300715] [PMID: 24243758]

[122] Bessette, A.; Hanan, G.S. Design, synthesis and photophysical studies of dipyrromethene-based materials: insights into their applications in organic photovoltaic devices. *Chem. Soc. Rev.,* **2014**, *43*(10), 3342-3405.
[http://dx.doi.org/10.1039/c3cs60411j] [PMID: 24577078]

[123] Li, F.; Yang, S.I.; Ciringh, Y.; Seth, J.; Martin, C.H., III; Singh, D.L.; Kim, D.; Birge, R.R.; Bocian, D.F.; Holten, D.; Lindsey, J.S. Design, synthesis, and photodynamics of light-harvesting arrays comprised of a porphyrin and one, two, or eight boron-dipyrrin accessory pigments. *J. Am. Chem. Soc.,* **1998**, *120*, 10001-10017.
[http://dx.doi.org/10.1021/ja9812047]

[124] Erten-Ela, S.; Yilmaz, M.D.; Icli, B.; Dede, Y.; Icli, S.; Akkaya, E.U. A panchromatic boradiazaindacene (BODIPY) sensitizer for dye-sensitized solar cells. *Org. Lett.,* **2008**, *10*(15), 3299-3302.
[http://dx.doi.org/10.1021/ol8010612] [PMID: 18588306]

[125] Rousseau, T.; Cravino, A.; Bura, T.; Ulrich, G.; Ziessel, R.; Roncali, J. BODIPY derivatives as donor materials for bulk heterojunction solar cells. *Chem. Commun. (Camb.),* **2009**, (13), 1673-1675.
[http://dx.doi.org/10.1039/b822770e] [PMID: 19294258]

[126] Kumaresan, D.; Thummel, R.P.; Bura, T.; Ulrich, G.; Ziessel, R. Color tuning in new metal-free organic sensitizers (Bodipys) for dye-sensitized solar cells. *Chemistry,* **2009**, *15*(26), 6335-6339.
[http://dx.doi.org/10.1002/chem.200900518] [PMID: 19462381]

[127] Bozdemir, O.A.; Büyükcakir, O.; Akkaya, E.U. Novel molecular building blocks based on the boradiazaindacene chromophore: applications in fluorescent metallosupramolecular coordination polymers. *Chemistry,* **2009**, *15*(15), 3830-3838.
[http://dx.doi.org/10.1002/chem.200802538] [PMID: 19229925]

[128] Algi, F.; Cihaner, A. An ambipolar low band gap material based on BODIPY and EDOT. *Org. Electron.,* **2009**, *10*, 453-458.
[http://dx.doi.org/10.1016/j.orgel.2009.01.010]

[129] Kolemen, S.; Cakmak, Y.; Erten-Ela, S.; Altay, Y.; Brendel, J.; Thelakkat, M.; Akkaya, E.U. Solid-state dye-sensitized solar cells using red and near-IR absorbing Bodipy sensitizers. *Org. Lett.,* **2010**, *12*(17), 3812-3815.
[http://dx.doi.org/10.1021/ol1014762] [PMID: 20704314]

[130] Kim, B.; Ma, B.; Donuru, V.R.; Liu, H.; Fréchet, J.M. Bodipy-backboned polymers as electron donor in bulk heterojunction solar cells. *Chem. Commun. (Camb.),* **2010**, *46*(23), 4148-4150.
[http://dx.doi.org/10.1039/b927350f] [PMID: 20390122]

[131] Araneda, J.F.; Piers, W.E.; Heyne, B.; Parvez, M.; McDonald, R. High Stokes shift anilido-pyridine boron difluoride dyes. *Angew. Chem. Int. Ed. Engl.,* **2011**, *50*(51), 12214-12217.
[http://dx.doi.org/10.1002/anie.201105228] [PMID: 22021120]

[132] Kolemen, S.; Bozdemir, O.A.; Cakmak, Y.; Barin, G.; Erten-Ela, S.; Marszalek, M.; Yum, J.H.; Zakeeruddin, S.M.; Nazeeruddin, M.K.; Gratzel, M.; Akkaya, E.U. Optimization of distyryl-Bodipy chromophores for efficient panchromatic sensitization in dye sensitized solar cells. *Chem. Sci. (Camb.),* **2011**, *2*, 949-954.
[http://dx.doi.org/10.1039/c0sc00649a]

[133] Duran-Sampedro, G.; Agarrabeitia, A.R.; Garcia-Moreno, I.; Costela, A.; Banuelos, J.; Arbeloa, T.; Arbeloa, I.L.; Chiara, J.L.; Ortiz, M.J. Chlorinated BODIPYs: surprisingly efficient and highly photostable laser dyes. *Eur. J. Org. Chem.,* **2012**, 6335-6350.
[http://dx.doi.org/10.1002/ejoc.201200946]

[134] Lefebvre, J.F.; Sun, X.Z.; Calladine, J.A.; George, M.W.; Gibson, E.A. Promoting charge-separation in p-type dye-sensitized solar cells using bodipy. *Chem. Commun. (Camb.),* **2014**, *50*(40), 5258-5260.
[http://dx.doi.org/10.1039/C3CC46133E] [PMID: 24129545]

[135] Khan, T.K.; Sheokand, P.; Agarwal, N. Synthesis and studies of aza-BODIPY-based pi-conjugates for organic electronic applications. *Eur. J. Org. Chem.,* **2014**, 1416-1422.
[http://dx.doi.org/10.1002/ejoc.201301300]

[136] Hilderbrand, S.A.; Weissleder, R. Near-infrared fluorescence: application to in vivo molecular imaging. *Curr. Opin. Chem. Biol.,* **2010**, *14*(1), 71-79.
[http://dx.doi.org/10.1016/j.cbpa.2009.09.029] [PMID: 19879798]

[137] Escobedo, J.O.; Rusin, O.; Lim, S.; Strongin, R.M. NIR dyes for bioimaging applications. *Curr. Opin. Chem. Biol.,* **2010**, *14*(1), 64-70.
[http://dx.doi.org/10.1016/j.cbpa.2009.10.022] [PMID: 19926332]

[138] Lavis, L.D.; Raines, R.T. Bright ideas for chemical biology. *ACS Chem. Biol.,* **2008**, *3*(3), 142-155.
[http://dx.doi.org/10.1021/cb700248m] [PMID: 18355003]

[139] Gomez-Hens, A.; Aguilar-Caballos, M.P. *Trends Analyt. Chem.,* **2004**, *23*, 127-136. [http://dx.doi.org/10.1016/S0165-9936(04)00305-X]

[140] Atilgan, S.; Ekmekci, Z.; Dogan, A.L.; Guc, D.; Akkaya, E.U. Water soluble distyryl-boradiazaindacenes as efficient photosensitizers for photodynamic therapy. *Chem. Commun. (Camb.),* **2006**, (42), 4398-4400. [http://dx.doi.org/10.1039/b612347c] [PMID: 17057856]

[141] Johnson, I.D.; Kang, H.C.; Haugland, R.P. Fluorescent membrane probes incorporating dipyrrometheneboron difluoride fluorophores. *Anal. Biochem.,* **1991**, *198*(2), 228-237. [http://dx.doi.org/10.1016/0003-2697(91)90418-S] [PMID: 1799206]

[142] Keller, R.C.; Silvius, J.R.; De Kruijff, B. Characterization of the resonance energy transfer couple coumarin-BODIPY and its possible applications in protein-lipid research. *Biochem. Biophys. Res. Commun.,* **1995**, *207*(2), 508-514. [http://dx.doi.org/10.1006/bbrc.1995.1217] [PMID: 7864836]

[143] Schade, S.Z.; Jolley, M.E.; Sarauer, B.J.; Simonson, L.G. BODIPY-α-casein, a pH-independent protein substrate for protease assays using fluorescence polarization. *Anal. Biochem.,* **1996**, *243*(1), 1-7. [http://dx.doi.org/10.1006/abio.1996.0475] [PMID: 8954519]

[144] Jones, L.J.; Upson, R.H.; Haugland, R.P.; Panchuk-Voloshina, N.; Zhou, M.; Haugland, R.P. Quenched BODIPY dye-labeled casein substrates for the assay of protease activity by direct fluorescence measurement. *Anal. Biochem.,* **1997**, *251*(2), 144-152. [http://dx.doi.org/10.1006/abio.1997.2259] [PMID: 9299009]

[145] Durand, J.; Kojic, L.; Wang, Y.; Lee, P.; Cynader, M.S.; Gu, Q. Confocal imaging of N-methyl-D-aspartate receptors in living cortical neurons. *Neuroscience,* **2000**, *97*(1), 11-23. [http://dx.doi.org/10.1016/S0306-4522(99)00595-3] [PMID: 10771336]

[146] Marmé, N.; Knemeyer, J-P.; Sauer, M.; Wolfrum, J. Inter- and intramolecular fluorescence quenching of organic dyes by tryptophan. *Bioconjug. Chem.,* **2003**, *14*(6), 1133-1139. [http://dx.doi.org/10.1021/bc0341324] [PMID: 14624626]

[147] Traverso, G.; Diehl, F.; Hurst, R.; Shuber, A.; Whitney, D.; Johnson, C.; Levin, B.; Kinzler, K.W.; Vogelstein, B. Multicolor *in vitro* translation. *Nat. Biotechnol.,* **2003**, *21*(9), 1093-1097. [http://dx.doi.org/10.1038/nbt857] [PMID: 12910244]

[148] Epand, R.F.; Sayer, B.G.; Epand, R.M. Induction of raft-like domains by a myristoylated NAP-22 peptide and its Tyr mutant. *FEBS J.,* **2005**, *272*(7), 1792-1803. [http://dx.doi.org/10.1111/j.1742-4658.2005.04612.x] [PMID: 15794765]

[149] Jang, H.G.; Park, M.; Wishnok, J.S.; Tannenbaum, S.R.; Wogan, G.N. Hydroxyl-specific fluorescence labeling of ABP-deoxyguanosine, PhIP-deoxyguanosine, and AFB1-formamidopyrimidine with BODIPY-FL. *Anal. Biochem.,* **2006**, *359*(2), 151-160. [http://dx.doi.org/10.1016/j.ab.2006.08.030] [PMID: 17081492]

[150] Deniz, E.; Isbasar, G.C.; Bozdemir, O.A.; Yildirim, L.T.; Siemiarczuk, A.; Akkaya, E.U. Bidirectional switching of near IR emitting boradiazaindacene fluorophores. *Org. Lett.,* **2008**, *10*(16), 3401-3403. [http://dx.doi.org/10.1021/ol801062h] [PMID: 18613696]

[151] Ziessel, R.; Ulrich, G.; Harriman, A.; Alamiry, M.A.; Stewart, B.; Retailleau, P. Solid-state gas sensors developed from functional difluoroboradiazaindacene dyes. *Chemistry,* **2009**, *15*(6), 1359-1369. [http://dx.doi.org/10.1002/chem.200801911] [PMID: 19115286]

[152] Myochin, T.; Hanaoka, K.; Komatsu, T.; Terai, T.; Nagano, T. Design strategy for a near-infrared fluorescence probe for matrix metalloproteinase utilizing highly cell permeable boron dipyrromethene. *J. Am. Chem. Soc.,* **2012**, *134*(33), 13730-13737. [http://dx.doi.org/10.1021/ja303931b] [PMID: 22830429]

[153] Stoddart, L.A.; Vernall, A.J.; Denman, J.L.; Briddon, S.J.; Kellam, B.; Hill, S.J. Fragment screening at adenosine-A(3) receptors in living cells using a fluorescence-based binding assay. *Chem. Biol.,* **2012**, *19*(9), 1105-1115. [http://dx.doi.org/10.1016/j.chembiol.2012.07.014] [PMID: 22999879]

[154] Vernall, A.J.; Stoddart, L.A.; Briddon, S.J.; Ng, H.W.; Laughton, C.A.; Doughty, S.W.; Hill, S.J.; Kellam, B. Conversion of a non-selective adenosine receptor antagonist into A3-selective high affinity fluorescent probes using peptide-based linkers. *Org. Biomol. Chem.,* **2013**, *11*(34), 5673-5682. [http://dx.doi.org/10.1039/c3ob41221k] [PMID: 23881285]

[155] Baker, J.G.; Adams, L.A.; Salchow, K.; Mistry, S.N.; Middleton, R.J.; Hill, S.J.; Kellam, B. Synthesis and characterization of high-affinity 4,4-difluoro-4-bora-3a,4a-diaza-s-indacene-labeled fluorescent ligands for human β-adrenoceptors. *J. Med. Chem.,* **2011**, *54*(19), 6874-6887. [http://dx.doi.org/10.1021/jm2008562] [PMID: 21870877]

[156] Ono, M.; Watanabe, H.; Kimura, H.; Saji, H. BODIPY-based molecular probe for imaging of cerebral β-amyloid plaques. *ACS Chem. Neurosci.,* **2012**, *3*(4), 319-324. [http://dx.doi.org/10.1021/cn3000058] [PMID: 22860198]

[157] Qi, X.; Jun, E.J.; Xu, L.; Kim, S-J.; Hong, J.S.; Yoon, Y.J.; Yoon, J. New BODIPY derivatives as OFF-ON fluorescent chemosensor and fluorescent chemodosimeter for Cu2+: cooperative selectivity enhancement toward Cu2+. *J. Org. Chem.,* **2006**, *71*(7), 2881-2884. [http://dx.doi.org/10.1021/jo052542a] [PMID: 16555847]

[158] Qi, X.; Kim, S.K.; Han, S.J.; Xu, L.; Jee, A.Y.; Kim, H.N.; Lee, C.; Kim, Y.; Lee, M.; Kim, S-J.; Yoon, J. Study on the BODIPY-triazine-based tripod fluorescent systems: various structures from similar procedure. *Supramol. Chem.,* **2009**, *21*, 455-464. [http://dx.doi.org/10.1080/10610270802283077]

[159] Zhao, Y.; Lv, X.; Liu, Y.; Liu, J.; Zhang, Y.; Shi, H.; Guo, W. The emission enhancement of the NIR distyryl Bodipy dyes by the indirect S0 → S2 excitation and their application towards a Hg2+ probe. *J. Mater. Chem.,* **2012**, *22*, 11475. [http://dx.doi.org/10.1039/c2jm30963g]

[160] Cao, J.; Zhao, C.; Feng, P.; Zhang, Y.; Zhu, W. A colorimetric and ratiometric NIR fluorescent turn-on fluoride chemodosimeter based on BODIPY derivatives: high selectivity *via* specific Si–O cleavage. *RSC Advances,* **2012**, *2*, 418-420. [http://dx.doi.org/10.1039/C1RA00942G]

[161] Bellier, Q.; Bouit, P-A.; Kamada, K.; Feneyrou, P.; Malmström, E.; Maury, O.; Andraud, C. Design of near-infrared dyes for nonlinear optics: toward optical limiting applications at telecommunication wavelengths. *Proc. SPIE,* **2009**, *7487*, 74870G.

[http://dx.doi.org/10.1117/12.831226]

[162] Bellier, Q.; Dalier, F.; Jeanneau, E.; Maury, O.; Andraud, C. Thiophene-substituted aza-bodipy as a strategic synthon for the design of near-infrared dyes. *New J. Chem.,* **2012**, *36*, 768-773. [http://dx.doi.org/10.1039/c2nj20943h]

[163] Warnan, J.; Buchet, F.; Pellegrin, Y.; Blart, E.; Odobel, F. Panchromatic trichromophoric sensitizer for dye-sensitized solar cells using antenna effect. *Org. Lett.,* **2011**, *13*(15), 3944-3947. [http://dx.doi.org/10.1021/ol2014686] [PMID: 21721515]

[164] Rousseau, T.; Cravino, A.; Ripaud, E.; Leriche, P.; Rihn, S.; De Nicola, A.; Ziessel, R.; Roncali, J. A tailored hybrid BODIPY-oligothiophene donor for molecular bulk heterojunction solar cells with improved performances. *Chem. Commun. (Camb.),* **2010**, *46*(28), 5082-5084. [http://dx.doi.org/10.1039/c0cc01144d] [PMID: 20559594]

[165] Bura, T.; Leclerc, N.; Fall, S.; Lévêque, P.; Heiser, T.; Retailleau, P.; Rihn, S.; Mirloup, A.; Ziessel, R. High-performance solution-processed solar cells and ambipolar behavior in organic field-effect transistors with thienyl-BODIPY scaffoldings. *J. Am. Chem. Soc.,* **2012**, *134*(42), 17404-17407. [http://dx.doi.org/10.1021/ja3072513] [PMID: 23035618]

CHAPTER 5

Dendrimers as Top Materials

Abstract: Dendrimers are monodispersed macromolecules, also known as cascade molecules, arborols or starburst polymers. The term dendrimers derives from the Greek language (*dendron* = tree, *meros* = part) and exploits two qualities that are frequently observed in many naturally occurring systems: globular structure and polyvalence. Herein is presented a conspectus on the field of dendrimers, including preparation and functionalization methods, as well as detailed examples on several promising applications with 291 references provided. Some synthetic details on the preparation of state-of-the-art compounds are given as well.

Keywords: Dendrimers, Drug delivery, Nanocarriers, Nanocatalysts, Starburst polymers.

INTRODUCTION

Dendrimers are also known as cascade molecules, arborols or starburst polymers. The term dendrimer (Greek: *dendron* = tree, *meros* = part) describes the architecture of this type of molecules. Dendrimers exploit two qualities that are frequently observed in many naturally occurring systems: globular structure and polyvalence [1 - 3].

Branched polymers have been known for many years, however, the control of their branching has always been considered an issue in polymer synthesis. Their are a class of polymers intermediate between linear and network polymers. In the early 1940's, observations of what appeared to be three-dimensional macromolecules were described theoretically and experimentally by Flory [4, 5], who received the Nobel Prize for Chemistry, due to his work on the design of linear and non-linear polymer chemistry in 1974.

The synthesis of polymers containing a controlled number and position of branches is a more recent phenomenon. The synthesis of both star and comb polymers have been reported during the last few decades [6]. Star polymers

Mário J. F. Calvete

possess a single focal point from which emanates a number of arms [7, 8], while comb polymers are produced when the arms are attached to a polymer backbone (Fig. **1**) [9, 10].

The exponential evolution of synthetic dendritic polymers has resulted in the creation of two classes of branched macromolecules, specifically dendrimers and hyperbranched polymers [11, 12]. The later class encompasses irregularly branched structures that have varying degrees of branching and are polydisperse in size and structure [13]. On the other hand, dendrimers are usually defined as perfectly branched, monodisperse macromolecules [1]. Dendrimers are normally synthesized using a repetitive sequence of several reactions, frequently coupled with purification actions. A dendrimer encompasses three distinctive regions: the core, the interior branching units and the terminal or end groups (Fig. **2**). The number of layers within a dendrimer is usually described as the generation (G) of the structure.

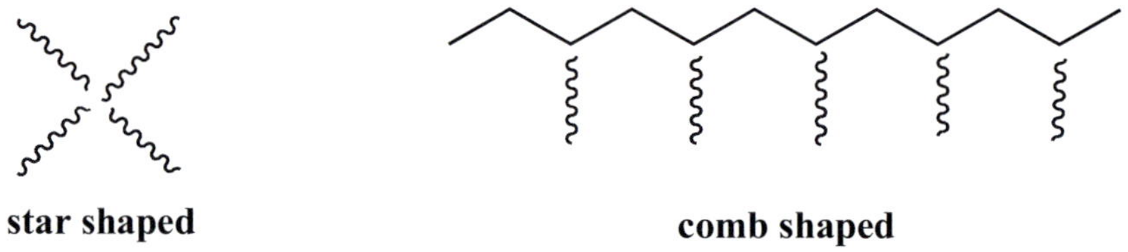

Fig. (1). Star and comb polymers.

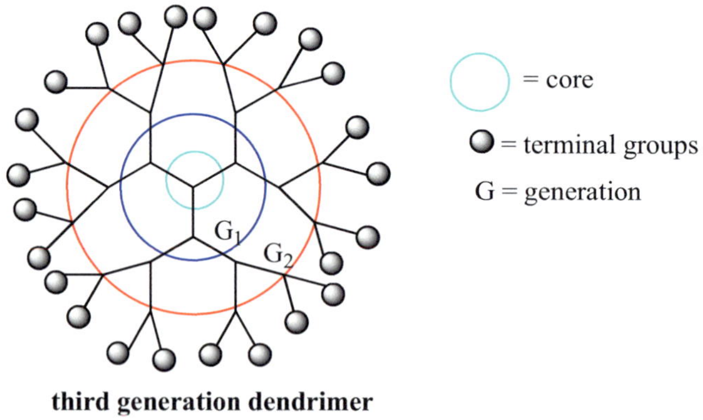

Fig. (2). Dendrimer schematics.

Dendrimers were first reported in 1978, named as "cascade molecules", by Vögtle [14], who acknowledged the analogy with a waterfall, as a cascade (poly(propylene imine) dendrimer) (Fig. **3**) [15]. The methodology for preparing these cascade molecules [16 - 18] consisted of two steps; i) Michael addition in which a diamine becomes a tetranitrile and ii) nitrile reduction to primary amines. This sequence of reactions could be taken repetitively to obtain regularly branched molecules bearing several arms. Regardless of the novelty, the growth on research on this type of molecules was hampered by synthetic and subsequent purification difficulties.

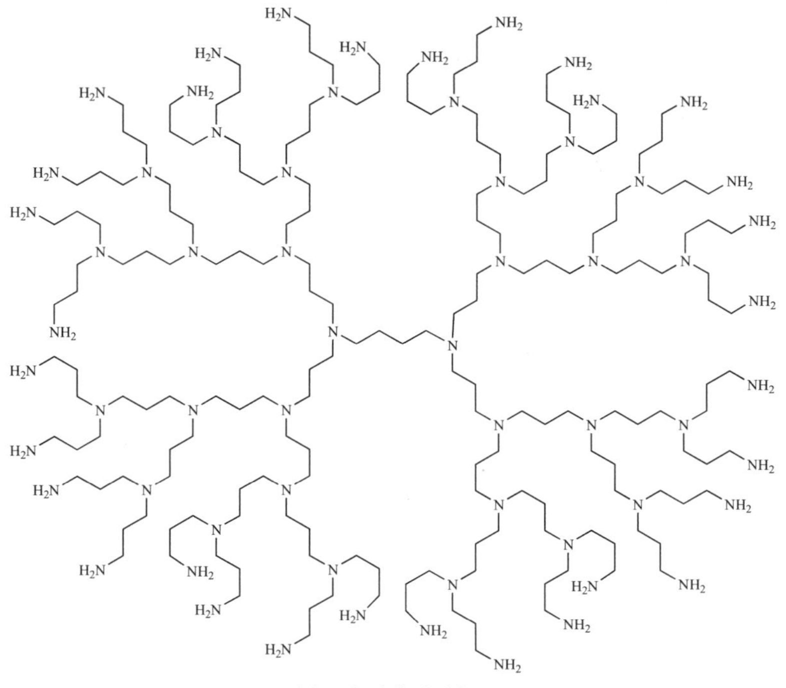

Fig. (3). Poly(propylene imine) dendrimer (PPI).

Later, in the 80's, progress continued with Denkewalter (polylysine [19]) and Maciejewski (dense cascade-like polymers [20]). It was not until 1985 that Tomalia synthesized the first branched "starburst" polyamidoamine (PAMAM) molecule (Fig. **4**) [21] and coined the name *dendrimers* [21 - 23], which were also the first commercially available ones (Starburst™ polymers) [24]. In the same year, Newkome [25] synthesized the so-called "arborols"; hydroxyl-terminated water soluble dendrimers (Fig. **5**).

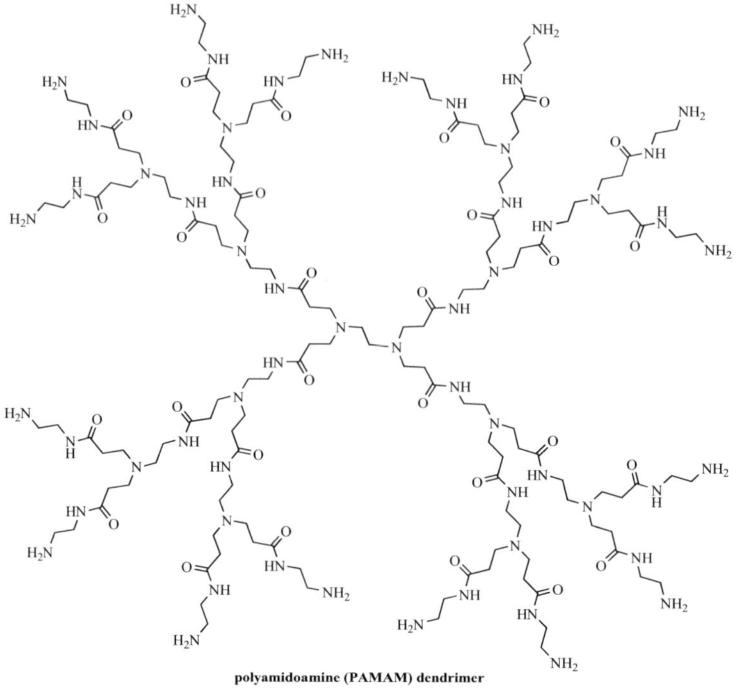

polyamidoamine (PAMAM) dendrimer

Fig. (4). Polyamidoamine dendrimer (PAMAM).

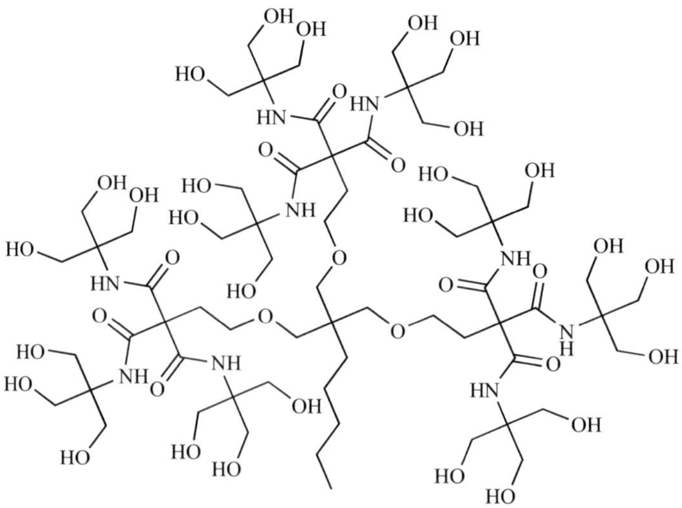

arborol dendrimer

Fig. (5). Arborol dendrimer.

The first polyarylether dendrimers were synthesized by Hawker and Fréchet [26], while Neenan and Miller [27] prepared aryl based dendrimers. The first dendrimers containing heteroatoms were prepared afterwards [28, 29], and Caminade and Majoral reported also the first dendrimers containing phosphorus atoms [30 - 32]. Shortly after, Müllen prepared the first shape persistent dendrimers, the polyphenylene dendrimers [33, 34].

SYNTHESIS OF DENDRIMER DERIVATIVES

Dendrimers are usually synthesized using a repetitive sequence of two reactions, comprising the addition of the monomer, corresponding to a structural increase of one generation, and another step involving the deprotection/activation of a function or functions [35, 36]. In sequence, the activation of the groups enables repetition of the two reactions resulting in the synthesis of dendritic molecules of the next generation. Two main synthetic strategies have been extensively used in the synthesis of dendrimers over the past three decades. These methods diverge in the direction in which the molecules are constructed and are termed accordingly as either the divergent route or the convergent route.

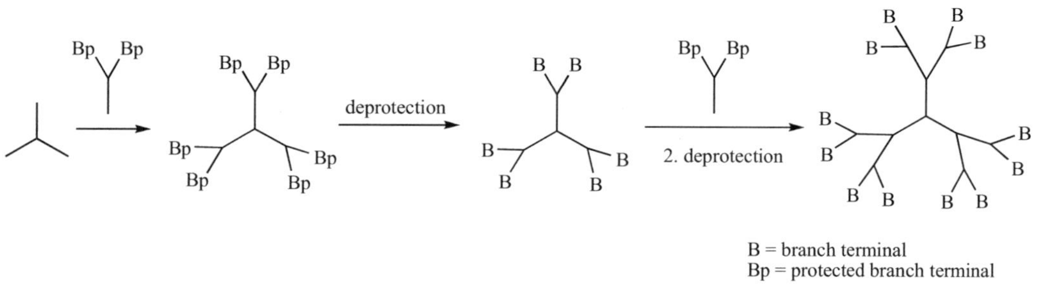

Scheme (1). Divergent synthesis of dendrimers.

The *divergent route* builds up the dendrimer, layer by layer, starting from a multifunctional core molecule diverging out towards the periphery (Scheme **1**). In the first step, an AB* monomer ($x = 2$ in this case) containing protected B groups (Bp) reacts with a trifunctional core to yield a first generation dendrimer with an unreactive (by protection) periphery. In the second step, deprotection or activation of these end groups forms a first generation dendrimer that is able to add the next layer. Replication of these steps yields then a second generation dendrimer with activated end groups [11, 37].

Using the divergent route has enabled the preparation of dendrimers of high generations; for example, the synthesis of the PAMAM series has been successful up to the tenth generation dendrimer (Scheme **2**); the ideal structure of which contains 3072 terminal groups [23]. However, the number of reactions required to occur on each molecule increases exponentially for successive generations in the divergent route and, consequently, at high generations the incomplete formation of the branches is a problem. These imperfect macromolecules and other impurities resulting from side reactions are generally impossible to remove because of their similarity to the desired perfectly branched dendrimer. Therefore, dendrimers of high generations probably contain defects and have some degree of polydispersity.

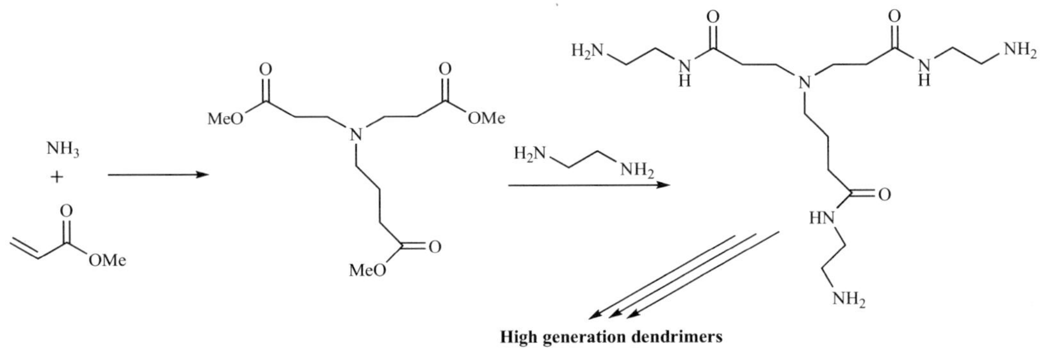

Scheme (2). PAMAM dendrimer synthesis.

In the preparation of PAMAM dendrimers both reactions occur rapidly and in good yields, using a large excess of ethylenediamine (80 equivalents to obtain generation 1, with 3 terminal amines). This large excess is needed to guarantee that reaction takes place at only one of the two amine groups. The synthesis of dendrimers up to generation 8 (384 end groups) were reported by Tomalia in 1987 [38]. The initial ideal perfectly branched structure with a known number of end groups revealed to be imprecise, and these materials possessed a significant number of defects [39, 40]. Regardless of this weakness, most dendrimers are prepared using a divergent manner.

Contrarily to the previous method, the *convergent route* to dendrimers assembles the macromolecule by starting at the structure periphery, moving toward the core. This is accomplished by the construction of dendrons (or dendritic wedges), which can be coupled to a multifunctional core molecule, producing dendrimers of different generations (Scheme **3**). The peak of a dendron is often referred to as the focal point and the function at this position can be either reactive (F) or protected (Fp), depending on the stage in the procedure. The first step is the addition of the branched monomer, which contains a protected focal point (Fp) to the first generation dendron with a reactive focal point (F). This yields a second generation structure with a protected focal point and after deprotection, the dendron can be coupled to a multifunctional core to produce a dendrimer of second generation.

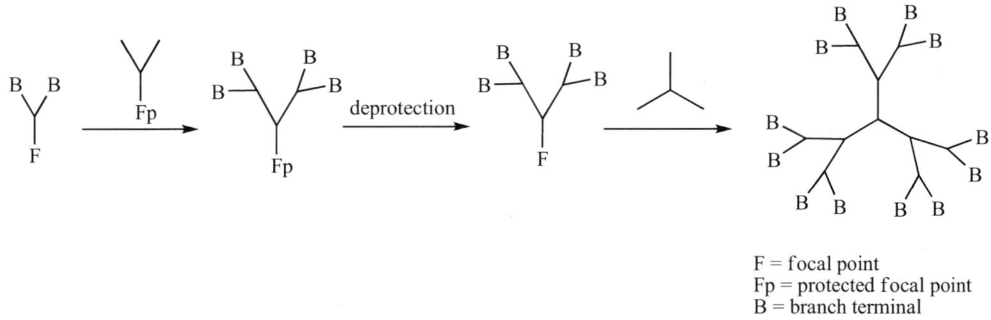

Scheme (3). Convergent dendrimer synthesis.

The convergent method was introduced for the first time by Frechet and Hawker, who reported a series of poly(benzyl ether) dendrimers (Scheme **4**) [26, 41, 42]. The molecule 3,5-dihydroxybenzyl alcohol was used as an AB2 monomer in the assembly of the dendrons. The first step was the reaction of benzyl bromide with the phenolic hydroxyl groups. The primary alcohol function at the focal point of the branched molecule was then activated by conversion to a benzyl bromide function. These two steps were repeated to prepare higher generation dendrons, coupled to multifunctional core molecule to yield the first generation dendrimer.

A similar series of poly(benzyl ester) dendrimers were also synthesized using the convergent route and the isolation of dendrimers of both series up to the fifth generation has been accomplished [41]. The main setback of this method is the difficulty in synthesizing higher generation structures, attributed to the steric demands involved in coupling two (or more) large molecules to functions in close immediacy on the same branching unit or core. Nevertheless, the small number of reactions required for each generational increase and easier purification allows isolation of more precisely monodisperse dendrimers than those isolated by the divergent route. In addition, the convergent method offers a precise location of functions within the dendrimers and high structural control [43], along with the possibility of preparation of asymmetric dendrimers [44 - 47].

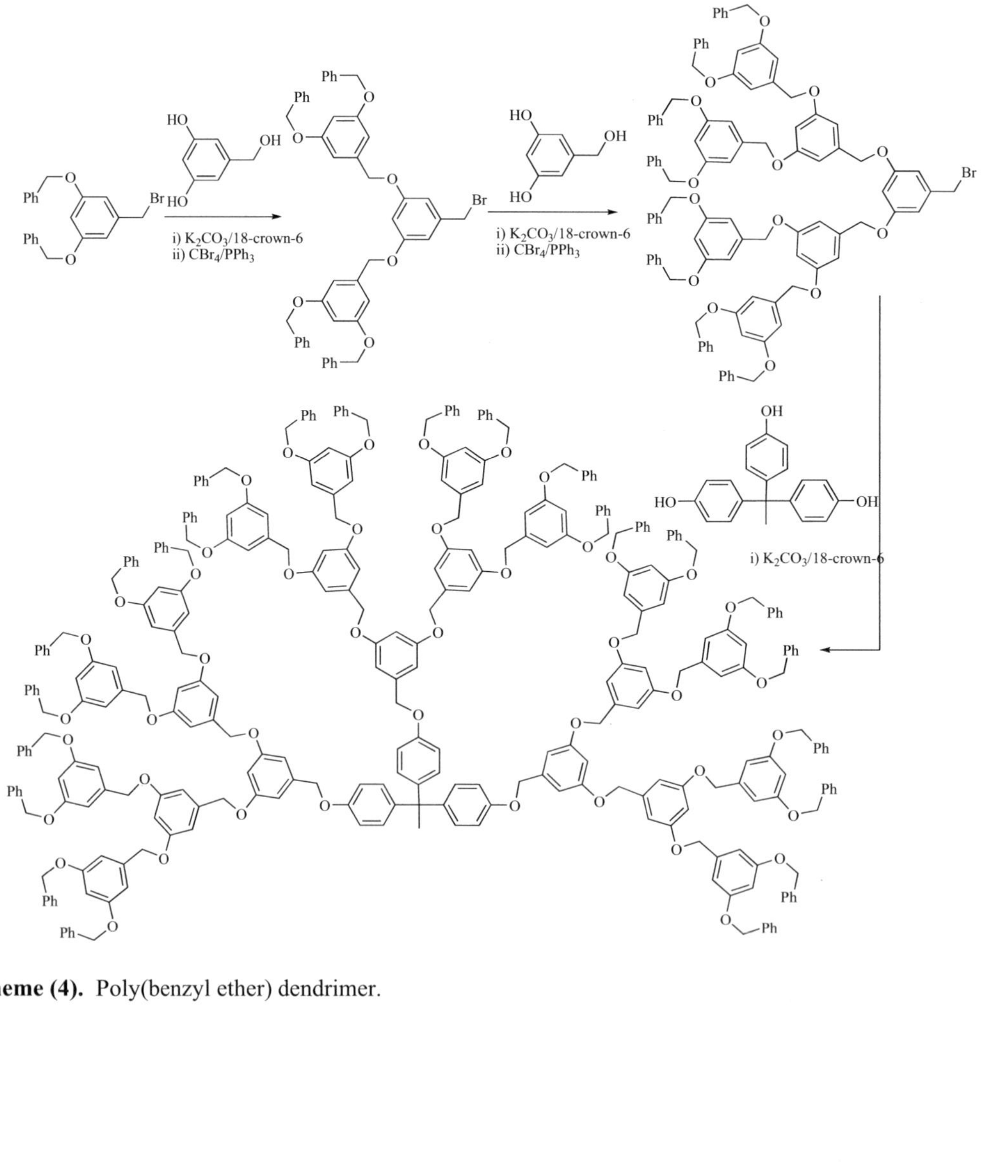

Scheme (4). Poly(benzyl ether) dendrimer.

Both convergent and divergent growths have associated difficulties [48]. The divergent approach may originate surface defects as the generation number increases, and use of excess reagent can lead to purification problems. The convergent approach suffers from steric constraints when the dendrons to be attached to the core become too large and is generally more synthetically challenging. Whichever method of dendrimer synthesis is used, it is always very time consuming, often difficult and hence expensive. Modifications of the convergent route and its combination with the divergent route results in dendrimers being synthesized using methods that reduce the number of steps [37], required for high generation structures. These methods are called accelerated syntheses and include the branched monomer approach [49, 50], in which an AB4 monomer is used, instead of an AB2 monomer and the orthogonal approach [51], which incorporates two different monomer units with two pairs of complementary coupling units, each pair reacting under conditions in which the other functional groups are inert. Therefore, deprotection or activation steps are bypassed, reducing the number of steps necessary to yield high generations. For instance, solid phase techniques [52] used in peptide synthesis have also been adapted for dendrimer synthesis, particularly PAMAM dendrimers, which were grown on the surface of amine functionalized resins [53]. The resin beads can be washed easily to remove the excess reagents and the PAMAM dendrimers can be cleaved from the resin after synthesis, allowing production of more homogenous materials.

APPLICATIONS OF DENDRIMERS

There are now many families of dendrimers, each with unique features like nanoscopic size, spheroidal surface, high branching, cavernous interior, among others. Exciting properties arise from tailoring its surface, interior and core. All the above mentioned dendrimers' properties make them suitable for a variety of technological uses, including in catalysis [3, 54 - 81], as liquid crystals [82 - 111], in light harvesting [48, 112 - 131], in guest-host interactions [23, 132 - 153], as sensors [154 - 166], for biological functions [24, 167 - 180], including drug delivery [181 - 209], gene delivery [210 - 231], therapeutics [232 - 247] and diagnostics [183, 236, 248 - 268].

Applications in Catalysis

The main issues in catalysis deal with catalyst efficiency, along with catalyst recovery, for economical and ecological. Selectivity factors, such as chemo-, regio-, stereo-, enantio- and diastereoselectivity, are other important issues, which are currently mostly optimized with homogeneous catalysts. However, most of the times, catalysts are difficult to remove from reaction media, due to their usual small sizes and amounts used.

In this context, catalysis using dendrimers has appeared as an interesting alternative, because homogeneous catalysts can be linked to the dendrimer's interior or periphery, keeping catalysts homogeneity. Moreover, such dendritic catalysts can be effortlessly recovered from the reaction media, by ultrafiltration or precipitation. van Leeuwen reported and patented polymerization of alkenes using star-shaped hexaphosphine-palladium catalysts in 1992 [65]. Later, several other examples of the use of dendrimers as catalysts were then reported, including the Cu(I)-catalyzed enantioselective styrene cyclopropanation and Rh(I)-catalyzed hydrogenation of acetamidocinnamic acid [54, 66, 67], the electrocatalytic reduction of CO_2 to CO [54], Ni(II)-catalyzed anti-Markovnikov Kharash addition of CCl_4 to methacrylate [64], decarboxylation and phosphonate hydrolysis [69].

Reetz pioneered the recycling of metallodendrimers, using poly(propylene)imine dendrimers (PPI), whose Pd complexes were used in the catalyzed the styrene and bromobenzene Heck reaction. The authors managed to recover the catalyst almost completely by precipitation and reused it, with comparable activity and selectivity [70]. Another quite important development was the introduction of membrane nanofiltration method, first reported by Kragl and Reetz [71]. These dendrimer-based catalysts were used for allylic substitution reactions and very high rates of retention (99.9%) resulted in total TON increases (by 6-fold).

Alper's group [72] reported in 1999 the immobilization of metallodendritic catalysts [72]. Alper synthesized PAMAM-dendronized silica gel support for RhI catalysts, which were then used in the supported hydroformylation of olefins. A breakthrough in dendrimer catalysis was the in catalysis by Crooks' group, in a scientific breakthrough, reported of dendrimer-encapsulated Pd nanoparticles using PAMAM dendrimers in 1999 [73], managing to use them in organic solvents, fluorous/organic biphasic solvents, supercritical CO_2 ($scCO_2$) or water.

The concept of protecting the catalyst active metal at the dendrimer center was also pursued by Suslick group [74]. Kinetic restrictions were a weakness, but the metal center protection and stability could be crucial. Suslick reported dendrimers manganese-porphyrin centered catalysts, providing enhanced selectivity. Years later, van Leeuwen also reported 1,1′-diphenylphosphinoferrocene-centered dendrimers offering increased Pd catalysts' stability [75]. Catalysis using encapsulated catalysts into dendrimers was also evaluated by Crooks, reporting catalysts based on Pd nanoparticles encapsulated into PAMAM dendrimer [76] and by Fréchet [77] in organic catalysis, using dendronized polymers containing 4-(dialkylamino) pyridine substituents, that were able to catalyze acylations using sterically hindered tertiary alcohols.

Applications as Liquid Crystals

Dendritic liquid crystals is a very broad subject [86]. Classic dendrimers, such as PPI, PAMAM [89 - 91] and carbosilane dendrimers [92 - 96] have been extensively studied for their potential applications as liquid crystalline materials.

One particular example is known for G1-G4 pyridine terminated PPI dendrimer H-bonding, in presence of 3-cholesteryloxycarbonyl propanoic acid [96], which formed birefringent glasses at room temperature and viscous smectic A phases at high temperatures (a characteristic from liquid crystals). The presence of chiral mesogenic calamitic groups at amphiphilic polyester dendrimers' ends also formed ferroelectric liquid crystals, displaying smectic A and C phases [97, 106]. Donnio-Guillon [98, 99] reported dendrimer with an octopus shape terminated with aliphatic chains, showing notable smectic A and B phases. A quite important number of publications then came out, reporting mesogenic structures designed with heterolytic and homolytic dendrimers based on octopus cores [3, 11, 48, 93, 96, 98, 99, 267, 269].

Moore also reported columnar discotic liquid crystals constituted by dendrimers having rigid cores, bearing oligo(ethylene oxide) chains as termini [106], while PPI and PAMAM dendrimers were tailored by 1,3,5-trioctyloxybenzoic, decanoic and 4-octylbenzoic acids termini, generating luminescent dendritic liquid crystals [107].

Applications in Light Harvesting

Dendrimers are normally solution processable and, using appropriate fine-tuning, one can direct their optoelectronic properties, which resembles small molecules. Additionally, their architecture allows controlling of the intermolecular interactions that govern device performance [116]. For instance, there are some reports of dendrimers bearing porphyrins [117 - 120]. Among them, propylene imine dendrimers **1** (Fig. **6**), containing porphyrins at their periphery were synthesized using divergent route, forming a dendrimer with porphyrin rings bound to the surface.

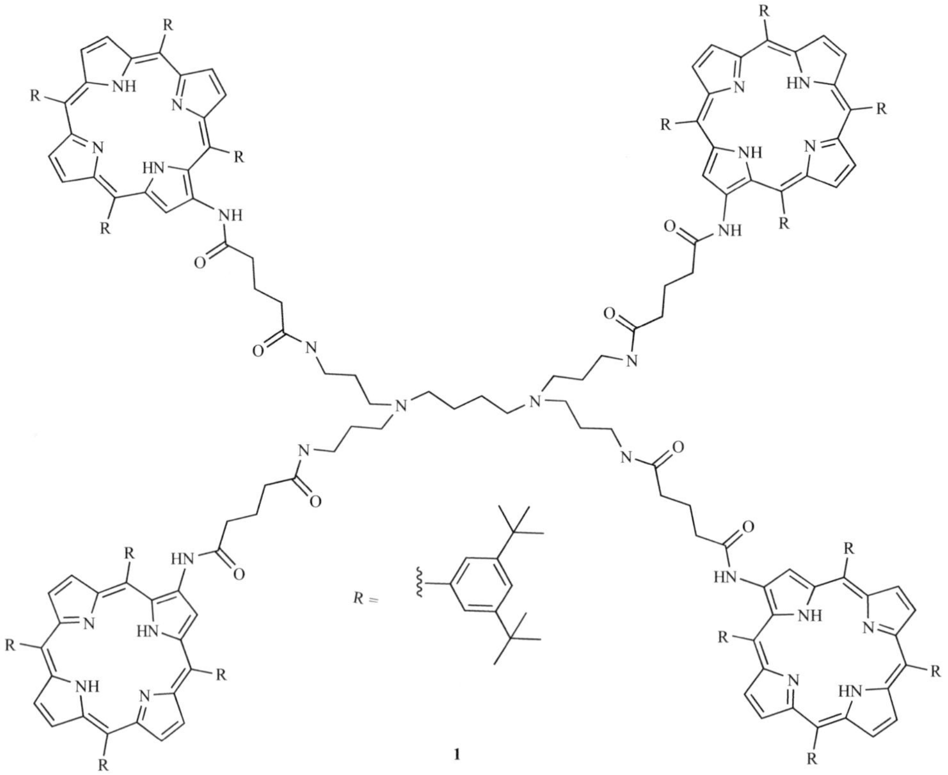

1

Fig. (6). Porphyrin substituted dendrimer for photovoltaics.

Dendrimer **1** exhibited higher absorptivity than that of nondendritic porphyrins, and the porphyrin-propylene imine dendrimers were blended with C_{60}, originating a bulk heterojuction solar cell displaying higher conversion efficiency values than those of devices having simple porphyrins. A maximum photocurrent conversion efficiency (PCE) value of 15% was observed in a Grätzel-like cell for the G3 dendrimer blend, derived from **1** (Fig. **6**), and an overall PCE $\eta = 0.32\%$ was attained.

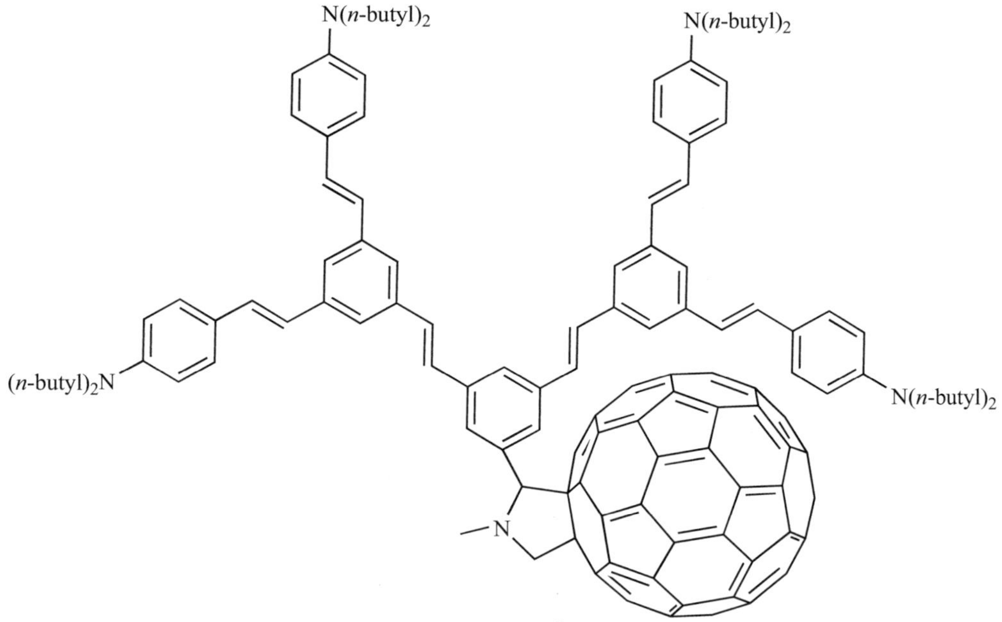

2

Fig. (7). Dendritic C60 based dendrimer for solar cell applications.

There have been few reports on conjugated arylalkene bearing dendrimers for solar cells so far. For instance, a solar cell material **2**, (Fig. **7**) was reported; however, no device performance was reported [121]. Dendritic thiophene materials were also investigated for solar cell applications. Highly branched (*e.g.***3** Fig. **8**) [123] and well-defined (*e.g.* **4a,b**, Fig. **8**) [122] materials have been prepared. While **3** was prepared using nickel-catalyzed couplings, dendrimers **4a,b** used Stille reaction. The material **4b** (n = 3) displayed enhanced hole mobility and improved absorption overlap with the solar spectrum, with corresponding photovoltaic cell PCBM (ITO/**4b**:PCBM/Al) attaining a power conversion efficiency η = 1.3% [124].

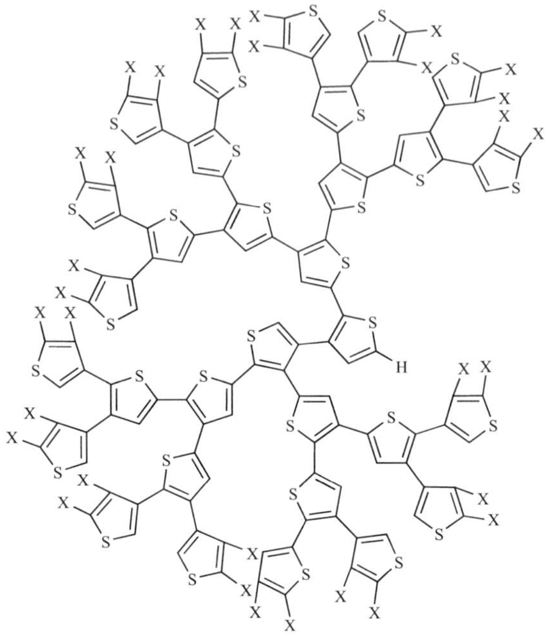

3 X = Br or *n*-hexyl

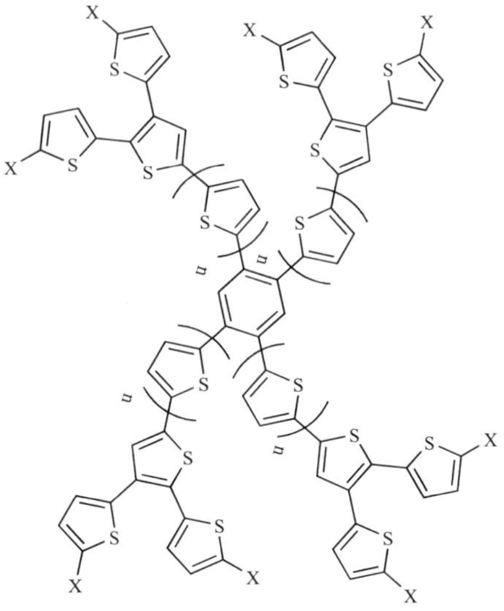

X = *n*-hexyl

4a: n = 1
4b: n = 3

Fig. (8). Dendritic thiophene compound for photovoltaic applications.

Several well-defined dendrimers containing thiophene arms with tris(thieno)-benzene or triphenylamine cores were also used in photovoltaic cells. The bilayer photovoltaic cell using **5** (Fig. **9**) exhibited a PCE η = 1.0% in a (ITO/**5**/C_{60}/Al) bilayer device [125], while a PCBM blended analogous material bearing perylene groups displayed η = 0.25% in a non optimized device (ITO/PEDOT:PSS/dendrimer:PCBM/Al) [126].

The branched thiophene **6** (Fig. **9**), (ITO/PEDOT:PSS/**6**/perylene/LiF/Al), exhibited a performance of η = 1.3%, at low incident light intensity [127]. Finally, single-layer photovoltaic cell **7** (Fig. **9**), (ITO/**7**:CdSe/Al) [128], showed PCE η = 0.29% at an illumination intensity of 0.14 mW/cm^2.

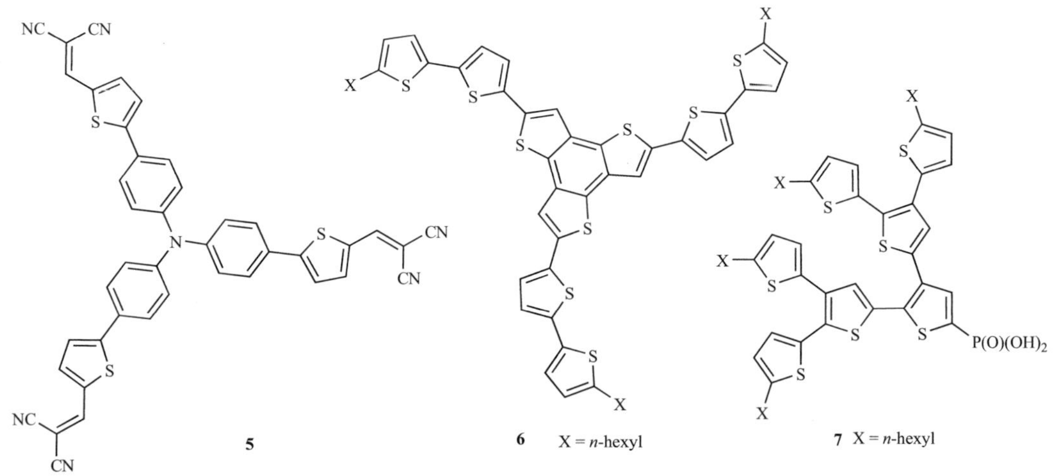

Fig. (9). Oligothiophene based dendritic solar cell material.

Applications in Guest-host Interactions

Since the influential Newkome's work [25], dendrimer guest encapsulation emerged as an important dendritic property [133]. Several authors performed encapsulation studies of neutral guests, involving molecules such as benzoic acid, pyridine, iodine or 2,6-dibromo-4-nitrophenol [137], Reichart's dye [138], along with other dyes [139, 140] and other molecules of photophysical interest (anthracene [141] and pyrene [142, 143]).

A high quantity of studies related to PAMAM dendrimers, due to their commercially availability, ease of encapsulation, ability to be functionalized with terminal groups and water-solubility. Most of the times, PAMAM dendrimers were functionalized through their terminal amine groups; however occasionally the PAMAM hosts possessed PEG or OH termini groups or even esters or acids [144]. Dendrimers with hydrophilic periphery and hydrophobic core showed micelle-type behavior, also acting as molecular containers [25, 145, 147 - 150, 173].

Applications as Sensors

The majority of all fluorescent dendritic materials are capable to be directly applied to sensing. For instance, Samuel [159] used a bisfluorene dendrimerdistributed compound in the sensing of 1,4-dinitrobenzene, a 2,4,6-trinitrotoluene (TNT) model explosive, by fluorescence quenching measurement of G1-dendrimer coupled with 2,2′-bis(9,9-di-*n* -hexylfluorene) core.

Most of dendritic sensors are based on electrochemistry or photophysics. For instance, gas sensors typically use composite dendrimer membranes and films. The thin films of Au nanoparticles containing dendrimers were prepared by layer-by-layer assemblies, being effectively tested in conductivity measurements for vapor sensing of 1-propanol, toluene and water [160]. Also PAMAM based membranes were used for CO_2 separation with high CO_2/N_2 selectivity [161, 162], and dendrimer composites containing carbon black could be used to detect volatile organic carboxylic acids and amines [163]. Additionally, a PAMAM dendrimer composite containing chitosan layered hyaluronic acid was reported for improved CO_2 separation, by enhancing the membrane's swelling degree [164]. *G* 0-*G* 8 PAMAM dendrimer films were evaluated in the sensing of volatile organic compounds [165], where it was demonstrated that G4 dendrimer was the optimal generation, due to being the smallest dendrimer spheroid with internal access.

Applications in Drug Delivery

The biomedical prospect for dendrimers was evident since their first early 1980s report [23]. The analogy and connection of dendrimers with biomolecules arise from their globular shapes and sizes, which are similar to bioassemblies such as hemoglobin (5.5 nm), lipid bilayer membranes (5.5 nm), insulin (4 nm) and DNA duplexes (2.4 nm) [188, 189].

Medicinal applications using dendrimers blasted off in the 1990s in antibodies mimicks, for sensing functionalities and immunological applications on DNA branches' selective recognition and to quantify and detect the AIDS virus [190]. Dendrimers can also be considered nanocarriers designed to enhance pharmacodynamics, pharmacokinetics, bioavailability of drugs *in vivo* and water solubility [191 - 193], taking advantages of dendrimer's capability to link with specific targeting groups (such as peptides, glycosides, folic acid and monoclonal antibodies).

The solubility of dendrimers is mainly ruled by their terminal groups. Hence, biologically suitable dendrimers with hydrophobic interiors and water-soluble termini have been designed, being able to encapsulate hydrophobic drugs by noncovalent bonds. Larger molecules rather adsorb at surface, while small molecules can be encapsulated in the dendrimer interior.

A seminal example regards the anti-inflammatory drug ibuprofen, which can complexate with G4-PAMAM dendrimer through carboxyl-amine electrostatic interactions [194, 195]. This bioconjugate penetrated A549 cells more rapidly, than ibuprofen itself. Similarly, the encapsulation of anticancer camptothecin in G4.5 carboxylate-terminated polyester dendrimer was also reported [196, 197] along with the encapsulation of indomethacin into PAMAM for transdermal delivery [198], while the solubilization of the hydrophobic drug nifedipine was enhanced by PAMAM's generation increase [199, 200]. Also successful results were obtained with aspirin, proicam, diflunisal, diclofenac, naproxen, ketoprofen, aceclofenac and flurbiprofen [201 - 208].

Among several water solubilization possibilities [270], poly(ethylene)glycol (PEG) emerged as a good choice to be used as dendrimeric component [181, 271]. Pioneered by Fréchet's group [181, 272], he reported the combination of biodegradable polyester dendrimers with PEGs for drug delivery therapeutics performing several solid tumors accumulation studies.

Water soluble melanine dendrimers are another particular dendrimer family, firstly synthesized by Simanek's group [273, 274] along with their functionalization using several water soluble groups, including PEG, phosphonate, carboxylate, guanidine, amine and sulfonate, evaluating them in *in vivo* cytotoxicity and hemolytic potential in mice, being the PEGylated examples, the best in terms of lethality or toxicity [273 - 275]. Park [276, 277] and Simanek [278] independently reported the application of G3, G4, and G5 polyglycerol dendrimers with the anticancer drug paclitaxel by 270-, 370- and 434-fold, respectively [276 - 278]. Tumor cells overexpressed several receptors, like specific peptide sequences, allowing their use as targets when they were linked to dendrimer-drug conjugates [271, 279, 280]. This vectorization strategy significantly reduced the drug distribution to normal cells, it improved drug efficiency, concomitantly with side effects reduction. For instance, the folate receptor is overexpressed by several cancers [281], and has been dendrimer conjugated by Tomalia [24] and Fréchet [282] for the first time. The former used MTX complexated G5-PAMAM-folate dendrimers [24], while the latter used polyether dendrimers conjugated with the drug MTX [282]. Also glycopeptide dendrimers show applicability in biomedicine, including as antiviral agents (including influenza and HIV), anticancer agents (immunotherapy and antiangiogenic) and antibacterial agents [283 - 289].

Gene therapy regards genetic defect corrections though DNA transfer (transfection) into cells [216, 217], for which dendrimers stability (higher than liposomes) accurate size design, generation, nature of termini groups and monodispersity are of acute importance [212, 218]. For instance, electrostatic interactions between the anionic DNA backbone's phosphate groups and the physiologically protonated primary amine terminated dendrimer yielded the formation of a dendrimer-DNA named "dendriplex" [219, 220]. The dendriplex bound to the cell membrane by electrostatic interaction of the carboxylate and phosphate membrane groups with its positive charges and then internalized into the cytoplasm upon endosomal uptake.

Low-generation PPI and PAMAM dendrimers are more suitable for efficient gene transfection than high-generation dendrimers, due to their lower cytotoxicity [221]. Thus, gene transfer has been carried out *in vivo* using this type of amine terminated, dendrimers to cure mice tumors [222 - 225, 290], including phosphorus dendrimers [222, 223] and carbosilane dendrimers [290]. Thomas [226] and Park [227] reported that the use of poly(propylene imine) dendrimers for prostate, ovarian and breast cancer oligonucleotide delivery. Protein transduction for intracellular protein was also carried out using lipidic dendrimers [228 - 230].

Applications in Therapeutics

Aida's group found that G3 poly(benzylether) dendrimers, having zinc porphyrin cores with carboxylate groups as termini, were found to be phototoxic against *in vitro* Lewis lung cells, when they were enclosed inside positively charged PEG-lysine block copolymers [238, 239]. To be used as vehicles for PDT, also

PEGylated tetraarylporphyrins have been included in vesicular carriers [240], as well as PEGylated PPI and PAMAM dendrimers [241].

The Kataoka group [238, 242 - 244] demonstrated that micelles loaded with dendrimer-photosensitizer conjugates displayed a higher antitumor effect than Photofrin® without any sign of normal tissue phototoxicity.

Applications in Disease Diagnostics

Pre-targeting of receptors is a practical approach in molecular image and therapeutics to reduce background toxicity or noise and improve selectivity. For instance, the medical diagnostics currently used modality, MRI, is based on the improvement of the quality of organ visualization by improving the longitudinal (T1) relaxation rate of H_2O protons by coordination to paramagnetic contrast agents that can be some lanthanide chelate complexes (*e.g.* GdIII), such as the generally used [GdIII(DTPA)] (DTPA) diethylenetriamine pentacetic acid, commercially known as Magnevist (Schering AG) and [GdIII(DOTA)] (DOTA) tetracarboxymethyl-1,4,7,10-tetraaza- cyclododecane).

The main problems of complexes having low molecular weight are the lack of discrimination between normal and diseased tissues body and short circulation times. Hence, PAMAM dendrimers' commercialization was critical for the MRI field when, in 1994, the groups of Tomalia and Lauterbur (2003 Nobel Prize in medicine) reported G6- and G2-PAMAM dendrimer-based Gd(III) chelates [255], displaying long blood circulation times (over 100 min) and first-rate blood vessel MRI images with *G* 6-PAMAM-[Gd(III)(DTPA)] upon rabbit intravenous injection. In the same way, ovarian tumor xenografts resulted in great MRI contrast improvement [256, 257].

The best results were obtained later with G9- and G10-PAMAM dendrimers, probably due to a possible relaxativity linear increase with the molecular weight. The introduction of PEG units significantly lowered liver retention after one week (from 40% without PEG) to 1-8%, while conjugation to avidin or monoclonal antibodies offered tumor-targeted MRI agents. **Gadomer 17**, a 24-Gd(III)-DTPA dendrimer containing trimesic acid core connected to *G* 2-polylysine dendrons possessing peripheral chelating-Gd(III) groups, was later reported, and even commercialized [188].

Many dendrimer-Gd(III) complexes for MRI have been developed since [183, 207, 258]. Besides lysine and PAMAM dendrimers, diaminobutane (DAB) and PPI dendrimers conjugates have also been used as MRI agents [259, 260], improving body clearance, when compared with PAMAM-based agents. Also the groups of Prasad and Fréchet reported the generation of cytotoxic singlet oxygen for PDT to subcutaneous tumors by fluorescence resonance energy transfer (FRET), using porphyrin sensitizers as dendritic cores [261 - 263].

Tridentate bis(pyridyl)amine terminated G5-G7-dendrimers was complexated with radioactive $^{99m}Tc(I)$, providing a dendrimer radiolabeled agent for single photon emission computed tomography in healthy adult Copenhagen rats in determining the biodistribution [264]. Fréchet [265] reported a biodegradable dendritic radiohalogen-based (^{125}I and ^{76}Br) PET nanoprobe targeted at $\alpha_v\beta_3$ integrin, a angiogenesis biological marker. This dendrimer enabled a 50-fold improvement of the $\alpha_v\beta_3$ integrin receptors binding affinity, when compared to a monovalent RGD.

In vivo biodistribution studies of ^{76}Br-labeled dendritic probes displayed high bioavailability and high nanoprobe accumulation targeted at $\alpha_v\beta_3$ integrins in angiogenic muscles, permitting a selective imaging. A G5-PAMAM dendrimer conjugated to a fluorescent probe (6-carboxytetramethylrhodamine succinic ester) and folic acid was also used to target xenograft tumors in mice, displaying tumor accumulation up to 673 ± 67 nM at 2 h, whereas the analogous conjugate without folic acid reached only 136 ± 28 nM in the same time [266].

The covalent inclusion of near-IR fluorophores in a dendrimer surrounded by a shell of polyethylene oxide originated dansyl fluorescence in dendrimers [267], which was also explored for intracellular localization and cellular uptake, monitored using confocal fluorescence microscopy, displaying increased nanoprobe stability with prolonged blood residence time and added resistance to enzymatic degradation, permitting to monitor *in vivo* fluorescence lifetime changes [268].

EXPERIMENTAL SECTION

Synthesis of Poly(Propylene Imine) Dendrimer (PPI) (Scheme 5)

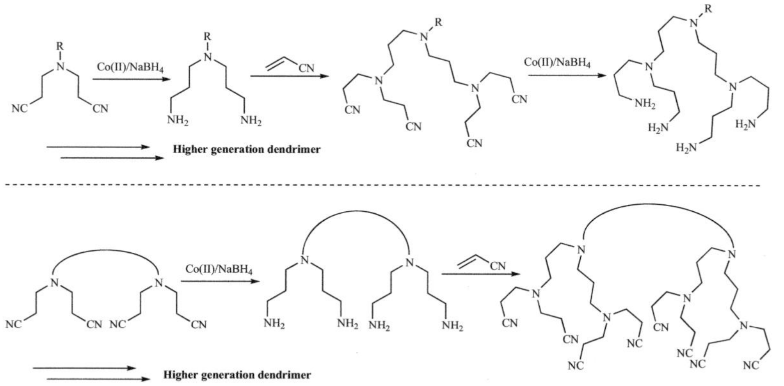

Scheme (5). Synthesis of poly(propylene imine) dendrimer. Procedure for total synthesis of dendrimer PPI can be directly found in Reference [14].

Synthesis of Melamine (Triazine) Dendrimers (Scheme 6)

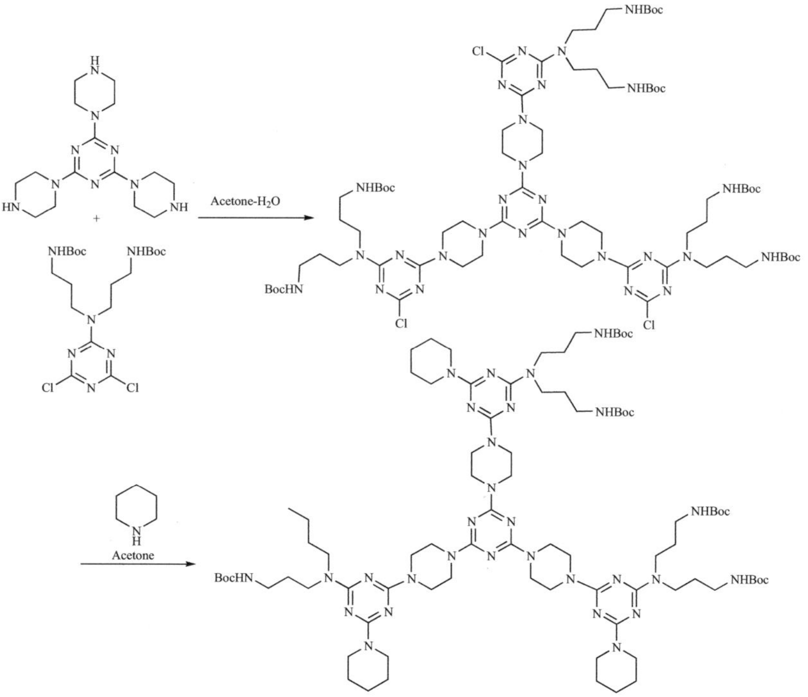

Scheme (6). Synthesis of melamine centered dendrimer. Procedure for total synthesis of dendrimer can be directly found in Reference [291].

CONCLUDING REMARKS

Dendrimers hold a stimulating possibility for chemists to generate macromolecular structures with an expressly customized function or several functions, exposing a multivalent surface, which is a promising motif to enhance a given functionality. Furthermore, the microenvironment in the core may be used to carry low-molecular substances, *e.g.* drugs, or may be useful to create altered properties of core-chromophores or fluorophores, among other possibilities. These selected properties will naturally render dendrimers to be multivalent in areas ranging from catalysis, energy, materials to biomedicine and biology.

Most of synthetic approaches to dendrimers are currently well developed; however, the study of the interactions between them and "guest molecules" still requires insight, naturally depending upon the desired purpose. Here lies the future for dendrimers, in applications that may lead to amplification of pre-existing properties in many applications.

REFERENCES

[1] Newkome, G.R.; Moorefield, C.N.; Vögtle, F. *Dendritic Molecules: Concepts, Syntheses, Perspectives*; VCH: Weinheim, Germany, **1996**. [http://dx.doi.org/10.1002/9783527614875]

[2] Pillai, O.; Panchagnula, R. Polymers in drug delivery. *Curr. Opin. Chem. Biol.,* **2001**, *5*(4), 447-451. [http://dx.doi.org/10.1016/S1367-5931(00)00227-1] [PMID: 11470609]

[3] Fréchet, J.M.; Tomalia, D.A. *Dendrimers and other Dendritic Polymers*; Wiley: Chichester, UK, **2002**.

[4] Flory, P.J. Thermodynamics of high polymer solutions. *J. Chem. Phys.,* **1941**, *9*, 660-661. [http://dx.doi.org/10.1063/1.1750971]

[5] Flory, P.J. Thermodynamics of high polymer solutions. *J. Chem. Phys.,* **1942**, *10*, 51-61. [http://dx.doi.org/10.1063/1.1723621]

[6] Hadjichristidis, N.; Pitsikalis, M.; Pispas, S.; Iatrou, H. Polymers with complex architecture by living anionic polymerization. *Chem. Rev.,* **2001**, *101*(12), 3747-3792. [http://dx.doi.org/10.1021/cr9901337] [PMID: 11740920]

[7] Schaefgen, J.R.; Flory, P.J. Synthesis of multichain polymers and investigation of their viscosities. *J. Am. Chem. Soc.,* **1948**, *70*, 2709-2718. [http://dx.doi.org/10.1021/ja01188a026]

[8] Morton, M.; Helminiak, T.E.; Gadkary, S.D.; Buecke, F. Preparation and properties of monodisperse branched polystyrene. *J. Polym. Sci. Part A: Chem.,* **1962**, *57*, 471-482.

[9] Decker, D. Etude en solution diluée de polymères à structure en "peigne". *Makromol. Chem.,* **1969**,

125, 136-160.
[http://dx.doi.org/10.1002/macp.1969.021250114]

[10] Falk, J.C.; Scott, R.J.; Hoeg, D.F.; Pendleton, J.F. New thermoplastic elastomers. styrene grafts on lithiated polydienes and their hydrogenated counterparts. *Rubber Chem. Technol.,* **1973**, *46*, 1044-1054.
[http://dx.doi.org/10.5254/1.3547404]

[11] Frechet, J.M.; Hawker, C.J. Synthesis and Properties of Dendrimers and Hyperbranched Polymers In: *Comprehensive Polymer Science*; , **1996**; p. 2nd Suppl.71.

[12] Gao, C.; Yan, D. Hyperbranched polymers: from synthesis to applications. *Prog. Polym. Sci.,* **2004**, *29*, 183-275.
[http://dx.doi.org/10.1016/j.progpolymsci.2003.12.002]

[13] Hult, A.; Johansson, M.; Malmstrom, E. Hyperbranched polymers. *Adv. Polym. Sci.,* **1999**, *143*, 1-34.
[http://dx.doi.org/10.1007/3-540-49780-3_1]

[14] Buhleier, E.; Wehner, W.; Vögtle, F. "Cascade"- and "Nonskid-Chain-like" syntheses of molecular cavity topologies. *Synthesis,* **1978**, 155-158.
[http://dx.doi.org/10.1055/s-1978-24702]

[15] Vögtle, F.; Richardt, G.; Werner, N. *Dendritische moleküle*; Teubner Verlag: Wiesbaden, **2007**.

[16] Feuerbacher, N.; Vogtle, F. Iterative synthesis in organic chemistry. *Top. Curr. Chem.,* **1998**, *197*, 1-18.
[http://dx.doi.org/10.1007/3-540-69779-9_1]

[17] Jacobi, P.A.; Liu, H. Iterative synthesis of semicorrins, tripyrrolines, and higher analogues. *J. Am. Chem. Soc.,* **1999**, *121*, 1958-1959.
[http://dx.doi.org/10.1021/ja983698d]

[18] Zhao, Y.M.; Tykwinski, R.R. Iterative synthesis and properties of cross-conjugated iso-polydiacetylene oligomers. *J. Am. Chem. Soc.,* **1999**, *121*, 458-459.
[http://dx.doi.org/10.1021/ja983288p]

[19] Denkewalter, R.G.; Kolc, J.F.; Lukasavage, W.J. . Preparation of lysine based macromolecular highly branched homogeneous compound. . U.S. Patent 4360646, 1979.

[20] Maciejewski, M. Concepts of trapping topologically by shell molecules. *J. Macromol. Sci. Part A Pure Appl. Chem.,* **1982**, *A17*, 689-707.
[http://dx.doi.org/10.1080/00222338208062416]

[21] Tomalia, D.A.; Baker, H.; Dewald, J.; Hall, M.; Kallos, G.; Martin, S.; Roeck, J.; Ryder, J.; Smith, P. A new class of polymers: starburst-dendritic macromolecules. *Polym. J.,* **1985**, *17*, 117-132.
[http://dx.doi.org/10.1295/polymj.17.117]

[22] Tomalia, D.A.; Baker, H.; Dewald, J.; Hall, M.; Kallos, G.; Martin, S.; Roeck, J.; Ryder, J.; Smith, P. Dendritic macromolecules: synthesis of starburst dendrimers. *Macromolecules,* **1986**, *19*, 2466-2468.
[http://dx.doi.org/10.1021/ma00163a029]

[23] Tomalia, D.A.; Naylor, A.M.; Goddard, W.A. III Starburst dendrimers: molecular-level control of size, shape, surface chemistry, topology, and flexibility from atoms to macroscopic matter. *Angew. Chem. Int. Ed. Engl.,* **1990**, *29*, 138-175.

[http://dx.doi.org/10.1002/anie.199001381]

[24] Esfand, R.; Tomalia, D.A. Poly(amidoamine) (PAMAM) dendrimers: from biomimicry to drug delivery and biomedical applications. *Drug Discov. Today,* **2001**, *6*(8), 427-436. [http://dx.doi.org/10.1016/S1359-6446(01)01757-3] [PMID: 11301287]

[25] Newkome, G.R.; Yao, Z-Q.; Baker, G.R.; Gupta, V.K. Micelles. Part 1. Cascade molecules: a new approach to micelles. A [27]-arborol. *J. Org. Chem.,* **1985**, *50*, 2003-2004. [http://dx.doi.org/10.1021/jo00211a052]

[26] Hawker, C.; Frechet, J.M. A new convergent approach to monodisperse dendritic macromolecules. *J. Chem. Soc. Chem. Commun.,* **1990**, 1010-1013. [http://dx.doi.org/10.1039/c39900001010]

[27] Miller, T.M.; Neenan, T.X. Convergent synthesis of monodisperse dendrimers based upon 1,3,5-trisubstituted benzenes. *Chem. Mater.,* **1990**, *2*, 346-349. [http://dx.doi.org/10.1021/cm00010a006]

[28] Rebrov, E.A.; Muzafarov, A.M.; Papkov, V.S.; Zhdanov, A.A. Dendritic polyorganosiloxanes. *Dokl. Akad. Nauk SSSR,* **1989**, *309*, 376-381.

[29] Rengan, K.; Engel, R. The synthesis of phosphonium cascade molecules. *J. Chem. Soc., Perkin Trans. 1,* **1991**, 987-990. [http://dx.doi.org/10.1039/p19910000987]

[30] Launay, N.; Caminade, A-M.; Lahana, R.; Majoral, J-P. A general synthetic strategy for neutral phosphorus-containing dendrimers. *Angew. Chem. Int. Ed. Engl.,* **1994**, *33*, 1589-1592. [http://dx.doi.org/10.1002/anie.199415891]

[31] Galliot, C.; Prevote, D.; Caminade, A.M.; Majoral, J.P. Polyaminophosphine containing dendrimers. syntheses and characterization. *J. Am. Chem. Soc.,* **1995**, *117*, 5470-5476. [http://dx.doi.org/10.1021/ja00125a006]

[32] Slany, M.; Bardaji, M.; Casanove, M.J.; Caminade, A.M.; Majoral, J.P.; Chaudret, B. Dendrimer surface chemistry. facile route to polyphosphines and their gold complexes. *J. Am. Chem. Soc.,* **1995**, *117*, 9764-9765. [http://dx.doi.org/10.1021/ja00143a023]

[33] Mihov, G.; Grebel-Koehler, D.; Lübbert, A.; Vandermeulen, G.W.; Herrmann, A.; Klok, H.A.; Müllen, K. Polyphenylene dendrimers as scaffolds for shape-persistent multiple peptide conjugates. *Bioconjug. Chem.,* **2005**, *16*(2), 283-293. [http://dx.doi.org/10.1021/bc049839k] [PMID: 15769081]

[34] Morgenroth, F.; Reuther, E.; Mullen, K. Polyphenylene dendrimers: from three-dimensional to two-dimensional structures. *Angew. Chem. Int. Ed. Engl.,* **1997**, *36*, 631-634. [http://dx.doi.org/10.1002/anie.199706311]

[35] Sowinska, M.; Urbanczyk-Lipkowska, Z. Advances in the chemistry of dendrimers. *New J. Chem.,* **2014**, *38*, 2168-2203. [http://dx.doi.org/10.1039/c3nj01239e]

[36] Newkome, G.R.; Shreiner, C. Dendrimers derived from 1 $\rightarrow$ 3 branching motifs. *Chem. Rev.,* **2010**, *110*(10), 6338-6442.

[http://dx.doi.org/10.1021/cr900341m] [PMID: 20666373]

[37] Matthews, O.A.; Shipway, A.N.; Stoddart, J.F. Dendrimers-Branching out from curiosities into new technologies. *Prog. Polym. Sci.,* **1998**, *23*, 1-56.
[http://dx.doi.org/10.1016/S0079-6700(97)00025-7]

[38] Tomalia, D.A.; Berry, V.; Hall, M.; Hedstrand, D.M. Starburst dendrimers. 4. Covalently fixed unimolecular assemblages reiminiscent of spheroidal micelles. *Macromolecules,* **1987**, *20*, 1164-1167.
[http://dx.doi.org/10.1021/ma00171a051]

[39] Hummelen, J.C.; van Dongen, J.L.; Meijer, E.W. Electrospray mass spectrometry of poly(propylene imine) dendrimers-the issue of dendritic purity or polydispersity. *Chemistry,* **1997**, *3*, 1489-1493.
[http://dx.doi.org/10.1002/chem.19970030916]

[40] Newkome, G.R.; Weis, C.D.; Moorefield, C.N.; Weis, I. Detection and functionalization of dendrimers possessing free carboxylic acid moieties. *Macromolecules,* **1997**, *30*, 2300-2304.
[http://dx.doi.org/10.1021/ma961716a]

[41] Hawker, C.J.; Frechet, J.M. Unusual macromolecular architectures: the convergent growth approach to dendritic polyesters and novel block copolymers. *J. Am. Chem. Soc.,* **1992**, *114*, 8405-8413.
[http://dx.doi.org/10.1021/ja00048a009]

[42] Hawker, C.J.; Frechet, J.M. Preparation of polymers with controlled molecular architecture. A new convergent approach to dendritic macromolecules. *J. Am. Chem. Soc.,* **1990**, *112*, 7638-7647.
[http://dx.doi.org/10.1021/ja00177a027]

[43] Wooley, K.L.; Hawker, C.J.; Frechet, J.M. Polymers with controlled molecular architecture: control of surface functionality in the synthesis of dendritic hyperbranched macromolecules using the convergent approach. *J. Chem. Soc., Perkin Trans. 1,* **1991**, 1059-1076.
[http://dx.doi.org/10.1039/p19910001059]

[44] Maraval, V.; Maraval, A.; Spataro, G.; Caminade, A.M.; Majoral, J.P.; Kim, D.H.; Knoll, W. Design of tailored multi-charged phosphorus surface-block dendrimers. *New J. Chem.,* **2006**, *30*, 1731-1736.
[http://dx.doi.org/10.1039/b610632n]

[45] Maraval, V.; Sebastian, R.M.; Ben, F.; Laurent, R.; Caminade, A.M.; Majoral, J.P. Varying topology of dendrimers-a new approach toward the synthesis of di-block dendrimers. *Eur. J. Inorg. Chem.,* **2001**, *7*, 1681-1691.
[http://dx.doi.org/10.1002/1099-0682(200107)2001:7<1681::AID-EJIC1681>3.0.CO;2-T]

[46] Wooley, K.L.; Hawker, C.J.; Frechet, J.M. Unsymmetrical three-dimensional macromolecules: preparation and characterization of strongly dipolar dendritic macromolecules. *J. Am. Chem. Soc.,* **1993**, *115*, 11496-11505.
[http://dx.doi.org/10.1021/ja00077a055]

[47] Yamamoto, K. Dendrimer complexes: Fine control of metal assembly in macromolecules. *J. Polym. Sci. A Polym. Chem.,* **2005**, *43*, 3719-3727.
[http://dx.doi.org/10.1002/pola.20925]

[48] Grayson, S.M.; Fréchet, J.M. Convergent dendrons and dendrimers: from synthesis to applications. *Chem. Rev.,* **2001**, *101*(12), 3819-3868.
[http://dx.doi.org/10.1021/cr990116h] [PMID: 11740922]

[49] Wooley, K.L.; Hawker, C.J.; Frechet, J.M. "branched-monomer approach" for the rapid synthesis of dendimers. *Angew. Chem. Int. Ed. Engl.,* **1994**, *33*, 82-85. [http://dx.doi.org/10.1002/anie.199400821]

[50] Walter, M.V.; Malkoch, M. Simplifying the synthesis of dendrimers: accelerated approaches. *Chem. Soc. Rev.,* **2012**, *41*(13), 4593-4609. [http://dx.doi.org/10.1039/c2cs35062a] [PMID: 22592560]

[51] Zeng, F.; Zimmermann, S.C. Rapid synthesis of dendrimers by an orthogonal coupling strategy. *J. Am. Chem. Soc.,* **1996**, *118*, 5326-5327. [http://dx.doi.org/10.1021/ja960317s]

[52] Merrifield, R.B. Solid phase peptide synthesis. I. The synthesis of a tetrapeptide. *J. Am. Chem. Soc.,* **1963**, *85*, 2149-2154. [http://dx.doi.org/10.1021/ja00897a025]

[53] Swali, V.; Wells, N.J.; Langley, G.J.; Bradley, M. Solid-phase dendrimer synthesis and the generation of super-high-loading resin beads for combinatorial chemistry. *J. Org. Chem.,* **1997**, *62*, 4902-4903. [http://dx.doi.org/10.1021/jo9708654]

[54] Brunner, H.; Dendrimer, I.V. Metal coordination, self assembly, catalysis. *J. Organomet. Chem.,* **1995**, *500*, 39-46. [http://dx.doi.org/10.1016/0022-328X(95)00507-M]

[55] Connelly, N.G.; Geiger, W.E. Chemical redox agents for organometallic chemistry. *Chem. Rev.,* **1996**, *96*(2), 877-910. [http://dx.doi.org/10.1021/cr940053x] [PMID: 11848774]

[56] Bosman, A.W.; Janssen, H.M.; Meijer, E.W. About dendrimers: structure, physical properties, and applications. *Chem. Rev.,* **1999**, *99*(7), 1665-1688. [http://dx.doi.org/10.1021/cr970069y] [PMID: 11849007]

[57] Cuadrado, I.; Moran, M.; Casado, C.M.; Alonso, B.; Losada, J. Organometallic dendrimers with transition metals. *Coord. Chem. Rev.,* **1999**, *193-195*, 395-445. [http://dx.doi.org/10.1016/S0010-8545(99)00036-3]

[58] Newkome, G.R.; He, E.; Moorefield, C.N.; Vögtle, F. *Dendrimers and Dendrons. Concepts, Synthesis, Applications,* **2001**.

[59] Fihri, A.; Bouhrara, M.; Nekoueishahraki, B.; Basset, J.M.; Polshettiwar, V. Nanocatalysts for Suzuki cross-coupling reactions. *Chem. Soc. Rev.,* **2011**, *40*(10), 5181-5203. [http://dx.doi.org/10.1039/c1cs15079k] [PMID: 21804997]

[60] Caminade, A.M.; Ouali, A.; Keller, M.; Majoral, J.P. Organocatalysis with dendrimers. *Chem. Soc. Rev.,* **2012**, *41*(11), 4113-4125. [http://dx.doi.org/10.1039/c2cs35030k] [PMID: 22517447]

[61] Zaera, F. Nanostructured materials for applications in heterogeneous catalysis. *Chem. Soc. Rev.,* **2013**, *42*(7), 2746-2762. [http://dx.doi.org/10.1039/C2CS35261C] [PMID: 23072831]

[62] Wörner, C.; Mülhaupt, R. Polynitrile- and polyamine-functional poly(trimethylene imine) dendrimers. *Angew. Chem. Int. Ed. Engl.,* **1993**, *32*, 1306-1308.

[http://dx.doi.org/10.1002/anie.199313061]

[63] Tomalia, D.A.; Dvornic, P.R. What promise for dendrimers? *Nature,* **1994**, *372*, 617-618. [http://dx.doi.org/10.1038/372617a0]

[64] Knapen, J.W.; van der Made, A.W.; de Wilde, J.C.; van Leeuwen, P.W.; Wijkens, P.; Grove, D.M.; van Koten, G. Homogenous catalysts based on silane dendrimers functionalized with arylnickel(II) complexes. *Nature,* **1994**, *372*, 659-663. [http://dx.doi.org/10.1038/372659a0]

[65] Kleij, R.A.; van Leeuwen, P.W.N.M; van der Made, A.W. Eur. Patent EP0456317, 1991.

[66] Brunner, H.; Fürst, J.; Ziegler, J. Enantioselektive katalyse: LXXXI. Optisch aktive zweischalenphosphine. *J. Organomet. Chem.,* **1993**, *454*, 87-94. [http://dx.doi.org/10.1016/0022-328X(93)83227-M]

[67] Brunner, H.; Fürst, J. Enantioselective catalysis, 85: optically active expanded chelate phosphines derived from 1,ω-Bis(dichlorophosphino)alkanes. *Tetrahedron,* **1994**, *50*, 4303-4310. [http://dx.doi.org/10.1016/S0040-4020(01)89367-4]

[68] Miedaner, A.; Curtis, C.J.; Barkley, R.M.; DuBois, D.L. Electrochemical reduction of CO2 catalyzed by small organophosphine dendrimers containing palladium. *Inorg. Chem.,* **1994**, *33*, 5482-5490. [http://dx.doi.org/10.1021/ic00102a022]

[69] Lee, J-J.; Ford, W.T.; Moore, J.A.; Li, Y. Reactivity of organic anions promoted by a quaternary ammonium ion dendrimer. *Macromolecules,* **1994**, *27*, 4632-4634. [http://dx.doi.org/10.1021/ma00094a033]

[70] Reetz, M-T.; Lohmer, G.; Schwickardi, R. Systhesis and catalytic activity of dendritic diphosphane metal complexes. *Angew. Chem. Int. Ed. Engl.,* **1997**, *36*, 1526-1529. [http://dx.doi.org/10.1002/anie.199715261]

[71] Brinkmann, N.; Giebel, D.; Lohmer, G.; Reetz, M.T.; Kragl, U. *J. Catal.,* **1999**, *183*, 163. [http://dx.doi.org/10.1006/jcat.1999.2462]

[72] Bourque, S.C.; Maltais, F.; Xiao, W.J.; Tardif, O.; Alper, H.; Manzer, L.E.; Arya, P. Hydroformylation reactions with rhodium-complexed dendrimers on silica. *J. Am. Chem. Soc.,* **1999**, *121*, 3035-3038. [http://dx.doi.org/10.1021/ja983764b]

[73] Zhao, M.; Crooks, R.M. Homogeneous hydrogenation catalysis with monodisperse, dendrimer-encapsulated Pd and Pt nanoparticles. *Angew. Chem. Int. Ed. Engl.,* **1999**, *38*, 364-366. [http://dx.doi.org/10.1002/(SICI)1521-3773(19990201)38:3<364::AID-ANIE364>3.0.CO;2-L]

[74] Bhyrappa, P.; Young, J.K.; Moore, J.S.; Suslick, K.S. Shape selective epoxidation of alkenes by metalloporphyrin-dendrimers. *J. Mol. Catal. Chem.,* **1996**, *113*, 109-116. [http://dx.doi.org/10.1016/S1381-1169(96)00161-6]

[75] Bhyrappa, P.; Young, J.K.; Moore, J.S.; Suslick, K.S. Dendrimer-metalloporphyrins: synthesis and catalysis. *J. Am. Chem. Soc.,* **1996**, *118*, 5708-5711. [http://dx.doi.org/10.1021/ja953474k]

[76] Leung, K.C.; Aricó, F.; Cantrill, S.J.; Stoddart, J.F. Template-directed dynamic synthesis of mechanically interlocked dendrimers. *J. Am. Chem. Soc.,* **2005**, *127*(16), 5808-5810. [http://dx.doi.org/10.1021/ja0501363] [PMID: 15839677]

[77] Liang, C.O.; Helms, B.; Hawker, C.J.; Fréchet, J.M. Dendronized cyclocopolymers with a radial gradient of polarity and their use to catalyze a difficult esterification. *Chem. Commun. (Camb.),* **2003**, *20*(20), 2524-2525.
[http://dx.doi.org/10.1039/b307425k] [PMID: 14594262]

[78] Gross, E.; Liu, J.H.; Toste, F.D.; Somorjai, G.A. Control of selectivity in heterogeneous catalysis by tuning nanoparticle properties and reactor residence time. *Nat. Chem.,* **2012**, *4*(11), 947-952.
[http://dx.doi.org/10.1038/nchem.1465] [PMID: 23089871]

[79] Laurier, K.G.; Poets, M.; Vermoortele, F.; De Cremer, G.; Martens, J.A.; UjI-i, H.; De Vos, D.E.; Hofkens, J.; Roeffaers, M.B. Photocatalytic growth of dendritic silver nanostructures as SERS substrates. *Chem. Commun. (Camb.),* **2012**, *48*(10), 1559-1561.
[http://dx.doi.org/10.1039/C1CC14727G] [PMID: 21984411]

[80] Isfahani, A.L.; Mohammadpoor-Baltork, I.; Mirkhani, V.; Khosropour, A.R.; Moghadam, M.; Tangestaninejad, S.; Kia, R. Palladium nanoparticles immobilized on nano-silica triazine dendritic polymer (Pd-np-nSTDP): an efficient and reusable catalyst for Suzuki-Miyaura cross-coupling and Heck reactions. *Adv. Synth. Catal.,* **2013**, *355*, 957-972.
[http://dx.doi.org/10.1002/adsc.201200707]

[81] Song, Y.J.; Lopez-De Jesus, Y.M.; Fanson, P.T.; Williams, C.T. Kinetic evaluation of direct NO decomposition and NO-CO reaction over dendrimer-derived bimetallic Ir-Au/Al2O3 catalysts. *Appl. Catal. B,* **2014**, *154*, 62-72.
[http://dx.doi.org/10.1016/j.apcatb.2014.01.065]

[82] Tschierske, C. Non-conventional liquid crystals - the importance of micro-segregation for self-organisation. *J. Mater. Chem.,* **1998**, *8*, 1485-1508.
[http://dx.doi.org/10.1039/a800946e]

[83] Kato, T.; Mizoshita, N.; Kishimoto, K. Functional liquid-crystalline assemblies: self-organized soft materials. *Angew. Chem. Int. Ed. Engl.,* **2005**, *45*(1), 38-68.
[http://dx.doi.org/10.1002/anie.200501384] [PMID: 16353263]

[84] Donnio, B.; Guillon, D. , **2006**.

[85] Marcos, M.; Martín-Rapún, R.; Omenat, A.; Serrano, J.L. Highly congested liquid crystal structures: dendrimers, dendrons, dendronized and hyperbranched polymers. *Chem. Soc. Rev.,* **2007**, *36*(12), 1889-1901.
[http://dx.doi.org/10.1039/b611123h] [PMID: 17982516]

[86] Donnio, B.; Buathong, S.; Bury, I.; Guillon, D. Liquid crystalline dendrimers. *Chem. Soc. Rev.,* **2007**, *36*(9), 1495-1513.
[http://dx.doi.org/10.1039/b605531c] [PMID: 17660881]

[87] Astruc, D.; Boisselier, E.; Ornelas, C. Dendrimers designed for functions: from physical, photophysical, and supramolecular properties to applications in sensing, catalysis, molecular electronics, photonics, and nanomedicine. *Chem. Rev.,* **2010**, *110*(4), 1857-1959.
[http://dx.doi.org/10.1021/cr900327d] [PMID: 20356105]

[88] Bauer, S.; Fischer, H.; Ringsdorf, H. Highly branched liquid crystalline polymers with chiral terminal groups. *Angew. Chem. Int. Ed. Engl.,* **1993**, *32*, 1589-1592.

[http://dx.doi.org/10.1002/anie.199315891]

[89] McKenna, M.D.; Barberá, J.; Marcos, M.; Serrano, J.L. Discotic liquid crystalline poly(propylene imine) dendrimers based on triphenylene. *J. Am. Chem. Soc.,* **2005**, *127*(2), 619-625. [http://dx.doi.org/10.1021/ja0455906] [PMID: 15643886]

[90] Martín-Rapún, R.; Marcos, M.; Omenat, A.; Barberá, J.; Romero, P.; Serrano, J.L. Ionic thermotropic liquid crystal dendrimers. *J. Am. Chem. Soc.,* **2005**, *127*(20), 7397-7403. [http://dx.doi.org/10.1021/ja042264h] [PMID: 15898788]

[91] Rueff, J.M.; Barbera, J.; Marcos, M.; Omenat, A.; Martin-Rapun, R.; Donnio, B.; Guillon, D.; Serrano, J.L. PAMAM- and DAB-derived dendromesogens: the plastic supermolecules. *Chem. Mater.,* **2006**, *18*, 249-254. [http://dx.doi.org/10.1021/cm048105e]

[92] Han, H.; Keith, C.; Lang, H.; Reddy, R.A.; Tschierke, C. First example of a third-generation liquid-crystalline carbosilane dendrimer with peripheral bent-core mesogenic units: understanding of "dark conglomerate phases". *Adv. Mater.,* **2006**, *18*, 2629-2633. [http://dx.doi.org/10.1002/adma.200600161]

[93] Saez, I.M.; Goodby, J.W. Supermolecular liquid crystals. *J. Mater. Chem.,* **2005**, *15*, 26-40. [http://dx.doi.org/10.1039/b413416h]

[94] Donnio, B.; Guillon, D. Liquid crystalline dendrimers and polypedes. *Adv. Polym. Sci.,* **2006**, *201*, 45-155. [http://dx.doi.org/10.1007/12_079]

[95] Bioko, N.I.; Lysachkov, A.I.; Ponomarenko, S.A.; Shibaev, V.P.; Richardson, R.M. Synthesis and comparative studies of carbosilane liquid crystalline dendrimers with chiral terminal mesogenic groups. *Colloid Polym. Sci.,* **2005**, *283*, 1155-1162. [http://dx.doi.org/10.1007/s00396-005-1313-6]

[96] Felekis, T.; Tsiourvas, D.; Tziveleka, L.; Paleos, C.M. Hydrogen-bonded liquid crystals derived from supramolecular complexes of pyridylated poly(propyleneimine) dendrimers and a cholesterol-based carboxylic acid. *Liq. Cryst.,* **2005**, *32*, 39-43. [http://dx.doi.org/10.1080/02678290412331320566]

[97] Percec, V.; Chu, P.W.; Ungar, G.; Zhou, J.P. Rational design of the first nonspherical dendrimer which displays calamitic nematic and smectic thermotropic liquid crystalline phases. *J. Am. Chem. Soc.,* **1995**, *117*, 11441-11454. [http://dx.doi.org/10.1021/ja00151a008]

[98] Gehringer, L.; Guillon, D.; Donnio, B. Liquid crystalline octopus: an alternative class of mesomorphic dendrimers. *Macromolecules,* **2003**, *36*, 5593-5601. [http://dx.doi.org/10.1021/ma034038i]

[99] Gehringer, L.; Bourgogne, C.; Guillon, D.; Donnio, B. Liquid-crystalline octopus dendrimers: block molecules with unusual mesophase morphologies. *J. Am. Chem. Soc.,* **2004**, *126*(12), 3856-3867. [http://dx.doi.org/10.1021/ja031506v] [PMID: 15038740]

[100] Deschenaux, R.; Serrano, E.; Levelu, A.M. Ferrocene-containing liquid-crystalline dendrimers: a novel family of mesomorphic macromolecules. *Chem. Commun. (Camb.),* **1997**, 1577-1578.

[http://dx.doi.org/10.1039/a702850d]

[101] Campidelli, S.; Pérez, L.; Rodriguez-Lopez, J.; Barbera, J.; Langa, F.; Deschenaux, R. Dendritic liquid-crystalline fullerene–ferrocene dyads. *Tetrahedron,* **2006**, *62*, 2115-2122. [http://dx.doi.org/10.1016/j.tet.2005.10.065]

[102] Campidelli, S.; Vazquez, E.; Milic, D.; Lenoble, J.; Atienza Castellanos, C.; Sarova, G.; Guldi, D.M.; Deschenaux, R.; Prato, M. Liquid-crystalline bisadducts of [60]fullerene. *J. Org. Chem.,* **2006**, *71*(20), 7603-7610. [http://dx.doi.org/10.1021/jo0609576] [PMID: 16995664]

[103] Campidelli, S.; Sevenac, M.; Scanu, D.; Descheneaux, R.; Vazquez, E.; Milic, D.; Prato, M.; Carano, M.; Marcaccio, M.; Paolucci, F.; Rahman, G.M.; Guldi, D.M. Photophysical, electrochemical, and mesomorphic properties of a liquid-crystalline [60]fullerene–peralkylated ferrocene dyad. *J. Mater. Chem.,* **2008**, *18*, 1504-1509. [http://dx.doi.org/10.1039/b716806c]

[104] Marcos, M.; Omenat, A.; Barbera, J.; Duran, F.; Serrano, J.L. Structural study of metallomesogens derived from tris-[2-(salicylideneamino)ethyl]amine. A molecular meccano. *J. Mater. Chem.,* **2004**, *14*, 3321-3327. [http://dx.doi.org/10.1039/b409361e]

[105] Terazzi, E.; Bocquet, B.; Campidelli, S.; Donnio, B.; Guillon, D.; Deschenaux, R.; Piguet, C. Encoding calamitic mesomorphism in thermotropic lanthanidomesogens. *Chem. Commun. (Camb.),* **2006**, (27), 2922-2924. [http://dx.doi.org/10.1039/b605253c] [PMID: 17007418]

[106] Pesak, D.J.; Moore, J.S. Columnar liquid crystals from shape-persistent dendritic molecules. *Angew. Chem. Int. Ed. Engl.,* **1997**, *36*, 1636-1639. [http://dx.doi.org/10.1002/anie.199716361]

[107] Antharjanam, P.K.; Sudhadevi, J.; Jaseer, M.; Ragi, K.N.; Prasad, E.J. Intrinsic luminescence properties of ionic liquid crystals based on PAMAM and PPI dendrimers. *Photochem. Photobiol. A: Chem.,* **2009**, *203*, 50-56. [http://dx.doi.org/10.1016/j.jphotochem.2008.12.023]

[108] Chen, F-L.; Jamieson, A.M.; Kawasumi, M.; Percec, V. Viscoelastic properties of dilute nematic mixtures containing cyclic and hyperbranched liquid crystal polymers dissolved in a nematic solvent. *J. Polym. Sci., B, Polym. Phys.,* **1995**, *33*, 1213-1223. [http://dx.doi.org/10.1002/polb.1995.090330807]

[109] Percec, V.; Johansson, G.; Ungar, G.; Zhou, J. Fluorophobic effect induces the self-assembly of semifluorinated tapered monodendrons containing crown ethers into supramolecular columnar dendrimers which exhibit a homeotropic hexagonal columnar liquid crystalline phase. *J. Am. Chem. Soc.,* **1996**, *118*, 9855-9866. [http://dx.doi.org/10.1021/ja9615738]

[110] Belaissaoui, A.; Saez, I.M.; Cowling, S.J.; Goodby, J.W. Chirally homogeneous and heterogeneous dendritic liquid crystals. *Macromolecules,* **2013**, *46*, 1268-1273. [http://dx.doi.org/10.1021/ma3026332]

[111] Yang, C.A.; Lu, Y.; Li, G.Y.; Huang, Z.B. Synthesis and phase Structures of dendronized polymers

with dendritic azobenzene side groups based on mesogen-jacketed liquid crystalline polymers. *Mol. Cryst. Liq. Cryst. (Phila. Pa.),* **2014**, *592*, 28-43. [http://dx.doi.org/10.1080/15421406.2013.837996]

[112] Thelakkat, M. Star-shaped, dendrimeric and polymeric triarylamines as photoconductors and hole transport materials for electro-optical applications. *Macromol. Mater. Eng.,* **2002**, *287*, 442-461. [http://dx.doi.org/10.1002/1439-2054(20020701)287:7<442::AID-MAME442>3.0.CO;2-H]

[113] Lo, S.C.; Burn, P.L. Development of dendrimers: macromolecules for use in organic light-emitting diodes and solar cells. *Chem. Rev.,* **2007**, *107*(4), 1097-1116. [http://dx.doi.org/10.1021/cr050136l] [PMID: 17385927]

[114] Hwang, S.H.; Moorefield, C.N.; Newkome, G.R. Dendritic macromolecules for organic light-emitting diodes. *Chem. Soc. Rev.,* **2008**, *37*(11), 2543-2557. [http://dx.doi.org/10.1039/b803932c] [PMID: 18949125]

[115] Schenning, A.; Martin, R.E.; Ito, M.; Diederich, F.; Boudon, C.; Gisselbrecht, J.P.; Gross, M. Dendritic rods with a poly(triacetylene) backbone: insulated molecular wires. *Chem. Commun. (Camb.),* **1998**, 1013-1014. [http://dx.doi.org/10.1039/a801269e]

[116] Soomro, S.A.; Benmouna, R.; Berger, R.; Meier, H. Preparation and photochemistry of dendrimers with isolated stilbene chromophores. *Eur. J. Org. Chem.,* **2005**, 3586-3593. [http://dx.doi.org/10.1002/ejoc.200500185]

[117] Li, W.S.; Jiang, D.L.; Suna, Y.; Aida, T. Cooperativity in chiroptical sensing with dendritic zinc porphyrins. *J. Am. Chem. Soc.,* **2005**, *127*(21), 7700-7702. [http://dx.doi.org/10.1021/ja0513335] [PMID: 15913359]

[118] Choi, M.S.; Aida, T.; Luo, H.; Araki, Y.; Ito, O. Fullerene-terminated dendritic multiporphyrin arrays: "dendrimer effects" on photoinduced charge separation. *Angew. Chem. Int. Ed. Engl.,* **2003**, *42*(34), 4060-4063. [http://dx.doi.org/10.1002/anie.200351460] [PMID: 12973770]

[119] Hasobe, T.; Kashiwagi, Y.; Absalom, M.A.; Sly, J.; Hosomizu, K.; Crossley, M.J.; Imahori, H.; Kamat, P.V.; Fukuzumi, S. Supramolecular photovoltaic cells using porphyrin dendrimers and fullerene. *Adv. Mater.,* **2004**, *16*, 975-979. [http://dx.doi.org/10.1002/adma.200306519]

[120] Hasobe, T.; Kamat, P.V.; Absalom, M.A.; Kashiwagi, Y.; Sly, J.; Crossley, M.J.; Hosomizu, K.; Imahori, H.; Fukuzumi, S. Supramolecular photovoltaic cells based on composite molecular nanoclusters: dendritic porphyrin and C60, porphyrin dimer and C60, and porphyrin-C60 dyad. *J. Phys. Chem. B,* **2004**, *108*, 12865-12872. [http://dx.doi.org/10.1021/jp048404r]

[121] Segura, J.L.; Giacalone, F.; Gómez, R.; Martín, N.; Guldi, D.M.; Luo, C.; Swartz, A.; Riedel, I.; Chirvase, D.; Parisi, J.; Dyakonov, V.; Sariciftci, N.S.; Padinger, F. Design, synthesis and photovoltaic properties of [60]fullerene based molecular materials. *Mater. Sci. Eng. C,* **2005**, *25*, 835-842. [http://dx.doi.org/10.1016/j.msec.2005.06.017]

[122] Mitchell, W.; Kopidakis, N.; Rumbles, G.; Ginley, D.S.; Shaheen, S.E. The synthesis and properties of solution processable phenyl cored thiophene dendrimers. *J. Mater. Chem.,* **2005**, *15*, 4518-4528.

[http://dx.doi.org/10.1039/b508683c]

[123] Xu, M.H.; Pu, L. Novel unsymmetrically hyperbranched polythiophenes with conjugation gradient. *Tetrahedron Lett.,* **2002**, *43*, 6347-6350.
[http://dx.doi.org/10.1016/S0040-4039(02)01367-9]

[124] Kopidakis, N.; Mitchell, W.J.; Van de Lagemaat, J.; Ginley, D.S.; Rumbles, G.; Shaheen, S.E.; Rance, W.L. Bulk heterojunction organic photovoltaic devices based on phenyl-cored thiophene dendrimers. *Appl. Phys. Lett.,* **2006**, *89*, 103524.
[http://dx.doi.org/10.1063/1.2337859]

[125] Roquet, S.; Cravino, A.; Leriche, P.; Alévêque, O.; Frère, P.; Roncali, J. Triphenylamine-thienylenevinylene hybrid systems with internal charge transfer as donor materials for heterojunction solar cells. *J. Am. Chem. Soc.,* **2006**, *128*(10), 3459-3466.
[http://dx.doi.org/10.1021/ja058178e] [PMID: 16522126]

[126] Cremer, J.; Bäuerle, P. Star-shaped perylene–oligothiophene–triphenylamine hybrid systems for photovoltaic applications. *J. Mater. Chem.,* **2006**, *16*, 874-884.
[http://dx.doi.org/10.1039/B515657B]

[127] Bettignies, R.; Nicolas, Y.; Blanchard, P.; Levillain, E.; Nunzi, J.M.; Roncali, J. Planarized star-shaped oligothiophenes as a new class of organic semiconductors for heterojunction solar cells. *Adv. Mater.,* **2003**, *15*, 1939-1943.
[http://dx.doi.org/10.1002/adma.200305168]

[128] Locklin, J.; Patton, D.; Deng, S.; Baba, A.; Millan, M.; Advincula, R.C. Conjugated oligothiophene-dendron-capped CdSe nanoparticles: synthesis and energy Transfer. *Chem. Mater.,* **2004**, *16*, 5187-5193.
[http://dx.doi.org/10.1021/cm048961q]

[129] Raja, S.; Satheeshkumar, C.; Rajakumar, P.; Ganesan, S.; Maruthamuthu, P. Influence of triazole dendritic additives in electrolytes on dye-sensitized solar cell (DSSC) performance. *J. Mater. Chem.,* **2011**, *21*, 7700-7704.
[http://dx.doi.org/10.1039/c1jm10334b]

[130] Hua, Y.; Jin, B.; Wang, H.D.; Zhu, X.J.; Wu, W.J.; Cheung, M.S.; Lin, Z.Y.; Wong, W.Y.; Wong, W.K. Bulky dendritic triarylamine-based organic dyes for efficient co-adsorbent-free dye-sensitized solar cells. *J. Power Sources,* **2013**, *237*, 195-203.
[http://dx.doi.org/10.1016/j.jpowsour.2013.03.018]

[131] Schuz, G.L.; Mastalerz, M.; Ma, C.Q.; Wienk, M.; Janssen, R.; Bauerle, P. Synthesis and photovoltaic performance of pyrazinoquinoxaline containing conjugated thiophene-based dendrimers and polymers. *Macromolecules,* **2013**, *46*, 2141-2151.
[http://dx.doi.org/10.1021/ma302404y]

[132] Zeng, F.; Zimmerman, S.C. Dendrimers in supramolecular chemistry: From molecular recognition to self-assembly. *Chem. Rev.,* **1997**, *97*(5), 1681-1712.
[http://dx.doi.org/10.1021/cr9603892] [PMID: 11851463]

[133] Baars, M.W.; Meijer, E.W. Host-guest chemistry of dendritic molecules. *Top. Curr. Chem.,* **2000**, *210*, 131-182.
[http://dx.doi.org/10.1007/3-540-46577-4_3]

[134] Kim, S.H.; Lamm, M.H. Multiscale modeling for host-guest chemistry of dendrimers in solution. *Polymers (Basel),* **2012**, *4*, 463-485.
[http://dx.doi.org/10.3390/polym4010463]

[135] Hu, J.; Xu, T.; Cheng, Y. NMR insights into dendrimer-based host-guest systems. *Chem. Rev.,* **2012**, *112*(7), 3856-3891.
[http://dx.doi.org/10.1021/cr200333h] [PMID: 22486250]

[136] Jansen, J.F.; de Brabander-van den Berg, E.M.; Meijer, E.W. Encapsulation of guest molecules into a dendritic box. *Science,* **1994**, *266*(5188), 1226-1229.
[http://dx.doi.org/10.1126/science.266.5188.1226] [PMID: 17810265]

[137] Beezer, A.E.; King, A.S.; Martin, I.K.; Mitchel, J.C.; Twyman, L.J.; Wain, C.F. Dendrimers as potential drug carriers; encapsulation of acidic hydrophobes within water soluble PAMAM derivatives. *Tetrahedron,* **2003**, *59*, 3873-3880.
[http://dx.doi.org/10.1016/S0040-4020(03)00437-X]

[138] Morgan, M.T.; Carnahan, M.A.; Immoos, C.E.; Ribeiro, A.A.; Finkelstein, S.; Lee, S.J.; Grinstaff, M.W. Dendritic molecular capsules for hydrophobic compounds. *J. Am. Chem. Soc.,* **2003**, *125*(50), 15485-15489.
[http://dx.doi.org/10.1021/ja0347383] [PMID: 14664594]

[139] Baars, M.W.; Kleppinger, R.; Koch, M.H.; Yeu, S.L.; Meijer, E.W. The Localization of Guests in Water-Soluble Oligoethyleneoxy-Modified Poly(propylene imine) Dendrimers This work was supported by the Netherlands Foundation for Chemical Research (CW), with financial aid from the Netherlands Organisation for Scientific Research (NWO). The authors thank Stefan Meskers, Harry Dekkers, Marcel van Genderen, and Rint Sijbesma for stimulating discussions. DSM Research is acknowleged for providing the poly(propyleneimine) dendrimers. M.B. and R.K. acknowledge financial support under the TMR/LSF program of the European Union (No. ERBFMGECT980134) to the EMBL Hamburg outstation. *Angew. Chem. Int. Ed. Engl.,* **2000**, *39*(7), 1285-1288.
[http://dx.doi.org/10.1002/(SICI)1521-3773(20000403)39:7<1285::AID-ANIE1285>3.0.CO;2-F] [PMID: 10767033]

[140] Chapman, T.; Hillyer, G.L.; Mahan, E.J.; Shaffer, K.A. Hydraamphiphiles: novel linear dendritic block copolymer surfactants. *J. Am. Chem. Soc.,* **1994**, *116*, 11195-11196.
[http://dx.doi.org/10.1021/ja00103a060]

[141] Hawker, C.J.; Wooley, K.L.; Fréchet, J.M. Unimolecular micelles and globular amphiphiles: dendritic macromolecules as novel recyclable solubilization agents. *J. Chem. Soc., Perkin Trans. 1,* **1993**, 1287-1297.
[http://dx.doi.org/10.1039/p19930001287]

[142] Pistolis, G.; Malliaris, A. Study of poly(propylene imine) dendrimers in water, by exciplex formation. *Langmuir,* **2002**, *18*, 246-251.
[http://dx.doi.org/10.1021/la010859+]

[143] Vutukuri, D.R.; Basu, S.; Thayumanavan, S. Dendrimers with both polar and apolar nanocontainer characteristics. *J. Am. Chem. Soc.,* **2004**, *126*(48), 15636-15637.
[http://dx.doi.org/10.1021/ja0449628] [PMID: 15571373]

[144] Gupta, U.; Agashe, H.B.; Asthana, A.; Jain, N.K. Dendrimers: novel polymeric nanoarchitectures for

solubility enhancement. *Biomacromolecules,* **2006**, *7*(3), 649-658.
[http://dx.doi.org/10.1021/bm050802s] [PMID: 16529394]

[145] Namazi, H.; Adeli, M. Dendrimers of citric acid and poly (ethylene glycol) as the new drug-delivery agents. *Biomaterials,* **2005**, *26*(10), 1175-1183.
[http://dx.doi.org/10.1016/j.biomaterials.2004.04.014] [PMID: 15451637]

[146] Smith, D.K. Dendritic supermolecules--towards controllable nanomaterials. *Chem. Commun. (Camb.),* **2006**, (1), 34-44.
[http://dx.doi.org/10.1039/B507416A] [PMID: 16353086]

[147] Liu, Y.; Bryantsev, V.S.; Diallo, M.S.; Goddard, W.A., III PAMAM dendrimers undergo pH responsive conformational changes without swelling. *J. Am. Chem. Soc.,* **2009**, *131*(8), 2798-2799.
[http://dx.doi.org/10.1021/ja8100227] [PMID: 19199433]

[148] Moorefield, C.N.; Jeong, K.U.; Hwang, S.H.; Li, S.; Cheng, S.Z.; Newkome, G.R. Dendrimer-metallomacrocycle composites: nanofiber formation by multi-ion pairing. *Adv. Mater.,* **2008**, *20*, 1381-1385.
[http://dx.doi.org/10.1002/adma.200701669]

[149] Kaanumalle, L.S.; Nithyanandhan, J.; Pattabiraman, M.; Jayaraman, N.; Ramamurthy, V. Water-soluble dendrimers as photochemical reaction media: chemical behavior of singlet and triplet radical pairs inside dendritic reaction cavities. *J. Am. Chem. Soc.,* **2004**, *126*(29), 8999-9006.
[http://dx.doi.org/10.1021/ja049492w] [PMID: 15264831]

[150] Yiyun, C.; Tongwen, X. Dendrimers as potential drug carriers. Part I. Solubilization of non-steroidal anti-inflammatory drugs in the presence of polyamidoamine dendrimers. *Eur. J. Med. Chem.,* **2005**, *40*(11), 1188-1192.
[http://dx.doi.org/10.1016/j.ejmech.2005.06.010] [PMID: 16153746]

[151] Kulkarni, A.; DeFrees, K.; Hyun, S.H.; Thompson, D.H. Pendant polymer:amino--cyclodextrin:siRNA guest:host nanoparticles as efficient vectors for gene silencing. *J. Am. Chem. Soc.,* **2012**, *134*(18), 7596-7599.
[http://dx.doi.org/10.1021/ja300690j] [PMID: 22545899]

[152] Liu, Y.; Yu, C.; Jin, H.; Jiang, B.; Zhu, X.; Zhou, Y.; Lu, Z.; Yan, D. A supramolecular Janus hyperbranched polymer and its photoresponsive self-assembly of vesicles with narrow size distribution. *J. Am. Chem. Soc.,* **2013**, *135*(12), 4765-4770.
[http://dx.doi.org/10.1021/ja3122608] [PMID: 23464832]

[153] Paolino, M.; Ennen, F.; Lamponi, S.; Cernescu, M.; Voit, B.; Cappelli, A.; Appelhans, D.; Komber, H. Cyclodextrin-adamantane host-guest interactions on the surface of biocompatible adamantyl-modified glycodendrimers. *Macromolecules,* **2013**, *46*, 3215-3227.
[http://dx.doi.org/10.1021/ma400352m]

[154] Snowden, T.S.; Anslyn, E.V. Anion recognition: synthetic receptors for anions and their application in sensors. *Curr. Opin. Chem. Biol.,* **1999**, *3*(6), 740-746.
[http://dx.doi.org/10.1016/S1367-5931(99)00034-4] [PMID: 10651521]

[155] Goddard, J.M.; Erickson, D. Bioconjugation techniques for microfluidic biosensors. *Anal. Bioanal. Chem.,* **2009**, *394*(2), 469-479.
[http://dx.doi.org/10.1007/s00216-009-2731-y] [PMID: 19280179]

[156] Andreescu, S.; Njagi, J.; Ispas, C.; Ravalli, M.T. JEM Spotlight: Applications of advanced nanomaterials for environmental monitoring. *J. Environ. Monit.,* **2009**, *11*(1), 27-40. [http://dx.doi.org/10.1039/B811063H] [PMID: 19137136]

[157] Wells, M.; Crooks, R.M. Interactions between organized, surface-confined monolayers and vapor-phase probe molecules. 10. Preparation and properties of chemically sensitive dendrimer surfaces. *J. Am. Chem. Soc.,* **1996**, *16*, 3988-3989. [http://dx.doi.org/10.1021/ja960097i]

[158] Valerio, C.; Fillaut, J.L.; Ruiz, J.; Guittard, J.; Blais, J.C.; Astruc, D. The dendritic effect in molecular recognition: Ferrocene dendrimers and their use as supramolecular redox sensors for the recognition of small inorganic anions. *J. Am. Chem. Soc.,* **1997**, *119*, 2588-2589. [http://dx.doi.org/10.1021/ja964127t]

[159] Richardson, S.; Barcena, H.S.; Turnbull, G.A.; Burn, P.L.; Samuel, I.D.W. Chemosensing of 1,4-dinitrobenzene using bisfluorene dendrimer distributed feedback lasers. *Appl. Phys. Lett,* **2009**, *95*, 063305/1.

[160] Krasteva, N.; Besnard, I.; Guse, B.; Bauer, R.E.; Müllen, K.; Yasuda, A.; Vossmeyer, T. Self-assembled gold nanoparticle/dendrimer composite Films for vapor sensing applications. *Nano Lett.,* **2002**, *2*, 551-555. [http://dx.doi.org/10.1021/nl020242s]

[161] Duan, S.; Kouketsu, T.; Kasama, S.; Yamada, K. Development of PAMAM dendrimer composite membranes for CO2 separation. *J. Membr. Sci.,* **2006**, *283*, 2-6. [http://dx.doi.org/10.1016/j.memsci.2006.06.026]

[162] Kouketsu, T.; Duan, S.; Kai, T.; Kasama, S.; Yamada, K. PAMAM dendrimer composite membrane for CO2 separation: Formation of a chitosan gutter layer. *J. Membr. Sci.,* **2007**, *287*, 51-59. [http://dx.doi.org/10.1016/j.memsci.2006.10.014]

[163] Gao, T.; Tillman, E.S.; Lewis, N.S. Detection and classification of volatile organic amines and carboxylic acids using arrays of carbon black-dendrimer composite vapor detectors. *Chem. Mater.,* **2005**, *17*, 2904-2911. [http://dx.doi.org/10.1021/cm049457o]

[164] Duan, S.; Chowdhury, F.A.; Kai, T.; Kazama, S.; Fujioka, Y. PAMAM dendrimer composite membrane for CO2 separation: addition of hyaluronic acid in gutter layer and application of novel hydroxyl PAMAM dendrimers. *Desalination,* **2008**, *234*, 278-285. [http://dx.doi.org/10.1016/j.desal.2007.09.095]

[165] Koo, B.W.; Song, C.K.; Kim, C. CO gas sensor based on a conducting dendrimers. *Sens. Actuators B Chem.,* **2001**, *77*, 432-436. [http://dx.doi.org/10.1016/S0925-4005(01)00750-X]

[166] Feng, G.; Liang, J.; Liu, B. Hyperbranched conjugated polyelectrolytes for biological sensing and imaging. *Macromol. Rapid Commun.,* **2013**, *34*(9), 705-715. [http://dx.doi.org/10.1002/marc.201200821] [PMID: 23471733]

[167] Lee, C.C.; MacKay, J.A.; Fréchet, J.M.; Szoka, F.C. Designing dendrimers for biological applications. *Nat. Biotechnol.,* **2005**, *23*(12), 1517-1526.

[http://dx.doi.org/10.1038/nbt1171] [PMID: 16333296]

[168] Wei, H.; Wang, E. Nanomaterials with enzyme-like characteristics (nanozymes): next-generation artificial enzymes. *Chem. Soc. Rev.,* **2013**, *42*(14), 6060-6093. [http://dx.doi.org/10.1039/c3cs35486e] [PMID: 23740388]

[169] Martínez, Á.; Ortiz Mellet, C.; García Fernández, J.M. Cyclodextrin-based multivalent glycodisplays: covalent and supramolecular conjugates to assess carbohydrate-protein interactions. *Chem. Soc. Rev.,* **2013**, *42*(11), 4746-4773. [http://dx.doi.org/10.1039/c2cs35424a] [PMID: 23340678]

[170] Jiang, D.L.; Aida, T. A dendritic iron porphyrin as a novel haematoprotein mimic: effects of the dendrimer cage on dioxygen-binding activity. *Chem. Commun. (Camb.),* **1996**, 1523-1524. [http://dx.doi.org/10.1039/cc9960001523]

[171] Dandiker, P.J.; Diederich, F.; Zingg, A.; Gisselbrecht, J.P.; Gross, M.; Louati, A.; Sanford, E. Dendrimers with porphyrin cores: synthetic models for globular heme proteins. *Helv. Chim. Acta,* **1997**, *80*, 1773-1801. [http://dx.doi.org/10.1002/hlca.19970800603]

[172] Collman, J.P.; Fu, L.; Zingg, A.; Diederich, F. Dioxygen and carbon monoxide dinding in dendritic iron(II)porphyrins. *Chem. Commun. (Camb.),* **1997**, 193-194. [http://dx.doi.org/10.1039/a607413h]

[173] Smith, D.K.; Diederich, F. Functional dendrimers: unique biological mimics. *Chemistry,* **1998**, *4*, 1353-1361. [http://dx.doi.org/10.1002/(SICI)1521-3765(19980807)4:8<1353::AID-CHEM1353>3.0.CO;2-0]

[174] Weyermann, P.; Gisselbrecht, J.P.; Boudon, C.; Diederich, F.; Gross, M. Dendritic iron porphyrins with tethered axial ligands: new model compounds for cytochromes. *Angew. Chem. Int. Ed. Engl.,* **1999**, *38*(21), 3215-3219. [http://dx.doi.org/10.1002/(SICI)1521-3773(19991102)38:21<3215::AID-ANIE3215>3.0.CO;2-S] [PMID: 10556907]

[175] Enomoto, M.; Aida, T. Self-assembly of a copper-ligating dendrimer that provides a new non-heme metalloprotein mimic: "Dendrimer effects" on stability of the bis(mu-oxo)dicopper(III) core. *J. Am. Chem. Soc.,* **1999**, *121*, 874-875. [http://dx.doi.org/10.1021/ja982282x]

[176] Tomalia, D.A.; Huang, B.; Swanson, D.R.; Brothers, H.M.; Klimash, J.W. Structure control within poly(amidoamine) dendrimers: size, shape and regio-chemical mimicry of globular proteins. *Tetrahedron,* **2003**, *59*, 3799-3813. [http://dx.doi.org/10.1016/S0040-4020(03)00430-7]

[177] Twyman, L.J.; Ge, Y. Porphyrin cored hyperbranched polymers as heme protein models. *Chem. Commun. (Camb.),* **2006**, (15), 1658-1660. [http://dx.doi.org/10.1039/b600831n] [PMID: 16583011]

[178] Twyman, L.J.; Ellis, A.; Gittins, P.J. Pyridine encapsulated hyperbranched polymers as mimetic models of haeme containing proteins, that also provide interesting and unusual porphyrin-ligand geometries. *Chem. Commun. (Camb.),* **2012**, *48*(1), 154-156. [http://dx.doi.org/10.1039/C1CC14396D] [PMID: 22039580]

[179] Wang, X.; Zhang, Y.; Li, T.; Tian, W.; Zhang, Q.; Cheng, Y. Generation 9 polyamidoamine dendrimer encapsulated platinum nanoparticle mimics catalase size, shape, and catalytic activity. *Langmuir,* **2013**, *29*(17), 5262-5270.
[http://dx.doi.org/10.1021/la3046077] [PMID: 23544351]

[180] Lamanna, G.; Smulski, C.R.; Chekkat, N.; Estieu-Gionnet, K.; Guichard, G.; Fournel, S.; Bianco, A. Multimerization of an apoptogenic TRAIL-mimicking peptide by using adamantane-based dendrons. *Chemistry,* **2013**, *19*(5), 1762-1768.
[http://dx.doi.org/10.1002/chem.201202415] [PMID: 23239456]

[181] Liu, M.; Kono, K.; Fréchet, J.M. Water-soluble dendritic unimolecular micelles: their potential as drug delivery agents. *J. Control. Release,* **2000**, *65*(1-2), 121-131.
[http://dx.doi.org/10.1016/S0168-3659(99)00245-X] [PMID: 10699276]

[182] Boas, U.; Heegaard, P.M. Dendrimers in drug research. *Chem. Soc. Rev.,* **2004**, *33*(1), 43-63.
[http://dx.doi.org/10.1039/b309043b] [PMID: 14737508]

[183] Svenson, S.; Tomalia, D.A. Dendrimers in biomedical applications--reflections on the field. *Adv. Drug Deliv. Rev.,* **2005**, *57*(15), 2106-2129.
[http://dx.doi.org/10.1016/j.addr.2005.09.018] [PMID: 16305813]

[184] Gillies, E.R.; Fréchet, J.M. Dendrimers and dendritic polymers in drug delivery. *Drug Discov. Today,* **2005**, *10*(1), 35-43.
[http://dx.doi.org/10.1016/S1359-6446(04)03276-3] [PMID: 15676297]

[185] Paleos, C.M.; Tsiourvas, D.; Sideratou, Z.; Tziveleka, L. Multifunctional dendritic drug delivery systems: design, synthesis, controlled and triggered release. *Curr. Top. Med. Chem.,* **2008**, *8*(14), 1204-1224.
[http://dx.doi.org/10.2174/156802608785848996] [PMID: 18855706]

[186] Sadekar, S.; Ghandehari, H. Transepithelial transport and toxicity of PAMAM dendrimers: implications for oral drug delivery. *Adv. Drug Deliv. Rev.,* **2012**, *64*(6), 571-588.
[http://dx.doi.org/10.1016/j.addr.2011.09.010] [PMID: 21983078]

[187] Twyman, L.J.; Beezer, A.E.; Esfand, R.; Hardy, M.J.; Mitchell, J.C. The synthesis of water soluble dendrimers, and their application as possible drug delivery systems. *Tetrahedron Lett.,* **1999**, *40*, 1743-1746.
[http://dx.doi.org/10.1016/S0040-4039(98)02680-X]

[188] Kreiter, R.; Kleij, A.W.; Klein Gebbink, R.J.; van Koten, G. Dendritic catalysts. *Top. Curr. Chem.,* **2001**, *217*, 163-199.
[http://dx.doi.org/10.1007/3-540-45003-3_5]

[189] van Heerbeek, R.; Kamer, P.C.; van Leeuwen, P.W.; Reek, J.N. Dendrimers as support for recoverable catalysts and reagents. *Chem. Rev.,* **2002**, *102*(10), 3717-3756.
[http://dx.doi.org/10.1021/cr0103874] [PMID: 12371900]

[190] Astruc, *Dendrimers and Nanoscience*; C. R. Chimie: Paris, **2003**, 6, p. 709.

[191] Allen, T.M.; Cullis, P.R. Drug delivery systems: entering the mainstream. *Science,* **2004**, *303*(5665), 1818-1822.
[http://dx.doi.org/10.1126/science.1095833] [PMID: 15031496]

[192] Tyshenko, M.G. Medical nanotechnology using genetic material and the need for precaution in design and risk assessments. *Int. J. Nanotechnol.,* **2008**, *5*, 116-123. [http://dx.doi.org/10.1504/IJNT.2008.016551]

[193] Cheng, Y.; Xu, T. The effect of dendrimers on the pharmacodynamic and pharmacokinetic behaviors of non-covalently or covalently attached drugs. *Eur. J. Med. Chem.,* **2008**, *43*(11), 2291-2297. [http://dx.doi.org/10.1016/j.ejmech.2007.12.021] [PMID: 18276038]

[194] Kolhe, P.; Misra, E.; Kannan, R.M.; Kannan, S.; Lieh-Lai, M. Drug complexation, *in vitro* release and cellular entry of dendrimers and hyperbranched polymers. *Int. J. Pharm.,* **2003**, *259*(1-2), 143-160. [http://dx.doi.org/10.1016/S0378-5173(03)00225-4] [PMID: 12787643]

[195] Kannan, S.; Kolhe, P.; Raykova, V.; Glibatec, M.; Kannan, R.M.; Lieh-Lai, M.; Bassett, D. Dynamics of cellular entry and drug delivery by dendritic polymers into human lung epithelial carcinoma cells. *J. Biomater. Sci. Polym. Ed.,* **2004**, *15*(3), 311-330. [http://dx.doi.org/10.1163/156856204322977201] [PMID: 15147164]

[196] Morgan, M.T.; Nakanishi, Y.; Kroll, D.J.; Griset, A.P.; Carnahan, M.A.; Wathier, M.; Oberlies, N.H.; Manikumar, G.; Wani, M.C.; Grinstaff, M.W. Dendrimer-encapsulated camptothecins: increased solubility, cellular uptake, and cellular retention affords enhanced anticancer activity *in vitro. Cancer Res.,* **2006**, *66*(24), 11913-11921. [http://dx.doi.org/10.1158/0008-5472.CAN-06-2066] [PMID: 17178889]

[197] Cheng, Y.; Li, M.; Xu, T. Potential of poly(amidoamine) dendrimers as drug carriers of camptothecin based on encapsulation studies. *Eur. J. Med. Chem.,* **2008**, *43*(8), 1791-1795. [http://dx.doi.org/10.1016/j.ejmech.2007.09.030] [PMID: 18215444]

[198] Chauhan, A.S.; Sridevi, S.; Chalasani, K.B.; Jain, A.K.; Jain, S.K.; Jain, N.K.; Diwan, P.V. Dendrimer-mediated transdermal delivery: enhanced bioavailability of indomethacin. *J. Control. Release,* **2003**, *90*(3), 335-343. [http://dx.doi.org/10.1016/S0168-3659(03)00200-1] [PMID: 12880700]

[199] Devarakonda, B.; Hill, R.A.; de Villiers, M.M. The effect of PAMAM dendrimer generation size and surface functional group on the aqueous solubility of nifedipine. *Int. J. Pharm.,* **2004**, *284*(1-2), 133-140. [http://dx.doi.org/10.1016/j.ijpharm.2004.07.006] [PMID: 15454304]

[200] Devarakonda, B.; Hill, R.A.; Liebenberg, W.; Brits, M.; de Villiers, M.M. Comparison of the aqueous solubilization of practically insoluble niclosamide by polyamidoamine (PAMAM) dendrimers and cyclodextrins. *Int. J. Pharm.,* **2005**, *304*(1-2), 193-209. [http://dx.doi.org/10.1016/j.ijpharm.2005.07.023] [PMID: 16198076]

[201] Na, M.; Yiyun, C.; Tongwen, X.; Yang, D.; Xiaomin, W.; Zhenwei, L.; Zhichao, C.; Guanyi, H.; Yunyu, S.; Longping, W. Dendrimers as potential drug carriers. Part II. Prolonged delivery of ketoprofen by *in vitro* and *in vivo* studies. *Eur. J. Med. Chem.,* **2006**, *41*(5), 670-674. [http://dx.doi.org/10.1016/j.ejmech.2006.01.001] [PMID: 16527374]

[202] Cheng, Y.; Qu, H.; Ma, M.; Xu, Z.; Xu, P.; Fang, Y.; Xu, T. Polyamidoamine (PAMAM) dendrimers as biocompatible carriers of quinolone antimicrobials: an *in vitro* study. *Eur. J. Med. Chem.,* **2007**, *42*(7), 1032-1038. [http://dx.doi.org/10.1016/j.ejmech.2006.12.035] [PMID: 17336426]

[203] Calabretta, M.K.; Kumar, A.; McDermott, A.M.; Cai, C. Antibacterial activities of poly(amidoamine) dendrimers terminated with amino and poly(ethylene glycol) groups. *Biomacromolecules,* **2007**, *8*(6), 1807-1811.
[http://dx.doi.org/10.1021/bm0701088] [PMID: 17511499]

[204] Cheng, Y.; Wang, J.; Rao, T.; He, X.; Xu, T. Pharmaceutical applications of dendrimers: promising nanocarriers for drug delivery. *Front. Biosci.,* **2008**, *13*, 1447-1471.
[http://dx.doi.org/10.2741/2774] [PMID: 17981642]

[205] Cho, K.; Wang, X.; Nie, S.; Chen, Z.G.; Shin, D.M. Therapeutic nanoparticles for drug delivery in cancer. *Clin. Cancer Res.,* **2008**, *14*(5), 1310-1316.
[http://dx.doi.org/10.1158/1078-0432.CCR-07-1441] [PMID: 18316549]

[206] Meyers, S.R.; Juhn, F.S.; Griset, A.P.; Luman, N.R.; Grinstaff, M.W. Anionic amphiphilic dendrimers as antibacterial agents. *J. Am. Chem. Soc.,* **2008**, *130*(44), 14444-14445.
[http://dx.doi.org/10.1021/ja806912a] [PMID: 18842041]

[207] Svenson, S. Dendrimers as versatile platform in drug delivery applications. *Eur. J. Pharm. Biopharm.,* **2009**, *71*(3), 445-462.
[http://dx.doi.org/10.1016/j.ejpb.2008.09.023] [PMID: 18976707]

[208] Gajbhiye, V.; Kumar, P.V.; Sharma, A.; Jain, N.K. Novel PEGylated PPI dendritic nanostructures for sustained delivery of anti-inflammatory agent. *Curr. Nanosci.,* **2008**, *4*, 267-277.
[http://dx.doi.org/10.2174/157341308785161136]

[209] Zou, J.; Zhang, F.; Zhang, S.; Pollack, S.F.; Elsabahy, M.; Fan, J.; Wooley, K.L. Poly(ethylene oxide)-block-polyphosphoester-graft-paclitaxel conjugates with acid-labile linkages as a pH-sensitive and functional nanoscopic platform for paclitaxel delivery. *Adv. Healthc. Mater.,* **2014**, *3*(3), 441-448.
[http://dx.doi.org/10.1002/adhm.201300235] [PMID: 23997013]

[210] Haesnler, J.; Szoka, F.C. Polymeric gene delivery: principles and applications. *Bioconjug. Chem.,* **1993**, *4*, 372.
[PMID: 8274523]

[211] Eichman, J.D.; Bielinska, A.U.; Kukowska-Latallo, J.F.; Baker, J.R., Jr The use of PAMAM dendrimers in the efficient transfer of genetic material into cells. *Pharm. Sci. Technol. Today,* **2000**, *3*(7), 232-245.
[http://dx.doi.org/10.1016/S1461-5347(00)00273-X] [PMID: 10884679]

[212] Dufès, C.; Uchegbu, I.F.; Schätzlein, A.G. Dendrimers in gene delivery. *Adv. Drug Deliv. Rev.,* **2005**, *57*(15), 2177-2202.
[http://dx.doi.org/10.1016/j.addr.2005.09.017] [PMID: 16310284]

[213] Pack, D.W.; Hoffman, A.S.; Pun, S.; Stayton, P.S. Design and development of polymers for gene delivery. *Nat. Rev. Drug Discov.,* **2005**, *4*(7), 581-593.
[http://dx.doi.org/10.1038/nrd1775] [PMID: 16052241]

[214] Mintzer, M.A.; Simanek, E.E. Nonviral vectors for gene delivery. *Chem. Rev.,* **2009**, *109*(2), 259-302.
[http://dx.doi.org/10.1021/cr800409e] [PMID: 19053809]

[215] Haensler, J.; Szoka, F.C., Jr Polyamidoamine cascade polymers mediate efficient transfection of cells in culture. *Bioconjug. Chem.,* **1993**, *4*(5), 372-379.

[http://dx.doi.org/10.1021/bc00023a012] [PMID: 8274523]

[216] Paleos, C.M.; Tziveleka, L-A.; Sideratou, Z.; Tsiourvas, D. Gene delivery using functional dendritic polymers. *Expert Opin. Drug Deliv.,* **2009**, *6*(1), 27-38.
[http://dx.doi.org/10.1517/17425240802607345] [PMID: 19236206]

[217] Gao, Y.; Gao, G.; He, Y.; Liu, T.; Qi, R. Recent advances of dendrimers in delivery of genes and drugs. *Mini Rev. Med. Chem.,* **2008**, *8*(9), 889-900.
[http://dx.doi.org/10.2174/138955708785132729] [PMID: 18691146]

[218] Kitchens, K.M.; El-Sayed, M.E.; Ghandehari, H. Transepithelial and endothelial transport of poly (amidoamine) dendrimers. *Adv. Drug Deliv. Rev.,* **2005**, *57*(15), 2163-2176.
[http://dx.doi.org/10.1016/j.addr.2005.09.013] [PMID: 16289433]

[219] Ramaswamy, C.; Sakthivel, T.; Wilderspin, A.F.; Florence, A.T. Dendriplexes and their characterisation. *Int. J. Pharm.,* **2003**, *254*(1), 17-21.
[http://dx.doi.org/10.1016/S0378-5173(02)00670-1] [PMID: 12615402]

[220] Caminade, A-M.; Turrin, C-O.; Majoral, J-P. Dendrimers and DNA: combinations of two special topologies for nanomaterials and biology. *Chemistry,* **2008**, *14*(25), 7422-7432.
[http://dx.doi.org/10.1002/chem.200800584] [PMID: 18537210]

[221] Zinselmeyer, B.H.; Mackay, S.P.; Schatzlein, A.G.; Uchegbu, I.F. The lower-generation polypropylenimine dendrimers are effective gene-transfer agents. *Pharm. Res.,* **2002**, *19*(7), 960-967.
[http://dx.doi.org/10.1023/A:1016458104359] [PMID: 12180548]

[222] Dufès, C.; Keith, W.N.; Bilsland, A.; Proutski, I.; Uchegbu, I.F.; Schätzlein, A.G. Synthetic anticancer gene medicine exploits intrinsic antitumor activity of cationic vector to cure established tumors. *Cancer Res.,* **2005**, *65*(18), 8079-8084.
[http://dx.doi.org/10.1158/0008-5472.CAN-04-4402] [PMID: 16166279]

[223] Maszewska, M.; Leclaire, J.; Cieslak, M.; Nawrot, B.; Okruszek, A.; Caminade, A-M.; Majoral, J-P. Water-soluble polycationic dendrimers with a phosphoramidothioate backbone: preliminary studies of cytotoxicity and oligonucleotide/plasmid delivery in human cell culture. *Oligonucleotides,* **2003**, *13*(4), 193-205.
[http://dx.doi.org/10.1089/154545703322460586] [PMID: 15000835]

[224] Guillot-Nieckowski, M.; Joester, D.; Stöhr, M.; Losson, M.; Adrian, M.; Wagner, B.; Kansy, M.; Heinzelmann, H.; Pugin, R.; Diederich, F.; Gallani, J.L. Self-assembly, DNA complexation, and pH response of amphiphilic dendrimers for gene transfection. *Langmuir,* **2007**, *23*(2), 737-746.
[http://dx.doi.org/10.1021/la0624891] [PMID: 17209628]

[225] Ewert, K.K.; Evans, H.M.; Zidovska, A.; Bouxsein, N.F.; Ahmad, A.; Safinya, C.R. A columnar phase of dendritic lipid-based cationic liposome-DNA complexes for gene delivery: hexagonally ordered cylindrical micelles embedded in a DNA honeycomb lattice. *J. Am. Chem. Soc.,* **2006**, *128*(12), 3998-4006.
[http://dx.doi.org/10.1021/ja055907h] [PMID: 16551108]

[226] Santhakumaran, L.M.; Thomas, T.; Thomas, T.J. Enhanced cellular uptake of a triplex-forming oligonucleotide by nanoparticle formation in the presence of polypropylenimine dendrimers. *Nucleic Acids Res.,* **2004**, *32*(7), 2102-2112.
[http://dx.doi.org/10.1093/nar/gkh526] [PMID: 15087489]

[227] Hwa Kim, S.; Hoon Jeong, J.; Chul Cho, K.; Wan Kim, S.; Gwan Park, T. Target-specific gene silencing by siRNA plasmid DNA complexed with folate-modified poly(ethylenimine). *J. Control. Release,* **2005**, *104*(1), 223-232.
[http://dx.doi.org/10.1016/j.jconrel.2005.02.006] [PMID: 15866348]

[228] Bayele, H.K.; Ramaswamy, C.; Wilderspin, A.F.; Srai, K.S.; Toth, I.; Florence, A.T. Protein transduction by lipidic peptide dendrimers. *J. Pharm. Sci.,* **2006**, *95*(6), 1227-1237.
[http://dx.doi.org/10.1002/jps.20606] [PMID: 16639724]

[229] Ewert, K.K.; Evans, H.M.; Bouxsein, N.F.; Safinya, C.R. Dendritic cationic lipids with highly charged headgroups for efficient gene delivery. *Bioconjug. Chem.,* **2006**, *17*(4), 877-888.
[http://dx.doi.org/10.1021/bc050310c] [PMID: 16848393]

[230] Takahashi, T.; Hirose, J.; Kojima, C.; Harada, A.; Kono, K. Synthesis of poly(amidoamine) dendron-bearing lipids with poly(ethylene glycol) grafts and their use for stabilization of nonviral gene vectors. *Bioconjug. Chem.,* **2007**, *18*(4), 1163-1169.
[http://dx.doi.org/10.1021/bc070014v] [PMID: 17569497]

[231] Liu, H.; Wang, Y.; Wang, M.; Xiao, J.; Cheng, Y. Fluorinated poly(propylenimine) dendrimers as gene vectors. *Biomaterials,* **2014**, *35*(20), 5407-5413.
[http://dx.doi.org/10.1016/j.biomaterials.2014.03.040] [PMID: 24709522]

[232] Duncan, R.; Malik, N.; Richardon, S.; Ferruti, P. Dendrimer nanocomposites in medicine. *Polymer Prepr.,* **1998**, *39*, 180.

[233] Stiriba, S.E.; Frey, H.; Haag, R. Dendritic polymers in biomedical applications: from potential to clinical use in diagnostics and therapy. *Angew. Chem. Int. Ed. Engl.,* **2002**, *41*(8), 1329-1334.
[http://dx.doi.org/10.1002/1521-3773(20020415)41:8<1329::AID-ANIE1329>3.0.CO;2-P] [PMID: 19750755]

[234] Bracci, L.; Falciani, C.; Lelli, B.; Lozzi, L.; Runci, Y.; Pini, A.; De Montis, M.G.; Tagliamonte, A.; Neri, P. Dendrimer in medicine and biotechnology. *J. Biol. Chem.,* **2003**, *278*, 46590.
[http://dx.doi.org/10.1074/jbc.M308615200] [PMID: 12972419]

[235] Haag, R.; Kratz, F. Polymer therapeutics: concepts and applications. *Angew. Chem. Int. Ed. Engl.,* **2006**, *45*(8), 1198-1215.
[http://dx.doi.org/10.1002/anie.200502113] [PMID: 16444775]

[236] Cheng, Y.; Zhao, L.; Li, Y.; Xu, T. Design of biocompatible dendrimers for cancer diagnosis and therapy: current status and future perspectives. *Chem. Soc. Rev.,* **2011**, *40*(5), 2673-2703.
[http://dx.doi.org/10.1039/c0cs00097c] [PMID: 21286593]

[237] Mintzer, M.A.; Grinstaff, M.W. Biomedical applications of dendrimers: a tutorial. *Chem. Soc. Rev.,* **2011**, *40*(1), 173-190.
[http://dx.doi.org/10.1039/B901839P] [PMID: 20877875]

[238] Ideta, R.; Tasaka, F.; Jang, W.D.; Nishiyama, N.; Zhang, G.D.; Harada, A.; Yanagi, Y.; Tamaki, Y.; Aida, T.; Kataoka, K. Nanotechnology-based photodynamic therapy for neovascular disease using a supramolecular nanocarrier loaded with a dendritic photosensitizer. *Nano Lett.,* **2005**, *5*(12), 2426-2431.
[http://dx.doi.org/10.1021/nl051679d] [PMID: 16351191]

[239] Jang, W.D.; Nishiyama, N.; Zhang, G.D.; Harada, A.; Jiang, D.L.; Kawauchi, S.; Morimoto, Y.; Kikuchi, M.; Koyama, H.; Aida, T.; Kataoka, K. Supramolecular nanocarrier of anionic dendrimer porphyrins with cationic block copolymers modified with polyethylene glycol to enhance intracellular photodynamic efficacy. *Angew. Chem. Int. Ed. Engl.,* **2005**, *44*(3), 419-423. [http://dx.doi.org/10.1002/anie.200461603] [PMID: 15624158]

[240] Kepszynski, M.; Nawalany, K.; Jachimska, B.; Romek, M.; Nowakowska, M. *Colloids Surf. Biointerfaces,* **2006**, *15*, 22. [http://dx.doi.org/10.1016/j.colsurfb.2006.02.008]

[241] Kojima, C.; Toi, Y.; Harada, A.; Kono, K. Preparation of poly(ethylene glycol)-attached dendrimers encapsulating photosensitizers for application to photodynamic therapy. *Bioconjug. Chem.,* **2007**, *18*(3), 663-670. [http://dx.doi.org/10.1021/bc060244u] [PMID: 17375896]

[242] Nishiyama, N.; Morimoto, Y.; Jang, W-D.; Kataoka, K. Design and development of dendrimer photosensitizer-incorporated polymeric micelles for enhanced photodynamic therapy. *Adv. Drug Deliv. Rev.,* **2009**, *61*(4), 327-338. [http://dx.doi.org/10.1016/j.addr.2009.01.004] [PMID: 19385091]

[243] Cabral, H.; Nakanishi, M.; Kumagai, M.; Jang, W-D.; Nishiyama, N.; Kataoka, K. A photo-activated targeting chemotherapy using glutathione sensitive camptothecin-loaded polymeric micelles. *Pharm. Res.,* **2009**, *26*(1), 82-92. [http://dx.doi.org/10.1007/s11095-008-9712-2] [PMID: 18758920]

[244] Nishiyama, N.; Nakagishi, Y.; Morimoto, Y.; Lai, P-S.; Miyazaki, K.; Urano, K.; Horie, S.; Kumagai, M.; Fukushima, S.; Cheng, Y.; Jang, W-D.; Kikuchi, M.; Kataoka, K. Enhanced photodynamic cancer treatment by supramolecular nanocarriers charged with dendrimer phthalocyanine. *J. Control. Release,* **2009**, *133*(3), 245-251. [http://dx.doi.org/10.1016/j.jconrel.2008.10.010] [PMID: 19000725]

[245] He, H.; Li, Y.; Jia, X.R.; Du, J.; Ying, X.; Lu, W.L.; Lou, J.N.; Wei, Y. PEGylated Poly(amidoamine) dendrimer-based dual-targeting carrier for treating brain tumors. *Biomaterials,* **2011**, *32*(2), 478-487. [http://dx.doi.org/10.1016/j.biomaterials.2010.09.002] [PMID: 20934215]

[246] Ly, T.U.; Tran, N.Q.; Hoang, T.K.; Phan, K.N.; Truong, H.N.; Nguyen, C.K. Pegylated dendrimer and its effect in fluorouracil loading and release for enhancing antitumor activity. *J. Biomed. Nanotechnol.,* **2013**, *9*(2), 213-220. [http://dx.doi.org/10.1166/jbn.2013.1479] [PMID: 23627047]

[247] Albertazzi, L.; Gherardini, L.; Brondi, M.; Sulis Sato, S.; Bifone, A.; Pizzorusso, T.; Ratto, G.M.; Bardi, G. *In vivo* distribution and toxicity of PAMAM dendrimers in the central nervous system depend on their surface chemistry. *Mol. Pharm.,* **2013**, *10*(1), 249-260. [http://dx.doi.org/10.1021/mp300391v] [PMID: 23163881]

[248] Fischer, M.; Vögtle, F. Dendrimers: from design to application-a progress report. *Angew. Chem. Int. Ed.,* **1999**, *38*, 884-905. [http://dx.doi.org/10.1002/(SICI)1521-3773(19990401)38:7<884::AID-ANIE884>3.0.CO;2-K]

[249] Frey, H.; Haag, R. Dendritic polyglycerol: a new versatile biocompatible-material. *J. Biotechnol.,* **2002**, *90*(3-4), 257-267.

[PMID: 12071228]

[250] Kobayashi, H.; Brechbiel, M.W. Dendrimer-based macromolecular MRI contrast agents: characteristics and application. *Mol. Imaging,* **2003**, *2*(1), 1-10. [http://dx.doi.org/10.1162/153535003765276237] [PMID: 12926232]

[251] Bawarski, W.E.; Chidlowsky, E.; Bharali, D.J.; Mousa, S.A. Emerging nanopharmaceuticals. *Nanomedicine (Lond.),* **2008**, *4*(4), 273-282. [PMID: 18640076]

[252] Wolinsky, J.B.; Grinstaff, M.W. Therapeutic and diagnostic applications of dendrimers for cancer treatment. *Adv. Drug Deliv. Rev.,* **2008**, *60*(9), 1037-1055. [http://dx.doi.org/10.1016/j.addr.2008.02.012] [PMID: 18448187]

[253] Hamoudeh, M.; Kamleh, M.A.; Diab, R.; Fessi, H. Radionuclides delivery systems for nuclear imaging and radiotherapy of cancer. *Adv. Drug Deliv. Rev.,* **2008**, *60*(12), 1329-1346. [http://dx.doi.org/10.1016/j.addr.2008.04.013] [PMID: 18562040]

[254] Wang, M.; Thanou, M. Targeting nanoparticles to cancer. *Pharmacol. Res.,* **2010**, *62*(2), 90-99. [http://dx.doi.org/10.1016/j.phrs.2010.03.005] [PMID: 20380880]

[255] Wiener, E.C.; Brechbiel, M.W.; Brothers, H.; Magin, R.L.; Gansow, O.A.; Tomalia, D.A.; Lauterbur, P.C. Dendrimer-based metal chelates: a new class of magnetic resonance imaging contrast agents. *Magn. Reson. Med.,* **1994**, *31*(1), 1-8. [http://dx.doi.org/10.1002/mrm.1910310102] [PMID: 8121264]

[256] Konda, S.D.; Aref, M.; Wang, S.; Brechbiel, M.; Wiener, E.C. Specific targeting of folate-dendrimer MRI contrast agents to the high affinity folate receptor expressed in ovarian tumor xenografts. *Biol. Med.,* **2001**, *12*, 104.

[257] Mantovani, L.T.; Miotti, S.; Ménard, S.; Canevari, S.; Raspagliesi, F.; Bottini, C.; Bottero, F.; Colnaghi, M.I. Folate binding protein distribution in normal tissues and biological fluids from ovarian carcinoma patients as detected by the monoclonal antibodies MOv18 and MOv19. *Eur. J. Cancer,* **1994**, *30A*(3), 363-369. [http://dx.doi.org/10.1016/0959-8049(94)90257-7] [PMID: 8204360]

[258] Tomalia, D.A.; Reyna, L.A.; Svenson, S. Dendrimers as multi-purpose nanodevices for oncology drug delivery and diagnostic imaging. *Biochem. Soc. Trans.,* **2007**, *35*(Pt 1), 61-67. [http://dx.doi.org/10.1042/BST0350061] [PMID: 17233602]

[259] Langereis, S.; de Lussanet, Q.G.; van Genderen, M.H.; Meijer, E.W.; Beets-Tan, R.G.; Griffioen, A.W.; van Engelshoven, J.M.; Backes, W.H. Evaluation of Gd(III)DTPA-terminated poly(propylene imine) dendrimers as contrast agents for MR imaging. *NMR Biomed.,* **2006**, *19*(1), 133-141. [http://dx.doi.org/10.1002/nbm.1015] [PMID: 16450331]

[260] Kobayashi, H.; Kawamoto, S.; Jo, S.K.; Bryant, H.L., Jr; Brechbiel, M.W.; Star, R.A. Macromolecular MRI contrast agents with small dendrimers: pharmacokinetic differences between sizes and cores. *Bioconjug. Chem.,* **2003**, *14*(2), 388-394. [http://dx.doi.org/10.1021/bc025633c] [PMID: 12643749]

[261] Dichtel, W.R.; Serin, J.M.; Edder, C.; Fréchet, J.M.; Matuszewski, M.; Tan, L.S.; Ohulchanskyy, T.Y.; Prasad, P.N. Singlet oxygen generation *via* two-photon excited FRET. *J. Am. Chem. Soc.,* **2004**,

126(17), 5380-5381.
[http://dx.doi.org/10.1021/ja031647x] [PMID: 15113208]

[262] Oar, M.A.; Serin, J.M.; Dichtel, W.R.; Fréchet, J.M. Photosensitization of singlet oxygen *via* two-photon-excited fluorescence resonance energy transfer in a water-soluble dendrimer. *Chem. Mater.,* **2005**, *17*, 2267-2275.
[http://dx.doi.org/10.1021/cm047825i]

[263] Oar, M.A.; Dichtel, W.R.; Serin, J.M.; Fréchet, J.M. Light-harvesting chromophores with metalated porphyrin cores for tuned photosensitization of singlet oxygen *via* two-photon excited FRET. *Chem. Mater.,* **2006**, *18*, 3682-3692.
[http://dx.doi.org/10.1021/cm0606070]

[264] Parrott, M.C.; Benhabbour, S.R.; Saab, C.; Lemon, J.A.; Parker, S.; Valliant, J.F.; Adronov, A. Synthesis, radiolabeling, and bio-imaging of high-generation polyester dendrimers. *J. Am. Chem. Soc.,* **2009**, *131*(8), 2906-2916.
[http://dx.doi.org/10.1021/ja8078175] [PMID: 19239268]

[265] Almutairi, A.; Rossin, R.; Shokeen, M.; Hagooly, A.; Ananth, A.; Capoccia, B.; Guillaudeu, S.; Abendschein, D.; Anderson, C.J.; Welch, M.J.; Fréchet, J.M. Biodegradable dendritic positron-emitting nanoprobes for the noninvasive imaging of angiogenesis. *Proc. Natl. Acad. Sci. USA,* **2009**, *106*(3), 685-690.
[http://dx.doi.org/10.1073/pnas.0811757106] [PMID: 19129498]

[266] Majoros, I.J.; Thomas, T.P.; Mehta, C.B.; Baker, J.R., Jr Poly(amidoamine) dendrimer-based multifunctional engineered nanodevice for cancer therapy. *J. Med. Chem.,* **2005**, *48*(19), 5892-5899.
[http://dx.doi.org/10.1021/jm0401863] [PMID: 16161993]

[267] Fuchs, S.; Otto, H.; Jehle, S.; Henklein, P.; Schlüter, A.D. Fluorescent dendrimers with a peptide cathepsin B cleavage site for drug delivery applications. *Chem. Commun. (Camb.),* **2005**, (14), 1830-1832.
[http://dx.doi.org/10.1039/b415651j] [PMID: 15795757]

[268] Friesen, D.T.; Shanker, R.; Crew, M.; Smithey, D.T.; Curatolo, W.J.; Nightingale, J.A. Hydroxypropyl methylcellulose acetate succinate-based spray-dried dispersions: an overview. *Mol. Pharm.,* **2008**, *5*(6), 1003-1019.
[http://dx.doi.org/10.1021/mp8000793] [PMID: 19040386]

[269] Kato, T.; Mizoshita, N.; Kishimoto, K. Functional liquid-crystalline assemblies: self-organized soft materials. *Angew. Chem. Int. Ed. Engl.,* **2005**, *45*(1), 38-68.
[http://dx.doi.org/10.1002/anie.200501384] [PMID: 16353263]

[270] Ringsdorf, H.; Schlarb, B.; Venzmer, J. Molecular architecture and function of polymeric oriented systems: models for the study of organization, surface recognition, and dynamics of biomembranes. *Angew. Chem. Int. Ed. Engl.,* **1988**, *27*, 113-158.
[http://dx.doi.org/10.1002/anie.198801131]

[271] Duncan, R. The dawning era of polymer therapeutics. *Nat. Rev. Drug Discov.,* **2003**, *2*(5), 347-360.
[http://dx.doi.org/10.1038/nrd1088] [PMID: 12750738]

[272] Liu, M.; Kono, K.; Fréchet, J.M. Water-soluble dendrimer–poly(ethylene glycol) starlike conjugates as potential drug carriers. *J. Polym. Sci. A Polym. Chem.,* **1999**, *37*, 3492-3503.

[http://dx.doi.org/10.1002/(SICI)1099-0518(19990901)37:17<3492::AID-POLA7>3.0.CO;2-0]

[273] Chen, H.T.; Neerman, M.F.; Parrish, A.R.; Simanek, E.E. Cytotoxicity, hemolysis, and acute *in vivo* toxicity of dendrimers based on melamine, candidate vehicles for drug delivery. *J. Am. Chem. Soc.,* **2004**, *126*(32), 10044-10048.
[http://dx.doi.org/10.1021/ja048548j] [PMID: 15303879]

[274] Neerman, M.F.; Zhang, W.; Parrish, A.R.; Simanek, E.E. *In vitro* and *in vivo* evaluation of a melamine dendrimer as a vehicle for drug delivery. *Int. J. Pharm.,* **2004**, *281*(1-2), 129-132.
[http://dx.doi.org/10.1016/j.ijpharm.2004.04.023] [PMID: 15288350]

[275] Lim, J.; Guo, Y.; Rostollan, C.L.; Stanfield, J.; Hsieh, J.T.; Sun, X.; Simanek, E.E. Biodistribution and pharmacokinetic studies of bone-targeting N-(2-hydroxypropyl)methacrylamide copolymer–alendronate conjugates. *Mol. Pharm.,* **2008**, *5*, 540-558.
[http://dx.doi.org/10.1021/mp8000292] [PMID: 18672950]

[276] Ooya, T.; Lee, J.; Park, K. Effects of ethylene glycol-based graft, star-shaped, and dendritic polymers on solubilization and controlled release of paclitaxel. *J. Control. Release,* **2003**, *93*(2), 121-127.
[http://dx.doi.org/10.1016/j.jconrel.2003.07.001] [PMID: 14636718]

[277] Ooya, T.; Lee, J.; Park, K. Hydrotropic dendrimers of generations 4 and 5: synthesis, characterization, and hydrotropic solubilization of paclitaxel. *Bioconjug. Chem.,* **2004**, *15*(6), 1221-1229.
[http://dx.doi.org/10.1021/bc049814l] [PMID: 15546187]

[278] Lim, J.; Simanek, E.E. Synthesis of water-soluble dendrimers based on melamine bearing 16 paclitaxel groups. *Org. Lett.,* **2008**, *10*(2), 201-204.
[http://dx.doi.org/10.1021/ol7024907] [PMID: 18088131]

[279] Wu, G.; Barth, R.F.; Yang, W.; Chatterjee, M.; Tjarks, W.; Ciesielski, M.J.; Fenstermaker, R.A. Site-specific conjugation of boron-containing dendrimers to anti-EGF receptor monoclonal antibody cetuximab (IMC-C225) and its evaluation as a potential delivery agent for neutron capture therapy. *Bioconjug. Chem.,* **2004**, *15*(1), 185-194.
[http://dx.doi.org/10.1021/bc0341674] [PMID: 14733599]

[280] Shukla, S.; Wu, G.; Chatterjee, M.; Yang, W.; Sekido, M.; Diop, L.A.; Müller, R.; Sudimack, J.J.; Lee, R.J.; Barth, R.F.; Tjarks, W. Synthesis and biological evaluation of folate receptor-targeted boronated PAMAM dendrimers as potential agents for neutron capture therapy. *Bioconjug. Chem.,* **2003**, *14*(1), 158-167.
[http://dx.doi.org/10.1021/bc025586o] [PMID: 12526705]

[281] Tekade, R.K.; Kumar, P.V.; Jain, N.K. Dendrimers in oncology: an expanding horizon. *Chem. Rev.,* **2009**, *109*(1), 49-87.
[http://dx.doi.org/10.1021/cr068212n] [PMID: 19099452]

[282] Kono, K.; Liu, M.; Fréchet, J.M. Design of dendritic macromolecules containing folate or methotrexate residues. *Bioconjug. Chem.,* **1999**, *10*(6), 1115-1121.
[http://dx.doi.org/10.1021/bc990082k] [PMID: 10563782]

[283] Rockendorf, N.; Lindhorst, T.K. Glycodendrimers. *Top. Curr. Chem.,* **2001**, *217*, 201-238.
[http://dx.doi.org/10.1007/3-540-45003-3_6]

[284] Roy, R.; Baek, M.G. Glycodendrimers: novel glycotope isosteres unmasking sugar coding. case study

with T-antigen markers from breast cancer MUC1 glycoprotein. *J. Biotechnol.,* **2002**, *90*(3-4), 291-309.
[PMID: 12071230]

[285] Turnbull, W.B.; Stoddart, J.F. Design and synthesis of glycodendrimers. *J. Biotechnol.,* **2002**, *90*(3-4), 231-255.
[PMID: 12071227]

[286] Bezouska, K. Design, functional evaluation and biomedical applications of carbohydrate dendrimers (glycodendrimers). *J. Biotechnol.,* **2002**, *90*(3-4), 269-290.
[PMID: 12071229]

[287] Lindhorst, T.K. Artificial multivalent sugar ligands to understand and manipulate carbohydrate-protein interactions. *Top. Curr. Chem.,* **2002**, *218*, 201-235.
[http://dx.doi.org/10.1007/3-540-45010-6_7]

[288] Roy, R. Decade of glycodendrimer chemistry. *Trends Glycosci. Glycotechnol.,* **2003**, *15*, 291-310.
[http://dx.doi.org/10.4052/tigg.15.291]

[289] Niederhafner, P.; Reinis, M.; Sebestík, J.; Jezek, J. Glycopeptide dendrimers, part III: a review. Use of glycopeptide dendrimers in immunotherapy and diagnosis of cancer and viral diseases. *J. Pept. Sci.,* **2008**, *14*(5), 556-587.
[http://dx.doi.org/10.1002/psc.1011] [PMID: 18275089]

[290] Bermejo, J.F.; Ortega, P.; Chonco, L.; Eritja, R.; Samaniego, R.; Müllner, M.; de Jesus, E.; de la Mata, F.J.; Flores, J.C.; Gomez, R.; Muñoz-Fernandez, A. Water-soluble carbosilane dendrimers: synthesis biocompatibility and complexation with oligonucleotides; evaluation for medical applications. *Chemistry,* **2007**, *13*(2), 483-495.
[http://dx.doi.org/10.1002/chem.200600594] [PMID: 17004291]

[291] Chouai, A.; Venditto, V.J.; Simanek, E.E.; Vanderplas, B.C.; Ragan, J.A. Large scale, green synthesis of a generation-1 melamine (triazine) dendrimer. *Org. Synth.,* **2009**, *86*, 151-160.
[http://dx.doi.org/10.15227/orgsyn.086.0151] [PMID: 20072717]

CHAPTER 6

Carbon Allotropes as Top Materials

Abstract: Carbon is the basic element of organic materials. Depending on how the carbon atoms are bonded, carbon-based systems can be divided into groups with different physical and chemical properties. As both the widely diamond and graphite carbon allotropes are of natural sources, the main interest here discussed will be regarding the other main existing forms of carbon arising from laboratory practice, which are the fullerenes, carbon nanotubes and graphene. Herein is presented a conspectus on the field of the above mentioned synthetic carbon allotropes, where emphasis is set on their preparation and functionalization methods, along with more detailed examples on several promising applications with 770 references provided. Some synthetic details on the preparation of state-of-the-art compounds are given as well.

Keywords: Capacitors, Carbon allotropes, Carbon nanotubes, Fullerenes, Functionalization of carbon nanotubes, Functionalization of fullerenes, Functionalization of graphene, Graphene, Graphene oxide, Guel cells, Sensors, Solar cells.

INTRODUCTION

Carbon is the basic element of organic materials. The great interest in carbon materials fundamentally originates from a variety of chemical bonding styles, which has never been seen in other elements [1, 2]. Each carbon atom possesses six electrons occupying $1s^2$, $2s^2$, and $2p^2$ atomic orbitals. As a result, a carbon atom may assume three different bonding configurations, sp, sp^2, and sp^3 [3]. Depending on how the carbon atoms are bonded, carbon-based systems can be divided into groups with different physical and chemical properties.

The specific carbon hybridization and its binding to surrounding atoms determine the type of allotrope assumed. sp^3 hybridized carbon possesses tetrahedral lattice (diamond), while sp^2 hybridized carbon may produce buckminsterfullerene, carbon nanotubes or graphite, depending on the conditions in which it is formed.

Mário J. F. Calvete

Graphite and diamond are known since primeval times, but it was only in the nineteenth century that the structural difference between diamond and graphite was clarified [4], while buckminsterfullerene and carbon nanotubes have only been discovered within the last twenty years, and are just beginning to be understood. Although the theoretical calculations about graphene properties existed for many decades and graphite thin film has been grown by all types of methods over decades, interest in this material was ignited by two independent reports, published in 2004. de Heer [5] demonstrated that ultrathin epitaxial graphite films grown on SiC with 2D electron gas behavior could be patterned *via* conventional lithographic techniques.

Each of the graphitic materials possesses different dimensionality (Fig **1**). The two-dimensional graphene can be considered as the "mother" of all other dimensionalities, *i.e* . fullerenes (0-dimension), carbon nanotubes (1-dimension) and graphite (3-dimensions) [6].

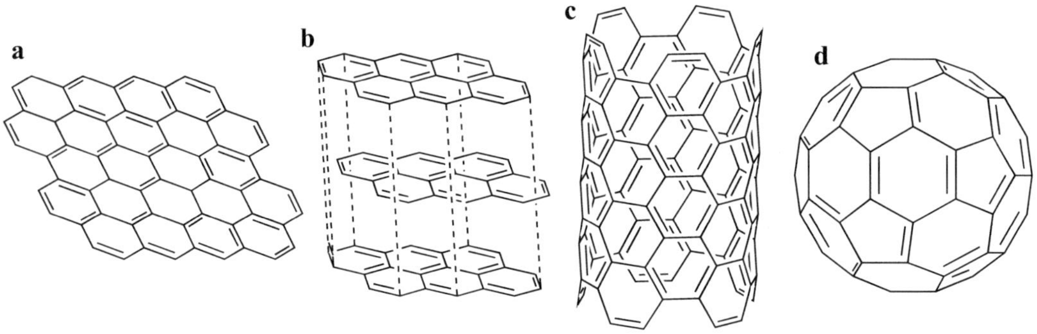

Fig. (1). (a) graphene; **(b)** graphite; **(c)** carbon nanotube; **(d)** fullerene.

As both diamond and graphite carbon allotropes are of natural sources, the main interest here discussed will be regarding the other main existing forms of carbon, which arise from laboratory practice, fullerenes, carbon nanotubes and graphene.

Buckminsterfullerene, C_{60}, was named after Buckminster Fuller, inventor of the geodesic dome), was discovered by Smalley, Kroto and Curl in 1985 [7], being awarded with the Nobel Prize in Chemistry in 1996 for those discoveries [8]. However, C_{60} has a pre-history. In 1970 Osawa, a Japanese chemist reported about a molecule that would be consisted of 60 carbon atoms [9]. It would be, he guessed, a spherical, stable molecule with soccer ball shape, later discussing its probable theoretical aromatic properties in more detail [10]. Also in theory, Gal'pern and Bochvar published a study in 1973 bringing up these structures [11].

About a decade later, Orville L. Chapman initiated its intents to synthesize C_{60} [12]. While Chapman and his colleagues were working on attempted syntheses of C_{60}, Smalley and his group were busy with a completely distinct subject, the study of atomic clusters. They used a self-made device in which laser energy was used for vaporization of an element sample, having no interest in carbon until a visitor, Kroto, suggested the use of their device to direct light on some carbon clusters detected by Smalley. When performing the experiment, the researchers observed an atypical concentration of carbon clusters, spontaneously self-assembled from hot nucleating carbon plasma [7].

After this acclaimed disclosure, the unambiguous proof of the soccer ball structure still remained opened. Several groups made crucial contributions [13 - 15], including theoreticians [16, 17], but the ultimate thrust was given by Krätschmer's group, who succeeded in the first extraction of macroscopic quantities of C_{60} [18 - 20], igniting the explosive expansion of Fullerene Science [21 - 25]. In their milestone work, they produced carbon soot by carbon evaporation in He gas atmosphere and found macroscopic quantities of C_{60} in the soot. They also found that fullerenes are soluble in aromatic solvents (*e.g.* benzene and toluene) and C_{60} crystals, named fullerite, could be grown from the solution as the solvent evaporated. These properties were used to separate C_{60}, C_{70} and other molecules from the other non-soluble forms of carbon products, using chromatography [26]. At last, nuclear magnetic resonance (NMR) study using the mass-produced C_{60} clearly demonstrated that the C_{60} actually has the soccer ball structure [27]. Later, in 1991, Haddon reported the first evidence that the fullerene solids could be doped by the alkali metals [28]. They grown thin films of C_{60} after first attaching electrical leads to the glass substrate and when exposing the film to a flux of alkali metal atoms, they found that the resistivity dropped with exposure time, reached a minimum, then increased and ultimately saturated. Among the properties of the doped fullerene solids, named fullerides, the most attention from the physics community was the appearance of superconductivity. For $Rb_1Cs_2C_{60}$, a rather high superconducting transition temperature *Tc* reached 33 K [29]. By using laser ablation technique with a target of La-impregnated low-density graphite, Heath succeeded in producing La@C_{60} [30]. This metallofullerene suggested a "super atom" concept in which the caged atom would donate (or accept) charge from the cage, providing a positive (or negative) core. Solids derived from such species are believed to offer intriguing possibilities showing promise as the building blocks of a host of new chemical species [31].

When fullerenes were revealed, the natural curiosity led scientists to other discoveries. The C_{60} relatives called carbon nanotubes are among them. These are quasi-one dimensional graphitic structures that exist either in the form of a single cylindrical shell or as concentric arrangements thereof. They are respectively known as single- and multi- walled carbon nanotubes (SWNTs and MWNTs). SWNTs can be seen as a formation of rolled graphene sheets in the shape of a cylinder, with 1-10 nm diameter [32].

MWNTs were found by Ijima, in 1991 [33]. In addition to the MWNT, the synthesis of SWNT with seamless single cylinder [34, 35] and ropes of closed packed SWNTs [36] were later reported. The research rate on SWNT was then naturally accelerated by the findings of the outstanding electrical transport properties of SWNTs. As their longitudinal conductivity is very sensitive to the detailed atomic structure and their diameter; a slight change of the diameter of the nanotube can result in the SWNT being either metallic or semiconducting [2, 37]. Such SWNT behavior reasoned the birth of nano-electronics and make SWNTs important parts at the cutting edge of nanotechnology [38, 39]. This research fever did not stop here and, subsequently to nanotubes, carbon onions [40] and carbon nanocapsules [41, 42] have also been discovered as carbonaceous nano-materials. Additionally, there are reports regarding the creation of single-walled carbon nanohorns [43], SWNTs encapsulating fullerenes or metallofullerene (carbon peapods) [44], and shortened multi-walled nanotube structures encapsulating fullerenes (Bucky shuttles) [45]. These novel carbonaceous nano-materials are expected to be key elements in the century of nanoscience and nanotechnology. However, the carbonaceous material which currently deserves most attention is graphene. Studies on the electronic structure of graphene date as early as the first half of the twentieth century [46], when scientists were largely motivated by graphite properties.

Although the theoretical calculations about graphene properties existed for many decades and graphite thin film has been grown by all types of methods over decades, interest in this material was ignited by two independent reports, published in 2004. de Heer [5] demonstrated that ultrathin epitaxial graphite films grown on SiC with 2D electron gas behavior could be patterned *via* conventional lithographic techniques. In the meantime, Geim [47] demonstrated that crystalline graphitic film with a few atoms thick exhibited a strong ambipolar electric field effect. The breakthrough for those experiments is the optimized method to prepare and identify thin layer graphite or graphene.

Graphene growth was verified [48] on several transition metals such as ruthenium [49], iridium [50], nickel [51] *via* hydrocarbon surface thermal decomposition or hydrocarbon surface segregation upon cooling from a metastable metal-carbon solid solution. One example of graphene chemical vapor deposition growth process on Cu foils is [52]: i) filling of vacuum system with hydrogen, heat it to 1000ºC and maintain a H_2 pressure of 40 mTorr under a 2 sccm (standard cubic centimeters per minute) flow; ii) heat Cu film up to 1000 ºC, and introduce 35 sccm of CH_4 for a desired period of time at a total pressure of 500 mTorr; iii) after exposure to CH_4, the furnace is cooled to room temperature.

SYNTHESIS OF CARBON BASED MATERIALS

Several methods have already been developed to synthesize nanotubes in considerable amounts, making CNTs more commercially viable, including chemical vapor deposition, laser ablation, arc discharge and plasma torch.

Nanotubes were accidentally discovered in 1991 [33]; however the first measurable carbon nanotube production only took place in 1992 [53], by sublimation due to high-discharge temperature, in up to 30% yield, producing SWNTs and MWNTs with lengths of up to 50 μm [54].

In the method of laser ablation, the graphite layers are vaporized by a pulsed laser, producing nanotubes on the reactor's cooler surfaces, by inert gas introduction into the reactor [55, 56]. This method yields primarily SWNTs, in up to 70% yield and manages to control the tubes' diameters, depending on the reaction temperature [54].

SWNTs can also be prepared using the method of thermal plasma, where the goal is to replicate the laser ablation and arc discharge established conditions, supplying the carbon necessary using a carbon-containing gas, managing to produce around 1 gram of nanotube material per minute [57 - 59].

Another method for nanotube preparation consists on chemical vapor deposition, in which a metal catalyst particle layer, usually of cobalt or nickel [60], iron, or a combination of them [61] is used to prepare the substrate. The substrate is then heated to 700 °C, after which two gases are introduced into the reactor: a process gas (such as ammonia, hydrogen or nitrogen) and a carbon-containing gas (such as ethanol, methane, acetylene or ethylene). Then the carbon-containing gas reacts at the catalyst surface, forming the nanotubes [62]. Chemical vapor deposition is the main carbon nanotube production method [63].

Exfoliated graphene can be obtained by mechanical exfoliation of thin graphitic layers from natural graphite or highly ordered pyrolitic graphite with scotch tape. This original method was described by Geim's group and is as follows [64]: A fresh surface of graphite is rubbed against another surface, leaving several flakes attached to it. The preliminary identification of single layer among wider flakes and other residues is done in an optical microscope and then the single layers become visible on top of an oxidized Si wafer (300nm SiO_2). Graphene made from this method can virtually be transferred to any substrate or no substrate at all. Suspended graphene was achieved by etching away SiO_2 underneath with wet chemistry [65]. Graphene on hexagonal boron nitride (h-BN) was also demonstrated by transferring graphene onto the top of a few layer h-BN [66].

Functionalization of Carbon Materials

Chemical functionalization of carbon allotropes is of primary importance, paving the way unique materials applications. Technically, the chemical and physical properties of fullerene, carbon nanotube and graphene are interconnected with each other, though stages of development on their functionalization yet vary considerably [67]. The fullerenes represent the first family of synthetic carbon allotropes and the most widely studied [68], while carbon nanotubes display the same application potential; however, its chemistry is yet lesser developed [69]. Many fundamental problems remain to be solved, although many functionalization protocols have been already reported. These include the highly demanding single helical SWNTs isolation or the highly selective semiconducting SWNTs functionalization, among several others. Lastly, graphene, the newer representative in the list of carbon-allotropes it is still at its infancy, and the study of its chemistry has just kicked off.

The fullerenes are quite reactive, due to the pyramidalization angles of the carbon atoms, which causes an enormous strain provoked by their spherical geometry [70]. For this reason, strain relief increases stability, highly favoring addition chemistry in fullerenes [71 - 74]. Similar to a fullerene, a non functionalized SWNT is chemically inert. However, misalignment of the π - orbitals and curvature-induced pyramidalization [75 - 80] of the carbon atoms provoke a localized strain, causing a higher reactivity than for a graphene sheet, even if less reactive than C_{60}.

Chemical functionalization permits for the synthesis of soluble C_{60} derivatives, maintaining the electronic properties of fullerenes. The chemistry of [60]fullerene is characteristic of electron-deficient alkenes, thus reacting with nucleophiles [81 - 83] and in many other reactions [84 - 95]. The term fulleroid is normally used to indicate a ring-open (methanoannulene) fullerene derivative, while methanofullerene is normally used to indicate a cyclopropane-fullerene derivative (ring-closed) [96, 97]. Also its [6, 6] bonds are good dienophiles, and cycloaddition reactions, such as [2+2], [3+2], and [4+2], play an important role in the preparation of C_{60} derivatives, from which a wide variety of cycloadducts have been prepared [98, 99].

Four-member rings fused to 6,6-ring intersections on C_{60} are the product of [2+2] cycloadditions. Tsuda [100, 101] and Hoke [102] independently reported the first [2+2] benzyne thermal addition to C_{60}. In addition, Schuster [101, 103 - 105] demonstrated [2+2] photochemical cycloadditions of 1,3-diones and cyclic enones, as well as dimer formation [106]. The first report on dimer formation was given by Komatsu [107, 108], when the researchers reacted C_{60} (Scheme **1**) using potassium cyanide, producing a dimer. However, when the reaction was carried out in solution only C_{60}-cyanated derivatives were obtained [109]. Likewise, Forman [110] and Tagmatarchis [111] have reported reactions of dimerization of C_{70}, as well as the synthesis of $C_{60} - C_{70}$, respectively.

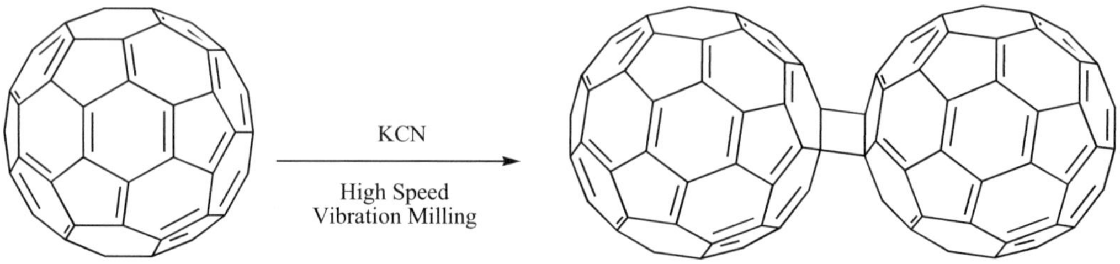

Scheme (1). Synthesis of substituted fulleropyrrolidines.

[3+2] cycloadditions can be used to form five-member rings fused to 6,6-junctions. Azomethine ylides are the most common reactive intermediates, being *in situ* generated by decarboxylation of immonium salt derived from the thermal condensation of amino acids with aldehydes (or ketones) or thermal aziridine ring opening (Scheme **2**). The first reported reaction under these conditions was made by Prato [112], who revealed a widely used strategy for C_{60} functionalization [113]. For example, 1,3-dipolar addition of azomethine ylide to C_{60} yielded fulleropyrrolidine at the 6,6-junction. The use of *N* -substituted glycines gives *N* -substituted fulleropyrrolidines, while f functionalized aldehydes leads to 2-substituted fulleropyrrolidines, (Scheme **2**). Furthermore, mono-fulleropy--rolidines can be synthesized by controlling the reaction conditions and reagent stoichiometry. Thus, different fullerene derivatives can be prepared using this methodology either by modifying a fulleropyrrolidine intermediate or by use of adequately azomethine ylides. Functionalized fullerene skeletons combined with further substituents or reactions also lead to materials based on fullerenes with unique properties [114 - 126].

The [6, 6] double bonds of [60]fullerene are excellent dienophiles. Thus, C_{60} can easily undergo reactions with different dienes by Diels–Alder cycloadditions. Depending on the reactivity of the diene, it can be accomplished by heating the reaction under reflux in a high boiling solvent or using microwave energy [127].

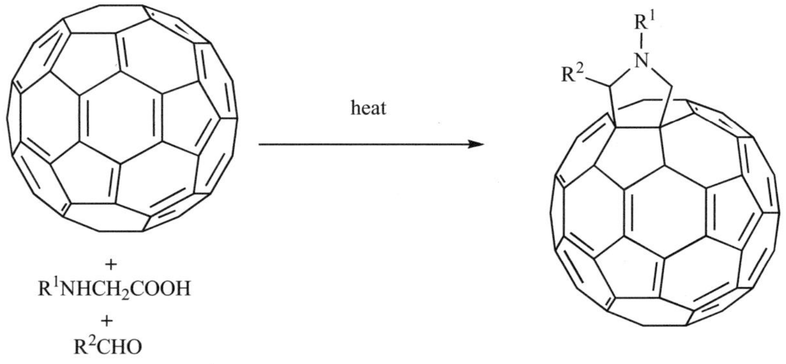

Scheme (2). Cycloaddition reactions of C_{60}.

The first example of a [4+2] cycloaddition reaction of C_{60} was carried out using cyclopentadiene [105] as the 1,4-diene system (Scheme **3**). Cyclopentadiene derivatives, such as methylcyclopentadiene and cyclopentadienone, afforded the corresponding monoadducts [128 - 132]. The dienophilic behavior of C_{60} in [4+2] cycloaddition reactions was also demonstrated in the reaction of C_{60} with anthracene (Scheme **3**) and other anthracene derivatives [133, 134].

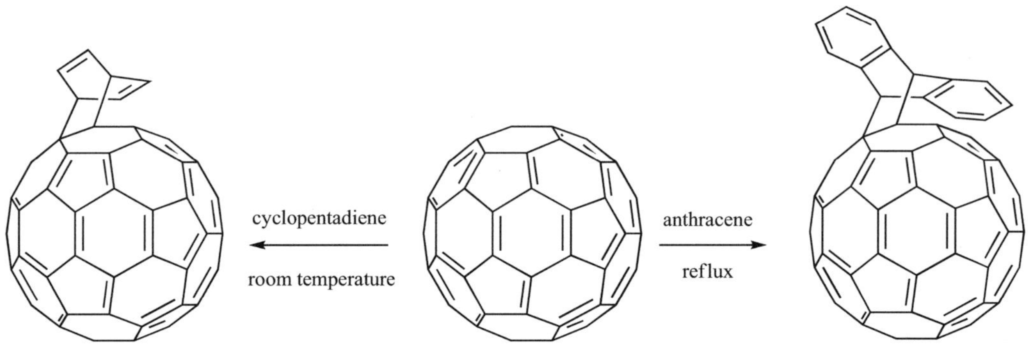

Scheme (3). Model structures of graphene oxide.

The production of C_{60} anions in a defined oxidation state was firstly reported in 1991 [135], by applying a suitable potential to prepare fulleride salts [135 - 140], which can also be used in the preparation of covalent organofullerene derivatives by anion quenching with electrophiles [141 - 146]. Fullerenes can also be reduced using electropositive metals such as alkali and alkaline earth metals [147 - 150], usually by titration of C_{60} suspensions [151 - 164]. Fullerene readily reacts with alkyl, phenyl or alkynyl Grignard and organolithium reagents, forming RC_{60}^- anions as primary intermediates [109, 165 - 169]. Other organolithium acetylides have been attached to C_{60} [170 - 172], along with several other electrophiles, such alkyl, cycloheptatrienyl, benzoyl- or vinylether-derivatives [169, 173]. Due to their high nucleophilicity, secondary and primary aliphatic amines can easily react with electron-deficient C_{60} [165 - 169, 174 - 177]. Furthermore, Kampe [178] reported that diamines such as piperazine, homopiperazine or *N,N'* - dimethylethylenediamine, reacted r with C_{60}, giving mono- and bisadducts as predominant products.

Following the first carbon nanotubes' report, by Ijima [33], interest on this nanomaterial emerged and has grown exponentially specially regarding their functionalization protocols [179 - 184]. However, nanotubes possess usually lower reactivity than fullerenes, requiring more demanding experimental conditions for the functionalization of carbon nanotubes. [185 - 187]. They can be chemically modified in order to induce solubility for integration in biological, inorganic and organic systems [188, 189]. Modification can occur by covalently attaching chemical fucntionalities by reactions with the nanotube's π-conjugated system, by noncovalent wrapping or adsorption of several functionalities around the nanotubes and by endohedral filling of the inner empty cavity (Fig. **2**).

The nanotubes' end-caps are highly reactive, when compared with their sidewalls [184, 190]. The sidewalls also contain defect sites such as *sp3* -hybridized defects, Stone-Walls defects, five- or seven- member rings in the carbon framework, holes (vacancies) in the nanotube lattice carbon framework damaging by oxidative conditions and open endings of the SWNT [184].

Covalent functionalization can be carried out at the nanotubes' sidewall defects or at their end-caps. This procedure can be carried out by reaction with some high chemically reactive molecules. For instance, halogenation of nanotubes [191], particularly fluorination, was one of the first topics on covalent functionalization of nanotubes [192, 193]. C-F bonds in fluorinated nanotubes are easier to break than in alkyl fluorides [194], consequently offering additional sites for functionalization, replacing the fluorine atoms by hydroxyl, alkyl or amino fucntionalities [191, 195 - 201]. Functionalization of nanotubes by hydrogenation has also been reported [202 - 206]. For instance, Pekker [202] promoted the reduction of pristine nanotubes using Birch reduction method (methanol and lithium in liquid ammonia). In addition, nanotubes can also be functionalized with H_2 by proton bombardment [206] or by using a glow discharge [203 - 205].

Cycloaddition reaction can also be successfully applied for nanotubes modification. Haddon reported for the first time [2+1] carbene cycloadditions to nanotubes [74, 207 - 210], through carbene generation *in situ* , with the reaction taking place on the side wall of the nanotubes. Also Hu promoted the reaction of dichlorocarbenes with nanotubes [210]. Hirsch [184, 211] carried out other studies showing that carbene nucleophilic addition to nanotubes, rather than cyclopropane derivatives, mainly gave zwitterionic 1:1 adducts. Additionally, nanotube thermal functionalization by nitrenes by [2+1] cycloaddition reaction was also promoted, yielding alkoxycarbonylaziridino-nanotubes [184, 211, 212 - 214].

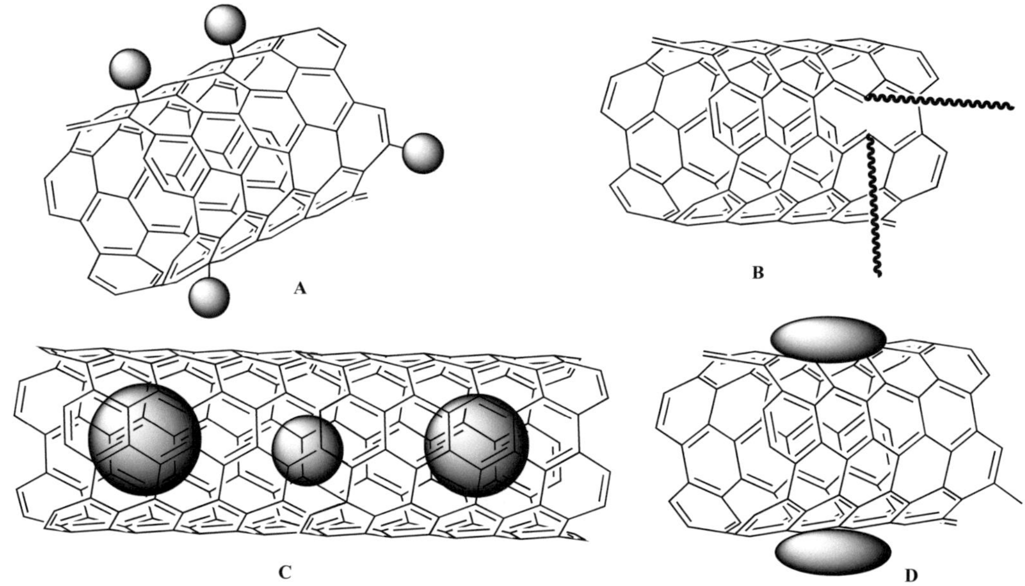

Fig. (2). Types of functionalization in SWNTs: **A)** covalent sidewall functionalization, **B)** defect-group functionalization, **C)** endohedral functionalization, and **D)** non-covalent exohedral functionalization.

Functionalization of nanotubes through application of azomethine ylides was first reported by Hirsch [215]. The azomethine ylides were thermally formed *in situ* by aldehyde/amino acid condensation, followed by 1,3-dipolar cycloaddition, yielding pyrrolidine-nanotube derivatives [215, 216].

Amidation/esterification reactions are another effective manner to modify nanotubes. The first step consists normally on performing a carboxylation reaction by using an oxidizing acid, or a mixture of concentrated sulfuric and nitric acids [217 - 220], a method usually used for raw nanotube purification [221 - 224]. This procedure induces tube caps opening, as well as hole formations on the sidewalls. Once the nanotube is functionalized with acid terminal groups, these serve to bind various types of chemical functionalities onto the defect sites and ends of the nanotubes. Nanotubes can also be precursors for subsequent chemical reactions, such as silanation, amination [225 - 230], thiolation [224], esterification [231] and polymer grafting [232, 233]. Nanotubes functionalized by covalent methods have several advantages which tend to the higher solubility in various organic solvents, due to the possible number of functional groups that can be tethered to their walls.

The greatest advantage of non-covalent functionalization is the lack of interference with the nanotube sidewalls' conjugated system, unaffecting the structural properties of the target material. Nanotubes are commonly non-covalently functionalized by surfactants, polymers and aromatic compounds, employing mostly hydrophobic or π-π stacking interactions. Aromatic molecules, such as pyrenes, porphyrins, phthalocyanines, and other molecules, usually interact with the nanotubes' sidewalls by means of π-π stacking interactions [234 - 236]. In addition, functionalization of nanotubes was also carried out using surfactant polymers, which diminished their surface tension, thus preventing nanotube aggregation [237 - 242].

Being the youngest carbon-based material, graphene has already attracted great attention, due to its exceptional performances and unique structural features. Additionally, graphene production is much cheaper than of other carbon-based nanomaterials. Thus, it is important to develop several aspects in graphene chemistry, particularly surface modification of graphene [243, 244].

Appropriate graphene chemical functionalization avoids single layer graphene agglomeration during solvent phase reduction, helping also to keep graphene inherent properties. Graphene oxide is widely used as a synthetic starting material for usable graphene. The most used method for graphene modification is called the Hummers method for graphene oxide preparation [245, 246]. It consists in oxidizing graphite with $NaNO_3$ and H_2SO_4 mixed solution. During the harsh oxidation and thermal expansion process, thin layers of graphite oxide are formed. Additional ultrasonication of solution can expand the graphite oxide flakes into single layer graphene oxide. Researchers have found that by exposing such graphite oxide layers in the saturated vapor of dimethylhydrazine [247], or through thermal annealing [248] the oxygen based groups in graphene oxide can be easily reduced and material conductivity can be enhanced.

The surfaces of graphene oxide sheets are extremely oxygenated, containing hydroxyl, epoxide, ketone, carboxylic and diol functional groups, leading to higher organic solvent and water solubility (Fig. **3**) [246, 249 - 254]. The presence of supplementary carboxyl and carbonyl groups situated at graphene oxide endings increases hydrophilicity, tolerating easy water swelling and dispersing [253 - 257].

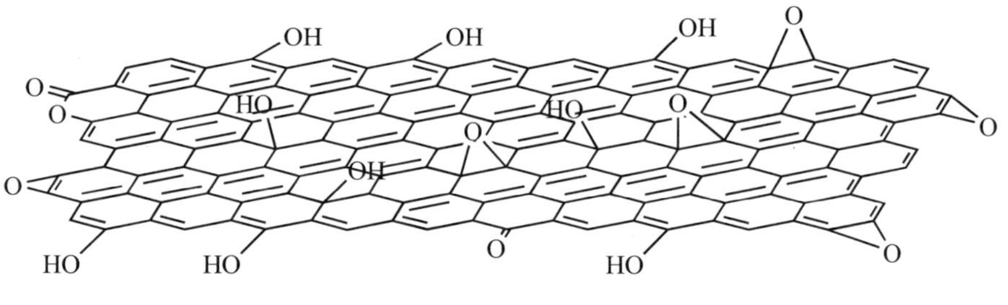

Fig. (3). Model structures of graphene oxide.

There are several manners to structurally modify graphene, which include non-covalent and covalent functionalization. The structural modification through covalent functionalization takes place on the surface and/or at the end of the sheets. Surface functionalization causes rehybridization of one or more sp^2 carbon atoms into the sp^3 configuration together with coincident loss of electronic conjugation, while graphene covalent modulation can be attained by cycloaddition, electrophilic substitution, condensation and nucleophilic substitution.

Cycloaddition reactions can be successfully applied for graphene modification. Among the possible addition reactions, the use of azomethine ylide 1,3-dipolar cycloaddition on the surface of graphene was applied for epitaxial graphene functionalization through azidotrimethylsilane cycloaddition [258]. After N_2 removal, graphene reacted with nitrene either by a biradical pathway or electrophilic [2+1] cycloaddition, forming the functionalized graphene. In a similar fashion, graphene sheets were also functionalized using *N* - methylpyrrolidone and pyridine as solvents [259, 260], or using cyclopropanated malonate as precursor [261]. Valiyaveettil has functionalized graphene sheets wrapped in surfactants using several alkylazides [262]. Materials bearing polar functional groups and longer alkyl chains showed improved dispersibility in toluene and acetone. Also polyacetylene is a suitable organic modulator for graphene following nitrene chemistry [263, 264], while Ma developed an efficient and simple method for the graphene sheets preparation under mild reaction conditions, *via* aryne cycloadditions [260, 265].

Electrophilic substitution reactions using graphene imply a hydrogen atom displacement by an electrophile. This procedure led to a *para* -nitro aniline diazonium salt grafting, onto the graphene surface [266]. Hence, Tour [267, 268] managed to prepare graphene soluble in organic solvents by aryl diazonium salt electrophilic substitution on the surfactant-wrapped graphene surface (Scheme **4**).

The same group also performed an edge-selective functionalization of graphene using 4-bromo-aniline diazonium salt [269], while Samulski reported a simple strategy for preparing sulfonated graphene [270], through sulfonation with the aryl diazonium salt of sulfanilic acid. Also Lu used polystyrene to covalently functionalize graphene [271]. The initiating molecules were bonded covalently to the surface of the graphene through addition of diazonium salt and consequent radical polymerization linked the polystyrene chains (82 wt.% grafting efficacy) to the graphene sheets.

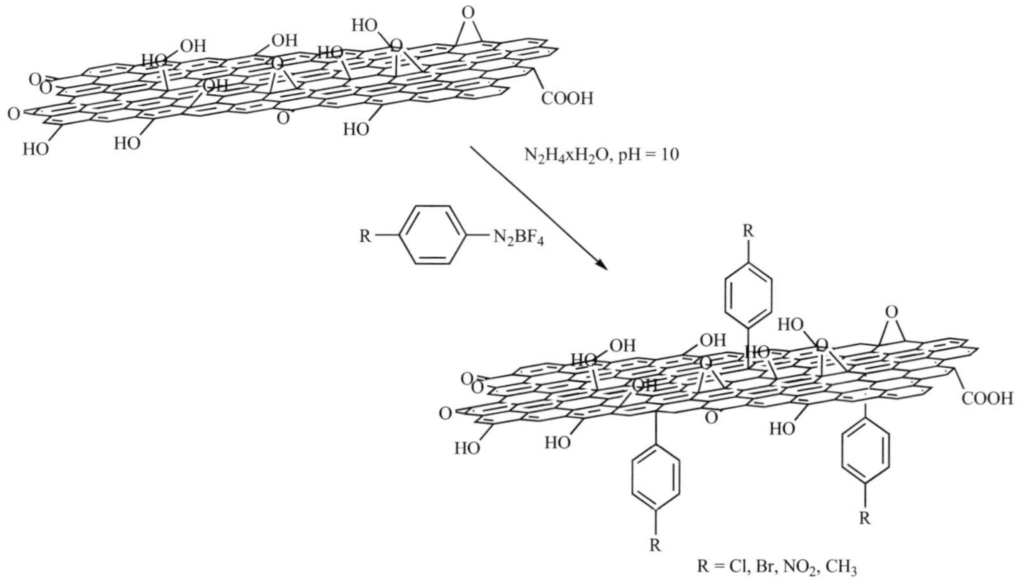

Scheme (4). Aryl diazonium salt electrophilic substitution.

Reactions of condensation on graphene surface occur with amine, isocyanate, and diisocyanate compounds, through amide and carbamate ester linkage formation. Nguyen and Ruoff used several isocyanate compounds for modifying the surface of graphene oxide [250, 272], under nitrogen atmosphere and the resultant functionalized graphene oxide could be dispersed in *N,N* -dimethyl formamide. Diisocyanate can be also used in graphene oxide cross-linking and functionalization [273]. Grafting diisocyanate on graphene oxide surface could be carried out by activation of the graphene oxide carboxylic functionality, using thionyl chloride originating an amide (Scheme **5**) [274]. Other authors have also reported graphene oxide functionalization using poly(ethylene glycol) terminated with amine groups [275]. The functionalization of graphene oxide occurred *via* carbodiimide-catalyzed amide formation.

Graphene modified with chitosan has been also prepared by microwave irradiation, using *N,N* -dimethylformamide as solvent [276]. The functionalization proceeded by reacting the amine groups of chitosan with the graphene oxide carboxylic groups, followed by hydrazine mediated reduction. Other studies reported that the dispersibility of graphene oxide in acidic aqueous media could be enhanced by graphene oxide amidation with chitosan [277, 278]. The graphene oxide sheets grafted with chitosan consisted of about 64 wt.% chitosan, originating good aqueous solubility and biocompatibility.

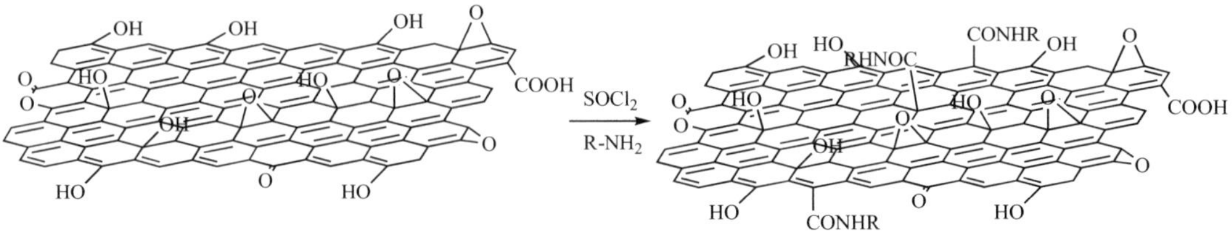

Scheme (5). Amidation of graphene.

Esterification of graphene oxide was also accomplished by reaction of the β-cyclodextrin hydroxyl groups with the carboxylic group, in presence of thionyl chloride [279]. A more straightforward method was also developed to prepare cyclodextrin (α,β,γ)-graphene inorganic–organic hybrid nanosheets, displaying higher stability/dispersibility in DMF, DMSO, water and ethanol [280]. Similar to β-cyclodextrin, modification of graphene surface would also be accomplished esterification of poly(vinyl alcohol) [281]. An amine-functionalized porphyrin was also used for the same purpose; however, *via* formation of an amide bond [282], significantly improving the dispersion stability and solubility of the material in organic solvents. Similarly, phthalocyanines [283], oligothiophene [284] poly(N-vinylcarbazole) [285] 2-amino-4,6-didodecylamino-1,3,5-triazine [286] have been also appended to graphene.

In nucleophilic substitution, the epoxy groups of graphene oxide are the main reactive sites and are straightforward reactions, occurring even at room temperature and in aqueous media.

For instance, Petridis and Dekany used several amino acids and primary amines for surface functionalization of graphite oxide [287]. For amines with small chains, grafting could be carried out at 25 °C, while the reaction had to be carried out at reflux when long alkyl chain amines were used (with eighteen carbons) [287, 288]. Wang [289] and Lee [290] have also used the same procedure to append dodecyl and octadecyl amines for the synthesis of amino functionalized graphene, showing that graphene modified with the longer chain alkyl amine was soluble in several organic solvents, like ethanol toluene, cyclohexane, tetrahydrofuran and DMF.

Other covalently functionalized graphene sheets were prepared by functionalization with 3-aminopropyltriethoxysilane [291], capable of being dispersed in DMF, DMSO, ethanol and water in a homogeneous way. Niu used ionic liquids with amine ends for graphene functionalization [292], which was carried out *via* nucleophilic epoxyde ring-opening reaction with the ionic liquid amine groups. Also biocompatible graphene sheets functionalized with poly--lysine were synthesized, using as solvent an alkaline solution [293]. The thickness of graphene– poly-L-lysine was determined to be ca. 3.6 nm. Govindaraju reported the Friedel-Crafts acylation of graphene oxide using ferrocene [294] and Chung used graphene oxide and *N* -methylpyrrolidone as starting materials in a solvothermal process [267, 295, 296].

Non-covalent interactions mainly involve electrostatic, hydrophobic and van der Waals forces. This type of functionalization can be accomplished through adsorption of aromatic molecules and surfactants, interaction with biomolecules and polymer wrapping. Nguyen and Ruoff reported for the first time the use of poly(sodium 4-styrenesulfonate) for non-covalently functionalize graphitic nanoplatelets [297], by *in situ* reduction after exfoliation of graphene oxide in the polymer presence. Also sulfonated polyaniline could be used similarly as a surface modifier [298], as can amine-terminated polymers [299]. Additionally, graphene materials can also be functionalized after the reduction of graphene oxide to graphene in aqueous medium using ammonia-hydrazine [299].

Non-covalent surface modification of graphene with PPI dendrimers was also accomplished [300], by reducing graphene oxide in the presence of the dendrimers, forming a stable aqueous dispersion in presence of excess of dendrimer molecules. Other authors [301] described a functionalized graphene flexible film, made from pyrenebutyric acid derivative and graphene oxide in aqueous medium. Due to the presence of the pyrenebutyrate, the material was highly soluble in water.

Davis also described the modulation of graphene sheets by taking advantage of the π-π interactions between graphene and poly(*N* -isopropylacrylamide [302]. The pyrene-terminated poly(N-isopropylacrylamide) was dissolved in water, and a graphene aqueous suspension was then added. Using this π-π interaction, several other functionalization of graphene materials were reported [303 - 309]. Also sodium dodecylbenzenesulfonate has been largely utilized for graphene surface modification. Tao [310], and Li [311] have prepared sodium dodecylbenzenesulfonate -modified graphene, which, by reduction produced surfactant-wrapped graphene sheets.

APPLICATIONS OF FULLERENES

Fullerene's outstanding properties such as remarkable electrical and thermal conductivity, flexibility and high mechanical strength make itself a potential candidate for various applications, *e.g.* in catalysis [312 - 316], including as catalyst for fuel cells [317 - 321], as electroactive materials [314, 322 - 328], solar cells [329 - 337], in thin films [338 - 344], in nonlinear optics, [345 - 350] and biological applications [351 - 357], including therapy [358 - 360] and imaging [361 - 366].

Applications in Catalysis

Catalysis is one of the multiple possible fullerene applications. There are three main types of catalysis in which fullerenes are used: (i) C_{60} as catalyst; (ii) C_{60} as a homogeneous catalyst ligand; and (iii) C_{60} as a heterogeneous catalyst support medium.

Impelled by fullerene ability to form and cleave strong bonds, besides participating in hydrogen transfer reactions, Hirshon [367] reported the use of materials based on fullerene for CH_4 conversion to higher hydrocarbon compounds. Before that, it had also been described the use of C_{60} as a transition metal ligand [368], suggesting that C_{60} can act as a η_2-C_{60} ligand. Furthermore, it was reported the efficiency of $RhH(CO)(C_{60})(PPh_3)_2$ as alkenes hydroformylation catalyst [369].

Later, Schlögl [316, 370, 371] studied some catalysts behavior supported on fullerene black (FB), displaying similar catalytic properties on hydrogenation of

CO and cyclohexenone to those of Ru_3C_{60} [316]. A limitation on C_{60} use as catalysis heterogeneous support is its sublimation temperature and relatively low surface area. To overcome such limitation, some authors dispersed the C_{60} moieties onto high surface area carrier, by chemisorption or physisorption, forming several oxide carriers [344, 372].

Grafting of fullerene to silica can be accomplished by the addition of an amine [373] or pyridyl osmium complexes as linkers [374] and hydrosilylation reactions [375]. A 0.8 weight % Pt catalyst silica supported and C_{60}-grafted was synthesized and tested in hydrogenation of cinnamaldehyde [376]. It was observed 89% selectivity to cinnamalcohol after 90% cinnamaldehyde conversion using the hybrid catalyst.

Applications as Electroactive Materials

The combination of C_{60} with other electroactive species is presently focus of strong research, since chemically modulated C_{60} are quite relevant in molecular electronic device design, particularly in artificial photosynthesis. Thus, several electron-rich groups were covalently conjugated to C_{60}, yielding a large set of dyads. Donor units used include aromatics (**1** and **2**) (Fig. **4**) [377, 378], porphyrins (**3**) Fig. (**4**) [379], ruthenium(II) tris(bipyridine) (**4**) (Fig. **4**) [380] and ferrocene Fig. (**5**) [381], which have been studied for photoinduced charge separation, where both electron transfer and energy processes between the acceptor and the donor (C_{60} moiety) have been demonstrated.

For instance, C_{60}singlet excited state intramolecular quenching from ferrocene electron transfer was observed in dyads **5-9** (Fig. **5**) [381], where the spacer nature played an significant role. Electronic transfer was observed in dyads **5-7**, while transient intramolecular exciplex formation was seen in materials **8** and **9**, whose spacer avoided recombination of charge.

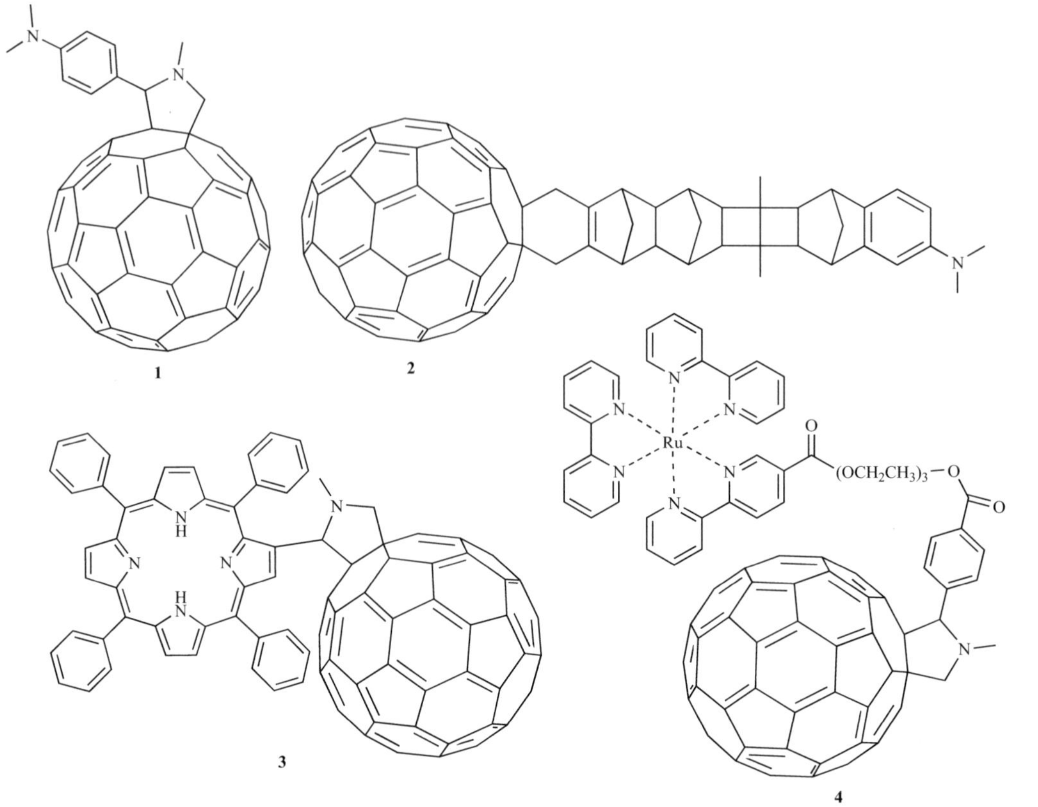

Fig. (4). Fullerene based compound for application in electronic devices.

The photoinduced electron transfer in C_{60} doped π-conjugated polymers has been demonstrated by some groups [382 - 384], toward the manufacturing of efficient photovoltaic devices. The authors reported a very fast electron transfer, with strong polymer photoluminescence quenching. However, the fullerene high crystallization tendency was a big disadvantage in using C_{60} in this type of polymer blends, resulting in poor homogeneity and low optical films quality. To overcome these issues, soluble C_{60} derivatives were then used, for example, the methanofullerenes **10** and **11** (Fig. **6**) [385, 386], showing that the enhanced fullerene derivative miscibility fulfilled the electro-optical devices prerequisite.

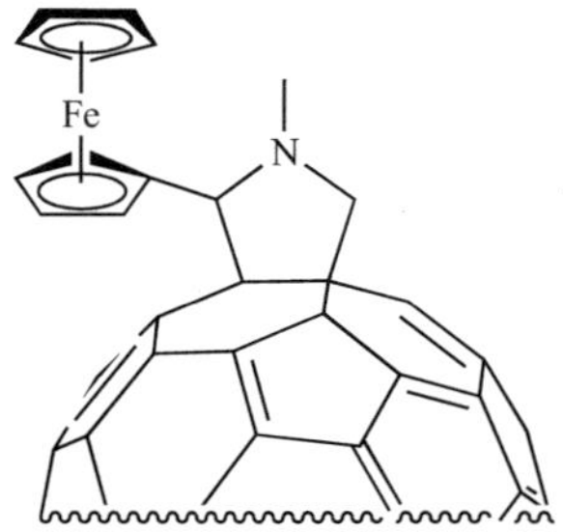

5

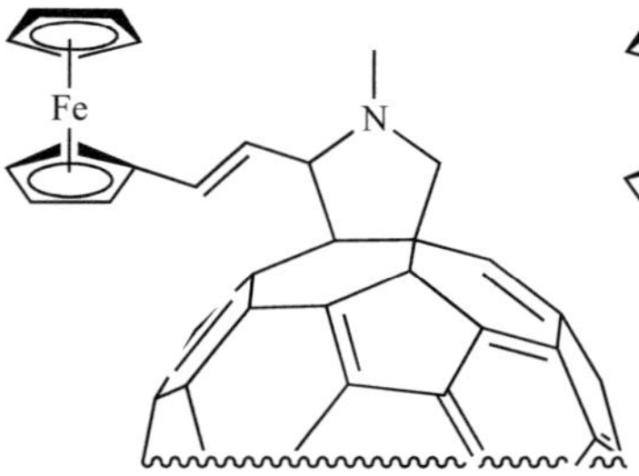

6

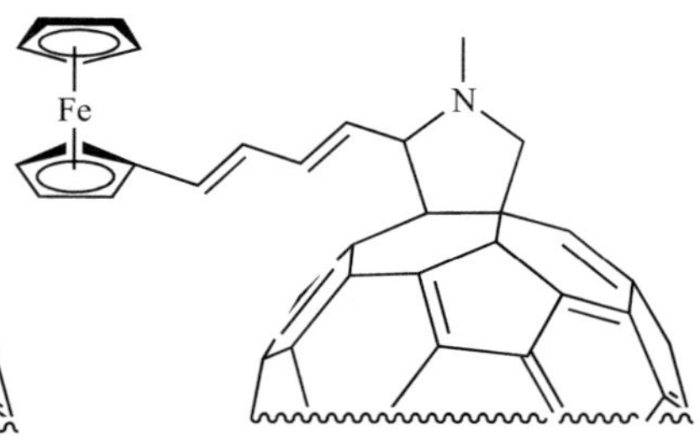

7

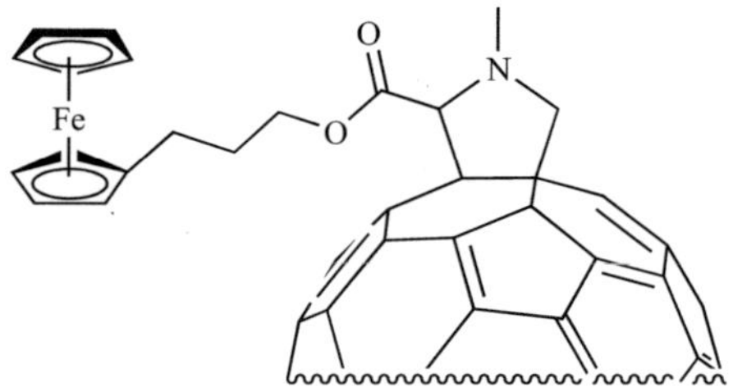

8

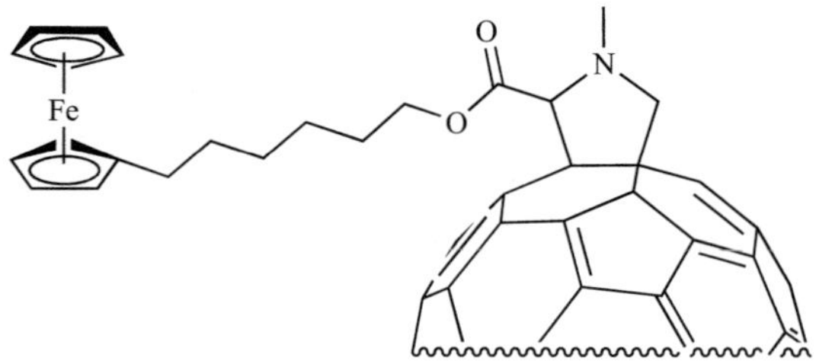

9

Fig. (5). Ferrocene substituted fullerenes.

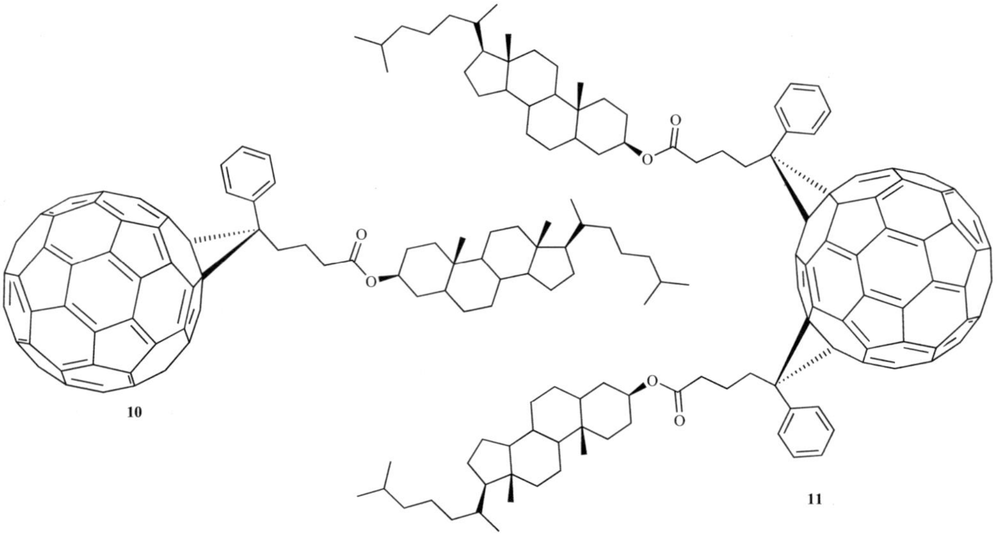

10

11

Fig. (6). Methanofullerene derivatives.

Applications in Organo-photovoltaics

A common photovoltaic cell requires three main components: an electrontransport agent (n-type), a hole-transport agent (p-type) and a light-absorber (dye). The development of n-type acceptors for efficient bulk heterojunction organic solar cells has been quite limited, reason why the most efficient organic photovoltaic cells still contain fullerene derivatives [387]. Significant progress has been made on the use of acceptors based on fullerene, since some fullerene weaknesses (for instance low LUMO energy levels) [331, 388] have been solved by suitable fullerene modifications.

Fullerenes are extensively used in organo photovoltaics mainly due to their high electron accepting tendency and high electron mobility [389, 390]. Commonly used fullerene derivatives processable in solution for bulk heterojunctions include 6, 6]-phenyl-C_{61}-butyric acid methyl ester (**12a**), and analogues with thiophene (**12b**) [391], fluorene (**12c**) [392] or triphenylamine (**12d**) Fig. (**7**) [392] moieties have also been reported, giving origin to devices with *ca* 4% efficiencies, when combined with electron donor poly(3-hexylthiophene). The use of dihydronaphthyl fullerene benzyl alcohol benzoic acid ester [393] or compound **13** [394] gave higher efficiency devices with $\eta = 4.5\%$. Nevertheless, a high open circuit voltage = 0.89 V was attained for a endohedral fullerene using used trimetallic nitride composite **14** Fig. (**7**) [395, 396]. Matsuo group reported silylmethyl[60]fullerene **15**, producing a device with PCE = 5.2% (PCE = power conversion efficiency) Fig. (**7**) [397, 398]. In 2010, Li reported the indene-C_{60} bisadduct **16** Fig. (**7**), providing an efficient device with 6.48% power conversion efficiency, when blended with poly(3-hexylthiophene) [399], which was the most efficient device reported to date with a poly(3-hexylthiophene) based polymer solar cell.

Applications in Thin Films

The fabrication of fullerene containing thin films is a subject of high attention, given an easy surface coating. For instance, Echegoyen prepared molecular monolayers of fullerene derivative **17** (Fig. **8**) [400], along with several other reports of self-assembled Langmuir film monolayers [339, 374, 401].

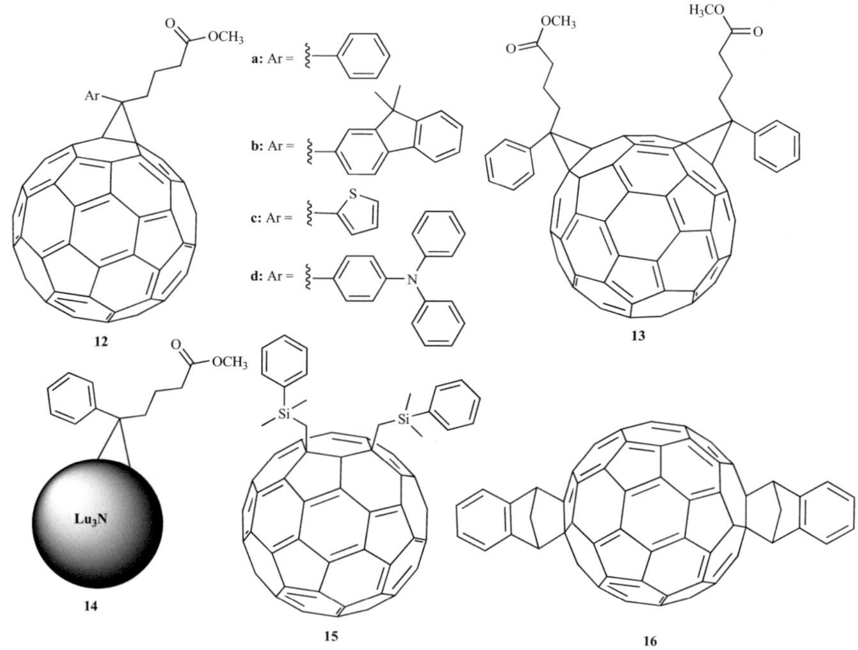

Fig. (7). Fullerene based derivatives for solar cell applications.

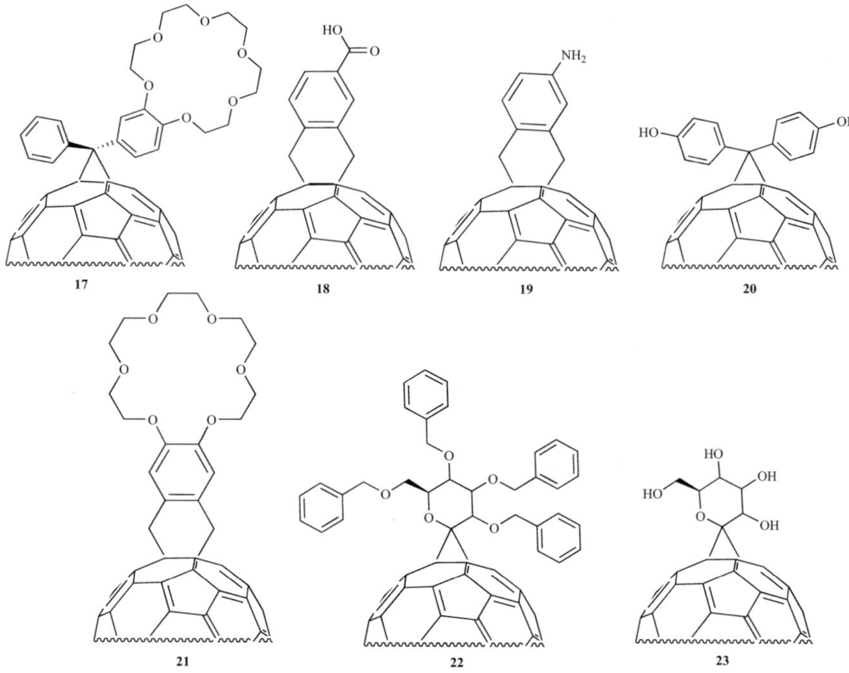

Fig. (8). C_{60} based compound for application in thin films.

The high hydrophobicity of the fullerene compounds is a significant problem in preparing fullerene Langmuir films. Thus, efforts have been taken on the preparation of fullerene derivatives having hydrophilic ends [402 - 410]. For instance, a broad investigation on the Langmuir behavior of several C_{60} derivatives was reported by Diederich and Ringsdorf [408]. The amphiphilic fullerene derivatives included carboxylates(**18**) and amines (**19**), bis-phenol (**20**), a crown-ether (**21**) and some protected (**22**) and deprotected (**23**) sugars (Fig. **8**).

Guldi and Prato also reported the deposition of [60] fullerene-donor dyad **24** Fig. (**9**) to modify ITO surfaces *via* layer-by-layer [409], using the positively charged Ru(III)-complex moiety in the dyad. Also Gulino and Fragalà prepared monolayers containing [60] fullerenes on silica surfaces by reaction of a [4-(chloromethyl)phenyl]silane-terminated silica surface with [60] fullerene derivative **25** (Fig. **9**) [410].

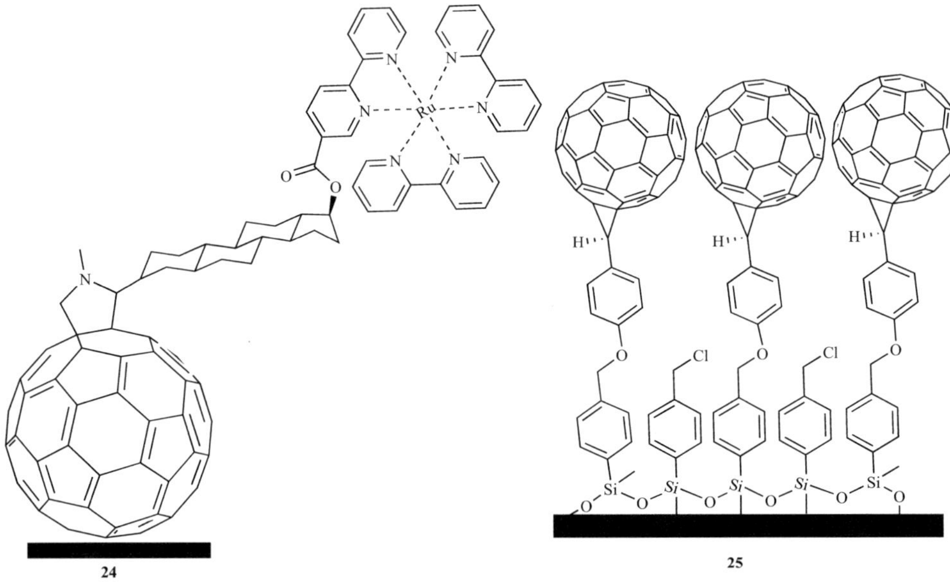

24 **25**

Fig. (9). Fullerene derivatives for thin film fabrication.

Applications in Nonlinear Optics

Nonlinear refraction and absorption are nonlinear optical features directly related. Generally, several mechanisms, both non-parametric and parametric, contribute to the nonlinearities and a consistent evaluation of the responsible mechanisms requires experiments involving different pulse lengths. Optical limiting consists of a sample transmittance decrease under fluence illumination or high intensity. Optical limiting is also regarded as the reverse of saturable absorption where increase in transmittance is seen at high illumination. As many organic molecules, fullerene derivatives show potential for optical limiting applications, exhibiting a sizeable nonlinear optical response.

Tutt and Kost [411] were the first to report the optical limiting properties of C_{60} in toluene solution using a nanosecond pulsed Nd:YAG laser at 532 nm, demonstrating that C_{60} showed superior optical limiting performance than the known optical limiting materials, at the time.

It was also observed that, upon photoexcitation, the C_{60} derivative **26** (Fig. **10**) suffered charge transfer with poly(3-octylthiophene) polymer in the film, forming a fullerene radical anion with high quantum efficiency. It was also observed that the fullerene-based film nonlinearity, at 760 nm, was much higher than that of each of the two components alone [412].

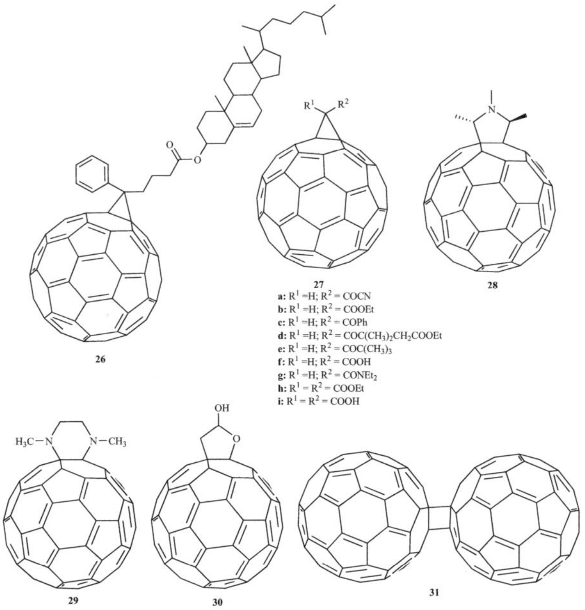

Fig. (10). Fullerene derivatives for nonlinear optics.

To compare their optical limiting properties, Sun performed a systematic investigation on several mono-functionalized fullerene derivatives (Fig. **10**), the methano-C_{60} derivatives **27a-i**, pyrrolidino-C_{60} derivative **28**, amino-C_{60} derivative **29**, C_{60} derivative **30** [413 - 415], and also C_{60} dimer **31** [413, 416].

Additionally, the incorporation of fullerene derivatives into glassy matrices *via* sol-gel processing was also reported, attempting to use the films thereby created for practical applications in solid state devices [417, 418]. For that, compounds **32-34**, incorporating siloxane groups, Fig. (**11**) were prepared, and were found to behave similarly as optical limiters in the solid state and in solution [419, 420].

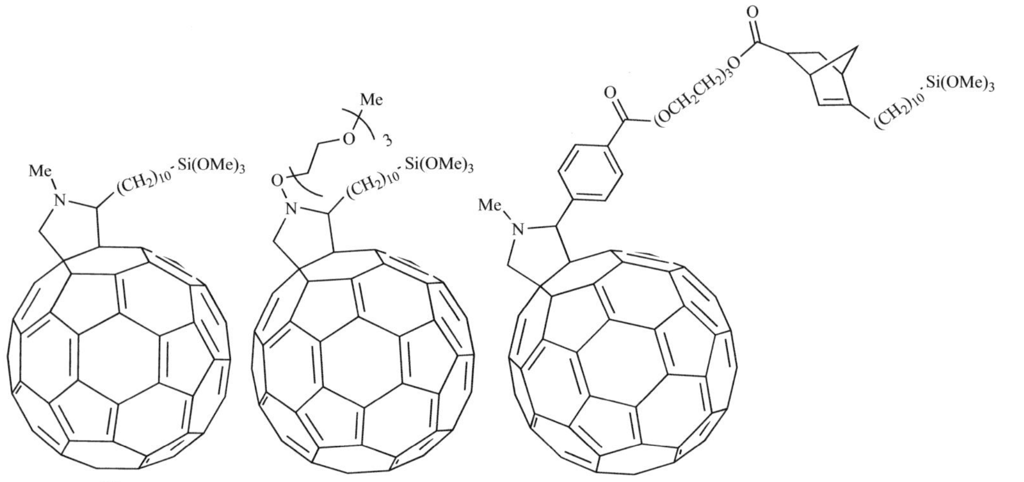

Fig. (11). Fullerene derivatives for NLO.

Applications in Biomedicine

There has been much interest in the potentialities of the biological activities of fullerenes. Even if unmodified C_{60}, in the form of nanocrystalline preparations present biological activity [421 - 424], most researchers aim to study modulated fullerenes in order to gain water solubility or in biologically compatible solvents, thus having higher versatility [425 - 427].

Photodynamic therapy (PDT) is a promising treatment approach for several diseases, particularly cancer. In PDT light, oxygen, and a photosensitizer are coupled, producing a selective therapeutic action. In fact, a photosensitizer of insignificant toxicity (in the dark) can accumulate preferentially in fast dividing cells. Then, a synchronized laser light beam is directed to unhealthy tissues. The drug is activated by light, inducing cytotoxicity. Soon after treatment, damaged cells become necrotic or suitably modified [428].

For instance, the highly soluble hexa(sulfobutyl)fullerenes **35** (Fig. **12**) were reported [429] for $^1O_2$ measurements, with results comparable to that of Photofrin®. Also Hamblin reported a study of the photosensitizing properties of several fullerenic compounds, bearing diserinol groups (**36-38**) (Fig. **12**) [430], or with quaternary pyrrolidinium groups (**39-41**) [431] (Fig. **12**). Ymakoshi and Iwamoto described a highly water soluble C_{60}-*N* -vinylpyrrolidine copolymers of type **42** as agents for photodynamic therapy [432], being considered the most water-soluble fullerene reported to date. Tabata also reported the use of polyethyleneglycol (PEG)-functionalized fullerene containing Gd^{3+} ions for MRI in combination with PDT [433].

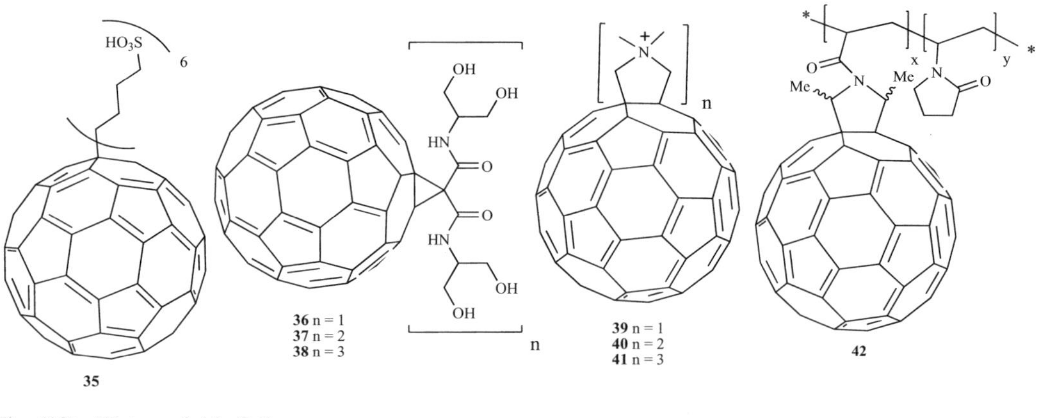

Fig. (12). Water soluble fullerenes.

DNA cleavage was probably the first applications of photoactivated fullerenes in biology [434, 435]. Miyata also studied fullerene biological activities under light, namely cell-toxicity, DNA-cleavage, hemolysis and mutagenicity [436].

Fullerene based compounds have also been used as photosensitizers for antimicrobial PDT. Hamblin [431] demonstrated that the soluble cationic functionalized fullerenes **39-41** Fig. (**12**), could be efficiently used for this purpose with various classes of microbial cells.

APPLICATIONS OF CARBON NANOTUBES

Carbon nanotubes (CNTs) and their outstanding properties, such as remarkable thermal and electrical conductivity and flexibility, along with high mechanical strength, make them potential candidates for several applications, including energy storage [437 - 445], fuel cells [446 - 449], batteries [450 - 456] and capacitors [457 - 462], in catalysis [463 - 470], as sensors [464, 471 - 483], in electronics [38, 39, 480, 484 - 495], in solar cells [496 - 504], and in biomedical applications [505 - 516].

Applications in Energy Storage

Hydrogen storage capability is a necessary requirement for fuel cell applications. Premature reports [517, 518] on the use of CNTs and graphitic nanofibers for storage of hydrogen claimed up to 67% weight storage capacities, triggering worldwide research on the subject. Since then, several experiments have been performed, using other methodologies; however, with replication of such high values [519]. In addition, has been reported that MWNTs could also have good hydrogen storage capacities [520 - 523]. Several carbon nanotube parameters account for these values, namely their structure (and defects), tubes arrangement ('ropes' and/or bundles), geometry, purification and pressure [524].

Since hydrogen uptake is proportional to the surface area (one layer adsorption), the uptake has a linearly dependence upon the nanotube diameter,. Within the nanotube, H_2 must either pass through an opening in the tube, or the nanotube wall. Naturally, a strong hydrogen uptake increase upon oxidative opening of the nanotubes was observed, leading to the hypothesis that most of H_2 internally stored in the tubes [525]. Higher hydrogen capacity is also obtained by increasing hydrogen pressure. However, pressure and temperature also influence carbon nanostructure's mechanical behavior. Smalley [526] reported a phase transition of SWNTs at 77 K, while other groups reported a tube breakdown, caused by excessive hydrogen adsorption, due to high repulsive energies [527, 528].

Applications in Fuel Cells

The fullerene principle applies to carbon nanotubes regarding application in fuel cells, which can be highly influenced by the physical properties of the carbon support [529 - 535]. Carbon nanotubes with high surface area and good crystallinity could provoke high Pt nanoparticles dispersion, facilitating electron transfer, hence, ensuing better device performance [533, 534].

According to theoretical predictions, SWNTs can be either semiconducting or metallic, depending on the tube helicity and diameter [536], while for MWNTs, scanning tunneling spectroscopy showed that conduction is mostly caused by the outer shell [537], normally larger than in single walled carbon nanotubes. Therefore, MWNT should possess a relatively high electrical conductivity.

Pristine carbon nanotubes are short on adequate binding sites for anchoring metal nanoparticles or metal ions, leading to metal nanoparticle's aggregation and poor dispersion, particularly at high loading conditions. This occurrence was reported by several authors, who found that, carbon black managed to immobilize high-loadings of metal nanoparticles, only less than 30 weight% Pt/multiwalled carbon nanotube catalysts could be obtained, due to aggregation [538 - 540]. As a consequence, carbon nanotube's functionalization is usually a requirement for further applications.

Sun [541] prepared Pt nanoparticles immobilized onto MWNT's, producing a high number of functional groups containing oxygen on carbon nanotube's surface, leading to monodispersed and highly dense Pt nanoparticles. For instance, Viswanathan [542] synthesized carbon nanotubes enclosing 87 wt% carbon nitrogen and impregnated Pt nanoclusters by using H_2PtCl_6, followed by H_2 flowing at 550°C to reduce Pt(II) to Pt(0). Zhang [543] loaded Pd nanoparticles on multiwalled carbon nanotubes. Moreover, Xu and Wu [544] reported an exhaustive comparison between MWNT and SWNT, regarding Pd nanoparticle loading.

Regarding the methanol electrooxidation reaction, Yan [545] showed that multi walled carbon nanotubes could outlast the commonly used catalyst. As a result of high corrosion resistance, multi walled carbon nanotubes have showed lower oxygen reduction reaction activity and lower Pt-surface area loss, when used as fuel cell catalyst supports.

Applications as Capacitors

Supercapacitors are energy storing devices with long cyclic stability and high power density and high capacity. They are also safe and environment friendly, operating in wide temperature ranges. Their applications contain industrial equipments, hybrid/electric vehicles and back-up power supplies [441, 546, 547].

Normally, supercapacitors are divided in two classes, according to their charge storage mechanism. One of the types of electrochemical capacitors is known as redox supercapacitors or pseudocapacitors, using transition metal oxides as electrode materials or electrical conducting polymers. The other type is known as electrical double layer capacitors (EDLC), operating by electrolyte charged ion adsorprtion-desorption onto electrodes with high surface area, as a double layer structure. The current EDLCs are manufactured by activation and pyrolysis of plant matter, petroleum sources, and wood [548, 549].

Synthetic carbons, which include carbon nanotubes (CNT), have high surface areas, controlled surface and bulk chemistry, good conductivity and tunable porous structure, allowing rapid transport of ions and better charge accumulation. Higher capacity is attained using outer carbon nanotubes walls bearing high number of defects [550], as well as by activation processes which increase the surface area. For instance, Lee observed large power density of 20 kW kg^{-1} with maximal specific capacitance of 180 F g^{-1}, upon heating carbon nanotubes to improve pore distribution and specific surface area [458]. Several groups have also performed carbon nanotube surface treatment with aqueous KOH-NaOH solution, nitric acid or ammonia, to obtain improved density of oxygen-containing groups [551 - 553]. For example, Niu performed nitric acid treatment on multi walled carbon nanotubes, obtaining a specific capacitance = 113 F g^{-1}, using as electrolyte a H_2SO_4 38 wt% solution [457, 554, 555]. Finally, carbon nanotubes can also be deposited on the collector of current as electrodes (Ni, Al, or other metals and alloys), increasing the carbon nanotube supercapacitor's energy storage properties [458, 551, 554, 556 - 558].

Applications in Catalysis

To date, carbon nanotube-based catalysts have been mostly used for liquid-phase (hydroformylation, hydrogenation) or gas-phase (ammonia decomposition, Fischer-Tropsch process) reactions, for electrocatalysis (fuel-cell electrodes) and for photocatalysis [559 - 567].

Pt nanoparticles supported on single walled carbon nanotube electrodes displayed enhanced catalytic properties for methanol oxidation and oxygen reduction reactions [568], where methanol oxidation currents showed to be ca. 10 times higher. In addition, Sheu used Pt nanoparticles electrodeposited on multi walled carbon nanotubes for electrocatalytic oxygen reduction reaction [569], observing noteworthy increase in oxygen reduction potential and current, showing also stability for methanol oxidation as well as improved electron transfer rates [570].

The use of Pt-Ru bimetallic catalysts to protect Pt from CO poisoning and to enhance catalytic efficiency is also established. Chen [571] studied the Pt-Ru electrodeposition on single walled carbon nanotubes, obtaining uniform distributions of catalyst nanoparticles (60-80 nm). Methanol oxidation reaction studies revealed a maximum current density obtained with a 4:3 Pt:Ru ratio, showing higher stability catalytic and activity than simple Pt nanoparticles.

Applications as Sensors

Single-walled carbon nanotubes show high potential for use in chemiresistive sensors (signal transduction changes in resistance are measured upon analyte desorption or adsorption) due to high surface area and high sensitivity, enabling the detection of lower concentrations than common sensors [481, 572 - 574].

Single walled carbon nanotubes are quite sensitive chemiresistive sensors, however they lack selectivity, a prerequisite for practical sensor applications. Consequently, the modification of single walled carbon nanotubes using metal nanoparticles has been examined, achieving enhanced selectivity, especially in hydrogen sensing [575 - 579].

Nanotube based composites can also be used in vapor sensing. The changes in electrical resistivity were measured and ascribed to conductive modification due to solvent absorption and/or polymer matrix swelling [580], while Dai [581] reported that, when subjected to optical or thermal exposure, alligned nanotube composite arrays showed similar changes in resistance.

Real time, highly sensitive biomolecular interactions monitoring has become a very important issue in modern science. The properties of CNTs, such as high conductivity or small size allow development of optical and electrochemical biosensors based on field-effect transistors. Over the last years, many reports on electrochemical biosensors based on carbon nanotubes appeared, used for biological detection [582 - 584]. For instance, poly(Nile blue A)-SWNT nanocomposites were used as biosensor for glucose, based on the electrocatalytic oxidation of reduced nicotinamide adenine dinucleotide phosphate (NADPH) [585]. This biosensor was capable to, at a very low potential value, electrocatalyze the NADPH oxidation.

The functionalization of carbon nanotubes with antigens or antibodies allowed the production of electrochemical immunosensors, used in the detection of small molecules [586]. A representative example was the reaction of clenbuterol with multi walled carbon nanotubes, using *N* -hydroxysulfo-succinimide and carbodiimide as crosslinkers. Also a tyrosinase biosensor was prepared by coating carbon nanotubes with Fe_3O_4 magnetic nanoparticle, being used for coliform detection with a flow-injection assay system [587].

Applications in Solar Cells

The employment of carbon nanotubes in organic photovoltaics includes the direct introduction of carbon nanotubes inside a photoactive bulky heterojunction film, the construction of transparent flexible hole collecting electrodes, the use of carbon nanotube thin films as hole extraction layers, or the synthesis of carbon nanotube-polymer nanocomposites. The main purpose for using carbon nanotubes in a bulky heterojunction is the substitution of the fullerenes, which are less stable to air, less conductive and more expensive, besides the combination of the unique charge transport characteristics of carbon nanotubes with suitable physical and chemical properties.

Some theoretical studies [588 - 590] show that single-walled carbon nanotubes and regioregular poly-3-(hexylthiophene) demonstrated that due to efficient spatial separation at the donor-acceptor interface, photoexcitation of poly---(hexylthiophene) resulted in long life electron carriers. Single-walled carbon nanotubes also induce organization in several conjugated polymers, therefore influencing their optical properties. For instance, it was reported that carbon nanotubes could act as a nano-substrate for regioregular poly-3-(hexylthiophene) [591], or that poly-3-(hexylthiophene) crystallinity increases with the single-walled carbon nanotube amount, demonstrating the enhancement of charge mobility in single-walled carbon nanotubes composites as well as exciton diffusion in polythiophene [592]. An efficient charge separation, similar to that of well known poly-3(hexylthiophene):fullerene heterojunctions can be attained with just 1% single-walled carbon nanotube content [592].

The most frequent strategy to integrate carbon nanotubes (SWNTs [593, 594] and MWNTs [595, 596]) within photoactive bulky heterojunctions is to induce the dispersion of CNTs in an electron donor polymer solution, followed by spin-coating on plastic or glass. Followed by inclusion on conductive ITO transparent electrode layer, 60-100 nm thick, covered by another layer of poly(3,4-ethylenedioxythiophene): poly(styrenesulfonate) (PEDOT:PSS). Following the method, Kymakis [597] prepared a bulky heterojunction solar cell using poly-3(hexylthiophene): single-walled carbon nanotubes (1% wt) as donor:acceptor couple, depositing a photoactive film with 100 nm thickness. Following thermal annealing, power conversion efficiency was increased to 0.22% for a 0.25 cm^2 cell. Lanzi also obtained a very efficient carbon nanotube based bulky heterojunction solar cell [598], using thiophene derivatives as donors, blended with single-walled carbon nanotubes as acceptor (3% wt), leading to solar cells with power conversion of 0.53%.

Applications in Biomedicine

Carbon nanotubes are able to penetrate various biological tissues and interact with cell membranes, due to their fair flexibility [599 - 602]. Therefore, both the toxicological [603] and pharmacological [604] profiles of carbon nanotubes have been the focus of undivided attention [605, 606].

SWNTs are poorly soluble in water due to their extremely strong π–π interactions. A surfactant can be an effective nanotube surface coating agent, because of the nanotube's interactions with the aliphatic chains present in the surfactan in aqueous solution. A surfactant provides both colloidal stability and steric repulsion to the dispersion, which is why surfactants as sodium dodecylbenzene sulfonate (SDBS) [608] and sodium dodecyl sulfate (SDS) [607] were already used for an efficient dispersion of single-walled carbon nanotubes (Fig. **13**).

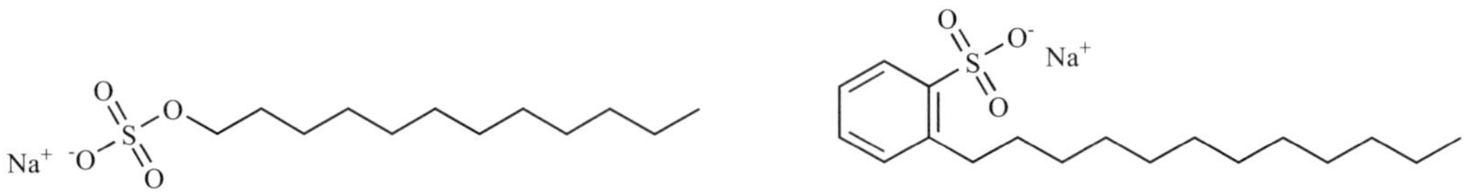

sodium dodecyl sulphate (SDS) **sodium dodecylbenzene sulphonate (SDBS)**

Fig. (13). Surfactants used to disperse carbon nanotubes.

SWNT dispersion in water can also be achieved using DNA, which forms stable complexes with CNTs, like for instance, alternating thymine (T) and guanine (G) nucleotides ($(GT)_n$- n = 10 to 45) DNA sequences [609 - 611].

Biocompatibility of carbon nanotubes usually depends on the surface modifications and additives on them [505]. For instance, pristine carbon nanotubes exhibit both *in vitro* and *in vivo* toxicity [612 - 617], suitable biochemical functionalization of carbon nanotubes can usually diminish their toxicity. For instance, single-walled carbon nanotubes containing phenyl-SO_3H and phenyl-$(COOH)_2$ provided improved cell viability, when compared with pristine single-walled carbon nanotubes [618]. Glycodendrimers have also been reported as homogeneous bioactive coatings for single-walled carbon nanotubes, also diminishing their cytotoxicity [619]. Also biotin-, fluorescein- and carboxyl-coated single-walled carbon nanotubes exhibited little toxicity to Jurkat T and HL60 cells [620].

Drug bioavailability can also be maximized by attaching it to carbon nanotubes due to the increased circulation time and solubility. Single-walled carbon nanotubes are considered good drug delivery carriers since they are quite innocuous materials, when compared, for instance, with quantum dots [621, 622], providing also a route for cytoplasm delivery due to their cell membranes penetration capability [623].

For instance, cisplatin, a platinum anticancer drug has been delivered to the tumor cells using the carrier **43**, represented in (Fig. **14**) [624]. In this system, the platinum (IV) complex contained a folic acid derivative and was linked to an amine-functionalized SWNT. The functionalized SWNT was able to deliver the complex into cancer cells through endocytosis, as it was observed using fluorescence microscopy, by fluorophore-labeled single-walled carbon nanotube localization, being the first conjugate in which both the delivery and targeting moieties were present in the same molecule.

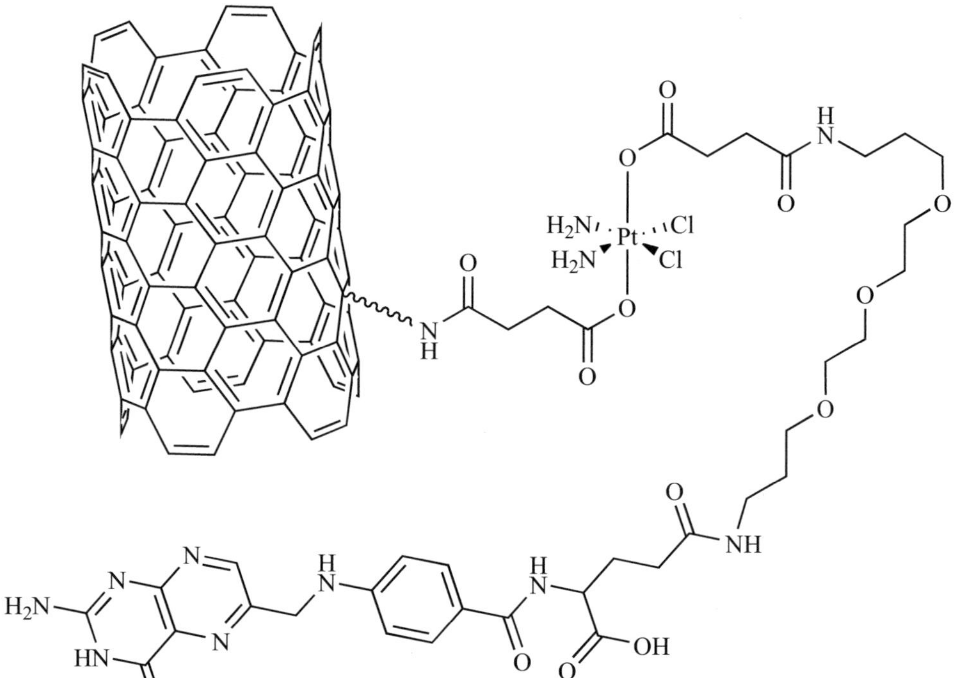

43

Fig. (14). Carbon nanotube based conjugate for drug delivery.

A biotin- single-walled carbon nanotube-linker-taxoid drug delivery vehicle was prepared [625] through covalent linkage of the nanotubes to the taxoid anticancer drug, being biotin and disulfide bond as the module for tumor-targeting. The conjugate was internalized mainly by endocytosis mediated by the receptor.

Furthermore, doxorubicin has been also adsorbed onto single-walled carbon nanotubes, *via* π-π stacking interaction [626, 627]. The drug loading in this single-walled carbon nanotubes-doxorubicin complex was found to be ~50-60 wt%, a surprisingly higher value than the ~8-10 wt% obtained with dendrimer drug carriers [628] and conventional liposomes [629], displaying high cancer cell cytotoxicity.

CNTs were also used as protein delivery molecular transporters both *in vivo* and *in vitro*. Strepavidin was linked to biotin substituted single-walled carbon nanotubes, and labeled with 5-(5-aminopentyl)thioureidyl fluorescein, for cellular uptake monitoring by human T cells (Jurkat) and human promyelocytic leukemia (HL60) cells [620].

Also paclitaxel drug was linked to PEG chains on single-walled carbon nanotubes, yielding conjugate **44** (Fig. **15**) [630]. The loading of paclitaxel on single-walled carbon nanotubes was observed to be ~150 per single-walled carbon nanotube with ~100 nm length. When intravenously administered to mice, tumor paclitaxel-SWNTs uptake was 10-fold higher than the free drug and PEGylated-paclitaxel.

APPLICATIONS OF GRAPHENE

The carbonaceous material which currently deserves most attention is graphene. A surge on applications involving this material has emerged. Its properties have rendered it applications in catalysis [631 - 634], in energy storage [635 - 646], as sensors [647 - 651], including biosensors [652 - 654], in electronics [6, 655 - 664], in solar cells [665 - 671], in biomedicine [275, 672 - 692].

Applications in Catalysis

In the efforts taken toward sustainable chemistry, carbon nanomaterials have been extensively studied in catalysis where, in particular, graphene has attracted increasing importance [633].

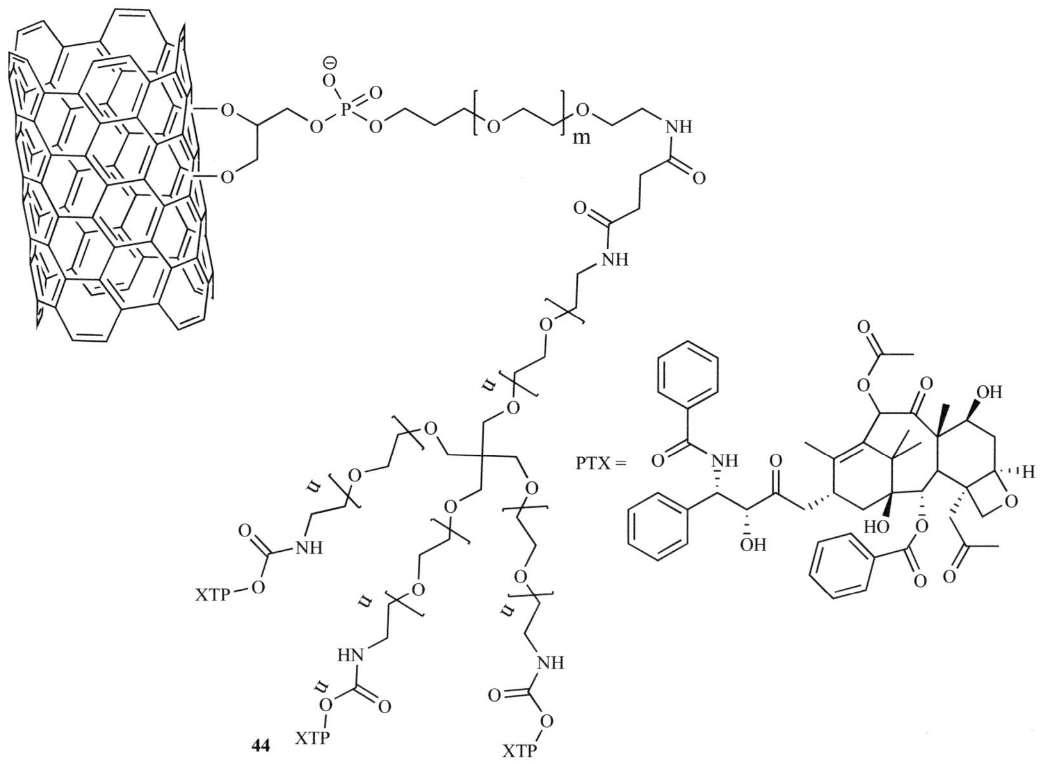

Fig. (15). Carbon nanotube based conjugate.

Bielawski and Dreyer reported the first use of metal-free catalysts based on graphene [693], who demonstrated that graphene oxide was a good catalysts for *cis* -stilbene and alcohol oxidation, displaying high selectivities and yields. Bielawski and co-workers have made important contributions for the understanding of graphene oxide catalytic applications [694], using graphene oxide (GO) in various reactions, namely ring opening lactam and cyclic lactone polymerization [695], polymerization of olefins [696], synthesis of poly(phenylene methylene) composites [697], Claisen-Schmidt condensation [698], C-H oxidation [699] and oxidation of sulfides and thiols [700]. Liu and Cao [701] also reported a GO based catalytic process for imine synthesis under neat and mild conditions, using O_2 as the oxidant, while Su [702] evaluated the GO activity on the oxidative coupling of imines to amines.

Electrocatalysis is a main graphene oxide catalytic area. For instance, the electrooxidation of uric acid, dopamine, acetaminophen and ascorbic acid, using graphene oxide was reported by Su's group [703], obtaining low high oxidation current and oxidation potential, ascribed to the oxygen functional groups and increased surface area of graphene oxide. Aside, Li also prepared reduced graphene oxide sheet films [704, 705] as electrode materials, possessing excellent electrocatalytic activity toward oxygen reduction, also exhibiting fast electron-transfer kinetics.

Further developments were accomplished; for instance, it has been shown that graphene oxide was a reusable acid catalyst for epoxides ring opening reactions [706], being also an effective and recyclable catalyst for room temperature aldehyde acetalyzation [707], oxidative dehydrogenation of propane [708], Friedel-Crafts addition of indoles to α,β-unsaturated ketones [711], synthesis of dipyrromethane and calyx[4]pyrrole [709] and aza-Michael addition of amines to activated alkenes [710].

C-H bond activation is conventionally catalyzed by organometallic complexes and transition metals [712]. Graphene derivatives, as cheaper metal-free catalysts, are suitable alternatives as catalysts for this reaction. For instance, *N* -graphene exhibited high selectivity and activity for aqueous phase arylalkane oxidation, yielding products with high value for biomedical applications [713]. Recently, Chen [714] reported the catalytic behavior of *N* -graphene in 4-nitrophenol reduction to 4-aminophenol. Furthermore, Xia [715] studied the electrocatalytic activity toward nitroaromatics reduction, observing a decreased over-potential, when compared with thermally reduced graphene. Wang [716] synthesized *N* -graphene using a heat treatment procedure and demonstrated its effective activity toward benzylic alcohol aerobic selective oxidation.

Applications in Energy Storage

There is an increase in graphene research for energy applications, which include hydrogen storage. For instance, Ruoff used graphene nanosheets produced from exfoliated graphite oxide for this purpose and observed an isosteric adsorption heat of 5.9 kJ mol^{-1}, a similar value to those of carbon materials possessing high surface area [717], while Rao observed a 3 wt% H_2 uptake using 100 bar pressure for exfoliated graphene oxide [718]. Creation of stacked graphene structures is necessary for increasing the hydrogen storage capacity. Theoretical studies have showed that a hydrogen layer could be stored between two 6 Å apart graphene sheets. If the separation distance is increased to 8 Å, theoretically, one could manage to achieve 5–6.5 wt% at 5 bar [719]. Indeed, Louie demonstrated that Ca predominantly adsorbed onto the armchair edge doped with boron and graphene zigzag edge, allowing up to 6 H_2 molecules to be adsorbed by each atom of Ca, displaying a 5 wt% hydrogen capacity [720].

Graphene possesses a theoretical surface area of 2675 m^2 g^{-1}, leading to a theoretical capacitance of 550 F g^{-1}. However, due to graphene sheet aggregation, capacitances of 100 F g^{-1} are usually attained [638, 721]. Wang and Chen prevented agglomeration by using a gas-solid reduction process, obtaining a maximal specific capacitance of 205 F g^{-1} [722], while other authors effectively used carbon nanotubes as spacers in between graphene sheets in a weight ratio graphene:nanotube of 9 to 1, also to avoid aggregation [723, 724], reaching specific capacitances of 326 F g^{-1} [724].

Applications as Sensors

Graphene is known for exhibiting high carrier density, extremely high carrier mobility and low intrinsic noise, anticipating a high signal-to-noise ratio in detection, combined with highly sensitive conductance to chemical and electrical environment alterations.

The first graphene sensor was used for gas detection, reported by Novoselov [725], by using as sensing element mechanically-exfoliated pristine graphene containing a few layers, used for NO_2 gas detection, till 1 ppb NO_2. A variety of graphene-based gas sensors have been reported afterward [726 - 728]. Nomani [729] verified that 6H-SiC (SiC = silicon carbide) substrates were superior to the conventional Si/SiO_2 ones, owing to the SiC C-face and graphene interaction, leading to less scattering events (therefore lower noise and higher conductivity) [729] and consequent NO_2 detection at a lower concentration (10 ppb) [730].

Sensors for explosive analytes are quite important and, for instance, dinitrotoluene is frequently used as a reporter of unstable trinitrotoluene. A dinitrotoluene sensor, based on a spin-coated reduced graphene oxide thin-film was reported [727]. A limit of detection with a value of 28 ppb was observed, a quite impressive result, since the vapor pressure of dinitrotoluene is 173 ppb at 25 ºC. Later, another reduced graphene oxide sensor was reported by Robinson [731], showing a clearly improved detection limit (0.1 ppb) for dinitrotoluene detection.

For instance, Shafiei reported a H_2 sensor using reduced graphene oxide thin-film [732]. Likewise, a reduced graphene oxide sensor was developed for H_2S detection, a poisoning gas, with a limit of detection of 2 ppm at 25 ºC [733]. Later, Chen [734] reported a NH_3 sensor, demonstrating that reduced graphene oxide operated in n-type mode, by application of an adequately positive gate voltage, attaining faster recovery and faster response than the biased at p-type mode. Thus, electronic sensors based on graphene have also been used in the detection of other vapors, including HCN [731], trimethylamine [735], I_2 [725], CH_4 [736] and ethanol [737].

Optical systems probably are the best strategy to develop nanoscale biosensors that can function inside cells. Graphene materials could display very low limits of detection for several vapors and gases [727, 738, 739]. Qazi [740] detected NO_2 at 60 ppb levels, upon electrical conductivity changes in thin graphite flakes. Novoselov verified single-molecule detection [725] using electrical sensors prepared from mechanically exfoliated graphene.

So far, the majority of biosensors used electrochemical reactions for biomolecule detection. For instance, Worden and Lee [741, 742] sensitively measured glucose concentration by using graphitic electrode materials. The electrodes were consisted of graphite films thermally exfoliated, dispersed on Nafion, a conductive polymer, and exhibited noteworthy reduction and oxidation currents in H_2O_2, while a gold electrode modified with Nafion showed insignificant electrochemical response. Shan used a different approach to detect glucose [743], by preparing polyvinylpyrrolidone protected reduced graphene oxide electrodes, attempting to obtain direct electron transfer between glucose oxidase and the electrode.

Other groups studied how reduced graphene oxide functional groups and edges

[744] or multilayer graphene nanoflakes edges [745] performed on electrochemical detection of serotonin and dopamine neurotransmitters. In 2008, Papakonstantinou grew multilayered graphene nanoflake films on silicone [745], used for electrochemically active biomolecule detection.

Berry and Mohanty reported the first electrical biosensors based on graphene, prepared by treating graphene oxide with ethylenediamine or nitrogenous plasma [746], and further electrostatically adsorbed onto silica surfaces.

They explored the graphene amine or graphene oxide functional groups to prepare biosensors, through graphene single-stranded oligonucleotides chemical grafting, followed by rhodamine green based fluorescence monitoring.

Later, Yang [653] used fluorescent tags on graphene oxide to sense proteins and DNA fragments in solution, obtaining 97% fluorescence signal quenching. As expected, subsequent complementary DNA strands addition caused hybridization, re-establishing 77% of initial fluorescence signal.

Applications in Solar Cells

Due to their multifunctionality, graphene materials have swiftly emerged in dye sensitized solar cells (DSSC) applications. Graphene materials can act as transparent conductors in DSSC photoanodes, both as the sensitizer itself and in the semiconducting layer. Electronically doped graphene derivatives might show potential as transparent electrodes. A graphene layer can absorb 2.3% of light; so, a maximum of 5 sheets can be traversed by light, for 90% transmittance [657]. Some authors provided several studies on transparent electrodes based on graphene materials, proposing a standardization strategy [747]. They have noted that limits of conductivity could be highly enhanced by rising the carrier concentration through electronic doping or substrate interactions. Tour reported graphene based transparent conducting films [748].

By photolithography, the authors prepared 5−10 μm wide and 100 nm thick metal nanogrids (Al, Au, and Cu), spaced 100-200 μm apart, followed by chemical vapor deposition (CVD) derived graphene grid coating, forming a continuous conductive network. Similarly, Huang [749] used graphene prepared by CVD directly on silica glass substrates.

Many groups have thrived to improve DSSCs semiconducting layer performance using graphene materials. Kim reported the first graphene derivative in a DSSC semiconducting layer [750], using a composite constituted by photocatalytically reduced graphene oxide and TiO_2 as blocking layer, obtaining PCE = 7.6%.

In various other studies similar approaches were used [751 - 753], also with improved efficiencies. For instance, Li [751] produced, by spin coating, thermally reduced graphene oxide thin layers, used as blocking layer, originating a cell with 8.1% efficiency, which was superior to a cell without a blocking layer (η = 7.2%), an improvement ascribed to photocurrent increase.

Some studies report the occurrence of efficient charge injection from graphene photosensitizers to TiO_2, however with poor performance [754 - 756]. Li [757] manufactured the first graphene-based photosensitizing DSSC, using graphene quantum dots, ranging from diameters of ~1 to 30 nm.

Applications in Biomedicine

Graphene in biomedicine was first reported by Dai [690]. Graphene oxide was functionalized with a branched PEG bearing amine endings, yielding physiologically highly stable PEGylated graphene (NGO-PEG). Then, through graphene surface π-π stacking, anticancer drugs with aromatic moieties were efficiently loaded for drug delivery [690].

Since then, several researchers also reported a number of interesting results. For instance, Zhang [758] investigated the targeted delivery of anticancer drugs (doxorubicin and camptothecin) by folic acid-functionalized graphene oxide, through hydrophobic interactions and π-π stacking. Results demonstrated that the conjugate showed remarkably higher cytotoxicity against target cells compared to amine graphene oxide loaded with only a single drug.

Using graphene oxide as the vehicle, co-operative drug and gene delivery was also investigated by Zhang [691], where a cationic polymer, polyethylenimine (PEI) was covalently linked to graphene oxide through amidation process. The formed PEI-graphene oxide facilitated loading siRNA (small interference RNA, a protein expression inhibitor) through π-π stacking and electrostatic adsorption.

Owing to the excellent biocompatibility, ready cellular uptake, flexible chemical modifications and unique optical properties, graphene oxide derivatives have also been explored for biological imaging. Liu [759] reported a NIR fluorescence dye labeled PEGylated nano-graphene to be tested in fluorescence imaging *in vivo* . The conjugate's behavior was studied using several xenograft tumor mouse models, displaying high uptake and good tumor photothermal ablation efficacy [759]. A later study further proved the superiority of nano-graphene over CNTs in the *in vitro* photothermal killing of U251 human glioma cells [760].

Nevertheless, it is a challenge to prepare fluorescent graphene because graphene efficiently quenches the fluorescence of fluorophores. Liu [759] conjugated nano-graphene sheets (NGS) with Cy7 *via* a bridge of polyethylene glycol. The *in vivo* fluorescence imaging revealed a high uptake of NGS in xenograft 4T1 murine breast cancer tumor, KB human epidermoid carcinoma tumor and U87MG human glioblastoma tumor mouse models. Using poly-acrylic acid and bovine serum albumin as bridge, fluorescein *o* -methacrylate [686] and CdTe quantum dots [761] were also linked to graphene and the complexes showed strong fluorescence *in vitro.*

MRI (magnetic resonance imaging) can also be used as a powerful *in vivo* non-invasive imaging modality that has been widely used in imaging applications [684]. For instance, in order to enhance biocompatibility and the cellular MRI, aminodextran coated Fe_3O_4 nanoparticles were immobilized onto graphene oxide [762]. The composites possessed low cytotoxicity, good physiological stability and efficiently internalization by Hela cells. When compared with isolated Fe_3O_4 nanoparticles, the Fe_3O_4-graphene oxide composites exhibited significantly increased cellular MRI signal.

PET (positron emission tomography) imaging is a nuclear medical imaging modality that quantitatively measures radioisotope concentrations *in vivo* [675]. For instance, Liu and Cao [685] conjugated graphene oxide with CD105 (a marker for tumor angiogenesis) antibody TRC105 and labeled it with ^{64}Cu for targeting *in vivo* and PET in breast cancer mouse model. *In vivo* PET imaging demonstrated the radiolabeled graphene oxide conjugates quickly accumulated in 4T1 murine breast tumor, demonstrating the utility of the imaging agent for CD105 targeting.

Besides chemotherapy and gene therapy, phototherapy is another therapeutic approach which may be used to treat a variety of diseases. Phototherapy, including photodynamic therapy (PDT) and photothermal therapy (PTT), can control disease by specific light irradiation [680]. In PTT an optical-absorbing agent generates heat under light irradiation, exposing biological tissues to temperature increase, promoting abnormal cell selective destruction [681]. Due to its strong NIR optical absorption, graphene has received close attention from the PTT field. For instance, Guo [763] developed a doxorubicin (DOX) DOX-loaded PEGylated nanographene oxide nanocomposite, which could deliver both the drug and heat to the tumor area, facilitating combined photothermal and chemotherapeutic treatment in a single system, in a superior fashion.

Liu [682] designed a chemically reduced iron oxide nanoparticle-graphene oxide nanocomposite probe to combine the capability of tumor bioimaging with PTT. Under the guidance of MRI, they found solid tumors were efficiently irradiation ablated with NIR laser (λ=808 nm), while Chen [764] synthesized a quantum-do--tagged chemically reduced graphene oxide nanocomposite that combined the capability of cell/tumor bright fluorescence bioimaging with photothermal therapy. The composite reduced the inherent toxicity of quantum dots and prevented fluorescence quenching by maintaining a precisely controlled spacer between the quantum dots and the reduced graphene oxide. Using folic acid, the composite could target MCF-7 cells selectively which, after irradiation at 808 nm, were killed by the generated heat from the nanoconjugate.

Different from PTT, PDT relies on the concept that free radicals or reactive oxygen species are produced upon irradiation of photosensitizers with a suitable light, producing irreversible cancer cell damage. Considering its solubility and biocompatibility, Zhou and Jiang [765] immobilized hypocrellin-A (HA, a perylenequinonoid hydrophobic non-porphyrin photodynamic antitumor drug) onto graphene oxide *via* the π-π stacking interaction, hydrophobic effect and hydrogen-bonding interactions, demonstrating that the nanocojugate could generate singlet oxygen, thus inducing significant cell death. Liu [766] managed to combine benefits of PDT and PTT. They loaded Chlorine6 (Ce6, a photosensitizer molecule) on PEG functionalized graphene oxide and delivered the complex to KB cells. It was found that a low power density of 808 nm laser would promote the delivery of Ce6 molecules by mild local heating because of the photothermal effect of graphene. Compared with Ce6 or graphene oxide-PEG-Ce6 complex without near-infrared laser, PDT efficacy against cancer cells was highly enhanced.

EXPERIMENTAL SECTION

Synthesis of N-methylpyrrolidine-C_{60} (Scheme 6)

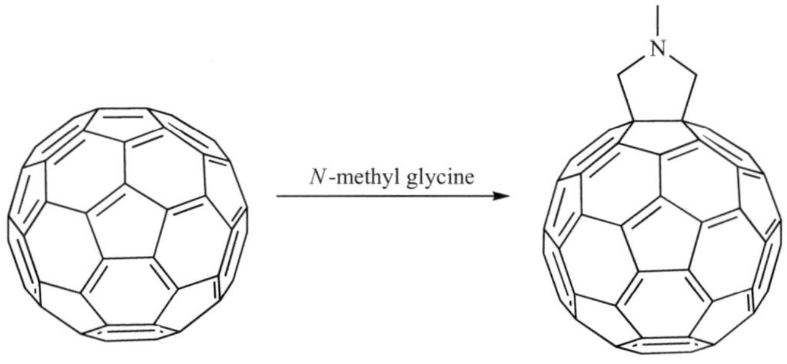

A

Scheme (6). Synthesis of N-methylpyrrolidine- C_{60}. Procedure for total synthesis of fullerene A can be directly found in Reference [112].

Preparation of N-benzyl-2-carbomethoxy Pyrrolidine-C_{60} (Scheme 7)

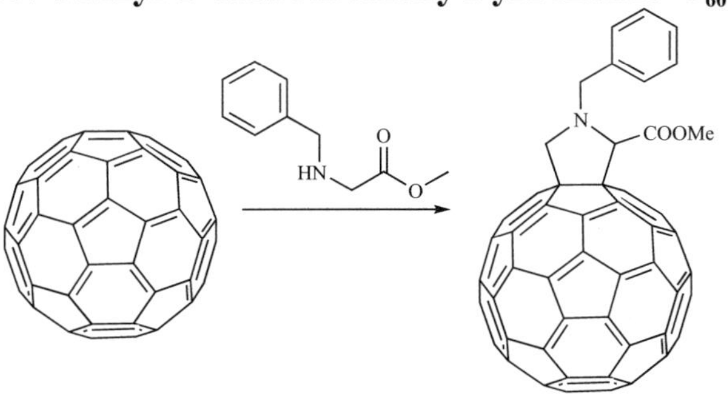

B

Scheme (7). Synthesis of N-benzyl-2-carbomethoxy pyrrolidine- C_{60}. Procedure for total synthesis of fullerene B can be directly found in Reference [112].

Preparation of Methano-$PC_{61}BM$ (Scheme 8)

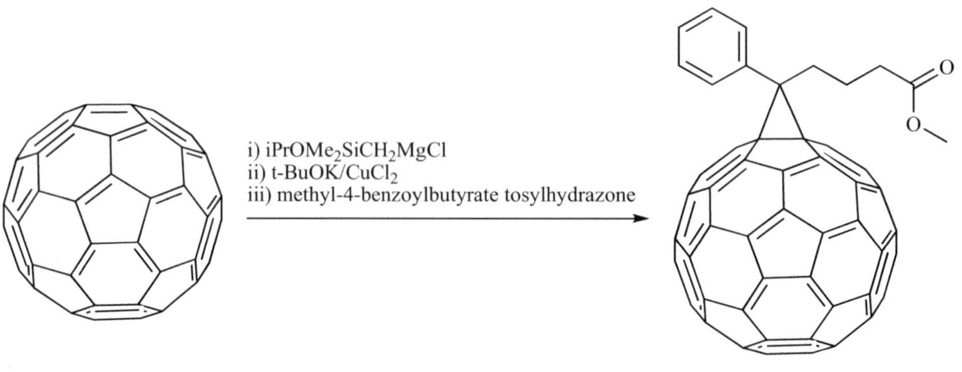

Scheme (8). Synthesis of methano-PC61BM derivative. Procedure for total synthesis of fullerene C can be directly found in Reference [767].

Synthesis of N-substituted Single-wall Carbon Nanotubes (Scheme 9)

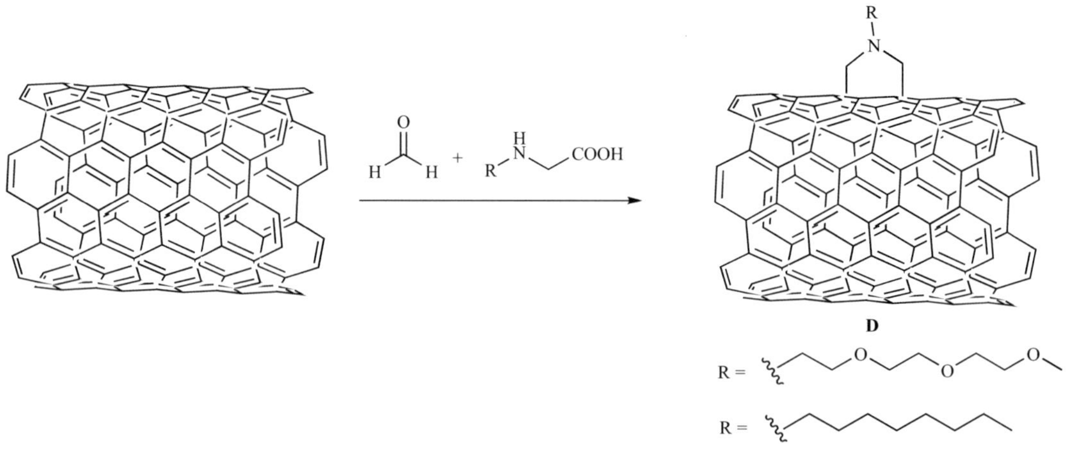

Scheme (9). Synthesis of N-substituted single-wall carbon nanotubes. Procedure for total synthesis of nanotube D can be directly found in Reference [215].

Synthesis of Amine Functionalized Carbon Nanotubes (Scheme 10)

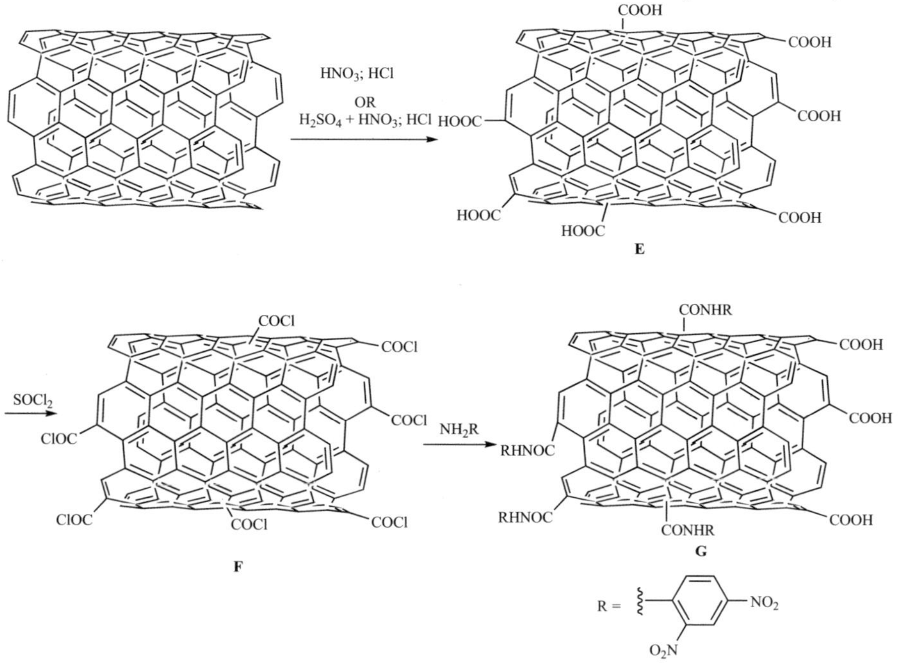

Scheme (10). Synthesis of amide substituted carbon nanotubes. Procedure for total synthesis of nanotube G can be directly found in Reference [768].

Functionalization of Carbon Nanotubes with Pyrene Derivatives (Scheme 11)

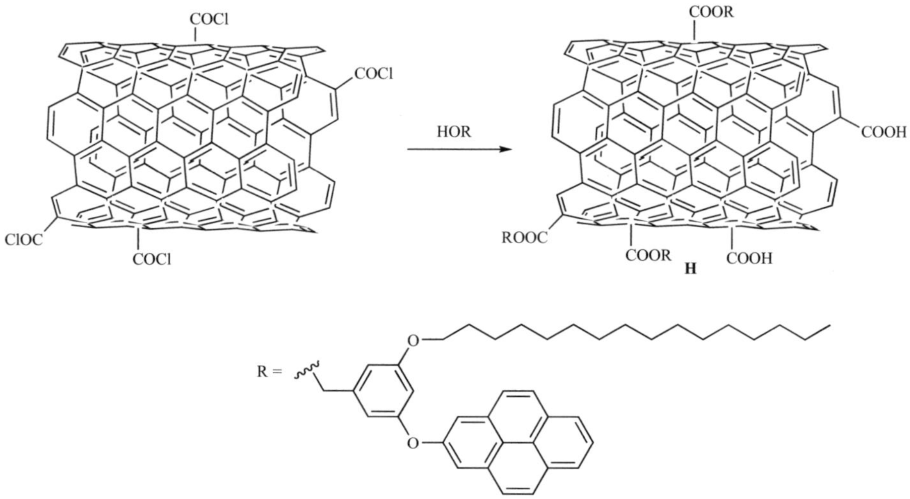

Scheme (11). Synthesis of pyrene functionalized carbon nanotubes. Procedure for total synthesis of nanotube H can be directly found in Reference [229].

CONCLUDING REMARKS

The incredible scientific interest originally generated by the finding of fullerenes has been quickly extended to other forms of carbon nanomaterials. Significant achievements in carbon science occurred swiftly with the discovery of carbon nanotubes by Ijima in 1991 and the isolation of hexagonal carbon sheets from graphite (graphene) by Geim and Novoselov in 2004.

Among the impressive applicability of carbon allotropes, their use in solar cells, energy storage and even in environmental cleaning and biology renders them both an astonishing record track and a bright future ahead. As far as any scientist imagination can travel, so can the possible applications for materials based on carbon. The possibility of widely varying the content of carbon allotrope-containing structures is probably the single possible direction for successful applications; however not less important than the establishment of definite structure-property relationships for any desired purpose, something has been only partially achieved so far. Standardization of procedures will definitely boost the understanding and appliance of carbon allotropes in vast areas of science.

REFERENCES

[1] Rao, A.M.; Dresselhaus, M.S. *Nanostructured Carbon for Advanced Applications*; Benedek, G.; Milani, P.; Ralchenko, V.G., Eds.; Kluwer Academic Publishers: Dordrecht, **2001**.

[2] Saito, R.; Dresselhaus, G.; Dresselhaus, M.S. *Physical Properties of Carbon Nanotubes*; Imperial College Press: London, **1998**.

[3] Robertson, J.; O'Reilly, E.P. Electronic and atomic structure of amorphous carbon. *Phys. Rev. B Condens. Matter,* **1987**, *35*(6), 2946-2957. [http://dx.doi.org/10.1103/PhysRevB.35.2946] [PMID: 9941778]

[4] Delhaes, P.; Issi, J.P.; Bonnamy, S.; Launois, P. Polymorphism and structure of carbons In: *Lecture Notes in Physics*; , **2006**; 677, p. 47.

[5] Berger, C.; Song, Z.; Li, T.; Li, X.; Ogbazghi, A.Y.; Feng, R.; Dai, Z.; Marchenkov, A.N.; Conrad, E.H.; First, P.N.; de Heer, W.A. Ultrathin epitaxial graphite: 2D electron gas properties and a route toward graphene-based nanoelectronics. *J. Phys. Chem. B,* **2004**, *108*, 19912-19916. [http://dx.doi.org/10.1021/jp040650f]

[6] Geim, A.K.; Novoselov, K.S. The rise of graphene. *Nat. Mater.,* **2007**, *6*(3), 183-191. [http://dx.doi.org/10.1038/nmat1849] [PMID: 17330084]

[7] Kroto, H.W.; Heath, J.R.; O'Brien, S.C.; Curl, R.F.; Smalley, R.E. C60: Buckminsterfullerene. *Nature,* **1985**, *318*, 162-163.

[http://dx.doi.org/10.1038/318162a0]

[8] Kroto, H. Symmetry, space, stars and C60. *Rev. Mod. Phys.*, **1997**, *69*, 703-722. [http://dx.doi.org/10.1103/RevModPhys.69.703]

[9] Osawa, E. *Kagaku,***1970**, *25* , 854-863 (in Japanese); b) Osawa, E. *Chem. Abstr.*, **1971**, *74*, 75698v.

[10] Yoshida, Z.; Osawa, E. *Aromaticity*; Kagakudojin: Kyoto, **1971**, pp. 174-178. (in Japanese)

[11] Bochvar, D.A.; Gal'pern, E.G. Hypothetical systems: carbodecahedron, s-icosahedrane and carbo--icosahedron. *Proc. Acad. Sci. USSR,* **1973**, *209*, 239-241.

[12] Russell, A.; Houk, K.N. Orville Chapman (1932-2004) Biographical Memoir , *National Academy of Sciences,* **2013**.

[13] Curl, R.F.; Smalley, R.E. Probing c60. *Science,* **1988**, *242*(4881), 1017-1022. [http://dx.doi.org/10.1126/science.242.4881.1017] [PMID: 17834045]

[14] Kroto, H. Space, stars, c60, and soot. *Science,* **1988**, *242*(4882), 1139-1145. [http://dx.doi.org/10.1126/science.242.4882.1139] [PMID: 17799730]

[15] Meijer, G.; Bethune, D.S. Mass spectroscopic confirmation of the presence of C60 in laboratory-produced carbon dust. *Chem. Phys. Lett.*, **1990**, *175*, 1-2. [http://dx.doi.org/10.1016/0009-2614(90)85507-9]

[16] Weltner, W., Jr; Van Zee, R.J. Carbon molecules, ions, and clusters. *Chem. Rev.*, **1989**, *89*, 1713-1747. [http://dx.doi.org/10.1021/cr00098a005]

[17] Kroto, H.W.; Allaf, W.; Balm, S.P. C60: Buckminsterfullerene. *Chem. Rev.*, **1991**, *91*, 1213-1235. [http://dx.doi.org/10.1021/cr00006a005]

[18] Krätschmer, W.; Lamb, L.D.; Fostiropoulos, K.; Huffman, D.R. Solid C60: a new form of carbon. *Nature,* **1990**, *347*, 354-358. [http://dx.doi.org/10.1038/347354a0]

[19] Krätschmer, W.; Fostiropoulos, K.; Huffman, D.R. The infrared and ultraviolet absorption spectra of laboratory-produced carbon dust: evidence for the presence of the C60 molecule. *Chem. Phys. Lett.*, **1990**, *170*, 167-170. [http://dx.doi.org/10.1016/0009-2614(90)87109-5]

[20] Krätschmer, W.; Fostiropoulos, K.; Huffman, D.R. *Dusty Objects in the Universe*; Bussoletti, E.; Vittone, A.A., Eds.; Kluwer: Dordrecht, **1990**.

[21] *The Fullerenes: New horizons for the chemistry, physics, and astrophysics of carbon*; Kroto, H.W.; Walton, D.R., Eds.; Cambridge University: Cambridge, England, **1993**.

[22] Hirsch, A. *The Chemistry of the Fullerenes*; Thieme: New York, **1994**. [http://dx.doi.org/10.1002/9783527619214]

[23] Baldwin, J. *BuckyWorks*; Wiley: New York, **1996**.

[24] Ball, P. *Designing the Molecular World*; Princeton University: Princeton, NJ, **1994**, pp. 13-53.

[25] Aldersey-Williams, H. *The most beautiful molecule in chemistry, an adventure in chemistry*; Aurum: London, **1995**.

[26] Katz, E.A. Fullerene thin films as photovoltaic material. In: *Nanostructured materials for solar energy conversion*; Sōga, T., Ed.; Elsevier, **2006**; pp. 361-443.
[http://dx.doi.org/10.1016/B978-044452844-5/50014-7]

[27] Taylor, R.; Hare, J.P.; Abdul-Sada, A.K.; Kroto, H.W. Isolation, separation and characterisation of the fullerenes C60 and C70: the third form of carbon *J. Chem. Soc., Chem. Comm.,* **1990**, 1423-1425.

[28] Haddon, R.C.; Hebard, A.F.; Rosseinsky, M.J.; Murphy, D.W.; Duclos, S.J.; Lyons, K.B.; Miller, B.; Rosamilia, J.M.; Fleming, R.M.; Kortan, A.R.; Glarum, S.H.; Makhija, A.V.; Muller, A.J.; Eick, R.H.; Zahurak, S.M.; Tycko, R.; Dabbagh, G.; Thiel, F.A. Conducting films of C60 and C70 by alkali-metal doping. *Nature,* **1991**, *350*, 320-322.
[http://dx.doi.org/10.1038/350320a0]

[29] Tanigaki, K.; Ebbesen, T.W.; Saito, S.; Mizuki, J.; Tsai, T.S.; Kubo, Y.; Kuroshima, S. Superconductivity at 33 K in CsxRbyC60. *Nature,* **1991**, *352*, 222-223.
[http://dx.doi.org/10.1038/352222a0]

[30] Heath, J.R.; O'Brien, S.C.; Zhang, Q.; Liu, Y.; Curl, R.F.; Kroto, H.W.; Smalley, R.E. Lanthanum complexes of spheroidal carbon shells. *J. Am. Chem. Soc.,* **1985**, *107*, 7779-7780.
[http://dx.doi.org/10.1021/ja00311a102]

[31] Weaver, J.H.; Poirier, D.M. *Solid Sate Physics 48*; Ehrenreich, H.; Spaepen, F., Eds.; Academic Press: San Diego, **1994**.

[32] Reich, S.; Thomsen, C.; Maultzsch, J. *Carbon nanotubes: basic concepts and physical properties*; Wiley-VCH: Weinheim, **2004**.

[33] Iijima, S. Helical microtubules of graphitic carbon. *Nature,* **1991**, *354*, 56-58.
[http://dx.doi.org/10.1038/354056a0]

[34] Iijima, S.; Ichihashi, T. Single-shell carbon nanotubes of 1-nm diameter. *Nature,* **1993**, *363*, 603-605.
[http://dx.doi.org/10.1038/363603a0]

[35] Bethune, D.S.; Kiang, C.H.; de Vries, M.S.; Gorman, G.; Savy, R.; Vazquez, J.; Beyers, R. Cobalt-catalysed growth of carbon nanotubes with single-atomic-layer walls. *Nature,* **1993**, *363*, 605-607.
[http://dx.doi.org/10.1038/363605a0]

[36] Guo, T.; Jin, C-M.; Smalley, R.E. Catalytic growth of single-walled nanotubes by laser vaporization. *Chem. Phys. Lett.,* **1995**, *243*, 49-54.
[http://dx.doi.org/10.1016/0009-2614(95)00825-O]

[37] Berber, S.; Kwon, Y-K.; Tomanek, D. Unusually high thermal conductivity of carbon nanotubes. *Phys. Rev. Lett.,* **2000**, *84*(20), 4613-4616.
[http://dx.doi.org/10.1103/PhysRevLett.84.4613] [PMID: 10990753]

[38] Tans, S.J.; Devoret, M.H.; Dai, H.J.; Thess, A.; Smalley, R.E.; Geerligs, L.J.; Dekker, C. Individual single-wall carbon nanotubes as quantum wires. *Nature,* **1997**, *386*, 474-477.
[http://dx.doi.org/10.1038/386474a0]

[39] Tans, S.J.; Verschueren, A.R.; Dekker, C. Room-temperature transistor based on a single carbon nanotube. *Nature,* **1998**, *393*, 49-52.
[http://dx.doi.org/10.1038/29954]

[40] Ugarte, D. Curling and closure of graphitic networks under electron-beam irradiation. *Nature,* **1992**, *359*(6397), 707-709.
[http://dx.doi.org/10.1038/359707a0] [PMID: 11536508]

[41] Ruoff, R.S.; Lorents, D.C.; Chan, B.; Malhotra, R.; Subramoney, S. Single crystal metals encapsulated in carbon nanoparticles. *Science,* **1993**, *259*(5093), 346-348.
[http://dx.doi.org/10.1126/science.259.5093.346] [PMID: 17832348]

[42] Tomita, M.; Saito, Y.; Hayashi, T. LaC2 encapsulated in graphite nano-particle. *Jpn. J. Appl. Phys.,* **1993**, *32*, L280-L282.
[http://dx.doi.org/10.1143/JJAP.32.L280]

[43] Iijima, S.; Yudasaka, M.; Yamada, R.; Bandow, S.; Suenaga, K.; Kokai, F.; Takahashi, K. Nano-aggregates of single-walled graphitic carbon nano-horns. *Chem. Phys. Lett.,* **1999**, *309*, 165-170.
[http://dx.doi.org/10.1016/S0009-2614(99)00642-9]

[44] Smith, B.W.; Monthioux, M.; Luzzi, D.E. Encapsulated C60 in carbon nanotubes. *Nature,* **1998**, *396*, 323-324.
[http://dx.doi.org/10.1038/24521]

[45] Kwon, Y-K.; Tomanek, D.; Iijima, S. "Bucky shuttle" memory device: synthetic approach and molecular dynamics simulations. *Phys. Rev. Lett.,* **1999**, *82*, 1470-1473.
[http://dx.doi.org/10.1103/PhysRevLett.82.1470]

[46] Wallace, P.R. The band theory of graphite. *Phys. Rev.,* **1947**, *71*, 622-634.
[http://dx.doi.org/10.1103/PhysRev.71.622]

[47] Novoselov, K.S.; Geim, A.K.; Morozov, S.V.; Jiang, D.; Zhang, Y.; Dubonos, S.V.; Grigorieva, I.V.; Firsov, A.A. Electric field effect in atomically thin carbon films. *Science,* **2004**, *306*(5696), 666-669.
[http://dx.doi.org/10.1126/science.1102896] [PMID: 15499015]

[48] Mattevi, C.O.; Kima, H.; Chhowalla, M. A review of chemical vapour deposition of graphene on copper. *J. Mater. Chem.,* **2010**, *21*, 3324-3334.
[http://dx.doi.org/10.1039/C0JM02126A]

[49] Sutter, P.W.; Flege, J-I.; Sutter, E.A. Epitaxial graphene on ruthenium. *Nat. Mater.,* **2008**, *7*(5), 406-411.
[http://dx.doi.org/10.1038/nmat2166] [PMID: 18391956]

[50] Coraux, J.; N'Diaye, A.T.; Busse, C.; Michely, T. Structural coherency of graphene on Ir(111). *Nano Lett.,* **2008**, *8*(2), 565-570.
[http://dx.doi.org/10.1021/nl0728874] [PMID: 18189442]

[51] Reina, A.; Jia, X.; Ho, J.; Nezich, D.; Son, H.; Bulovic, V.; Dresselhaus, M.S.; Kong, J. Large area, few-layer graphene films on arbitrary substrates by chemical vapor deposition. *Nano Lett.,* **2009**, *9*(1), 30-35.
[http://dx.doi.org/10.1021/nl801827v] [PMID: 19046078]

[52] Li, X.; Cai, W.; An, J.; Kim, S.; Nah, J.; Yang, D.; Piner, R.; Velamakanni, A.; Jung, I.; Tutuc, E.; Banerjee, S.K.; Colombo, L.; Ruoff, R.S. Large-area synthesis of high-quality and uniform graphene films on copper foils. *Science,* **2009**, *324*(5932), 1312-1314.
[http://dx.doi.org/10.1126/science.1171245] [PMID: 19423775]

[53] Ebbesen, T.W.; Ajayan, P.M. Large-scale synthesis of carbon nanotubes. *Nature,* **1992**, *358*, 220-222. [http://dx.doi.org/10.1038/358220a0]

[54] Collins, P.G.; Avouris, P. Nanotubes for electronics. *Sci. Am.,* **2000**, *283*(6), 62-69. [http://dx.doi.org/10.1038/scientificamerican1200-62] [PMID: 11103460]

[55] Guo, T.; Nikolaev, P.; Rinzler, A.G.; Tomanek, D.; Colbert, D.T.; Smalley, R.E. Self-assembly of tubular fullerenes. *J. Phys. Chem.,* **1995**, *99*, 10694-10697. [http://dx.doi.org/10.1021/j100027a002]

[56] Guo, T.; Nikolaev, P.; Thess, A.; Colbert, D.; Smalley, R. Catalytic growth of single-walled nanotubes by laser vaporization. *Chem. Phys. Lett.,* **1995**, *243*, 49-54. [http://dx.doi.org/10.1016/0009-2614(95)00825-O]

[57] Smiljanic, O.; Stansfield, B.L.; Dodelet, J-P.; Serventi, A.; Désilets, S. Gas-phase synthesis of SWNT by an atmospheric pressure plasma jet. *Chem. Phys. Lett.,* **2002**, *356*, 189-193. [http://dx.doi.org/10.1016/S0009-2614(02)00132-X]

[58] Smiljanic, O. Method and apparatus for producing single-wall carbon nanotubes US Patent 7591989, 2009.

[59] José-Yacamán, M.; Miki-Yoshida, M.; Rendón, L.; Santiesteban, J.G. Catalytic growth of carbon microtubules with fullerene structure. *Appl. Phys. Lett.,* **1993**, *62*, 202-204. [http://dx.doi.org/10.1063/1.109315]

[60] Inami, N.; Mohamed, M.A.; Shikoh, E.; Fujiwara, A. Synthesis-condition dependence of carbon nanotube growth by alcohol catalytic chemical vapor deposition method. *Sci. Technol. Adv. Mater.,* **2007**, *8*, 292-296. [http://dx.doi.org/10.1016/j.stam.2007.02.009]

[61] Ishigami, N.; Ago, H.; Imamoto, K.; Tsuji, M.; Iakoubovskii, K.; Minami, N. Crystal plane dependent growth of aligned single-walled carbon nanotubes on sapphire. *J. Am. Chem. Soc.,* **2008**, *130*(30), 9918-9924. [http://dx.doi.org/10.1021/ja8024752] [PMID: 18597459]

[62] Naha, S.; Puri, I.K. A model for catalytic growth of carbon nanotubes. *J. Phys. D Appl. Phys.,* **2008**, *41*, 065304. [http://dx.doi.org/10.1088/0022-3727/41/6/065304]

[63] Kumar, M.; Ando, Y. Chemical vapor deposition of carbon nanotubes: a review on growth mechanism and mass production. *J. Nanosci. Nanotechnol.,* **2010**, *10*(6), 3739-3758. [http://dx.doi.org/10.1166/jnn.2010.2939] [PMID: 20355365]

[64] Novoselov, K.S.; Jiang, D.; Schedin, F.; Booth, T.J.; Khotkevich, V.V.; Morozov, S.V.; Geim, A.K. Two-dimensional atomic crystals. *Proc. Natl. Acad. Sci. USA,* **2005**, *102*(30), 10451-10453. [http://dx.doi.org/10.1073/pnas.0502848102] [PMID: 16027370]

[65] Bolotin, K.I.; Sikes, K.J.; Jiang, Z.; Klima, M.; Fudenberg, G.; Hone, J.; Kim, P.; Stormer, H.L. Ultrahigh electron mobility in suspended graphene. *Solid State Commun.,* **2008**, *146*, 351-355. [http://dx.doi.org/10.1016/j.ssc.2008.02.024]

[66] Xue, J.; Sanchez-Yamagishi, J.; Bulmash, D.; Jacquod, P.; Deshpande, A.; Watanabe, K.; Taniguchi, T.; Jarillo-Herrero, P.; LeRoy, B.J. Scanning tunnelling microscopy and spectroscopy of ultra-flat

graphene on hexagonal boron nitride. *Nat. Mater.,* **2011**, *10*(4), 282-285.
[http://dx.doi.org/10.1038/nmat2968] [PMID: 21317900]

[67] Hirsch, A. The era of carbon allotropes. *Nat. Mater.,* **2010**, *9*(11), 868-871.
[http://dx.doi.org/10.1038/nmat2885] [PMID: 20966925]

[68] Wilson, S.R.; Schuster, D.I.; Nuber, B.; Meier, M.S.; Maggini, M.; Prato, M.; Taylor, R. *Fullerenes: chemistry, physics, and technology in Organic Chemistry of Fullerenes*; Kadish, K. M.; Ruoff, R.S., Eds.; John Wiley & Sons: New York, **2000**.

[69] Jeon, I-Y.; Chang, D.W.; Kumar, N.A.; Baek, J-B. Functionalization of carbon nanotubes. In: *Carbon Nanotubes - Polymer Nanocomposites*; Yellampalli, S., Ed.; InTech: Rijeka, Croatia, **2011**.
[http://dx.doi.org/10.5772/18396]

[70] Haddon, R.C. Chemistry of the fullerenes: the manifestation of strain in a class of continuous aromatic molecules. *Science,* **1993**, *261*(5128), 1545-1550.
[http://dx.doi.org/10.1126/science.261.5128.1545] [PMID: 17798112]

[71] Grant, P. Superconductivity. Up on the C60 elevator. *Nature,* **2001**, *413*(6853), 264-265.
[http://dx.doi.org/10.1038/35095149] [PMID: 11565014]

[72] Dagotto, E. Superconductivity. The race to beat the cuprates. *Science,* **2001**, *293*(5539), 2410-2411.
[http://dx.doi.org/10.1126/science.1065711] [PMID: 11577227]

[73] Taylor, R.D.; Walton, M.R. The chemistry of fullerenes. *Nature,* **1993**, *363*, 685-693.
[http://dx.doi.org/10.1038/363685a0]

[74] Haddon, R.C.; Fang, S.; Rao, A.M.; Eklund, P.C.; Lee, W.H.; Dickey, E.C.; Grulke, E.A.; Pendergrass, J.C.; Chavan, A.; Haley, B.E.; Smalley, R.E. Chemical attachment of organic functional groups to single-walled Carbon nanotube material. *J. Mater. Res.,* **1998**, *13*, 2423-2431.
[http://dx.doi.org/10.1557/JMR.1998.0337]

[75] Hamon, M.A.; Itkis, M.E.; Niyogi, S.; Alvaraez, T.; Kuper, C.; Menon, M.; Scott, L.T.; Bratcher, M.S.; Hagen, S. Synthesis and characterization of a C36H12 fullerene subunit. *J. Am. Chem. Soc.,* **1996**, *118*, 8743-8744.
[http://dx.doi.org/10.1021/ja9621511]

[76] Hamon, M.A.; Itkis, M.E.; Niyogi, S.; Alvaraez, T.; Kuper, C.; Menon, M.; Haddon, R.C. Effect of rehybridization on the electronic structure of single-walled carbon nanotubes. *J. Am. Chem. Soc.,* **2001**, *123*(45), 11292-11293.
[http://dx.doi.org/10.1021/ja0109702] [PMID: 11697973]

[77] Haddon, R.C. Measure of nonplanarity in conjugated organic molecules: which structurally characterized molecule displays the highest degree of pyramidalization? *J. Am. Chem. Soc.,* **1990**, *112*, 3385-3389.
[http://dx.doi.org/10.1021/ja00165a020]

[78] Rabideau, P.W.; Sygula, A. Buckybowls: polynuclear aromatic hydrocarbons related to the Buckminsterfullerene surface. *Acc. Chem. Res.,* **1996**, *29*, 235-242.
[http://dx.doi.org/10.1021/ar950197d]

[79] Weedon, B.R.; Haddon, R.C.; Spielmann, H.P.; Meier, M.S. Fulleroid addition regiochemistry is driven by π-orbital misalignment. *J. Am. Chem. Soc.,* **1999**, *121*, 335-340.

[http://dx.doi.org/10.1021/ja983269f]

[80] Bodwell, G.J.; Bridson, J.N.; Houghton, T.J.; Kennedy, J.W.; Mannion, M.R. 1,7-Dioxa[7](2,7)pyrenophane: the pyrene moiety is more bent than that of C70. *Chemistry,* **1999**, *5*, 1823-1827.
[http://dx.doi.org/10.1002/(SICI)1521-3765(19990604)5:6<1823::AID-CHEM1823>3.0.CO;2-I]

[81] Bingel, C. Cyclopropanierung von Fullerenen. *Chem. Ber.,* **1993**, *126*, 1957.
[http://dx.doi.org/10.1002/cber.19931260829]

[82] Iwashita, A.; Matsuo, Y.; Nakamura, E. AlCl(3)-mediated mono-, di-, and trihydroarylation of [60]fullerene. *Angew. Chem. Int. Ed. Engl.,* **2007**, *46*(19), 3513-3516.
[http://dx.doi.org/10.1002/anie.200700062] [PMID: 17385815]

[83] Matsuo, Y.; Muramatsu, A.; Tahara, K.; Koide, M.; Nakamura, E. Synthesis of 6,9,12,15,18-pentamethyl-1,6,9,12,15,18-hexahydro(c60-ih)[5,6]fullerene. *Org. Synth.,* **2006**, *83*, 80-87.
[http://dx.doi.org/10.15227/orgsyn.083.0080]

[84] Yamada, M.; Akasaka, T.; Nagase, S. Carbene additions to fullerenes. *Chem. Rev.,* **2013**, *113*(9), 7209-7264.
[http://dx.doi.org/10.1021/cr3004955] [PMID: 23773169]

[85] Talyzin, A.V.; Tsybin, Y.O.; Purcell, J.M.; Schaub, T.M.; Shulga, Y.M.; Noréus, D.; Sato, T.; Dzwilewski, A.; Sundqvist, B.; Marshall, A.G. Reaction of hydrogen gas with C60 at elevated pressure and temperature: hydrogenation and cage fragmentation. *J. Phys. Chem. A,* **2006**, *110*(27), 8528-8534.
[http://dx.doi.org/10.1021/jp0557971] [PMID: 16821837]

[86] Chiang, L.Y.; Swirczewski, J.W.; Hsu, C.S.; Chowdhury, S.K.; Cameron, S.; Creegan, K. Multi-hydroxy additions onto C60 fullerene molecules. *J. Chem. Soc. Chem. Commun.,* **1992**, 1791.
[http://dx.doi.org/10.1039/c39920001791]

[87] Chiang, L.Y.; Upasani, R.B.; Swirczewski, J.W.; Soled, S. Evidence of hemiketals incorporated in the structure of fullerols derived from aqueous acid chemistry. *J. Am. Chem. Soc.,* **1993**, *115*, 5453-5457.
[http://dx.doi.org/10.1021/ja00066a014]

[88] Li, J.; Takeuchi, A.; Ozawa, M.; Li, X.; Saigo, K.; Kitazawa, K. C60 fullerol formation catalysed by quaternary ammonium hydroxides. *J. Chem. Soc. Chem. Commun.,* **1993**, 1784.
[http://dx.doi.org/10.1039/c39930001784]

[89] Wang, S.; He, P.; Zhang, J.M.; Jiang, H.; Zhu, S.Z. Novel and efficient synthesis of water-soluble [60]fullerenol by solvent-free reaction. *Synth. Commun.,* **2005**, *35*, 1803-1808.
[http://dx.doi.org/10.1081/SCC-200063958]

[90] Zhang, G.; Liu, Y.; Liang, D.; Gan, L.; Li, Y. Facile synthesis of isomerically pure fullerenols and formation of spherical aggregates from C60(OH)8. *Angew. Chem. Int. Ed. Engl.,* **2010**, *49*(31), 5293-5295.
[http://dx.doi.org/10.1002/anie.201001280] [PMID: 20575126]

[91] Kokubo, K.; Matsubayashi, K.; Tategaki, H.; Takada, H.; Oshima, T. Facile synthesis of highly water-soluble fullerenes more than half-covered by hydroxyl groups. *ACS Nano,* **2008**, *2*(2), 327-333.
[http://dx.doi.org/10.1021/nn700151z] [PMID: 19206634]

[92] Jia, J.; Wu, H.S.; Xu, X.H.; Zhang, X.M.; Jiao, H. Fused five-membered rings determine the stability

of C60F60. *J. Am. Chem. Soc.,* **2008**, *130*(12), 3985-3988.
[http://dx.doi.org/10.1021/ja0781590] [PMID: 18311972]

[93] Tsuda, M.; Ishida, T.; Nogami, T.; Kurono, S.; Ohashi, M. C61Cl2. Synthesis and characterization of dichlorocarbene adducts of C60. *Tetrahedron Lett.,* **1993**, *34*, 6911-6912.
[http://dx.doi.org/10.1016/S0040-4039(00)91828-8]

[94] Tzirakis, M.D.; Orfanopoulos, M. Radical reactions of fullerenes: from synthetic organic chemistry to materials science and biology. *Chem. Rev.,* **2013**, *113*(7), 5262-5321.
[http://dx.doi.org/10.1021/cr300475r] [PMID: 23570603]

[95] Morton, J.R.; Preston, K.F.; Krusic, P.J.; Hill, S.A.; Wasserman, E. ESR studies of the reaction of alkyl radicals with fullerene (C60). *J. Phys. Chem.,* **1992**, *96*, 3576-3578.
[http://dx.doi.org/10.1021/j100188a006]

[96] Prato, M.; Lucchini, V.; Maggini, M.; Stimpfl, E.; Scorrano, G.; Eiermann, M.; Suzuki, T.; Wudl, F. Energetic preference in 5,6 and 6,6 ring junction adducts of C60: fulleroids and methanofullerenes. *J. Am. Chem. Soc.,* **1993**, *115*, 8479-8480.
[http://dx.doi.org/10.1021/ja00071a080]

[97] Vogel, E. Recent advances in the chemistry of bridged annulenes. *Pure Appl. Chem.,* **1982**, *54*, 1015-1039.
[http://dx.doi.org/10.1351/pac198254051015]

[98] Yurovskaya, M.A.; Trushkov, I.V. Cycloaddition to buckminsterfullerene C60: advancements and future prospects *Russ. Chem. Bull., Int. Ed.,* **2002**, *51*, 367-443.

[99] Śliwa, W. Cycloaddition reactions of fullerenes. *Fullerene Science and Technology,* **1995**, *3*, 243-281.
[http://dx.doi.org/10.1080/153638X9508543782]

[100] Tsuda, M.; Ishida, T.; Nogami, T.; Kurono, S.; Ohashi, M. Addition reaction of benzyne to C60. *Chem. Lett.,* **1992**, *21*, 2333-2334.
[http://dx.doi.org/10.1246/cl.1992.2333]

[101] Wilson, S.R.; Kaprinidis, N.; Wu, Y.; Schuster, D.I. A new reaction of fullerenes: [2+2]-photocycloaddition of enones. *J. Am. Chem. Soc.,* **1993**, *115*, 8495-8496.
[http://dx.doi.org/10.1021/ja00071a088]

[102] Hoke, S.H., II; Molstad, J.; Dilettato, D.; Jay, M.J.; Carlson, D.; Kahr, B.; Cooks, R.G. Reaction of fullerenes and benzyne. *J. Org. Chem.,* **1992**, *57*, 5069-5071.
[http://dx.doi.org/10.1021/jo00045a012]

[103] Wilson, S.R.; Wu, Y.; Kaprinidis, N.A.; Schuster, D.I.; Welch, C.J. Resolution of enantiomers of cis- and trans-fused C60-enone [2+2] photoadducts. *J. Org. Chem.,* **1993**, *58*, 6548-6549.
[http://dx.doi.org/10.1021/jo00076a007]

[104] Schuster, D.I.; Cao, J.; Kaprinidis, N.; Wu, Y.; Jensen, A.W.; Lu, Q.; Wang, H.; Wilson, S.R. [2 + 2] Photocycloaddition of cyclic enones to C60. *J. Am. Chem. Soc.,* **1996**, *118*, 5639-5647.
[http://dx.doi.org/10.1021/ja9534009]

[105] Jensen, A.W.; Khong, A.; Saunders, M.; Wilson, S.R.; Schuster, D.I. Photocycloaddition of cyclic 1,3-diones to C60. *J. Am. Chem. Soc.,* **1997**, *119*, 7303-7307.
[http://dx.doi.org/10.1021/ja962509t]

[106] Segura, J.L.; Martin, N. [60]Fullerene dimers. *Chem. Soc. Rev.,* **2000**, *29*, 13-25. [http://dx.doi.org/10.1039/a903716k]

[107] Wang, G-W.; Komatsu, K.; Murata, Y.; Shiro, M. Synthesis and X-ray structure of dumb-bell-shaped C120. *Nature,* **1997**, *387*, 583-585. [http://dx.doi.org/10.1038/42439]

[108] Komatsu, K.; Wang, G-W.; Murata, Y.; Tanaka, T.; Fujiwara, K. Mechanochemical synthesis and characterization of the fullerene dimer C120. *J. Org. Chem.,* **1998**, *63*, 9358-9366. [http://dx.doi.org/10.1021/jo981319t]

[109] Keshavarz, M-K.; Knight, B.; Srdanov, G.; Wudl, F. Cyanodihydrofullerenes and dicyanodihydrofullerene: the first polar solid based on C60. *J. Am. Chem. Soc.,* **1995**, *117*, 11371-11372. [http://dx.doi.org/10.1021/ja00150a048]

[110] Forman, G.S.; Tagmatarchis, N.; Shinohara, H. Novel synthesis and characterization of five isomers of (C(70))(2) fullerene dimers. *J. Am. Chem. Soc.,* **2002**, *124*(2), 178-179. [http://dx.doi.org/10.1021/ja0168662] [PMID: 11782162]

[111] Tagmatarchis, N.; Forman, G.S.; Taninaka, A.; Shinohara, H. Cross-fullerene dimers [(C60)(C70)]: solid-state synthesis, characterization and mechanism. *Synlett,* **2002**, 235-238. [http://dx.doi.org/10.1055/s-2002-19784]

[112] Maggini, M.; Scorrano, G.; Prato, M. Addition of azomethine ylides to C60: synthesis, characterization, and functionalization of fullerene pyrrolidines. *J. Am. Chem. Soc.,* **1993**, *115*, 9798-9799. [http://dx.doi.org/10.1021/ja00074a056]

[113] Prato, M.; Maggini, M. Fulleropyrrolidines: A family of full-fledged fullerene derivatives. *Acc. Chem. Res.,* **1998**, *31*, 519-526. [http://dx.doi.org/10.1021/ar970210p]

[114] Wang, P.; Metzger, R.M.; Chen, B. Stable monolayers of fullerene derivatives. *Thin Solid Films,* **1998**, *327-329*, 96-99. [http://dx.doi.org/10.1016/S0040-6090(98)00596-3]

[115] Guo, Z.; Li, Y.; Xu, J.; Mao, Z.; Wu, Y.; Zhu, D. The first crown ether-bearing [60]fullero-pyrrolidine: synthesis and regulation of absorption spectrum by ions binding. *Synth. Commun.,* **1998**, *28*, 1957-1962. [http://dx.doi.org/10.1080/00397919808007169]

[116] Murakami, H.; Shirakusa, M.; Sagara, T.; Nakashima, N. Synthesis, aggregate structure and electrochemical properties of a water-soluble fullerene-bearing ammonium amphiphile. *Chem. Lett.,* **1999**, *28*, 815-816. [http://dx.doi.org/10.1246/cl.1999.815]

[117] Ge, Z.X.; Li, Y.L.; Guo, Z.X.; Shi, Z.Q.; Zhu, D.B. Synthesis of a new crown ether-bearing [60]fulleropyrrolidine containing a benzothiazolium styryl dye. *Tetrahedron Lett.,* **1999**, *40*, 5759-5762. [http://dx.doi.org/10.1016/S0040-4039(99)01124-7]

[118] Guo, Z.; Yan, J.; Li, Y.; Ge, Z.; Zhu, D. AMI study of a novel crowned [60]fulleropyrrolidine. *Synth. Met.,* **1999**, *102*, 1567-1568.
[http://dx.doi.org/10.1016/S0379-6779(98)00738-3]

[119] Leo, L.; Mele, G.; Rosso, G.; Valli, L.; Vasapollo, G.; Guldi, D.M.; Mascolo, G. Interfacial properties of substituted fulleropyrrolidines on the water surface. *Langmuir,* **2000**, *16*, 4599-4606.
[http://dx.doi.org/10.1021/la9912038]

[120] Zhang, S.; Gan, L.; Huang, C. Photoelectric response of pyrrolidinofullerene and bis-pyrrolidinofullerene LB films. *Chem. Phys. Lett.,* **2000**, *331*, 143-148.
[http://dx.doi.org/10.1016/S0009-2614(00)01185-4]

[121] Tomberli, V.; Da Ros, T.; Bosi, S.; Prato, M. Synthetic approaches towards the preparation of water-soluble fulleropyrrolidines. *Carbon,* **2000**, *38*, 1551-1555.
[http://dx.doi.org/10.1016/S0008-6223(99)00285-7]

[122] Bosi, S.; Da Ros, T.; Castellano, S.; Banfi, E.; Prato, M. Antimycobacterial activity of ionic fullerene derivatives. *Bioorg. Med. Chem. Lett.,* **2000**, *10*(10), 1043-1045.
[http://dx.doi.org/10.1016/S0960-894X(00)00159-1] [PMID: 10843212]

[123] Kordatos, K.; Da Ros, T.; Bosi, S.; Vázquez, E.; Bergamin, M.; Cusan, C.; Pellarini, F.; Tomberli, V.; Baiti, B.; Pantarotto, D.; Georgakilas, V.; Spalluto, G.; Prato, M. Novel versatile fullerene synthons. *J. Org. Chem.,* **2001**, *66*(14), 4915-4920.
[http://dx.doi.org/10.1021/jo015608k] [PMID: 11442425]

[124] Kang, S.H.; Ma, H.; Kang, M.S.; Kim, K.S.; Jen, A.K.; Zareie, M.H.; Sarikaya, M. Ordered self-assembly and electronic behavior of C60-anthrylphenylacetylene hybrid. *Angew. Chem. Int. Ed. Engl.,* **2004**, *43*(12), 1512-1516.
[http://dx.doi.org/10.1002/anie.200353001] [PMID: 15022220]

[125] Maggini, M.; Scorrano, G.; Bianco, A.; Toniolo, C.; Sijbesma, R.P.; Wudl, F.; Prato, M. Addition reactions of C60 leading to fulleroprolines. *J. Chem. Soc. Chem. Commun.,* **1994**, 305-306.
[http://dx.doi.org/10.1039/c39940000305]

[126] Bianco, A.; Gasparrini, F.; Maggini, M.; Misiti, D.; Polese, A.; Prato, M.; Scorrano, G.; Toniolo, C.; Villani, C. Molecular recognition by a silica-bound fullerene derivative. *J. Am. Chem. Soc.,* **1997**, *119*, 7550-7554.
[http://dx.doi.org/10.1021/ja970502l]

[127] Langa, F.; de la Cruz, P.; de la Hoz, A.; Diaz-Ortiz, A.; Diez-Barra, E. Microwave irradiation: more than just a method for accelerating reactions. *Contemp. Org. Synth.,* **1997**, •••, 373-386.
[http://dx.doi.org/10.1039/co9970400373]

[128] Rotello, V.M.; Howard, J.B.; Yadav, T.; Conn, M.M.; Viani, E.; Giovane, L.M.; Lafleur, A.L. Isolation of fullerene products from flames: Structure and synthesis of the C60-cyclopentadiene adduct. *Tetrahedron Lett.,* **1993**, *34*, 1561-1562.
[http://dx.doi.org/10.1016/0040-4039(93)85006-I]

[129] Tsuda, M.; Ishida, T.; Nogami, T.; Kurono, S.; Ohashi, M. Isolation and characterization of Diels–Alder adducts of C60 with anthracene and cyclopentadiene. *J. Chem. Soc. Chem. Commun.,* **1993**, 1296-1298.

[http://dx.doi.org/10.1039/c39930001296]

[130] Meidine, M.F.; Avent, A.G.; Darwish, A.D.; Kroto, H.W.; Ohashi, O.; Taylor, R.; Walton, D.R. Pentamethylcyclopentadiene adducts of [60]- and [70]- fullerene. *J. Chem. Soc., Perkin Trans. 2,* **1994**, 1189-1193.
[http://dx.doi.org/10.1039/p29940001189]

[131] Takeshita, H.; Liu, J-F.; Kato, N.; Mori, A.; Isobe, R. Highly thermally-stable Diels-Alder adducts of [60]fullerene with 2-cycloalkenones and their acetals. *Chem. Lett.,* **1995**, *24*, 377-378.
[http://dx.doi.org/10.1246/cl.1995.377]

[132] Wilson, S.R.; Yurchenko, M.E.; Schuster, D.I.; Khong, A.; Saunders, M. Synthesis of fluorous fullerene adducts: reversible solubilization of fullerenes in perfluorinated solvents. *J. Org. Chem.,* **2000**, *65*(9), 2619-2623.
[http://dx.doi.org/10.1021/jo991008y] [PMID: 10808431]

[133] Schluter, J.A.; Seaman, J.M.; Taha, S.; Cohen, H.; Lykke, K.R.; Wang, H.H.; Williams, J.M. Synthesis, purification, and characterization of the 1: 1 addition product of C60 and anthracene. *J. Chem. Soc. Chem. Commun.,* **1993**, 972-974.
[http://dx.doi.org/10.1039/c39930000972]

[134] Komatsu, K.; Murata, Y.; Sugita, N.; Takeuchi, K.; Wan, T.S. Use of naphthalene as a solvent for selective formation of the 1:1 diels-alder adduct of C60 with anthracene. *Tetrahedron Lett.,* **1993**, *34*, 8473-8476.
[http://dx.doi.org/10.1016/S0040-4039(00)61362-X]

[135] Allemand, P.M.; Srdanov, G.; Koch, A.; Khemani, K.; Wudl, F.; Rubin, Y.; Diederich, F.; Alvarez, M.M.; Anz, S.J.; Whetten, R.L. The unusual electron spin resonance of fullerene C60 anion radical. *J. Am. Chem. Soc.,* **1991**, *113*, 2780-2781.
[http://dx.doi.org/10.1021/ja00007a078]

[136] Becker, U.; Denninger, G.; Dyakonov, V.; Gotschy, B.; Klos, H.; Rösler, G.; Hirsch, A. C60-tetraphenylphosphonium+: an organic alloy. *Europhys. Lett.,* **1993**, *21*, 267-272.
[http://dx.doi.org/10.1209/0295-5075/21/3/003]

[137] Moriyama, H.; Kobayashi, H.; Kobayashi, A.; Watanabe, T. Electrocrystallization and ESR spectra of the single crystal [N(P(C6H5)3)2].C60. *J. Am. Chem. Soc.,* **1993**, *115*, 1185-1187.
[http://dx.doi.org/10.1021/ja00056a075]

[138] Penicaud, A.; Perez-Benitez, A.; Gleason, R.; Munoz, V.E.; Escudero, P.R. Electrocrystallizing C60: synthesis, single crystal x-ray structure, and magnetic (ESR, SQUID) characterization of [(C6H5)4P]2[C60][I]x. *J. Am. Chem. Soc.,* **1993**, *115*, 10392-10393.
[http://dx.doi.org/10.1021/ja00075a078]

[139] Foss, C.A.; Feldheim, D.L.; Lawson, D.R.; Dorhout, P.K.; Elliott, C.M.; Martin, C.R.; Parkinson, B.A. Transition metal chelate-fulleride compounds: electrocrystallization of semiconducting [Ru(bpy)3](C60)2. *J. Electrochem. Soc.,* **1993**, *140*, L84-L86.
[http://dx.doi.org/10.1149/1.2221762]

[140] Miller, B.; Rosamilia, J.M. Electrochemical crystallization of tetraphenylphosphonium and tetraphenylarsonium fullerides. *J. Chem. Soc., Faraday Trans.,* **1993**, *89*, 273-276.
[http://dx.doi.org/10.1039/ft9938900273]

[141] Caron, C.; Subramanian, R.; D'Souza, F.; Kim, J.; Kutner, W.; Jones, M.T.; Kadish, K.M. Selective electrosynthesis of dimethylfullerene [(CH3)2C60]: a novel method for the controlled functionalization of fullerenes. *J. Am. Chem. Soc.,* **1993**, *115*, 8505-8506. [http://dx.doi.org/10.1021/ja00071a093]

[142] Allard, E.; Delaunay, J.; Cheng, F.; Cousseau, J.; Ordúna, J.; Garín, J. Novel C60-based building blocks derived from C60($^{2-}$) anion. *Org. Lett.,* **2001**, *3*(22), 3503-3506. [http://dx.doi.org/10.1021/ol0101786] [PMID: 11678693]

[143] Allard, E.; Delaunay, J.; Cousseau, J. C60 $^{2-}$ chemistry: C60 adducts bearing two ester, carbonyl, or alcohol groups. *Org. Lett.,* **2003**, *5*(13), 2239-2242. [http://dx.doi.org/10.1021/ol034545k] [PMID: 12816418]

[144] Fukuzumi, S.; Suenobu, T.; Hirasaka, T.; Arakawa, R.; Kadish, K.M. Formation of C60 adducts with two different alkyl groups *via* combination of electron transfer and SN2 reactions. *J. Am. Chem. Soc.,* **1998**, *120*, 9220-9227. [http://dx.doi.org/10.1021/ja9815430]

[145] Ederle, Y.; Mathis, C. Carbanions on grafted C60 as initiators for anionic polymerization. *Macromolecules,* **1997**, *30*, 4262-4267. [http://dx.doi.org/10.1021/ma9700643]

[146] Wang, Z.; Meier, M.S. Monoalkylation of C60 and C70 with Zn and active alkyl bromides. *J. Org. Chem.,* **2003**, *68*(8), 3043-3048. [http://dx.doi.org/10.1021/jo020563o] [PMID: 12688770]

[147] Reed, C.A.; Bolskar, R.D. Discrete fulleride anions and fullerenium cations. *Chem. Rev.,* **2000**, *100*(3), 1075-1120. [http://dx.doi.org/10.1021/cr980017o] [PMID: 11749258]

[148] Yildirim, T.; Zhou, O.; Fischer, J.E. Intercalation compounds of fullerenes I: Synthesis, characterization and solid state properties. *Physics and Chemistry of Materials with Low-Dimensional Structures,* **2000**, *23*, 23-66. [http://dx.doi.org/10.1007/978-94-011-4038-6_2]

[149] Yildirim, T.; Zhou, O.; Fischer, J.E. Intercalation compounds of fullerenes II: Structure and superconductivity of alkali metal fullerides. *Physics and Chemistry of Materials with Low-Dimensional Structures,* **2000**, *23*, 67-133. [http://dx.doi.org/10.1007/978-94-011-4038-6_3]

[150] Yildirim, T.; Zhou, O.; Fischer, J.E. Intercalation compounds of fullerenes III: Other fullerides and intercalated nanotubes. *Physics and Chemistry of Materials with Low-Dimensional Structures,* **2000**, *23*, 249-289. [http://dx.doi.org/10.1007/978-94-011-4038-6_7]

[151] Fullagar, W.K.; Gentle, I.R.; Heath, G.A.; White, J.W. Reversible alkali-metal reduction of C60 in liquid ammonia; first observation of near-infrared spectrum of C5–60. *J. Chem. Soc. Chem. Commun.,* **1993**, 525-527. [http://dx.doi.org/10.1039/c39930000525]

[152] Margadonna, S.; Prassides, K. Recent advances in fullerene superconductivity. *J. Solid State Chem.,*

2002, *168*, 639-652.
[http://dx.doi.org/10.1006/jssc.2002.9762]

[153] Gunnarsson, O. Superconductivity in fullerides. *Rev. Mod. Phys.,* **1997**, *69*, 575-606.
[http://dx.doi.org/10.1103/RevModPhys.69.575]

[154] Iwasa, Y.; Takenobu, T. Superconductivity, Mott–Hubbard states, and molecular orbital order in intercalated fullerides. *J. Phys. Cond. Mater.,* **2003**, *15*, R495-R520.
[http://dx.doi.org/10.1088/0953-8984/15/13/202]

[155] Chen, Y.; Stepniak, F.; Weaver, J.H.; Chibante, L.P.; Smalley, R.E. Fullerides of alkaline-earth metals. *Phys. Rev. B Condens. Matter,* **1992**, *45*(15), 8845-8848.
[http://dx.doi.org/10.1103/PhysRevB.45.8845] [PMID: 10000744]

[156] Kortan, A.R.; Kopylov, N.; Glarum, S.; Gyorgy, E.M.; Ramirez, A.P.; Fleming, R.M.; Thiel, F.A.; Haddon, R.C. Superconductivity at 8.4 K in calcium-doped C60. *Nature,* **1992**, *355*, 529-531.
[http://dx.doi.org/10.1038/355529a0]

[157] Kortan, A.R.; Kopylov, N.; Glarum, S.; Gyorgy, E.M.; Ramirez, A.P.; Fleming, R.M.; Zhou, O.; Thiel, F.A.; Trevor, P.L.; Haddon, R.C. Superconductivity in barium fulleride. *Nature,* **1992**, *360*, 566-568.
[http://dx.doi.org/10.1038/360566a0]

[158] Kortan, A.R.; Kopylov, N.; Ozdas, E.; Ramirez, A.P.; Fleming, R.M.; Haddon, R.C. Strontium doped fullerite compounds. *Chem. Phys. Lett.,* **1994**, *223*, 501-505.
[http://dx.doi.org/10.1016/0009-2614(94)00495-1]

[159] Bänitz, M.; Heinze, M.; Lüders, K.; Werner, H.; Schlögl, R.; Weiden, M.; Sparn, G.; Steglich, F. Superconductivity of Ba doped C60-susceptibility results and upper critical field. *Solid State Commun.,* **1995**, *96*, 539-544.
[http://dx.doi.org/10.1016/0038-1098(95)00498-X]

[160] Brown, C.M.; Taga, S.; Gogia, B.; Kordatos, K.; Margadonna, S.; Prassides, K.; Iwasa, Y.; Tanigaki, K.; Fitch, A.N.; Pattison, P. Structural and electronic properties of the noncubic superconducting fullerides A′4C60 (A′=Ba,Sr). *Phys. Rev. Lett.,* **1999**, *83*, 2258-2261.
[http://dx.doi.org/10.1103/PhysRevLett.83.2258]

[161] Iwasa, Y.; Hayashi, H.; Furudate, T.; Mitani, T. Superconductivity in K3Ba3C60. *Phys. Rev. B Condens. Matter,* **1996**, *54*(21), 14960-14962.
[http://dx.doi.org/10.1103/PhysRevB.54.14960] [PMID: 9985546]

[162] Iwasa, Y.; Kawaguchi, M.; Iwasaki, H.; Mitani, T.; Wada, N.; Hasegawa, T. Superconducting and normal-state properties of nonavalent fullerides. *Phys. Rev. B,* **1998**, *57*, 13395-13398.
[http://dx.doi.org/10.1103/PhysRevB.57.13395]

[163] Wertheim, G.K.; Buchanan, D.N.; Rowe, J.E. Electronic structure of barium fullerides. *Chem. Phys. Lett.,* **1993**, *206*, 193-196.
[http://dx.doi.org/10.1016/0009-2614(93)85540-5]

[164] Boulas, P.; Subramanian, R.; Kutner, W.; Jones, M.T.; Kadish, K.M. Facile preparation of the C60 monoanion in aprotic solvents. *J. Electrochem. Soc.,* **1993**, *140*, L130-L132.
[http://dx.doi.org/10.1149/1.2220841]

[165] Fagan, P.J.; Krusic, P.J.; Evans, D.H.; Lerke, S.A.; Johnston, E. Synthesis, chemistry, and properties

of a monoalkylated buckminsterfullerene derivative, tert-BuC60 anion. *J. Am. Chem. Soc.,* **1992**, *114*, 9697-9699.
[http://dx.doi.org/10.1021/ja00050a081]

[166] Murata, Y.; Komatsu, K.; Wan, T.S. The reaction of [60]fullerene with lithium fluorenide: Formation of a novel 1,4-adduct of [60]fullerene. *Tetrahedron Lett.,* **1996**, *37*, 7061-7064.
[http://dx.doi.org/10.1016/0040-4039(96)01552-3]

[167] Nagashima, H.; Terasaki, H.; Kimura, E.; Nakajima, K.; Itoh, K. Silylmethylations of C60 with Grignard reagents: selective synthesis of HC60CH2SiMe2Y and C60(CH2SiMe2Y)2 with selection of solvents. *J. Org. Chem.,* **1994**, *59*, 1246-1248.
[http://dx.doi.org/10.1021/jo00085a007]

[168] Nagashima, H.; Terasaki, H.; Saito, Y.; Jinno, K.; Itoh, K. Chlorosilanes and silyl triflates containing C60 as a partial structure. A versatile synthetic entry linking the C60 moieties with alcohols, phenols, and silica. *J. Org. Chem.,* **1995**, *60*, 4966-4967.
[http://dx.doi.org/10.1021/jo00121a004]

[169] Murata, Y.; Motoyama, K.; Komatsu, K.; Wan, T.S. Synthesis, properties, and reactions of a stable carbanion derived from alkynyldihydrofullerene: 1-octynyl-C60 carbanion. *Tetrahedron,* **1996**, *52*, 5077-5090.
[http://dx.doi.org/10.1016/0040-4020(96)00115-9]

[170] Okamura, H.; Murata, Y.; Minoda, M.; Komatsu, K.; Miyamoto, T.; Wan, T.S. Synthesis of disubstituted 1,2-dihydro[60]fullerenes with well-defined structure by addition of 1-alkoxy-1-chloroethanes to 2-(1-octynyl)-1,2-dihydro[60]fulleren-1-ide ion. *J. Org. Chem.,* **1996**, *61*, 8500-8502.
[http://dx.doi.org/10.1021/jo961320x]

[171] Kunieda, R.; Fujitsuka, M.; Ito, O.; Ito, M.; Murata, Y.; Komatsu, K. Photochemical and photophysical properties of C60 dendrimers studied by laser flash photolysis. *J. Phys. Chem. B,* **2002**, *106*, 7193-7199.
[http://dx.doi.org/10.1021/jp0256995]

[172] Murata, Y.; Ito, M.; Komatsu, K. Synthesis and properties of novel fullerene derivatives having dendrimer units and the fullerenyl anions generated therefrom. *J. Mater. Chem.,* **2002**, *12*, 2009-2020.
[http://dx.doi.org/10.1039/b201334g]

[173] Timmerman, P.; Anderson, H.L.; Faust, R.; Nierengarten, J-F.; Habicher, T.; Seiler, P.; Diederich, F. Fullerene-acetylene hybrids: Towards a novel class of molecular carbon allotropes. *Tetrahedron,* **1996**, *52*, 4925-4947.
[http://dx.doi.org/10.1016/0040-4020(96)00102-0]

[174] Seshadri, R.; Govindaraj, A.; Nagarajan, R.; Pradeep, T.; Rao, C.N. Addition of amines and halogens to fullerenes C60 and C70. *Tetrahedron Lett.,* **1992**, *33*, 2069-2070.
[http://dx.doi.org/10.1016/0040-4039(92)88144-T]

[175] Balch, A.L.; Cullison, B.; Fawcett, W.R.; Ginwalla, A.S.; Olmstead, M.M.; Winkler, K. Structural and electrochemical characterization of the products of the C60–piperazine reaction. *J. Chem. Soc. Chem. Commun.,* **1995**, 2287-2288.
[http://dx.doi.org/10.1039/c39950002287]

[176] Schick, G.; Kampe, K-D.; Hirsch, A. Reaction of [60]fullerene with morpholine and piperidine: preferred 1,4-additions and fullerene dimer formation. *J. Chem. Soc. Chem. Commun.,* **1995**, 2023-2024.
[http://dx.doi.org/10.1039/c39950002023]

[177] Skiebe, A.; Hirsch, A.; Klos, H.; Gotschy, B. [DBU] C60. Spin pairing in a fullerene salt. *Chem. Phys. Lett.,* **1994**, *220*, 138-140.
[http://dx.doi.org/10.1016/0009-2614(94)00124-3]

[178] Kampe, K-D.; Egger, N.; Vogel, M. Diamino and tetraamino derivatives of Buckminsterfullerene C60. *Angew. Chem. Int. Ed. Engl.,* **1993**, *32*, 1174-1176.
[http://dx.doi.org/10.1002/anie.199311741]

[179] *Physical and Chemical Properties of Carbon Nanotubes*; Suzuki, S, Ed.; InTechOpen: Rijeka, Croatia, **2013**.

[180] Strano, M.S.; Dyke, C.A.; Usrey, M.L.; Barone, P.W.; Allen, M.J.; Shan, H.; Kittrell, C.; Hauge, R.H.; Tour, J.M.; Smalley, R.E. Electronic structure control of single-walled carbon nanotube functionalization. *Science,* **2003**, *301*(5639), 1519-1522.
[http://dx.doi.org/10.1126/science.1087691] [PMID: 12970561]

[181] Li, H.; Zhou, B.; Lin, Y.; Gu, L.; Wang, W.; Fernando, K.A.; Kumar, S.; Allard, L.F.; Sun, Y-P. Selective interactions of porphyrins with semiconducting single-walled carbon nanotubes. *J. Am. Chem. Soc.,* **2004**, *126*(4), 1014-1015.
[http://dx.doi.org/10.1021/ja037142o] [PMID: 14746457]

[182] Niyogi, S.; Hamon, M.A.; Hu, H.; Zhao, B.; Bhowmik, P.; Sen, R.; Itkis, M.E.; Haddon, R.C. Chemistry of single-walled carbon nanotubes. *Acc. Chem. Res.,* **2002**, *35*(12), 1105-1113.
[http://dx.doi.org/10.1021/ar010155r] [PMID: 12484799]

[183] Tasis, D.; Tagmatarchis, N.; Georgakilas, V.; Prato, M. Soluble carbon nanotubes. *Chemistry,* **2003**, *9*(17), 4000-4008.
[http://dx.doi.org/10.1002/chem.200304800] [PMID: 12953186]

[184] Hirsch, A. Functionalization of single-walled carbon nanotubes. *Angew. Chem. Int. Ed. Engl.,* **2002**, *41*(11), 1853-1859.
[http://dx.doi.org/10.1002/1521-3773(20020603)41:11<1853::AID-ANIE1853>3.0.CO;2-N] [PMID: 19750614]

[185] Bahr, J.L.; Tour, J.M. Covalent chemistry of single-wall carbon nanotubes. *J. Mater. Chem.,* **2002**, *12*, 1952-1958.
[http://dx.doi.org/10.1039/b201013p]

[186] Dyke, C.A.; Tour, J.M. Overcoming the insolubility of carbon nanotubes through high degrees of sidewall functionalization. *Chemistry,* **2004**, *10*(4), 812-817.
[http://dx.doi.org/10.1002/chem.200305534] [PMID: 14978808]

[187] Dyke, C.A.; Tour, J.M. Covalent functionalization of single-walled carbon nanotubes for materials applications. *J. Phys. Chem. A,* **2004**, *108*, 11151-11159.
[http://dx.doi.org/10.1021/jp046274g]

[188] *TechOpen*; Marulanda, J.M, Ed.; Croatia: Rijeka, **2010**.

[189] Sun, Y-P.; Fu, K.; Lin, Y.; Huang, W. Functionalized carbon nanotubes: properties and applications. *Acc. Chem. Res.,* **2002**, *35*(12), 1096-1104.
[http://dx.doi.org/10.1021/ar010160v] [PMID: 12484798]

[190] Sinnott, S.B. Chemical functionalization of carbon nanotubes. *J. Nanosci. Nanotechnol.,* **2002**, *2*(2), 113-123.
[http://dx.doi.org/10.1166/jnn.2002.107] [PMID: 12908295]

[191] Unger, E.; Graham, A.; Kreupl, F.; Liebau, M.; Hoenlein, W. Electrochemical functionalization of multi-walled carbon nanotubes for solvation and purification. *Curr. Appl. Phys.,* **2002**, *2*, 107-111.
[http://dx.doi.org/10.1016/S1567-1739(01)00072-4]

[192] Prato, M.; Kostarelos, K.; Bianco, A.; Bianco, A. Functionalized carbon nanotubes in drug design and discovery. *Acc. Chem. Res.,* **2008**, *41*(1), 60-68.
[http://dx.doi.org/10.1021/ar700089b] [PMID: 17867649]

[193] Mickelson, E.T.; Huffman, C.B.; Rinzler, A.G.; Smalley, R.E.; Hauge, R.H.; Margrave, J.L. Fluorination of single-wall carbon nanotubes. *Chem. Phys. Lett.,* **1998**, *296*, 188-194.
[http://dx.doi.org/10.1016/S0009-2614(98)01026-4]

[194] Kelly, K.F.; Chiang, I.W.; Mickelson, E.T.; Hauge, R.H.; Margrave, J.L.; Wang, X.; Scuseria, G.E.; Radloff, C.; Halas, N.J. Insight into the mechanism of sidewall functionalization of single-walled nanotubes: an STM study. *Chem. Phys. Lett.,* **1993**, *313*, 445-450.
[http://dx.doi.org/10.1016/S0009-2614(99)00973-2]

[195] Touhara, H.; Inahara, J.; Mizuno, T.; Yokoyama, Y.; Okanao, S.; Yanagiuch, K.; Mukopadhyay, I.; Kawasaki, S.; Okino, F.; Shirai, H.; Xu, W.H.; Kyotani, T.; Tomita, A. Fluorination of cup-stacked carbon nanotubes, structure and properties. *Fluorine Chem.,* **2002**, *114*, 181-188.
[http://dx.doi.org/10.1016/S0022-1139(02)00026-X]

[196] Dettlaff-Weglikowska, U.; Skakalova, V.; Meyer, J.; Cech, J.; Mueller, B.G.; Roth, S. Effect of fluorination on electrical properties of single walled carbon nanotubes and C60 peapods in networks. *Curr. Appl. Phys.,* **2007**, *7*, 42-46.
[http://dx.doi.org/10.1016/j.cap.2005.08.003]

[197] Valentini, L.; Puglia, D.; Carniato, F.; Boccaleri, E.; Marchese, L.; Kenny, J.M. Use of plasma fluorinated single-walled carbon nanotubes for the preparation of nanocomposites with epoxy matrix. *Comp. Sci. Tech,* **2008**, *68*, 1008-1014.
[http://dx.doi.org/10.1016/j.compscitech.2007.07.011]

[198] Khabashesku, V.N.; Billups, W.E.; Margrave, J.L. Fluorination of single-wall carbon nanotubes and subsequent derivatization reactions. *Acc. Chem. Res.,* **2002**, *35*(12), 1087-1095.
[http://dx.doi.org/10.1021/ar020146y] [PMID: 12484797]

[199] Stevens, J.L.; Huang, A.Y.; Peng, H.; Chiang, I.W.; Khabashesku, V.N.; Margrave, J.L. Sidewall amino-functionalization of SWNTs through fluorination and subsequent reactions with terminal diamines. *Nano Lett.,* **2003**, *3*, 331.
[http://dx.doi.org/10.1021/nl025944w]

[200] Boul, P.J.; Liu, J.; Mickelson, E.T.; Huffman, C.B.; Ericson, L.M.; Chiang, I.W.; Smith, K.A.; Colbert, D.T.; Hauge, R.H.; Margrave, J.L.; Smalley, R.E. Reversible sidewall functionalization of buckytubes.

Chem. Phys. Lett., **1999**, *310*, 367-372.
[http://dx.doi.org/10.1016/S0009-2614(99)00713-7]

[201] Saini, R.K.; Chiang, I.W.; Peng, H.; Smalley, R.E.; Billups, W.E.; Hauge, R.H.; Margrave, J.L. Covalent sidewall functionalization of single wall carbon nanotubes. *J. Am. Chem. Soc.,* **2003**, *125*(12), 3617-3621.
[http://dx.doi.org/10.1021/ja021167q] [PMID: 12643724]

[202] Pekker, S.; Salvetat, J-P.; Jakab, E.; Bonard, J-M.; Forro, L. Hydrogenation of carbon nanotubes and graphite in liquid ammonia. *J. Phys. Chem. B,* **2001**, *105*, 7938-7943.
[http://dx.doi.org/10.1021/jp010642o]

[203] Khare, B.N.; Meyyappan, M.; Cassell, A.M.; Nguyen, C.V.; Han, J. Functionalization of carbon nanotubes using atomic hydrogen from a glow discharge. *Nano Lett.,* **2002**, *2*, 73-77.
[http://dx.doi.org/10.1021/nl015646j]

[204] Khare, B.N.; Meyyappan, M.; Kralj, J.; Wilhite, Pv; Sisay, M.; Imanaka, H.; Koehne, J.; Baushchlicher, C.W. A glow-discharge approach for functionalization of carbon nanotubes. *Appl. Phys. Lett.,* **2002**, *81*, 5237-5239.
[http://dx.doi.org/10.1063/1.1533859]

[205] Kim, K.S.; Bae, D.J.; Kim, J.R.; Park, K.A.; Lim, S.C.; Kim, J-J.; Choi, W.B.; Park, C.Y.; Lee, Y.H. Modification of electronic structures of a carbon nanotube by hydrogen functionalization. *Adv. Mater.,* **2002**, *14*, 1818-1821.
[http://dx.doi.org/10.1002/adma.200290008]

[206] Khare, B.; Meyyappan, M.; Moore, M.H.; Wilhite, P.; Imanaka, H.; Chen, B. Proton irradiation of carbon nanotubes. *Nano Lett.,* **2003**, *3*, 643-646.
[http://dx.doi.org/10.1021/nl034058y]

[207] Chen, J.; Hamon, M.A.; Hu, H.; Chen, Y.; Rao, A.M.; Eklund, P.C.; Haddon, R.C. Solution properties of single-walled carbon nanotubes. *Science,* **1998**, *282*(5386), 95-98.
[http://dx.doi.org/10.1126/science.282.5386.95] [PMID: 9756485]

[208] Lee, W.H.; Kim, S.J.; Lee, W.J.; Lee, J.G.; Haddon, R.C.; Reucroft, P.J. X-ray photoelectron spectroscopic studies of surface modified single-walled carbon nanotube material. *Appl. Surf. Sci.,* **2001**, *181*, 121-127.
[http://dx.doi.org/10.1016/S0169-4332(01)00381-6]

[209] Kamaras, K.; Itkis, M.E.; Hu, H.; Zhao, B.; Haddon, R.C. Covalent bond formation to a carbon nanotube metal. *Science,* **2003**, *301*(5639), 1501-1501.
[http://dx.doi.org/10.1126/science.1088083] [PMID: 12970557]

[210] Hu, H.; Zhao, B.; Hamon, M.A.; Kamaras, K.; Itkis, M.E.; Haddon, R.C. Sidewall functionalization of single-walled carbon nanotubes by addition of dichlorocarbene. *J. Am. Chem. Soc.,* **2003**, *125*(48), 14893-14900.
[http://dx.doi.org/10.1021/ja0356737] [PMID: 14640666]

[211] Holzinger, M.; Vostrowsky, O.; Hirsch, A.; Hennrich, F.; Kappes, M.; Weiss, R.; Jellen, F. Sidewall functionalization of carbon nanotubes. *Angew. Chem. Int. Ed. Engl.,* **2001**, *40*(21), 4002-4005.
[http://dx.doi.org/10.1002/1521-3773(20011105)40:21<4002::AID-ANIE4002>3.0.CO;2-8] [PMID: 12404474]

[212] Holzinger, M.; Abraham, J.; Whelan, P.; Graupner, R.; Ley, L.; Hennrich, F.; Kappes, M.; Hirsch, A. Functionalization of single-walled carbon nanotubes with (R-)oxycarbonyl nitrenes. *J. Am. Chem. Soc.,* **2003**, *125*(28), 8566-8580.
[http://dx.doi.org/10.1021/ja029931w] [PMID: 12848565]

[213] Holzinger, M.; Steinmetz, J.; Samaille, D.; Glerup, M.; Paillet, M.; Bernier, P.; Ley, L.; Graupner, R. [2+1] cycloaddition for cross-linking SWCNTs. *Carbon,* **2004**, *42*, 941-947.
[http://dx.doi.org/10.1016/j.carbon.2003.12.019]

[214] Moghaddam, M.J.; Taylor, S.; Gao, M.; Huang, S.; Dai, L.; McCall, M.J. Highly efficient binding of DNA on the sidewalls and tips of carbon nanotubes using photochemistry. *Nano Lett.,* **2004**, *4*, 89-93.
[http://dx.doi.org/10.1021/nl034915y]

[215] Georgakilas, V.; Kordatos, K.; Prato, M.; Guldi, D.M.; Holzinger, M.; Hirsch, A. Organic functionalization of carbon nanotubes. *J. Am. Chem. Soc.,* **2002**, *124*(5), 760-761.
[http://dx.doi.org/10.1021/ja016954m] [PMID: 11817945]

[216] Tagmatarchis, N.; Prato, M.J. Functionalization of carbon nanotubes *via* 1,3-dipolar cycloadditions. *J. Mater. Chem.,* **2004**, *14*, 437-439.
[http://dx.doi.org/10.1039/b314039c]

[217] Yang, B.X.; Shi, J.H.; Pramoda, K.P.; Goh, S.H. Enhancement of stiffness, strength, ductility and toughness of poly(ethylene oxide) using phenoxy-grafted multiwalled carbon nanotubes. *Nanotechnology,* **2007**, *18*, 125606.
[http://dx.doi.org/10.1088/0957-4484/18/12/125606]

[218] Goh, H.W.; Goh, S.H.; Xu, G.Q.; Pramoda, K.P.; Zhang, W.D. Dynamic mechanical behavior of in situ functionalized multi-walled carbon nanotube/phenoxy resin composite. *Chem. Phys. Lett.,* **2003**, *373*, 277-283.
[http://dx.doi.org/10.1016/S0009-2614(03)00621-3]

[219] Ramanathan, T.; Fisher, F.T.; Ruoff, R.S.; Brinson, L.C. Amino-functionalized carbon nanotubes for binding to polymers and biological systems. *Chem. Mater.,* **2005**, *17*, 1290-1295.
[http://dx.doi.org/10.1021/cm048357f]

[220] Esumi, K.; Ishigami, M.; Nakajima, A.; Sawada, K.; Honda, H. Chemical treatment of carbon nanotubes. *Carbon,* **1996**, *34*, 279-281.
[http://dx.doi.org/10.1016/0008-6223(96)83349-5]

[221] Mapkar, J.A.; Iyer, G.; Coleman, M.R. Functionalization of carbon nanofibers with elastomeric block copolymer using carbodiimide chemistry. *Appl. Surf. Sci.,* **2009**, *255*, 4806-4813.
[http://dx.doi.org/10.1016/j.apsusc.2008.11.068]

[222] Wang, Z.; Liu, Q.; Zhu, H.; Liu, H.; Chen, Y.; Yang, M. Dispersing multi-walled carbon nanotubes with water–soluble block copolymers and their use as supports for metal nanoparticles. *Carbon,* **2007**, *45*, 285-292.
[http://dx.doi.org/10.1016/j.carbon.2006.09.025]

[223] Ivanov, V.; Fonseca, A.; Nagy, J.B.; Lucas, A.; Lambin, P.; Bernaerts, D.; Zhang, X.B. Catalytic production and purification of nanotubules having fullerene-scale diameters. *Carbon,* **1995**, *33*, 1727-1738.

[http://dx.doi.org/10.1016/0008-6223(95)00132-1]

[224] Liu, J.; Rinzler, A.G.; Dai, H.; Hafner, J.H.; Bradley, R.K.; Boul, P.J.; Lu, A.; Iverson, T.; Shelimov, K.; Huffman, C.B.; Rodriguez-Macias, F.; Shon, Y.S.; Lee, T.R.; Colbert, D.T.; Smalley, R.E. Fullerene pipes. *Science,* **1998**, *280*(5367), 1253-1256. [http://dx.doi.org/10.1126/science.280.5367.1253] [PMID: 9596576]

[225] Sun, Y-P.; Huang, W.; Lin, Y.; Fu, K.; Kitaygorodskiy, A.; Riddle, L.A.; Yu, Y.J.; Carroll, L. Soluble dendron-functionalized carbon nanotubes: preparation, characterization, and properties. *Chem. Mater.,* **2001**, *13*, 2864-2869. [http://dx.doi.org/10.1021/cm010069l]

[226] Sun, Y-P.; Zhou, B.; Henbest, K.; Fu, K.F.; Huang, W.J.; Lin, Y.; Taylor, S.; Carroll, D.L. Luminescence anisotropy of functionalized carbon nanotubes in solution. *Chem. Phys. Lett.,* **2002**, *351*, 349-353. [http://dx.doi.org/10.1016/S0009-2614(01)01404-X]

[227] Lin, Y.; Taylor, S.; Huang, W.; Sun, Y-P. Characterization of fractions from repeated functionalization reactions of carbon nanotubes. *J. Phys. Chem. B,* **2003**, *107*, 914-919. [http://dx.doi.org/10.1021/jp020701d]

[228] Fu, K.; Li, H.; Zhou, B.; Kitaygorodskiy, A.; Allard, L.F.; Sun, Y-P. Deuterium attachment to carbon nanotubes in deuterated water. *J. Am. Chem. Soc.,* **2004**, *126*(14), 4669-4675. [http://dx.doi.org/10.1021/ja030312c] [PMID: 15070384]

[229] Martin, R.B.; Qu, L.; Lin, Y.; Harruff, B.A.; Bunker, C.E.; Gord, J.R.; Allard, L.F.; Sun, Y-P. Functionalized carbon nanotubes with tethered pyrenes: synthesis and photophysical properties. *J. Phys. Chem. B,* **2004**, *108*, 11447-11453. [http://dx.doi.org/10.1021/jp048984s]

[230] Fu, K.; Huang, W.; Lin, Y.; Riddle, L.A.; Carroll, D.L.; Sun, Y-P. Defunctionalization of functionalized carbon nanotubes. *Nano Lett.,* **2001**, *1*, 439-441. [http://dx.doi.org/10.1021/nl010040g]

[231] Hamon, M.A.; Hui, H.; Bhowmik, P. Ester-functionalized soluble single-walled carbon nanotubes. *Appl. Phys., A Mater. Sci. Process.,* **2002**, *74*, 333-338. [http://dx.doi.org/10.1007/s003390201281]

[232] Sano, M.; Kamino, A.; Okamura, J.; Shinkai, S. Self-organization of PEO-graft-single-walled carbon nanotubes in solutions and Langmuir–Blodgett films. *Langmuir,* **2001**, *17*, 5125-5128. [http://dx.doi.org/10.1021/la010126p]

[233] Kong, H.; Gao, C.; Yan, D. Controlled functionalization of multiwalled carbon nanotubes by in situ atom transfer radical polymerization. *J. Am. Chem. Soc.,* **2004**, *126*(2), 412-413. [http://dx.doi.org/10.1021/ja0380493] [PMID: 14719907]

[234] Chen, R.J.; Zhang, Y.; Wang, D.; Dai, H. Noncovalent sidewall functionalization of single-walled carbon nanotubes for protein immobilization. *J. Am. Chem. Soc.,* **2001**, *123*(16), 3838-3839. [http://dx.doi.org/10.1021/ja010172b] [PMID: 11457124]

[235] Hecht, D.S.; Ramirez, R.J.; Briman, M.; Artukovic, E.; Chichak, K.S.; Stoddart, J.F.; Grüner, G. Bioinspired detection of light using a porphyrin-sensitized single-wall nanotube field effect transistor.

Nano Lett., **2006**, *6*(9), 2031-2036.
[http://dx.doi.org/10.1021/nl061231s] [PMID: 16968021]

[236] Hu, L.; Zhao, Y-L.; Ryu, K.; Zhou, C.; Stoddart, J.F.; Gruner, G. Light-induced charge transfer in pyrene/CdSe-SWNT hybrids. *Adv. Mater.,* **2008**, *20*, 939-946.
[http://dx.doi.org/10.1002/adma.200701125]

[237] Vaisman, L.; Marom, G.; Wagner, H.D. Dispersions of surface-modified carbon nanotubes in water-soluble and water-insoluble polymers. *Adv. Funct. Mater.,* **2006**, *16*, 357-363.
[http://dx.doi.org/10.1002/adfm.200500142]

[238] Islam, M.F.; Rojas, E.; Bergey, D.M.; Johnson, A.T.; Yodh, A.G. High weight fraction surfactant solubilization of single-wall Carbon nanotubes in water. *Nano Lett.,* **2003**, *3*, 269-273.
[http://dx.doi.org/10.1021/nl025924u]

[239] Yu, J.; Grossiord, N.; Koning, C.E.; Loos, J. Controlling the dispersion of multi-wall carbon nanotubes in aqueous surfactant solution. *Carbon,* **2007**, *45*, 618-623.
[http://dx.doi.org/10.1016/j.carbon.2006.10.010]

[240] Whitsitt, E.A.; Barron, A.R. Silica coated single walled carbon nanotubes. *Nano Lett.,* **2003**, *3*, 775-778.
[http://dx.doi.org/10.1021/nl034186m]

[241] Kim, T.H.; Doe, C.; Kline, S.R.; Choi, S.M. Water-redispersible isolated single-walled carbon nanotubes fabricated by in situ polymerization of micelles. *Adv. Mater.,* **2007**, *19*, 929-933.
[http://dx.doi.org/10.1002/adma.200601991]

[242] Klumpp, C.; Kostarelos, K.; Prato, M.; Bianco, A. **2006**.

[243] Kuila, T.; Bose, S.; Mishra, A.K.; Khanra, P.; Kim, N.H.; Lee, J.H. Chemical functionalization of graphene and its applications. *Prog. Mater. Sci.,* **2012**, *57*, 1061-1105.
[http://dx.doi.org/10.1016/j.pmatsci.2012.03.002]

[244] Hirsch, A.; Englert, J.M.; Hauke, F. Wet chemical functionalization of graphene. *Acc. Chem. Res.,* **2013**, *46*(1), 87-96.
[http://dx.doi.org/10.1021/ar300116q] [PMID: 22946482]

[245] Hummers, W.S.; Offeman, R.E. Preparation of graphitic oxide. *J. Am. Chem. Soc.,* **1958**, *80*, 1339-1339.
[http://dx.doi.org/10.1021/ja01539a017]

[246] Park, S.; Ruoff, R.S. Chemical methods for the production of graphenes. *Nat. Nanotechnol.,* **2009**, *4*(4), 217-224.
[http://dx.doi.org/10.1038/nnano.2009.58] [PMID: 19350030]

[247] Eda, G.; Fanchini, G.; Chhowalla, M. Large-area ultrathin films of reduced graphene oxide as a transparent and flexible electronic material. *Nat. Nanotechnol.,* **2008**, *3*(5), 270-274.
[http://dx.doi.org/10.1038/nnano.2008.83] [PMID: 18654522]

[248] Ganguly, A.; Sharma, S.; Papakonstantinou, P.; Hamilton, J. Probing the thermal deoxygenation of graphene oxide using high-resolution in situ X-ray-based spectroscopies. *J. Phys. Chem. C,* **2011**, *115*, 17009-17019.
[http://dx.doi.org/10.1021/jp203741y]

[249] Dreyer, D.R.; Park, S.; Bielawski, C.W.; Ruoff, R.S. The chemistry of graphene oxide. *Chem. Soc. Rev.,* **2010**, *39*(1), 228-240.
[http://dx.doi.org/10.1039/B917103G] [PMID: 20023850]

[250] Stankovich, S.; Dikin, D.A.; Dommett, G.H.; Kohlhaas, K.M.; Zimney, E.J.; Stach, E.A.; Piner, R.D.; Nguyen, S.T.; Ruoff, R.S. Graphene-based composite materials. *Nature,* **2006**, *442*(7100), 282-286.
[http://dx.doi.org/10.1038/nature04969] [PMID: 16855586]

[251] Dikin, D.A.; Stankovich, S.; Zimney, E.J.; Piner, R.D.; Dommett, G.H.; Evmenenko, G.; Nguyen, S.T.; Ruoff, R.S. Preparation and characterization of graphene oxide paper. *Nature,* **2007**, *448*(7152), 457-460.
[http://dx.doi.org/10.1038/nature06016] [PMID: 17653188]

[252] Paredes, J.I.; Villar-Rodil, S.; Martínez-Alonso, A.; Tascón, J.M. Graphene oxide dispersions in organic solvents. *Langmuir,* **2008**, *24*(19), 10560-10564.
[http://dx.doi.org/10.1021/la801744a] [PMID: 18759411]

[253] Lerf, A.; He, H.; Forster, M.; Klinowski, J. Structure of graphite oxide revisited. *J. Phys. Chem. B,* **1998**, *102*, 4477-4482.
[http://dx.doi.org/10.1021/jp9731821]

[254] Gao, W.; Alemany, L.B.; Ci, L.; Ajayan, P.M. New insights into the structure and reduction of graphite oxide. *Nat. Chem.,* **2009**, *1*(5), 403-408.
[http://dx.doi.org/10.1038/nchem.281] [PMID: 21378895]

[255] Nethravathi, C.; Rajamathi, J.T.; Ravishankar, N.; Shivakumara, C.; Rajamathi, M. Graphite oxide-intercalated anionic clay and its decomposition to graphene-inorganic material nanocomposites. *Langmuir,* **2008**, *24*(15), 8240-8244.
[http://dx.doi.org/10.1021/la8000027] [PMID: 18597500]

[256] Szabo, T.; Szeri, A.; Dekany, I. Composite graphitic nanolayers prepared by self-assembly between finely dispersed graphite oxide and a cationic polymer. *Carbon,* **2005**, *43*, 87-94.
[http://dx.doi.org/10.1016/j.carbon.2004.08.025]

[257] Szabo, T.; Berkesi, O.; Forgo, P.; Josepovits, K.; Sanakis, Y.; Petridis, D.; Dékány, I. Evolution of surface functional groups in a series of progressively oxidized graphite oxides. *Chem. Mater.,* **2006**, *18*, 2740-2749.
[http://dx.doi.org/10.1021/cm060258+]

[258] Choi, J.; Kim, K.J.; Kim, B.; Lee, H.; Kim, S. Covalent functionalization of epitaxial graphene by azidotrimethylsilane. *J. Phys. Chem. C,* **2009**, *113*, 9433-9435.
[http://dx.doi.org/10.1021/jp9010444]

[259] Georgakilas, V.; Bourlinos, A.B.; Zboril, R.; Steriotis, T.A.; Dallas, P.; Stubos, A.K.; Trapalis, C. Organic functionalisation of graphenes. *Chem. Commun. (Camb.),* **2010**, *46*(10), 1766-1768.
[http://dx.doi.org/10.1039/b922081j] [PMID: 20177643]

[260] Quintana, M.; Spyrou, K.; Grzelczak, M.; Browne, W.R.; Rudolf, P.; Prato, M. Functionalization of graphene *via* 1,3-dipolar cycloaddition. *ACS Nano,* **2010**, *4*(6), 3527-3533.
[http://dx.doi.org/10.1021/nn100883p] [PMID: 20503982]

[261] Economopoulos, S.P.; Rotas, G.; Miyata, Y.; Shinohara, H.; Tagmatarchis, N. Exfoliation and

chemical modification using microwave irradiation affording highly functionalized graphene. *ACS Nano,* **2010**, *4*(12), 7499-7507.
[http://dx.doi.org/10.1021/nn101735e] [PMID: 21080708]

[262] Vadukumpully, S.; Gupta, J.; Zhang, Y.; Xu, G.Q.; Valiyaveettil, S. Functionalization of surfactant wrapped graphene nanosheets with alkylazides for enhanced dispersibility. *Nanoscale,* **2011**, *3*(1), 303-308.
[http://dx.doi.org/10.1039/C0NR00547A] [PMID: 21052576]

[263] Xu, X.; Luo, Q.; Lv, W.; Dong, Y.; Lin, Y.; Yang, Q.; Shen, A.; Pang, D.; Hu, J.; Qin, J.; Li, Z. Functionalization of graphene sheets by polyacetylene: convenient synthesis and enhanced emission. *Macromol. Chem. Phys.,* **2011**, *212*, 768-773.
[http://dx.doi.org/10.1002/macp.201000608]

[264] He, H.; Gao, C. General approach to individually dispersed, highly soluble, and conductive graphene nanosheets functionalized by nitrene chemistry. *Chem. Mater.,* **2010**, *22*, 5054-5064.
[http://dx.doi.org/10.1021/cm101634k]

[265] Zhong, X.; Jin, J.; Li, S.; Niu, Z.; Hu, W.; Li, R.; Ma, J. Aryne cycloaddition: highly efficient chemical modification of graphene. *Chem. Commun. (Camb.),* **2010**, *46*(39), 7340-7342.
[http://dx.doi.org/10.1039/c0cc02389b] [PMID: 20820532]

[266] Bekyarova, E.; Itkis, M.E.; Ramesh, P.; Berger, C.; Sprinkle, M.; de Heer, W.A.; Haddon, R.C. Chemical modification of epitaxial graphene: spontaneous grafting of aryl groups. *J. Am. Chem. Soc.,* **2009**, *131*(4), 1336-1337.
[http://dx.doi.org/10.1021/ja8057327] [PMID: 19173656]

[267] Lomeda, J.R.; Doyle, C.D.; Kosynkin, D.V.; Hwang, W.F.; Tour, J.M. Diazonium functionalization of surfactant-wrapped chemically converted graphene sheets. *J. Am. Chem. Soc.,* **2008**, *130*(48), 16201-16206.
[http://dx.doi.org/10.1021/ja806499w] [PMID: 18998637]

[268] Zhu, Y.; Higginbotham, A.L.; Tour, J.M. Covalent functionalization of surfactant-wrapped graphene nanoribbons. *Chem. Mater.,* **2009**, *21*, 5284-5291.
[http://dx.doi.org/10.1021/cm902939n]

[269] Sun, Z.; Kohama, S.I.; Zhang, Z.; Lomeda, J.R.; Tour, J.M. Soluble graphene through edge-selective functionalization. *Nano Res.,* **2010**, *3*, 117-125.
[http://dx.doi.org/10.1007/s12274-010-1016-2]

[270] Si, Y.; Samulski, E.T. Synthesis of water soluble graphene. *Nano Lett.,* **2008**, *8*(6), 1679-1682.
[http://dx.doi.org/10.1021/nl080604h] [PMID: 18498200]

[271] Fang, M.; Wang, K.; Lu, H.; Yang, Y.; Nutt, S. Covalent polymer functionalization of graphenenanosheets and mechanical properties of composites. *J. Mater. Chem.,* **2009**, *19*, 7098-7105.
[http://dx.doi.org/10.1039/b908220d]

[272] Stankovich, S.; Piner, R.D.; Nguyen, S.T.; Ruoff, R.S. Synthesis and exfoliation of isocyanate-treated graphene oxide nanoplatelets. *Carbon,* **2006**, *44*, 3342-3347.
[http://dx.doi.org/10.1016/j.carbon.2006.06.004]

[273] Zhang, D.D.; Zua, S.Z.; Hana, B.H. Inorganic–organic hybrid porous materials based on graphite

oxide sheets. *Carbon,* **2009**, *47*, 2993-3000.
[http://dx.doi.org/10.1016/j.carbon.2009.06.052]

[274] Niyogi, S.; Bekyarova, E.; Itkis, M.E.; McWilliams, J.L.; Hamon, M.A.; Haddon, R.C. Solution properties of graphite and graphene. *J. Am. Chem. Soc.,* **2006**, *128*(24), 7720-7721.
[http://dx.doi.org/10.1021/ja060680r] [PMID: 16771469]

[275] Liu, Z.; Robinson, J.T.; Sun, X.; Dai, H. PEGylated nanographene oxide for delivery of water-insoluble cancer drugs. *J. Am. Chem. Soc.,* **2008**, *130*(33), 10876-10877.
[http://dx.doi.org/10.1021/ja803688x] [PMID: 18661992]

[276] Hu, H.; Wang, X.; Wang, J.; Liu, F.; Zhang, M.; Xu, C. Microwave-assisted covalent modification of graphene nanosheets with chitosan and its electrorheological characteristics. *Appl. Surf. Sci.,* **2011**, *257*, 2637-2642.
[http://dx.doi.org/10.1016/j.apsusc.2010.10.035]

[277] Rana, V.K.; Choi, M.C.; Kong, J.Y.; Kim, G.Y.; Kim, M.J.; Kim, S.H.; Mishra, S.; Singh, R.P.; Ha, C-S. Synthesis and drug-delivery behavior of chitosanfunctionalized graphene oxide hybrid nanosheets. *Macromol. Mater. Eng.,* **2011**, *296*, 131-140.
[http://dx.doi.org/10.1002/mame.201000307]

[278] Bao, H.; Pan, Y.; Ping, Y.; Sahoo, N.G.; Wu, T.; Li, L.; Li, J.; Gan, L.H. Chitosan-functionalized graphene oxide as a nanocarrier for drug and gene delivery. *Small,* **2011**, *7*(11), 1569-1578.
[http://dx.doi.org/10.1002/smll.201100191] [PMID: 21538871]

[279] Xu, C.; Wang, X.; Wang, J.; Hu, H.; Wan, L. Synthesis and photoelectrical properties of b-cyclodextrin functionalized graphene materials with high bio-recognition capability. *Chem. Phys. Lett.,* **2010**, *498*, 162-167.
[http://dx.doi.org/10.1016/j.cplett.2010.08.060]

[280] Guo, Y.; Guo, S.; Ren, J.; Zhai, Y.; Dong, S.; Wang, E. Cyclodextrin functionalized graphene nanosheets with high supramolecular recognition capability: synthesis and host-guest inclusion for enhanced electrochemical performance. *ACS Nano,* **2010**, *4*(7), 4001-4010.
[http://dx.doi.org/10.1021/nn100939n] [PMID: 20583782]

[281] Salavagione, H.J.; Gomez, M.A.; Martinez, G. Polymeric modification of graphene through esterification of graphite oxide and poly(vinyl alcohol). *Macromolecules,* **2009**, *42*, 6331-6334.
[http://dx.doi.org/10.1021/ma900845w]

[282] Xu, Y.; Liu, Z.; Zhang, X.; Wang, Y.; Tian, J.; Huang, Y.; Ma, Y.; Zhang, X.; Chen, Y. A graphene hybrid material covalently functionalized with porphyrin: synthesis and optical limiting property. *Adv. Mater.,* **2009**, *21*, 1275-1279.
[http://dx.doi.org/10.1002/adma.200801617]

[283] Zhu, J.; Li, Y.; Chen, Y. Graphene oxide covalently functionalized with zinc phthalocyanine for broadband optical limiting. *Carbon,* **2011**, *49*, 1900-1905.
[http://dx.doi.org/10.1016/j.carbon.2011.01.014]

[284] Liu, Y.; Zhou, J.; Zhang, X. Synthesis, characterization and optical limiting property of covalently oligothiophene-functionalized graphene material. *Carbon,* **2009**, *47*, 3113-3121.
[http://dx.doi.org/10.1016/j.carbon.2009.07.027]

[285] Zhang, B.; Chen, Y.; Zhuang, X. Poly(N-vinylcarbazole) chemically modified graphene oxide. *J. Polym. Sci. A Polym. Chem.,* **2010**, *48*, 2642-2649.
[http://dx.doi.org/10.1002/pola.24047]

[286] Tang, X.; Li, W.; Yu, Z. Enhanced thermal stability in graphene oxide covalently functionalized with 2-amino-4,6-didodecylamino-1,3,5-triazine. *Carbon,* **2011**, *49*, 1258-1265.
[http://dx.doi.org/10.1016/j.carbon.2010.11.044]

[287] Bourlinos, A.B.; Gournis, D.; Petridis, D.; Szabo, T.; Szeri, A.; Dekany, I. Graphite oxide: chemical reduction to graphite and surface modification with primary aliphatic amines and amino acids. *Langmuir,* **2003**, *19*, 6050-6055.
[http://dx.doi.org/10.1021/la026525h]

[288] Matsuo, Y.; Miyabe, T.; Fukutsuka, T.; Sugie, Y. Preparation and characterization of alkylamine-intercalated graphite oxides. *Carbon,* **2007**, *45*, 1005-1012.
[http://dx.doi.org/10.1016/j.carbon.2006.12.023]

[289] Wang, G.; Shen, X.; Wang, B.; Yao, J.; Park, J. Synthesis and characterization of hydrophilic and organophilic graphene nanosheets. *Carbon,* **2009**, *47*, 1359-1364.
[http://dx.doi.org/10.1016/j.carbon.2009.01.027]

[290] Kuila, T.; Bose, S.; Hong, C.E.; Uddin, M.E.; Khanra, P.; Kim, N.H.; Lee, J.H. Preparation of functionalized graphene/linear low density polyethylene composites by a solution mixing method. *Carbon,* **2011**, *49*, 1033-1037.
[http://dx.doi.org/10.1016/j.carbon.2010.10.031]

[291] Yang, H.; Li, F.; Shan, C.; Han, D.; Zhang, Q.; Niu, L.; Ivaska, A. Covalent functionalization of chemically converted graphene sheets *via* silane and its reinforcement. *J. Mater. Chem.,* **2009**, *19*, 4632-4638.
[http://dx.doi.org/10.1039/b901421g]

[292] Yang, H.; Shan, C.; Li, F.; Han, D.; Zhang, Q.; Niu, L. Covalent functionalization of polydisperse chemically-converted graphene sheets with amine-terminated ionic liquid. *Chem. Commun. (Camb.),* **2009**, (26), 3880-3882.
[http://dx.doi.org/10.1039/b905085j] [PMID: 19662239]

[293] Shan, C.; Yang, H.; Han, D.; Zhang, Q.; Ivaska, A.; Niu, L. Water-soluble graphene covalently functionalized by biocompatible poly-L-lysine. *Langmuir,* **2009**, *25*(20), 12030-12033.
[http://dx.doi.org/10.1021/la903265p] [PMID: 19769343]

[294] Avinash, M.B.; Subrahmanyam, K.S.; Sundarayya, Y.; Govindaraju, T. Covalent modification and exfoliation of graphene oxide using ferrocene. *Nanoscale,* **2010**, *2*(9), 1762-1766.
[http://dx.doi.org/10.1039/c0nr00024h] [PMID: 20820707]

[295] Pham, V.H.; Cuong, T.V.; Hur, S.H.; Oh, E.; Kim, E.J.; Shin, E.W.; Chung, J.S. Chemical functionalization of graphene sheets by solvothermal reduction of a graphene oxide suspension in N-methyl-2-pyrrolidone. *J. Mater. Chem.,* **2011**, *21*, 3371-3377.
[http://dx.doi.org/10.1039/C0JM02790A]

[296] Jin, Z.; Lomeda, J.R.; Price, B.K.; Lu, W.; Zhu, Y.; Tour, J.M. Mechanically assisted exfoliation and functionalization of thermally converted graphene sheets. *Chem. Mater.,* **2009**, *21*, 3045-3047.

[http://dx.doi.org/10.1021/cm901601g]

[297] Stankovich, S.; Piner, R.D.; Chen, X.; Wu, N.; Nguyen, S.T.; Ruoff, R.S. Stable aqueous dispersions of graphitic nanoplatelets *via* the reduction of exfoliated graphite oxide in the presence of poly(sodium 4-styrenesulfonate). *J. Mater. Chem.,* **2006**, *16*, 155-158. [http://dx.doi.org/10.1039/B512799H]

[298] Bai, H.; Xu, Y.; Zhao, L.; Li, C.; Shi, G. Non-covalent functionalization of graphene sheets by sulfonated polyaniline. *Chem. Commun. (Camb.),* **2009**, (13), 1667-1669. [http://dx.doi.org/10.1039/b821805f] [PMID: 19294256]

[299] Choi, E.Y.; Han, T.H.; Hong, J.; Kim, J.E.; Lee, S.H.; Kim, H.W.; Kim, S.O. Noncovalent functionalization of graphene with end-functional polymers. *J. Mater. Chem.,* **2010**, *20*, 1907-1912. [http://dx.doi.org/10.1039/b919074k]

[300] Yoon, S.; In, I. Solubilization of reduced graphene in water through noncovalent interaction with dendrimers. *Chem. Lett.,* **2010**, *39*, 1160-1161. [http://dx.doi.org/10.1246/cl.2010.1160]

[301] Xu, Y.; Bai, H.; Lu, G.; Li, C.; Shi, G. Flexible graphene films *via* the filtration of water-soluble noncovalent functionalized graphene sheets. *J. Am. Chem. Soc.,* **2008**, *130*(18), 5856-5857. [http://dx.doi.org/10.1021/ja800745y] [PMID: 18399634]

[302] Liu, J.; Yang, W.; Tao, L.; Li, D.; Boyer, C.; Davis, T.P. Thermosensitive graphene nanocomposites formed using pyrene-terminal polymers made by RAFT polymerization. *J. Polym. Sci. A Polym. Chem.,* **2010**, *48*, 425-433. [http://dx.doi.org/10.1002/pola.23802]

[303] Ghosh, A.; Rao, K.V.; George, S.J.; Rao, C.N. Noncovalent functionalization, exfoliation, and solubilization of graphene in water by employing a fluorescent coronene carboxylate. *Chemistry,* **2010**, *16*(9), 2700-2704. [http://dx.doi.org/10.1002/chem.200902828] [PMID: 20108284]

[304] Yang, H.; Zhang, Q.; Shan, C.; Li, F.; Han, D.; Niu, L. Stable, conductive supramolecular composite of graphene sheets with conjugated polyelectrolyte. *Langmuir,* **2010**, *26*(9), 6708-6712. [http://dx.doi.org/10.1021/la100365z] [PMID: 20229996]

[305] Liu, H.; Gao, J.; Xue, M.; Zhu, N.; Zhang, M.; Cao, T. Processing of graphene for electrochemical application: noncovalently functionalize graphene sheets with water-soluble electroactive methylene green. *Langmuir,* **2009**, *25*(20), 12006-12010. [http://dx.doi.org/10.1021/la9029613] [PMID: 19764798]

[306] Kamada, S.; Nomoto, H.; Fukuda, K.; Fukawa, T.; Shirai, H.; Kimura, M. Noncovalent wrapping of chemically modified graphene with p-conjugated disk-like molecules. *Colloid Polym. Sci.,* **2011**, *289*, 925-932. [http://dx.doi.org/10.1007/s00396-011-2414-z]

[307] Kodali, V.K.; Scrimgeour, J.; Kim, S.; Hankinson, J.H.; Carroll, K.M.; de Heer, W.A.; Berger, C.; Curtis, J.E. Nonperturbative chemical modification of graphene for protein micropatterning. *Langmuir,* **2011**, *27*(3), 863-865. [http://dx.doi.org/10.1021/la1033178] [PMID: 21182241]

[308] Li, F.; Bao, Y.; Chai, J.; Zhang, Q.; Han, D.; Niu, L. Synthesis and application of widely soluble graphene sheets. *Langmuir,* **2010**, *26*(14), 12314-12320.
[http://dx.doi.org/10.1021/la101534n] [PMID: 20536161]

[309] Malig, J.; Jux, N.; Kiessling, D.; Cid, J.J.; Vázquez, P.; Torres, T.; Guldi, D.M. Towards tunable graphene/phthalocyanine-PPV hybrid systems. *Angew. Chem. Int. Ed. Engl.,* **2011**, *50*(15), 3561-3565.
[http://dx.doi.org/10.1002/anie.201003364] [PMID: 21391293]

[310] Chang, H.; Wang, G.; Yang, A.; Tao, X.; Liu, X.; Shen, Y.; Zheng, Z. A transparent, flexible, low-temperature, and solution-processible graphene composite electrode. *Adv. Funct. Mater.,* **2010**, *20*, 2893-2902.
[http://dx.doi.org/10.1002/adfm.201000900]

[311] Zeng, Q.; Cheng, J.; Tang, L.; Liu, X.; Liu, Y.; Li, J.; Jiang, J. Self-assembled graphene–enzyme hierarchical nanostructures for electrochemical biosensing. *Adv. Funct. Mater.,* **2010**, *20*, 3366-3372.
[http://dx.doi.org/10.1002/adfm.201000540]

[312] Coq, B.; Planeix, J.M.; Brotons, V. Fullerene-based materials as new support media in heterogeneous catalysis by metals. *Appl. Catal. A Gen.,* **1998**, *173*, 175-183.
[http://dx.doi.org/10.1016/S0926-860X(98)00177-X]

[313] Goldshleger, N.F. Fullerenes and fullerene-based materials in catalysis. *Fullerene Sci. Techn.,* **2001**, *9*, 255-280.
[http://dx.doi.org/10.1081/FST-100104493]

[314] Sherigara, B.S.; Kutner, W.; D'Souza, F. Electrocatalytic properties and sensor applications of fullerenes and carbon nanotubes. *Electroanal.,* **2003**, *15*, 753-772.
[http://dx.doi.org/10.1002/elan.200390094]

[315] Malhotra, R.; McMillen, D.F.; Tse, D.S.; Lorents, D.C.; Ruoff, R.S.; Keegan, D.M. Hydrogen-transfer reactions catalyzed by fullerenes. *Energy Fuels,* **1993**, *7*, 685-686.
[http://dx.doi.org/10.1021/ef00041a020]

[316] Braun, Th.; Wohlers, M.; Belz, T.; Nowitzke, G.; Wortmann, G.; Uchida, Y.; Pfänder, N.; Schlögl, R. Fullerene-based ruthenium catalysts: a novel approach for anchoring metal to carbonaceous supports. I. Structure. *Catal. Lett.,* **1997**, *43*, 167-173.
[http://dx.doi.org/10.1023/A:1018998808208]

[317] Koprinarov, N.; Konstantinova, M.; Pchelarov, G.; Marinov, M. *Fuel cell technologies: state and perspectives in fullerene structures suitable for fuel cells electrodes. NATO Science Series, Sammes, N.; Smirnova, A*; Vasylyev, O., Ed.; Springer, Netherlands, **2005**, Vol. 202, pp. 81-95.

[318] Loutfy, R.O.; Loutfy, K.; Pugazhendhi, P. Advances in Fuel Cell Technology. *Adv. Mater. Process.,* **2008**, •••, 35-38.

[319] Gabriel, M.A.; Deutsch, T.; Franco, A.A. Fullerene-based materials as catalysts for fuel cells. *ECS Trans.,* **2010**, *25*, 1-6.

[320] Gust, D.; Moore, T.A.; Moore, A.L. Solar fuels *via* artificial photosynthesis. *Acc. Chem. Res.,* **2009**, *42*(12), 1890-1898.
[http://dx.doi.org/10.1021/ar900209b] [PMID: 19902921]

[321] Dai, L.; Chang, D.W.; Baek, J.B.; Lu, W. Carbon nanomaterials for advanced energy conversion and storage. *Small,* **2012**, *8*(8), 1130-1166.
[http://dx.doi.org/10.1002/smll.201101594] [PMID: 22383334]

[322] Synowczyk, A.W.; Heinze, J. *Electronic Properties of Fullerenes. in Springer Series in Solid-State Sciences,* **1993**, *117*, 73-77.
[http://dx.doi.org/10.1007/978-3-642-85049-3_11]

[323] Prato, M. [60] Fullerene chemistry for materials science applications. *J. Mater. Chem.,* **1997**, *7*, 1097-1109.
[http://dx.doi.org/10.1039/a700080d]

[324] Diederich, F.; Gomez-Lopez, M. Supramolecular fullerene chemistry. *Chem. Soc. Rev.,* **1999**, *28*, 263-277.
[http://dx.doi.org/10.1039/a804248i]

[325] Imahori, H.; Sakata, Y. Fullerenes as novel acceptors in photosynthetic electron transfer. *Eur. J. Org. Chem.,* **1999**, 2445-2457.
[http://dx.doi.org/10.1002/(SICI)1099-0690(199910)1999:10<2445::AID-EJOC2445>3.0.CO;2-G]

[326] Guldi, D.M.; Prato, M. Excited-state properties of C(60) fullerene derivatives. *Acc. Chem. Res.,* **2000**, *33*(10), 695-703.
[http://dx.doi.org/10.1021/ar990144m] [PMID: 11041834]

[327] Guldi, D.M. Fullerene-porphyrin architectures; photosynthetic antenna and reaction center models. *Chem. Soc. Rev.,* **2002**, *31*(1), 22-36.
[http://dx.doi.org/10.1039/b106962b] [PMID: 12108980]

[328] El-Khouly, M.E.; Ito, O.; Smith, P.M.; D'Souza, F. Intermolecular and supramolecular photoinduced electron transfer processes of fullerene-porphyrin/phthalocyanine systems. *J. Photochem. Photobiol. Photochem. Rev.,* **2004**, *5*, 79-104.
[http://dx.doi.org/10.1016/j.jphotochemrev.2004.01.003]

[329] Hoppe, H.; Sariciftci, N.S. Organic solar cells: An overview. *J. Mater. Res.,* **2004**, *19*, 1924-1945.
[http://dx.doi.org/10.1557/JMR.2004.0252]

[330] Spanggaard, H.; Krebs, F.C. A brief history of the development of organic and polymeric photovoltaics. *Sol. Energy Mater. Sol. Cells,* **2004**, *83*, 125-146.
[http://dx.doi.org/10.1016/j.solmat.2004.02.021]

[331] Scharber, M.C.; Wuhlbacher, D.; Koppe, M.; Denk, P.; Waldauf, C.; Heeger, A.J.; Brabec, C.L. Design rules for donors in bulk-heterojunction solar cells - towards 10#x0025; energy-conversion efficiency. *Adv. Mater.,* **2006**, *18*, 789-794.
[http://dx.doi.org/10.1002/adma.200501717]

[332] Günes, S.; Neugebauer, H.; Sariciftci, N.S. Conjugated polymer-based organic solar cells. *Chem. Rev.,* **2007**, *107*(4), 1324-1338.
[http://dx.doi.org/10.1021/cr050149z] [PMID: 17428026]

[333] Thompson, B.C.; Frechet, J.M. Organic photovoltaics-polymer-fullerene composite solar cells. *Angew. Chem. Int. Ed.,* **2008**, *47*, 58-77.
[http://dx.doi.org/10.1002/anie.200702506]

[334] Dennler, G.; Scharber, M.C.; Brabec, C.J. Polymer-fullerene bulk-heterojunction solar cells. *Adv. Mater.,* **2009**, *21*, 1323-1338.
[http://dx.doi.org/10.1002/adma.200801283]

[335] Mishra, A.; Bäuerle, P. Small molecule organic semiconductors on the move: promises for future solar energy technology. *Angew. Chem. Int. Ed. Engl.,* **2012**, *51*(9), 2020-2067.
[http://dx.doi.org/10.1002/anie.201102326] [PMID: 22344682]

[336] Yang, X.H.; Uddin, A. Effect of thermal annealing on P3HT:PCBM bulk-heterojunction organic solar cells: A critical review. *Renew. Sustain. Energy Rev.,* **2014**, *30*, 324-336.
[http://dx.doi.org/10.1016/j.rser.2013.10.025]

[337] Dang, M.T.; Hirsch, L.; Wantz, G.; Wuest, J.D. Controlling the morphology and performance of bulk heterojunctions in solar cells. Lessons learned from the benchmark poly(3-hexylthiophene):[6,6--phenyl-C61-butyric acid methyl ester system. *Chem. Rev.,* **2013**, *113*(5), 3734-3765.
[http://dx.doi.org/10.1021/cr300005u] [PMID: 23347135]

[338] Gurav, A.; Kodas, T.; Pluym, T.; Xiong, Y. Aerosol processing materials. *Aerosol Sci. Technol.,* **1993**, *19*, 411-452.
[http://dx.doi.org/10.1080/02786829308959650]

[339] Mirkin, C.A.; Caldwell, W.B. Thin film, fullerene-based materials. *Tetrahedron,* **1996**, *52*, 5113-5130.
[http://dx.doi.org/10.1016/0040-4020(96)00118-4]

[340] Tsukruk, V.V. Assembly of supramolecular polymers in ultrathin films. *Prog. Polym. Sci.,* **1997**, *22*, 247-311.
[http://dx.doi.org/10.1016/S0079-6700(96)00005-6]

[341] Nicole, L.; Boissiere, C.; Grosso, D.; Quach, A.; Sanchez, C. Mesostructured hybrid organic-inorganic thin films. *J. Mater. Chem.,* **2005**, *15*, 35-36.
[http://dx.doi.org/10.1039/b506072a]

[342] Bonifazi, D.; Enger, O.; Diederich, F. Supramolecular [60]fullerene chemistry on surfaces. *Chem. Soc. Rev.,* **2007**, *36*(2), 390-414.
[http://dx.doi.org/10.1039/B604308A] [PMID: 17264939]

[343] Dou, L.; You, J.; Hong, Z.; Xu, Z.; Li, G.; Street, R.A.; Yang, Y. 25th anniversary article: a decade of organic/polymeric photovoltaic research. *Adv. Mater.,* **2013**, *25*(46), 6642-6671.
[http://dx.doi.org/10.1002/adma.201302563] [PMID: 24105687]

[344] Hebard, A.F.; Rosseinsky, M.J.; Haddon, R.C.; Murphy, D.W.; Glarum, S.H.; Palstra, T.T.; Ramirez, A.P.; Kortan, A.R. Superconductivity at 18K in potassium doped C-60. *Nature,* **1991**, *350*, 600-601.
[http://dx.doi.org/10.1038/350600a0]

[345] Tutt, L.W.; Boggess, T.F. A review of optical limiting machanisms and devices using organics, fullerenes, semiconductors and other materials. *Prog. Quantum Electron.,* **1993**, *17*, 299-338.
[http://dx.doi.org/10.1016/0079-6727(93)90004-S]

[346] Nalwa, H.S. Organic materials for 3rd order nonlinear optics. *Adv. Mater.,* **1993**, *5*, 341-358.
[http://dx.doi.org/10.1002/adma.19930050504]

[347] Sun, Y.P.; Riggs, J.E. Organic and inorganic optical limiting materials. From fullerenes to nanoparticles. *Int. Rev. Phys. Chem.,* **1999**, *18*, 43-90.
[http://dx.doi.org/10.1080/014423599230008]

[348] Prato, M. *Fullerene materials* . in Fullerenes and related structures, Book Series. *Top. Curr. Chem.,* **1999**, *199*, 173-187.
[http://dx.doi.org/10.1007/3-540-68117-5_5]

[349] Hollins, R.C. Materials for optical limiters. *Curr. Opin. Solid State Mater. Sci.,* **1999**, *4*, 189-196.
[http://dx.doi.org/10.1016/S1359-0286(99)00009-1]

[350] Lopez, A.M.; Mateo-Alonso, A.; Prato, M. Materials chemistry of fullerene C-60 derivatives. *J. Mater. Chem.,* **2011**, *21*, 1305-1318.
[http://dx.doi.org/10.1039/C0JM02386H]

[351] Da Ros, T.; Prato, M. Medicinal chemistry with fullerenes and fullerene derivatives. *Chem. Commun. (Camb.),* **1999**, 663-669.
[http://dx.doi.org/10.1039/a809495k]

[352] Tabata, Y.; Ikada, Y. Biological functions of fullerene. *Pure Appl. Chem.,* **1999**, *71*, 2047-2053.
[http://dx.doi.org/10.1351/pac199971112047]

[353] Tagmatarchis, N.; Shinohara, H. Fullerenes in medicinal chemistry and their biological applications. *Mini Rev. Med. Chem.,* **2001**, *1*(4), 339-348.
[PMID: 12369961]

[354] Bianco, A.; Da Ros, T.; Prato, M.; Toniolo, C. Fullerene-based amino acids and peptides. *J. Pept. Sci.,* **2001**, *7*(4), 208-219.
[http://dx.doi.org/10.1002/psc.313] [PMID: 11354464]

[355] Bosi, S.; Da Ros, T.; Spalluto, G.; Prato, M. Fullerene derivatives: an attractive tool for biological applications. *Eur. J. Med. Chem.,* **2003**, *38*(11-12), 913-923.
[http://dx.doi.org/10.1016/j.ejmech.2003.09.005] [PMID: 14642323]

[356] Bakry, R.; Vallant, R.M.; Najam-ul-Haq, M.; Rainer, M.; Szabo, Z.; Huck, C.W.; Bonn, G.K. Medicinal applications of fullerenes. *Int. J. Nanomedicine,* **2007**, *2*(4), 639-649.
[PMID: 18203430]

[357] Anilkumar, P.; Lu, F.; Cao, L.; Luo, P.G.; Liu, J.H.; Sahu, S.; Tackett, K.N.; Wang, Y.; Sun, Y.P. Fullerenes for applications in biology and medicine. *Curr. Med. Chem.,* **2011**, *18*(14), 2045-2059.
[http://dx.doi.org/10.2174/092986711795656225] [PMID: 21517770]

[358] Mroz, P.; Tegos, G.P.; Gali, H.; Wharton, T.; Sarna, T.; Hamblin, M.R. Photodynamic therapy with fullerenes. *Photochem. Photobiol. Sci.,* **2007**, *6*(11), 1139-1149.
[http://dx.doi.org/10.1039/b711141j] [PMID: 17973044]

[359] Yano, S.; Hirohara, S.; Obata, M.; Hagiya, Y.; Ogura, S.; Ikeda, A.; Kataoka, H.; Tanaka, M.; Joh, T. Current states and future views in photodynamic therapy. *J. Photochem. Photobiol. Photochem. Rev.,* **2011**, *12*, 46-67.
[http://dx.doi.org/10.1016/j.jphotochemrev.2011.06.001]

[360] Chistyakov, V.A.; Smirnova, Y.O.; Prazdnova, E.V.; Soldatov, A.V. Possible mechanisms of fullerene C□□ antioxidant action. *BioMed Res. Int.,* **2013**, *2013*, 821498.

[http://dx.doi.org/10.1155/2013/821498] [PMID: 24222918]

[361] Pandey, R.K.; Goswami, L.N.; Chen, Y.; Gryshuk, A.; Missert, J.R.; Oseroff, A.; Dougherty, T.J. Nature: a rich source for developing multifunctional agents. Tumor-imaging and photodynamic therapy. *Lasers Surg. Med.,* **2006**, *38*(5), 445-467. [http://dx.doi.org/10.1002/lsm.20352] [PMID: 16788930]

[362] Sitharaman, B.; Wilson, L.J. Gadofullerenes and gadonanotubes: A new paradigm for high-performance magnetic resonance imaging contrast agent probes. *J. Biomed. Nanotechnol.,* **2007**, *3*, 342-352. [http://dx.doi.org/10.1166/jbn.2007.043]

[363] Chen, Z.; Mao, R.; Liu, Y. Fullerenes for cancer diagnosis and therapy: preparation, biological and clinical perspectives. *Curr. Drug Metab.,* **2012**, *13*(8), 1035-1045. [http://dx.doi.org/10.2174/138920012802850128] [PMID: 22380017]

[364] Chen, Z.; Ma, L.; Liu, Y.; Chen, C. Applications of functionalized fullerenes in tumor theranostics. *Theranostics,* **2012**, *2*(3), 238-250. [http://dx.doi.org/10.7150/thno.3509] [PMID: 22509193]

[365] Mody, V.V.; Nounou, M.I.; Bikram, M. Novel nanomedicine-based MRI contrast agents for gynecological malignancies. *Adv. Drug Deliv. Rev.,* **2009**, *61*(10), 795-807. [http://dx.doi.org/10.1016/j.addr.2009.04.020] [PMID: 19427886]

[366] Braun, K.; Dunsch, L.; Pipkorn, R.; Bock, M.; Baeuerle, T.; Yang, S.; Waldeck, W.; Wiessler, M. Gain of a 500-fold sensitivity on an intravital MR contrast agent based on an endohedral gadolinium-cluster-fullerene-conjugate: a new chance in cancer diagnostics. *Int. J. Med. Sci.,* **2010**, *7*(3), 136-146. [http://dx.doi.org/10.7150/ijms.7.136] [PMID: 20567614]

[367] Hirshon, A.S.; Wu, H-J.; Wilson, R.B.; Malhotra, R. Investigation of fullerene-based catalysts for methane activation. *J. Phys. Chem.,* **1995**, *99*, 17483-17486. [http://dx.doi.org/10.1021/j100049a001]

[368] Fagan, P.J.; Calabrese, J.C.; Malone, B. Metal complexes of buckminsterfullerene (C60). *Acc. Chem. Res.,* **1992**, *25*, 134-142. [http://dx.doi.org/10.1021/ar00015a006]

[369] Claridge, J.B.; Douthwaite, R.E.; Green, M.L.; Lago, R.M.; Tsang, S.C.; York, A.P. Studies on a new catalyst for the hydroformylation of alkenes using C60 as a ligand. *J. Mol. Catal. Chem.,* **1994**, *89*, 113-120. [http://dx.doi.org/10.1016/0304-5102(93)E0328-E]

[370] Braun, Th.; Wohlers, M.; Belz, T.; Schlögl, R. Fullerene-based ruthenium catalysts: a novel approach for anchoring metal to carbonaceous supports. II. Hydrogenation activity. *Catal. Lett.,* **1997**, *43*, 175-180. [http://dx.doi.org/10.1023/A:1018950925046]

[371] Datta, A.; Khwaja, H.I.; Kelkar, R.Y.; Saple, A.R.; Datta, M. Characterization and catalytic activity of palladium dispersed on Kratschmer–Huffman (K–H) soot. *Chem. Commun. (Camb.),* **1996**, 851-852. [http://dx.doi.org/10.1039/cc9960000851]

[372] Zhang, Y.H.; Liu, R.; Teng, Y.M.; Fang, E.T.; Huang, Z.Y.; Zeng, X.K. The support and

characterization of C60 and MoO3 on Al2O3. *React. Kinet. Catal. Lett.,* **1996**, *58*, 91-96. [http://dx.doi.org/10.1007/BF02071110]

[373] Chen, K.; Caldwell, W.B.; Mirkin, C.A. Fullerene self-assembly onto (MeO)3Si(CH2)3NH2-modified oxide surfaces. *J. Am. Chem. Soc.,* **1993**, *115*, 1193-1194. [http://dx.doi.org/10.1021/ja00056a080]

[374] Chupa, J.A.; Xu, S.; Fischetti, R.F.; Strongin, R.M.; McCauley, J.P.; Smith, A.B.; Blasie, J.K.; Peticolas, L.J.; Bean, J.C. A monolayer of C60 tethered to the surface of an inorganic substrate: assembly and structure. *J. Am. Chem. Soc.,* **1993**, *115*, 4383-4384. [http://dx.doi.org/10.1021/ja00063a075]

[375] Planeix, J.M.; Coq, B.; de Meânorval, L-C.; Medina, P. Grafting [60]fullerene on high surface area silica. *Chem. Commun. (Camb.),* **1996**, 2087-2088. [http://dx.doi.org/10.1039/cc9960002087]

[376] Coq, B.; Brotons, V.; Planeix, J.M.; de Meânorval, L-C.; Dutartre, R. Platinum supported on [60]fullerene-grafted silica as a new potential catalyst for hydrogenation. *J. Catal.,* **1998**, *176*, 358-364. [http://dx.doi.org/10.1006/jcat.1998.2040]

[377] Williams, R.M.; Zwier, J.M.; Verhoeven, J.W. Photoinduced intramolecular electron transfer in a bridged C60 (acceptor)-aniline (donor) system; photophysical properties of the first "active" fullerene diad. *J. Am. Chem. Soc.,* **1995**, *117*, 4093-4099. [http://dx.doi.org/10.1021/ja00119a025]

[378] Williams, R.M.; Koeberg, M.; Lawson, J.M.; An, Y-Z.; Rubin, Y.; Paddon-Row, M.N.; Verhoeven, J.W. Photoinduced electron transfer to C60 across extended 3- and 11-bond hydrocarbon bridges: creation of a long-lived charge-separated state. *J. Org. Chem.,* **1996**, *61*, 5055-5062. [http://dx.doi.org/10.1021/jo960678q]

[379] Kuciauskas, D.; Lin, S.; Seely, G.R.; Moore, A.L.; Moore, T.A.; Gust, D.; Drovetskaya, T.; Reed, C.A.; Boyd, P.D. Energy and photoinduced electron transfer in porphyrin−fullerene dyads. *J. Phys. Chem.,* **1996**, *100*, 15926-15932. [http://dx.doi.org/10.1021/jp9612745]

[380] Sariciftci, N.S.; Wudl, F.; Heeger, A.J.; Maggini, M.; Scorrano, G.; Prato, M.; Bourassa, J.; Ford, P.C. Photoinduced electron transfer and long lived charge separation in a donor-bridge-acceptor supramolecular 'diad' consisting of ruthenium(II) tris(bipyridine) functionalized C60. *Chem. Phys. Lett.,* **1995**, *247*, 510. [http://dx.doi.org/10.1016/S0009-2614(95)01276-1]

[381] Guldi, D.M.; Maggini, M.; Scorrano, G.; Prato, M. Intramolecular electron transfer in fullerene/ferrocene based donor−bridge−acceptor dyads. *J. Am. Chem. Soc.,* **1997**, *119*, 974-980. [http://dx.doi.org/10.1021/ja960070z]

[382] Wang, Y. Photoconductivity of fullerene-doped polymers. *Nature,* **1992**, *356*, 585-587. [http://dx.doi.org/10.1038/356585a0]

[383] Sariciftci, N.S.; Smilowitz, L.; Heeger, A.J.; Wudl, F. Photoinduced electron transfer from a conducting polymer to buckminsterfullerene. *Science,* **1992**, *258*(5087), 1474-1476. [http://dx.doi.org/10.1126/science.258.5087.1474] [PMID: 17755110]

[384] Morita, S.; Zakidov, A.A.; Yoshino, K. Doping effect of buckminsterfullerene in conducting polymer: Change of absorption spectrum and quenching of luminescence. *Solid State Commun.,* **1992**, *82*, 249-252.
[http://dx.doi.org/10.1016/0038-1098(92)90636-N]

[385] Janssen, R.A.; Hummelen, J.C.; Lee, K.; Pakbaz, K.; Sariciftci, N.S.; Heeger, A.J.; Wudl, F. Photoinduced electron transfer from π-conjugated polymers onto Buckminsterfullerene, fulleroids, and methanofullerenes. *J. Chem. Phys.,* **1995**, *103*, 788-793.
[http://dx.doi.org/10.1063/1.470110]

[386] Yu, G.; Gao, J.; Hummelen, J.C.; Wudl, F.; Heeger, A.J. Polymer photovoltaic cells: enhanced efficiencies *via* a network of internal donor-acceptor heterojunctions. *Science,* **1995**, *270*, 1789-1791.
[http://dx.doi.org/10.1126/science.270.5243.1789]

[387] Brabec, C.J.; Gowrisanker, S.; Halls, J.J.; Laird, D.; Jia, S.; Williams, S.P. Polymer-fullerene bulk-heterojunction solar cells. *Adv. Mater.,* **2010**, *22*(34), 3839-3856.
[http://dx.doi.org/10.1002/adma.200903697] [PMID: 20717982]

[388] Bundgaard, E.; Krebs, F.C. Low band gap polymers for organic photovoltaics. *Sol. Energy Mater. Sol. Cells,* **2007**, *91*, 954-985.
[http://dx.doi.org/10.1016/j.solmat.2007.01.015]

[389] Mihailetchi, V.D.; Xie, H.; de Boer, B.; Koster, L.J.; Blom, P.W. Charge transport and photocurrent generation in poly(3-hexylthiophene): methanofullerene bulk-heterojunction solar cells. *Adv. Funct. Mater.,* **2006**, *16*, 699-708.
[http://dx.doi.org/10.1002/adfm.200500420]

[390] Huang, J.S.; Li, G.; Yang, Y. Influence of composition and heat-treatment on the charge transport properties of poly(3-hexylthiophene) and [6,6]-phenyl C61-butyric acid methyl ester blends. *Appl. Phys. Lett.,* **2005**, *87*, 112105.
[http://dx.doi.org/10.1063/1.2045554]

[391] Choi, J.H.; Son, K.I.; Kim, T.; Kim, K.; Ohkubo, K.; Fukuzumi, S. Thienyl-substituted methanofullerene derivatives for organic photovoltaic cells. *J. Mater. Chem.,* **2010**, *20*, 475-482.
[http://dx.doi.org/10.1039/B916597E]

[392] Zhang, Y.; Yip, H-L.; Acton, O.; Hau, S.K.; Huang, F.; Jen, A.K. A simple and effective way of achieving highly efficient and thermally stable bulk-heterojunction polymer solar cells using amorphous fullerene derivatives as electron acceptor. *Chem. Mater.,* **2009**, *21*, 2598-2602.
[http://dx.doi.org/10.1021/cm9009282]

[393] Backer, S.A.; Sivula, K.; Kavulak, D.F.; Frechet, J.M. High efficiency organic photovoltaics incorporating a new Family of soluble fullerene derivatives. *Chem. Mater.,* **2007**, *19*, 2927-2929.
[http://dx.doi.org/10.1021/cm070893v]

[394] Lenes, M.; Wetzelaer, G-J.; Kooistra, F.B.; Veenstra, S.C.; Hummelen, J.C.; Blom, P.W. Fullerene bisadducts for enhanced open-circuit voltages and efficiencies in polymer solar cells. *Adv. Mater.,* **2008**, *20*, 2116-2119.
[http://dx.doi.org/10.1002/adma.200702438]

[395] Ross, R.B.; Cardona, C.M.; Guldi, D.M.; Sankaranarayanan, S.G.; Reese, M.O.; Kopidakis, N.; Peet,

J.; Walker, B.; Bazan, G.C.; Van Keuren, E.; Holloway, B.C.; Drees, M. Endohedral fullerenes for organic photovoltaic devices. *Nat. Mater.,* **2009**, *8*(3), 208-212. [http://dx.doi.org/10.1038/nmat2379] [PMID: 19202546]

[396] Ross, R.B.; Cardona, C.M.; Swain, F.B.; Guldi, D.M.; Reese, M.O.; Sankaranarayanan, S.G.; van Keuren, E.; Holloway, B.C.; Drees, M. Tuning conversion efficiency in metallo endohedral fullerene-based organic photovoltaic devices. *Adv. Funct. Mater.,* **2009**, *19*, 2332-2337. [http://dx.doi.org/10.1002/adfm.200900214]

[397] Matsuo, Y.; Iwashita, A.; Abe, Y.; Li, C-Z.; Matsuo, K.; Hashiguchi, M.; Nakamura, E. Regioselective synthesis of 1,4-di(organo)[60]fullerenes through DMF-assisted monoaddition of silylmethyl Grignard reagents and subsequent alkylation reaction. *J. Am. Chem. Soc.,* **2008**, *130*(46), 15429-15436. [http://dx.doi.org/10.1021/ja8041299] [PMID: 18950174]

[398] Matsuo, Y.; Sato, Y.; Niinomi, T.; Soga, I.; Tanaka, H.; Nakamura, E. Columnar structure in bulk heterojunction in solution-processable three-layered p-i-n organic photovoltaic devices using tetrabenzoporphyrin precursor and silylmethyl[60]fullerene. *J. Am. Chem. Soc.,* **2009**, *131*(44), 16048-16050. [http://dx.doi.org/10.1021/ja9048702] [PMID: 19886695]

[399] Zhao, G.; He, Y.; Li, Y. 6.5% Efficiency of polymer solar cells based on poly(3-hexylthiophene) and indene-C(60) bisadduct by device optimization. *Adv. Mater.,* **2010**, *22*(39), 4355-4358. [http://dx.doi.org/10.1002/adma.201001339] [PMID: 20589774]

[400] Arias, F.; Godínez, L.A.; Wilson, S.R.; Kaifer, A.E.; Echegoyen, L. Interfacial hydrogen bonding. self-assembly of a monolayer of a fullerene-crown ether derivative on gold surfaces derivatized with an ammonium-terminated alkanethiolate. *J. Am. Chem. Soc.,* **1996**, *118*, 6086-6087. [http://dx.doi.org/10.1021/ja960445g]

[401] Shi, X.; Caldwell, W.B.; Chen, K.; Mirkin, C.A. A well-defined surface-confinable fullerene: monolayer self-assembly on Au(111). *J. Am. Chem. Soc.,* **1994**, *116*, 11598-11599. [http://dx.doi.org/10.1021/ja00104a062]

[402] Maggini, M.; Karlsson, A.; Pasimeni, L.; Scorrano, G.; Prato, M.; Valli, L. Synthesis of N-acylated fulleropyrrolidines: New materials for the preparation of Langmuir-Blodgett films containing fullerenes. *Tetrahedron Lett.,* **1994**, *35*, 2985-2988. [http://dx.doi.org/10.1016/S0040-4039(00)76678-0]

[403] Maggini, M.; Pasimeni, L.; Prato, M.; Scorrano, G.; Valli, L. Incorporation of an acyl group in fulleropyrrolidines: effects on Langmuir monolayers. *Langmuir,* **1994**, *10*, 4164-4166. [http://dx.doi.org/10.1021/la00023a042]

[404] Maliszewskyi, N.C.; Heiney, P.A.; Jones, D.H.; Strongin, R.M.; Cichy, M.A.; Smith, A.B., III Langmuir films of fullerene C60, fullerene epoxide C60O, and dihydrofulleroid C61H2. *Langmuir,* **1993**, *9*, 1439-1441. [http://dx.doi.org/10.1021/la00030a001]

[405] Matsumoto, M.; Tachibana, H.; Azumi, R.; Tanaka, M.; Nakamura, T.; Yunome, G.; Abe, M.; Yamago, S.; Nakamura, E. Langmuir-Blodgett film of amphiphilic C60 carboxylic acid. *Langmuir,* **1995**, *11*, 660-665. [http://dx.doi.org/10.1021/la00002a051]

[406] Patel, H.M.; Didymus, J.M.; Wang, K.K.; Hirsch, A.; Skiebe, A.; Lamparth, I.; Mann, S. Fullerates: interaction of divalent metal ions with Langmuir monolayers and multilayers of mono-substituted C60–malonic acid. *Chem. Commun. (Camb.),* **1996**, 611-612.
[http://dx.doi.org/10.1039/cc9960000611]

[407] Ravaine, S.; Le Peq, F.; Mingotaud, C.; Delhaes, P.; Hummelen, J.C.; Wudl, F.; Patterson, L.K. Synthesis, physical characterizations, and Langmuir films of new methanofullerenes. *J. Phys. Chem.,* **1995**, *99*, 9551-9557.
[http://dx.doi.org/10.1021/j100023a037]

[408] Jonas, U.; Cardullo, F.; Belik, P.; Diederich, F.; Gügel, A.; Harth, E.; Herrmann, A.; Isaacs, L.; Müllen, K.; Ringsdorf, H.; Thilgen, C.; Uhlmann, P.; Vasella, A.; Waldra, C.A.; Walter, M. Synthesis of a fullerene[60] cryptate and systematic Langmuir – Blodgett and thin-film investigations of amphiphilic fullerene derivatives. *Chemistry,* **1995**, *1*, 243-21.
[http://dx.doi.org/10.1002/chem.19950010408]

[409] Luo, C.; Guldi, D.M.; Maggini, M.; Menna, E.; Mondini, S.; Kotov, N.A.; Prato, M. Stepwise assembled photoactive films containing donor-linked fullerenes. *Angew. Chem. Int. Ed.,* **2000**, *39*, 3905-3909.
[http://dx.doi.org/10.1002/1521-3773(20001103)39:21<3905::AID-ANIE3905>3.0.CO;2-Q]

[410] Gulino, A.; Bazzano, S.; Condorelli, G.G.; Giuffrida, S.; Mineo, P.; Satriano, C.; Scamporrino, E.; Ventimiglia, G.; Vitalini, D.; Fragalà, I. Engineered silica surfaces with an assembled C60 fullerene monolayer. *Chem. Mater.,* **2005**, *17*, 1079-1084.
[http://dx.doi.org/10.1021/cm048861k]

[411] Tutt, L.W.; Kost, A. Optical limiting performance of C60 and C70 solutions. *Nature,* **1992**, *356*, 225-226.
[http://dx.doi.org/10.1038/356225a0]

[412] Cha, M.; Sariciftci, N.S.; Heeger, A.J.; Hummelen, J.C.; Wudl, F. Enhanced nonlinear absorption and optical limiting in semiconducting polymer/methanofullerene charge transfer films. *Appl. Phys. Lett.,* **1995**, *67*, 3850-3852.
[http://dx.doi.org/10.1063/1.115294]

[413] Ma, B.; Riggs, J.E.; Sun, Y-P. Photophysical and nonlinear absorptive optical limiting properties of [60]fullerene dimer and poly[60]fullerene polymer. *J. Phys. Chem. B,* **1998**, *102*, 5999-6009.
[http://dx.doi.org/10.1021/jp981715f]

[414] Sun, Y-P.; Lawson, G.E.; Riggs, J.E.; Ma, B.; Wang, N.; Moton, D.K. Photophysical and nonlinear optical properties of [60]fullerene derivatives. *J. Phys. Chem. A,* **1998**, *102*, 5520-5528.
[http://dx.doi.org/10.1021/jp9801121]

[415] Ma, B.; Bunker, C.E.; Guduru, R.; Zhang, X.; Sun, Y-P. Quantitative spectroscopic studies of the photoexcited state properties of methano- and pyrrolidino-[60]fullerene derivatives. *J. Phys. Chem. A,* **1997**, *101*, 5626-5632.
[http://dx.doi.org/10.1021/jp971067k]

[416] Sun, Y-P.; Ma, B.; Bunker, C.E. Photoinduced intramolecular $n-\pi^*$ electron transfer in aminofullerene derivatives. *J. Phys. Chem. A,* **1998**, *102*, 7580-7590.
[http://dx.doi.org/10.1021/jp982591+]

[417] Maggini, M.; Scorrano, G.; Prato, M.; Brusatin, G.; Innocenzi, P.; Guglielmi, M.; Renier, A.; Signorini, R.; Meneghetti, M.; Bozio, R. C60 derivatives embedded in sol-gel silica films. *Adv. Mater.,* **1995**, *7*, 404-406.
[http://dx.doi.org/10.1002/adma.19950070414]

[418] Signorini, R.; Zerbetto, M.; Meneghetti, M.; Bozio, R.; Maggini, M.; De Faveri, C.; Prato, M.; Scorrano, G. Optical limiting properties of soluble fullerene derivatives for incorporation in sol–gel materials. *Chem. Commun. (Camb.),* **1996**, 1891-1892.
[http://dx.doi.org/10.1039/cc9960001891]

[419] Maggini, M.; De Faveri, C.; Scorrano, G.; Prato, M.; Brusatin, G.; Guglielmi, M.; Meneghetti, M.; Signorini, R.; Bozio, R. Synthesis and optical-limiting behavior of hybrid inorganic–organic materials from the sol–gel Processing of organofullerenes. *Chemistry,* **1999**, *5*, 2501-2510.
[http://dx.doi.org/10.1002/(SICI)1521-3765(19990903)5:9<2501::AID-CHEM2501>3.0.CO;2-8]

[420] Brustain, G.; Signorini, R. Linear and nonlinear optical properties of fullerenes in solid state materials. *J. Mater. Chem.,* **2002**, *12*, 1964-1977.
[http://dx.doi.org/10.1039/b202399g]

[421] Levi, N.; Hantgan, R.R.; Lively, M.O.; Carroll, D.L.; Prasad, G.L. C60-fullerenes: detection of intracellular photoluminescence and lack of cytotoxic effects. *J. Nanobiotechnology,* **2006**, *4*, 14.
[http://dx.doi.org/10.1186/1477-3155-4-14] [PMID: 17169152]

[422] Belgorodsky, B.; Fadeev, L.; Kolsenik, J.; Gozin, M. Formation of a soluble stable complex between pristine C60-fullerene and a native blood protein. *ChemBioChem,* **2006**, *7*(11), 1783-1789.
[http://dx.doi.org/10.1002/cbic.200600237] [PMID: 16933357]

[423] Gharbi, N.; Pressac, M.; Hadchouel, M.; Szwarc, H.; Wilson, S.R.; Moussa, F. [60]fullerene is a powerful antioxidant *in vivo* with no acute or subacute toxicity. *Nano Lett.,* **2005**, *5*(12), 2578-2585.
[http://dx.doi.org/10.1021/nl051866b] [PMID: 16351219]

[424] Sayes, C.M.; Gobin, A.M.; Ausman, K.D.; Mendez, J.; West, J.L.; Colvin, V.L. Nano-C60 cytotoxicity is due to lipid peroxidation. *Biomaterials,* **2005**, *26*(36), 7587-7595.
[http://dx.doi.org/10.1016/j.biomaterials.2005.05.027] [PMID: 16005959]

[425] Nakamura, E.; Isobe, H. Functionalized fullerenes in water. The first 10 years of their chemistry, biology, and nanoscience. *Acc. Chem. Res.,* **2003**, *36*(11), 807-815.
[http://dx.doi.org/10.1021/ar030027y] [PMID: 14622027]

[426] Pantarotto, D.; Tagmatarchis, N.; Bianco, A.; Prato, M. Synthesis and biological properties of fullerene-containing amino acids and peptides. *Mini Rev. Med. Chem.,* **2004**, *4*(7), 805-814.
[PMID: 15379647]

[427] Bagno, A.; Claeson, S.; Maggini, M.; Martini, M.L.; Prato, M.; Scorrano, G. [60]Fullerene as a substituent. *Chemistry,* **2002**, *8*(5), 1015-1023.
[http://dx.doi.org/10.1002/1521-3765(20020301)8:5<1015::AID-CHEM1015>3.0.CO;2-Q] [PMID: 11891887]

[428] Castano, A.P.; Demidova, T.N.; Hamblin, M.R. Mechanisms in photodynamic therapy: Part three-Photosensitizer pharmacokinetics, biodistribution, tumor localization and modes of tumor destruction. *Photodiagn. Photodyn. Ther.,* **2005**, *2*(2), 91-106.

[http://dx.doi.org/10.1016/S1572-1000(05)00060-8] [PMID: 25048669]

[429] Yu, C.; Canteenwala, T.; El-Khouly, M.E.; Araki, Y.; Pritzker, K.; Ito, O.; Wilson, B.C.; Chiang, L.Y. Efficiency of singlet oxygen production from self-assembled nanospheres of molecular micelle-like photosensitizers FC4S. *J. Mater. Chem.,* **2005**, *15*, 1857-1864. [http://dx.doi.org/10.1039/b500369e]

[430] Mroz, P.; Pawlak, A.; Satti, M.; Lee, H.; Wharton, T.; Gali, H.; Sarna, T.; Hamblin, M.R. Functionalized fullerenes mediate photodynamic killing of cancer cells: Type I *versus* Type II photochemical mechanism. *Free Radic. Biol. Med.,* **2007**, *43*(5), 711-719. [http://dx.doi.org/10.1016/j.freeradbiomed.2007.05.005] [PMID: 17664135]

[431] Tegos, G.P.; Demidova, T.N.; Arcila-Lopez, D.; Lee, H.; Wharton, T.; Gali, H.; Hamblin, M.R. Cationic fullerenes are effective and selective antimicrobial photosensitizers. *Chem. Biol.,* **2005**, *12*(10), 1127-1135. [http://dx.doi.org/10.1016/j.chembiol.2005.08.014] [PMID: 16242655]

[432] Iwamoto, Y.; Yamakoshi, Y. A highly water-soluble C60-NVP copolymer: a potential material for photodynamic therapy. *Chem. Commun. (Camb.),* **2006**, (46), 4805-4807. [http://dx.doi.org/10.1039/b614305a] [PMID: 17345735]

[433] Liu, J.; Ohta, S.; Sonoda, A.; Yamada, M.; Yamamoto, M.; Nitta, N.; Murata, K.; Tabata, Y. Preparation of PEG-conjugated fullerene containing Gd3+ ions for photodynamic therapy. *J. Control. Release,* **2007**, *117*(1), 104-110. [http://dx.doi.org/10.1016/j.jconrel.2006.10.008] [PMID: 17156882]

[434] Tokuyama, H.; Yamago, S.; Nakamura, E. Photoinduced biochemical activity of fullerene carboxylic acid. *J. Am. Chem. Soc.,* **1993**, *115*, 7918-7919. [http://dx.doi.org/10.1021/ja00070a064]

[435] Nakanishi, I.; Fukuzumi, S.; Konishi, T.; Ohkubo, K.; Fujitsuka, M.; Ito, O.; Miyata, N. DNA cleavage *via* electron transfer from NADH to molecular oxygen photosensitized by c-cyclodextrin-bicapped C60. In: *Fullerenes for the New Millennium*; Kamat, P.V.; Guldi, D.M.; Kadish, D.M., Eds.; The Electrochemical Society: Pennigton, NJ, **2001**; Vol. 11, pp. 138-151.

[436] Yamakoshi, Y.; Sueyoshi, S.; Miyata, N. [Biological activity of photoexcited fullerene]. *Kokuritsu Iyakuhin Shokuhin Eisei Kenkyusho Hokoku,* **1999**, *117*(117), 50-60. [PMID: 10859936]

[437] Ding, R.G.; Finnerty, J.J.; Zhu, Z.H.; Yan, Z.F.; Lu, G.Q. *Encyclopedia of Nanoscience and Nanotechnology*; Nalwa, H.S., Ed.; American Science Publishers: New York, **2004**, Vol. 10, pp. 1-21.

[438] Aricò, A.S.; Bruce, P.; Scrosati, B.; Tarascon, J.M.; van Schalkwijk, W. Nanostructured materials for advanced energy conversion and storage devices. *Nat. Mater.,* **2005**, *4*(5), 366-377. [http://dx.doi.org/10.1038/nmat1368] [PMID: 15867920]

[439] Strobel, R.; Garche, J.; Moseley, P.T.; Jorissen, L.; Wolf, G. Hydrogen storage by carbon materials. *J. Power Sources,* **2006**, *159*, 781-801. [http://dx.doi.org/10.1016/j.jpowsour.2006.03.047]

[440] Wang, L.F.; Yang, R.T. New sorbents for hydrogen storage by hydrogen spillover - a review. *Energy Environ. Sci.,* **2008**, *1*, 268-279.

[http://dx.doi.org/10.1039/b807957a]

[441] Liu, C.; Li, F.; Ma, L.P.; Cheng, H.M. Advanced materials for energy storage. *Adv. Mater.,* **2010**, *22*(8), E28-E62.
[http://dx.doi.org/10.1002/adma.200903328] [PMID: 20217798]

[442] Jena, P. Materials for hydrogen storage: past, present, and future. *J. Phys. Chem. Lett.,* **2011**, *2*, 206-211.
[http://dx.doi.org/10.1021/jz1015372]

[443] Simon, P.; Gogotsi, Y. Capacitive energy storage in nanostructured carbon-electrolyte systems. *Acc. Chem. Res.,* **2013**, *46*(5), 1094-1103.
[http://dx.doi.org/10.1021/ar200306b] [PMID: 22670843]

[444] Zhang, Q.; Uchaker, E.; Candelaria, S.L.; Cao, G. Nanomaterials for energy conversion and storage. *Chem. Soc. Rev.,* **2013**, *42*(7), 3127-3171.
[http://dx.doi.org/10.1039/c3cs00009e] [PMID: 23455759]

[445] De Volder, M.F.; Tawfick, S.H.; Baughman, R.H.; Hart, A.J. Carbon nanotubes: present and future commercial applications. *Science,* **2013**, *339*(6119), 535-539.
[http://dx.doi.org/10.1126/science.1222453] [PMID: 23372006]

[446] Liu, H.S.; Song, C.J.; Zhang, L.; Zhang, J.J.; Wang, H.J.; Wilkinson, D.P. A review of anode catalysis in the direct methanol fuel cell. *J. Power Sources,* **2006**, *155*, 95-110.
[http://dx.doi.org/10.1016/j.jpowsour.2006.01.030]

[447] Dicks, A.L. The role of carbon in fuel cells. *J. Power Sources,* **2006**, *156*, 128-141.
[http://dx.doi.org/10.1016/j.jpowsour.2006.02.054]

[448] Lee, K.; Zhang, J.J.; Wang, H.J.; Wilkinson, D.P. Progress in the synthesis of carbon nanotube- and nanofiber-supported Pt electrocatalysts for PEM fuel cell catalysis. *J. Appl. Electrochem.,* **2006**, *36*, 507-522.
[http://dx.doi.org/10.1007/s10800-006-9120-4]

[449] Antolini, E. Carbon supports for low-temperature fuel cell catalysts. *Appl. Catal. B,* **2009**, *88*, 1-24.
[http://dx.doi.org/10.1016/j.apcatb.2008.09.030]

[450] Simon, P.; Gogotsi, Y. Materials for electrochemical capacitors. *Nat. Mater.,* **2008**, *7*(11), 845-854.
[http://dx.doi.org/10.1038/nmat2297] [PMID: 18956000]

[451] Manthiram, A.; Fu, Y.; Su, Y.S. Challenges and prospects of lithium-sulfur batteries. *Acc. Chem. Res.,* **2013**, *46*(5), 1125-1134.
[http://dx.doi.org/10.1021/ar300179v] [PMID: 23095063]

[452] Zhao, J.; Buldum, A.; Han, J.; Ping, L.J. First-principles study of Li-intercalated carbon nanotube ropes. *Phys. Rev. Lett.,* **2000**, *85*(8), 1706-1709.
[http://dx.doi.org/10.1103/PhysRevLett.85.1706] [PMID: 10970594]

[453] Chen, W.X.; Lee, J.Y.; Liu, Z. The nanocomposites of carbon nanotube with Sb and SnSb0.5 as Li-ion battery anodes. *Carbon,* **2003**, *41*, 959-966.
[http://dx.doi.org/10.1016/S0008-6223(02)00425-6]

[454] Morris, R.S.; Dixon, B.G.; Gennett, T.; Raffaelle, R.; Heben, M.J. High-energy, rechargeable Li-ion battery based on carbon nanotube technology. *J. Power Sources,* **2004**, *138*, 277-280.

[http://dx.doi.org/10.1016/j.jpowsour.2004.06.014]

[455] Bruce, P.G.; Freunberger, S.A.; Hardwick, L.J.; Tarascon, J.M. Li-O2 and Li-S batteries with high energy storage. *Nat. Mater.,* **2012**, *11*(1), 19-29. [http://dx.doi.org/10.1038/nmat3191] [PMID: 22169914]

[456] Wu, H.; Chan, G.; Choi, J.W.; Ryu, I.; Yao, Y.; McDowell, M.T.; Lee, S.W.; Jackson, A.; Yang, Y.; Hu, L.; Cui, Y. Stable cycling of double-walled silicon nanotube battery anodes through solid-electrolyte interphase control. *Nat. Nanotechnol.,* **2012**, *7*(5), 310-315. [http://dx.doi.org/10.1038/nnano.2012.35] [PMID: 22447161]

[457] Pandolfo, A.G.; Hollenkamp, A.F. Carbon properties and their role in supercapacitors. *J. Power Sources,* **2006**, *157*, 11-27. [http://dx.doi.org/10.1016/j.jpowsour.2006.02.065]

[458] An, K.H.; Kim, W.S.; Park, Y.S.; Choi, Y.C.; Lee, S.M.; Chung, D.C.; Bae, D.J.; Lim, S.C.; Lee, Y.H. Supercapacitors using single-walled carbon nanotube electrodes. *Adv. Mater.,* **2001**, *13*, 497-500. [http://dx.doi.org/10.1002/1521-4095(200104)13:7<497::AID-ADMA497>3.0.CO;2-H]

[459] Chen, J.H.; Li, W.Z.; Wang, D.Z.; Yang, S.X.; Wen, J.G.; Ren, Z.F. Electrochemical characterization of carbon nanotubes as electrode in electrochemical double-layer capacitors. *Carbon,* **2002**, *40*, 1193-1197. [http://dx.doi.org/10.1016/S0008-6223(01)00266-4]

[460] Futaba, D.N.; Hata, K.; Yamada, T.; Hiraoka, T.; Hayamizu, Y.; Kakudate, Y.; Tanaike, O.; Hatori, H.; Yumura, M.; Iijima, S. Shape-engineerable and highly densely packed single-walled carbon nanotubes and their application as super-capacitor electrodes. *Nat. Mater.,* **2006**, *5*(12), 987-994. [http://dx.doi.org/10.1038/nmat1782] [PMID: 17128258]

[461] Xie, X.; Gao, L. Characterization of a manganese dioxide/carbon nanotube composite fabricated using an in situ coating method. *Carbon,* **2007**, *45*, 2365-2373. [http://dx.doi.org/10.1016/j.carbon.2007.07.014]

[462] Bavio, M.A.; Acosta, G.G.; Kessler, T. Synthesis and characterization of polyaniline and polyaniline - carbon nanotubes nanostructures for electrochemical supercapacitors. *J. Power Sources,* **2014**, *245*, 475-481. [http://dx.doi.org/10.1016/j.jpowsour.2013.06.119]

[463] Serp, P.; Corrias, M.; Kalck, P. Carbon nanotubes and nanofibers in catalysis. *Appl. Catal. A Gen.,* **2003**, *253*, 337-358. [http://dx.doi.org/10.1016/S0926-860X(03)00549-0]

[464] Vairavapandian, D.; Vichchulada, P.; Lay, M.D. Preparation and modification of carbon nanotubes: review of recent advances and applications in catalysis and sensing. *Anal. Chim. Acta,* **2008**, *626*(2), 119-129. [http://dx.doi.org/10.1016/j.aca.2008.07.052] [PMID: 18790113]

[465] Serp, P.; Castillejos, E. Catalysis in carbon nanotubes. *ChemCatChem,* **2010**, *2*, 41-47. [http://dx.doi.org/10.1002/cctc.200900283]

[466] Pan, X.; Bao, X. The effects of confinement inside carbon nanotubes on catalysis. *Acc. Chem. Res.,* **2011**, *44*(8), 553-562.

[http://dx.doi.org/10.1021/ar100160t] [PMID: 21707038]

[467] Zhu, J.; Holmen, A.; Chen, D. Carbon nanomaterials in catalysis: proton affinity, chemical and electronic properties, and their catalytic consequences. *ChemCatChem,* **2013**, *5*, 378-401. [http://dx.doi.org/10.1002/cctc.201200471]

[468] Yu, H.; Peng, F.; Tan, J.; Hu, X.; Wang, H.; Yang, J.; Zheng, W. Selective catalysis of the aerobic oxidation of cyclohexane in the liquid phase by carbon nanotubes. *Angew. Chem. Int. Ed. Engl.,* **2011**, *50*(17), 3978-3982. [http://dx.doi.org/10.1002/anie.201007932] [PMID: 21433233]

[469] Chen, Z.; Guan, Z.; Li, M.; Yang, Q.; Li, C. Enhancement of the performance of a platinum nanocatalyst confined within carbon nanotubes for asymmetric hydrogenation. *Angew. Chem. Int. Ed. Engl.,* **2011**, *50*(21), 4913-4917. [http://dx.doi.org/10.1002/anie.201006870] [PMID: 21370365]

[470] Cao, R.; Thapa, R.; Kim, H.; Xu, X.; Gyu Kim, M.; Li, Q.; Park, N.; Liu, M.; Cho, J. Promotion of oxygen reduction by a bio-inspired tethered iron phthalocyanine carbon nanotube-based catalyst. *Nat. Commun.,* **2013**, *4*, 2076. [http://dx.doi.org/10.1038/ncomms3076] [PMID: 23797710]

[471] Swager, T.M. The molecular wire approach to sensory signal amplification. *Acc. Chem. Res.,* **1998**, *31*, 201-207. [http://dx.doi.org/10.1021/ar9600502]

[472] Zhao, Q.; Gan, Z.H.; Zhuang, Q.K. Electrochemical sensors based on carbon nanotubes. *Electroanal.,* **2002**, *14*, 1609-1613. [http://dx.doi.org/10.1002/elan.200290000]

[473] Lin, Y.; Yantasee, W.; Wang, J. Carbon nanotubes (CNTs) for the development of electrochemical biosensors. *Front. Biosci.,* **2005**, *10*, 492-505. [http://dx.doi.org/10.2741/1545] [PMID: 15574386]

[474] Wang, J. Carbon-nanotube based electrochemical biosensors: A review. *Electroanal.,* **2005**, *17*, 7-14. [http://dx.doi.org/10.1002/elan.200403113]

[475] Trojanowicz, M. Analytical applications of carbon nanotubes: a review. *TRAC- . Trends Analyt. Chem.,* **2006**, *25*, 480-489. [http://dx.doi.org/10.1016/j.trac.2005.11.008]

[476] Balasubramanian, K.; Burghard, M. Biosensors based on carbon nanotubes. *Anal. Bioanal. Chem.,* **2006**, *385*(3), 452-468. [http://dx.doi.org/10.1007/s00216-006-0314-8] [PMID: 16568294]

[477] Li, C.; Thostenson, E.T.; Chou, T.W. Sensors and actuators based on carbon nanotubes and their composites: A review. *Compos. Sci. Technol.,* **2008**, *68*, 1227-1249. [http://dx.doi.org/10.1016/j.compscitech.2008.01.006]

[478] Kauffman, D.R.; Star, A. Carbon nanotube gas and vapor sensors. *Angew. Chem. Int. Ed. Engl.,* **2008**, *47*(35), 6550-6570. [http://dx.doi.org/10.1002/anie.200704488] [PMID: 18642264]

[479] Yang, W.; Ratinac, K.R.; Ringer, S.P.; Thordarson, P.; Gooding, J.J.; Braet, F. Carbon nanomaterials

in biosensors: should you use nanotubes or graphene? *Angew. Chem. Int. Ed. Engl.,* **2010**, *49*(12), 2114-2138.
[http://dx.doi.org/10.1002/anie.200903463] [PMID: 20187048]

[480] Collins, P.G.; Bradley, K.; Ishigami, M.; Zettl, A. Extreme oxygen sensitivity of electronic properties of carbon nanotubes. *Science,* **2000**, *287*(5459), 1801-1804.
[http://dx.doi.org/10.1126/science.287.5459.1801] [PMID: 10710305]

[481] Kong, J.; Franklin, N.R.; Zhou, C.; Chapline, M.G.; Peng, S.; Cho, K.; Dai, H. Nanotube molecular wires as chemical sensors. *Science,* **2000**, *287*(5453), 622-625.
[http://dx.doi.org/10.1126/science.287.5453.622] [PMID: 10649989]

[482] Shahmiri, M.R.; Bahari, A.; Karimi-Maleh, H.; Hosseinzadeh, R.; Mirnia, N. Ethynylferrocene-NiO/MWCNT nanocomposite modified carbon paste electrode as a novel voltammetric sensor for simultaneous determination of glutathione and acetaminophen. *Sens. Actuators B Chem.,* **2013**, *177*, 70-77.
[http://dx.doi.org/10.1016/j.snb.2012.10.098]

[483] Jung, D.; Han, M.E.; Lee, G.S. pH-sensing characteristics of multi-walled carbon nanotube sheet. *Mater. Lett.,* **2014**, *116*, 57-60.
[http://dx.doi.org/10.1016/j.matlet.2013.10.095]

[484] Dresselhaus, M.S.; Eklund, P.C. Phonons in carbon nanotubes. *Adv. Phys.,* **2000**, *49*, 705-814.
[http://dx.doi.org/10.1080/000187300413184]

[485] Thostenson, E.T.; Ren, Z.F.; Chou, T.W. Advances in the science and technology of carbon nanotubes and their composites: a review. *Compos. Sci. Technol.,* **2001**, *61*, 1899-1912.
[http://dx.doi.org/10.1016/S0266-3538(01)00094-X]

[486] Li, Y.; Qian, F.; Xiang, J.; Lieber, C.M. Nanowire electronic and optoelectronic devices. *Mater. Today,* **2006**, *9*, 18-27.
[http://dx.doi.org/10.1016/S1369-7021(06)71650-9]

[487] Avouris, P.; Chen, Z.; Perebeinos, V. Carbon-based electronics. *Nat. Nanotechnol.,* **2007**, *2*(10), 605-615.
[http://dx.doi.org/10.1038/nnano.2007.300] [PMID: 18654384]

[488] Allen, B.L.; Kichambare, P.D.; Star, A. Carbon nanotube field-effect-transistor-based biosensors. *Adv. Mater.,* **2007**, *19*, 1439-1451.
[http://dx.doi.org/10.1002/adma.200602043]

[489] Hu, L.; Hecht, D.S.; Grüner, G. Carbon nanotube thin films: fabrication, properties, and applications. *Chem. Rev.,* **2010**, *110*(10), 5790-5844.
[http://dx.doi.org/10.1021/cr9002962] [PMID: 20939616]

[490] Park, S.; Vosguerichian, M.; Bao, Z. A review of fabrication and applications of carbon nanotube film-based flexible electronics. *Nanoscale,* **2013**, *5*(5), 1727-1752.
[http://dx.doi.org/10.1039/c3nr33560g] [PMID: 23381727]

[491] Wang, C.; Takei, K.; Takahashi, T.; Javey, A. Carbon nanotube electronics--moving forward. *Chem. Soc. Rev.,* **2013**, *42*(7), 2592-2609.
[http://dx.doi.org/10.1039/C2CS35325C] [PMID: 23229523]

[492] Ebbcsen, T.W.; Lezec, H.J.; Hiura, H.; Bennett, J.W.; Ghaemi, H.F.; Thio, T. Electrical conductivity of individual carbon nanotubes. *Nature,* **1996**, *382*, 54-56. [http://dx.doi.org/10.1038/382054a0]

[493] Yao, Z.; Postma, H.W.; Balents, L.; Dekker, C. Carbon nanotube intramolecular junctions. *Nature,* **1999**, *402*, 273-276. [http://dx.doi.org/10.1038/46241]

[494] Behabtu, N.; Young, C.C.; Tsentalovich, D.E.; Kleinerman, O.; Wang, X.; Ma, A.W.; Bengio, E.A.; ter Waarbeek, R.F.; de Jong, J.J.; Hoogerwerf, R.E.; Fairchild, S.B.; Ferguson, J.B.; Maruyama, B.; Kono, J.; Talmon, Y.; Cohen, Y.; Otto, M.J.; Pasquali, M. Strong, light, multifunctional fibers of carbon nanotubes with ultrahigh conductivity. *Science,* **2013**, *339*(6116), 182-186. [http://dx.doi.org/10.1126/science.1228061] [PMID: 23307737]

[495] Shulaker, M.M.; Hills, G.; Patil, N.; Wei, H.; Chen, H.Y.; Wong, H.S.; Mitra, S. Carbon nanotube computer. *Nature,* **2013**, *501*(7468), 526-530. [http://dx.doi.org/10.1038/nature12502] [PMID: 24067711]

[496] Granqvist, C.G. Transparent conductors as solar energy materials: A panoramic review. *Sol. Energy Mater. Sol. Cells,* **2007**, *91*, 1529-1598. [http://dx.doi.org/10.1016/j.solmat.2007.04.031]

[497] Hains, A.W.; Liang, Z.; Woodhouse, M.A.; Gregg, B.A. Molecular semiconductors in organic photovoltaic cells. *Chem. Rev.,* **2010**, *110*(11), 6689-6735. [http://dx.doi.org/10.1021/cr9002984] [PMID: 20184362]

[498] Cataldo, S.; Salice, P.; Menna, P.; Pignataro, B. Carbon nanotubes and organic solar cells. *Energy Environ. Sci.,* **2012**, *5*, 5919-5940. [http://dx.doi.org/10.1039/C1EE02276H]

[499] Chen, T.; Qiu, L.; Yang, Z.; Peng, H. Novel solar cells in a wire format. *Chem. Soc. Rev.,* **2013**, *42*(12), 5031-5041. [http://dx.doi.org/10.1039/c3cs35465b] [PMID: 23467786]

[500] Ago, H.; Petritsch, K.; Shaffer, M.S.; Windle, A.H.; Friend, R.H. Composites of carbon nanotubes and conjugated polymers for photovoltaic devices. *Adv. Mater.,* **1999**, *11*, 1281-1285. [http://dx.doi.org/10.1002/(SICI)1521-4095(199910)11:15<1281::AID-ADMA1281>3.0.CO;2-6]

[501] Jung, Y.; Li, X.; Rajan, N.K.; Taylor, A.D.; Reed, M.A. Record high efficiency single-walled carbon nanotube/silicon p-n junction solar cells. *Nano Lett.,* **2013**, *13*(1), 95-99. [http://dx.doi.org/10.1021/nl3035652] [PMID: 23237412]

[502] Yang, Z.; Liu, M.; Zhang, C.; Tjiu, W.W.; Liu, T.; Peng, H. Carbon nanotubes bridged with graphene nanoribbons and their use in high-efficiency dye-sensitized solar cells. *Angew. Chem. Int. Ed. Engl.,* **2013**, *52*(14), 3996-3999. [http://dx.doi.org/10.1002/anie.201209736] [PMID: 23401014]

[503] Yang, L.; Leung, W.W. Electrospun TiO2 nanorods with carbon nanotubes for efficient electron collection in dye-sensitized solar cells. *Adv. Mater.,* **2013**, *25*(12), 1792-1795. [http://dx.doi.org/10.1002/adma.201204256] [PMID: 23355287]

[504] Golobostanfard, M.R.; Abdizadeh, H. Hierarchical porous titania/carbon nanotube nanocomposite

photoanode synthesized by controlled phase separation for dye sensitized solar cell. *Sol. Energy Mater. Sol. Cells,* **2014**, *120*, 295-302.
[http://dx.doi.org/10.1016/j.solmat.2013.09.021]

[505] Yang, W.; Thordarson, P.; Gooding, J.J.; Ringer, S.P.; Braet, F. Carbon nanotubes for biological and biomedical applications. *Nanotechnology,* **2007**, *18*, 412001.
[http://dx.doi.org/10.1088/0957-4484/18/41/412001]

[506] Liang, F.; Chen, B. A review on biomedical applications of single-walled carbon nanotubes. *Curr. Med. Chem.,* **2010**, *17*(1), 10-24.
[http://dx.doi.org/10.2174/092986710789957742] [PMID: 19941481]

[507] *Carbon nanotubes for biomedical applications in Springer Series: Carbon Nanostructures. Klingeler, R*; Sim, R.B., Ed.; Springer-Verlag: Berlin, **2011**.

[508] Vardharajula, S.; Ali, S.Z.; Tiwari, P.M.; Eroğlu, E.; Vig, K.; Dennis, V.A.; Singh, S.R. Functionalized carbon nanotubes: biomedical applications. *Int. J. Nanomedicine,* **2012**, *7*, 5361-5374.
[PMID: 23091380]

[509] Roldo, M.; Fatouros, D.G. Biomedical applications of carbon nanotubes. *Annu. Rep. Prog. Chem. Sect. C: Phys. Chem.,* **2013**, *109*, 10-35.

[510] Ding, R.G.; Lu, G.Q.; Yan, Z.F.; Wilson, M.A. Recent advances in the preparation and utilization of carbon nanotubes for hydrogen storage. *J. Nanosci. Nanotechnol.,* **2001**, *1*(1), 7-29.
[http://dx.doi.org/10.1166/jnn.2001.012] [PMID: 12914026]

[511] Kim, O-K.; Je, J.; Baldwin, J.W.; Kooi, S.; Pehrsson, P.E.; Buckley, L.J. Solubilization of single-wall carbon nanotubes by supramolecular encapsulation of helical amylose. *J. Am. Chem. Soc.,* **2003**, *125*(15), 4426-4427.
[http://dx.doi.org/10.1021/ja029233b] [PMID: 12683805]

[512] Staii, C.; Johnson, A.T., Jr; Chen, M.; Gelperin, A. DNA-decorated carbon nanotubes for chemical sensing. *Nano Lett.,* **2005**, *5*(9), 1774-1778.
[http://dx.doi.org/10.1021/nl051261f] [PMID: 16159222]

[513] Kam, N.W.; O'Connell, M.; Wisdom, J.A.; Dai, H. Carbon nanotubes as multifunctional biological transporters and near-infrared agents for selective cancer cell destruction. *Proc. Natl. Acad. Sci. USA,* **2005**, *102*(33), 11600-11605.
[http://dx.doi.org/10.1073/pnas.0502680102] [PMID: 16087878]

[514] Pai, P.; Nair, K.; Jamade, S.; Shah, R.; Ekshinge, V.; Jadhav, N. Pharmaceutical applications of carbon tubes and nanohorns. *Curr. Pharma Res. J.,* **2006**, *1*, 11-15.

[515] Gheith, M.K.; Pappas, T.C.; Liopo, A.V.; Sinani, V.A.; Shim, B.S.; Motamedi, M.; Wicksted, J.P.; Kotov, N.A. Stimulation of neural cells by lateral currents in conductive layer-by-layer films of single-walled carbon nanotubes. *Adv. Mater.,* **2006**, *18*, 2975-2975.
[http://dx.doi.org/10.1002/adma.200600878]

[516] Barone, P.W.; Strano, M.S. Reversible control of carbon nanotube aggregation for a glucose affinity sensor. *Angew. Chem. Int. Ed. Engl.,* **2006**, *45*(48), 8138-8141.
[http://dx.doi.org/10.1002/anie.200603138] [PMID: 17099921]

[517] Chambers, A.; Park, C.; Baker, R.T.; Rodriguez, N.M. Hydrogen storage in graphite nanofibers. *J.*

Phys. Chem. B, **1998**, *102*, 4253-4256.
[http://dx.doi.org/10.1021/jp980114l]

[518] Dillon, A.C.; Jones, K.M.; Bekkedahl, T.A.; Kiang, C.H.; Bethune, D.S.; Heben, M.J. Storage of hydrogen in single-walled carbon nanotubes. *Nature,* **1997**, *386*, 377-379.
[http://dx.doi.org/10.1038/386377a0]

[519] Wang, Q.Y.; Johnson, J.K. Computer simulations of hydrogen adsorption on graphite nanofibers. *J. Phys. Chem. B,* **1999**, *103*, 277-281.
[http://dx.doi.org/10.1021/jp9839100]

[520] Lee, H.; Kang, Y.S.; Kim, S.H.; Lee, J.Y. Hydrogen desorption properties of multiwall carbon nanotubes with closed and open structures. *Appl. Phys. Lett.,* **2002**, *80*, 577-579.
[http://dx.doi.org/10.1063/1.1446208]

[521] Lueking, A.D.; Yang, R.T. Hydrogen storage in carbon nanotubes: Residual metal content and pretreatment temperature. *AIChE J.,* **2003**, *49*, 1556-1568.
[http://dx.doi.org/10.1002/aic.690490619]

[522] Li, X.; Zhu, H.; Ci, L.; Xu, C.; Mao, Z.; Wei, B.; Ji, L.; Wu, D. Hydrogen uptake by graphitized multi-walled carbon nanotubes under moderate pressure and at room temperature. *Carbon,* **2002**, *39*, 2077-2079.
[http://dx.doi.org/10.1016/S0008-6223(01)00183-X]

[523] Hou, P.; Yang, Q.; Bai, S.; Xu, S.; Liu, M.; Cheng, H. Bulk storage capacity of hydrogen in purified multiwalled carbon nanotubes. *J. Phys. Chem. B,* **2002**, *106*, 963-966.
[http://dx.doi.org/10.1021/jp0136112]

[524] Rzepka, M.; Lamp, P.; de la Casa-lillo, M.A. Physisorption of hydrogen on microporous carbon and carbon nanotubes. *J. Phys. Chem. B,* **1998**, *102*, 10894-10898.
[http://dx.doi.org/10.1021/jp9829602]

[525] Dillon, A.C.; Heben, M.J. Hydrogen storage using carbon adsorbents: past, present and future. *Appl. Phys., A Mater. Sci. Process.,* **2001**, *72*, 133-142.
[http://dx.doi.org/10.1007/s003390100788]

[526] Ye, Y.; Ahn, C.C.; Whitam, C.; Fultz, B.; Liu, L.; Rinzler, A.G.; Colbert, D.; Smith, K.A.; Smalley, R.E. Hydrogen adsorption and cohesive energy of single-walled carbon nanotubes. *Appl. Phys. Lett.,* **1999**, *74*, 2307-2309.
[http://dx.doi.org/10.1063/1.123833]

[527] Lee, S.M.; Lee, Y.H. Hydrogen storage in single-walled carbon nanotubes. *Appl. Phys. Lett.,* **2000**, *76*, 2877-2879.
[http://dx.doi.org/10.1063/1.126503]

[528] Park, C.; Anderson, P.E.; Chambers, A.; Tan, S.D.; Hidalgo, R.; Rodriguez, N.M. Further studies of the interaction of hydrogen with graphite nanofibers. *J. Phys. Chem. B,* **1999**, *103*, 10572-10581.
[http://dx.doi.org/10.1021/jp990500i]

[529] Zhang, W.H.; Shi, J.L.; Wang, L.Z.; Yan, D.S. Preparation and characterization of ZnO clusters inside mesoporous silica. *Chem. Mater.,* **2000**, *12*, 1408-1413.
[http://dx.doi.org/10.1021/cm990740a]

[530] Bessel, C.A.; Laubernds, K.; Rodriguez, N.M.; Baker, R.T. Graphite nanofibers as an electrode for fuel cell applications. *J. Phys. Chem. B,* **2001**, *105*, 1115-1118. [http://dx.doi.org/10.1021/jp003280d]

[531] Steigerwalt, E.S.; Deluga, G.A.; Lukehart, C.M. Pt–Ru/carbon fiber nanocomposites: synthesis, characterization, and performance as anode catalysts of direct methanol fuel cells. A search for exceptional performance. *J. Phys. Chem. B,* **2002**, *106*, 760-766. [http://dx.doi.org/10.1021/jp012707t]

[532] Liu, Y.C.; Qiu, X.P.; Huang, Y.Q.; Zhu, W.T. Methanol electro-oxidation on mesocarbon microbead supported Pt catalysts. *Carbon,* **2002**, *40*, 2375-2380. [http://dx.doi.org/10.1016/S0008-6223(02)00115-X]

[533] Takasu, Y.; Kawaguchi, T.; Sugimoto, W.; Murakami, Y. Effects of the surface area of carbon support on the characteristics of highly-dispersed Pt-Ru particles as catalysts for methanol oxidation. *Electrochim. Acta,* **2003**, *48*, 3861-3868. [http://dx.doi.org/10.1016/S0013-4686(03)00521-8]

[534] Park, K.W.; Sung, Y.E.; Han, S.; Yun, Y.; Hyeon, T. Origin of the enhanced catalytic activity of carbon nanocoil-supported PtRu alloy electrocatalysts. *J. Phys. Chem. B,* **2004**, *108*, 939-944. [http://dx.doi.org/10.1021/jp0368031]

[535] Hull, R.V.; Li, L.; Xing, Y.; Chusuei, C.C. Pt Nanoparticle binding on functionalized multiwalled carbon nanotubes. *Chem. Mater.,* **2006**, *18*, 1780-1788. [http://dx.doi.org/10.1021/cm0518978]

[536] Saito, R.; Fujita, M.; Dresselhaus, G.; Dresselhaus, M. Electronic structure of chiral graphene tubules. *Appl. Phys. Lett.,* **1992**, *60*, 2204-2206. [http://dx.doi.org/10.1063/1.107080]

[537] Kasumov, A.Y.; Bouchiat, H.; Reulet, B.; Stephan, O.; Khodos, I.I.; Gorbatov, Y.B.; Colliex, C. *Europhys. Lett.,* **1998**, *43*, 89. [http://dx.doi.org/10.1209/epl/i1998-00324-1]

[538] Kim, C.; Kim, Y.J.; Kim, Y.A.; Yanagisawa, T.; Park, K.C.; Endo, M.; Dresselhaus, M.S. High performance of cup-stacked-type carbon nanotubes as a Pt–Ru catalyst support for fuel cell applications. *J. Appl. Phys.,* **2004**, *96*, 5903-5905. [http://dx.doi.org/10.1063/1.1804242]

[539] Li, W.; Liang, C.; Zhou, W.; Qiu, J.; Li, H.; Sun, G.; Xin, Q. Homogeneous and controllable Pt particles deposited on multi-wall carbon nanotubes as cathode catalyst for direct methanol fuel cells. *Carbon,* **2004**, *42*, 436-439. [http://dx.doi.org/10.1016/j.carbon.2003.10.033]

[540] Matsumoto, T.; Komatsu, T.; Arai, K.; Yamazaki, T.; Kijima, M.; Shimizu, H.; Takasawa, Y.; Nakamura, J. Reduction of Pt usage in fuel cell electrocatalysts with carbon nanotube electrodes. *Chem. Commun. (Camb.),* **2004**, (7), 840-841. [http://dx.doi.org/10.1039/b400607k] [PMID: 15045090]

[541] Saha, M.S.; Li, R.; Sun, X. High loading and monodispersed Pt nanoparticles on multiwalled carbon nanotubes for high performance proton exchange membrane fuel cells. *J. Power Sources,* **2008**, *177*,

314-322.
[http://dx.doi.org/10.1016/j.jpowsour.2007.11.036]

[542] Maiyalagan, T.; Viswanathan, B.; Varadaraju, U.V. Nitrogen containing carbon nanotubes as supports for Pt – Alternate anodes for fuel cell applications. *Electrochem. Commun.,* **2005**, *7*, 905-912.
[http://dx.doi.org/10.1016/j.elecom.2005.07.007]

[543] Yang, S.; Zhang, X.; Mi, H.; Ye, X. Pd nanoparticles supported on functionalized multi-walled carbon nanotubes (MWCNTs) and electrooxidation for formic acid. *J. Power Sources,* **2008**, *175*, 26-32.
[http://dx.doi.org/10.1016/j.jpowsour.2007.09.080]

[544] Wu, G.; Xu, B.Q. Carbon nanotube supported Pt electrodes for methanol oxidation: A comparison between multi- and single-walled carbon nanotubes. *J. Power Sources,* **2007**, *174*, 148-158.
[http://dx.doi.org/10.1016/j.jpowsour.2007.08.024]

[545] Wang, X.; Li, W.; Chen, Z.; Waje, M.; Yan, Y. Durability investigation of carbon nanotube as catalyst support for proton exchange membrane fuel cell. *J. Power Sources,* **2006**, *158*, 154-159.
[http://dx.doi.org/10.1016/j.jpowsour.2005.09.039]

[546] Kim, J.H.; Zhu, K.; Yan, Y.; Perkins, C.L.; Frank, A.J. Microstructure and pseudocapacitive properties of electrodes constructed of oriented NiO-TiO2 nanotube arrays. *Nano Lett.,* **2010**, *10*(10), 4099-4104.
[http://dx.doi.org/10.1021/nl102203s] [PMID: 20873847]

[547] Korenblit, Y.; Rose, M.; Kockrick, E.; Borchardt, L.; Kvit, A.; Kaskel, S.; Yushin, G. High-rate electrochemical capacitors based on ordered mesoporous silicon carbide-derived carbon. *ACS Nano,* **2010**, *4*(3), 1337-1344.
[http://dx.doi.org/10.1021/nn901825y] [PMID: 20180559]

[548] Burke, A.F. Batteries and ultracapacitors for electric, hybrid, and fuel cell vehicles. *Proc. IEEE,* **2007**, *95*, 806-820.
[http://dx.doi.org/10.1109/JPROC.2007.892490]

[549] Obreja, V.V. On the performance of supercapacitors with electrodes based on carbon nanotubes and carbon activated material-A review. *Physica E,* **2008**, *40*, 2596-2605.
[http://dx.doi.org/10.1016/j.physe.2007.09.044]

[550] Lota, G.; Fic, K.; Frackowiak, E. Carbon nanotubes and their composites in electrochemical applications. *Energy Environ. Sci.,* **2011**, *4*, 1592-1605.
[http://dx.doi.org/10.1039/c0ee00470g]

[551] Yoon, B.J.; Jeong, S.H.; Lee, K.H.; Kim, H.S.; Park, C.G.; Han, J.H. Electrical properties of electrical double layer capacitors with integrated carbon nanotube electrodes. *Chem. Phys. Lett.,* **2004**, *388*, 170-174.
[http://dx.doi.org/10.1016/j.cplett.2004.02.071]

[552] Huang, Z.P.; Wu, J.W.; Ren, Z.F.; Wang, J.H.; Siegal, M.P.; Provencio, P.N. Growth of highly oriented carbon nanotubes by plasma-enhanced hot filament chemical vapor deposition. *Appl. Phys. Lett.,* **1998**, *73*, 3845-3847.
[http://dx.doi.org/10.1063/1.122912]

[553] Frackowiak, E.; Beguin, F. Carbon materials for the electrochemical storage of energy in capacitors.

Carbon, **2001**, *39*, 937-950.
[http://dx.doi.org/10.1016/S0008-6223(00)00183-4]

[554] Niu, C.M.; Sichel, E.K.; Hoch, R.; Moy, D.; Tennent, H. High power electrochemical capacitors based on carbon nanotube electrodes. *Appl. Phys. Lett.,* **1997**, *70*, 1480-1482.
[http://dx.doi.org/10.1063/1.118568]

[555] Azais, P.; Duclaux, L.; Florian, P.; Massiot, D.; Lillo-Rodenas, M.A.; Linares-Solano, A.; Peres, J.P.; Jehoulet, C.; Beguin, F. Causes of supercapacitors ageing in organic electrolyte. *J. Power Sources,* **2007**, *171*, 1046-1053.
[http://dx.doi.org/10.1016/j.jpowsour.2007.07.001]

[556] Du, C.S.; Yeh, J.; Pan, N. High power density supercapacitors using locally aligned carbon nanotube electrodes. *Nanotechnology,* **2005**, *16*, 350-353.
[http://dx.doi.org/10.1088/0957-4484/16/4/003]

[557] Gao, L.J.; Peng, A.P.; Wang, Z.Y.; Zhang, H.; Shi, Z.J.; Gu, Z.N.; Cao, G.P.; Ding, B.Z. Growth of aligned carbon nanotube arrays on metallic substrate and its application to supercapacitors. *Solid State Commun.,* **2008**, *146*, 380-383.
[http://dx.doi.org/10.1016/j.ssc.2008.03.034]

[558] Emmenegger, C.; Bonard, J.M.; Mauron, P.; Sudan, P.; Lepora, A.; Grobety, B.; Zuttel, A.; Schlapbach, L. Synthesis of carbon nanotubes over Fe catalyst on aluminium and suggested growth mechanism. *Carbon,* **2003**, *41*, 539-547.
[http://dx.doi.org/10.1016/S0008-6223(02)00362-7]

[559] Kang, J.; Zhang, S.; Zhang, Q.; Wang, Y. Ruthenium nanoparticles supported on carbon nanotubes as efficient catalysts for selective conversion of synthesis gas to diesel fuel. *Angew. Chem. Int. Ed. Engl.,* **2009**, *48*(14), 2565-2568.
[http://dx.doi.org/10.1002/anie.200805715] [PMID: 19248073]

[560] Vu, H.; Gonçalves, F.; Philippe, R.; Lamouroux, E.; Corrias, M.; Kihn, Y.; Plee, D.; Kalck, P.; Serp, P. Bimetallic catalysis on carbon nanotubes for the selective hydrogenation of cinnamaldehyde. *J. Catal.,* **2006**, *240*, 18-22.
[http://dx.doi.org/10.1016/j.jcat.2006.03.003]

[561] Yin, S.F.; Zhang, Q.H.; Xu, B.Q.; Zhu, W-X.; Ng, C-F.; Au, C-T. Investigation on the catalysis of COx-free hydrogen generation from ammonia. *J. Catal.,* **2004**, *224*, 384-396.
[http://dx.doi.org/10.1016/j.jcat.2004.03.008]

[562] Bambagioni, V.; Bianchini, C.; Marchionni, A.; Filippi, J.; Vizza, F.; Teddy, J.; Serp, P.; Zhiani, M. Pd and Pt–Ru anode electrocatalysts supported on multi-walled carbon nanotubes and their use in passive and active direct alcohol fuel cells with an anion-exchange membrane (alcohol = methanol, ethanol, glycerol). *J. Power Sources,* **2009**, *190*, 241-251.
[http://dx.doi.org/10.1016/j.jpowsour.2009.01.044]

[563] Oresmaa, L.; Moreno, M.A.; Jakonen, M.; Suvanto, S.; Auca, M. Catalytic activity of linear chain ruthenium carbonyl polymer [Ru(CO)4]n in 1-hexene hydroformylation. *Appl. Catal. A Gen.,* **2009**, *353*, 113-116.
[http://dx.doi.org/10.1016/j.apcata.2008.10.028]

[564] Wang, W.D.; Serp, P.; Kalck, P.; Silva, C.G.; Faria, J.L. Preparation and characterization of

nanostructured MWCNT-TiO2 composite materials for photocatalytic water treatment applications. *Mater. Res. Bull.,* **2008**, *43*, 958-967.
[http://dx.doi.org/10.1016/j.materresbull.2007.04.032]

[565] Giordano, R.; Serp, P.; Kalck, P.; Kihn, Y.; Schreiber, J.; Marhic, C.; Duvail, J-L. Preparation of rhodium catalysts supported on carbon nanotubes by a surface mediated organometallic reaction. *Eur. J. Inorg. Chem.,* **2003**, 610-617.
[http://dx.doi.org/10.1002/ejic.200390083]

[566] Luo, J.Z.; Gao, L.Z.; Leung, Y.L.; Au, C.T. The decomposition of NO on CNTs and 1 wt% Rh/CNTs. *Catal. Lett.,* **2000**, *66*, 91-97.
[http://dx.doi.org/10.1023/A:1019035220233]

[567] Liu, Z-J.; Xu, Z.; Yuan, Z-Y.; Lu, D.; Chen, W.; Zhou, W. Cyclohexanol dehydrogenation over Co/carbon nanotube catalysts and the effect of promoter K on performance. *Catal. Lett.,* **2001**, *72*, 203-206.
[http://dx.doi.org/10.1023/A:1009040210044]

[568] Girishkumar, G.; Vinodgopal, K.; Kamat, P.V. Carbon nanostructures in portable fuel cells: single-walled carbon nanotube electrodes for methanol oxidation and oxygen reduction. *J. Phys. Chem. B,* **2004**, *108*, 19960-19966.
[http://dx.doi.org/10.1021/jp046872v]

[569] Cui, H.F.; Ye, J.S.; Zhang, W.D.; Wang, J.; Sheu, F.S. Electrocatalytic reduction of oxygen by a platinum nanoparticle/carbon nanotube composite electrode. *J. Electroanal. Chem.,* **2005**, *577*, 295-302.
[http://dx.doi.org/10.1016/j.jelechem.2004.12.004]

[570] Ye, J.S.; Cui, H.F.; Wen, Y.; Zhang, W.D.; Xu, G.Q.; Sheu, F.S. Electrodeposition of platinum nanoparticles on multi-walled carbon nanotubes for electrocatalytic oxidation of methanol. *Mikrochim. Acta,* **2006**, *152*, 267-275.
[http://dx.doi.org/10.1007/s00604-005-0446-0]

[571] He, Z.B.; Chen, J.H.; Liu, D.Y.; Zhou, H.H.; Kuang, Y.F. Electrodeposition of Pt–Ru nanoparticles on carbon nanotubes and their electrocatalytic properties for methanol electrooxidation. *Diamond Related Materials,* **2004**, *13*, 1764-1770.
[http://dx.doi.org/10.1016/j.diamond.2004.03.004]

[572] Collins, P.G.; Bradley, K.; Ishigami, M.; Zettl, A. Extreme oxygen sensitivity of electronic properties of carbon nanotubes. *Science,* **2000**, *287*(5459), 1801-1804.
[http://dx.doi.org/10.1126/science.287.5459.1801] [PMID: 10710305]

[573] Vichchulada, P.; Zhang, Q.; Lay, M.D. Recent progress in chemical detection with single-walled carbon nanotube networks. *Analyst (Lond.),* **2007**, *132*(8), 719-723.
[http://dx.doi.org/10.1039/b618824a] [PMID: 17646869]

[574] Li, J.; Lu, Y.; Ye, Q.; Cinke, M.; Han, J.; Meyyappan, M. Carbon nanotube sensors for gas and organic vapor detection. *Nano Lett.,* **2003**, *3*, 929-933.
[http://dx.doi.org/10.1021/nl034220x]

[575] Kong, J.; Chapline, M.G.; Dai, H.J. Functionalized carbon nanotubes for molecular hydrogen sensors. *Adv. Mater.,* **2001**, *13*, 1384-1386.

[http://dx.doi.org/10.1002/1521-4095(200109)13:18<1384::AID-ADMA1384>3.0.CO;2-8]

[576] Mubeen, S.; Zhang, T.; Yoo, B.; Deshusses, M.A.; Myung, N.V. Palladium nanoparticles decorated single-walled carbon nanotube hydrogen sensor. *J. Phys. Chem. C,* **2007**, *111*, 6321-6327. [http://dx.doi.org/10.1021/jp067716m]

[577] Sayago, I.; Terrado, E.; Lafuente, E.; Horrillo, M.C.; Maser, W.K.; Benito, A.M.; Navarro, R.; Urriolabeitia, E.P.; Martinez, M.T.; Gutierrez, J. Hydrogen sensors based on carbon nanotubes thin films. *Synth. Met.,* **2005**, *148*, 15-19. [http://dx.doi.org/10.1016/j.synthmet.2004.09.013]

[578] Kumar, M.K.; Ramaprabhu, S. Nanostructured Pt functionlized multiwalled carbon nanotube based hydrogen sensor. *J. Phys. Chem. B,* **2006**, *110*(23), 11291-11298. [http://dx.doi.org/10.1021/jp0611525] [PMID: 16771399]

[579] Sun, Y.; Wang, H.H.; Xia, M. Single-walled carbon nanotubes modified with Pd nanoparticles: unique building blocks for high-performance, flexible hydrogen sensors. *J. Phys. Chem. C,* **2008**, *112*, 1250-1259. [http://dx.doi.org/10.1021/jp076965n]

[580] Yoon, H.; Xie, J.N.; Abraham, J.K.; Varadan, V.K.; Ruffin, P.B. Passive wireless sensors using electrical transition of carbon nanotube junctions in polymer matrix. *Smart Mater. Struct.,* **2006**, *15*, S14-S20. [http://dx.doi.org/10.1088/0964-1726/15/1/004]

[581] Wei, C.; Dai, L.; Roy, A.; Tolle, T.B. Multifunctional chemical vapor sensors of aligned carbon nanotube and polymer composites. *J. Am. Chem. Soc.,* **2006**, *128*(5), 1412-1413. [http://dx.doi.org/10.1021/ja0570335] [PMID: 16448087]

[582] Gooding, J.J. Nanostructuring electrodes with carbon nanotubes: A review on electrochemistry and applications for sensing. *Electrochim. Acta,* **2005**, *50*, 3049-3060. [http://dx.doi.org/10.1016/j.electacta.2004.08.052]

[583] McCreery, R.L. Advanced carbon electrode materials for molecular electrochemistry. *Chem. Rev.,* **2008**, *108*(7), 2646-2687. [http://dx.doi.org/10.1021/cr068076m] [PMID: 18557655]

[584] Katz, E.; Willner, I. Biomolecule-functionalized carbon nanotubes: applications in nanobioelectronics. *ChemPhysChem,* **2004**, *5*(8), 1084-1104. [http://dx.doi.org/10.1002/cphc.200400193] [PMID: 15446731]

[585] Du, P.; Wu, P.; Cai, C. A glucose biosensor based on electrocatalytic oxidation of NADPH at single-walled carbon nanotubes functionalized with poly(nile blue A). *J. Electroanal. Chem.,* **2008**, *624*, 21-26. [http://dx.doi.org/10.1016/j.jelechem.2008.07.031]

[586] He, P.; Wang, Z.; Zhang, L.; Yang, W. Development of a label-free electrochemical immunosensor based on carbon nanotube for rapid determination of clenbuterol. *Food Chem.,* **2009**, *112*, 707-714. [http://dx.doi.org/10.1016/j.foodchem.2008.05.116]

[587] Cheng, Y.; Liu, Y.; Huang, J.; Li, K.; Xian, Y.; Zhang, W.; Jin, L. Amperometric tyrosinase biosensor based on Fe3O4 nanoparticles-coated carbon nanotubes nanocomposite for rapid detection of

coliforms. *Electrochim. Acta,* **2009**, *54*, 2588-2594.
[http://dx.doi.org/10.1016/j.electacta.2008.10.072]

[588] Schuettfort, T.; Nish, A.; Nicholas, R.J. Observation of a type II heterojunction in a highly ordered polymer-carbon nanotube nanohybrid structure. *Nano Lett.,* **2009**, *9*(11), 3871-3876.
[http://dx.doi.org/10.1021/nl902081t] [PMID: 19795839]

[589] Ferguson, A.J.; Blackburn, J.L.; Holt, J.M.; Kopidakis, N.; Tenent, R.C.; Barnes, T.M.; Heben, M.J.; Rumbles, G. Photoinduced energy and charge transfer in P3HT:SWNT composites. *J. Phys. Chem. Lett.,* **2010**, *1*, 2406-2411.
[http://dx.doi.org/10.1021/jz100768f]

[590] Kanai, Y.; Grossman, J.C. Role of semiconducting and metallic tubes in P3HT/carbon-nanotube photovoltaic heterojunctions: density functional theory calculations. *Nano Lett.,* **2008**, *8*(3), 908-912.
[http://dx.doi.org/10.1021/nl0732777] [PMID: 18290634]

[591] Schuettfort, T.; Snaith, H.J.; Nish, A.; Nicholas, R.J. Synthesis and spectroscopic characterization of solution processable highly ordered polythiophene-carbon nanotube nanohybrid structures. *Nanotechnology,* **2010**, *21*(2), 025201.
[http://dx.doi.org/10.1088/0957-4484/21/2/025201] [PMID: 19955610]

[592] Geng, J.; Zeng, T. Influence of single-walled carbon nanotubes induced crystallinity enhancement and morphology change on polymer photovoltaic devices. *J. Am. Chem. Soc.,* **2006**, *128*(51), 16827-16833.
[http://dx.doi.org/10.1021/ja065035z] [PMID: 17177433]

[593] Miller, A.J.; Hatton, R.A.; Silva, S.R. Water-soluble multiwall-carbon-nanotube-polythiophene composite for bilayer photovoltaics. *Appl. Phys. Lett.,* **2006**, *89*, 123115.
[http://dx.doi.org/10.1063/1.2356115]

[594] Kymakis, E.; Alexandrou, I.; Amaratunga, G.A. High open-circuit voltage photovoltaic devices from carbon-nanotube-polymer composites. *J. Appl. Phys.,* **2003**, *93*, 1764-1768.
[http://dx.doi.org/10.1063/1.1535231]

[595] Kymakis, E.; Amaratunga, G.A. Single-wall carbon nanotube/conjugated polymer photovoltaic devices. *Appl. Phys. Lett.,* **2002**, *80*, 112-114.
[http://dx.doi.org/10.1063/1.1428416]

[596] Kazaoui, S.; Minami, N.; Nalini, B.; Kim, Y.; Hara, K. Near-infrared photoconductive and photovoltaic devices using single-wall carbon nanotubes in conductive polymer films. *J. Appl. Phys.,* **2005**, *98*, 084314.
[http://dx.doi.org/10.1063/1.2113419]

[597] Kymakis, E.; Koudoumas, E.; Franghiadakis, I.; Amaratunga, G.A. Post-fabrication annealing effects in polymer-nanotube photovoltaic cells. *J. Phys. D Appl. Phys.,* **2006**, *39*, 1058-1062.
[http://dx.doi.org/10.1088/0022-3727/39/6/010]

[598] Lanzi, M.; Paganin, L.; Caretti, D. New photoactive oligo- and poly-alkylthiophenes. *Polymer (Guildf.),* **2008**, *49*, 4942-4948.
[http://dx.doi.org/10.1016/j.polymer.2008.09.013]

[599] Khodakovskaya, M.V.; de Silva, K.; Nedosekin, D.A.; Dervishi, E.; Biris, A.S.; Shashkov, E.V.;

Galanzha, E.I.; Zharov, V.P. Complex genetic, photothermal, and photoacoustic analysis of nanoparticle-plant interactions. *Proc. Natl. Acad. Sci. USA,* **2011**, *108*(3), 1028-1033. [http://dx.doi.org/10.1073/pnas.1008856108] [PMID: 21189303]

[600] Mahmood, M.; Casciano, D.A.; Mocan, T.; Iancu, C.; Xu, Y.; Mocan, L.; Iancu, D.T.; Dervishi, E.; Li, Z.; Abdalmuhsen, M.; Biris, A.R.; Ali, N.; Howard, P.; Biris, A.S. Cytotoxicity and biological effects of functional nanomaterials delivered to various cell lines. *J. Appl. Toxicol.,* **2010**, *30*(1), 74-83. [http://dx.doi.org/10.1002/jat.1475] [PMID: 19760634]

[601] Zhang, Y.; Ali, S.F.; Dervishi, E.; Xu, Y.; Li, Z.; Casciano, D.; Biris, A.S. Cytotoxicity effects of graphene and single-wall carbon nanotubes in neural phaeochromocytoma-derived PC12 cells. *ACS Nano,* **2010**, *4*(6), 3181-3186. [http://dx.doi.org/10.1021/nn1007176] [PMID: 20481456]

[602] Zhang, Y.; Xu, Y.; Li, Z.; Chen, T.; Lantz, S.M.; Howard, P.C.; Paule, M.G.; Slikker, W., Jr; Watanabe, F.; Mustafa, T.; Biris, A.S.; Ali, S.F. Mechanistic toxicity evaluation of uncoated and PEGylated single-walled carbon nanotubes in neuronal PC12 cells. *ACS Nano,* **2011**, *5*(9), 7020-7033. [http://dx.doi.org/10.1021/nn2016259] [PMID: 21866971]

[603] Ali-Boucetta, H.; Al-Jamal, K.T.; Kostarelos, K. Cytotoxic assessment of carbon nanotube interaction with cell cultures. *Methods Mol. Biol.,* **2011**, *726*, 299-312. [http://dx.doi.org/10.1007/978-1-61779-052-2_19] [PMID: 21424457]

[604] Higgins, P.; Dawson, J.; Walters, M. Nanomedicine: Nanotubes reduce stroke damage. *Nat. Nanotechnol.,* **2011**, *6*(2), 83-84. [http://dx.doi.org/10.1038/nnano.2011.5] [PMID: 21278751]

[605] Ge, C.; Du, J.; Zhao, L.; Wang, L.; Liu, Y.; Li, D.; Yang, Y.; Zhou, R.; Zhao, Y.; Chai, Z.; Chen, C. Binding of blood proteins to carbon nanotubes reduces cytotoxicity. *Proc. Natl. Acad. Sci. USA,* **2011**, *108*(41), 16968-16973. [http://dx.doi.org/10.1073/pnas.1105270108] [PMID: 21969544]

[606] Wang, X.; Xia, T.; Ntim, S.A.; Ji, Z.; George, S.; Meng, H.; Zhang, H.; Castranova, V.; Mitra, S.; Nel, A.E. Quantitative techniques for assessing and controlling the dispersion and biological effects of multiwalled carbon nanotubes in mammalian tissue culture cells. *ACS Nano,* **2010**, *4*(12), 7241-7252. [http://dx.doi.org/10.1021/nn102112b] [PMID: 21067152]

[607] Nish, A.; Nicholas, R.J. Temperature induced restoration of fluorescence from oxidised single-walled carbon nanotubes in aqueous sodium dodecylsulfate solution. *Phys. Chem. Chem. Phys.,* **2006**, *8*(30), 3547-3551. [http://dx.doi.org/10.1039/b604045d] [PMID: 16871344]

[608] Matarredona, O.; Rhoads, H.; Li, Z.; Harwell, J.H.; Balzano, L.; Resasco, D.E. Dispersion of single-walled carbon nanotubes in aqueous solutions of the anionic surfactant NaDDBS. *J. Phys. Chem. B,* **2003**, *107*, 13357-13367. [http://dx.doi.org/10.1021/jp0365099]

[609] Tu, X.; Zheng, M. A DNA-based approach to the carbon nanotube sorting problem. *Nano Res.,* **2008**, *1*, 185-194. [http://dx.doi.org/10.1007/s12274-008-8022-7]

[610] Kim, S.N.; Kuang, Z.; Grote, J.G.; Farmer, B.L.; Naik, R.R. Enrichment of (*6* ,5) single wall carbon

nanotubes using genomic DNA. *Nano Lett.,* **2008**, *8*(12), 4415-4420. [http://dx.doi.org/10.1021/nl802332v] [PMID: 19053794]

[611] Zhang, L.; Zaric, S.; Tu, X.; Wang, X.; Zhao, W.; Dai, H. Assessment of chemically separated carbon nanotubes for nanoelectronics. *J. Am. Chem. Soc.,* **2008**, *130*(8), 2686-2691. [http://dx.doi.org/10.1021/ja7106492] [PMID: 18251484]

[612] Barlow, P.G.; Donaldson, K.; MacCallum, J.; Clouter, A.; Stone, V. Serum exposed to nanoparticle carbon black displays increased potential to induce macrophage migration. *Toxicol. Lett.,* **2005**, *155*(3), 397-401. [http://dx.doi.org/10.1016/j.toxlet.2004.11.006] [PMID: 15649623]

[613] Cui, D.; Tian, F.; Ozkan, C.S.; Wang, M.; Gao, H. Effect of single wall carbon nanotubes on human HEK293 cells. *Toxicol. Lett.,* **2005**, *155*(1), 73-85. [http://dx.doi.org/10.1016/j.toxlet.2004.08.015] [PMID: 15585362]

[614] Jia, G.; Wang, H.; Yan, L.; Wang, X.; Pei, R.; Yan, T.; Zhao, Y.; Guo, X. Cytotoxicity of carbon nanomaterials: single-wall nanotube, multi-wall nanotube, and fullerene. *Environ. Sci. Technol.,* **2005**, *39*(5), 1378-1383. [http://dx.doi.org/10.1021/es048729l] [PMID: 15787380]

[615] Lam, C.W.; James, J.T.; McCluskey, R.; Hunter, R.L. Pulmonary toxicity of single-wall carbon nanotubes in mice 7 and 90 days after intratracheal instillation. *Toxicol. Sci.,* **2004**, *77*(1), 126-134. [http://dx.doi.org/10.1093/toxsci/kfg243] [PMID: 14514958]

[616] Manna, S.K.; Sarkar, S.; Barr, J.; Wise, K.; Barrera, E.V.; Jejelowo, O.; Rice-Ficht, A.C.; Ramesh, G.T. Single-walled carbon nanotube induces oxidative stress and activates nuclear transcription factor-kappaB in human keratinocytes. *Nano Lett.,* **2005**, *5*(9), 1676-1684. [http://dx.doi.org/10.1021/nl0507966] [PMID: 16159204]

[617] Monteiro-Riviere, N.A.; Nemanich, R.J.; Inman, A.O.; Wang, Y.Y.; Riviere, J.E. Multi-walled carbon nanotube interactions with human epidermal keratinocytes. *Toxicol. Lett.,* **2005**, *155*(3), 377-384. [http://dx.doi.org/10.1016/j.toxlet.2004.11.004] [PMID: 15649621]

[618] Sayes, C.M.; Liang, F.; Hudson, J.L.; Mendez, J.; Guo, W.; Beach, J.M.; Moore, V.C.; Doyle, C.D.; West, J.L.; Billups, W.E.; Ausman, K.D.; Colvin, V.L. Functionalization density dependence of single-walled carbon nanotubes cytotoxicity *in vitro. Toxicol. Lett.,* **2006**, *161*(2), 135-142. [http://dx.doi.org/10.1016/j.toxlet.2005.08.011] [PMID: 16229976]

[619] Wu, P.; Chen, X.; Hu, N.; Tam, U.C.; Blixt, O.; Zettl, A.; Bertozzi, C.R. Biocompatible carbon nanotubes generated by functionalization with glycodendrimers. *Angew. Chem. Int. Ed. Engl.,* **2008**, *47*(27), 5022-5025. [http://dx.doi.org/10.1002/anie.200705363] [PMID: 18509843]

[620] Shi Kam, N.W.; Jessop, T.C.; Wender, P.A.; Dai, H. Nanotube molecular transporters: internalization of carbon nanotube-protein conjugates into Mammalian cells. *J. Am. Chem. Soc.,* **2004**, *126*(22), 6850-6851. [http://dx.doi.org/10.1021/ja0486059] [PMID: 15174838]

[621] Liu, Z.; Davis, C.; Cai, W.; He, L.; Chen, X.; Dai, H. Circulation and long-term fate of functionalized, biocompatible single-walled carbon nanotubes in mice probed by Raman spectroscopy. *Proc. Natl. Acad. Sci. USA,* **2008**, *105*(5), 1410-1415.

[http://dx.doi.org/10.1073/pnas.0707654105] [PMID: 18230737]

[622] Lacerda, L.; Herrero, M.A.; Venner, K.; Bianco, A.; Prato, M.; Kostarelos, K. Carbon-nanotube shape and individualization critical for renal excretion. *Small,* **2008**, *4*(8), 1130-1132. [http://dx.doi.org/10.1002/smll.200800323] [PMID: 18666166]

[623] Kam, N.W.; Liu, Z.; Dai, H. Carbon nanotubes as intracellular transporters for proteins and DNA: an investigation of the uptake mechanism and pathway. *Angew. Chem. Int. Ed. Engl.,* **2006**, *45*(4), 577-581. [http://dx.doi.org/10.1002/anie.200503389] [PMID: 16345107]

[624] Dhar, S.; Liu, Z.; Thomale, J.; Dai, H.; Lippard, S.J. Targeted single-wall carbon nanotube-mediated Pt(IV) prodrug delivery using folate as a homing device. *J. Am. Chem. Soc.,* **2008**, *130*(34), 11467-11476. [http://dx.doi.org/10.1021/ja803036e] [PMID: 18661990]

[625] Chen, J.; Chen, S.; Zhao, X.; Kuznetsova, L.V.; Wong, S.S.; Ojima, I. Functionalized single-walled carbon nanotubes as rationally designed vehicles for tumor-targeted drug delivery. *J. Am. Chem. Soc.,* **2008**, *130*(49), 16778-16785. [http://dx.doi.org/10.1021/ja805570f] [PMID: 19554734]

[626] Ali-Boucetta, H.; Al-Jamal, K.T.; McCarthy, D.; Prato, M.; Bianco, A.; Kostarelos, K. Multiwalled carbon nanotube-doxorubicin supramolecular complexes for cancer therapeutics. *Chem. Commun. (Camb.),* **2008**, (4), 459-461. [http://dx.doi.org/10.1039/B712350G] [PMID: 18188467]

[627] Liu, Z.; Sun, X.; Nakayama-Ratchford, N.; Dai, H. Supramolecular chemistry on water-soluble carbon nanotubes for drug loading and delivery. *ACS Nano,* **2007**, *1*(1), 50-56. [http://dx.doi.org/10.1021/nn700040t] [PMID: 19203129]

[628] Lee, C.C.; Gillies, E.R.; Fox, M.E.; Guillaudeu, S.J.; Fréchet, J.M.; Dy, E.E.; Szoka, F.C. A single dose of doxorubicin-functionalized bow-tie dendrimer cures mice bearing C-26 colon carcinomas. *Proc. Natl. Acad. Sci. USA,* **2006**, *103*(45), 16649-16654. [http://dx.doi.org/10.1073/pnas.0607705103] [PMID: 17075050]

[629] Storm, G.; ten Kate, M.T.; Working, P.K.; Bakker-Woudenberg, I.A. Doxorubicin entrapped in sterically stabilized liposomes: effects on bacterial blood clearance capacity of the mononuclear phagocyte system. *Clin. Cancer Res.,* **1998**, *4*(1), 111-115. [PMID: 9516959]

[630] Liu, Z.; Chen, K.; Davis, C.; Sherlock, S.; Cao, Q.; Chen, X.; Dai, H. Drug delivery with carbon nanotubes for *in vivo* cancer treatment. *Cancer Res.,* **2008**, *68*(16), 6652-6660. [http://dx.doi.org/10.1158/0008-5472.CAN-08-1468] [PMID: 18701489]

[631] Wang, Y.; Wang, X.; Antonietti, M. Polymeric graphitic carbon nitride as a heterogeneous organocatalyst: from photochemistry to multipurpose catalysis to sustainable chemistry. *Angew. Chem. Int. Ed. Engl.,* **2012**, *51*(1), 68-89. [http://dx.doi.org/10.1002/anie.201101182] [PMID: 22109976]

[632] Zhang, N.; Zhang, Y.; Xu, Y.J. Recent progress on graphene-based photocatalysts: current status and future perspectives. *Nanoscale,* **2012**, *4*(19), 5792-5813. [http://dx.doi.org/10.1039/c2nr31480k] [PMID: 22907128]

[633] Machado, B.F.; Serp, P. Graphene-based materials for catalysis. *Catal. Sci. Technol.,* **2012**, *2*, 54-75. [http://dx.doi.org/10.1039/C1CY00361E]

[634] Kong, X.K.; Chen, C.L.; Chen, Q.W. Doped graphene for metal-free catalysis. *Chem. Soc. Rev.,* **2014**, *43*(8), 2841-2857. [http://dx.doi.org/10.1039/c3cs60401b] [PMID: 24500122]

[635] Zhu, C.; Dong, S. Recent progress in graphene-based nanomaterials as advanced electrocatalysts towards oxygen reduction reaction. *Nanoscale,* **2013**, *5*(5), 1753-1767. [http://dx.doi.org/10.1039/c2nr33839d] [PMID: 23364753]

[636] Zhu, J.; Yang, D.; Yin, Z.; Yan, Q.; Zhang, H. Graphene and graphene-based materials for energy storage applications. *Small,* **2014**, *10*(17), 3480-3498. [http://dx.doi.org/10.1002/smll.201303202] [PMID: 24431122]

[637] Sun, Y.Q.; Wu, Q.O.; Shi, G.Q. Graphene based new energy materials. *Energy Environ. Sci.,* **2011**, *4*, 1113-1132. [http://dx.doi.org/10.1039/c0ee00683a]

[638] Stoller, M.D.; Park, S.; Zhu, Y.; An, J.; Ruoff, R.S. Graphene-based ultracapacitors. *Nano Lett.,* **2008**, *8*(10), 3498-3502. [http://dx.doi.org/10.1021/nl802558y] [PMID: 18788793]

[639] Qu, L.; Liu, Y.; Baek, J.B.; Dai, L. Nitrogen-doped graphene as efficient metal-free electrocatalyst for oxygen reduction in fuel cells. *ACS Nano,* **2010**, *4*(3), 1321-1326. [http://dx.doi.org/10.1021/nn901850u] [PMID: 20155972]

[640] Zhou, Y.G.; Chen, J.J.; Wang, F.B.; Sheng, Z.H.; Xia, X.H. A facile approach to the synthesis of highly electroactive Pt nanoparticles on graphene as an anode catalyst for direct methanol fuel cells. *Chem. Commun. (Camb.),* **2010**, *46*(32), 5951-5953. [http://dx.doi.org/10.1039/c0cc00394h] [PMID: 20601996]

[641] Yin, S.; Zhang, Y.; Kong, J.; Zou, C.; Li, C.M.; Lu, X.; Ma, J.; Boey, F.Y.; Chen, X. Assembly of graphene sheets into hierarchical structures for high-performance energy storage. *ACS Nano,* **2011**, *5*(5), 3831-3838. [http://dx.doi.org/10.1021/nn2001728] [PMID: 21510618]

[642] Liang, Y.; Li, Y.; Wang, H.; Zhou, J.; Wang, J.; Regier, T.; Dai, H. Co□O□ nanocrystals on graphene as a synergistic catalyst for oxygen reduction reaction. *Nat. Mater.,* **2011**, *10*(10), 780-786. [http://dx.doi.org/10.1038/nmat3087] [PMID: 21822263]

[643] Zhang, S.; Shao, Y.Y.; Liao, H.G.; Liu, J.; Aksay, I.A.; Yin, G.P.; Lin, Y.H. Graphene decorated with PtAu alloy nanoparticles: facile synthesis and promising application for formic acid oxidation. *Chem. Mater.,* **2011**, *23*, 1079-1081. [http://dx.doi.org/10.1021/cm101568z]

[644] David, L.; Bhandavat, R.; Kulkarni, G.; Pahwa, S.; Zhong, Z.; Singh, G. Synthesis of graphene films by rapid heating and quenching at ambient pressures and their electrochemical characterization. *ACS Appl. Mater. Interfaces,* **2013**, *5*(3), 546-552. [http://dx.doi.org/10.1021/am301782h] [PMID: 23268553]

[645] Hsu, H.C.; Shown, I.; Wei, H.Y.; Chang, Y.C.; Du, H.Y.; Lin, Y.G.; Tseng, C.A.; Wang, C.H.; Chen,

L.C.; Lin, Y.C.; Chen, K.H. Graphene oxide as a promising photocatalyst for CO2 to methanol conversion. *Nanoscale,* **2013**, *5*(1), 262-268.
[http://dx.doi.org/10.1039/C2NR31718D] [PMID: 23160369]

[646] Zhu, S.; Li, T. Hydrogenation-assisted graphene origami and its application in programmable molecular mass uptake, storage, and release. *ACS Nano,* **2014**, *8*(3), 2864-2872.
[http://dx.doi.org/10.1021/nn500025t] [PMID: 24564284]

[647] Shao, Y.Y.; Wang, J.; Wu, H.; Liu, J.; Aksay, I.A.; Lin, Y.H. Graphene based electrochemical sensors and biosensors: a review. *Electroanal.,* **2010**, *22*, 1027-1036.
[http://dx.doi.org/10.1002/elan.200900571]

[648] Pumera, M.; Ambrosi, A.; Bonanni, A.; Chng, E.L.; Poh, H.L. Graphene for electrochemical sensing and biosensing. *TRAC- . Trends Analyt. Chem.,* **2010**, *29*, 954-965.
[http://dx.doi.org/10.1016/j.trac.2010.05.011]

[649] Liu, Y.; Dong, X.; Chen, P. Biological and chemical sensors based on graphene materials. *Chem. Soc. Rev.,* **2012**, *41*(6), 2283-2307.
[http://dx.doi.org/10.1039/C1CS15270J] [PMID: 22143223]

[650] Bae, S.H.; Lee, Y.; Sharma, B.K.; Lee, H.J.; Kim, J.H.; Ahn, J.H. Graphene-based transparent strain sensor. *Carbon,* **2013**, *51*, 236-242.
[http://dx.doi.org/10.1016/j.carbon.2012.08.048]

[651] Yuan, W.; Liu, A.; Huang, L.; Li, C.; Shi, G. High-performance NO2 sensors based on chemically modified graphene. *Adv. Mater.,* **2013**, *25*(5), 766-771.
[http://dx.doi.org/10.1002/adma.201203172] [PMID: 23139053]

[652] Kuila, T.; Bose, S.; Khanra, P.; Mishra, A.K.; Kim, N.H.; Lee, J.H. Recent advances in graphene-based biosensors. *Biosens. Bioelectron.,* **2011**, *26*(12), 4637-4648.
[http://dx.doi.org/10.1016/j.bios.2011.05.039] [PMID: 21683572]

[653] Lu, C.H.; Yang, H.H.; Zhu, C.L.; Chen, X.; Chen, G.N. A graphene platform for sensing biomolecules. *Angew. Chem. Int. Ed. Engl.,* **2009**, *48*(26), 4785-4787.
[http://dx.doi.org/10.1002/anie.200901479] [PMID: 19475600]

[654] Wang, L.; Hua, E.; Liang, M.; Ma, C.; Liu, Z.; Sheng, S.; Liu, M.; Xie, G.; Feng, W. Graphene sheets, polyaniline and AuNPs based DNA sensor for electrochemical determination of BCR/ABL fusion gene with functional hairpin probe. *Biosens. Bioelectron.,* **2014**, *51*, 201-207.
[http://dx.doi.org/10.1016/j.bios.2013.07.049] [PMID: 23962707]

[655] Wu, J.; Pisula, W.; Müllen, K. Graphenes as potential material for electronics. *Chem. Rev.,* **2007**, *107*(3), 718-747.
[http://dx.doi.org/10.1021/cr068010r] [PMID: 17291049]

[656] Neto, A.H.; Guinea, F.; Peres, N.M.; Novoselov, K.S.; Geim, A.K. The electronic properties of graphene. *Rev. Mod. Phys.,* **2009**, *81*, 109-162.
[http://dx.doi.org/10.1103/RevModPhys.81.109]

[657] Bonaccorso, F.; Sun, Z.; Hasan, T.; Ferrari, A.C. Graphene photonics and optoelectronics. *Nat. Photonics,* **2010**, *4*, 611-622.
[http://dx.doi.org/10.1038/nphoton.2010.186]

[658] Sun, Y.; Gao, S.; Xie, Y. Atomically-thick two-dimensional crystals: electronic structure regulation and energy device construction. *Chem. Soc. Rev.,* **2014**, *43*(2), 530-546. [http://dx.doi.org/10.1039/C3CS60231A] [PMID: 24122032]

[659] Britnell, L.; Gorbachev, R.V.; Geim, A.K.; Ponomarenko, L.A.; Mishchenko, A.; Greenaway, M.T.; Fromhold, T.M.; Novoselov, K.S.; Eaves, L. Resonant tunnelling and negative differential conductance in graphene transistors. *Nat. Commun.,* **2013**, *4*, 1794-1799. [http://dx.doi.org/10.1038/ncomms2817] [PMID: 23653206]

[660] Li, X.; Wang, X.; Zhang, L.; Lee, S.; Dai, H. Chemically derived, ultrasmooth graphene nanoribbon semiconductors. *Science,* **2008**, *319*(5867), 1229-1232. [http://dx.doi.org/10.1126/science.1150878] [PMID: 18218865]

[661] Schwierz, F. Graphene transistors. *Nat. Nanotechnol.,* **2010**, *5*(7), 487-496. [http://dx.doi.org/10.1038/nnano.2010.89] [PMID: 20512128]

[662] Liao, L.; Lin, Y.C.; Bao, M.; Cheng, R.; Bai, J.; Liu, Y.; Qu, Y.; Wang, K.L.; Huang, Y.; Duan, X. High-speed graphene transistors with a self-aligned nanowire gate. *Nature,* **2010**, *467*(7313), 305-308. [http://dx.doi.org/10.1038/nature09405] [PMID: 20811365]

[663] Mueller, T.; Xia, F.N.; Avouris, P. Graphene photodetectors for high-speed optical communications. *Nat. Photonics,* **2010**, *4*, 297-301. [http://dx.doi.org/10.1038/nphoton.2010.40]

[664] Liu, M.; Yin, X.; Ulin-Avila, E.; Geng, B.; Zentgraf, T.; Ju, L.; Wang, F.; Zhang, X. A graphene-based broadband optical modulator. *Nature,* **2011**, *474*(7349), 64-67. [http://dx.doi.org/10.1038/nature10067] [PMID: 21552277]

[665] Yip, H.L.; Jen, A.K. Recent advances in solution-processed interfacial materials for efficient and stable polymer solar cells. *Energy Environ. Sci.,* **2012**, *5*, 5994-6011. [http://dx.doi.org/10.1039/c2ee02806a]

[666] Yin, Z.; Zhu, J.; He, Q.; Cao, X.; Tan, C.; Chen, H.; Yan, Q.; Zhan, H. Graphene-based materials for Solar cell applications. *Adv. Energy Mater.,* **2014**, *4*, 1300574. [http://dx.doi.org/10.1002/aenm.201300574]

[667] Roy-Mayhew, J.D.; Aksay, I.A. Graphene materials and their use in dye-sensitized solar cells. *Chem. Rev.,* **2014**, *114*(12), 6323-6348. [http://dx.doi.org/10.1021/cr400412a] [PMID: 24814731]

[668] Mukhopadhyay, P. *Graphite, graphene and their polymer nanocomposites*; Taylor & Francis Group: Boca Raton, Florida, **2013**, pp. 202-213.

[669] Li, X.; Xie, D.; Park, H.; Zeng, T.H.; Wang, K.; Wei, J.; Zhong, M.; Wu, D.; Kong, J.; Zhu, H. Anomalous behaviors of graphene transparent conductors in graphene–silicon heterojunction solar cells. *Adv. Energy Mater.,* **2013**, *3*, 1029-1034. [http://dx.doi.org/10.1002/aenm.201300052]

[670] Li, X.; Xie, D.; Park, H.; Zhu, M.; Zeng, T.H.; Wang, K.; Wei, J.; Wu, D.; Kong, J.; Zhu, H.; Kong, J.; Zhu, H. Ion doping of graphene for high-efficiency heterojunction solar cells. *Nanoscale,* **2013**, *5*(5), 1945-1948. [http://dx.doi.org/10.1039/c2nr33795a] [PMID: 23358527]

[671] Wang, J.T.; Ball, J.M.; Barea, E.M.; Abate, A.; Alexander-Webber, J.A.; Huang, J.; Saliba, M.; Mora-Sero, I.; Bisquert, J.; Snaith, H.J.; Nicholas, R.J. Low-temperature processed electron collection layers of graphene/TiO2 nanocomposites in thin film perovskite solar cells. *Nano Lett.,* **2014**, *14*(2), 724-730.
[http://dx.doi.org/10.1021/nl403997a] [PMID: 24341922]

[672] Yang, L.; Zhang, L.; Webster, T.J. Carbon nanostructures for orthopedic medical applications. *Nanomedicine (Lond.),* **2011**, *6*(7), 1231-1244.
[http://dx.doi.org/10.2217/nnm.11.107] [PMID: 21929458]

[673] Feng, L.; Liu, Z. Graphene in biomedicine: opportunities and challenges. *Nanomedicine (Lond.),* **2011**, *6*(2), 317-324.
[http://dx.doi.org/10.2217/nnm.10.158] [PMID: 21385134]

[674] Shen, H.; Zhang, L.; Liu, M.; Zhang, Z. Biomedical applications of graphene. *Theranostics,* **2012**, *2*(3), 283-294.
[http://dx.doi.org/10.7150/thno.3642] [PMID: 22448195]

[675] Zhang, Y.; Nayak, T.R.; Hong, H.; Cai, W. Graphene: a versatile nanoplatform for biomedical applications. *Nanoscale,* **2012**, *4*(13), 3833-3842.
[http://dx.doi.org/10.1039/c2nr31040f] [PMID: 22653227]

[676] Bitounis, D.; Ali-Boucetta, H.; Hong, B.H.; Min, D-H.; Kostarelos, K. Prospects and challenges of graphene in biomedical applications. *Adv. Mater.,* **2013**, *25*(16), 2258-2268.
[http://dx.doi.org/10.1002/adma.201203700] [PMID: 23494834]

[677] Bianco, A. Graphene: safe or toxic? The two faces of the medal. *Angew. Chem. Int. Ed. Engl.,* **2013**, *52*(19), 4986-4997.
[http://dx.doi.org/10.1002/anie.201209099] [PMID: 23580235]

[678] Broude, N.E. Stem-loop oligonucleotides: a robust tool for molecular biology and biotechnology. *Trends Biotechnol.,* **2002**, *20*(6), 249-256.
[http://dx.doi.org/10.1016/S0167-7799(02)01942-X] [PMID: 12007493]

[679] Xu, M.; Fujita, D.; Hanagata, N. Perspectives and challenges of emerging single-molecule DNA sequencing technologies. *Small,* **2009**, *5*(23), 2638-2649.
[http://dx.doi.org/10.1002/smll.200900976] [PMID: 19904762]

[680] Yang, K.; Feng, L.; Shi, X.; Liu, Z. Nano-graphene in biomedicine: theranostic applications. *Chem. Soc. Rev.,* **2013**, *42*(2), 530-547.
[http://dx.doi.org/10.1039/C2CS35342C] [PMID: 23059655]

[681] Li, M.; Yang, X.; Ren, J.; Qu, K.; Qu, X. Using graphene oxide high near-infrared absorbance for photothermal treatment of Alzheimer's disease. *Adv. Mater.,* **2012**, *24*(13), 1722-1728.
[http://dx.doi.org/10.1002/adma.201104864] [PMID: 22407491]

[682] Yang, K.; Hu, L.; Ma, X.; Ye, S.; Cheng, L.; Shi, X.; Li, C.; Li, Y.; Liu, Z. Multimodal imaging guided photothermal therapy using functionalized graphene nanosheets anchored with magnetic nanoparticles. *Adv. Mater.,* **2012**, *24*(14), 1868-1872.
[http://dx.doi.org/10.1002/adma.201104964] [PMID: 22378564]

[683] Frangioni, J.V. *In vivo* near-infrared fluorescence imaging. *Curr. Opin. Chem. Biol.,* **2003**, *7*(5), 626-

634.
[http://dx.doi.org/10.1016/j.cbpa.2003.08.007] [PMID: 14580568]

[684] Pohlmann, A.; Cantow, K.; Hentschel, J.; Arakelyan, K.; Ladwig, M.; Flemming, B.; Hoff, U.; Persson, P.B.; Seeliger, E.; Niendorf, T. Linking non-invasive parametric MRI with invasive physiological measurements (MR-PHYSIOL): towards a hybrid and integrated approach for investigation of acute kidney injury in rats. *Acta Physiol. (Oxf.),* **2013**, *207*(4), 673-689.
[http://dx.doi.org/10.1111/apha.12065] [PMID: 23336404]

[685] Hong, H.; Yang, K.; Zhang, Y.; Engle, J.W.; Feng, L.; Yang, Y.; Nayak, T.R.; Goel, S.; Bean, J.; Theuer, C.P.; Barnhart, T.E.; Liu, Z.; Cai, W. *In vivo* targeting and imaging of tumor vasculature with radiolabeled, antibody-conjugated nanographene. *ACS Nano,* **2012**, *6*(3), 2361-2370.
[http://dx.doi.org/10.1021/nn204625e] [PMID: 22339280]

[686] Gollavelli, G.; Ling, Y.C. Multi-functional graphene as an *in vitro* and *in vivo* imaging probe. *Biomaterials,* **2012**, *33*(8), 2532-2545.
[http://dx.doi.org/10.1016/j.biomaterials.2011.12.010] [PMID: 22206596]

[687] Peng, J.; Gao, W.; Gupta, B.K.; Liu, Z.; Romero-Aburto, R.; Ge, L.; Song, L.; Alemany, L.B.; Zhan, X.; Gao, G.; Vithayathil, S.A.; Kaipparettu, B.A.; Marti, A.A.; Hayashi, T.; Zhu, J.J.; Ajayan, P.M. Graphene quantum dots derived from carbon fibers. *Nano Lett.,* **2012**, *12*(2), 844-849.
[http://dx.doi.org/10.1021/nl2038979] [PMID: 22216895]

[688] Tetsuka, H.; Asahi, R.; Nagoya, A.; Okamoto, K.; Tajima, I.; Ohta, R.; Okamoto, A. Optically tunable amino-functionalized graphene quantum dots. *Adv. Mater.,* **2012**, *24*(39), 5333-5338.
[http://dx.doi.org/10.1002/adma.201201930] [PMID: 22833282]

[689] Wang, Y.; Li, Z.; Weber, T.J.; Hu, D.; Lin, C-T.; Li, J.; Lin, Y. In situ live cell sensing of multiple nucleotides exploiting DNA/RNA aptamers and graphene oxide nanosheets. *Anal. Chem.,* **2013**, *85*(14), 6775-6782.
[http://dx.doi.org/10.1021/ac400858g] [PMID: 23758346]

[690] Sun, X.; Liu, Z.; Welsher, K.; Robinson, J.T.; Goodwin, A.; Zaric, S.; Dai, H. Nano-graphene oxide for cellular imaging and drug delivery. *Nano Res.,* **2008**, *1*(3), 203-212.
[http://dx.doi.org/10.1007/s12274-008-8021-8] [PMID: 20216934]

[691] Zhang, L.; Lu, Z.; Zhao, Q.; Huang, J.; Shen, H.; Zhang, Z. Enhanced chemotherapy efficacy by sequential delivery of siRNA and anticancer drugs using PEI-grafted graphene oxide. *Small,* **2011**, *7*(4), 460-464.
[http://dx.doi.org/10.1002/smll.201001522] [PMID: 21360803]

[692] Yang, K.; Wan, J.; Zhang, S.; Zhang, Y.; Lee, S.T.; Liu, Z. *In vivo* pharmacokinetics, long-term biodistribution, and toxicology of PEGylated graphene in mice. *ACS Nano,* **2011**, *5*(1), 516-522.
[http://dx.doi.org/10.1021/nn1024303] [PMID: 21162527]

[693] Dreyer, D.R.; Jia, H.P.; Bielawski, C.W. Graphene oxide: a convenient carbocatalyst for facilitating oxidation and hydration reactions. *Angew. Chem. Int. Ed. Engl.,* **2010**, *49*(38), 6813-6816.
[http://dx.doi.org/10.1002/anie.201003238] [PMID: 20602388]

[694] Dreyer, D.R.; Bielawski, C.W. Carbocatalysis: Heterogeneous carbons finding utility in synthetic chemistry. *Chem. Sci. (Camb.),* **2011**, *2*, 1233-1240.
[http://dx.doi.org/10.1039/c1sc00035g]

[695] Dreyer, D.R.; Jarvis, K.A.; Ferreira, P.J.; Bielawski, C.W. Graphite oxide as a carbocatalyst for the preparation of fullerene-reinforced polyester and polyamide nanocomposites. *Polym. Chem.,* **2012**, *3*, 757-766.
[http://dx.doi.org/10.1039/c2py00545j]

[696] Dreyer, D.R.; Bielawski, C.W. Graphite oxide as an olefin polymerization carbocatalyst: applications in electrochemical double Layer capacitors. *Adv. Funct. Mater.,* **2012**, *22*, 3247-3253.
[http://dx.doi.org/10.1002/adfm.201103152]

[697] Dreyer, D.R.; Jarvis, K.A.; Ferreira, P.J.; Bielawski, C.W. Graphite oxide as a dehydrative polymerization catalyst: a one-step synthesis of carbon-reinforced poly(phenylene methylene) composites. *Macromolecules,* **2011**, *44*, 7659-7667.
[http://dx.doi.org/10.1021/ma201306x]

[698] Jia, H.P.; Dreyer, D.R.; Bielawski, C.W. Graphite oxide as an auto-tandem oxidation-hydration–aldol coupling catalyst. *Adv. Synth. Catal.,* **2011**, *353*, 528-532.
[http://dx.doi.org/10.1002/adsc.201000748]

[699] Jia, H.P.; Dreyer, D.R.; Bielawski, C.W. C–H oxidation using graphite oxide. *Tetrahedron,* **2011**, *67*, 4431-4434.
[http://dx.doi.org/10.1016/j.tet.2011.02.065]

[700] Dreyer, D.R.; Jia, H.P.; Todd, A.D.; Geng, J.; Bielawski, C.W. Graphite oxide: a selective and highly efficient oxidant of thiols and sulfides. *Org. Biomol. Chem.,* **2011**, *9*(21), 7292-7295.
[http://dx.doi.org/10.1039/c1ob06102j] [PMID: 21909587]

[701] Huang, H.; Huang, J.; Liu, Y.M.; He, H.Y.; Cao, Y.; Fan, K.N. Graphite oxide as an efficient and durable metal-free catalyst for aerobic oxidative coupling of amines to imines. *Green Chem.,* **2012**, *14*, 930-934.
[http://dx.doi.org/10.1039/c2gc16681j]

[702] Su, C.; Acik, M.; Takai, K.; Lu, J.; Hao, S-J.; Zheng, Y.; Wu, P.; Bao, Q.; Enoki, T.; Chabal, Y.J.; Loh, K.P. Probing the catalytic activity of porous graphene oxide and the origin of this behaviour. *Nat. Commun.,* **2012**, *3*, 1298-1306.
[http://dx.doi.org/10.1038/ncomms2315] [PMID: 23250428]

[703] Zeng, F.; Sun, Z.; Sang, X.; Diamond, D.; Lau, K.T.; Liu, X.; Su, D.S. In Situ one-step electrochemical preparation of graphene oxide nanosheet-modified electrodes for biosensors. *ChemSusChem,* **2011**, *4*(11), 1587-1591.
[http://dx.doi.org/10.1002/cssc.201100319] [PMID: 21953723]

[704] Tang, L.; Wang, Y.; Li, Y.; Feng, H.; Lu, J.; Li, J. Preparation, structure, and electrochemical properties of reduced graphene sheet films. *Adv. Funct. Mater.,* **2009**, *19*, 2782-2789.
[http://dx.doi.org/10.1002/adfm.200900377]

[705] Chen, D.; Feng, H.; Li, J. Graphene oxide: preparation, functionalization, and electrochemical applications. *Chem. Rev.,* **2012**, *112*(11), 6027-6053.
[http://dx.doi.org/10.1021/cr300115g] [PMID: 22889102]

[706] Dhakshinamoorthy, A.; Alvaro, M.; Concepción, P.; Fornés, V.; Garcia, H. Graphene oxide as an acid catalyst for the room temperature ring opening of epoxides. *Chem. Commun. (Camb.),* **2012**, *48*(44),

5443-5445.
[http://dx.doi.org/10.1039/c2cc31385e] [PMID: 22534622]

[707] Dhakshinamoorthy, A.; Alvaro, M.; Puche, M.; Fornes, V.; Garcia, H. Graphene oxide as catalyst for the acetalization of aldehydes at room temperature. *ChemCatChem,* **2012**, *4*, 2026-2030.
[http://dx.doi.org/10.1002/cctc.201200461]

[708] Tang, S.; Cao, Z. Site-dependent catalytic activity of graphene oxides towards oxidative dehydrogenation of propane. *Phys. Chem. Chem. Phys.,* **2012**, *14*(48), 16558-16565.
[http://dx.doi.org/10.1039/c2cp41343d] [PMID: 22801590]

[709] Singh Chauhan, S.M.; Mishra, S. Use of graphite oxide and graphene oxide as catalysts in the synthesis of dipyrromethane and calix[4]pyrrole. *Molecules,* **2011**, *16*(9), 7256-7266.
[http://dx.doi.org/10.3390/molecules16097256] [PMID: 21869753]

[710] Verma, S.; Mungse, H.P.; Kumar, N.; Choudhary, S.; Jain, S.L.; Sain, B.; Khatri, O.P. Graphene oxide: an efficient and reusable carbocatalyst for aza-Michael addition of amines to activated alkenes. *Chem. Commun. (Camb.),* **2011**, *47*(47), 12673-12675.
[http://dx.doi.org/10.1039/c1cc15230k] [PMID: 22039588]

[711] Kumar, A.V.; Rao, K.R. Recyclable graphite oxide catalyzed Friedel–Crafts addition of indoles to á,â-unsaturated ketones. *Tetrahedron Lett.,* **2011**, *52*, 5188-5191.
[http://dx.doi.org/10.1016/j.tetlet.2011.08.002]

[712] Labinger, J.A.; Bercaw, J.E. Understanding and exploiting C-H bond activation. *Nature,* **2002**, *417*(6888), 507-514.
[http://dx.doi.org/10.1038/417507a] [PMID: 12037558]

[713] Gao, Y.; Hu, G.; Zhong, J.; Shi, Z.; Zhu, Y.; Su, D.S.; Wang, J.; Bao, X.; Ma, D. Nitrogen-doped sp2-hybridized carbon as a superior catalyst for selective oxidation. *Angew. Chem. Int. Ed. Engl.,* **2013**, *52*(7), 2109-2113.
[http://dx.doi.org/10.1002/anie.201207918] [PMID: 23307693]

[714] Kong, X.; Sun, Z.; Chen, M.; Chen, C.; Chen, Q. Metal-free catalytic reduction of 4-nitrophenol to 4-aminophenol by N-doped graphene. *Energy Environ. Sci.,* **2013**, *6*, 3260-3266.
[http://dx.doi.org/10.1039/c3ee40918j]

[715] Chen, T.W.; Xu, J.Y.; Sheng, Z.H.; Wang, K.; Wang, F.B.; Liang, T.M.; Xia, X.H. Enhanced electrocatalytic activity of nitrogen-doped graphene for the reduction of nitro explosives. *Electrochem. Commun.,* **2012**, *16*, 30-33.
[http://dx.doi.org/10.1016/j.elecom.2011.12.016]

[716] Long, J.; Xie, X.; Xu, J.; Gu, Q.; Che, L.; Wang, X. Nitrogen-doped graphene nanosheets as metal-free catalysts for aerobic selective oxidation of benzylic alcohols. *ACS Catal.,* **2012**, *2*, 622-631.
[http://dx.doi.org/10.1021/cs3000396]

[717] Srinivas, G.; Zhu, Y.W.; Piner, R.; Skipper, N.; Ellerby, M.; Ruoff, R. Synthesis of graphene-like nanosheets and their hydrogen adsorption capacity. *Carbon,* **2010**, *48*, 630-635.
[http://dx.doi.org/10.1016/j.carbon.2009.10.003]

[718] Subrahmanyam, K.S.; Vivekchand, S.R.; Govindaraj, A.; Rao, C.N. A study of graphenes prepared by different methods: characterization, properties and solubilization. *J. Mater. Chem.,* **2008**, *18*, 1517-

1523.
[http://dx.doi.org/10.1039/b716536f]

[719] Pumera, M. Graphene-based nanomaterials for energy storage. *Energy Environ. Sci.,* **2011**, *4*, 668-674.
[http://dx.doi.org/10.1039/C0EE00295J]

[720] Lee, H.; Ihm, J.; Cohen, M.L.; Louie, S.G. Calcium-decorated graphene-based nanostructures for hydrogen storage. *Nano Lett.,* **2010**, *10*(3), 793-798.
[http://dx.doi.org/10.1021/nl902822s] [PMID: 20104855]

[721] Yu, A.P.; Roes, I.; Davies, A.; Chen, Z.W. Ultrathin, transparent, and flexible graphene films for supercapacitor application. *Appl. Phys. Lett.,* **2010**, *96*, 253105.
[http://dx.doi.org/10.1063/1.3455879]

[722] Wang, Y.; Shi, Z.Q.; Huang, Y.; Ma, Y.F.; Wang, C.Y.; Chen, M.M.; Chen, Y.S. Supercapacitor devices based on graphene materials. *J. Phys. Chem. C,* **2009**, *113*, 13103-13107.
[http://dx.doi.org/10.1021/jp902214f]

[723] Cheng, Q.; Tang, J.; Ma, J.; Zhang, H.; Shinya, N.; Qin, L.C. Graphene and carbon nanotube composite electrodes for supercapacitors with ultra-high energy density. *Phys. Chem. Chem. Phys.,* **2011**, *13*(39), 17615-17624.
[http://dx.doi.org/10.1039/c1cp21910c] [PMID: 21887427]

[724] Yang, S.Y.; Chang, K.H.; Tien, H.W.; Lee, Y.F.; Li, S.M.; Wang, Y.S.; Wang, J.Y.; Ma, C.C.; Hu, C.C. Design and tailoring of a hierarchical graphene-carbon nanotube architecture for supercapacitors. *J. Mater. Chem.,* **2011**, *21*, 2374-2380.
[http://dx.doi.org/10.1039/C0JM03199B]

[725] Schedin, F.; Geim, A.K.; Morozov, S.V.; Hill, E.W.; Blake, P.; Katsnelson, M.I.; Novoselov, K.S. Detection of individual gas molecules adsorbed on graphene. *Nat. Mater.,* **2007**, *6*(9), 652-655.
[http://dx.doi.org/10.1038/nmat1967] [PMID: 17660825]

[726] Lu, G.; Ocola, L.E.; Chen, J. Reduced graphene oxide for room-temperature gas sensors. *Nanotechnology,* **2009**, *20*(44), 445502.
[http://dx.doi.org/10.1088/0957-4484/20/44/445502] [PMID: 19809107]

[727] Fowler, J.D.; Allen, M.J.; Tung, V.C.; Yang, Y.; Kaner, R.B.; Weiller, B.H. Practical chemical sensors from chemically derived graphene. *ACS Nano,* **2009**, *3*(2), 301-306.
[http://dx.doi.org/10.1021/nn800593m] [PMID: 19236064]

[728] Dua, V.; Surwade, S.P.; Ammu, S.; Agnihotra, S.R.; Jain, S.; Roberts, K.E.; Park, S.; Ruoff, R.S.; Manohar, S.K. All-organic vapor sensor using inkjet-printed reduced graphene oxide. *Angew. Chem. Int. Ed. Engl.,* **2010**, *49*(12), 2154-2157.
[http://dx.doi.org/10.1002/anie.200905089] [PMID: 20187049]

[729] Nomani, M.W.; Shishir, R.; Qazi, M.; Diwan, D.; Shields, V.B.; Spencer, M.G.; Tompa, G.S.; Sbrockey, N.M.; Koley, G. Highly sensitive and selective detection of NO2 using epitaxial graphene on 6H-SiC. *Sens. Actuators B Chem.,* **2010**, *150*, 301-307.
[http://dx.doi.org/10.1016/j.snb.2010.06.069]

[730] Ko, G.; Kim, H.Y.; Ahn, J.; Park, Y.M.; Lee, K.Y.; Kim, J. Graphene-based nitrogen dioxide gas

sensors. *Curr. Appl. Phys.,* **2010**, *10*, 1002-1004.
[http://dx.doi.org/10.1016/j.cap.2009.12.024]

[731] Robinson, J.T.; Perkins, F.K.; Snow, E.S.; Wei, Z.; Sheehan, P.E. Reduced graphene oxide molecular sensors. *Nano Lett.,* **2008**, *8*(10), 3137-3140.
[http://dx.doi.org/10.1021/nl8013007] [PMID: 18763832]

[732] Shafiei, M.; Spizzirri, P.G.; Arsat, R.; Yu, J.; du Plessis, J.; Dubin, S.; Kaner, R.B. Kalantar-Zadeh, K. Wlodarski, W. Platinum/graphene nanosheet/SiC contacts and their application for hydrogen gas sensing. *J. Phys. Chem. C,* **2010**, *114*, 13796-13801.
[http://dx.doi.org/10.1021/jp104459s]

[733] Cuong, T.V.; Pham, V.H.; Chung, J.S.; Shin, E.W.; Yoo, D.H.; Hahn, S.H.; Huh, J.S.; Rue, G.H.; Kim, E.J.; Hur, S.H.; Kohl, P.A. Solution-processed ZnO-chemically converted graphene gas sensor. *Mater. Lett.,* **2010**, *64*, 2479-2482.
[http://dx.doi.org/10.1016/j.matlet.2010.08.027]

[734] Lu, G.; Yu, K.; Ocola, L.E.; Chen, J. Ultrafast room temperature NH3 sensing with positively gated reduced graphene oxide field-effect transistors. *Chem. Commun. (Camb.),* **2011**, *47*(27), 7761-7763.
[http://dx.doi.org/10.1039/c1cc12658j] [PMID: 21643620]

[735] Dan, Y.; Lu, Y.; Kybert, N.J.; Luo, Z.; Johnson, A.T. Intrinsic response of graphene vapor sensors. *Nano Lett.,* **2009**, *9*(4), 1472-1475.
[http://dx.doi.org/10.1021/nl8033637] [PMID: 19267449]

[736] Chu, B.H.; Nicolosi, J.; Lo, C.F.; Strupinski, W.; Pearton, S.J.; Ren, F. Effect of coated platinum thickness on hydrogen detection sensitivity of graphene-based sensors. *Electrochem. Solid-State Lett.,* **2011**, *14*, K43-K45.
[http://dx.doi.org/10.1149/1.3589250]

[737] Yi, J.; Lee, J.M.; Il-Park, W. Vertically aligned ZnO nanorods and graphene hybrid architectures for high-sensitive flexible gas sensors. *Sens. Actuators B Chem.,* **2011**, *155*, 264-269.
[http://dx.doi.org/10.1016/j.snb.2010.12.033]

[738] Arsat, R.; Breedon, M.; Shafiei, M.; Spizziri, P.G.; Gilje, S.; Kaner, R.B.; Kalantar-Zadeh, K.; Wlodarski, W. Graphene-like nano-sheets for surface acoustic wave gas sensor applications. *Chem. Phys. Lett.,* **2009**, *467*, 344-347.
[http://dx.doi.org/10.1016/j.cplett.2008.11.039]

[739] Lu, G.H.; Ocola, L.E.; Chen, J.H. Gas detection using low-temperature reduced graphene oxide sheets. *Appl. Phys. Lett.,* **2009**, *94*, 083111.
[http://dx.doi.org/10.1063/1.3086896]

[740] Qazi, M.; Vogt, T.; Koley, G. Trace gas detection using nanostructured graphite layers. *Appl. Phys. Lett.,* **2007**, *91*, 233101.
[http://dx.doi.org/10.1063/1.2820387]

[741] Lu, J.; Drzal, L.T.; Worden, R.M.; Lee, I. Simple fabrication of a highly sensitive glucose biosensor using enzymes immobilized in exfoliated graphite nanoplatelets nafion membrane. *Chem. Mater.,* **2007**, *19*, 6240-6246.
[http://dx.doi.org/10.1021/cm702133u]

[742] Lu, J.; Do, I.; Drzal, L.T.; Worden, R.M.; Lee, I. Nanometal-decorated exfoliated graphite nanoplatelet based glucose biosensors with high sensitivity and fast response. *ACS Nano,* **2008**, *2*(9), 1825-1832. [http://dx.doi.org/10.1021/nn800244k] [PMID: 19206421]

[743] Shan, C.; Yang, H.; Song, J.; Han, D.; Ivaska, A.; Niu, L. Direct electrochemistry of glucose oxidase and biosensing for glucose based on graphene. *Anal. Chem.,* **2009**, *81*(6), 2378-2382. [http://dx.doi.org/10.1021/ac802193c] [PMID: 19227979]

[744] Alwarappan, S.; Erdem, A.; Liu, C.; Li, C.Z. Probing the electrochemical properties of graphene nanosheets for biosensing applications. *J. Phys. Chem. C,* **2009**, *113*, 8853-8857. [http://dx.doi.org/10.1021/jp9010313]

[745] Shang, N.G.; Papakonstantinou, P.; McMullan, M.; Chu, M.; Tamboulis, A.; Potenza, A.; Dhesi, S.S.; Marchetto, H. Catalyst-free efficient growth, orientation and biosensing properties of multilayer graphene nanoflake films with sharp edge planes. *Adv. Funct. Mater.,* **2008**, *18*, 3506-3514. [http://dx.doi.org/10.1002/adfm.200800951]

[746] Mohanty, N.; Berry, V. Graphene-based single-bacterium resolution biodevice and DNA transistor: interfacing graphene derivatives with nanoscale and microscale biocomponents. *Nano Lett.,* **2008**, *8*(12), 4469-4476. [http://dx.doi.org/10.1021/nl802412n] [PMID: 19367973]

[747] De, S.; Coleman, J.N. Are there fundamental limitations on the sheet resistance and transmittance of thin graphene films? *ACS Nano,* **2010**, *4*(5), 2713-2720. [http://dx.doi.org/10.1021/nn100343f] [PMID: 20384321]

[748] Zhu, Y.; Sun, Z.; Yan, Z.; Jin, Z.; Tour, J.M. Rational design of hybrid graphene films for high-performance transparent electrodes. *ACS Nano,* **2011**, *5*(8), 6472-6479. [http://dx.doi.org/10.1021/nn201696g] [PMID: 21774533]

[749] Bi, H.; Sun, S.; Huang, F.; Xie, X.; Jiang, M. Direct growth of few-layer graphene films on SiO2 substrates and their photovoltaic applications. *J. Mater. Chem.,* **2012**, *22*, 411-416. [http://dx.doi.org/10.1039/C1JM14778A]

[750] Kim, S.R.; Parvez, M.K.; Chhowalla, M. UV-reduction of graphene oxide and its application as an interfacial layer to reduce the back-transport reactions in dye-sensitized solar cells. *Chem. Phys. Lett.,* **2009**, *483*, 124-127. [http://dx.doi.org/10.1016/j.cplett.2009.10.066]

[751] Chen, T.; Hu, W.; Song, J.; Guai, G.H.; Li, C.M. Interface functionalization of photoelectrodes with graphene for high performance dye-sensitized solar cells. *Adv. Funct. Mater.,* **2012**, *22*, 5245-5250. [http://dx.doi.org/10.1002/adfm.201201126]

[752] Durantini, J.; Boix, P.P.; Gervaldo, M.; Morales, G.M.; Otero, L.; Bisquert, J.; Barea, E.M. Photocurrent enhancement in dye-sensitized photovoltaic devices with titania–graphene composite electrodes. *J. Electroanal. Chem.,* **2012**, *683*, 43. [http://dx.doi.org/10.1016/j.jelechem.2012.07.032]

[753] Tang, B.; Hu, G. Two kinds of graphene-based composites for photoanode applying in dye-sensitized solar cell. *J. Power Sources,* **2012**, *220*, 95-102. [http://dx.doi.org/10.1016/j.jpowsour.2012.07.093]

[754] Williams, K.J.; Nelson, C.A.; Yan, X.; Li, L-S.; Zhu, X. Hot electron injection from graphene quantum dots to TiO_2. *ACS Nano,* **2013**, *7*(2), 1388-1394. [http://dx.doi.org/10.1021/nn305080c] [PMID: 23347000]

[755] Tielrooij, K.J.; Song, J.C.; Jensen, S.A.; Centeno, A.; Pesquera, A.; Zurutuza-Elorza, A.; Bonn, M.; Levitov, L.S.; Koppens, F.H. Photoexcitation cascade and multiple hot-carrier generation in graphene. *Nat. Phys.,* **2013**, *9*, 248-252. [http://dx.doi.org/10.1038/nphys2564]

[756] Long, R.; English, N.J.; Prezhdo, O.V. Photo-induced charge separation across the graphene-TiO2 interface is faster than energy losses: a time-domain ab initio analysis. *J. Am. Chem. Soc.,* **2012**, *134*(34), 14238-14248. [http://dx.doi.org/10.1021/ja3063953] [PMID: 22880690]

[757] Yan, X.; Cui, X.; Li, B.; Li, L-S. Large, solution-processable graphene quantum dots as light absorbers for photovoltaics. *Nano Lett.,* **2010**, *10*(5), 1869-1873. [http://dx.doi.org/10.1021/nl101060h] [PMID: 20377198]

[758] Zhang, L.; Xia, J.; Zhao, Q.; Liu, L.; Zhang, Z. Functional graphene oxide as a nanocarrier for controlled loading and targeted delivery of mixed anticancer drugs. *Small,* **2010**, *6*(4), 537-544. [http://dx.doi.org/10.1002/smll.200901680] [PMID: 20033930]

[759] Yang, K.; Zhang, S.; Zhang, G.; Sun, X.; Lee, S.T.; Liu, Z. Graphene in mice: ultrahigh *in vivo* tumor uptake and efficient photothermal therapy. *Nano Lett.,* **2010**, *10*(9), 3318-3323. [http://dx.doi.org/10.1021/nl100996u] [PMID: 20684528]

[760] Markovic, Z.M.; Harhaji-Trajkovic, L.M.; Todorovic-Markovic, B.M.; Kepić, D.P.; Arsikin, K.M.; Jovanović, S.P.; Pantovic, A.C.; Dramićanin, M.D.; Trajkovic, V.S. *In vitro* comparison of the photothermal anticancer activity of graphene nanoparticles and carbon nanotubes. *Biomaterials,* **2011**, *32*(4), 1121-1129. [http://dx.doi.org/10.1016/j.biomaterials.2010.10.030] [PMID: 21071083]

[761] Chen, M.L.; Liu, J-W.; Hu, B.; Chen, M-L.; Wang, J-H. Conjugation of quantum dots with graphene for fluorescence imaging of live cells. *Analyst (Lond.),* **2011**, *136*(20), 4277-4283. [http://dx.doi.org/10.1039/c1an15474e] [PMID: 21879034]

[762] Chen, W.; Yi, P.; Zhang, Y.; Zhang, L.; Deng, Z.; Zhang, Z. Composites of aminodextran-coated Fe3O4 nanoparticles and graphene oxide for cellular magnetic resonance imaging. *ACS Appl. Mater. Interfaces,* **2011**, *3*(10), 4085-4091. [http://dx.doi.org/10.1021/am2009647] [PMID: 21882840]

[763] Zhang, W.; Guo, Z.; Huang, D.; Liu, Z.; Guo, X.; Zhong, H. Synergistic effect of chemo-photothermal therapy using PEGylated graphene oxide. *Biomaterials,* **2011**, *32*(33), 8555-8561. [http://dx.doi.org/10.1016/j.biomaterials.2011.07.071] [PMID: 21839507]

[764] Hu, S.H.; Chen, Y-W.; Hung, W-T.; Chen, I-W.; Chen, S-Y. Quantum-dot-tagged reduced graphene oxide nanocomposites for bright fluorescence bioimaging and photothermal therapy monitored in situ. *Adv. Mater.,* **2012**, *24*(13), 1748-1754. [http://dx.doi.org/10.1002/adma.201104070] [PMID: 22422734]

[765] Zhou, L.; Wang, W.; Tang, J.; Zhou, J-H.; Jiang, H-J.; Shen, J. Graphene oxide noncovalent

photosensitizer and its anticancer activity *in vitro. Chemistry,* **2011**, *17*(43), 12084-12091. [http://dx.doi.org/10.1002/chem.201003078] [PMID: 21915922]

[766] Tian, B.; Wang, C.; Zhang, S.; Feng, L.; Liu, Z. Photothermally enhanced photodynamic therapy delivered by nano-graphene oxide. *ACS Nano,* **2011**, *5*(9), 7000-7009. [http://dx.doi.org/10.1021/nn201560b] [PMID: 21815655]

[767] Li, C-Z.; Chien, S-C.; Yip, H-L.; Chueh, C-C.; Chen, F-C.; Matsuo, Y.; Nakamura, E.; Jen, A.K. Facile synthesis of a 56π-electron 1,2-dihydromethano-[60]PCBM and its application for thermally stable polymer solar cells. *Chem. Commun. (Camb.),* **2011**, *47*(36), 10082-10084. [http://dx.doi.org/10.1039/c1cc14446d] [PMID: 21833380]

[768] Wang, Y.; Iqbal, Z.; Malhotra, S.V. Functionalization of carbon nanotubes with amines and enzymes. *Chem. Phys. Lett.,* **2005**, *402*, 96-101. [http://dx.doi.org/10.1016/j.cplett.2004.11.099]

CHAPTER 7

Organic Frameworks as Top Materials

Abstract: Organic frameworks are interconnected networks of channels, created in order to obtain stable rigid one-, two-, or three-dimensional pore structures, having well-ordered structures and uniform pore sizes. Hence, herein is presented a conspectus on the field of organic frameworks, namely covalent-organic frameworks (COF) and metal-organic frameworks (MOF), including their preparation, as well as detailed examples on several promising applications with 300 references provided. Some synthetic details on the preparation of some state-of-the-art compounds are given as well.

Keywords: COF, Covalent organic framework, Energy storage, Gas adsorption, Metal organic framework, MOF, Separation technologies, Porous catalysts.

INTRODUCTION

Porous solids are classified as microporous, mesoporous and macroporous, according to the size of the pores [1]. Solids with a pore size of 2 nm or below are known as microporous [2]. The mesoporous solids are in the range of 2 nm – 50 nm and above 50 nm are known as macroporous [3, 4]. For instance, zeolites, which are crystalline aluminosilicates, are the main example of microporous materials [5, 6]. Usually zeolites occur as natural minerals, but a considerable number of synthetic zeolites might also be synthesized and are used in various applications [7 - 10]. Regardless of their widespread use, zeolites have several potential drawbacks that hamper their utility as porous solids. Those drawbacks include syntheses that can be difficult to control, a limited number of structural and channel architectures that are available, and crystalline structures based on covalently-bonded networks of atoms that cannot be modified easily to vary the structures, topologies or properties of channels without altering the structure of the zeolite. The micropore presence ($d < 2$ nm) leads to quite high surface areas (>1000 m^2/g). Gas adsorption (usually nitrogen at 77 K) is commonly used to measure the surface areas in polymer networks and are determined using

Mário J. F. Calvete

Brunauer–Emmett–Teller (BET) [11] or Langmuir [12] theories, the most frequently used approach the BET theory [13].

In order to obtain an accessible stable pore structure, an interconnected channels network should be formed, using rigid building blocks to avoid collapsing the networks, being reported years ago [14]. Rigidity is frequently ascribed to aromatic monomers, either linked by other rigid groups or directly linked together [15, 16]. Porous polymer networks can be of two types: crystalline and amorphous networks. Amorphous networks can possess a broad pore size distribution and a structure statistically disordered, while crystalline networks possess structures that are well-ordered and pore sizes that are uniform [17]. This chapter will be dealing with examples of the latter category, which for instance include covalent organic frameworks (COFs) or metal-organic frameworks (MOFs). MOFs can be seen as inorganic zeolites' organic analogs, where oxygen atoms are swapped by metal ion linking rigid organic ligands. The resulting crystalline solids are comprised of rigid frameworks of molecules coordinated to metal ions in one, two and three dimensional open networks (Fig. **1**), providing high porosity for the crystalline structure. MOFs modulation depends on the ligand's chemical structure and the properties of the connecting metals [18].

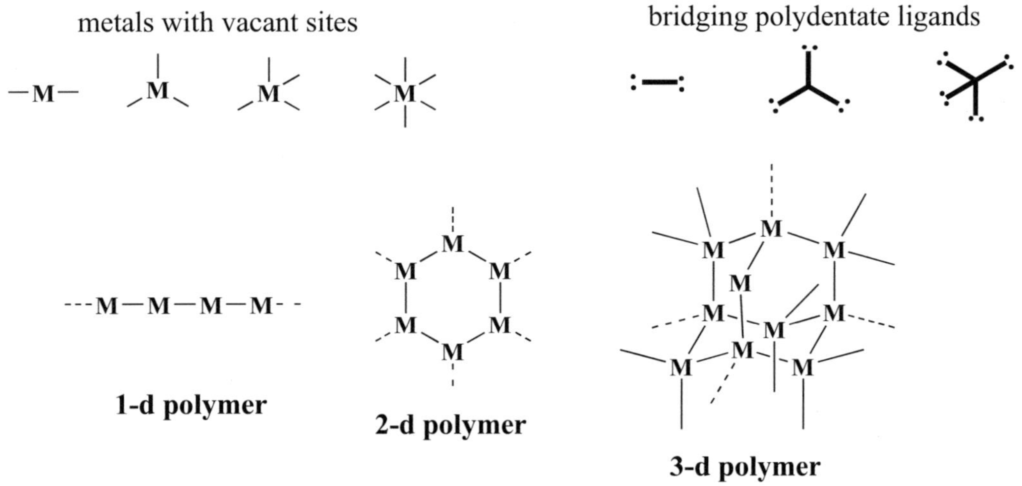

Fig. (1). Modular principle for the formation of coordination polymers.

In a large amount of cases, MOFs also exhibit permanent porosity and high thermal stability to above 300 °C, high dimensions and posses tunable properties (*e.g.*, hydrophobicity, exposed functionality, reactivity), channels can be controlled by synthetic modification of the ligand either before or after the MOF preparation [19 - 25].

In literature we can find several derived acronyms: IRMOF (IsoReticular MOF), MMOF (Microporous MOF), ZIF (Zeolitic Imidazole Framework), PCP (Porous Coordination Polymers), MOP (Metal Organic Polyhedra), CP (Coordination Polymers) and PMOF (Porous Metal Organic Framework) [26 - 31]. To organize and describe MOFs structures, a nomenclature system has been created and can be found at the Reticular Chemistry Structure Resource [32, 33].

The name coordination polymer was used for the first time in 1916 [34], but it was impossible to demonstrate the veracity of the structure at that time. Later in 1936, the well-known Prussian Blue compounds, based on Fe-CN-Fe linkages, were introduced in literature, however without assuming there three-dimensional coordination framework [35], which was not known at that time. Only in the end of the 1980's-early 1990's, new reports were made by Robson [36 - 38]. Immediately after, materials with polymeric structures based on organic bridging ligands and metal ions have been reported by Moore [39, 40], Yaghi [41] and Zawarotko [42].

The first sanctioned "MOF", denoted as MOF-5, was synthesized by the group of Yaghi in 1999 [22]. MOF-5 consists of Zn_4O tetraheda clusters connected with each other by 1,4- benzenedicarboxylate organic linkers. Since then, this seminal paper triggered the field of organic frameworks, and a large number of reports appeared in the literature [27, 43 - 52]. Among the several remarkable structures have been reported, MOF-177 [52] and MIL-101 [51] are exemplary, having surface areas of 5640 m2/g and 5900 m2/g respectively. Nowadays, many other mesoporous MOFs with very high surface areas have been reported [51, 53 - 56].

Most MOFs are constructed from metal carboxylates bearing divalent (*e.g.* Cu^{2+}, Zn^{2+}) or trivalent (*e.g.* Cr^{3+}, Al^{3+}) cations, interlinked by different organic fragments. The rigid building block assembly process is called reticular synthesis

[57]. In general, the chemical property and the volume of the resultant pores can be determined by the chemical functionality and length of the organic spacer. The ability to accept different organic linkers as well as many cations allows a vast number of possibilities in the creation of MOFs with high mechanical and thermal stability.

Closely related, another class of porous materials was reported by Yaghi, denominated COF-1 and COF-5 [58]. These molecules were prepared by exclusive assembly from elements like H, B, C and O, providing highly porous materials. Shortly after, the same group also prepared other 2D COFs, denominated COF-6, COF-8, COF-10 [59] and COF-202 [60], having pores with diameter sizes varying between 6 and 34 Å. COFs are stable against organic solvents and air, possess high thermal stability, have very low density and very large surface area [61 - 63]. Analogously to MOF's synthesis, reticular chemistry can also be used to assure pore size control in the resultant materials [58].

SYNTHESIS OF ORGANIC FRAMEWORKS

As mentioned above, this chapter will mainly focus on synthetic approaches related to the preparation of organic frameworks. For instance, there are several methods to synthesize MOFs [46]. The most frequently used synthetic approaches are: hydro- and solvothermal synthesis, microwave synthesis, electrochemical synthesis and high-throughput synthesis.

The first method is the considered the classical methodology, in which MOFs are synthesized using as metal sources well soluble salts, *e.g.* metal nitrates, acetates or sulfates, together with the organic linker (mostly carboxylic acids or pyridyl-based linkers) in a solvent or mixture of solvents [64]. Hydrothermal synthesis involves water as solvent, while solvothermal synthesis regards the use of organic solvents (*e.g.* alcohols, dialkylformamides and pyridine) [44] where its choice depends on its ability to dissolve the organic ligand. In both cases the reaction takes place in a Teflon-lined stainless autoclave in which an autogeneous pressure is built up. The self-assembly of the frameworks is mostly subjective to temperature, metal-salt concentration, reactants solubility in the solvent and pH of the solution [65]. The operational temperature varies between 80-260 °C, during a time scale of several hours to several days [66]. For this reason the classical synthesis is somewhat slow in comparison to other synthetic methodologies (microwave, electrochemical and high-throughput synthesis). A valid alternative refers to ionothermal synthesis, where ionic liquids or eutectic mixtures are used as solvents. Usually high vapor pressures are generated in solvothermal methods but the most fascinating feature of ionic liquids is that they possess a very low vapor pressure which in turn is more safe than solvothermal methods [67], beside possessing excellent solvating properties and high thermal stabilities [68].

The use of microwave irradiation reveals to be a suitable alternative for MOF synthesis, due to the rapid synthesis and dielectric uniform heating [69]. Thus, microwave heating allows short crystallization times, narrow particle size distribution, high yields and selectivities and simple morphology control [70 - 75]. The first synthesis of MOFs by microwave irradiation was reported in 2005, with the preparation of Cr-MIL-100 [76], in just 4 hours at 220 °C, in a yield comparable to the conventional synthesis method of 4 days. Since then, the syntheses of several other metal (III) MOFs have been revisited or prepared for the first time by microwave methods [77 - 84]. In addition, a large variety MOFs were also prepared by using microwave irradiation [85 - 88]. Among them, MOF-5 and HKUST-1 are the most studied ones so far [89 - 95].

Another method for MOF synthesis refers to electrochemical synthesis. The first report appeared in 2006 and was carried out by researchers at BASF for the synthesis of HKUST-130 [96]. Bulk copper plates were placed as anodes in an electrochemical cell, dissolving the carboxylate linker in methanol, using a copper cathode. The great advantage is that there were no anions present in solution that could block the pores [64, 96]. Later, using the same method, the same group of BASF researchers have prepared Zn-EZIF (electrochemically made ZIF) with 2-methylimidazole as linker [97]. Almost coincidentally, Ameloot modified the electrochemically produced HKUST-1, to enable the preparation of thin films, by applying an anodic voltage to the Cu-electrode. Cu(II) ions were introduced into the solution containing the carboxylate linker and methyltributylammonium methyl sulfate as a conduction salt. By changing the applied voltage, the crystal size could be controlled in a remarkably fast way, and the coatings can be produced continuously what makes them appropriate in industry. However, one main problem was identified, which was some pore blockage due organic linker and methyltributylammonium methyl sulfate coprecipitatation with the produced MOF.

As said before, MOF synthesis depends largely on its reaction parameters (temperature, pH, concentration of the reactants, *etc*). Regrettably, each parameter dependence increases with the number of studied parameters, raising exponentially the necessary number of experiments. In such cases, high throughput methods can be implemented. This method can be characterized by a complete set of parallel reactors and low reagent consumption, allowing proficient discovery of new components and fast optimization of synthetic settings [98]. Most of high throughput studies on MOFs were carried out to clarify the role of certain parameters, such as temperature, pH, concentration and reaction time [99 - 103]. Additionally, high throughput experiments were performed to discover new MOFs and hence the optimization of its synthetic methods [98, 104, 105].

MOFs are typically obtained from a ''one pot'' synthesis [21], hence without possibility to adjust properties during material formation. One way to do that is to prepare the functionalized networks by performing reactions on preformed MOFs. This transformation is the so-called ''Post Synthetic Modification'' which can be defined as chemical derivatization of MOFs after their formation [106]. This approach enables tailoring the pore architecture of the preformed MOF for a specific purpose, *i.e* .''tuning'' the material, facilitates the incorporation of diverse functional groups into MOFs which cannot endure the initial synthesis conditions (*e.g.* pH and temperature), avoiding competing between functional groups and donor groups or bridging ligands, allowing its introduction using the post synthetic method [21, 107, 108].

The main (most times the only) difference in the synthesis of COFs, when compared to the synthesis of COFs, lies in the substitution of metals salts by linkers, which typically interact with the building units to form covalent bonds. Ratios and solvent combinations are important issues in complementing formation of the frameworks and their crystallization, in COF synthesis. For instance, solvent combinations such as dimethylacetamide-*o* -dichlorobenzene [109], THF-methanol [110] and dioxane-mesitylene [58] have been utilized in boroxine and boronate ester linked COFs preparation. In addition, imine-linked COFs can be prepared using dioxane-aqueous acetic acid solvent combination [111], while aqueous acetic acid-dioxane-mesitylene was efficient for COFs linked with hydrazone [112]. Simultaneously, proper temperature is necessary to guarantee reaction reversibility. In general, COFs can be synthesized at 85-120 ºC, naturally depending on building block chemical reactivity.

APPLICATIONS OF ORGANIC FRAMEWORKS

Only application of organic frameworks, including MOFs and COFs will be here discussed. Initially the main interest of MOFs was gas storage [25, 52, 113 - 118]. However, now several other applications such as gas purification/separation [64, 65, 119 - 158], (also COFs [24, 62, 63, 159 - 178], catalysis [44, 51, 114, 117, 179 - 221], sensing [222 - 227] and as conducting materials [109, 228 - 239] are of interest.

Applications in Gas Purification/separation

The exhaustive usage of carbon-containing fuels produces terrific quantities of CO_2. Thus, the development of effective materials for gas separation is very important. During last decade, MOFs appeared as new materials for such purpose. MOFs feature significantly high porosities, functional sites that can be readily immobilized in their pore surfaces, over 6000 $m^2 g^{-1}$ of BET surface areas and microporous channels/spaces that can be rationally tuned. CO_2 is the main gas responsible for the greenhouse effect, due to industrial development and economic growth, and became a prime environmental issue. For this purpose, the exploration of materials with high separation capabilities and adsorption of CO_2 over other gases is critical.

The low pressure separation of CO_2 from N_2 (or CH_4) is the main technique for post-combustion outlet gas, while pre-combustion capture and separation mainly entail high pressure CO_2/H_2 separation. Thus, the selective capture of CO_2 over N_2 or CH_4 primarily involves CO_2/N_2 separation (postcombustion) and CO_2/CH_4 separation (purification of natural gas), for which the higher CO_2 quadrupole moment leads to stronger pore surface interactions in MOFs [240].

HKUST-1 is probably the most studied MOFs, bearing open metal sites, for CO_2 adsorption [241]. This material displayed high CO_2/N_2 and CO_2/CH_4 selectivities with modest adsorption enthalpy (30-35 $kJmol^{-1}$) [242]. The strong electrostatic interactions between open Cu^{2+} sites and CO_2molecules (O=C=O•••Cu^{2+}) were probably the reason for high selectivity [243, 244]. Particularly, Snurr tested the **HKUST-1** framework containing 4 wt% water and observed significant selectivities for CO_2/N_2 and CO_2/CH_4 [245].

The set of **M-MOF-74** (M= Zn, Mn, Ni, Co, Mg) is another well-known open metal site group of materials with high densities [246]. For instance, Matzger [247] reported, under ambient conditions, high CO_2 adsorption capacities, while Snurr showed a CO_2 uptake for **Mg-MOF-74** of 260 mg g^{-1}, representing the highest volumetric and gravimetric adsorption capacity of any MOF at 0.1 bar [248], and was based on their M-O bond length, in the order Mg-O (1.969 Å) < Ni-O (2.003 Å) < Co-O (2.031 Å) < Zn-O (2.083 Å). A similar enhancement in CO_2 uptake relative to N_2 was also reported [249, 250], for which a secondary amine (*N,N* -dimethylethylenediamine-mmen) was introduced in the **Mg_2(dobpdc)** (dobpdc = 4,4'-dioxidobiphenyl-3,3'-dicarboxylate) framework, producing amino-functionalized **mmen-Mg_2(dobpdc)**.

Zaworotko showed that the pore sizes in the **SIFSIX-3-Zn,SIFSIX-2-Cu**, and **SIFSIX-2-Cu-i** set could be modulated [251, 252]. The doubly interpenetrated **SIFSIX-2-Cu-i** framework formed narrower channels (5.15 by 5.15 Å) than that of **SIFSIX-2-Cu** (13.05 Å by 13.05 Å). Even with the surface area reduction, when compared with **SIFSIX-2-Cu**, the CO_2/CH_4 and CO_2/N_2 ratio selectivity displayed by **SIFSIX-2-Cu-i** showed to be 10 times higher, which was ascribed to the smaller pores of **SIFSIX-2-Cu-i**, when compared with **SIFSIX-2-Cu**, ensuing in stronger interactions between guest CO_2 molecules and the pore surfaces. Furthermore, **SIFSIX-2-Zn** with 3.84 Å pore dimension, displayed higher selectivity than **SIFSIX-2-Cu-i**, representing one of highest selectivities of CO_2 over N_2 or CH_4 so far achieved.

Contrasting with CO_2/N_2 and CO_2/CH_4 separations, precombustion CO_2 capture over H_2 is easier to perform, under high pressure, due to the greater difference in the quadrupole moment and polarizability between CO_2 and H_2 molecules [253].

For instance, Long selected five porous MOFs, such as **$Be_{12}(OH)_{12}(btb)_4$ (Be-BTB)**, **Co(bdp)** (bdp=1,4-benzenedipyrazolate), **$Zn_4O(btb)_2$** (**MOF-177**; btb=1,3,5-benzenetribenzoate), and **Mg-MOF-74**, **$H_3(Cu_4Cl)_3(BTTri)_8$ (Cu-BTTri)**, testing their properties on CO_2 and H_2 separation [254]. The material **Mg_2(dobdc)** and **Cu-BTTri**, which possessed higher open metal site densities, exhibited the highest selectivities under an 80:20 H_2/CO_2 mixture. Additionally, the framework **Cu_3(TDPAT)** (**Cu-TDPAT**; TDPAT=2,4,6-tris(3,5- dicarboxy lpheny lamino)-1,3,5-triazine) [255] also displayed significant performance for the separation of CO_2/H_2 mixtures at pressures up to 7 MPa.

Light C_1 to C_3 hydrocarbons are very important petrochemical raw chemicals, for which high quality and purity is required for an efficient utilization. Chen [256] reported the synthesis of **$Zn_2(pba)_2(bdc)$** (**UTSA-36a**; pba = 4-(4-pyridyl) benzoic, bdc = 1,4-benzenedicarboxylate), an interpenetrated framework with 3.1 to 4.8 Å pore sizes, and tested the selective separation of C_2H_2, C_2H_4 and C_2H_6 from CH_4 by adsorption. This material only displayed fair separation of C2 hydrocarbons over CH_4.

Later, other microporous MOFs (**UTSA-33a** [257] and **UTSA-34b** [258]) also were evaluated for C_2-C_1 separation. Under equimolar mixtures of C_2H_2, C_2H_4, C_2H_6 and CH_4 both frameworks displayed high selectivity for C_2 hydrocarbon separation from CH_4 at room temperature. Particularly, **UTSA-34b** exhibited the highest penetration of all C_2 hydrocarbons, when compared with **UTSA-33a**, which was ascribed to the strong interactions between open Cu^{2+} sites within the framework and C_2 hydrocarbons.

Also the O_2 separation from N_2 in air is a quite significant industrial process. Thus, Long reported the synthesis and evaluation of rigid **$Cr_3(btc)_2$** (btc = 1,4-benzenetricarboxylate), observing high selectivity in the room temperature adsorption of O_2 over N_2 [259]. Gas adsorption isotherms showed higher adsorption for O_2 (11 wt%) than for N_2 (0.58 wt%) at 25 ºC and 1 bar.; The same authors also reported the evaluation of the framework **Fe_2(dobdc) (Fe-MOF-74)**, in O_2/N_2 separation revealing also high selective capability, owing to its high density of open Fe^{2+} sites [260].

Applications in Catalysis

Even with the industrial approval of numerous homogeneous asymmetric organometallic catalysts, they frequently experience several limitations including tiresome separation/recycling. Process improvement is expected, with the use of the corresponding heterogeneous catalysts, by allowing several advantages over the homogeneous ones, including easy separation, improved handling and process control, minimization of metal traces in the product and efficient recycling. Many MOFs advantages, such as high density of active catalytic centers, relatively easy immobilization when compared to other heterogeneous systems, crystalline nature facilitating elucidation of structural details and high level of porosity, can make MOFs valuable heterogeneous asymmetric catalysts.

Kim [261] reported **POST-1**, the first homochiral MOF, used in asymmetric transesterification reactions, displaying high activity, with 77% conversion; however only with 8% enantiomeric excess (ee). This influential work prompted the interest in heterogeneous asymmetric catalysis and rational chiral ligand design. Lin reported the cyanosilylation of aldehydes as the first asymmetric catalysis metal-ion promoted at the framework nodes, with 55-69% conversions (less than 5% ee) [262]. Later, the same authors designed homochiral MOFs using chiral ligand BINOL as a building block, for diethylzinc additions to aldehydes, with > 99% conversion and remarkable 80-93% ee [187]. In 2006, Nguyen and Hupp [189] prepared a homochiral MOF using Mn(salen), for evaluation in asymmetric epoxidation reactions with 71% conversion and 82% ee. In 2009, Kim and co-workers used a new post synthetic modification methodology to prepare homochiral catalytically active MOFs, by the incorporation of organocatalyst units into non chiral frameworks [263]. These catalysts were then evaluated in aldol condensation reactions obtaining 58-91% conversions and 29-76% ee. Duan and co-workers [264] reported another chiral MOF, very active in aldehyde asymmetric cyanosilylation having conversions higher than 95% and high ee, up to 98%. In 2011, Telfer described the construction of the chiral framework, using organocatalysts-based ligands as bridging units [265].

Pd nanoparticle's use for formation of C-C bonds became a major chemical attainment [266, 267]. Inducing preferential formation of metal nanoparticles inside MOFs is a quite important issue that has been seldom addressed. Ferey [195] used amines (3-aminopropylsilane and ethylenediamine) as linking agents, in order to induce a privileged Pd nanoparticles loading inside the **MIL-101** pores. The catalyst was then used for benzaldehyde Knoevenagel condensation with ethyl cyanoacetate at 80 ºC. Aminopropylsilane-**MIL-101** gave conversion 96% and 99% selectivity, while ethylenediamine-**MIL-101** displayed conversion of 98% and 99% selectivity. On the other hand, the use of sole MIL-101 displayed just 32% conversion, however with similar selectivity.

MOF-5 has also been used as a Pd support [114] in the hydrogenation of styrene, for which the use of **Pd/MOF-5** as catalyst yielded ~85% ethylbenzene after 8 h reaction. In addition, all **Pd/MOF-5** catalysts were highly active in 1-octene hydrogenation, in 99.8% after 9 h. Later, a **Pd/MIL-101** catalyst was evaluated in ethylene and styrene hydrogenation, using acetylene and H_2 [193], leading to a complete ethylbenzene formation after 7 h, while, **Pd/MOF-5** yielded 80% hydrogenated product. **Pd/MIL-101** has been also used in heterogeneous Suzuki-Miyaura coupling reaction between 4-chloroanisole and phenylboronic acid [203], giving superior 4-phenylanisole (82%) yield in the presence of tetrabutylammonium bromide.

Besides Pd, also Au Au NPs immobilized in MOFs have emerged as an important topic on the homogeneous/heterogeneous catalysis interface. Haruta reported the incorporation of Au clusters on **MOF-5**, **Cu_3(btc)$_2$**(btc = 1,3,5-benzenetricarboxylate) and **MIL-53(Al)**, using Me_2Au(acac) (acac = acetylacetonate) as precursor [196], using the catalysts in benzyl alcohol and 1-phenylethanol aerobic oxidation.

Fischer [202] reported the loading of naked gold in zeolite imidazolate cavities, specifically **ZIF-8** and **ZIF-90 (Zn(ica)$_2$**; ica = imidazolate-2-carboxyaldehyde), aiming a Au nanoparticles dispersion control. **ZIF-8** and **ZIF-90** were inactive in benzyl alcohol oxidation, while **Au/ZIF-8** exhibited 81% conversion, with methyl benzoate selectivity of 98% [202]. Gold nanoparticles have also been incorporated into **MIL-101** [201]. Benzylic alcohols were efficiently oxidized in more than 95% conversion and 99% selectivity for aldehydes. For the same reason, Ru nanoparticles were employed as hydrogenation catalysts. An effective Ru nanoparticles immobilization route with MOF **La(BTC)•6H_2O**, in supercritical CO_2-methanol solution at 200 ºC, using $RuCl_3$•3H_2O as a precursor, was reported [268], and used for benzene and cyclohexene hydrogenation. Cyclohexene was almost completely converted into cyclohexane in 30 minutes, at 60 ºC, with 2 MPa H_2 pressure, using a 5000 substrate/Ru molar ratio, attaining a TOF = 10 000 h^{-1}. Benzene to cyclohexane hydrogenation required 2 hours reaction time, at 60 ºC with 6 MPa H_2 pressure using the same molar ratio, reaching TOF = 2475 [268]. **MOF-177** has been also used for Pt nanoparticle loading (43 wt%) [117]. Most of the Pt nanoparticles possessed 2-3 nm size and were used in allylic and benzylic alcohol aerobic oxidation, with TON values higher than 500.

Applications in Sensing

Regarding sensing applications, luminescent MOFs are frequently compared with other organic conjugate polymers, being good hosts for several analytes, enabling a sensitive detection [223, 269].

Detection of life-threatening materials is critical in societal safety, where the use of luminescent MOF sensors for explosive detection is probably the most promising application. In 2009, Li described, for the first time, the use of porous luminescent MOFs for detection of explosives [270, 271]. Upon **MOF-111$Zn_2(bpdc)_2$(bpee)2DMF**, bpdc = 4,4'-biphenyldicarboxylate; bpee = 1,2-bis(4-pyridyl)ethylene) room temperature dinitrotoluene vapors exposure (0.18 ppm) and 2,3-dimethyl-2,3-dinitrobutane (2.7 ppm), the authors observed a significant red shifted fluorescence quenching of more than 80% within 10 s.

Chen and Qian [272] also used **$Eu_2(bdc)_3(H_2O)_2(H_2O)_2$** (bdc = 1,4-benzenedicarboxylic acid) for dinitrotoluene and trinitrotoluene detection in ethanol, while Shi [273] studied the photoluminescence quenching of **$Zn_3(tdpat)(H_2O)_3$** upon nitrobenzene addition in methanol. Chen and Du [274] studied the photoluminescence response of **$In_2L_{29}NH_2(CH_3)_{22}$ $(DMF)_4(H_2O)_{16}$** (L_{29} = tetrakis(3,5-dicarboxyphenoxy)methylmethane) by several electron rich or deficient analytes addition.

Dinca and coworkers also reported volatile organic compound (VOC) sensing at high temperature [275]. Both **Mg(dhbdc)** (H_2DHBDC2− = 2,5-

dihydroxybenzene-1,4-dicarboxylate) and **Zn_2(tcpe)** (TCPE = tetrakis(4-carboxyphenyl)ethylene) MOFs were ammonia selective sensors at 100°C. In another example, Li and coworkers [276] demonstrated that lanthanide luminescent MOF $Eu_2(L_{33})_3$(H2O)4·3DMF (L_{33} = 20,50-bis(methoxymethyl--1,10:40,100-terphenyl-4,400-dicarboxylate) could function as vapor phase "turn-on" DMF detector, upon increase in the intensity of emission from the water-exchanged framework.

Other VOCs and other small molecules, such as H_2S, H_2O and CO_2 can also be successfully detected by luminescent MOFs. For instance, luminescent MOF **$MVMn_2Cu_3(mpba)_3(H_2O)_3\cdot 20H_2O$** could detect water, through increase of photoluminescence incremental loading of water, and emission shifting, upon CO_2 exposure [277]. Additionally, luminescent MOFs can also act as detectors for significant biologic molecules. For instance, Chen and Qian [278] utilized MOF **$Eu_2(fma)_2(ox)(H_2O)_4{\bullet}4H_2O$**, for effective turn-on detection of dipicolinic acid, an important biologically expressed compound. Duan reported the detection of nitric oxide (NO) in physiological media using **Cu-tca** (H_3tca = tricarboxytriphenyl amine) [279]. A weak blue emission was shown by Cu-MOF, upon 350 nm excitation, but room temperature incorporation of 1 mM in aqueous solution of NO inside the **Cu-tca** activated a considerable enhancement in MOF's emission, leading to a fluorescence intensity increase in 700-fold.

Applications as Conducting Materials

Transport or ion conductivity of solid compounds depends on the structures, which can be either amorphous or crystalline. Noteworthy properties of the conductive ions are mobility, charge and concentration, for which the control of the spatial distribution and dynamic behavior of target ions in solids is critical. Most inorganic compounds (*e.g.* metal oxides) exhibit ion conductivity above 400 °C, while organic polymers below 200 °C. On the other hand, MOFs work in the intermediate range, with additional important properties, including photoluminescence and magnetism.

A vast array of different MOF structures was evaluated for their ion conductivity, mostly H^+. The first examples of measured conductivies for porous materials occurred in 2003 by the group of Kitagawa [280 - 282], who prepared porous polymers of the type R_2dtoaCu (R = H, C_2H_4OH, C_3H_6OH and C_2H_5) (dtoa = dithiooxamide), reaching conductivities in the range of 10^{-6} - 10^{-5} S cm^{-1}. The same group had later reported the evaluation of MOF **$(NH_4)_2(adp)Zn_2(ox)_{33}$•$3H_2O$** [283], with an impressive conductivity of 8 x 10^{-3} S cm^{-1}, where adp = adipic acid and ox = oxalate, followed by MOF **M(dhbq)$_3$•nH_2O** (M = Mg, Mn, Ni, Zn and H_2(dhbq) = 2,5-dihydroxy-1,4-benzoquinone) with conductivity of 10^{-4} S cm^{-1} [284].

Other MOFs were also reported, displaying conductivities in the range 10^{-6} -10^{-5} S cm^{-1}, always in moisture conditions [285 - 292]. The MOF which showed highest conductivity in moisturized conditions was reported by Ponomareva [293], who synthesized MOF **$Cr_3F(H_2O)_2O(14bdc)_3$•H_2SO_4**, displaying 1 x 10^{-2} S cm^{-1} conductivity.

Conductivity measurements carried out in anhydrous conditions were also reported by other authors, namely using MOF **Na_3(2,4,6-trihydroxy-1,3,5-benzenetrisulfonate)•0.45(124triH)$_2$**, (124tri = 1,2,4-triazolate) with conductivity = 5 x 10^{-4} S cm^{-1} [294]. Additionally, other groups prepared MOFs His@Al(μ_2-OH)(1,4-bdc) (His = histamine, 1,4-bdc = 1,4-benzenedicarboxylate) [295] and Tz@β-PCMOF-2 (Tz = triazole) [294], displaying conductivities of 1.7 x 10^{-3} S cm^{-1} and 5 x 10^{-4} S cm^{-1}. Long [296] also reported Li^+ ion conductive MOF **$Mg_2(dobdc)_3$•0.35LiO^{i}Pr•0.25LiBF_4•EC•DEC** (dobdc = 1,4-dioxido-2,5-benzene dicarboxylate, DEC = diethyl carbonate, EC = ethylene carbonate), showing a conductivity of 1 x 10^{-4} S cm^{-1}

EXPERIMENTAL SECTION

Synthesis of HKUST-1 MOF

Procedure for total synthesis of HKUST-1 MOF can be directly found in Reference [297].

Synthesis of UMCM-2 MOF

Procedure for total synthesis of UMCM-2 MOF can be directly found in

Reference [56].

Microwave Synthesis of IRMOF2 MOF

Procedure for total synthesis of IRMOF2 MOF can be directly found in Reference [298].

Synthesis of IRMOF3 MOF

Procedure for total synthesis of IRMOF3 MOF can be directly found in Reference [298].

Microwave Synthesis of MOF-5

Procedure for total synthesis of MOF-5 can be directly found in Reference [87].

Synthesis of MIL-101 COF

Procedure for total synthesis of MIL-101 COF can be directly found in Reference [78].

Synthesis of COF-1

Procedure for total synthesis of COF-1 can be directly found in Reference [299].

Synthesis of COF-6

Procedure for total synthesis of COF-6 can be directly found in Reference [300].

Synthesis of COF-8

Procedure for total synthesis of COF-8 can be directly found in Reference [300].

Synthesis of COF-202

Procedure for total synthesis of COF-202 can be directly found in Reference [60].

CONCLUDING REMARKS

15 years and more than 5000 organic frameworks later, divided between metal organic (MOF) and covalent organic frameworks (COF), research in the field

apparently knows no bounds, representing one of today's hottest fields of research.

The structures and structural properties strongly pivot on the different preparation methods. Several successful methodologies have been used to date, achieving precise control on composite morphology. In the future, more facile and versatile production strategies will be probably envisaged, which predicts a continual progress on the field.

Although many challenges still exist, the opportunities to apply organic framework structures are limited only by imagination and ability to synthesize well-designed structures for a wide array of topics that highlight the importance and diversity of this chemistry, as well as its impact on other fields.

REFERENCES

[1] Sing, K.S.; Everett, D.H.; Haul, R.A.; Moscou, L.; Pierotti, R.A.; Rouquerol, J.; Siemieniewska, T. Reporting physisorption data for gas solid systems with special reference to the determination of surface-area and porosity (recommendations 1984). *Pure Appl. Chem.,* **1985**, *57*, 603-619. [http://dx.doi.org/10.1351/pac198557040603]

[2] Davis, M.E. Ordered porous materials for emerging applications. *Nature,* **2002**, *417*(6891), 813-821. [http://dx.doi.org/10.1038/nature00785] [PMID: 12075343]

[3] Inagaki, S.; Fukushima, Y.; Kuroda, K. Synthesis of highly ordered mesoporous materials from a layered polysilicate. *J. Chem. Soc. Chem. Commun.,* **1993**, 680-682. [http://dx.doi.org/10.1039/c39930000680]

[4] Imhof, A.; Pine, D.J. Ordered macroporous materials by emulsion templating. *Nature,* **1997**, *389*, 948-950. [http://dx.doi.org/10.1038/40105]

[5] Cundy, C.S.; Cox, P.A. The hydrothermal synthesis of zeolites: history and development from the earliest days to the present time. *Chem. Rev.,* **2003**, *103*(3), 663-702. [http://dx.doi.org/10.1021/cr020060i] [PMID: 12630849]

[6] Nishi, K.; Thompson, R.W. Synthesis of classical zeolites. In: *Handbook of porous solids*; Schuth, F.; Sing, K.S.; Weitkamp, F., Eds.; Wiley-VCH Verlag: Weinheim, **2002**; pp. 736-814. [http://dx.doi.org/10.1002/9783527618286.ch18a]

[7] Cappellazzo, O.; Cao, G.; Messina, G.; Morbidelli, M. Kinetics of shape-selective xylene isomerization over a ZSM-5 catalyst. *Ind. Eng. Chem. Res.,* **1991**, *30*, 2280-2287. [http://dx.doi.org/10.1021/ie00058a006]

[8] Degnan, T.F.; Chitnis, G.K.; Schipper, P.H. History of ZSM-5 fluid catalytic cracking additive development at Mobil. *Microporous Mesoporous Mater.,* **2000**, *35-36*, 245-252. [http://dx.doi.org/10.1016/S1387-1811(99)00225-5]

[9] Wilson, S.T.; Lok, B.M.; Messina, C.A.; Cannan, T.R.; Flanigen, E.M. Aluminophosphate molecular sieves: a new class of microporous crystalline inorganic solids. *J. Am. Chem. Soc.,* **1982**, *104*, 1146-1147.
[http://dx.doi.org/10.1021/ja00368a062]

[10] Silva, M.; Fernandes, A.; Bebiano, S.S.; Calvete, M.J.; Ribeiro, M.F.; Burrows, H.D.; Pereira, M.M. Size and ability do matter! Influence of acidity and pore size on the synthesis of hindered halogenated meso-phenyl porphyrins catalysed by porous solid oxides. *Chem. Commun. (Camb.),* **2014**, *50*(50), 6571-6573.
[http://dx.doi.org/10.1039/c4cc01100g] [PMID: 24643723]

[11] Brunauer, S.; Emmett, P.H.; Teller, E. Adsorption of gases in multimolecular layers. *J. Am. Chem. Soc.,* **1938**, *60*, 309.
[http://dx.doi.org/10.1021/ja01269a023]

[12] Langmuir, I. The adsorption of gases on plane surface of glass, mica and platinum. *J. Am. Chem. Soc.,* **1918**, *40*, 1361-1403.
[http://dx.doi.org/10.1021/ja02242a004]

[13] Walton, K.S.; Snurr, R.Q. Applicability of the BET method for determining surface areas of microporous metal-organic frameworks. *J. Am. Chem. Soc.,* **2007**, *129*(27), 8552-8556.
[http://dx.doi.org/10.1021/ja071174k] [PMID: 17580944]

[14] Aharoni, S.M.; Edwards, S.F. Rigid polymer networks. *Adv. Polym. Sci.,* **1994**, *118*, 1-231.
[http://dx.doi.org/10.1007/3-540-58340-8_1]

[15] Jiang, J.X.; Su, F.; Trewin, A.; Wood, C.D.; Campbell, N.L.; Niu, H.; Dickinson, C.; Ganin, A.Y.; Rosseinsky, M.J.; Khimyak, Y.Z.; Cooper, A.I. Conjugated microporous poly(aryleneethynylene) networks. *Angew. Chem. Int. Ed. Engl.,* **2007**, *46*(45), 8574-8578.
[http://dx.doi.org/10.1002/anie.200701595] [PMID: 17899616]

[16] Dawson, R.; Su, F.B.; Niu, H.J.; Wood, C.D.; Jones, J.T.; Khimyak, Y.Z.; Cooper, A.I. Mesoporous poly(phenylenevinylene) networks. *Macromolecules,* **2008**, *41*, 1591-1593.
[http://dx.doi.org/10.1021/ma702411b]

[17] Stöckel, E.; Wu, X.; Trewin, A.; Wood, C.D.; Clowes, R.; Campbell, N.L.; Jones, J.T.; Khimyak, Y.Z.; Adams, D.J.; Cooper, A.I. High surface area amorphous microporous poly(aryleneethynylene) networks using tetrahedral carbon- and silicon-centred monomers. *Chem. Commun. (Camb.),* **2009**, (2), 212-214.
[http://dx.doi.org/10.1039/B815044C] [PMID: 19099072]

[18] James, S.L. Metal-organic frameworks. *Chem. Soc. Rev.,* **2003**, *32*(5), 276-288.
[http://dx.doi.org/10.1039/b200393g] [PMID: 14518181]

[19] Eddaoudi, M.; Kim, J.; Rosi, N.; Vodak, D.; Wachter, J.; O'Keeffe, M.; Yaghi, O.M. Systematic design of pore size and functionality in isoreticular MOFs and their application in methane storage. *Science,* **2002**, *295*(5554), 469-472.
[http://dx.doi.org/10.1126/science.1067208] [PMID: 11799235]

[20] Zhang, X.; Llabrés i Xamena, F.X.; Corma, A. Gold(III) – metal organic framework bridges the gap between homogeneous and heterogeneous gold catalysts. *J. Catal.,* **2009**, *265*, 155-160.

[http://dx.doi.org/10.1016/j.jcat.2009.04.021]

[21] Wang, Z.; Cohen, S.M. Postsynthetic modification of metal-organic frameworks. *Chem. Soc. Rev.,* **2009**, *38*(5), 1315-1329.
[http://dx.doi.org/10.1039/b802258p] [PMID: 19384440]

[22] Li, H.; Eddaoudi, M.; O'Keeffe, M.; Yaghi, O.M. Design and synthesis of an exceptionally stable and highly porous metal-organic framework. *Nature,* **1999**, *402*, 276-278.
[http://dx.doi.org/10.1038/46248]

[23] Chae, H.K.; Siberio-Pérez, D.Y.; Kim, J.; Go, Y.; Eddaoudi, M.; Matzger, A.J.; O'Keeffe, M.; Yaghi, O.M. A route to high surface area, porosity and inclusion of large molecules in crystals. *Nature,* **2004**, *427*(6974), 523-527.
[http://dx.doi.org/10.1038/nature02311] [PMID: 14765190]

[24] Furukawa, H.; Ko, N.; Go, Y.B.; Aratani, N.; Choi, S.B.; Choi, E.; Yazaydin, A.O.; Snurr, R.Q.; O'Keeffe, M.; Kim, J.; Yaghi, O.M. Ultrahigh porosity in metal-organic frameworks. *Science,* **2010**, *329*(5990), 424-428.
[http://dx.doi.org/10.1126/science.1192160] [PMID: 20595583]

[25] Murray, L.J.; Dincă, M.; Long, J.R. Hydrogen storage in metal-organic frameworks. *Chem. Soc. Rev.,* **2009**, *38*(5), 1294-1314.
[http://dx.doi.org/10.1039/b802256a] [PMID: 19384439]

[26] Yaghi, O.M.; Li, H.L.; Davis, C.; Richardson, D.; Groy, T.L. Synthetic strategies, structure patterns, and emerging properties in the chemistry of modular porous solids. *Acc. Chem. Res.,* **1998**, *31*, 474-484.
[http://dx.doi.org/10.1021/ar970151f]

[27] Kitagawa, S.; Kitaura, R.; Noro, S. Functional porous coordination polymers. *Angew. Chem. Int. Ed. Engl.,* **2004**, *43*(18), 2334-2375.
[http://dx.doi.org/10.1002/anie.200300610] [PMID: 15114565]

[28] Rosseinsky, M.J. Recent developments in metal–organic framework chemistry: design, discovery, permanent porosity and flexibility. *Microporous Mesoporous Mater.,* **2004**, *73*, 15-30.
[http://dx.doi.org/10.1016/j.micromeso.2003.05.001]

[29] Tranchemontagne, D.J.; Ni, Z.; O'Keeffe, M.; Yaghi, O.M. Reticular chemistry of metal-organic polyhedra. *Angew. Chem. Int. Ed. Engl.,* **2008**, *47*(28), 5136-5147.
[http://dx.doi.org/10.1002/anie.200705008] [PMID: 18528833]

[30] Eddaoudi, M.; Moler, D.B.; Li, H.; Chen, B.; Reineke, T.M.; O'Keeffe, M.; Yaghi, O.M. Modular chemistry: secondary building units as a basis for the design of highly porous and robust metal-organic carboxylate frameworks. *Acc. Chem. Res.,* **2001**, *34*(4), 319-330.
[http://dx.doi.org/10.1021/ar000034b] [PMID: 11308306]

[31] Ockwig, N.W.; Delgado-Friedrichs, O.; O'Keeffe, M.; Yaghi, O.M. Reticular chemistry: occurrence and taxonomy of nets and grammar for the design of frameworks. *Acc. Chem. Res.,* **2005**, *38*(3), 176-182.
[http://dx.doi.org/10.1021/ar020022l] [PMID: 15766236]

[32] Tranchemontagne, D.J.; Mendoza-Cortés, J.L.; O'Keeffe, M.; Yaghi, O.M. Secondary building units,

nets and bonding in the chemistry of metal-organic frameworks. *Chem. Soc. Rev.,* **2009**, *38*(5), 1257-1283.
[http://dx.doi.org/10.1039/b817735j] [PMID: 19384437]

[33] *http://rcsr.anu.edu.au,* **2014**. (last access: 18th October)

[34] Shibata, Y. Polymer of coordination. In: *J. Coll. Sci*; Imperial Univ: Tokyo, **1916**; 37, pp. 1-31.

[35] Keggin, J.F.; Miles, F.D. Structures and formulæ of the prussian blues and related compounds. *Nature,* **1936**, *3466*, 577-578.
[http://dx.doi.org/10.1038/137577a0]

[36] Hoskins, B.F.; Robson, R. Infinite polymeric frameworks consisting of three dimensionally linked rod-like segments. *J. Am. Chem. Soc.,* **1989**, *111*, 5962-5964.
[http://dx.doi.org/10.1021/ja00197a079]

[37] Hoskins, B.F.; Robson, R. Design and construction of a new class of scaffolding-like materials comprising infinite polymeric frameworks of 3D-linked molecular rods. A reappraisal of the zinc cyanide and cadmium cyanide structures and the synthesis and structure of the diamond-related frameworks [N(CH3)4][CuIZnII(CN)4] and CuI [4,4',4",4'"-tetracy anotetraphenyl methane] BF4.xC6H5NO2. *J. Am. Chem. Soc.,* **1990**, *112*, 1546-1554.
[http://dx.doi.org/10.1021/ja00160a038]

[38] Abrahams, B.F.; Hoskins, B.F.; Michail, D.M.; Robson, R. Assembly of porphyrin building blocks into network structures with large channels. *Nature,* **1994**, *369*, 727-729.
[http://dx.doi.org/10.1038/369727a0]

[39] Venkataraman, D.; Gardner, G.B.; Lee, S.; Moore, J.S. Zeolite-like behavior of a coordination network. *J. Am. Chem. Soc.,* **1995**, *117*, 11600-11601.
[http://dx.doi.org/10.1021/ja00151a034]

[40] Gardner, G.B.; Venkataraman, D.; Moore, J.S.; Lee, S. Spontaneous assembly of a hinged coordination network. *Nature,* **1995**, *374*, 792-795.
[http://dx.doi.org/10.1038/374792a0]

[41] Yaghi, O.M.; Li, G.M.; Li, H.L. Selective binding and removal of guests in a microporous metal–organic framework. *Nature,* **1995**, *378*, 703-706.
[http://dx.doi.org/10.1038/378703a0]

[42] Subramanian, S.; Zaworotko, M.J. Porous solids by design: [Zn(4,4′-bpy)2(SiF6)]n·xDMF, a single framework octahedral coordination polymer with large square channels. *Angew. Chem. Int. Ed. Engl.,* **1995**, *34*, 2127-2129.
[http://dx.doi.org/10.1002/anie.199521271]

[43] Rowsell, J.L.; Yaghi, O.M. Metal–organic frameworks: a new class of porous materials. *Microp. Mesopor. Mater.,* **2004**, *73*, 3-14.
[http://dx.doi.org/10.1016/j.micromeso.2004.03.034]

[44] Férey, G. Hybrid porous solids: past, present, future. *Chem. Soc. Rev.,* **2008**, *37*(1), 191-214.
[http://dx.doi.org/10.1039/B618320B] [PMID: 18197340]

[45] Zhu, Q-L.; Xu, Q. Metal-organic framework composites. *Chem. Soc. Rev.,* **2014**, *43*(16), 5468-5512.
[http://dx.doi.org/10.1039/C3CS60472A] [PMID: 24638055]

[46] Stock, N.; Biswas, S. Synthesis of metal-organic frameworks (MOFs): routes to various MOF topologies, morphologies, and composites. *Chem. Rev.,* **2012**, *112*(2), 933-969. [http://dx.doi.org/10.1021/cr200304e] [PMID: 22098087]

[47] Lee, Y-R.; Kim, J.; Ahn, W-S. Korean Synthesis of metal-organic frameworks: A mini review. *Chem. Eng. J.,* **2013**, *30*, 1667-1680.

[48] Feng, X.; Ding, X.; Jiang, D. Covalent organic frameworks. *Chem. Soc. Rev.,* **2012**, *41*(18), 6010-6022. [http://dx.doi.org/10.1039/c2cs35157a] [PMID: 22821129]

[49] Thomas, A. Functional materials: from hard to soft porous frameworks. *Angew. Chem. Int. Ed. Engl.,* **2010**, *49*(45), 8328-8344. [http://dx.doi.org/10.1002/anie.201000167] [PMID: 20949565]

[50] Dawson, R.; Cooper, A.I.; Adams, D.J. Nanoporous organic polymer networks. *Prog. Polym. Sci.,* **2012**, *37*, 530-563. [http://dx.doi.org/10.1016/j.progpolymsci.2011.09.002]

[51] Férey, G.; Mellot-Draznieks, C.; Serre, C.; Millange, F.; Dutour, J.; Surblé, S.; Margiolaki, I. A chromium terephthalate-based solid with unusually large pore volumes and surface area. *Science,* **2005**, *309*(5743), 2040-2042. [http://dx.doi.org/10.1126/science.1116275] [PMID: 16179475]

[52] Wong-Foy, A.G.; Matzger, A.J.; Yaghi, O.M. Exceptional H_2 saturation uptake in microporous metal-organic frameworks. *J. Am. Chem. Soc.,* **2006**, *128*(11), 3494-3495. [http://dx.doi.org/10.1021/ja058213h] [PMID: 16536503]

[53] Lebedev, O.I.; Millange, F.; Serre, C.; van Tendeloo, G.; Ferey, G. First direct imaging of giant pores of the metal–organic framework MIL-101. *Chem. Mater.,* **2005**, *17*, 6525-6527. [http://dx.doi.org/10.1021/cm051870o]

[54] Klein, N.; Senkovska, I.; Gedrich, K.; Stoeck, U.; Henschel, A.; Mueller, U.; Kaskel, S. A mesoporous metal-organic framework. *Angew. Chem. Int. Ed. Engl.,* **2009**, *48*(52), 9954-9957. [http://dx.doi.org/10.1002/anie.200904599] [PMID: 19937878]

[55] Xuan, W.; Zhu, C.; Liu, Y.; Cui, Y. Mesoporous metal-organic framework materials. *Chem. Soc. Rev.,* **2012**, *41*(5), 1677-1695. [http://dx.doi.org/10.1039/C1CS15196G] [PMID: 22008884]

[56] Koh, K.; Wong-Foy, A.G.; Matzger, A.J. A porous coordination copolymer with over 5000 m2/g BET surface area. *J. Am. Chem. Soc.,* **2009**, *131*(12), 4184-4185. [http://dx.doi.org/10.1021/ja809985t] [PMID: 19267435]

[57] Yaghi, O.M.; O'Keeffe, M.; Ockwig, N.W.; Chae, H.K.; Eddaoudi, M.; Kim, J. Reticular synthesis and the design of new materials. *Nature,* **2003**, *423*(6941), 705-714. [http://dx.doi.org/10.1038/nature01650] [PMID: 12802325]

[58] Côté, A.P.; Benin, A.I.; Ockwig, N.W.; O'Keeffe, M.; Matzger, A.J.; Yaghi, O.M. Porous, crystalline, covalent organic frameworks. *Science,* **2005**, *310*(5751), 1166-1170. [http://dx.doi.org/10.1126/science.1120411] [PMID: 16293756]

[59] Côté, A.P.; El-Kaderi, H.M.; Furukawa, H.; Hunt, J.R.; Yaghi, O.M. Reticular synthesis of microporous and mesoporous 2D covalent organic frameworks. *J. Am. Chem. Soc.,* **2007**, *129*(43), 12914-12915.
[http://dx.doi.org/10.1021/ja0751781] [PMID: 17918943]

[60] Hunt, J.R.; Doonan, C.J.; LeVangie, J.D.; Côté, A.P.; Yaghi, O.M. Reticular synthesis of covalent organic borosilicate frameworks. *J. Am. Chem. Soc.,* **2008**, *130*(36), 11872-11873.
[http://dx.doi.org/10.1021/ja805064f] [PMID: 18707184]

[61] El-Kaderi, H.M.; Hunt, J.R.; Mendoza-Cortés, J.L.; Côté, A.P.; Taylor, R.E.; O'Keeffe, M.; Yaghi, O.M. Designed synthesis of 3D covalent organic frameworks. *Science,* **2007**, *316*(5822), 268-272.
[http://dx.doi.org/10.1126/science.1139915] [PMID: 17431178]

[62] Han, S.S.; Furukawa, H.; Yaghi, O.M.; Goddard, W.A., III Covalent organic frameworks as exceptional hydrogen storage materials. *J. Am. Chem. Soc.,* **2008**, *130*(35), 11580-11581.
[http://dx.doi.org/10.1021/ja803247y] [PMID: 18683924]

[63] Furukawa, H.; Yaghi, O.M. Storage of hydrogen, methane, and carbon dioxide in highly porous covalent organic frameworks for clean energy applications. *J. Am. Chem. Soc.,* **2009**, *131*(25), 8875-8883.
[http://dx.doi.org/10.1021/ja9015765] [PMID: 19496589]

[64] Czaja, A.U.; Trukhan, N.; Müller, U. Industrial applications of metal-organic frameworks. *Chem. Soc. Rev.,* **2009**, *38*(5), 1284-1293.
[http://dx.doi.org/10.1039/b804680h] [PMID: 19384438]

[65] Kuppler, R.J.; Timmons, D.J.; Fang, Q.R.; Li, J.R.; Makal, T.A.; Young, M.D.; Yuan, D.Q.; Zhao, D.; Zhuang, W.J.; Zhou, H.C. Potential applications of metal-organic frameworks. *Coord. Chem. Rev.,* **2009**, *253*, 3042-3066.
[http://dx.doi.org/10.1016/j.ccr.2009.05.019]

[66] Qiu, S.L.; Zhu, G.S. Molecular engineering for synthesizing novel structures of metal–organic frameworks with multifunctional properties. *Coord. Chem. Rev.,* **2009**, *253*, 2891-2911.
[http://dx.doi.org/10.1016/j.ccr.2009.07.020]

[67] Parnham, E.R.; Morris, R.E. Ionothermal synthesis of zeolites, metal-organic frameworks, and inorganic-organic hybrids. *Acc. Chem. Res.,* **2007**, *40*(10), 1005-1013.
[http://dx.doi.org/10.1021/ar700025k] [PMID: 17580979]

[68] Wasserscheid, P.; Keim, W. Ionic liquids-new "solutions" for transition metal catalysis. *Angew. Chem. Int. Ed. Engl.,* **2000**, *39*(21), 3772-3789.
[http://dx.doi.org/10.1002/1521-3773(20001103)39:21<3772::AID-ANIE3772>3.0.CO;2-5] [PMID: 11091453]

[69] Klinowski, J.; Paz, F.A.; Silva, P.; Rocha, J. Microwave-assisted synthesis of metal-organic frameworks. *Dalton Trans.,* **2011**, *40*(2), 321-330.
[http://dx.doi.org/10.1039/C0DT00708K] [PMID: 20963251]

[70] Park, S.E.; Chang, J.S.; Hwang, Y.K.; Kim, D.S.; Jhung, S.H.; Hwang, J.S. Supramolecular interactions and morphology control in microwave synthesis of nanoporous materials. *Catal. Surv. Asia,* **2004**, *8*, 91-110.

[http://dx.doi.org/10.1023/B:CATS.0000026990.25778.a8]

[71] Jhung, S.H.; Chang, J.S.; Hwang, J.S.; Park, S.E. Selective formation of SAPO-5 and SAPO-34 molecular sieves with microwave irradiation and hydrothermal heating. *Microporous Mesoporous Mater.,* **2003**, *64*, 33-39.
[http://dx.doi.org/10.1016/S1387-1811(03)00501-8]

[72] Jhung, S.H.; Lee, J.H.; Yoon, J.W.; Hwang, J.S.; Park, S.E.; Chang, J.S. Selective crystallization of CoAPO-34 and VAPO-5 molecular sieves under microwave irradiation in an alkaline or neutral condition. *Microporous Mesoporous Mater.,* **2005**, *80*, 147-152.
[http://dx.doi.org/10.1016/j.micromeso.2004.11.013]

[73] Kang, K.K.; Park, C.H.; Ahn, W.S. Microwave preparation of a titanium-substituted mesoporous molecular sieve. *Catal. Lett.,* **1999**, *59*, 45-49.
[http://dx.doi.org/10.1023/A:1019004101326]

[74] Jhung, S.H.; Chang, J.S.; Hwang, Y.K.; Park, S.E. Crystal morphology control of AFI type molecular sieves with microwave irradiation. *J. Mater. Chem.,* **2004**, *14*, 280-285.
[http://dx.doi.org/10.1039/b309142b]

[75] Hwang, Y.K.; Chang, J.S.; Park, S.E.; Kim, D.S.; Kwon, Y.U.; Jhung, S.H.; Hwang, J.S.; Park, M.S. Microwave fabrication of MFI zeolite crystals with a fibrous morphology and their applications. *Angew. Chem. Int. Ed. Engl.,* **2005**, *44*(4), 556-560.
[http://dx.doi.org/10.1002/anie.200461403] [PMID: 15523677]

[76] Jhung, S.H.; Lee, J.H.; Chang, J.S. Microwave synthesis of a nanoporous hybrid material, chromium trimesate. *Bull. Korean Chem. Soc.,* **2005**, *26*, 880-881.
[http://dx.doi.org/10.5012/bkcs.2005.26.6.880]

[77] Horcajada, P.; Chalati, T.; Serre, C.; Gillet, B.; Sebrie, C.; Baati, T.; Eubank, J.F.; Heurtaux, D.; Clayette, P.; Kreuz, C.; Chang, J.S.; Hwang, Y.K.; Marsaud, V.; Bories, P.N.; Cynober, L.; Gil, S.; Férey, G.; Couvreur, P.; Gref, R. Porous metal-organic-framework nanoscale carriers as a potential platform for drug delivery and imaging. *Nat. Mater.,* **2010**, *9*(2), 172-178.
[http://dx.doi.org/10.1038/nmat2608] [PMID: 20010827]

[78] Jhung, S.H.; Lee, J.H.; Yoon, J.W.; Serre, C.; Ferey, G.; Chang, J.S. Microwave synthesis of chromium terephthalate MIL-101 and its benzene sorption ability. *Adv. Mater.,* **2007**, *19*, 121-124.
[http://dx.doi.org/10.1002/adma.200601604]

[79] Taylor-Pashow, K.M.; Della Rocca, J.; Xie, Z.; Tran, S.; Lin, W. Postsynthetic modifications of iron-carboxylate nanoscale metal-organic frameworks for imaging and drug delivery. *J. Am. Chem. Soc.,* **2009**, *131*(40), 14261-14263.
[http://dx.doi.org/10.1021/ja906198y] [PMID: 19807179]

[80] Centrone, A.; Yang, Y.; Speakman, S.; Bromberg, L.; Rutledge, G.C.; Hatton, T.A. Growth of metal-organic frameworks on polymer surfaces. *J. Am. Chem. Soc.,* **2010**, *132*(44), 15687-15691.
[http://dx.doi.org/10.1021/ja106381x] [PMID: 20945899]

[81] Centrone, A.; Harada, T.; Speakman, S.; Hatton, T.A. Facile synthesis of vanadium metal-organic frameworks and their magnetic properties. *Small,* **2010**, *6*(15), 1598-1602.
[http://dx.doi.org/10.1002/smll.201000773] [PMID: 20623532]

[82] Ahnfeldt, T.; Moellmer, J.; Guillerm, V.; Staudt, R.; Serre, C.; Stock, N. High-throughput and time-resolved energy-dispersive X-ray diffraction (EDXRD) study of the formation of CAU-1-(OH)2: microwave and conventional heating. *Chemistry,* **2011**, *17*(23), 6462-6468. [http://dx.doi.org/10.1002/chem.201003708] [PMID: 21538608]

[83] Silva, P.; Valente, A.A.; Rocha, J.; Paz, F.A. Fast microwave synthesis of a microporous lanthanide–organic framework. *Cryst. Growth Des.,* **2010**, *10*, 2025-2028. [http://dx.doi.org/10.1021/cg900884d]

[84] Haque, E.; Khan, N.A.; Park, J.H.; Jhung, S.H. Synthesis of a metal-organic framework material, iron terephthalate, by ultrasound, microwave, and conventional electric heating: a kinetic study. *Chemistry,* **2010**, *16*(3), 1046-1052. [http://dx.doi.org/10.1002/chem.200902382] [PMID: 20014080]

[85] Kim, D.O.; Park, J.; Ahn, G.R.; Jeon, H.J.; Kim, J.S.; Kim, D.W.; Jung, M.S.; Lee, S.W.; Shin, S.H. Synthesis of MOF having functional side group. *Inorg. Chim. Acta,* **2011**, *370*, 76-81. [http://dx.doi.org/10.1016/j.ica.2011.01.030]

[86] Jhung, S.H.; Lee, J.H.; Forster, P.M.; Férey, G.; Cheetham, A.K.; Chang, J.S. Microwave synthesis of hybrid inorganic-organic porous materials: phase-selective and rapid crystallization. *Chemistry,* **2006**, *12*(30), 7899-7905. [http://dx.doi.org/10.1002/chem.200600270] [PMID: 16871506]

[87] Lee, J.S.; Halligudi, S.B.; Jang, N.H.; Hwang, D.W.; Chang, J.S.; Hwang, Y.K. Microwave synthesis of a porous metal-organic framework, nickel(II) dihydroxyterephthalate and its catalytic properties in oxidation of cyclohexene. *Bull. Korean Chem. Soc.,* **2010**, *31*, 1489-1495. [http://dx.doi.org/10.5012/bkcs.2010.31.6.1489]

[88] Jung, D.W.; Yang, D.A.; Kim, J.; Kim, J.; Ahn, W.S. Facile synthesis of MOF-177 by a sonochemical method using 1-methyl-2-pyrrolidinone as a solvent. *Dalton Trans.,* **2010**, *39*(11), 2883-2887. [http://dx.doi.org/10.1039/b925088c] [PMID: 20200716]

[89] Choi, J.Y.; Kim, J.; Jhung, S.H.; Kim, H.K.; Chang, J.S.; Chae, H.K. Microwave synthesis of a porous metal-organic framework, zinc terephthalate MOF-5. *Bull. Korean Chem. Soc.,* **2006**, *27*, 1523-1524. [http://dx.doi.org/10.5012/bkcs.2006.27.10.1523]

[90] Ni, Z.; Masel, R.I. Rapid production of metal-organic frameworks *via* microwave-assisted solvothermal synthesis. *J. Am. Chem. Soc.,* **2006**, *128*(38), 12394-12395. [http://dx.doi.org/10.1021/ja0635231] [PMID: 16984171]

[91] Choi, J.S.; Son, W.J.; Kim, J.; Ahn, W.S. Metal–organic framework MOF-5 prepared by microwave heating: Factors to be considered. *Microporous Mesoporous Mater.,* **2008**, *116*, 727-731. [http://dx.doi.org/10.1016/j.micromeso.2008.04.033]

[92] Xiang, Z.H.; Cao, D.P.; Shao, X.H.; Wang, W.C.; Zhang, J.W.; Wu, W.Z. Facile preparation of high-capacity hydrogen storage metal-organic frameworks: A combination of microwave-assisted solvothermal synthesis and supercritical activation. *Chem. Eng. Sci.,* **2010**, *65*, 3140-3146. [http://dx.doi.org/10.1016/j.ces.2010.02.005]

[93] Seo, Y.K.; Hundal, G.; Jang, I.T.; Hwang, Y.K.; Jun, C.H.; Chang, J.S. Microwave synthesis of hybrid inorganic–organic materials including porous Cu3(BTC)2 from Cu(II)-trimesate mixture. *Microporous*

Mesoporous Mater., **2009**, *119*, 331-337.
[http://dx.doi.org/10.1016/j.micromeso.2008.10.035]

[94] Schlesinger, M.; Schulze, S.; Hietschold, M.; Mehring, M. Evaluation of synthetic methods for microporous metal–organic frameworks exemplified by the competitive formation of [Cu2(btc)3(H2O)3] and. *Microporous Mesoporous Mater.,* **2010**, *132*, 121-127. [Cu2(btc)(OH)(H2O)].
[http://dx.doi.org/10.1016/j.micromeso.2010.02.008]

[95] Khan, N.A.; Haque, E.; Jhung, S.H. Rapid syntheses of a metal-organic framework material Cu3(BTC)2(H2O)3 under microwave: a quantitative analysis of accelerated syntheses. *Phys. Chem. Chem. Phys.,* **2010**, *12*(11), 2625-2631.
[http://dx.doi.org/10.1039/b921558a] [PMID: 20200739]

[96] Mueller, U.; Schubert, M.; Teich, F.; Puetter, H.; Schierle-Arndt, K.; Pastre, J. Metal–organic frameworks-prospective industrial applications. *J. Mater. Chem.,* **2006**, *16*, 626-636.
[http://dx.doi.org/10.1039/B511962F]

[97] Mueller, U.; Richter, I.; Schubert, M. Porous metal organic framework based on pyrroles and pyridinones . *WO Patent 2007131955A1,* **2007**.

[98] Sumida, K.; Horike, S.; Kaye, S.S.; Herm, Z.R.; Queen, W.L.; Brown, C.M.; Grandjean, F.; Long, G.J.; Dailly, A.; Long, J.R. Hydrogen storage and carbon dioxide capture in an iron-based sodalite-type metal–organic framework (Fe-BTT) discovered *via* high-throughput methods. *Chem. Sci. (Camb.),* **2010**, *1*, 184-191.
[http://dx.doi.org/10.1039/c0sc00179a]

[99] Forster, P.M.; Stock, N.; Cheetham, A.K. A high-throughput investigation of the role of pH, temperature, concentration, and time on the synthesis of hybrid inorganic-organic materials. *Angew. Chem. Int. Ed. Engl.,* **2005**, *44*(46), 7608-7611.
[http://dx.doi.org/10.1002/anie.200501766] [PMID: 16245378]

[100] Bauer, S.; Serre, C.; Devic, T.; Horcajada, P.; Marrot, J.; Férey, G.; Stock, N. High-throughput assisted rationalization of the formation of metal organic frameworks in the Iron(III) aminoterephthalate solvothermal system. *Inorg. Chem.,* **2008**, *47*(17), 7568-7576.
[http://dx.doi.org/10.1021/ic800538r] [PMID: 18681423]

[101] Ahnfeldt, T.; Gunzelmann, D.; Loiseau, T.; Hirsemann, D.; Senker, J.; Férey, G.; Stock, N. Synthesis and modification of a functionalized 3D open-framework structure with MIL-53 topology. *Inorg. Chem.,* **2009**, *48*(7), 3057-3064.
[http://dx.doi.org/10.1021/ic8023265] [PMID: 19245258]

[102] Biemmi, E.; Christian, S.; Stock, N.; Bein, T. High-throughput screening of synthesis parameters in the formation of the metal-organic frameworks MOF-5 and HKUST-1. *Microporous Mesoporous Mater.,* **2009**, *117*, 111-117.
[http://dx.doi.org/10.1016/j.micromeso.2008.06.040]

[103] Sonnauer, A.; Hoffmann, F.; Fröba, M.; Kienle, L.; Duppel, V.; Thommes, M.; Serre, C.; Férey, G.; Stock, N. Giant pores in a chromium 2,6-naphthalenedicarboxylate open-framework structure with MIL-101 topology. *Angew. Chem. Int. Ed. Engl.,* **2009**, *48*(21), 3791-3794.
[http://dx.doi.org/10.1002/anie.200805980] [PMID: 19191364]

[104] Banerjee, R.; Phan, A.; Wang, B.; Knobler, C.; Furukawa, H.; O'Keeffe, M.; Yaghi, O.M. High-throughput synthesis of zeolitic imidazolate frameworks and application to CO2 capture. *Science,* **2008**, *319*(5865), 939-943.
[http://dx.doi.org/10.1126/science.1152516] [PMID: 18276887]

[105] Maniam, P.; Stock, N. Investigation of porous Ni-based metal-organic frameworks containing paddle-wheel type inorganic building units *via* high-throughput methods. *Inorg. Chem.,* **2011**, *50*(11), 5085-5097.
[http://dx.doi.org/10.1021/ic200381f] [PMID: 21539354]

[106] Cohen, S.M. Postsynthetic methods for the functionalization of metal-organic frameworks. *Chem. Rev.,* **2012**, *112*(2), 970-1000.
[http://dx.doi.org/10.1021/cr200179u] [PMID: 21916418]

[107] Burrows, A.D.; Frost, C.G.; Mahon, M.F.; Richardson, C. Post-synthetic modification of tagged metal-organic frameworks. *Angew. Chem. Int. Ed. Engl.,* **2008**, *47*(44), 8482-8486.
[http://dx.doi.org/10.1002/anie.200802908] [PMID: 18825761]

[108] Doonan, C.J.; Morris, W.; Furukawa, H.; Yaghi, O.M. Isoreticular metalation of metal-organic frameworks. *J. Am. Chem. Soc.,* **2009**, *131*(27), 9492-9493.
[http://dx.doi.org/10.1021/ja903251e] [PMID: 19534523]

[109] Ding, X.; Guo, J.; Feng, X.; Honsho, Y.; Guo, J.; Seki, S.; Maitarad, P.; Saeki, A.; Nagase, S.; Jiang, D. Synthesis of metallophthalocyanine covalent organic frameworks that exhibit high carrier mobility and photoconductivity. *Angew. Chem. Int. Ed. Engl.,* **2011**, *50*(6), 1289-1293.
[http://dx.doi.org/10.1002/anie.201005919] [PMID: 21290495]

[110] Tilford, R.W.; Gemmill, W.R.; zur Loye, H.C.; Lavigne, J.J. Facile synthesis of a highly crystalline, covalently linked porous boronate network. *Chem. Mater.,* **2006**, *18*, 5296-5301.
[http://dx.doi.org/10.1021/cm061177g]

[111] Uribe-Romo, F.J.; Hunt, J.R.; Furukawa, H.; Klöck, C.; O'Keeffe, M.; Yaghi, O.M. A crystalline imine-linked 3-D porous covalent organic framework. *J. Am. Chem. Soc.,* **2009**, *131*(13), 4570-4571.
[http://dx.doi.org/10.1021/ja8096256] [PMID: 19281246]

[112] Uribe-Romo, F.J.; Doonan, C.J.; Furukawa, H.; Oisaki, K.; Yaghi, O.M. Crystalline covalent organic frameworks with hydrazone linkages. *J. Am. Chem. Soc.,* **2011**, *133*(30), 11478-11481.
[http://dx.doi.org/10.1021/ja204728y] [PMID: 21721558]

[113] Hamilton, C.W.; Baker, R.T.; Staubitz, A.; Manners, I. B-N compounds for chemical hydrogen storage. *Chem. Soc. Rev.,* **2009**, *38*(1), 279-293.
[http://dx.doi.org/10.1039/B800312M] [PMID: 19088978]

[114] Sabo, M.; Henschel, A.; Frode, H.; Klemm, E.; Kaskel, S. Solution infiltration of palladium into MOF-5: synthesis, physisorption and catalytic properties. *J. Mater. Chem.,* **2007**, *17*, 3827-3832.
[http://dx.doi.org/10.1039/b706432b]

[115] Cheon, Y.E.; Suh, M.P. Enhanced hydrogen storage by palladium nanoparticles fabricated in a redox-active metal-organic framework. *Angew. Chem. Int. Ed. Engl.,* **2009**, *48*(16), 2899-2903.
[http://dx.doi.org/10.1002/anie.200805494] [PMID: 19294708]

[116] Suh, M.P.; Park, H.J.; Prasad, T.K.; Lim, D-W. Hydrogen storage in metal-organic frameworks. *Chem.*

Rev., **2012**, *112*(2), 782-835.
[http://dx.doi.org/10.1021/cr200274s] [PMID: 22191516]

[117] Proch, S.; Herrmannsdörfer, J.; Kempe, R.; Kern, C.; Jess, A.; Seyfarth, L.; Senker, J. Pt@MOF-177: synthesis, room-temperature hydrogen storage and oxidation catalysis. *Chemistry,* **2008**, *14*(27), 8204-8212.
[http://dx.doi.org/10.1002/chem.200801043] [PMID: 18666269]

[118] Lim, D-W.; Yoon, J.W.; Ryu, K.Y.; Suh, M.P. Magnesium nanocrystals embedded in a metal-organic framework: hybrid hydrogen storage with synergistic effect on physi- and chemisorption. *Angew. Chem. Int. Ed. Engl.,* **2012**, *51*(39), 9814-9817.
[http://dx.doi.org/10.1002/anie.201206055] [PMID: 22961817]

[119] Figueroa, J.D.; Fout, T.; Plasynski, S.; McIlvried, H.; Srivastava, R.D. Advances in CO(2) capture technology - The US Department of Energy's carbon sequestration program. *Int. J. Greenh. Gas Control,* **2008**, *2*, 9-20.
[http://dx.doi.org/10.1016/S1750-5836(07)00094-1]

[120] Choi, S.; Drese, J.H.; Jones, C.W. Adsorbent materials for carbon dioxide capture from large anthropogenic point sources. *ChemSusChem,* **2009**, *2*(9), 796-854.
[http://dx.doi.org/10.1002/cssc.200900036] [PMID: 19731282]

[121] Kuppler, R.J.; Timmons, D.J.; Fang, Q.R.; Li, J.R.; Makal, T.A.; Young, M.D.; Yuan, D.Q.; Zhao, D.; Zhuang, W.J.; Zhou, H.C. Potential applications of metal-organic frameworks. *Coord. Chem. Rev.,* **2009**, *253*, 3042-3066.
[http://dx.doi.org/10.1016/j.ccr.2009.05.019]

[122] Chen, B.; Xiang, S.; Qian, G. Metal-organic frameworks with functional pores for recognition of small molecules. *Acc. Chem. Res.,* **2010**, *43*(8), 1115-1124.
[http://dx.doi.org/10.1021/ar100023y] [PMID: 20450174]

[123] Phan, A.; Doonan, C.J.; Uribe-Romo, F.J.; Knobler, C.B.; O'Keeffe, M.; Yaghi, O.M. Synthesis, structure, and carbon dioxide capture properties of zeolitic imidazolate frameworks. *Acc. Chem. Res.,* **2010**, *43*(1), 58-67.
[http://dx.doi.org/10.1021/ar900116g] [PMID: 19877580]

[124] Keskin, S.; van Heest, T.M.; Sholl, D.S. Can metal-organic framework materials play a useful role in large-scale carbon dioxide separations? *ChemSusChem,* **2010**, *3*(8), 879-891.
[http://dx.doi.org/10.1002/cssc.201000114] [PMID: 20730980]

[125] Lin, X.; Champness, N.R.; Schröder, M. Hydrogen, methane and carbon dioxide adsorption in metal-organic framework materials. *Top. Curr. Chem.,* **2010**, *293*, 35-76.
[http://dx.doi.org/10.1007/128_2009_21] [PMID: 21618742]

[126] Bae, Y.S.; Snurr, R.Q. Development and evaluation of porous materials for carbon dioxide separation and capture. *Angew. Chem. Int. Ed. Engl.,* **2011**, *50*(49), 11586-11596.
[http://dx.doi.org/10.1002/anie.201101891] [PMID: 22021216]

[127] Li, J.R.; Sculley, J.; Zhou, H.C. Metal-organic frameworks for separations. *Chem. Rev.,* **2012**, *112*(2), 869-932.
[http://dx.doi.org/10.1021/cr200190s] [PMID: 21978134]

[128] Li, B.; Wang, H.; Chen, B. Microporous metal-organic frameworks for gas separation. *Chem. Asian J.,* **2014**, *9*(6), 1474-1498.
[http://dx.doi.org/10.1002/asia.201400031] [PMID: 24668618]

[129] Yamaura, H.; Moriya, K.; Miura, N.; Yamazoe, N. Mechanism of sensitivity promotion in CO sensor using indium oxide and cobalt oxide. *Sens. Actuators B Chem.,* **2000**, *65*, 39-41.
[http://dx.doi.org/10.1016/S0925-4005(99)00456-6]

[130] Dybtsev, D.N.; Chun, H.; Yoon, S.H.; Kim, D.; Kim, K. Microporous manganese formate: a simple metal-organic porous material with high framework stability and highly selective gas sorption properties. *J. Am. Chem. Soc.,* **2004**, *126*(1), 32-33.
[http://dx.doi.org/10.1021/ja038678c] [PMID: 14709045]

[131] Landon, P.; Ferguson, J.; Solsona, B.E.; Garcia, T.; Carley, A.F.; Herzing, A.A.; Kiely, C.J.; Golunski, S.E.; Hutchings, G.J. Selective oxidation of CO in the presence of H2, H2O and CO2 *via* gold for use in fuel cells. *Chem. Commun. (Camb.),* **2005**, (27), 3385-3387.
[http://dx.doi.org/10.1039/b505295p] [PMID: 15997272]

[132] Chen, B.; Liang, C.; Yang, J.; Contreras, D.S.; Clancy, Y.L.; Lobkovsky, E.B.; Yaghi, O.M.; Dai, S. A microporous metal-organic framework for gas-chromatographic separation of alkanes. *Angew. Chem. Int. Ed. Engl.,* **2006**, *45*(9), 1390-1393.
[http://dx.doi.org/10.1002/anie.200502844] [PMID: 16425335]

[133] Pan, L.; Olson, D.H.; Ciemnolonski, L.R.; Heddy, R.; Li, J. Separation of hydrocarbons with a microporous metal-organic framework. *Angew. Chem. Int. Ed. Engl.,* **2006**, *45*(4), 616-619.
[http://dx.doi.org/10.1002/anie.200503503] [PMID: 16355421]

[134] Pan, L.; Parker, B.; Huang, X.; Olson, D.H.; Lee, J.; Li, J. Zn(tbip) (H2tbip= 5-tert-butyl isophthalic acid): a highly stable guest-free microporous metal organic framework with unique gas separation capability. *J. Am. Chem. Soc.,* **2006**, *128*(13), 4180-4181.
[http://dx.doi.org/10.1021/ja057667b] [PMID: 16568964]

[135] Wang, B.; Côté, A.P.; Furukawa, H.; O'Keeffe, M.; Yaghi, O.M. Colossal cages in zeolitic imidazolate frameworks as selective carbon dioxide reservoirs. *Nature,* **2008**, *453*(7192), 207-211.
[http://dx.doi.org/10.1038/nature06900] [PMID: 18464739]

[136] Bastin, L.; Barcia, P.S.; Hurtado, E.J.; Silva, J.A.; Rodrigues, A.E.; Chen, B.L. A microporous metal-organic framework for separation of CO2/N-2 and CO2/CH4 by fixed-bed adsorption. *J. Phys. Chem. C,* **2008**, *112*, 1575-1581.
[http://dx.doi.org/10.1021/jp077618g]

[137] Bae, Y.S.; Mulfort, K.L.; Frost, H.; Ryan, P.; Punnathanam, S.; Broadbelt, L.J.; Hupp, J.T.; Snurr, R.Q. Separation of CO2 from CH4 using mixed-ligand metal-organic frameworks. *Langmuir,* **2008**, *24*(16), 8592-8598.
[http://dx.doi.org/10.1021/la800555x] [PMID: 18616225]

[138] Bae, Y.S.; Farha, O.K.; Spokoyny, A.M.; Mirkin, C.A.; Hupp, J.T.; Snurr, R.Q. Carborane-based metal-organic frameworks as highly selective sorbents for CO(2) over methane. *Chem. Commun. (Camb.),* **2008**, (35), 4135-4137.
[http://dx.doi.org/10.1039/b805785k] [PMID: 18802507]

[139] Bae, Y.S.; Farha, O.K.; Hupp, J.T.; Snurr, R.Q. Enhancement of CO2/N-2 selectivity in a metal-organic framework by cavity modification. *J. Mater. Chem.,* **2009**, *19*, 2131-2134. [http://dx.doi.org/10.1039/b900390h]

[140] El-Shall, M.S.; Abdelsayed, V.; Khder, A.E.; Hassan, H.M.; El-Kaderi, H.M.; Reich, T.E. Metallic and bimetallic nanocatalysts incorporated into highly porous coordination polymer MIL-101. *J. Mater. Chem.,* **2009**, *19*, 7625-7631. [http://dx.doi.org/10.1039/b912012b]

[141] Jiang, H-L.; Liu, B.; Akita, T.; Haruta, M.; Sakurai, H.; Xu, Q. Au@ZIF-8: CO oxidation over gold nanoparticles deposited to metal-organic framework. *J. Am. Chem. Soc.,* **2009**, *131*(32), 11302-11303. [http://dx.doi.org/10.1021/ja9047653] [PMID: 19637919]

[142] Britt, D.; Furukawa, H.; Wang, B.; Glover, T.G.; Yaghi, O.M. Highly efficient separation of carbon dioxide by a metal-organic framework replete with open metal sites. *Proc. Natl. Acad. Sci. USA,* **2009**, *106*(49), 20637-20640. [http://dx.doi.org/10.1073/pnas.0909718106] [PMID: 19948967]

[143] Banerjee, R.; Furukawa, H.; Britt, D.; Knobler, C.; O'Keeffe, M.; Yaghi, O.M. Control of pore size and functionality in isoreticular zeolitic imidazolate frameworks and their carbon dioxide selective capture properties. *J. Am. Chem. Soc.,* **2009**, *131*(11), 3875-3877. [http://dx.doi.org/10.1021/ja809459e] [PMID: 19292488]

[144] Venna, S.R.; Carreon, M.A. Highly permeable zeolite imidazolate framework-8 membranes for CO2/CH4 separation. *J. Am. Chem. Soc.,* **2010**, *132*(1), 76-78. [http://dx.doi.org/10.1021/ja909263x] [PMID: 20014839]

[145] Jiang, H-L.; Singh, S.K.; Yan, J-M.; Zhang, X-B.; Xu, Q. Liquid-phase chemical hydrogen storage: catalytic hydrogen generation under ambient conditions. *ChemSusChem,* **2010**, *3*(5), 541-549. [http://dx.doi.org/10.1002/cssc.201000023] [PMID: 20379965]

[146] Gu, X.; Lu, Z-H.; Jiang, H-L.; Akita, T.; Xu, Q. Synergistic catalysis of metal-organic framework-immobilized Au-Pd nanoparticles in dehydrogenation of formic acid for chemical hydrogen storage. *J. Am. Chem. Soc.,* **2011**, *133*(31), 11822-11825. [http://dx.doi.org/10.1021/ja200122f] [PMID: 21761819]

[147] Jia, C-J.; Schwickardi, M.; Weidenthaler, C.; Schmidt, W.; Korhonen, S.; Weckhuysen, B.M.; Schüth, F. Co3O4-SiO2 nanocomposite: a very active catalyst for CO oxidation with unusual catalytic behavior. *J. Am. Chem. Soc.,* **2011**, *133*(29), 11279-11288. [http://dx.doi.org/10.1021/ja2028926] [PMID: 21696181]

[148] Wang, C.; deKrafft, K.E.; Lin, W. Pt nanoparticles@photoactive metal-organic frameworks: efficient hydrogen evolution *via* synergistic photoexcitation and electron injection. *J. Am. Chem. Soc.,* **2012**, *134*(17), 7211-7214. [http://dx.doi.org/10.1021/ja300539p] [PMID: 22486151]

[149] Horiuchi, Y.; Toyao, T.; Saito, M.; Mochizuki, K.; Iwata, M.; Higashimura, H.; Anpo, M.; Matsuoka, M. Visible-light-promoted photocatalytic hydrogen production by using an amino-functionalized Ti(IV) metal–organic framework. *J. Phys. Chem. C,* **2012**, *116*, 20848-20853. [http://dx.doi.org/10.1021/jp3046005]

[150] Yadav, M.; Xu, Q. Liquid-phase chemical hydrogen storage materials. *Energy Environ. Sci.,* **2012**, *5*, 9698-9725.
[http://dx.doi.org/10.1039/c2ee22937d]

[151] Aijaz, A.; Karkamkar, A.; Choi, Y.J.; Tsumori, N.; Rönnebro, E.; Autrey, T.; Shioyama, H.; Xu, Q. Immobilizing highly catalytically active Pt nanoparticles inside the pores of metal-organic framework: a double solvents approach. *J. Am. Chem. Soc.,* **2012**, *134*(34), 13926-13929.
[http://dx.doi.org/10.1021/ja3043905] [PMID: 22888976]

[152] Li, P-Z.; Aranishi, K.; Xu, Q. ZIF-8 immobilized nickel nanoparticles: highly effective catalysts for hydrogen generation from hydrolysis of ammonia borane. *Chem. Commun. (Camb.),* **2012**, *48*(26), 3173-3175.
[http://dx.doi.org/10.1039/c2cc17302f] [PMID: 22343827]

[153] Singh, A.K.; Xu, Q. Synergistic catalysis over bimetallic alloy nanoparticles. *ChemCatChem,* **2013**, *5*, 652-676.
[http://dx.doi.org/10.1002/cctc.201200591]

[154] Dalebrook, A.F.; Gan, W.; Grasemann, M.; Moret, S.; Laurenczy, G. Hydrogen storage: beyond conventional methods. *Chem. Commun. (Camb.),* **2013**, *49*(78), 8735-8751.
[http://dx.doi.org/10.1039/c3cc43836h] [PMID: 23964360]

[155] Zhu, Q-L.; Li, J.; Xu, Q. Immobilizing metal nanoparticles to metal-organic frameworks with size and location control for optimizing catalytic performance. *J. Am. Chem. Soc.,* **2013**, *135*(28), 10210-10213.
[http://dx.doi.org/10.1021/ja403330m] [PMID: 23805877]

[156] Singh, A.K.; Xu, Q. Addition of sulfonic acids to terminal alkynes catalyzed by a rhodium complex: ligand concentration-controlled reaction selectivity. *ChemCatChem,* **2013**, *5*, 3000-3004.
[http://dx.doi.org/10.1002/cctc.201300233]

[157] Wu, R.; Qian, X.; Zhou, K.; Liu, H.; Yandian, B.; Wei, J.; Zhu, H.; Huang, Y. Highly dispersed Au nanoparticles immobilized on Zr-based metal–organic frameworks as heterostructured catalyst for CO oxidation. *J. Mater. Chem. A Mater. Energy Sustain.,* **2013**, *1*, 14294-14299.
[http://dx.doi.org/10.1039/c3ta13114a]

[158] Zhu, Q-L.; Tsumori, N.; Xu, Q. Sodium hydroxide-assisted growth of uniform Pd nanoparticles on nanoporous carbon MSC-30 for efficient and complete dehydrogenation of formic acid under ambient conditions. *Chem. Sci. (Camb.),* **2014**, *5*, 195-199.
[http://dx.doi.org/10.1039/C3SC52448E]

[159] Cavenati, S.; Grande, C.A.; Rodrigues, A.E. *J. Chem. Eng. Data,* **2004**, *49*, 1095-1101.
[http://dx.doi.org/10.1021/je0498917]

[160] Millward, A.R.; Yaghi, O.M. Metal-organic frameworks with exceptionally high capacity for storage of carbon dioxide at room temperature. *J. Am. Chem. Soc.,* **2005**, *127*(51), 17998-17999.
[http://dx.doi.org/10.1021/ja0570032] [PMID: 16366539]

[161] Garberoglio, G. Computer simulation of the adsorption of light gases in covalent organic frameworks. *Langmuir,* **2007**, *23*(24), 12154-12158.
[http://dx.doi.org/10.1021/la701736m] [PMID: 17956137]

[162] Klontzas, E.; Tylianakis, E.; Froudakis, G.E. Hydrogen storage in 3D covalent organic frameworks. A multiscale theoretical investigation. *J. Phys. Chem. C,* **2008**, *112*, 9095-9098.
[http://dx.doi.org/10.1021/jp711326g]

[163] Babarao, R.; Jiang, J.W. Exceptionally high CO2storage in covalent-organic frameworks: Atomistic simulation study. *Energy Environ. Sci.,* **2008**, *1*, 139-143.
[http://dx.doi.org/10.1039/b805473h]

[164] Klontzas, E.; Tylianakis, E.; Froudakis, G.E. Hydrogen storage in lithium-functionalized 3-D covalent-organic framework materials. *J. Phys. Chem. C,* **2009**, *113*, 21253-21257.
[http://dx.doi.org/10.1021/jp907241y]

[165] Srepusharawoot, P.; Scheicher, R.H.; Araujo, C.M.; Blomqvist, A.; Pinsook, U.; Ahuja, R. Ab Initio study of molecular hydrogen adsorption in covalent organic framework-1. *J. Phys. Chem. C,* **2009**, *113*, 8498-8504.
[http://dx.doi.org/10.1021/jp809167b]

[166] Assfour, B.; Seifert, G. Hydrogen adsorption sites and energies in 2D and 3D covalent organic frameworks. *Chem. Phys. Lett.,* **2010**, *489*, 86-91.
[http://dx.doi.org/10.1016/j.cplett.2010.02.046]

[167] Klontzas, E.; Tylianakis, E.; Froudakis, G.E. Designing 3D COFs with enhanced hydrogen storage capacity. *Nano Lett.,* **2010**, *10*(2), 452-454.
[http://dx.doi.org/10.1021/nl903068a] [PMID: 20050693]

[168] Li, F.; Zhao, J.; Johansson, B.; Sun, L. Improving hydrogen storage properties of covalent organic frameworks by substitutional doping. *Int. J. Hydrogen Energy,* **2010**, *35*, 266-271.
[http://dx.doi.org/10.1016/j.ijhydene.2009.10.061]

[169] Assfour, B.; Seifert, G. Adsorption of hydrogen in covalent organic frameworks: Comparison of simulations and experiments. *Microporous Mesoporous Mater.,* **2010**, *133*, 59-65.
[http://dx.doi.org/10.1016/j.micromeso.2010.04.015]

[170] Mendoza-Cortés, J.L.; Han, S.S.; Furukawa, H.; Yaghi, O.M.; Goddard, W.A., III Adsorption mechanism and uptake of methane in covalent organic frameworks: theory and experiment. *J. Phys. Chem. A,* **2010**, *114*(40), 10824-10833.
[http://dx.doi.org/10.1021/jp1044139] [PMID: 20845983]

[171] Doonan, C.J.; Tranchemontagne, D.J.; Glover, T.G.; Hunt, J.R.; Yaghi, O.M. Exceptional ammonia uptake by a covalent organic framework. *Nat. Chem.,* **2010**, *2*(3), 235-238.
[http://dx.doi.org/10.1038/nchem.548] [PMID: 21124483]

[172] Lan, J.; Cao, D.; Wang, W.; Smit, B. Doping of alkali, alkaline-earth, and transition metals in covalent-organic frameworks for enhancing CO2 capture by first-principles calculations and molecular simulations. *ACS Nano,* **2010**, *4*(7), 4225-4237.
[http://dx.doi.org/10.1021/nn100962r] [PMID: 20568707]

[173] Choi, Y.J.; Choi, J.H.; Choi, K.M.; Kang, J.K. Covalent organic frameworks for extremely high reversible CO2 uptake capacity: a theoretical approach. *J. Mater. Chem.,* **2011**, *21*, 1073-1078.
[http://dx.doi.org/10.1039/C0JM02891F]

[174] Mendoza-Cortes, J.L.; Pascal, T.A.; Goddard, W.A., III Design of covalent organic frameworks for

methane storage. *J. Phys. Chem. A,* **2011**, *115*(47), 13852-13857.
[http://dx.doi.org/10.1021/jp209541e] [PMID: 21992457]

[175] Tylianakis, E.; Klontzas, E.; Froudakis, G.E. Multi-scale theoretical investigation of hydrogen storage in covalent organic frameworks. *Nanoscale,* **2011**, *3*(3), 856-869.
[http://dx.doi.org/10.1039/c0nr00505c] [PMID: 21218227]

[176] Kim, D.; Jung, D.H.; Kim, K-H.; Guk, H.; Han, S.S.; Choi, K.; Choi, S-H. Pillared covalent organic frameworks with balanced volumetric and gravimetric hydrogen uptake. *J. Phys. Chem. C,* **2012**, *116*, 1479-1484.
[http://dx.doi.org/10.1021/jp2080622]

[177] Lan, J.; Cao, D.; Wang, W. High uptakes of methane in Li-doped 3D covalent organic frameworks. *Langmuir,* **2010**, *26*(1), 220-226.
[http://dx.doi.org/10.1021/la9020383] [PMID: 20038169]

[178] Keskin, S. Adsorption, diffusion, and separation of CH4/H2 mixtures in covalent organic frameworks: molecular simulations and theoretical predictions. *J. Phys. Chem. C,* **2012**, *116*, 1772-1779.
[http://dx.doi.org/10.1021/jp209804x]

[179] Ma, L.; Abney, C.; Lin, W. Enantioselective catalysis with homochiral metal-organic frameworks. *Chem. Soc. Rev.,* **2009**, *38*(5), 1248-1256.
[http://dx.doi.org/10.1039/b807083k] [PMID: 19384436]

[180] Lee, J.; Farha, O.K.; Roberts, J.; Scheidt, K.A.; Nguyen, S.T.; Hupp, J.T. Metal-organic framework materials as catalysts. *Chem. Soc. Rev.,* **2009**, *38*(5), 1450-1459.
[http://dx.doi.org/10.1039/b807080f] [PMID: 19384447]

[181] Farrusseng, D.; Aguado, S.; Pinel, C. Metal-organic frameworks: opportunities for catalysis. *Angew. Chem. Int. Ed. Engl.,* **2009**, *48*(41), 7502-7513.
[http://dx.doi.org/10.1002/anie.200806063] [PMID: 19691074]

[182] Corma, A.; García, H.; Llabrés i Xamena, F.X. Engineering metal organic frameworks for heterogeneous catalysis. *Chem. Rev.,* **2010**, *110*(8), 4606-4655.
[http://dx.doi.org/10.1021/cr9003924] [PMID: 20359232]

[183] Yoon, M.; Srirambalaji, R.; Kim, K. Homochiral metal-organic frameworks for asymmetric heterogeneous catalysis. *Chem. Rev.,* **2012**, *112*(2), 1196-1231.
[http://dx.doi.org/10.1021/cr2003147] [PMID: 22084838]

[184] Dhakshinamoorthy, A.; Garcia, H. Catalysis by metal nanoparticles embedded on metal-organic frameworks. *Chem. Soc. Rev.,* **2012**, *41*(15), 5262-5284.
[http://dx.doi.org/10.1039/c2cs35047e] [PMID: 22695806]

[185] Dybtsev, D.N.; Chun, H.; Kim, K. Rigid and flexible: a highly porous metal-organic framework with unusual guest-dependent dynamic behavior. *Angew. Chem. Int. Ed. Engl.,* **2004**, *43*(38), 5033-5036.
[http://dx.doi.org/10.1002/anie.200460712] [PMID: 15384114]

[186] Schlichte, K.; Kratzke, T.; Kaskel, S. Improved synthesis, thermal stability and catalytic properties of the metal-organic framework compound CU3(BTC)(2). *Microporous Mesoporous Mater.,* **2004**, *73*, 81-88.
[http://dx.doi.org/10.1016/j.micromeso.2003.12.027]

[187] Wu, C.D.; Hu, A.; Zhang, L.; Lin, W. A homochiral porous metal-organic framework for highly enantioselective heterogeneous asymmetric catalysis. *J. Am. Chem. Soc.,* **2005**, *127*(25), 8940-8941. [http://dx.doi.org/10.1021/ja052431t] [PMID: 15969557]

[188] Park, K.S.; Ni, Z.; Côté, A.P.; Choi, J.Y.; Huang, R.; Uribe-Romo, F.J.; Chae, H.K.; O'Keeffe, M.; Yaghi, O.M. Exceptional chemical and thermal stability of zeolitic imidazolate frameworks. *Proc. Natl. Acad. Sci. USA,* **2006**, *103*(27), 10186-10191. [http://dx.doi.org/10.1073/pnas.0602439103] [PMID: 16798880]

[189] Cho, S.H.; Ma, B.; Nguyen, S.T.; Hupp, J.T.; Albrecht-Schmitt, T.E. A metal-organic framework material that functions as an enantioselective catalyst for olefin epoxidation. *Chem. Commun. (Camb.),* **2006**, (24), 2563-2565. [http://dx.doi.org/10.1039/b600408c] [PMID: 16779478]

[190] Dybtsev, D.N.; Nuzhdin, A.L.; Chun, H.; Bryliakov, K.P.; Talsi, E.P.; Fedin, V.P.; Kim, K. A homochiral metal-organic material with permanent porosity, enantioselective sorption properties, and catalytic activity. *Angew. Chem. Int. Ed. Engl.,* **2006**, *45*(6), 916-920. [http://dx.doi.org/10.1002/anie.200503023] [PMID: 16385607]

[191] Hasegawa, S.; Horike, S.; Matsuda, R.; Furukawa, S.; Mochizuki, K.; Kinoshita, Y.; Kitagawa, S. Three-dimensional porous coordination polymer functionalized with amide groups based on tridentate ligand: selective sorption and catalysis. *J. Am. Chem. Soc.,* **2007**, *129*(9), 2607-2614. [http://dx.doi.org/10.1021/ja067374y] [PMID: 17288419]

[192] Wu, C.D.; Lin, W. Heterogeneous asymmetric catalysis with homochiral metal-organic frameworks: network-structure-dependent catalytic activity. *Angew. Chem. Int. Ed. Engl.,* **2007**, *46*(7), 1075-1078. [http://dx.doi.org/10.1002/anie.200602099] [PMID: 17183496]

[193] Henschel, A.; Gedrich, K.; Kraehnert, R.; Kaskel, S. Catalytic properties of MIL-101. *Chem. Commun. (Camb.),* **2008**, (35), 4192-4194. [http://dx.doi.org/10.1039/b718371b] [PMID: 18802526]

[194] Schröder, F.; Esken, D.; Cokoja, M.; van den Berg, M.W.; Lebedev, O.I.; Van Tendeloo, G.; Walaszek, B.; Buntkowsky, G.; Limbach, H.H.; Chaudret, B.; Fischer, R.A. Ruthenium nanoparticles inside porous [Zn4O(bdc)3] by hydrogenolysis of adsorbed [Ru(cod)(cot)]: a solid-state reference system for surfactant-stabilized ruthenium colloids. *J. Am. Chem. Soc.,* **2008**, *130*(19), 6119-6130. [http://dx.doi.org/10.1021/ja078231u] [PMID: 18402452]

[195] Hwang, Y.K.; Hong, D.Y.; Chang, J.S.; Jhung, S.H.; Seo, Y.K.; Kim, J.; Vimont, A.; Daturi, M.; Serre, C.; Férey, G. Amine grafting on coordinatively unsaturated metal centers of MOFs: consequences for catalysis and metal encapsulation. *Angew. Chem. Int. Ed. Engl.,* **2008**, *47*(22), 4144-4148. [http://dx.doi.org/10.1002/anie.200705998] [PMID: 18435442]

[196] Ishida, T.; Nagaoka, M.; Akita, T.; Haruta, M. Deposition of gold clusters on porous coordination polymers by solid grinding and their catalytic activity in aerobic oxidation of alcohols. *Chemistry,* **2008**, *14*(28), 8456-8460. [http://dx.doi.org/10.1002/chem.200800980] [PMID: 18752232]

[197] Horike, S.; Dinca, M.; Tamaki, K.; Long, J.R. Size-selective Lewis acid catalysis in a microporous metal-organic framework with exposed Mn2+ coordination sites. *J. Am. Chem. Soc.,* **2008**, *130*(18),

5854-5855.
[http://dx.doi.org/10.1021/ja800669j] [PMID: 18399629]

[198] Alkordi, M.H.; Liu, Y.; Larsen, R.W.; Eubank, J.F.; Eddaoudi, M. Zeolite-like metal-organic frameworks as platforms for applications: on metalloporphyrin-based catalysts. *J. Am. Chem. Soc.,* **2008**, *130*(38), 12639-12641.
[http://dx.doi.org/10.1021/ja804703w] [PMID: 18759392]

[199] Ma, L.; Falkowski, J.M.; Abney, C.; Lin, W. A series of isoreticular chiral metal-organic frameworks as a tunable platform for asymmetric catalysis. *Nat. Chem.,* **2010**, *2*(10), 838-846.
[http://dx.doi.org/10.1038/nchem.738] [PMID: 20861899]

[200] Park, Y.K.; Choi, S.B.; Nam, H.J.; Jung, D-Y.; Ahn, H.C.; Choi, K.; Furukawa, H.; Kim, J. Catalytic nickel nanoparticles embedded in a mesoporous metal-organic framework. *Chem. Commun. (Camb.),* **2010**, *46*(18), 3086-3088.
[http://dx.doi.org/10.1039/c000775g] [PMID: 20379607]

[201] Liu, H.; Liu, Y.; Li, Y.; Tang, Z.; Jiang, H. Metal–organic framework supported gold nanoparticles as a highly active heterogeneous catalyst for aerobic oxidation of alcohols. *J. Phys. Chem. C,* **2010**, *114*, 13362-13369.
[http://dx.doi.org/10.1021/jp105666f]

[202] Esken, D.; Turner, S.; Lebedev, O.I.; van Tendeloo, G.; Fischer, R.A. Au@ZIFs: stabilization and encapsulation of cavity-size matching gold clusters inside functionalized zeolite imidazolate frameworks, ZIFs. *Chem. Mater.,* **2010**, *22*, 6393-6401.
[http://dx.doi.org/10.1021/cm102529c]

[203] Yuan, B.; Pan, Y.; Li, Y.; Yin, B.; Jiang, H. A highly active heterogeneous palladium catalyst for the Suzuki-Miyaura and Ullmann coupling reactions of aryl chlorides in aqueous media. *Angew. Chem. Int. Ed. Engl.,* **2010**, *49*(24), 4054-4058.
[http://dx.doi.org/10.1002/anie.201000576] [PMID: 20432496]

[204] Pan, Y.; Yuan, B.; Li, Y.; He, D. Multifunctional catalysis by Pd@MIL-101: one-step synthesis of methyl isobutyl ketone over palladium nanoparticles deposited on a metal-organic framework. *Chem. Commun. (Camb.),* **2010**, *46*(13), 2280-2282.
[http://dx.doi.org/10.1039/b922061e] [PMID: 20234932]

[205] Huang, Y.; Lin, Z.; Cao, R. Palladium nanoparticles encapsulated in a metal-organic framework as efficient heterogeneous catalysts for direct C2 arylation of indoles. *Chemistry,* **2011**, *17*(45), 12706-12712.
[http://dx.doi.org/10.1002/chem.201101705] [PMID: 21956646]

[206] Huang, Y.; Zheng, Z.; Liu, T.; Lü, J.; Lin, Z.; Li, H.; Cao, R. Palladium nanoparticles supported on amino functionalized metal-organic frameworks as highly active catalysts for the Suzuki–Miyaura cross-coupling reaction. *Catal. Commun.,* **2011**, *14*, 27-31.
[http://dx.doi.org/10.1016/j.catcom.2011.07.004]

[207] Ding, S.Y.; Gao, J.; Wang, Q.; Zhang, Y.; Song, W.G.; Su, C.Y.; Wang, W. Construction of covalent organic framework for catalysis: Pd/COF-LZU1 in Suzuki-Miyaura coupling reaction. *J. Am. Chem. Soc.,* **2011**, *133*(49), 19816-19822.
[http://dx.doi.org/10.1021/ja206846p] [PMID: 22026454]

[208] Hermannsdörfer, J.; Kempe, R. Selective palladium-loaded MIL-101 catalysts. *Chemistry,* **2011**, *17*(29), 8071-8077.
[http://dx.doi.org/10.1002/chem.201101004] [PMID: 21678515]

[209] Jiang, H-L.; Akita, T.; Ishida, T.; Haruta, M.; Xu, Q. Synergistic catalysis of Au@Ag core-shell nanoparticles stabilized on metal-organic framework. *J. Am. Chem. Soc.,* **2011**, *133*(5), 1304-1306.
[http://dx.doi.org/10.1021/ja1099006] [PMID: 21214205]

[210] Hermannsdörfer, J.; Friedrich, M.; Miyajima, N.; Albuquerque, R.Q.; Kümmel, S.; Kempe, R. Ni/Pd@MIL-101: synergistic catalysis with cavity-conform Ni/Pd nanoparticles. *Angew. Chem. Int. Ed. Engl.,* **2012**, *51*(46), 11473-11477.
[http://dx.doi.org/10.1002/anie.201205078] [PMID: 23065962]

[211] Lu, G.; Li, S.; Guo, Z.; Farha, O.K.; Hauser, B.G.; Qi, X.; Wang, Y.; Wang, X.; Han, S.; Liu, X.; DuChene, J.S.; Zhang, H.; Zhang, Q.; Chen, X.; Ma, J.; Loo, S.C.; Wei, W.D.; Yang, Y.; Hupp, J.T.; Huo, F. Imparting functionality to a metal-organic framework material by controlled nanoparticle encapsulation. *Nat. Chem.,* **2012**, *4*(4), 310-316.
[http://dx.doi.org/10.1038/nchem.1272] [PMID: 22437717]

[212] Kuo, C-H.; Tang, Y.; Chou, L-Y.; Sneed, B.T.; Brodsky, C.N.; Zhao, Z.; Tsung, C-K. Yolk-shell nanocrystal@ZIF-8 nanostructures for gas-phase heterogeneous catalysis with selectivity control. *J. Am. Chem. Soc.,* **2012**, *134*(35), 14345-14348.
[http://dx.doi.org/10.1021/ja306869j] [PMID: 22901021]

[213] Sun, Z.; Li, G.; Liu, L.; Liu, H.O. Au nanoparticles supported on Cr-based metal-organic framework as bimetallic catalyst for selective oxidation of cyclohexane to cyclohexanone and cyclohexanol. *Catal. Commun.,* **2012**, *27*, 200-205.
[http://dx.doi.org/10.1016/j.catcom.2012.07.017]

[214] Huang, Y.; Gao, S.; Liu, T.; Lü, J.; Lin, X.; Li, H.; Cao, R. Palladium nanoparticles supported on mixed-linker metal–organic frameworks as highly active catalysts for Heck reactions. *Chempluschem,* **2012**, *77*, 106-112.
[http://dx.doi.org/10.1002/cplu.201100021]

[215] Huang, Y.; Ma, T.; Huang, P.; Wu, D.; Lin, Z.; Cao, R.; Direct, C. BOND]H bBond arylation of indoles with aryl boronic acids catalyzed by palladium nanoparticles encapsulated in mesoporous metal–organic framework. *ChemCatChem,* **2013**, *5*, 1877-1883.
[http://dx.doi.org/10.1002/cctc.201200957]

[216] Zhang, L.; Su, Z.; Jiang, F.; Zhou, Y.; Xu, W.; Hong, M. Catalytic palladium nanoparticles supported on nanoscale MOFs: a highly active catalyst for Suzuki–Miyaura cross-coupling reaction. *Tetrahedron,* **2013**, *69*, 9237-9244.
[http://dx.doi.org/10.1016/j.tet.2013.08.059]

[217] Wang, P.; Zhao, J.; Li, X.; Yang, Y.; Yang, Q.; Li, C. Assembly of ZIF nanostructures around free Pt nanoparticles: efficient size-selective catalysts for hydrogenation of alkenes under mild conditions. *Chem. Commun. (Camb.),* **2013**, *49*(32), 3330-3332.
[http://dx.doi.org/10.1039/c3cc39275a] [PMID: 23505631]

[218] Ke, F.; Zhu, J.; Qiu, L-G.; Jiang, X. Controlled synthesis of novel Au@MIL-100(Fe) core-shell nanoparticles with enhanced catalytic performance. *Chem. Commun. (Camb.),* **2013**, *49*(13), 1267-

1269.
[http://dx.doi.org/10.1039/C2CC33964A] [PMID: 23135003]

[219] Zahmakiran, M. Iridium nanoparticles stabilized by metal organic frameworks (IrNPs@ZIF-8): synthesis, structural properties and catalytic performance. *Dalton Trans.,* **2012**, *41*(41), 12690-12696.
[http://dx.doi.org/10.1039/c2dt31779f] [PMID: 22961286]

[220] Pan, H.; Li, X.; Zhang, D.; Guan, Y.; Wu, P. Pt nanoparticles entrapped in mesoporous metal–organic frameworks MIL-101 as an efficient and recyclable catalyst for the asymmetric hydrogenation of α-ketoesters. *J. Mol. Catal. Chem.,* **2013**, *377*, 108-114.
[http://dx.doi.org/10.1016/j.molcata.2013.04.025]

[221] Long, J.; Liu, H.; Wu, S.; Liao, S.; Li, Y. Selective oxidation of saturated hydrocarbons using Au–Pd alloy nanoparticles supported on metal–organic frameworks. *ACS Catal.,* **2013**, *3*, 647-654.
[http://dx.doi.org/10.1021/cs300754k]

[222] Achmann, S.; Hagen, G.; Kita, J.; Malkowsky, I.M.; Kiener, C.; Moos, R. Metal-organic frameworks for sensing applications in the gas phase. *Sensors (Basel),* **2009**, *9*(3), 1574-1589.
[http://dx.doi.org/10.3390/s90301574] [PMID: 22573973]

[223] Kreno, L.E.; Leong, K.; Farha, O.K.; Allendorf, M.; Van Duyne, R.P.; Hupp, J.T. Metal-organic framework materials as chemical sensors. *Chem. Rev.,* **2012**, *112*(2), 1105-1125.
[http://dx.doi.org/10.1021/cr200324t] [PMID: 22070233]

[224] Chang, S.; Ko, H.; Singamaneni, S.; Gunawidjaja, R.; Tsukruk, V.V. Nanoporous membranes with mixed nanoclusters for Raman-based label-free monitoring of peroxide compounds. *Anal. Chem.,* **2009**, *81*(14), 5740-5748.
[http://dx.doi.org/10.1021/ac900537d] [PMID: 19518140]

[225] Panarin, A.Y.; Chirvony, V.S.; Kholostov, K.I.; Turpin, P.Y.; Terekhov, S.N. Formation of SERS-active silver structures on the surface of mesoporous silicon. *J. Appl. Spectrosc.,* **2009**, *76*, 280-287.
[http://dx.doi.org/10.1007/s10812-009-9175-1]

[226] Houk, R.J.; Jacobs, B.W.; El Gabaly, F.; Chang, N.N.; Talin, A.A.; Graham, D.D.; House, S.D.; Robertson, I.M.; Allendorf, M.D. Silver cluster formation, dynamics, and chemistry in metal-organic frameworks. *Nano Lett.,* **2009**, *9*(10), 3413-3418.
[http://dx.doi.org/10.1021/nl901397k] [PMID: 19757817]

[227] Sugikawa, K.; Nagata, S.; Furukawa, Y.; Kokado, K.; Sada, K. Stable and functional gold nanorod composites with a metal–organic framework crystalline shell. *Chem. Mater.,* **2013**, *25*, 2565-2570.
[http://dx.doi.org/10.1021/cm302735b]

[228] Horike, S.; Umeyama, D.; Kitagawa, S. Ion conductivity and transport by porous coordination polymers and metal-organic frameworks. *Acc. Chem. Res.,* **2013**, *46*(11), 2376-2384.
[http://dx.doi.org/10.1021/ar300291s] [PMID: 23730917]

[229] Yoon, M.; Suh, K.; Natarajan, S.; Kim, K. Proton conduction in metal-organic frameworks and related modularly built porous solids. *Angew. Chem. Int. Ed. Engl.,* **2013**, *52*(10), 2688-2700.
[http://dx.doi.org/10.1002/anie.201206410] [PMID: 23345157]

[230] Wan, S.; Guo, J.; Kim, J.; Ihee, H.; Jiang, D. A belt-shaped, blue luminescent, and semiconducting covalent organic framework. *Angew. Chem. Int. Ed. Engl.,* **2008**, *47*(46), 8826-8830.

[http://dx.doi.org/10.1002/anie.200803826] [PMID: 18830952]

[231] Wan, S.; Guo, J.; Kim, J.; Ihee, H.; Jiang, D. A photoconductive covalent organic framework: self-condensed arene cubes composed of eclipsed 2D polypyrene sheets for photocurrent generation. *Angew. Chem. Int. Ed. Engl.,* **2009**, *48*(30), 5439-5442.
[http://dx.doi.org/10.1002/anie.200900881] [PMID: 19434640]

[232] Spitler, E.L.; Dichtel, W.R. Lewis acid-catalysed formation of two-dimensional phthalocyanine covalent organic frameworks. *Nat. Chem.,* **2010**, *2*(8), 672-677.
[http://dx.doi.org/10.1038/nchem.695] [PMID: 20651731]

[233] Wan, S.; Gandara, F.; Asano, A.; Furukawa, H.; Saeki, A.; Dey, S.K.; Liao, L.; Ambrogio, M.W.; Botros, Y.Y.; Duan, X.F.; Seki, S.; Stoddart, J.F.; Yaghi, O.M. Covalent organic frameworks with high charge carrier mobility. *Chem. Mater.,* **2011**, *23*, 4094-4097.
[http://dx.doi.org/10.1021/cm201140r]

[234] Patwardhan, S.; Kocherzhenko, A.A.; Grozema, F.C.; Siebbeles, L.D. Delocalization and mobility of charge carriers in covalent organic frameworks. *J. Phys. Chem. C,* **2011**, *115*, 11768-11772.
[http://dx.doi.org/10.1021/jp202399u]

[235] Ding, X.; Chen, L.; Honsho, Y.; Feng, X.; Saengsawang, O.; Guo, J.; Saeki, A.; Seki, S.; Irle, S.; Nagase, S.; Parasuk, V.; Jiang, D. An n-channel two-dimensional covalent organic framework. *J. Am. Chem. Soc.,* **2011**, *133*(37), 14510-14513.
[http://dx.doi.org/10.1021/ja2052396] [PMID: 21863859]

[236] Feng, X.; Liu, L.; Honsho, Y.; Saeki, A.; Seki, S.; Irle, S.; Dong, Y.; Nagai, A.; Jiang, D. High-rate charge-carrier transport in porphyrin covalent organic frameworks: switching from hole to electron to ambipolar conduction. *Angew. Chem. Int. Ed. Engl.,* **2012**, *51*(11), 2618-2622.
[http://dx.doi.org/10.1002/anie.201106203] [PMID: 22290932]

[237] Spitler, E.L.; Colson, J.W.; Uribe-Romo, F.J.; Woll, A.R.; Giovino, M.R.; Saldivar, A.; Dichtel, W.R. Lattice expansion of highly oriented 2D phthalocyanine covalent organic framework films. *Angew. Chem. Int. Ed. Engl.,* **2012**, *51*(11), 2623-2627.
[http://dx.doi.org/10.1002/anie.201107070] [PMID: 22223402]

[238] Feng, X.; Chen, L.; Honsho, Y.; Saengsawang, O.; Liu, L.; Wang, L.; Saeki, A.; Irle, S.; Seki, S.; Dong, Y.; Jiang, D. An ambipolar conducting covalent organic framework with self-sorted and periodic electron donor-acceptor ordering. *Adv. Mater.,* **2012**, *24*(22), 3026-3031.
[http://dx.doi.org/10.1002/adma.201201185] [PMID: 22576320]

[239] Kundu, T.; Sahoo, S.C.; Banerjee, R. Alkali earth metal (Ca, Sr, Ba) based thermostable metal-organic frameworks (MOFs) for proton conduction. *Chem. Commun. (Camb.),* **2012**, *48*(41), 4998-5000.
[http://dx.doi.org/10.1039/c2cc31135f] [PMID: 22509491]

[240] Chowdhury, P.; Bikkina, C.; Gumma, S. Gas adsorption properties of the chromium-based metal organic framework MIL-101. *J. Phys. Chem. C,* **2009**, *113*, 6616-6621.
[http://dx.doi.org/10.1021/jp811418r]

[241] Chui, S.S.; Lo, S.M.; Charmant, J.P.; Orpen, A.G.; Williams, I.D. A chemically functionalizable nanoporous material. *Science,* **1999**, *283*(5405), 1148-1150.
[http://dx.doi.org/10.1126/science.283.5405.1148] [PMID: 10024237]

[242] Liang, Z.; Marshall, M.; Chaffee, A.L. CO2 adsorption-based separation by metal organic framework (Cu-BTC) *versus* zeolite (13X). *Energy Fuels,* **2009**, *23*, 2785-2789. [http://dx.doi.org/10.1021/ef800938e]

[243] Bordiga, S.; Regli, L.; Bonino, F.; Groppo, E.; Lamberti, C.; Xiao, B.; Wheatley, P.S.; Morris, R.E.; Zecchina, A. Adsorption properties of HKUST-1 toward hydrogen and other small molecules monitored by IR. *Phys. Chem. Chem. Phys.,* **2007**, *9*(21), 2676-2685. [http://dx.doi.org/10.1039/b703643d] [PMID: 17627311]

[244] Martín-Calvo, A.; García-Pérez, E.; Manuel Castillo, J.; Calero, S. Molecular simulations for adsorption and separation of natural gas in IRMOF-1 and Cu-BTC metal-organic frameworks. *Phys. Chem. Chem. Phys.,* **2008**, *10*(47), 7085-7091. [http://dx.doi.org/10.1039/b807470d] [PMID: 19039342]

[245] Yazaydın, A.; Benin, A.I.; Faheem, S.A.; Jakubczak, P.; Low, J.J.; Willis, R.R.; Snurr, R.Q. Enhanced CO2 adsorption in metal-organic frameworks *via* occupation of open-metal sites by coordinated water molecules. *Chem. Mater.,* **2009**, *21*, 1425-1430. [http://dx.doi.org/10.1021/cm900049x]

[246] Rosi, N.L.; Kim, J.; Eddaoudi, M.; Chen, B.; O'Keeffe, M.; Yaghi, O.M. Rod packings and metal-organic frameworks constructed from rod-shaped secondary building units. *J. Am. Chem. Soc.,* **2005**, *127*(5), 1504-1518. [http://dx.doi.org/10.1021/ja045123o] [PMID: 15686384]

[247] Caskey, S.R.; Wong-Foy, A.G.; Matzger, A.J. Dramatic tuning of carbon dioxide uptake *via* metal substitution in a coordination polymer with cylindrical pores. *J. Am. Chem. Soc.,* **2008**, *130*(33), 10870-10871. [http://dx.doi.org/10.1021/ja8036096] [PMID: 18661979]

[248] Yazaydın, A.O.; Snurr, R.Q.; Park, T-H.; Koh, K.; Liu, J.; Levan, M.D.; Benin, A.I.; Jakubczak, P.; Lanuza, M.; Galloway, D.B.; Low, J.J.; Willis, R.R. Screening of metal-organic frameworks for carbon dioxide capture from flue gas using a combined experimental and modeling approach. *J. Am. Chem. Soc.,* **2009**, *131*(51), 18198-18199. [http://dx.doi.org/10.1021/ja9057234] [PMID: 19954193]

[249] McDonald, T.M.; Lee, W.R.; Mason, J.A.; Wiers, B.M.; Hong, C.S.; Long, J.R. Capture of carbon dioxide from air and flue gas in the alkylamine-appended metal-organic framework mmen-Mg2(dobpdc). *J. Am. Chem. Soc.,* **2012**, *134*(16), 7056-7065. [http://dx.doi.org/10.1021/ja300034j] [PMID: 22475173]

[250] Planas, N.; Dzubak, A.L.; Poloni, R.; Lin, L-C.; McManus, A.; McDonald, T.M.; Neaton, J.B.; Long, J.R.; Smit, B.; Gagliardi, L. The mechanism of carbon dioxide adsorption in an alkylamine-functionalized metal-organic framework. *J. Am. Chem. Soc.,* **2013**, *135*(20), 7402-7405. [http://dx.doi.org/10.1021/ja4004766] [PMID: 23627764]

[251] Burd, S.D.; Ma, S.; Perman, J.A.; Sikora, B.J.; Snurr, R.Q.; Thallapally, P.K.; Tian, J.; Wojtas, L.; Zaworotko, M.J. Highly selective carbon dioxide uptake by [Cu(bpy-n)2(SiF6)] (bpy-1 = 4,4′-bipyridine; bpy-2 = 1,2-bis(4-pyridyl)ethene). *J. Am. Chem. Soc.,* **2012**, *134*(8), 3663-3666. [http://dx.doi.org/10.1021/ja211340t] [PMID: 22316279]

[252] Nugent, P.; Belmabkhout, Y.; Burd, S.D.; Cairns, A.J.; Luebke, R.; Forrest, K.; Pham, T.; Ma, S.;

Space, B.; Wojtas, L.; Eddaoudi, M.; Zaworotko, M.J. Porous materials with optimal adsorption thermodynamics and kinetics for CO2 separation. *Nature,* **2013**, *495*(7439), 80-84. [http://dx.doi.org/10.1038/nature11893] [PMID: 23446349]

[253] Li, J-R.; Kuppler, R.J.; Zhou, H-C. Selective gas adsorption and separation in metal-organic frameworks. *Chem. Soc. Rev.,* **2009**, *38*(5), 1477-1504. [http://dx.doi.org/10.1039/b802426j] [PMID: 19384449]

[254] Herm, Z.R.; Swisher, J.A.; Smit, B.; Krishna, R.; Long, J.R. Metal-organic frameworks as adsorbents for hydrogen purification and precombustion carbon dioxide capture. *J. Am. Chem. Soc.,* **2011**, *133*(15), 5664-5667. [http://dx.doi.org/10.1021/ja111411q] [PMID: 21438585]

[255] Wu, H.; Yao, K.; Zhu, Y.; Li, B.; Shi, Z.; Krishna, R.; Li, J. Cu-TDPAT, an rht-type dual-functional metal–organic framework offering significant potential for use in H2 and natural gas purification processes operating at high pressures. *J. Phys. Chem. C,* **2012**, *116*, 16609-16618. [http://dx.doi.org/10.1021/jp3046356]

[256] Das, M.C.; Xu, H.; Xiang, S.; Zhang, Z.; Arman, H.D.; Qian, G.; Chen, B. A new approach to construct a doubly interpenetrated microporous metal-organic framework of primitive cubic net for highly selective sorption of small hydrocarbon molecules. *Chemistry,* **2011**, *17*(28), 7817-7822. [http://dx.doi.org/10.1002/chem.201100350] [PMID: 21611990]

[257] He, Y.; Zhang, Z.; Xiang, S.; Fronczek, F.R.; Krishna, R.; Chen, B. A microporous metal-organic framework for highly selective separation of acetylene, ethylene, and ethane from methane at room temperature. *Chemistry,* **2012**, *18*(2), 613-619. [http://dx.doi.org/10.1002/chem.201102734] [PMID: 22162259]

[258] He, Y.; Zhang, Z.; Xiang, S.; Wu, H.; Fronczek, F.R.; Zhou, W.; Krishna, R.; O'Keeffe, M.; Chen, B. High separation capacity and selectivity of C2 hydrocarbons over methane within a microporous metal-organic framework at room temperature. *Chemistry,* **2012**, *18*(7), 1901-1904. [http://dx.doi.org/10.1002/chem.201103927] [PMID: 22262665]

[259] Murray, L.J.; Dinca, M.; Yano, J.; Chavan, S.; Bordiga, S.; Brown, C.M.; Long, J.R. Highly-selective and reversible O2 binding in Cr3(1,3,5-benzenetricarboxylate)2. *J. Am. Chem. Soc.,* **2010**, *132*(23), 7856-7857. [http://dx.doi.org/10.1021/ja1027925] [PMID: 20481535]

[260] Bloch, E.D.; Murray, L.J.; Queen, W.L.; Chavan, S.; Maximoff, S.N.; Bigi, J.P.; Krishna, R.; Peterson, V.K.; Grandjean, F.; Long, G.J.; Smit, B.; Bordiga, S.; Brown, C.M.; Long, J.R. Selective binding of O2 over N2 in a redox-active metal-organic framework with open iron(II) coordination sites. *J. Am. Chem. Soc.,* **2011**, *133*(37), 14814-14822. [http://dx.doi.org/10.1021/ja205976v] [PMID: 21830751]

[261] Seo, J.S.; Whang, D.; Lee, H.; Jun, S.I.; Oh, J.; Jeon, Y.J.; Kim, K. A homochiral metal-organic porous material for enantioselective separation and catalysis. *Nature,* **2000**, *404*(6781), 982-986. [http://dx.doi.org/10.1038/35010088] [PMID: 10801124]

[262] Evans, O.R.; Ngo, H.L.; Lin, W. Chiral porous solids based on lamellar lanthanide phosphonates. *J. Am. Chem. Soc.,* **2001**, *123*(42), 10395-10396. [http://dx.doi.org/10.1021/ja0163772] [PMID: 11603994]

[263] Banerjee, M.; Das, S.; Yoon, M.; Choi, H.J.; Hyun, M.H.; Park, S.M.; Seo, G.; Kim, K. Postsynthetic modification switches an achiral framework to catalytically active homochiral metal-organic porous materials. *J. Am. Chem. Soc.,* **2009**, *131*(22), 7524-7525. [http://dx.doi.org/10.1021/ja901440g] [PMID: 19438178]

[264] Dang, D.; Wu, P.; He, C.; Xie, Z.; Duan, C. Homochiral metal-organic frameworks for heterogeneous asymmetric catalysis. *J. Am. Chem. Soc.,* **2010**, *132*(41), 14321-14323. [http://dx.doi.org/10.1021/ja101208s] [PMID: 20879732]

[265] Lun, D.J.; Waterhouse, G.I.; Telfer, S.G. A general thermolabile protecting group strategy for organocatalytic metal-organic frameworks. *J. Am. Chem. Soc.,* **2011**, *133*(15), 5806-5809. [http://dx.doi.org/10.1021/ja202223d] [PMID: 21443248]

[266] Djakovitch, L.; Koehler, K.; de Vries, J.G. *The role of palladium nanoparticles as catalysts for carbon-carbon coupling reactions in nanoparticles and catalysis*; Astruc, D., Ed.; Wiley-VCH: Weinheim, **2008**.

[267] Moreno-Mañas, M.; Pleixats, R. Formation of carbon--carbon bonds under catalysis by transition-metal nanoparticles. *Acc. Chem. Res.,* **2003**, *36*(8), 638-643. [http://dx.doi.org/10.1021/ar020267y] [PMID: 12924961]

[268] Zhao, Y.; Zhang, J.; Song, J.; Li, J.; Liu, J.; Wu, T.; Zhang, P.; Han, B. Ru nanoparticles immobilized on metal–organic framework nanorods by supercritical CO2-methanol solution: highly efficient catalyst. *Green Chem.,* **2011**, *13*, 2078. [http://dx.doi.org/10.1039/c1gc15340d]

[269] Allendorf, M.D.; Bauer, C.A.; Bhakta, R.K.; Houk, R.J. Luminescent metal-organic frameworks. *Chem. Soc. Rev.,* **2009**, *38*(5), 1330-1352. [http://dx.doi.org/10.1039/b802352m] [PMID: 19384441]

[270] Lan, A.; Li, K.; Wu, H.; Olson, D.H.; Emge, T.J.; Ki, W.; Hong, M.; Li, J. A luminescent microporous metal-organic framework for the fast and reversible detection of high explosives. *Angew. Chem. Int. Ed. Engl.,* **2009**, *48*(13), 2334-2338. [http://dx.doi.org/10.1002/anie.200804853] [PMID: 19180622]

[271] Lan, A.; Li, K.; Wu, H.; Kong, L.; Nijem, N.; Olson, D.H.; Emge, T.J.; Chabal, Y.J.; Langreth, D.C.; Hong, M.; Li, J. RPM3: a multifunctional microporous MOF with recyclable framework and high H2 binding energy. *Inorg. Chem.,* **2009**, *48*(15), 7165-7173. [http://dx.doi.org/10.1021/ic9002115] [PMID: 19722690]

[272] Xu, H.; Liu, F.; Cui, Y.; Chen, B.; Qian, G. A luminescent nanoscale metal-organic framework for sensing of nitroaromatic explosives. *Chem. Commun. (Camb.),* **2011**, *47*(11), 3153-3155. [http://dx.doi.org/10.1039/c0cc05166g] [PMID: 21271003]

[273] Ma, D.; Li, B.; Zhou, X.; Zhou, Q.; Liu, K.; Zeng, G.; Li, G.; Shi, Z.; Feng, S. A dual functional MOF as a luminescent sensor for quantitatively detecting the concentration of nitrobenzene and temperature. *Chem. Commun. (Camb.),* **2013**, *49*(79), 8964-8966. [http://dx.doi.org/10.1039/c3cc44546a] [PMID: 23963022]

[274] Xue, Y-S.; He, Y.; Zhou, L.; Chen, F-J.; Xu, Y.; Du, H-B.; You, X-Z.; Chen, B. A photoluminescent microporous metal organic anionic framework for nitroaromatic explosive sensing. *J. Mater. Chem. A*

Mater. Energy Sustain., **2013**, *1*, 4525-4530.
[http://dx.doi.org/10.1039/c3ta01118f]

[275] Shustova, N.B.; Cozzolino, A.F.; Reineke, S.; Baldo, M.; Dincă, M. Selective turn-on ammonia sensing enabled by high-temperature fluorescence in metal-organic frameworks with open metal sites. *J. Am. Chem. Soc.,* **2013**, *135*(36), 13326-13329.
[http://dx.doi.org/10.1021/ja407778a] [PMID: 23981174]

[276] Li, Y.; Zhang, S.; Song, D. A luminescent metal-organic framework as a turn-on sensor for DMF vapor. *Angew. Chem. Int. Ed. Engl.,* **2013**, *52*(2), 710-713.
[http://dx.doi.org/10.1002/anie.201207610] [PMID: 23165755]

[277] Ferrando-Soria, J.; Khajavi, H.; Serra-Crespo, P.; Gascon, J.; Kapteijn, F.; Julve, M.; Lloret, F.; Pasán, J.; Ruiz-Pérez, C.; Journaux, Y.; Pardo, E. Highly selective chemical sensing in a luminescent nanoporous magnet. *Adv. Mater.,* **2012**, *24*(41), 5625-5629.
[http://dx.doi.org/10.1002/adma.201201846] [PMID: 22887721]

[278] Xu, H.; Rao, X.; Gao, J.; Yu, J.; Wang, Z.; Dou, Z.; Cui, Y.; Yang, Y.; Chen, B.; Qian, G. A luminescent nanoscale metal-organic framework with controllable morphologies for spore detection. *Chem. Commun. (Camb.),* **2012**, *48*(59), 7377-7379.
[http://dx.doi.org/10.1039/c2cc32346j] [PMID: 22627669]

[279] Wu, P.; Wang, J.; He, C.; Zhang, X.; Wang, Y.; Liu, T.; Duan, C. Luminescent metal-organic frameworks for selectively sensing nitric oxide in an aqueous solution and in living cCells. *Adv. Funct. Mater.,* **2012**, *22*, 1698-1703.
[http://dx.doi.org/10.1002/adfm.201102157]

[280] Kitagawa, H.; Nagao, Y.; Fujishima, M.; Ikeda, R.; Kanda, S. Highly proton-conductive copper coordination polymer, H2dtoaCu (H2dtoa = dithiooxamide anion). *Inorg. Chem. Commun.,* **2003**, *6*, 346-348.
[http://dx.doi.org/10.1016/S1387-7003(02)00749-9]

[281] Nagao, Y.; Fujishima, M.; Ikeda, R.; Kanda, S.; Kitagawa, H. Highly proton-conductive μcopper coordination polymers. *Synth. Met.,* **2003**, *133*, 431-432.
[http://dx.doi.org/10.1016/S0379-6779(02)00282-5]

[282] Nagao, Y.; Ikeda, R.; Iijima, K.; Kubo, T.; Nakasuji, K.; Kitagawa, H. A new proton-conductive copper coordination polymer, (HOC3H6)2dtoaCu (dtoa = dithiooxamide). *Synth. Met.,* **2003**, *135-136*, 283-284.
[http://dx.doi.org/10.1016/S0379-6779(02)00914-1]

[283] Sadakiyo, M.; Yamada, T.; Kitagawa, H. Rational designs for highly proton-conductive metal-organic frameworks. *J. Am. Chem. Soc.,* **2009**, *131*(29), 9906-9907.
[http://dx.doi.org/10.1021/ja9040016] [PMID: 19621952]

[284] Yamada, T.; Morikawa, S.; Kitagawa, H. Structures and proton conductivity of one-dimensional M(dhbq)•nH2O (M = Mg, Mn, Co, Ni, and Zn, H2(dhbq)=2,5-dihydroxy-1,4-benzoquinone) promoted by connected hydrogen-bond networks with absorbed water. *Bull. Chem. Soc. Jpn.,* **2010**, *83*, 42-48.
[http://dx.doi.org/10.1246/bcsj.20090216]

[285] Taylor, J.M.; Mah, R.K.; Moudrakovski, I.L.; Ratcliffe, C.I.; Vaidhyanathan, R.; Shimizu, G.K. Facile proton conduction *via* ordered water molecules in a phosphonate metal-organic framework. *J. Am.*

Chem. Soc., **2010**, *132*(40), 14055-14057.
[http://dx.doi.org/10.1021/ja107035w] [PMID: 20857972]

[286] Sahoo, S.C.; Kundu, T.; Banerjee, R. Helical water chain mediated proton conductivity in homochiral metal-organic frameworks with unprecedented zeolitic unh-topology. *J. Am. Chem. Soc.,* **2011**, *133*(44), 17950-17958.
[http://dx.doi.org/10.1021/ja2078637] [PMID: 21919488]

[287] Chen, W.X.; Xu, H.R.; Zhuang, G.L.; Long, L.S.; Huang, R.B.; Zheng, L.S. Temperature-dependent conductivity of Emim+ (Emim+ = 1-ethyl-3-methyl imidazolium) confined in channels of a metal-organic framework. *Chem. Commun. (Camb.),* **2011**, *47*(43), 11933-11935.
[http://dx.doi.org/10.1039/c1cc14702a] [PMID: 21975950]

[288] Jeong, N.C.; Samanta, B.; Lee, C.Y.; Farha, O.K.; Hupp, J.T. Coordination-chemistry control of proton conductivity in the iconic metal-organic framework material HKUST-1. *J. Am. Chem. Soc.,* **2012**, *134*(1), 51-54.
[http://dx.doi.org/10.1021/ja2110152] [PMID: 22148287]

[289] Panda, T.; Kundu, T.; Banerjee, R. Self-assembled one dimensional functionalized metal-organic nanotubes (MONTs) for proton conduction. *Chem. Commun. (Camb.),* **2012**, *48*(44), 5464-5466.
[http://dx.doi.org/10.1039/c2cc31527k] [PMID: 22538292]

[290] Pardo, E.; Train, C.; Gontard, G.; Boubekeur, K.; Fabelo, O.; Liu, H.; Dkhil, B.; Lloret, F.; Nakagawa, K.; Tokoro, H.; Ohkoshi, S.; Verdaguer, M. High proton conduction in a chiral ferromagnetic metal-organic quartz-like framework. *J. Am. Chem. Soc.,* **2011**, *133*(39), 15328-15331.
[http://dx.doi.org/10.1021/ja206917z] [PMID: 21913689]

[291] Shigematsu, A.; Yamada, T.; Kitagawa, H. Wide control of proton conductivity in porous coordination polymers. *J. Am. Chem. Soc.,* **2011**, *133*(7), 2034-2036.
[http://dx.doi.org/10.1021/ja109810w] [PMID: 21284399]

[292] Ohkoshi, S.; Nakagawa, K.; Tomono, K.; Imoto, K.; Tsunobuchi, Y.; Tokoro, H. High proton conductivity in prussian blue analogues and the interference effect by magnetic ordering. *J. Am. Chem. Soc.,* **2010**, *132*(19), 6620-6621.
[http://dx.doi.org/10.1021/ja100385f] [PMID: 20420388]

[293] Ponomareva, V.G.; Kovalenko, K.A.; Chupakhin, A.P.; Dybtsev, D.N.; Shutova, E.S.; Fedin, V.P. Imparting high proton conductivity to a metal-organic framework material by controlled acid impregnation. *J. Am. Chem. Soc.,* **2012**, *134*(38), 15640-15643.
[http://dx.doi.org/10.1021/ja305587n] [PMID: 22958118]

[294] Hurd, J.A.; Vaidhyanathan, R.; Thangadurai, V.; Ratcliffe, C.I.; Moudrakovski, I.L.; Shimizu, G.K. Anhydrous proton conduction at 150 °C in a crystalline metal-organic framework. *Nat. Chem.,* **2009**, *1*(9), 705-710.
[http://dx.doi.org/10.1038/nchem.402] [PMID: 21124357]

[295] Umeyama, D.; Horike, S.; Inukai, M.; Hijikata, Y.; Kitagawa, S. Confinement of mobile histamine in coordination nanochannels for fast proton transfer. *Angew. Chem. Int. Ed. Engl.,* **2011**, *50*(49), 11706-11709.
[http://dx.doi.org/10.1002/anie.201102997] [PMID: 21989965]

[296] Wiers, B.M.; Foo, M.L.; Balsara, N.P.; Long, J.R. A solid lithium electrolyte *via* addition of lithium

isopropoxide to a metal-organic framework with open metal sites. *J. Am. Chem. Soc.,* **2011**, *133*(37), 14522-14525.
[http://dx.doi.org/10.1021/ja205827z] [PMID: 21877685]

[297] Lin, K-S.; Adhikari, A.K.; Ku, C-N.; Chiang, C-L.; Kuo, H. Synthesis and characterization of porous HKUST-1 metal organic frameworks for hydrogen storage. *Int. J. Hydrogen Energy,* **2012**, *37*, 13865-13871.
[http://dx.doi.org/10.1016/j.ijhydene.2012.04.105]

[298] Ni, Z.; Masel, R.I. Rapid production of metal-organic frameworks *via* microwave-assisted solvothermal synthesis. *J. Am. Chem. Soc.,* **2006**, *128*(38), 12394-12395.
[http://dx.doi.org/10.1021/ja0635231] [PMID: 16984171]

[299] Yang, S-T.; Kim, J.; Cho, H-Y.; Kim, S.; Ahn, W-S. Facile synthesis of covalent organic frameworks COF-1 and COF-5 by sonochemical method. *RSC Advances,* **2012**, *2*, 10179-10181.
[http://dx.doi.org/10.1039/c2ra21531d]

[300] Côté, A.P.; El-Kaderi, H.M.; Furukawa, H.; Hunt, J.R.; Yaghi, O.M. Reticular synthesis of microporous and mesoporous 2D covalent organic frameworks. *J. Am. Chem. Soc.,* **2007**, *129*(43), 12914-12915.
[http://dx.doi.org/10.1021/ja0751781] [PMID: 17918943]

CHAPTER 8

Nanoparticles as Top Materials

Abstract: Nanoparticles are particles more than one dimension smaller than 1 micron. Theoretically, any metal is a candidate to produce metal nanoparticles. However, certain metals are more prone and advantageous for metal nanoparticle synthesis. Hence, herein is presented a conspectus on the field of metal nanoparticles, including the preparation and functionalization of the most representative types of metal nanoparticles, as well as detailed examples on several promising applications with 646 references provided. Some synthetic details on the preparation of some state-of-the-art compounds are given as well.

Keywords: Biomedicinal applications, Drug delivery, Gold nanoparticles, Magnetic nanoparticles, Metal nanoparticles, Nanocatalysts, Quantum dots.

INTRODUCTION

Nanomaterials generally refer to materials on the nanometer scale, usually in the 1-100 nm range, which exhibit superior properties compared to their bulk counterparts. This aspect of nanotechnology is closely related to all science disciplines, particularly in the field of chemistry. In nanochemistry, chemists use a chemical approach and the tool of synthetic chemistry to assemble individual atoms or molecules into nanomaterials in a size and shape - controlled manner [1 - 5].With the knowledge of coordination chemistry, the surface properties of nanomaterials can be modified to suit the applications in various fields. Of particular interest is utilizing the large active surface area provided by nanoparticles in organic transformation reaction and organic synthesis [1, 6 - 9]. Nanoparticles are particles as small as ~0.2 nm possessing high surface energy, spatial confinement and reduced imperfections and large fraction of surface atoms.

Faraday reported, for the first time, the optical properties of nanometer-scale metals in 1857 [10], where he described a method for preparing aqueous gold colloids by electrical discharge. Maxwell-Garnett attempted to theoretically

Mário J. F. Calvete

describe the colors of metal nanoparticles in 1904, using the theory of Lorenz and Rayleigh [11]. Later, Mie performed a thorough theoretical study on the optical properties of metal particles [12], applicable to spheres of any size.

In 1960, the famous talk of Richard Feynman [13] allured scientists for the nanotechnology field, with his premonitory words "*there is plenty of room at the bottom*" and then, in the 1970s, the term *nanotechnology* was created [14] and remarkable progress and new discoveries in the "nanoworld" followed – often resulting in a Nobel Prizes for the respective scientists. Since late 1970's and early 1980's, several groups became interested in "the neglected dimension between atoms or molecules and bulk materials" [15 - 18], followed by some convergence of perspectives in the 1990's. The use of nanoparticles in the preparation of bigger structures is yet in its early years. Pileni established the metal nanoparticles ability in producing ordered lattices [19], while Schiffrin reported the synthesis of highly ordered lattices with two dimensions [20].

Theoretically, any metal is a candidate to produce metal nanoparticles. However, certain metals are more prone and advantageous for the synthesis of metal nanoparticles. Among them, gold nanoparticles have been highly studied [21]. Other metal nanoparticles include the ones based on silver, iron, copper, cobalt, nickel, platinum, palladium, rhodium and aluminum [22 - 32]. Also various nanometric metal oxides have been prepared, the most common ones, being titanium dioxide, zinc oxide and silica. Other metal oxides are also produced, including iron, cerium, zirconium, copper, nickel, aluminium, yttrium, barium, manganese, antimony and oxides [22 - 24].

One of the most studied types of nanoparticles is the magnetic nanoparticle type. Superparamagnetic nanoparticles are small (typically 5-500 nanometers) iron, cobalt or nickel based crystals, characterized by possessing a large magnetic momentum in the presence of a static external magnetic field [33 - 36].

Last but definitely not least, quantum dots have emerged, during the past years, as an exciting field of research, commonly being combinations of periodic table Group II and IV or Group III and V elements and are nanometer-sized semiconductor nanocrystals or crystals of inorganic semiconductor. Quantum dots possess exclusive electronic and optical properties with 1-10 nm of diameter. They can, for instance, absorb ultraviolet or white light, with emission at a precise wavelength a few nanoseconds later [37].

For instance, there is a significant fluorescence dependence of the wavelength on the material composition and nanocrystal dimensions, which is consequence from using materials with different band gaps: CdSe (blue), InP (green) and InAs (red). The fluorescence emission fine-tuning is also controlled by the quantum dots size.

Quantum dots exist since more than 2000 years ago, when the Greeks and Romans used these materials as cosmetics [38]. In the beginning of the 20th century, red-yellow colors could be obtained by inclusion of CdSe and CdS into silicates [39]. In 1982, Efros [40] postulated that by changing CdS_xSe_{1-x} stoichiometry or size, the effect of quantum size could control the glass color.

Later, Brus [41] discussed the color change in semiconductor colloidal solutions. This booming research field has flourished, with theoretical and experimental research [42 - 50], also paying attention to commercialization efforts [51, 52].

Nanoparticles can be classified by their morphology, composition, dimensionality, agglomeration and uniformity [53]. It is useful to classify nanoparticles based on their number of dimensions: i) 1D nanomaterials have one dimension and are characteristically surface coatings or thin films (monolayers); ii) 2D nanomaterials are two-dimensional nanomaterials.

These include nanopore filters used for small particle separation and filtration or 2D nanostructured films, with nanostructures firmly attached to a substrate; iii) 3D nanomaterials have three dimensions, including free nanoparticles and colloids.

SYNTHESIS OF NANOPARTICLES

Metal Nanoparticles

Nanoparticles may be produced by various methods (Fig. **1**). Some of them are bottom-up methods (chemical methods), *i.e.* construction from smaller synthons, and involve precursor decomposition for atom formation, followed by atomic aggregation or the metal ion reduction.

Chemical methods usually furnish nanoparticles with narrow size distribution [54]. The other preparation methods are denominated top-down methods (physical methods), involving rupture of larger materials into nanoparticles [55, 56] and involve bulk metal subdivision, including pulverization of bulk metal in ball mills, mechanical crushing, arc discharge between metal electrodes or other size reducing mechanisms [57], producing usually large and wide size distributed metal nanoparticles. The "bottom-up" approach allows the control of the growth rate at different crystal faces giving nanomaterials of various structures. As a result, the "bottom-up" approach is more desirable and adopted by chemists in nanomaterials synthesis [1 - 5].

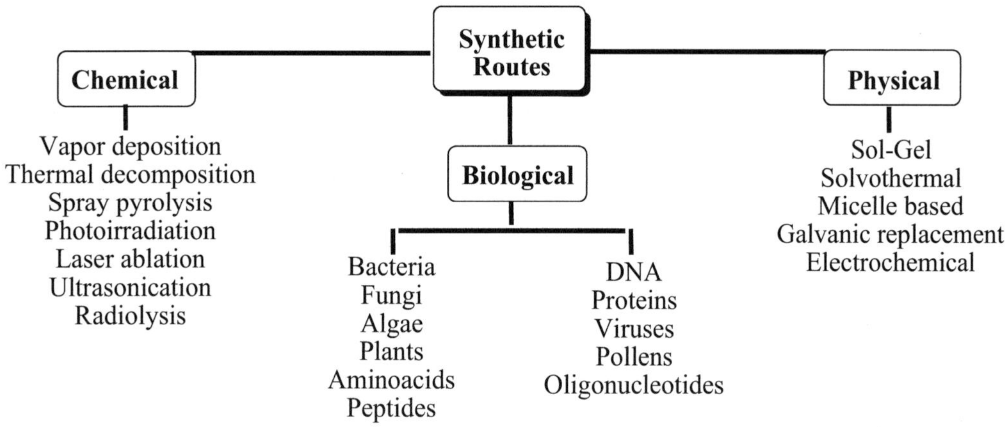

Fig. (1). Possible routes for metal nanoparticle synthesis.

The synthetic procedures can be classified mostly into liquid-phase (or solution-based) and gas-phase (or vapor-based) synthesis. Gas-phase synthesis of nanomaterials mainly involves the vapor-liquid-solid growth and chemical vapor deposition techniques [58]. Examples of gas-phase synthesis are Ge nanowires [59] and CoS nanowires [60] For instance, Li proposed a general synthesis of metal sulfide preparation, having one-dimensional nanostructure, by atmospheric pressure chemical vapor deposition [61]. These gas-phase synthetic experiments always require special handling and complicated instruments. The nanocrystals are also obtained in low quantity and productivity [62 - 64]. Even though, liquid-phase synthesis of nanoparticles is more expedient to handle-no special apparatus is required and vast efforts have made by chemists to develop liquid-phase synthesis of nanoparticles [1 - 5, 65 - 71].

The main two processes for preparation of metal nanoparticles are the precipitation method and hydrothermal/solvothermal synthesis. Precipitation method is a widespread synthetic strategy for the synthesis of nanomaterials, which involve mainly nucleation and growth steps. Initially, chemical reactions such as reduction, oxidation or hydrolysis give products of low solubility, resulting in a supersaturated condition. Nucleation begins and the nuclei continue to grow into the desired nanomaterials [66]. Ostwald ripening also contributes to the growth of nanomaterials by consuming the smaller particles [72].

The growth step is typically diffusion-limited and both concentration gradient and temperature are important factors governing the growth rate, in order to synthesize monodispersed nanoparticles [73]. The formed nanomaterials can become stable either by steric or electrostatic repulsion, preventing further agglomeration (Fig. **2**). Electrostatic stabilization is obtained by the adsorption of cations or anions (*e.g.* halides, citrates, carboxylates) onto the surface of nanomaterials [66, 74], while steric stabilization is usually achieved by capping surfactants, polymers or oligomers onto the surface of nanomaterials [74, 75].

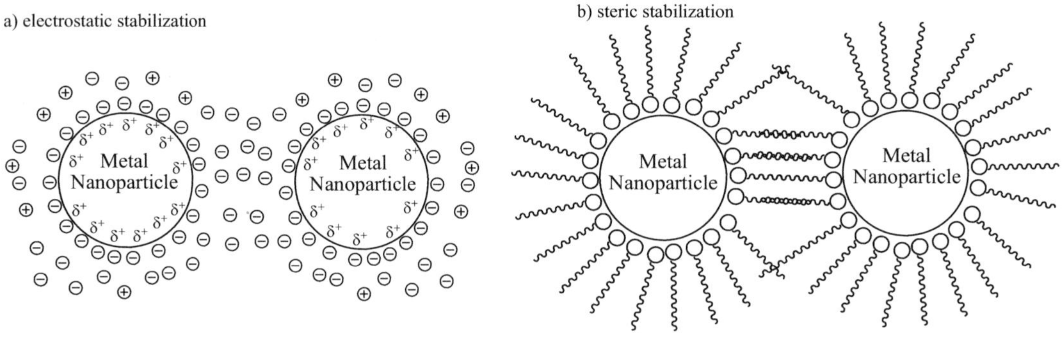

Fig. (2). Possible types of nanoparticle stabilization: **a)** electrostatic and **b)** steric.

To give an idea of the importance of the method, it has been frequently employed in the production of metal oxide and metal nanoparticles. The most prominent precipitation method is the citrate reduction used for the preparation of Au nanoparticles, reported by Turkevich [76], in which the formation of wine-red Au nanoparticles suspension results from boiling an aqueous solution of sodium citrate and $HAuCl_4$ for about ten minutes results. Citrate works both as reducing agent for the reduction $AuCl_4^-$, and stabilizer to prevent Au nanoparticles from aggregation, method that has been later improved by Frens [77].

Citrate reduction can also be applied to the synthesis of platinum [78] and silver nanoparticles [79 - 82]. Other milder reducing agents such as potassium bitartrate [83] and ascorbic acid [84, 85] have also been employed in the synthesis of palladium [83 - 85], platinum [83], silver [83 - 85] and gold nanoparticles [83]. However, stronger reducing agents such as borohydride ion (BH_4^-) or hydrazine hydrate (N_2H_4•H_2O) are needed reduce first-row transition metals, which have low reduction potentials (*e.g.* Ni^{2+} and Co^{2+}). Using these reducing agents metal nanoparticles such as that of cobalt [86, 87], nickel [86 - 89], copper [86, 87, 90], rhodium [91 - 93], palladium [94 - 96], platinum [93 - 96], silver [95, 97 - 99] and gold [93, 95, 100] were successfully synthesized. Also organic reducing agents such as trialkylborohydride and solvated magnesium have stronger reducing ability and can be used to synthesize Fe [101 - 103], Co [101 - 103] and Ni nanoparticles [103]. Alcohols, such as methanol, ethanol, 2-propanol, ethylene glycol and 1,2-propanediol, are also good reducing agents, simultaneously acting as solvents. This way, most transition metal (Ni, Cu, Ru, Rh, Pd, Pt, Ag and Au) nanoparticles can be synthesized in refluxing alcoholic solution [104 - 111].

Similarly, metal oxide nanoparticles are synthesized by the addition of precipitating agents, which provide a basic medium for amorphous metal precursors, such as metal hydroxides, metal carbonates or metal bicarbonates. Subsequent thermal treatment (*e.g.* reflux, calcination, annealing) of the metal precursors gives metal oxide nanoparticles, for which the most used precipitating agents are potassium, sodium and ammonium hydroxides, for instance in the synthesis of Sb_2O_3 nanoparticles [112].

Hydrothermal and solvothermal synthesis are usually employed in the synthesis of metal oxide nanomaterials, predominantly anisotropic nanowires [113 - 119]. Typically, an aqueous reaction mixture is heated in a sealed Teflon-lined stainless autoclave, resulting in an elevated temperature and pressure of the reaction medium is named hydrothermal synthesis, while the process is described as solvothermal synthesis, when organic solvents are used. The high temperature and pressure offer a near-supercritical solvent medium, so the solubility and reactivity of the metal salts can be hugely increased. One significant advantage of hydrothermal synthesis is the better crystal growth of nanoparticles obtainable by near-supercritical conditions [66, 120]

Magnetic Nanoparticles

The main source of magnetic nanoparticles comes from iron. The use of fine iron particles has been investigated for several decades, however not knowing the exact dimensions of the particles [121]. The theory demonstrating the iron particles magnetism was only presented and fundamented in the 1950s-1960s, [122 - 124]. Iron nanoparticles research has experienced a large growth in the last two decades, with particular significance for magnetic nanoparticles formed by iron oxide. There are three types of iron oxides: wustite FeO, hematite Fe_2O_3 and magnetite Fe_3O_4, but only the last two are used for magnetic nanoparticles production. The magnetite is the most high iron content material (about 72.5%) and for this reason it is widely used in industries dealing with metals. It is one of minerals with the more intense magnetic properties existing in nature. It is a mixture of both oxidation of iron: Fe^{2+} and Fe^{3+}, its molecular weight is 231.54 and in nature it can be found in metamorphic and igneous rocks.

There are several methods for production of magnetic nanoparticles [125, 126], including iron oxides, *e.g.* γ-Fe_2O_3 and Fe_3O_4 [127 - 129], alloys, namely FePt and $CoPt_3$ [130, 131], spinel-type ferromagnets, such as $CoFe_2O_4$, $MgFe_2O_4$ and $MnFe_2O_4$ [132, 133] and pure metals, such as Co and Fe [134, 135]. Several quite well-accepted methods include micelle synthesis, thermal decomposition and/or reduction, co-precipitation, laser pyrolysis techniques and hydrothermal synthesis. Since exhaustive discussion of each method would exceed the current purpose, simple topical considerations will be here made on the most representative methods.

Co-precipitation is probably the most spread method for iron oxide nanoparticle preparation (either γ-Fe_2O_3 or Fe_3O_4), using Fe^{2+}/Fe^{3+} aqueous salt solutions in presence of base, at elevated temperature or room temperature, but always under inert atmosphere. The magnetic nanoparticle's shape, size and composition depends highly on the used salts (nitrates, sulfates, chlorides), on the media ionic strength, on the reaction temperature, on the Fe^{2+}/Fe^{3+} ratio and on the pH value. The magnetite nanoparticles quality is completely reproducible, provided that the synthetic parameters are all set and fixed. Nevertheless, the main experimental problem in Fe_3O_4 synthesis by co-precipitation is obtaining narrow particle size distribution, as the nanoparticles have a propensity to be polydisperse. A short nucleation burst and successive controlled slow growth is critical for the production of monodisperse particles.

To aid in the preparation of the magnetic nanoparticles, addition of stabilizers can induce the desired nucleation burst and slow controlled growth. For instance, 4-10 nm sized magnetite nanoparticles can stabilize in a 1 wt% polyvinylalcohol aqueous solution [136, 137]. Size-tuned maghemite nanoparticles were also produced by magnetite formation, in an alkaline medium, in presence of citric acid trisodium salt, followed by iron(III) nitrate oxidation at 90ºC [137]. Adjusting the metal ions molar ratio (Fe^{2+} and Fe^{3+}) and citrate, particle sizes ranging 2-8 nm were obtained. Moreover, the consequence of hydroxy carboxylate and carboxylate ions on iron oxide formation has been extensively investigated [138 - 143], where recent studies provided evidence of Fe_3O_4 stabilization, upon use of oleic acid [142, 143].

Small sized monodisperse magnetic nanocrystals with can be prepared in presence of stabilizing surfactants, by thermal decomposition, using high-boiling solvents [133, 144, 145]. The organometallic salts comprise metal cuproferronates [MxCupx] (M=metal ion; Cup=N-nitrosophenylhydroxylamine, $C_6H_5N(NO)O^-$) [146], metal carbonyls or metal acetylacetonates, [M(acac)$_n$], (M= Cr; Ni, Co, Mn, Fe; n=2 or 3, acac=acetylacetonate) [147], or using hexadecylamine[148], oleic acid [144, 149] and fatty acids [150] as surfactants.

Monodisperse nanoparticles having controlled sizes and shape are typically coated using hydrocarbons with long chains, of several types of materials, such as dendrimers, lipids, dextran, proteins, poly(ethylene)glycol, chitosan, poly(vinylpyrrolidone), poly(vinyl alcohol), poly(ethylene-*co* -vinyl acetate) or poly(lactic-*co* -glycolic acid) [151 - 156]. Other molecules, such as silanes [157], dopamine [158 - 160] or bifunctional 2,3-dimercaptosuccinic acids, [161], among others [162] were also studied in functionalization of nanoparticles (Schemes **1** and **2**).

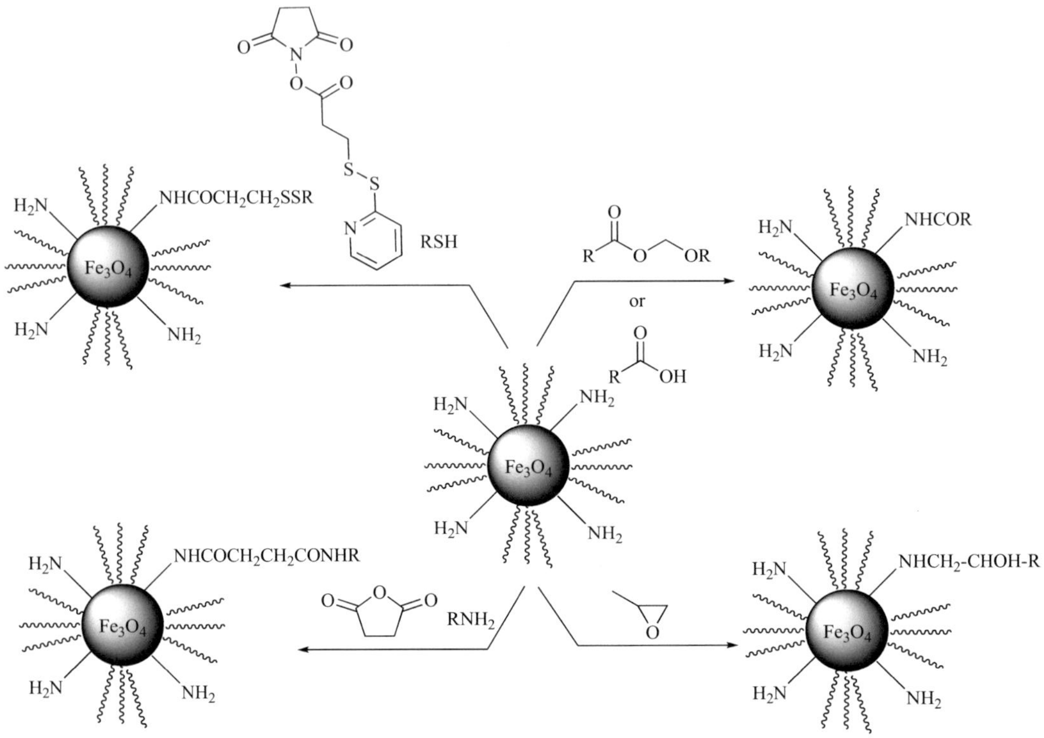

Scheme (1). Magnetic iron nanoparticle functionalization.

Magnetic nanoparticles have attracted significant importance in several fields, mostly due to their capacity to act, when properly functionalized, as quasi-organic compounds, with emphasis in biological applications [163 - 170], catalysis [34, 162, 171 - 175] and environmental remediation [176 - 179].

Quantum Dots

Various methods are used to prepare quantum dots, usually falling in two types: bottom-up or top-down approaches [180, 181]. Using the latter approaches, bulk semiconductors can be made thinner, forming quantum dots. Wet chemical etching, reactive-ion etching and electron beam lithography are frequently used to obtain ~30 nm diameter quantum dots.

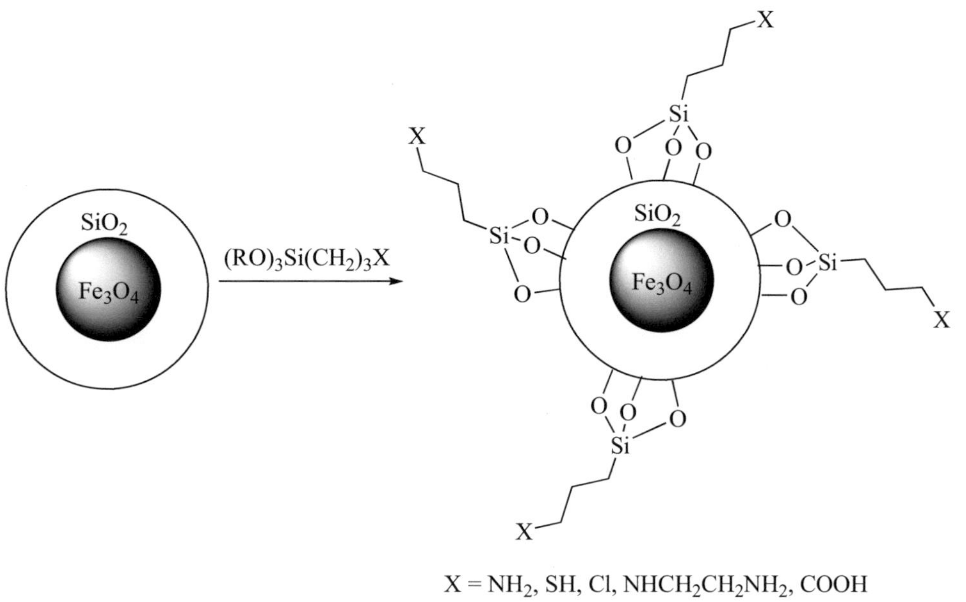

Scheme (2). Silane magnetic nanoparticle functionalization.

The dry etching process, however, is the most common method in quantum dot nanofabrication: by introduction in a etching chamber of a reactive gas, followed by application of a radio frequency voltage, creating plasma that ruptures the molecules of gas into reactive particles, then striking the surface, forming a product is volatile enough to cause patterned samples the etching [182 - 184].

Several techniques of self-assembly were used, by the bottom-up approach, being divided into vapor-phase and wet-chemical methods. The less used aggregation of gaseous monomers, liquid metal ion sources, molecular beam epitaxy, or sputtering can be sorted as vapor-phase methods [185 - 189], while sol-gel, microemulsion, electrochemistry and hot-solution decomposition are commonly denominated as wet-chemical methods [190 - 199]. Other synthesis methods for quantum dot preparation can be also used, including sonic waves or microwaves [200, 201]. Hydrothermal synthesis methods [202 - 204] have also been used to obtain quantum dots, using from inorganic salt crystallization from aqueous solutions, controlled by temperature and pressure.

Surface modification and stabilization of quantum dots are the main steps for the regulation of their functionality, as well as stability. For instance, is the desired application is biological, one has to take in consideration that (i) most of the synthesized quantum dots are water-insoluble and must be modified to become hydrophilic, (ii) uncoated quantum dots are very reactive and may inadvertently interact with nonspecific biomacromolecules, (iii) quantum dots are mainly made from heavy metal and inflict undesired toxic effects when applied in cell/animal models [205].

Generally, hydrophilic polymers are suitable quantum dot substituents for solubilization, like polyethylene glycol polymer (PEGylation). In addition, dihydrolipoic acid, dendrimers and multidentate phosphine polymers can be attached to the quantum dots surface, becoming more hydrophilic (Scheme **3**) [206 - 210]. Other amphiphilic polymers, triblock copolymer and acrylic acid can also be used for solubilization/stabilizing of quantum dots even though these coating protocols may inevitably enlarge the overall size of these nano-assemblies [206].

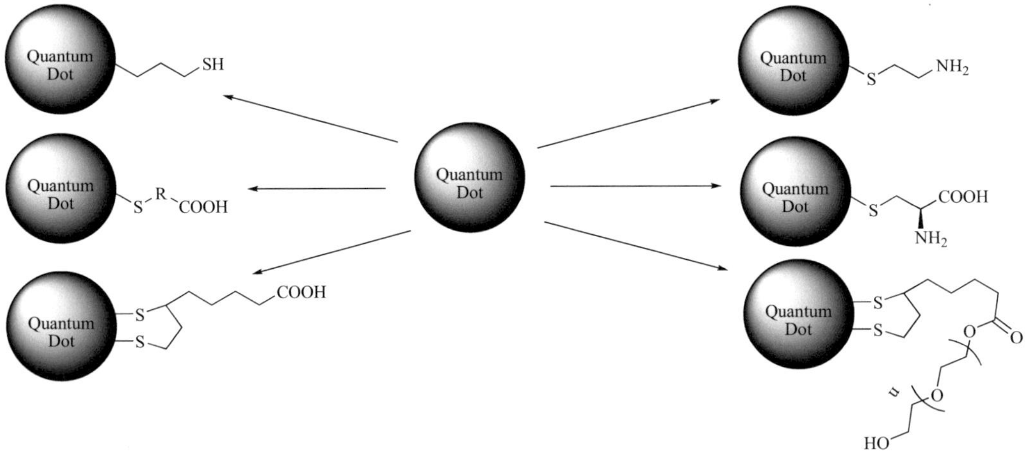

Scheme (3). Quantum dot functionalization.

APPLICATIONS OF METAL NANOPARTICLES

Nanomaterials have been extensively investigated in a wide range of applications, because of their extremely small size and their potential usefulness, namely energy storage [211 - 239], water purification [240 - 252], sensors [253 - 264], in solar cells [265 - 281] and biological applications, including drug delivery [282 - 306] and theranostics [285, 288, 307 - 330].

Applications in Energy Storage

Functional Pt and Pt-based nanomaterials provide highly active fuel cell catalysts, being the most efficient electrocatalysts, catalyzing anodic oxidations of small molecules and cathodic oxygen reduction reactions.

Small molecule's oxidations and oxygen reduction reactions normally require small size spherical nanoparticles (~3 nm) as electrocatalysts, due to their high volume-to-surface-area ratios. In addition, reports were given toward the importance of shape on Pt nanoparticle electrocatalytic activity [331 - 334]. For instance, Wang prepared Pt nanocubes [331, 332], observing that Pt nanocubes with 7 nm size presented higher activity than other Pt nanoparticles in H_2SO_4 solution, due to different sulfate ion adsorption on the nanocube's faces [331]. Another important finding was reported by the same group [333], where single-crystalline Pt nanospheres exhibited very high stability and specific activity for small organic fuels electrooxidation, when compared with commercial C/Pt catalysts. Nevertheless, Pt mass activity was relatively poorer, due to the big size of the Pt nanospheres. Other architectures were then developed, and nowadays other nanostructures have emerged, such as Pt hollow spheres [335], Pt nanotube [336, 337], Pt nanowires [338, 339], Pt nanoboxes [340] and Pt dendrites [341, 342]. For instance, Yan [336] demonstrated that Pt nanotubes prepared from Ag nanowires as sacrificial templates displayed durability higher and electrocatalytic activity than platinum black and a commercial catalyst, due to lower vulnerability to dissolution and aggregation. The same authors later generated porous Pt nanotubes using a similar strategy to enhance Pt nanotube's activity [337]. They confirmed that Pt nanotubes were more efficient in enhancing the activity of methanol oxidation reactions and oxygen reduction reactions. Dodelet and Sun [339] also prepared single-crystal Pt nanowires developed on carbon black nanospheres at room temperature, observing much higher catalytic activity for oxygen reduction reactions than with the C/Pt catalyst. In addition, Dong reported a hollow Pt nanosphere, which also enhanced the electroactivity of oxygen reduction and methanol oxidation reactions in H_2SO_4 solutions [335].

Bimetallic nanoelectrocatalysts in core/shell or alloy forms have gained much attention. Transition metal core/shell M/Pt nanoparticles are nowadays extensively investigated bimetallic systems, having two types of nanostructures. One type consists of the so-called Pt monolayer (Pt layers placed on metallic NP cores). For instance, Dong's group [343] designed Pt-coated Au-NP monolayer films, which displayed high electrocatalytic activity. Also Park and Yoo [344] synthesized bimetallic core/shell NR arrays, as well as Adzic [345, 346], who found that Pd_9Au_1 alloy and $AuNi_{0.5}Fe$ NP electrocatalysts proved advantageous for increasing the stability and activity in oxygen reduction reactions. Xia [347] reported Pd-Pt bimetallic nanodendrites possessing high surface area. These nanoparticles required no special treatment, being 2.5 times more active in oxygen reduction reactions than commercial Pt/C catalysts. Afterward, Dong [348] reported Pt-on-Pd bimetallic nanowires by wet-chemical strategy, being the shells made of dendritic Pt, exhibiting electrocatalytic activity on methanol reduction reactions. In fuel cell reactions, Pt-on-Au dendritic bimetallic nanoparticles were better suited due to higher chemical stability and durability of Au when compared with Pd [349]. Leaning on this, the same group [350] synthesized Pt-on-Au bimetallic dendritic nanoparticles at room temperature, in a one-step in high yield.

Ding's [351] reported the fabrication of bimetallic Pt/Au nanotubes, using a galvanic replacement reaction between Ag and $HAuCl_4$ or H_2PtCl_6, demonstrating with enhanced tolerance to CO poisoning in the formic acid oxidation. Chen [352, 353] reported the composition dependence of several FePt nanoparticles in oxygen reduction reactions, with activities following the trend $Fe_{63}Pt_{37} < Fe_{58}Pt_{42} < Fe_{54}Pt_{46} < Fe_{42}Pt_{58} << Fe_{15}Pt_{85} \leq Pt$.

Since nanoparticle's shape is a significant subject, some research has been directed to the design of Pt-(Mn, Ni, Co, Cu, Fe) nanocubes [354 - 356], all demonstrating activity, higher than commercialized catalysts. For example, Murray [356] verified higher activities in electrocatalytic oxidation for formic acid, oxygen and methanol, when cubic Mn-Pt nanocubes were used, when compared with the commercial Pt catalysts. Additionally, Fang [357] found that in

Pt_3Ni nanoctahedra activity in oxygen reduction reactions, in $HClO_4$ solution, was much better than when using similar size nanocubes. Ding [358, 359] also designed a synthetic methodology for nanotubular bimetallic mesoporous nanocatalysts (Pd/Cu and Pt/Cu), also showing enhanced electrocatalytic properties.

Also the utilization of trimetallic nanoelectrocatalysts has emerged recently, for their symbiotic increase of stability and catalytic activity. For instance, Sun reported the synthesis of Au/FePt, Pd/Au/FePt and Pd/FePt NPs nanoparticles [360 - 362] and used them for oxygen reduction reactions.

Applications in Water Purification

Heavy metal exposure is known to be a risk for living beings [363, 364]. Nowadays, various heavy metal removal methods are used, including electrochemical technologies chemical precipitation, adsorption, ion exchange, and membrane filtration [365, 366].

The use of nanoparticles is very prospective for water-purification due to their large surface areas, electronic, catalytic and optical properties [367]. For instance, titanium dioxide nanoparticles were proven to be quite capable photocatalysts for water purification [368]. These nanomaterials serve both as reductive and oxidative catalysts for inorganic and organic pollutants, usually in the presence of ultraviolet light, as shown by several authors [369, 370]. They successfully reported the use of TiO_2 nanoparticles both in toxic metal ion reduction of (*e.g.* Pt^{2+} and Ag^{+} Cr^{6+}) and in organic compound degradation (*e.g.* chlorinated benzenes and alkanes, furans dioxins, etc.) under UV light irradiation and in aqueous solutions. Ashasi [371] prepared *N* -doped TiO_2 nanoparticles for visible light-activated methylene blue photodegradation, while Choi and Bae [372] prepared TiO_2 nanoparticles modified by Pt deposits and Ru sensitizers, which considerably increased the carbon tetrachloride and trichloroacetate reductive dehalogenation rate under visible light, in aqueous solutions. It was also reported a multiple metal removal (Cu, Zn, Ni, Pb, Cd) from a pH = 8 solution using TiO_2 nanoparticles [373], displaying higher adsorption rate than bulk TiO_2 particles.

Detoxification of inorganic and organic pollutants in aqueous solutions was also carried out using zero valent bimetallic Fe(0) and iron Fe(0) nanoparticles [374 - 376]. Nanosized manganese oxide particles have also been used for sorption of anionic and cationic pollutants, including phosphates [377], arsenates [378] and heavy metal ions [379].

Nanostructured ZnO could also be used effectively for heavy metal removal [380], while others [381] produced nanometric ZnO powder. Compared with two TiO_2 powders and others powders, the ZnO displayed higher rate of removal of Cu^{2+}, with a capacity of ca. 1600 mg/g for Cu(II) ion adsorption.

Applications as Sensors

Noble metal nanoparticles are materials very extensively studied for high sensitive/selective platforms for electrochemical detection of target molecules [382]. Some authors reported several functional hybrid metal-based nanomaterials for detection, including high-density Pt nanoparticle/polyaniline nanofiber hybrids [383], Pt nanoparticle/silica coaxial nanocable supported Au/carbon nanotube hybrids [384], TiO_2 nanosphere supported high-density Pt/Au hybrid nanoparticles [385] and Au nanoparticle hybrids [386]. For instance, Hou and You [382] employed carbon nanofibers loaded with Pd-nanoparticles as improved materials for direct NADH and H_2O_2 sensing at low potentials, while Zhu [387] synthesized Au nanoparticle/C nanosphere hybrids, conjugated with horseradish peroxidase-labeled antibody, developed using self-assembly, and then used for protein assay sensing, as an amplified label. In 2010, Chen [388] reported a very selective and sensitive biosensor using graphene oxide sheets which were thermally reduced conjugated with a Au NP-antibody, for protein assay ultrasensitive detection, exhibiting a high detection limit of *ca.* 13 pM.

Also colorimetric sensors are quite appealing, owing to their low cost, high sensitivity and reading simplicity. Due to their extremely high ε (10^8-10^{10} M^{-1} cm^{-1}) in the visible region, metal nanoparticles meet these demands, since dispersed gold nanoparticles solutions are reddish while aggregated gold nanoparticles solutions are usually purple or bluish [389]. For instance, Mirkin stated to use Au NPs-thiol-DNA conjugates [390, 391], leading to a variety of new ultrasensitive detecting colorimetric sensors for polynucleotides [391], enzyme activity [392,

393], DNA [394, 395], proteins [396], and metal ions [263, 397 - 399]. He [400] further studied the sensing of potential triple helix specific DNA binders using colorimetric bioassays from sensing platform Au nanoparticles functionalized with DNA. Later, the same authors [393] used the same sensing platform for endonuclease activity screening and measuring inhibitors relative inhibitory strength, by monitoring kinetics of Au nanoparticles-DNA dissociation.

Rothberg [401] described the different adsorbing susceptibilities on colloidal gold nanoparticles solutions of non-labeled oligonucleotides. Following this seminal work, several groups explored diverse strategies for detection of DNA [402], metallic ions [403 - 405], proteins [406] and small molecules [407, 408] using Au nanoparticles modified with unlabeled DNA as sensing platforms.

Other platforms could also be engineered based on Au nanoparticles linked to other functional molecules. Some important examples include Au nanoparticles for detection of trinitrotoluene, through color change provoked by interactions acceptor-donor between primary amines and trinitrotoluene on Au nanoparticles surface [409], or melamine detection through triple hydrogen-bonding recognition induced color change, between a cyanuric acid derivative and melamine [410], Hg^{2+} ion detection on mercaptopropionic acid-modified gold nanoparticles [411], Ca^{2+} colorimetric detection by Au nanoparticles functionalized with calsequestrin [412] and Pb^{2+} ion detection using gallic acid functionalized Au nanoparticles [413].

Silver nanoparticles are also good candidates for optical sensing because, similarly to Au nanoparticles, they exhibit a distance dependent color, but having higher extinction coefficient for similar sized Ag nanoparticles [414]. Ag nanoparticles functionalization with DNA and functional molecules has been used to detect some important compounds. For instance, a sensitive colorimetric assay using unmodified silver nanoparticles for detection of enzymatic reactions was reported in 2008 [415]. The authors could readily detect adenosine diphosphate phosphorylation and adenosine triphosphate dephosphorylation by Ag nanoparticle change of color. Jiao and Yang [416] also used Ag nanoparticles and unmodified DNA to prepare a colorimetric strategy, distinguishing homoadenine ligand binding. The ligand's presence had driven the adenine deoxynucleoside aside from the Ag nanoparticle surface, with subsequent nanoparticle aggregation upon salt addition, giving a yellow-to-brown color change. Su and Kanjanawarut [417] used charged neutral peptide nucleic acid (PNA) and unmodified Ag nanoparticles as probes for DNA colorimetric assay since, citrate protected Ag nanoparticles suffered direct aggregation, in the presence of PNA molecules.

Unlike Ag and Au nanoparticles, Pd and Pt nanoparticles do not have typical visible range surface plasmon resonance absorption, hampering their use as colorimetric sensors. However, Pt and Pd nanoparticles can potentially mimic peroxidase or oxidase [418], AgM (M = Pd, Pt, Au) alloy nanoparticles, catalyzing, the oxidation of typical horseradish peroxidase substrates in the presence of H_2O_2. Later, Li [419] found that Au nanoparticles bearing a positive charge also possessed peroxidase-like activity, catalyzing the H_2O_2 peroxidase oxidation.

Also fluorescent sensors can be also used, mostly due to higher sensitivity and multiplicity, when compared with other optical absorbance methods. Currently, metal nanoparticle-based fluorescent sensors can be of different sources. For instance, Wang [420] prepared Ag nanoparticles stabilized with oligonucleotides for Hg^{2+} ion sensing, displaying high selectivity and low detection limit. Additionally, other functional stabilized metal nanoparticles such as bovine serum albumin-Au nanoparticles [421], alkanethiol protected gold nanoparticles [422], lysozyme-stabilized gold fluorescent nanoparticles [423] were also synthesized and used for water Hg^{2+}detection. Fluorescent sensors based on metal nanoparticles have also been used for other molecule detection, including Cu^{2}+ [424], CN− [425], cysteine [426] and single-nucleotide mutation [427].

In parallel, metal nanoparticle's high quenching ability through electron transfer/non-radiative energy processes has been also utilized. For instance Dong [428] used Au nanoparticles protected with a NIR dye for sensitive detection of thiols in human plasma. Additionally, fluorescent conjugated molecules-stabilized gold nanoparticles were used for detection of cysteine [429] and cyanine [430], and other immunoassays [431].

Applications in Solar Cells

Nanomaterials appeared as the next best light energy harvesting assemblies synthons. Inorganic and organic hybrid structures exhibiting enhanced efficiency and selectivity have been designed [432 - 434], establishing the base for next generation energy conversion devices [435 - 437].

There are two main strategies to exploit nanostructures for fabricating solar devices: photosynthetic mimicking using donor-acceptor clusters and assemblies and, the most promising, solar cells based on semiconductor nanostructures. Hence, organized organic-inorganic nanohybrids can be developed through assembly of organic molecule monolayers bearing isothiocyanates, silanes, amines and thiols onto metal nanoparticle 3D surface [438, 439].

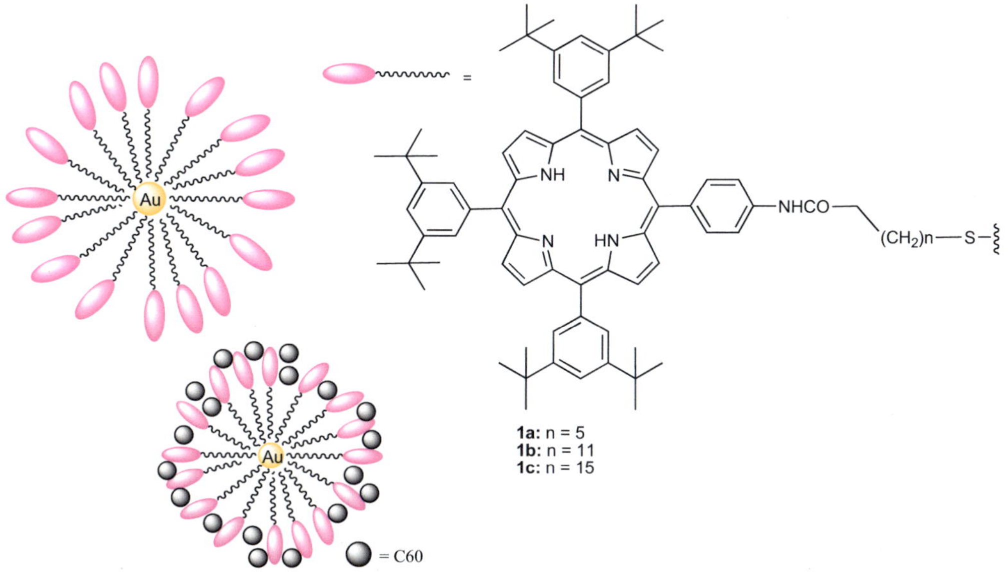

Fig. (3). Examples of porphyrin-functionalized gold nanoparticles.

Recent developments in assembling organized nanostructures for solar cell performance improving were endeavored. For example, Fukuzumi reported hybrids of type **1**, constituted by spherical gold nanoclusters protected with porphyrin monolayers, and were used as light harvesting antennae [440] Fig. (**3**), exhibiting efficient capability, while other authors improved the nanostructure assembly, by aggregating C_{60} to gold nanoclusters [441 - 443], further preparing composite molecular clusters consisting of **1** and C_{60} molecules [444, 445], stable at room temperature, reaching overall PCE = 1.5% at 11.2 mW cm^{-2} input power.

Similarly, fullerene-functionalized gold clusters **2** (Fig. **4**) were also evaluated for their energy transfer processes when in excited state [446]. The fullerenethiol emission spectrum displayed a maximum at 710 nm in the absence of Au nanoparticles, corresponding to its singlet excited state [447], while singlet excited of **2** was deactivated by intersystem crossing, generating its triplet excited state. Furthermore, some authors observed that pyrene-linked Au nanoparticles **3** Fig. (**5**) exhibited clearly lower quantum yields, when compared to sole pyrenethiol [448]. The quenching mechanism was then studied by time-resolved absorption, establishing that, using pulsed laser irradiation (337 nm) of nanoparticle **3**, the authors observed electronic transfer between pyrene the and gold nanoparticle, demonstrating Au nanoparticle storing and transporting abilities.

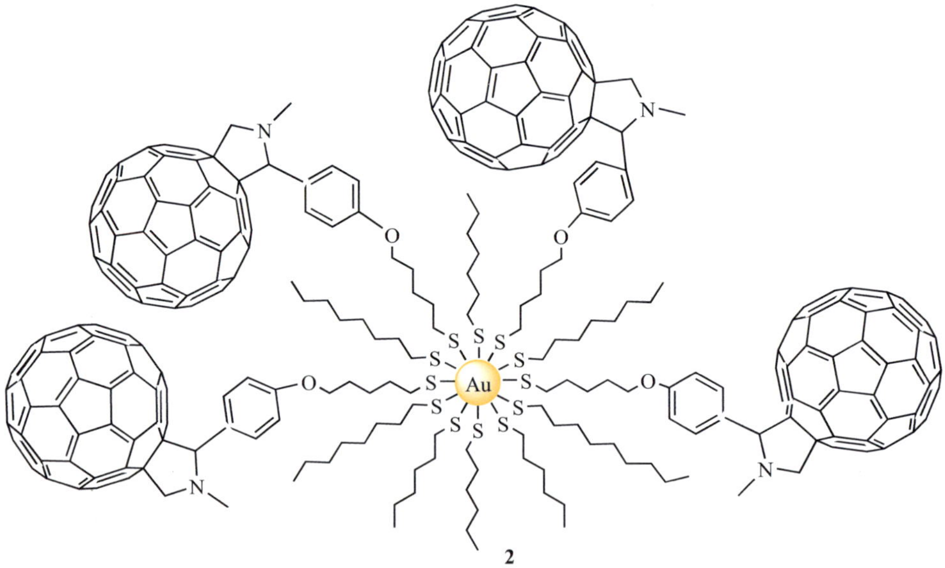

2

Fig. (4). C60 functionalized gold nanoparticles.

Some authors prepared polypeptide based porphyrin octamer **4**, which was used to fabricate solar cell fullerene-porphyrin assemblies [449]. The OTE/SnO_2/AA + C60)*m* system displayed an overall PCE = 1.3% at 3.4 mW cm^{-2} input power (Fig.

6) [450].

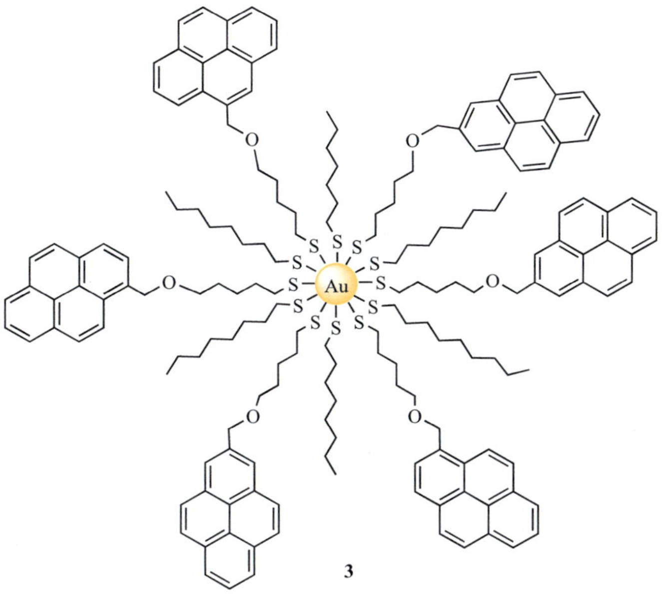

3

Fig. (5). Pyrene functionalized gold nanoparticles.

Applications in Biomedicine

The nanometer size of nanoparticles, makes them biologically relevant as they have the ability to permeate cellular membranes and interact with important biomolecules, such as DNA, proteins, and viruses. For instance, metal nanostructures, such as silver and gold, have allured much consideration. Their stability allows simple surface functionalization with various biological and organic molecules, through noncovalent and covalent interactions, allowing for not only the replacement of potentially toxic initial stabilizing agents, such as cetyl trimethylammonium bromide, an antiseptic agent, but also allows for specific biological targeting applications.

Noncovalent modifications might occur through electrostatic interactions, hydrophobic entrapment and van der Walls forces, while covalent modifications utilize direct chemical attachment, linker molecules, or click chemistry. Functionalization of silver and gold nanoparticles with biomolecules, such as DNA, peptides and antibodies as been widely accomplished using electrostatic interactions. This type of functionalization depends on the attraction between two oppositely charged species, the nanoparticle surface and the biomolecule of interest. This way, the biomolecule is not exposed to harsh chemical modifications. In 2001, Rotello and coworkers used this approach to functionalize cationic gold clusters with DNA using the negatively charged phosphate backbone of DNA [451], whose stability was confirmed through UV-Vis spectroscopy and dynamic light scattering.

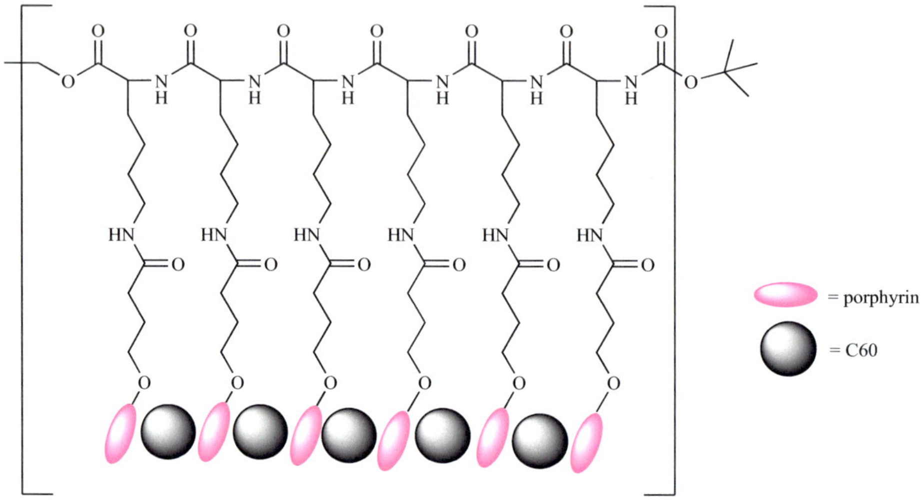

4

Fig. (6). Polypeptide functionalized porphyrin octamer.

El-Sayed group [452 - 454] has also used electrostatic interactions to functionalize Au nanoparticles and Au nanorods with antibodies for selective targeting and photothermal destruction of cancer cells. Yang [455] studied the electrostatic interactions to immobilize an antineurogenin 1 (anti-ngn 1) antibody on mercaptosuccinic acid (MSA) stabilized Ag nanoparticles, which were created between antibody cationic amine groups and the MSA-Ag nanoparticles anionic carboxylic groups, permitting label-free detection of the synthetic peptide fragment of ngn1.

In contrast to noncovalent interactions, covalent interactions usually offer superior stability and reproducibility of molecular functionalization. For instance, the bond energy between Au and S is 44 kcal mol^{-1} [456, 457]. Taking it in consideration, thiolated biopolymers such as PEG and polyoxazoline have been anchored to gold nanoparticle surfaces through Au-S bonds to increase the nanoparticle's stability in physiological medium and to avoid serum protein nonspecific binding in *in vivo* systems [458, 459]. Metal-S bond formation has also been used to hold oligonucleotides to nanoparticle surfaces. Several groups demonstrated oligonucleotide-functionalized Au nanoparticles [390, 460], this approach being further extended to various biomedical applications such as hepatitis C virus detection, intracellular gene regulation, and mercuric ion detection [263, 461, 462]. Targeting nuclear localizing signal peptides and arginine-glycine-aspartic acid (RGD) have also been conjugated to gold and silver nanoparticle surfaces through the metal-S bond to promote particle internalization by human oral squamous cell cells and their succeeding nuclear localization [463 - 465], through cysteine residues located within the peptides and the nanoparticle surface.

Bifunctionalized nanoparticle conjugate systems utilize linkers to covalently attach therapeutic drugs and biomolecules, reducing undesirable structural changes that may arise from direct interaction. Commonly used linkers for biomedical applications include oligoethylene glycol (OEG) and PEG polymers modified with amine, carboxylic, isothiocyanate and maleimide functional groups, permitting conventional coupling strategies, such as carbodiimide-mediated esterification and amidation [466]. The linker molecules could also be modified to contain thiol or disulfide groups to promote their exchange with capping agents at the nanoparticle surface, as reported by Gong [467], with a successful covalent attachment of doxorubicin (DOX) to a hydrophobic inner layer of a folic acid amphiphilic Au nanoparticle through a pH-sensitive hydrazone linkage, displaying selective uptake and DOX release to breast cancer cells. Ying and coworkers [468] also used covalently attached DOX to Ag nanoparticles through binding using NHS-PEG-NHS linker to create a DNA biosensor, while azide-alkyne Huisgen cycloaddition (Click chemistry), which involves the use of a copper catalyst, has also been utilized for nanoparticle functionalization and related applications [469, 470].

As stated previously, the easily accessible surface chemistry of Au nanoparticles functionalization with a variety of biomolecules makes them ideal candidates also as drug delivery platforms. Their drug delivery use was first reported by

Tamarkin, Paciotti, and coworkers [471], where Au nanoparticles were functionalized with polyethylene glycol (PEG) and tumor necrosis factor (TNF) *via* dative covalent attachment, exhibiting colon carcinoma (MC-38) tumor-specific accumulation and minimal systemic toxicity and, particularly enhanced antitumor efficacy at a lower dose than native TNF. The accomplishment made through application of this particular drug delivery platform has opened up many opportunities for the delivery of many other anticancer agents, specifically targeting tumor characteristics in a variety of cancer types. For instance, targeting cancer cell overexpressed folate receptors was carried out using methotrexate (MTX), which also exhibits anticancer activity [472]. MTX was conjugated to 13 nm Au nanoparticles at high loading capacity through electrostatic Interactions, highly enhancing MTX accumulation in Lewis lung carcinoma (LL2) tumors and subsequent antitumor activity. Another tumor characteristic that can be targeted with Au nanoparticles drug delivery systems are hormone receptors. El-Sayed group developed drug delivery systems taking advantage of the upregulated estrogen receptor in breast cancer [473] and targeting the upregulated androgen receptors in prostate cancer [474]. This ability to enhance chemotherapeutic efficacy in treatment-resistant cancers is one of the primary targets for developing new drug delivery systems [475]. Wang and coworkers [476] demonstrated enhanced doxorubicin (DOX) drug efficacy in treatment-resistant breast cancer cells just by attaching modified DOX to the 30 nm Au nanoparticles through a pH-sensitive linker.

Owing to their characteristic localized surface plasmon resonance, Au nanoparticles can absorb light with resonant frequency, but in addition to the strongly enhanced light energy decaying *via* radiative processes (*e.g.* scattering), energy can also decay *via* nonradiative processes. The localized surface plasmon resonance allows for electron coherent oscillation within the metallic lattice, and these electrons collide (electron-electron relaxation) on a femtosecond timescale, producing electrons, which interact with the phonons in the metallic lattice (electron–phonon relaxation) on a picosecond timescale, transferring their energy to the phonons and thus raising the temperature of the metallic lattice by about 30 °C. The metallic lattice "cools off" *via* slower phonon-phonon interactions and subsequent dissipation of heat to the surrounding medium [453, 477, 478]. The ability for these particles to heat the surrounding medium upon excitation, along with the sensitivity of cancer cells to hyperthermia, can be used in the cancer treatment method photothermal therapy (PTT) [479]. Plasmonic Au nanoparticles are promising candidates in this treatment method, as they can be systemically delivered to tumor sites through enhanced retention and permeability or specifically attached to cancer cells *via* molecular targeting [480] and then, through external radiation, subsequently cause temperature increases, inducing apoptosis [321, 481].

The first reports on PTT were given by Lin and coworkers [482], successfully showing the destruction of lymphocyte cells after attachment with antibody-conjugated Au nanoparticles and subsequent nanosecond-pulsed visible laser exposure. El-Sayed and coworkers [453] later used a similar strategy of attaching antibody-conjugated Au nanoparticles to human oral squamous carcinoma cell (HSC-3) cells, using a continuous wave visible laser to induce cell death. Gold nanoshells (silica-gold core-shell nanoparticles) were also used in PTT [483, 484]. These gold nanoshells were functionalized with polyethylene glycol (PEG). Upon NIR laser exposure, the photon energy absorbed by the gold nanoshells was translated to thermal energy, so much so that temperature increases of up to 37 °C were recorded intratumorally, resulting in complete tumor ablation in tumors specifically bearing the gold nanoshell photothermal contrast agent. Another photothermally significant gold nanoparticle that was used in PTT was an Au nanocomposite, first developed by Xia and coworkers in 2007 [485]. The Au nanocomposites were first modified with an antibody to promote specific binding to breast cancer (SK-BR-3) cells, resulting in the photothermal destruction of the cancer cells labeled with these hollow gold nanostructures, upon NIR laser pulse exposure. Li and coworkers [486, 487], shortly after, also applied cancer-specific targeting Au nanocomposites *in vivo* as photothermal contrast agents, being able to demonstrate the specific accumulation of Au nanocomposites at tumor sites, as well as their external activation as photothermal contrast agents by continuous wave NIR laser exposure, resulting in tumor destruction.

Borghs and coworkers synthesized branched Au nanoparticles and modified them with nanobodies specific for breast and ovarian cancer (SKOV3) cell surface binding [488], demonstrating that, upon NIR radiation, the Au nanoparticleslabeled cancer cells were destroyed.

Like their Au counterparts, Ag nanostructures have also been extensively used in therapeutic applications [489], where antimicrobial activity is one of the most investigated applications. The recent increase in antibiotic-resistant bacteria, due to antibiotic excessive use, has caused a need for microbiocide alternatives. Hence, many groups have begun investigating the antimicrobial activity of Ag nanoparticles [490, 491]. In 2004, Salopek-Sondi and Sondi [492] demonstrated Ag nanoparticle antimicrobial activity against *E. coli* , a gram-negative bacterium. TEM and SEM images revealed that agar, containing Ag nanoparticles, resulted in a concentration dependent growth inhibition, while media containing Ag nanoparticles only delayed *E. coli* growth.

The influence on *E. coli* growth was attributed to membrane damage in the form of pits. Yacaman [493] further demonstrated the antimicrobial activity of Ag nanoparticles on gram-negative bacteria and that growth inhibition was both size and concentration-dependent. At 75 μg/mL, there was no significant growth observed. Shape dependence on antimicrobial activity has also been studied, by Song [494], who prepared spherical, rod, and triangular-shaped Ag nanoparticles and assessed their antimicrobial activity against *E. coli* . Similar to previous works, the different shaped Ag nanoparticles were incorporated into agar plates as well as Luria Branti (LB) media. It was seen that bacteria grown on agar plates showed complete growth inhibition, while LB media only resulted in a growth delay.

The main outcome of this study was the disclosure that antimicrobial activity was related to the Ag nanoparticle shape. Triangular Ag nanoparticle showed the highest activity, followed by spheres and rods. These studies afforded the fundamentals for Ag nanoparticles antimicrobial activity, and since then, several groups have expanded upon these works [495 - 497].

Although Ag nanoparticles have been incorporated into many wound dressings due to their antimicrobial activity, they also can promote wound healing. In 2007, Tian *et al.* [498] reported that Ag nanoparticles topically administered to murine burn wounds accelerated healing by ~10 days and reduced scar appearance, as a result of the Ag nanoparticles' antimicrobial activity as well as a reduction in local and systemic inflammation.

The promotion of wound healing also was found to be dose-dependent, by Wong and coworkers [499], who demonstrated that during wound healing, Ag nanoparticles ranging from 5 to 15 nm were able to induce different responses in keratinocyte and fibroblast skin cells.

The accelerated wound closure was ascribed to the improved rate in fibroblast differentiation and keratinocyte proliferation into myofibroblasts. Furthermore, in 2011, the same group [500] reported the progress in the tensile properties of the repaired skin. This improvement was linked to the ability of Ag nanoparticles to control collagen deposition and direct collagen alignment and spatial arrangement.

APPLICATIONS OF MAGNETIC NANOPARTICLES

Biological Applications

When designing magnetic nanoparticles (MNP) for *in vivo* applications, two key issues must be accounted: surface and size functionality. Superparamagnetic iron oxide nanoparticles (SPIOs) size influences significantly *in vivo* biodistribution, reason why 10-40 nm particles, are most favorable for extended blood circulation [501].

MRI is a useful modality for medical diagnostics [302, 502 - 504]. Nevertheless, the main lies in their *in vivo* behavior. Detection of most magnetic nanoparticles is not very efficient, since they are nearly randomly absorbed by the reticulo-endothelial system [505], namely the lymph nodes, liver and spleen [506]. However, when coupled with specific receptor guide functionalities, these particles can effectively produce clear images, as for instance Jun [507] reported, when coupling Herceptin, a specifically binding antibody with $MnFe_2O_4$ nanoparticles on ovarian or breast cancer cell's surfaces.

Fluorescent organic dyes are usually conjugated with MNP for both combining MRI and optical properties in one nanocomposite [508 - 512]. For instance, dextran coated magnetic nanoparticles can be further conjugated with fluorescein isothiocyanate and HIV-Tat peptide [513]. Magnetic nanoparticles could also be tailored with phospholipids bearing PEG units substituted with Tat peptide and Texas Red [514], while PEG coated functionalized iron oxide magnetic nanoparticles were conjugated with chlorotoxin (a peptide) and Cy5.5 [515].

In another example, copolymerization of a Eu_3 complex, tetraethyl orthosilicate and 3-aminopropyltriethoxysilane in the presence of polyvinylpyrrolidone-stabilized Fe_3O_4 nanoparticles yielded optically active iron oxide/silica multifunctional magnetic nanoparticles [516], while coating Fe_3O_4 nanoparticles with polyacrylic acid, followed by a NIR dye grafting [517].

PET is another important imaging tool [518] where ^{11}C, ^{18}F, ^{64}Cu, ^{68}Ga and ^{124}I are used as tracers, being a very sensitive technique, but gives lower resolution than MRI, reason why combination of two techniques might allow obtaining highly resolved and highly sensitive images. For instance, a ^{64}Cu complex (**5**, Fig. **7**) was grafted onto magnetic nanoparticles dextran coating [519].

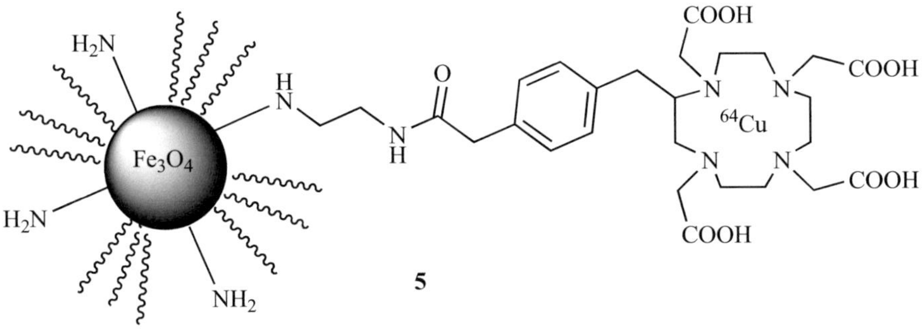

5

Fig. (7). Coated magnetic nanoparticles bearing chelated bifunctional ligands.

PET/MRI dual probes were also prepared by coupling cyclic arginine-glycin--aspartic acid peptides and ^{64}Cu to magnetic nanoparticles coated with poly(aspartic acid) (**6**, Fig.**8**) [520]. Also $MnFe_2O_4$ coated with serum albumin along with ^{124}I was used as MRI/PET probe [521].

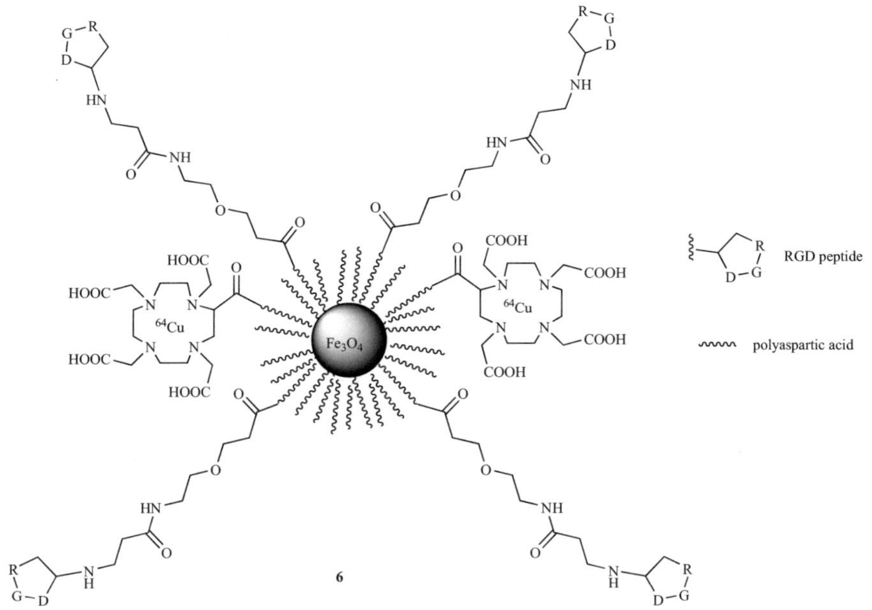

Fig. (8). PET/MRI magnetic nanoparticle based dual probes.

Other biological applications include drug delivery [34, 163, 522 - 525], bioseparations [526 - 530] and therapeutics [531 - 533]. For instance, by external magnetic field application, delivery of particles to designated targets was possible [33, 522, 523]. As an example, Banerjee [525] grafted cyclodextrin molecules on magnetic nanoparticles modified with gum arabic affording a magnetic drug carrier at the specified dose and elimination time.

Also nanometer size magnetic particles were widely studied in purification and separation of biomolecules and cells in physiological processes [526 - 528]. For instance, Sen [529] reported a mesoporous silica-magnetite composite for sheared salmon sperm DNA extraction, while Haynes [530] prepared ovalbumin bound $Fe_3O_4@Al_2O_3$ MNPs for *Escherichia coli* affinity capture.

Superparamagnetic iron oxide nanoparticles allow magnetic energy transfer as heat, which can also be used in hyperthermia therapeutics, due to tumor cell temperature increase, being considered one the key approaches in the future for cancer therapy [531 - 535].

Applications in Catalysis

The main goal in catalysis research is to provide a facilitated reutilization and recovery of quasi-homogeneous catalysts by covalent heterogeneous support [536]. Recently, the use magnetic nanoparticles as catalyst's support have been extensively studied, attempting to overcome heterogeneous catalysis limitations. Thus, several reactions catalyzed by transition metals using magnetic nanoparticles with grafted catalytic sites have been recently reported, including hydroformylation [537], hydrogenation [538, 539] polymerization [540], carbon-carbon cross-coupling reactions [541, 542], enzymatic carboxylate resolution [543], ester hydrolysis [544] and miscellaneous reactions [545].

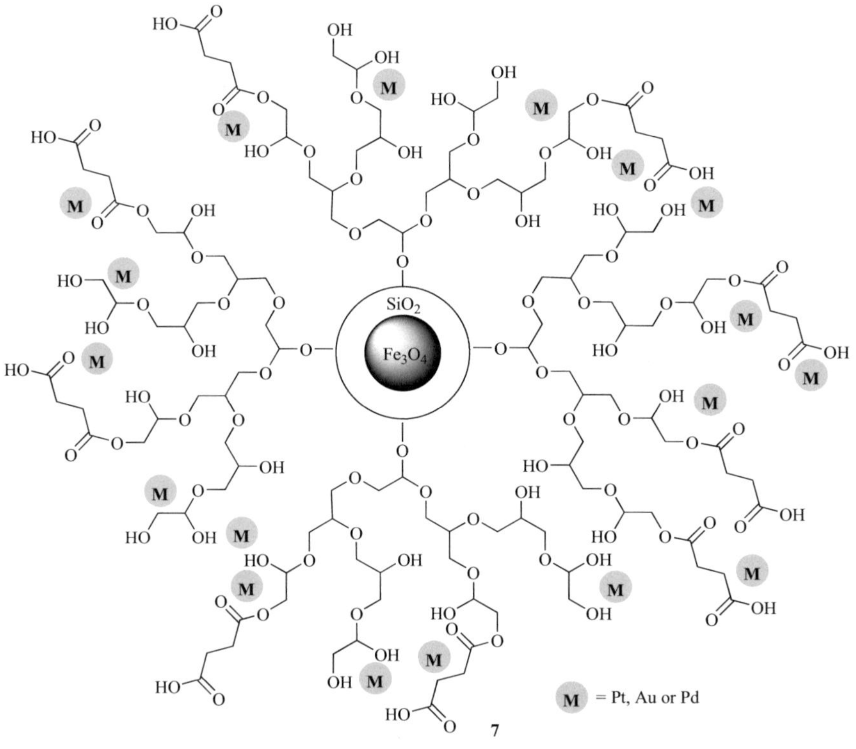

7

Fig. (9). Dendrimer based magnetic nanoparticle for metal catalyst encapsulation.

Some groups have reported the preparation of several metal nanocatalysts, based on Au, Pd and Pt, directly grown on polymer nanotube magnetic hybrids or Fe_3O_4/SiO_2 surfaces, with high-loading capacity and high dispersibility (**7**, Fig. **9**)

[546 - 548], usually observing high catalytic activity and recyclability in Heck coupling, alcohol oxidation and nitrophenol reduction.

Ying used SiO_2/Fe_2O_3 nanocomposites **8a,b** (Fig. **10**) as magnetic supports for the production of high dispersed stable Pd nanoclusters [549], using it for nitrobenzene hydrogenation, observing excellent reusability and reactivity (99% yield after 6 runs), which was ascribed to the presence of the amine ligands on SiO_2/Fe_2O_3, as agglomeration suppressors, during the hydrogenation reaction.

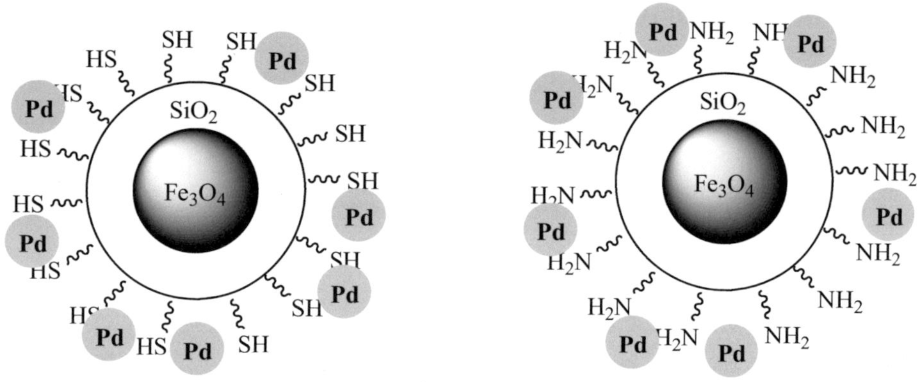

8a **8b**

Fig. (10). Magnetic nanoparticle for metallic palladium encapsulation.

Glorius [550] reported the synthesis of enantiomerically pure Pd nanoparticles on Fe_3O_4, using *N* -heterocyclic carbenes (**9**, Fig. **11**), acting also as anti-leaching stabilizers, and used in asymmetric ketones R-arylation reactions, with high enantiomeric excess (up to 85% ee) and great catalyst recyclability.

Connon reported a 4-*N,N* -dialkylaminopyridine-catalyst supported onto magnetite **10** (Fig. **11**) [551], and used in several organic reactions, namely the acetylation of 1-phenylethanol, achieving high catalytic activity at room temperature.

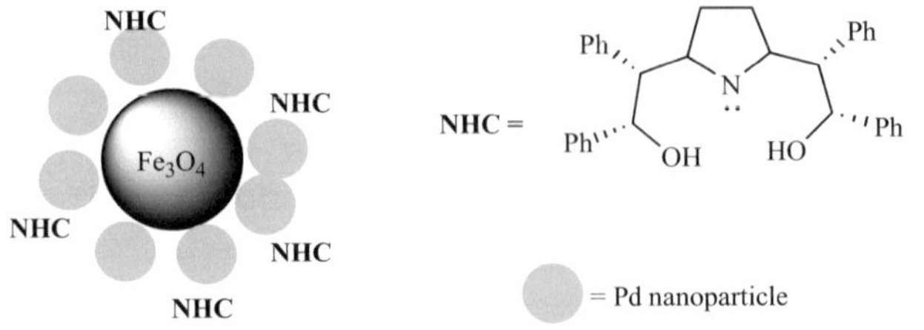

9

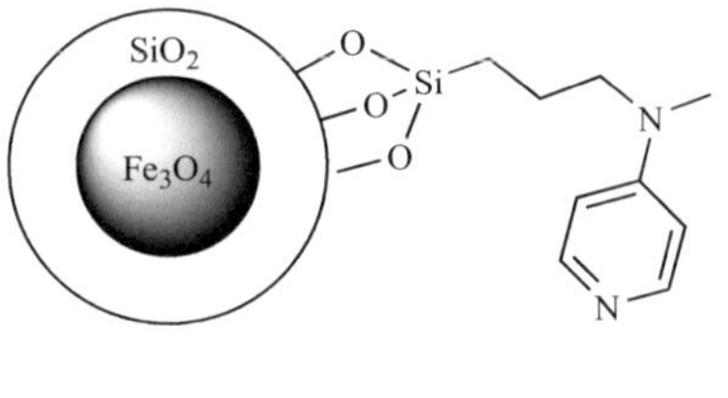

10

Fig. (11). Functionalized magnetic nanoparticles as catalysts.

Applications in Environmental Remediation

Iron nanoparticles can be considered cost-effective alternatives for the conventional water treatment technologies, [552, 554]. Several modified iron nanoparticles have already been synthesized to enhance remediation efficiency in removing high organic compound concentrations, existing for instance in many industrial wastewater streams [553].

For instance, polyacrylic acid-coated magnetic nanoparticles were used for recovery of lipase *Candida rugosa* [555] and methylene blue [556] from aqueous solutions, while cetyltrimethylammonium bromide coated iron oxide nanoparticles were reported as efficient adsorbents for amaranth from aqueous solutions [557]. Fe_3O_4 nanoparticles conjugated with chitosan by carbodiimide activation were also produced, being used for acid dye removal [558]. Also, Hajaligol [559] evaluated Fe_2O_3 nanoparticles both as oxidants and catalysts for carbon monoxide removal. Heavy metals have been also removed for wastewaters by the use of amino functionalized iron oxide nanoparticles [560, 561], which proved to be quite effective in their removal.

Xu [562] reported the use of metal nanoparticles modified with bisphosphate **(11)** (Fig. **12**) for radioactive metal toxins (*e.g.* , UO_2^{2+}) blood removal, showing quite high efficiency, while Jung [563] synthesized magnetic silica nanoparticles functionalized with BODIPY units for high selectivity/high affinity Pb^{2+} removal from human blood and water.

APPLICATIONS OF QUANTUM DOTS

Quantum dots are nanometer scale light-emitting particles, being currently an exciting field of research that has emerged, with use as fluorescent labels for several purposes, including biomedicine applications [564 - 570], optoelectronics [571 - 578] and solar cells [579 - 584].

Applications in Biomedicine

Quantum dot's emission wavelength can be adjusted by slightly altering its size. This property is a major advantage as they can be optimized for a specific study and/or can be used in multidisciplinary experiments [585]. Quantum dots are more photostable than standard fluorescent reporters, being also very bright owing to their high quantum yield [586].

For instance, there are several commercial quantum dots, possessing an extensive range of emission wavelengths that have already been conjugated with various tumor directed agents for use in cancer molecular imaging [587 - 590].

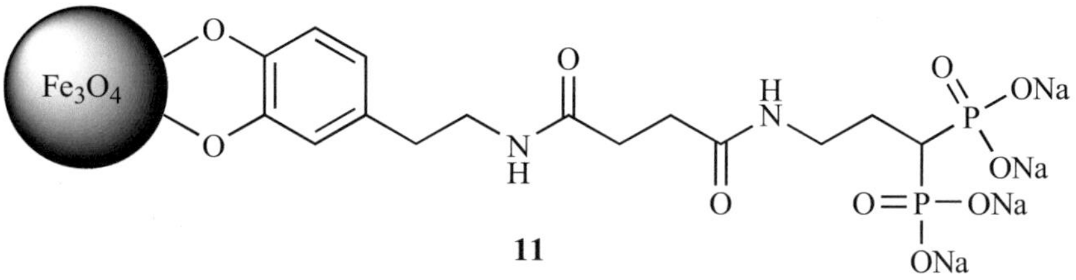

11

Fig. (12). Bisphosphate-modified metal nanoparticles.

Maysinger [591] subcutaneously injected mice CdTe and CdSe quantum dots and scanned their brain for fluorescence, detectable using *in vivo* imaging techniques for long times after intracortical injection, allowing conjugate identification and location with neuronal internalization.

By conjugating quantum dot with transferring, Chan [592] demonstrated the utility of NIR reporters for non-invasive applications or *in vivo* deep tissue analysis, which minimized light scattering and absorption by endogenous tissues.

Bawendi [593] prepared several small core sized InAs/ZnSe quantum dots (less than 10 nm), between 750 and 920 nm, permitting visualization within the interstitial fluid in rats upon conjugation of quantum dots with DHLA.

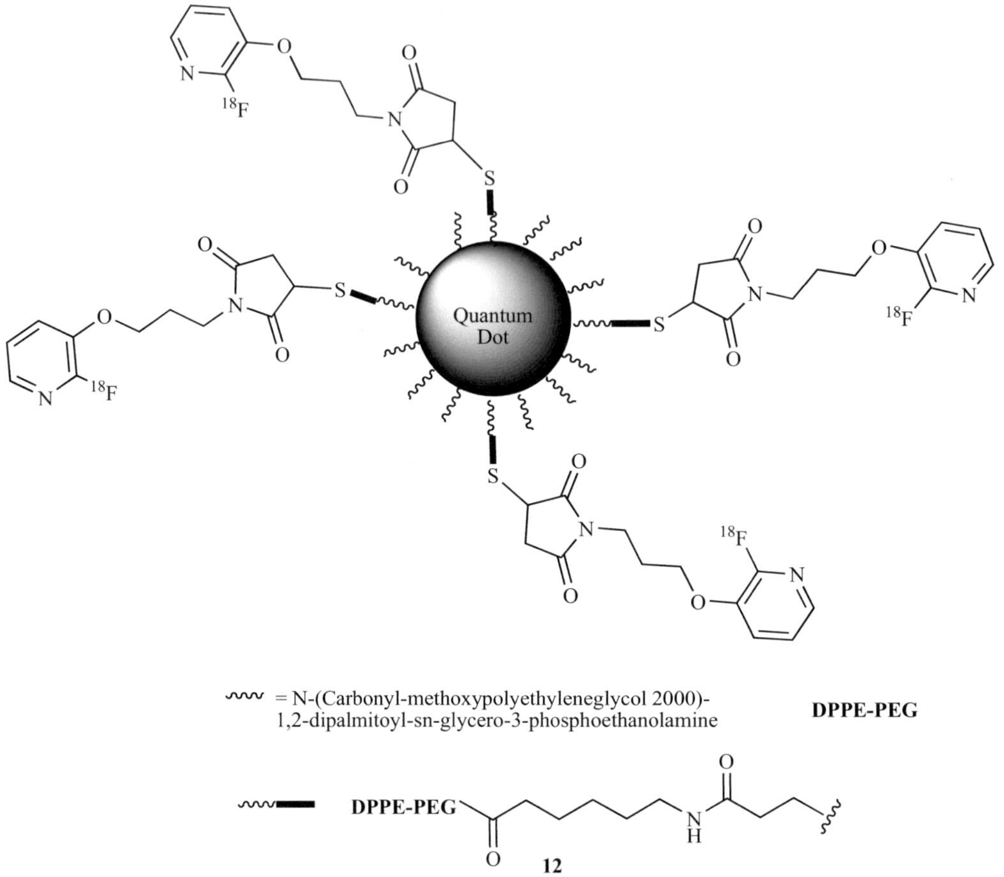

12

Fig. (13). Bimodal imaging probes based on encapsulated CdSe/ZnS QDs in fluorine-18 labeled phospholipids micelles.

In other reports, the quantum dots have been used in MRI applications [594 - 596]. For instance, Yang and Santra [595] prepared water soluble quantum dots, functionalized with Gd-silica coated CdS:Mn/ZnS, observing enhanced magnetic resonance signal intensity. Yong and Prasad [597] reported non-Cd-based quantum dots as nontoxic highly efficient bioimaging agents, for targeting and visualization of pancreatic cancer cells. In this approach, InP/ZnS quantum dots were functionalized with mercaptosuccinic acid and further conjugated with pancreatic cancer specific mAbs, and effectively used for *in vivo* and *in vitro* targeted bioimaging. Moreover, Dubertret and Tavitian [598] encapsulated CdSe/ZnS QDs in F-18 labeled phospholipidic micelles (**12**, Fig. **13**), resulting in bimodal imaging probes in combined *in vivo* fibered confocal fluorescence microscopy and positron emission tomography.

Quantum dots can also be used for cellular signaling, by studying cell surface protein membrane dynamics, using them as single protein fluorescent tags, for which has been reported by several groups [48, 599 - 601]. Chu investigated the nerve growth factor (NGF) signals' retrograde axonal transport [602], using streptavidin-conjugated quantum dots with biotinylated NGF to track the movement of NGF in live neurons in real time by total-internal-reflection fluorescence microscopy. In 2003, Dahan [603] prepared quantum dots conjugated with an antibody to track individual glycine receptors in neuronal cells by fluorescent microscopy. Bawendi, on the other hand, combined CdSe quantum dots with a fluorescent dye, using a PEG linker for enhanced stability and solubility (**13**, Fig. **14**) [604].

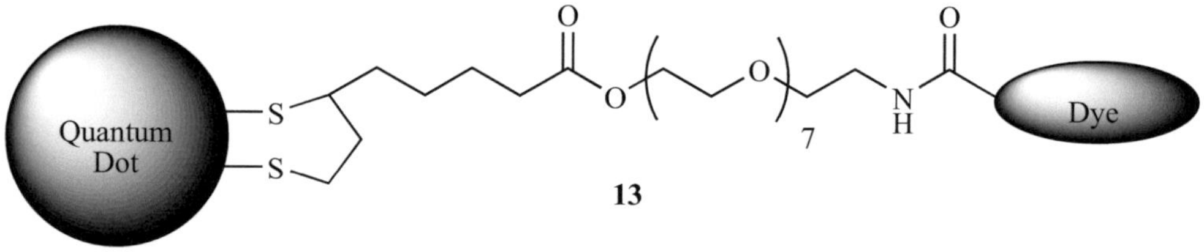

13

Fig. (14). Dehydrolipoic acid functionalized quantum dots.

Akiyoshi reported the preparation of monodisperse hybrid nanoparticles of quantum dots through mixing with cholesterol-bearing nanogels, which were able to efficiently internalize into various human cells [605]. Also Mansson and Montellius reported the labeling of individual isolated biotinylated F-actin fibres by streptavidin-coated quantum dots, in which smaller percentage of labelled filaments were motile in comparison with an organic fluorophore such as Alexa488 [606], while Lidke [607] also reported the visualization of tyrosine kinase-dependent signaling by using quantum dots functionalized with epidermal growth factor (EGF-QD), which proved to be highly potent and specific in EGF receptor (erbB1) activation and binding.

Enormous efforts have been devoted to achieve safe quantum dots with visible and NIR stable and strong photoluminescence. So far, CdSe and CdTe core-only quantum dots and CdSe/ZnCdS, CdSe/ZnS and CdTe/CdSe core/shell quantum dots were extensively studied; nevertheless the safety of these nanocrystals has been a big concern for *in vivo* uses despite great achievements for *in vitro* cell based applications [608, 609].

The development of new generation cadmium-free quantum dots [610] or addition of silica coating to cadmium QDs [611] can help addressing some toxicity issues. Other quantum dots limitations include their aggregation proneness as well as the available functionalization methods, which can be challenging [612, 613].

Applications in Optoelectronics

High photoluminescence efficiency and bandgap modulability make quantum dots very appealing for several optoelectronic applications, including photonic crystals and polymer films/gels. For instance, Yang [614] prepared lead sulfide (PbS) quantum dots in copolymer matrices constituted by methacrylate and styrene, while other examples appeared, using polymethyl methacrylate [615] and polyacrylonitrile [616].

Silver sulfide quantum dots have also been prepared *in situ* using polyvinylpyrrolidone films [617]. Yang and Wang [618] adopted a different methodology, first synthesizing HgCdTe alloy quantum dots using an aqueous route, while Knappenberger, Jr [619] have synthesized HgTe quantum dots in methanol at 5ºC using G5 and G7 PAMAM debdrimers.

Layer-by-layer deposition is another method for quantum dots/polymer films formation is known as [620], which uses immersion of the substrates in ionic charged quantum dot solutions and polyelectrolytes [621, 622] and HgTe polyelectrolyte multilayer/quantum dot deposition on polystyrene sphere surfaces [623].

An important step toward several types of commercial optoelectronic and electronic devices lies in assembling quantum dots into larger scale assemblies [624 - 628]. A number of quantum dots deposition techniques are now well established, including optical amplifiers light sources and optoelectronic switches [572], photo-imaging components and IR photodetectors [629] and field effect transistors [630].

Applications in Solar Cells

The first quantum dot based solar cells (QDSSCs) were fabricated in order to replace organometallic dyes in DSSCs. Even if quantum dots offered great modulability in band alignment engineering and higher photostability, the most important disadvantage remains finding an appropriate electrolyte redox pair that prevents the quantum dots from degradation. Instead of using the commonly used I^-/I^{3-} couple, CdS-CdSe quantum dots have been combined with the S^{2-}/S_x^{2-} polysulfide pair; however inefficient with narrow bandgap quantum dots such as PbS [631].

For that reason Grätzel [632, 633] used an alternative redox couple based on cobalt complex (Co^{2+}-Co^{3+}) for their QDSSCs and DSSC.

Both Parkinson [634] and Wang [635] have combined CdS and PbS in a QDSSC with a polysulfide electrolyte and mesoporous SnO_2, passivated with a layer of ZnS to increase PbS stability in air, reporting a power conversion efficiency of 2.23%. Teng [636] reported an efficiency of 4.2% using a TiO_2-$CuInS_2$ quantum dot-ZnS-CdS photoanode, a CuS counter electrode structure and a polysulfide electrolyte.

When preparing layered QDSSCs, ethanedithiol linker is often used. Seok [637] used this linker to synthesize PbS quantum dot-coated mesoporous TiO_2 layers, where the quantum dot multilayers were deposited within TiO_2 by alternating PbS quantum dot and dilute ethanedithiol solutions in spin coating cycles. The device was completed upon adding a spin coated hole transport layer (poly(3-

hexylthiophene), and a poly(3,4-ethylenedioxythiophene) polystyrene sulfonate (PEDOT:PSS) layer, followed by an evaporated gold top contact.

The QDSSc displayed power conversion efficiency (PCE) of 2.9%, under 1 sun illumination. Additionally, Grätzel [638] joined nanoplatelet layers of TiO_2 and PbS quantum dots, forming a depleted heterojunction solar cell, displaying a PCE of 4.7%.

Sandwiching an Al electrode between Ag and LiF improved device lifetimes, with efficiency values of 2% being obtained. Using PbS quantum dot cells with similar electrode structures, Tao [639] and Loi [640] obtained PCE values of 3.8% and 3.9%, respectively.

The same authors [641] also studied a Schottky cell (ITO/PEDOT/PbSe/Al) containing PbSe quantum dots, obtaining efficiencies of 4.75% for 2.3 nm PbSe quantum dots, representing one of the best power conversion efficiencies.

The highest efficiency for a QDSSC is currently 6%, as reported by Sargent's group [642]. Their FTO-TiO_2-PbS-Au-Ag heterostructure comprised PbS quantum dots were passivated with an inorganic layer. PbS quantum dots were subsequently treated with a solution of $CdCl_2$•tetradecylphosphonic acid-oleic acid, in order to enhance the PbS quantum dot surface layer with cadmium ions.

EXPERIMENTAL SECTION

Synthesis of Functionalized Au Nanoparticles with Doxorubicin (Scheme 4)

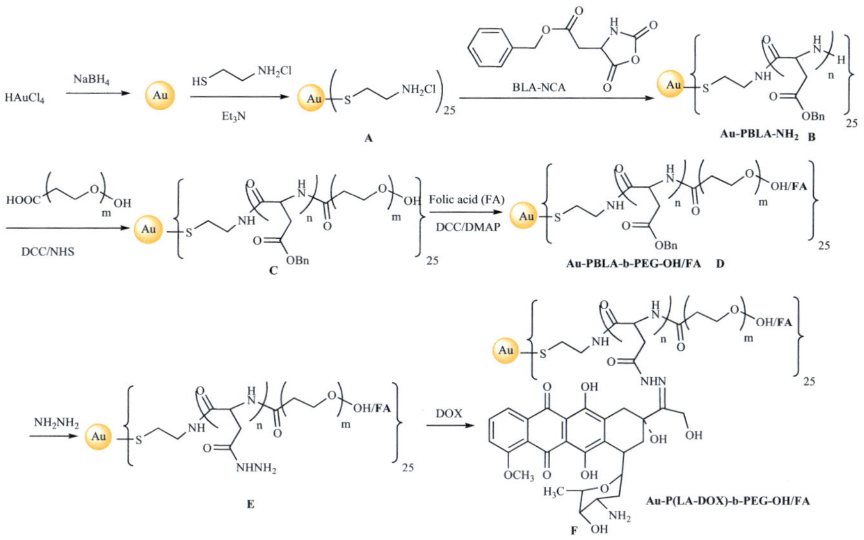

Scheme (4). Amine functionalized Au nanoparticles with doxorubicin. Procedure for total synthesis of gold nanoparticles F can be directly found in Reference [467].

Synthesis of Self-assembled Porphyrin Monolayers (Scheme 5)

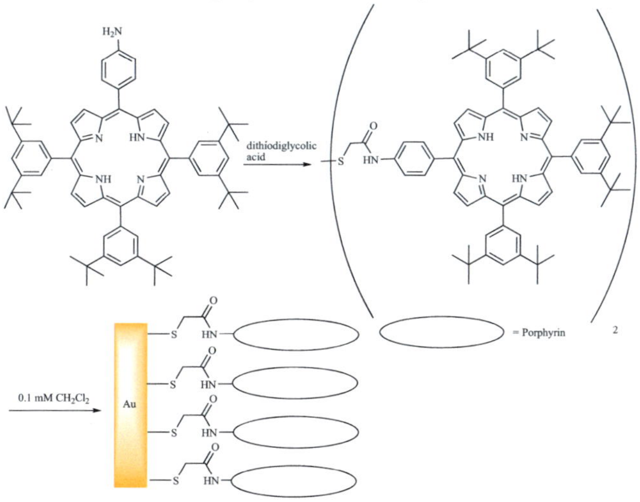

Scheme (5). Self-assembled monolayers of porphyrins on gold electrodes. Procedure for total synthesis of gold nanoparticles can be directly found in Reference [643].

Synthesis of Iron Based Magnetic Nanoparticles (Scheme 6)

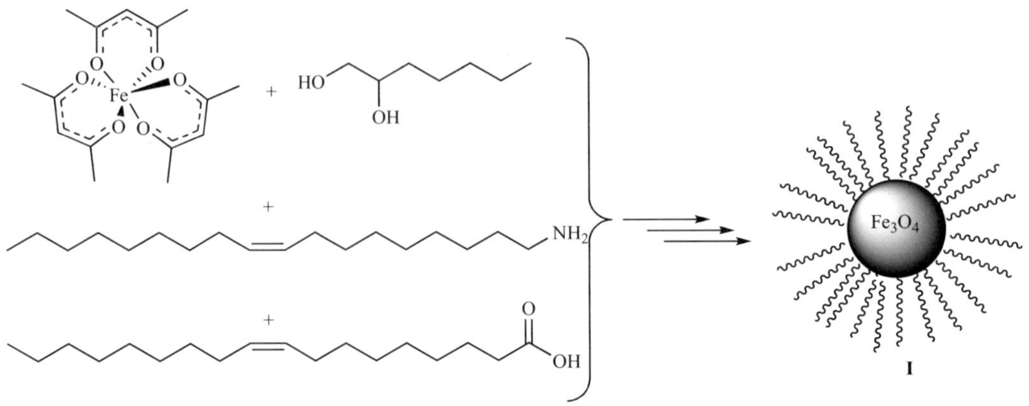

I

Scheme (6). Iron based magnetic nanoparticle synthesis. Procedure for total synthesis of magnetic nanoparticles can be directly found in Reference [144].

Synthesis of Hydrophilic Iron Magnetic Nanoparticles (Scheme 7)

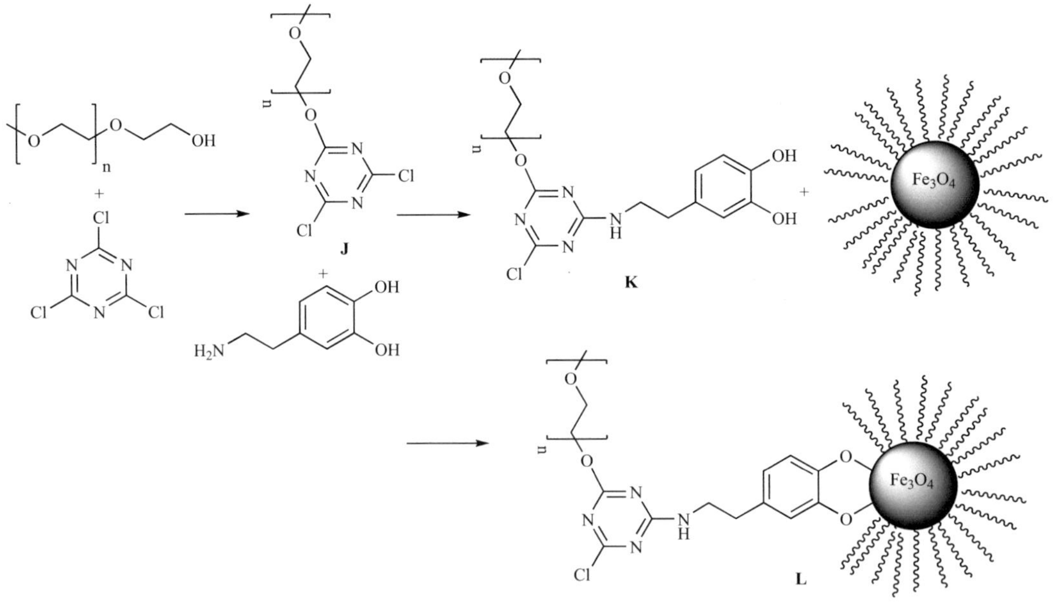

Scheme (7). Triazine functionalized iron magnetic hydrophilic nanoparticles. Procedure for total synthesis of magnetic nanoparticles L can be directly found in Reference [160].

Synthesis of cRGD-peptide Conjugated Near-infrared CdTe/ZnSe Core-shell Quantum Dots (Scheme 8)

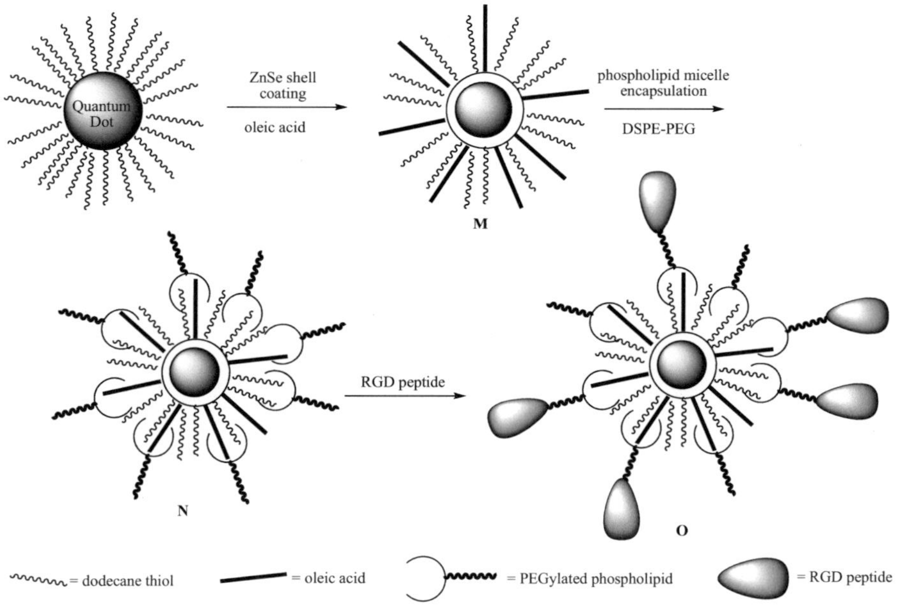

Scheme (8). cRGD-peptide conjugated CdTe/ZnSe quantum dots. Procedure for total synthesis of quantum dots O can be directly found in Reference [645].

Synthesis of Peptide Conjugated Quantum Dots (Scheme 9)

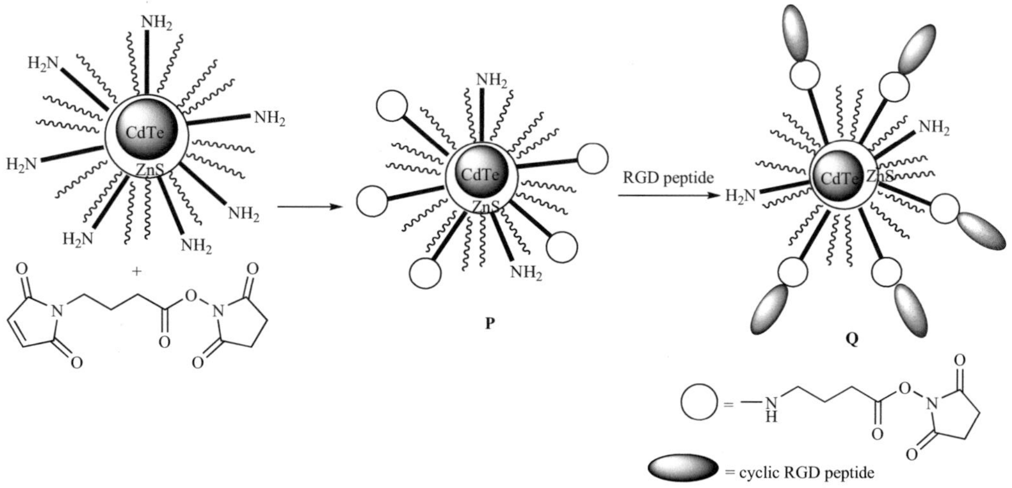

Scheme (9). Peptide conjugated quantum dots. Procedure for total synthesis of quantum dots Q can be directly found in Reference [646].

CONCLUDING REMARKS

Nanoparticles and their applications are probably some of the most fascinating areas of research for the next decade. Their inorganic core, with their nearly endless synthetic possibilities, makes these materials promising, regarding the use of organic chemistry for post-functionalization. The merging of inorganic chemistry with organic chemistry, in this particular case, can promote enthusiastic findings in several areas, such as biomedicine, catalysis, optoelectronics and in materials science, provided that standard procedures are applied, establishing structure-property relationships, in order to achieve the desired promising materials that this area of research should offer.

REFERENCES

[1] *The chemistry of nanomaterials: synthesis, properties and applications*; Cheetham, A.K.; Rao, C. N. R.; Muller, A., Eds.; Wiley-VCH: Weinheim, **2004**, Vol. 1, .

[2] Ozin, G.A.; Arsenault, A.C. *Nanochemistry: a chemical approach to nanomaterials*; Royal Society of Chemistry: Cambridge, **2005**.

[3] Sergeev, G.B. *Nanochemistry*; Elsevier: Amsterdam, **2006**.

[4] *Nanomaterials, chemistry: recent developments and new directions*; Cheetham, A.K.; Rao, C.N.R.; Müller, A., Eds.; Wiley-VCH: Weinheim, **2007**.

[5] *Nanomaterials and nanochemistry*; Lahmani, M.; Bréchignac, C.; Houdy, P., Eds.; Springer: Berlin, **2007**.

[6] Roucoux, A.; Schulz, J.; Patin, H. Reduced transition metal colloids: a novel family of reusable catalysts? *Chem. Rev.,* **2002**, *102*(10), 3757-3778.
[http://dx.doi.org/10.1021/cr010350j] [PMID: 12371901]

[7] Astruc, D.; Lu, F.; Aranzaes, J.R. Nanoparticles as recyclable catalysts: the frontier between homogeneous and heterogeneous catalysis. *Angew. Chem. Int. Ed. Engl.,* **2005**, *44*(48), 7852-7872.
[http://dx.doi.org/10.1002/anie.200500766] [PMID: 16304662]

[8] *Nanoparticles and catalysis*; Astruc, D., Ed.; Wiley-VCH: Weinheim, **2008**.

[9] Zhu, Y.; Lee, C.N.; Kemp, R.A.; Hosmane, N.S.; Maguire, J.A. Latest developments in the catalytic application of nanoscaled neutral group 8-10 metals. *Chem. Asian J.,* **2008**, *3*(4), 650-662.
[http://dx.doi.org/10.1002/asia.200800005] [PMID: 18348146]

[10] Faraday, M. Experimental relations of gold (and other metals) to light. *Philos. Trans. R. Soc. Lond.,* **1857**, *147*, 145-181.
[http://dx.doi.org/10.1098/rstl.1857.0011]

[11] Maxwell-Garnett, J.C. Colours in metal glasses and in metallic films. *Phil. Trans. Royal Soc. A,* **1904**, *203*, 385-420.
[http://dx.doi.org/10.1098/rsta.1904.0024]

[12] Mie, G. Sättigungsstrom und Stromkurve einer schlecht leitenden Flüssigkeit. *Ann. Phys.,* **1908**, *331*, 597-614.
[http://dx.doi.org/10.1002/andp.19083310810]

[13] Feynman, R. There is plenty of room at the bottom (talk tanscript). *Caltech Engineering and Science,* **1960**, *23*, 22-36.

[14] *On the Basic Concept of Nano-Technology Proceedings of the International Conference on Production Engineering*; Taniguchi, N., Ed.; Japan Society of Precision Engineering: Tokyo, **1974**. Part II.

[15] Henglein, A. Small-particle research: physicochemical properties of extremely small colloidal metal and semiconductor particles. *Chem. Rev.,* **1989**, *89*, 1861-1873.
[http://dx.doi.org/10.1021/cr00098a010]

[16] Henglein, A. Mechanism of reactions on colloidal microelectrodes and size quantization effects. *Top. Curr. Chem.,* **1988**, *143*, 113-180.
[http://dx.doi.org/10.1007/BFb0018073]

[17] Gratzel, M. *Energy resources through photochemistry and catalysis*; Academic Press: New York, **1983**.

[18] Schmid, G. Large clusters and colloids. Metals in the embryonic state. *Chem. Rev.,* **1992**, *92*, 1709-1727.
[http://dx.doi.org/10.1021/cr00016a002]

[19] Keily, C.J.; Fink, J.; Brust, M.; Bethell, D.; Schiffrin, D.J. Spontaneous ordering of bimodal ensembles of nanoscopic gold clusters. *Nature,* **1998**, *396*, 444-446.
[http://dx.doi.org/10.1038/24808]

[20] Pileni, M.P. Nanosized particles made in colloidal assemblies. *Langmuir,* **1997**, *13*, 3266-3276.
[http://dx.doi.org/10.1021/la960319q]

[21] Zhao, P.X.; Li, N.; Astruc, D. State of the art in gold nanoparticle synthesis. *Coord. Chem. Rev.,* **2013**, *257*, 638-665.
[http://dx.doi.org/10.1016/j.ccr.2012.09.002]

[22] *Clusters and Colloids. From theory to Applications*; Schmid, G., Ed.; VCH: Wienheim, **1994**.
[http://dx.doi.org/10.1002/9783527616077]

[23] Toshima, N.; Yonezawa, T. Bimetallic nanoparticles-novel materials for chemical and physical applications. *New J. Chem.,* **1998**, *22*, 1179-1201.
[http://dx.doi.org/10.1039/a805753b]

[24] Toshima, N.; Shiraishi, Y. Catalysis by metallic colloids. In: *Encyclopedia of Surface and Colloid Science*; Hubbard, A.T., Ed.; Marcel Dekker: New York, **2002**; pp. 879-886.

[25] Xiong, Y.; Chen, J.; Wiley, B.; Xia, Y.; Yin, Y.; Li, Z-Y. Size-dependence of surface plasmon resonance and oxidation for Pd nanocubes synthesized *via* a seed etching process. *Nano Lett.,* **2005**, *5*(7), 1237-1242.
[http://dx.doi.org/10.1021/nl0508826] [PMID: 16178217]

[26] Wiley, B.J.; Im, S.H.; Li, Z.Y.; McLellan, J.; Siekkinen, A.; Xia, Y. Maneuvering the surface plasmon

resonance of silver nanostructures through shape-controlled synthesis. *J. Phys. Chem. B,* **2006**, *110*(32), 15666-15675.
[http://dx.doi.org/10.1021/jp0608628] [PMID: 16898709]

[27] Thomas, S.; Nair, S.K.; Jamal, E.M.; Al-Harthi, S.H.; Varma, M.R.; Anantharaman, M.R. Size-dependent surface plasmon resonance in silver silica nanocomposites. *Nanotechnology,* **2008**, *19*(7), 075710.
[http://dx.doi.org/10.1088/0957-4484/19/7/075710] [PMID: 21817658]

[28] Yeshchenko, O.A.; Dmitruk, I.M.; Alexeenko, A.A.; Yu, M.; Losytskyy, A.V.; Kotko, A.; Pinchuk, O. Size-dependent surface-plasmon-enhanced photoluminescence from silver nanoparticles embedded in silica. *Phys. Rev. B,* **2009**, *79*, 235438-235439.
[http://dx.doi.org/10.1103/PhysRevB.79.235438]

[29] Link, S.; El-Sayed, M.A. Size and temperature dependence of the plasmon absorption of colloidal gold nanoparticles. *J. Phys. Chem. B,* **1999**, *103*, 4212-4217.
[http://dx.doi.org/10.1021/jp984796o]

[30] Hu, M.; Chen, J.; Li, Z-Y.; Au, L.; Hartland, G.V.; Li, X.; Marquez, M.; Xia, Y. Gold nanostructures: engineering their plasmonic properties for biomedical applications. *Chem. Soc. Rev.,* **2006**, *35*(11), 1084-1094.
[http://dx.doi.org/10.1039/b517615h] [PMID: 17057837]

[31] Ghosh, S.K.; Pal, T. Interparticle coupling effect on the surface plasmon resonance of gold nanoparticles: from theory to applications. *Chem. Rev.,* **2007**, *107*(11), 4797-4862.
[http://dx.doi.org/10.1021/cr0680282] [PMID: 17999554]

[32] El-Brolossy, T.A.; Abdallah, T.; Mohamed, M.B.; Abdallah, S.; Easawi, K.; Negm, S.; Talaat, H. Shape and size dependence of the surface plasmon resonance of gold nanoparticles studied by photoacoustic technique. *Eur. Phys. J. Spec. Top.,* **2008**, *153*, 361-364.
[http://dx.doi.org/10.1140/epjst/e2008-00462-0]

[33] Jeong, U.; Teng, X.; Wang, Y.; Yang, H.; Xia, Y. Superparamagnetic colloids: controlled synthesis and niche applications. *Adv. Mater.,* **2007**, *19*, 33-60.
[http://dx.doi.org/10.1002/adma.200600674]

[34] Lu, A-H.; Salabas, E.L.; Schüth, F. Magnetic nanoparticles: synthesis, protection, functionalization, and application. *Angew. Chem. Int. Ed. Engl.,* **2007**, *46*(8), 1222-1244.
[http://dx.doi.org/10.1002/anie.200602866] [PMID: 17278160]

[35] Mathew, D.S.; Juang, R-S. An overview of the structure and magnetism of spinel ferrite nanoparticles and their synthesis in microemulsions. *Chem. Eng. J.,* **2007**, *129*, 51-65.
[http://dx.doi.org/10.1016/j.cej.2006.11.001]

[36] Morrish, A.H. *The Physical Principles of Magnetism*; Wiley: New York, **1965**.

[37] Burda, C.; Chen, X.; Narayanan, R.; El-Sayed, M.A. Chemistry and properties of nanocrystals of different shapes. *Chem. Rev.,* **2005**, *105*(4), 1025-1102.
[http://dx.doi.org/10.1021/cr030063a] [PMID: 15826010]

[38] Walter, P.; Welcomme, E.; Hallégot, P.; Zaluzec, N.J.; Deeb, C.; Castaing, J.; Veyssière, P.; Bréniaux, R.; Lévêque, J.L.; Tsoucaris, G. Early use of PbS nanotechnology for an ancient hair dyeing formula.

Nano Lett., **2006**, *6*(10), 2215-2219.
[http://dx.doi.org/10.1021/nl061493u] [PMID: 17034086]

[39] Rocksby, H.P. Color of selenium ruby glasses. *J. Soc. Glass Technol.,* **1932**, *16*, 171-181.

[40] Efros, A.L.; Efros, A.L. Interband absorption of light in a semiconductor sphere. *Sov. Phys. Semicond.,* **1982**, *16*, 772-775.

[41] Rossetti, R.; Ellison, J.L.; Gibson, J.M.; Brus, L.E. Size effects in the excited electronic states of small colloidal CdS crystallites. *J. Chem. Phys.,* **1984**, *80*, 4464-4469.
[http://dx.doi.org/10.1063/1.447228]

[42] Henglein, A. Small-particle research−physicochemical properties of extremely small colloidal metal and semiconductor particles. *Chem. Rev.,* **1989**, *89*, 1861-1873.
[http://dx.doi.org/10.1021/cr00098a010]

[43] Sundaram, M.; Chalmers, S.A.; Hopkins, P.F.; Gossard, A.C. New quantum structures. *Science,* **1991**, *254*(5036), 1326-1335.
[http://dx.doi.org/10.1126/science.254.5036.1326] [PMID: 17773602]

[44] Cingolani, R.; Ploog, K. Frequency and density dependent radiative recombination processes in III-V semiconductor quantum-wells and superlattices. *Adv. Phys.,* **1991**, *40*, 535-623.
[http://dx.doi.org/10.1080/00018739100101522]

[45] Alivisatos, A.P. Semiconductor clusters, nanocrystals, and quantum dots. *Science,* **1996**, *271*, 933-937.
[http://dx.doi.org/10.1126/science.271.5251.933]

[46] Zhang, L.J.; Shen, X.C.; Liang, H.; Guo, S.; Liang, Z.H. Hot-injection synthesis of highly luminescent and monodisperse CdS nanocrystals using thioacetamide and cadmium source with proper reactivity. *J. Colloid Interface Sci.,* **2010**, *342*(2), 236-242.
[http://dx.doi.org/10.1016/j.jcis.2009.10.030] [PMID: 19932894]

[47] Kuçur, E.; Boldt, F.M.; Cavaliere-Jaricot, S.; Ziegler, J.; Nann, T. Quantitative analysis of cadmium selenide nanocrystal concentration by comparative techniques. *Anal. Chem.,* **2007**, *79*(23), 8987-8993.
[http://dx.doi.org/10.1021/ac0715064] [PMID: 17973353]

[48] Xie, R.; Peng, X. Synthesis of Cu-doped InP nanocrystals (d-dots) with ZnSe diffusion barrier as efficient and color-tunable NIR emitters. *J. Am. Chem. Soc.,* **2009**, *131*(30), 10645-10651.
[http://dx.doi.org/10.1021/ja903558r] [PMID: 19588970]

[49] Moreels, I.; Fritzinger, B.; Martins, J.C.; Hens, Z. Surface chemistry of colloidal PbSe nanocrystals. *J. Am. Chem. Soc.,* **2008**, *130*(45), 15081-15086.
[http://dx.doi.org/10.1021/ja803994m] [PMID: 18928251]

[50] Michalet, X.; Pinaud, F.F.; Bentolila, L.A.; Tsay, J.M.; Doose, S.; Li, J.J.; Sundaresan, G.; Wu, A.M.; Gambhir, S.S.; Weiss, S. Quantum dots for live cells, *in vivo* imaging, and diagnostics. *Science,* **2005**, *307*(5709), 538-544.
[http://dx.doi.org/10.1126/science.1104274] [PMID: 15681376]

[51] Graham-Rowe, D. From dots to devices. *Nat. Photonics,* **2009**, *3*, 307-309.
[http://dx.doi.org/10.1038/nphoton.2009.79]

[52] Coe-Sullivan, S. Quantum dot developments. *Nat. Photonics,* **2009**, *3*, 315-316. [http://dx.doi.org/10.1038/nphoton.2009.83]

[53] Hahn, H. Gas phase synthesis of nanocrystalline materials. *Nanostruct. Mater.,* **1997**, *9*, 3-12. [http://dx.doi.org/10.1016/S0965-9773(97)00013-5]

[54] Toshima, N. Reactions in homogeneous solutions. In: *Fine Particle. Synthesis, Characterization, and Mechanism of Growth*; Sugimoto, T., Ed.; Marcel Dekker: New York, **2000**; pp. 430-459.

[55] Chen, R.; So, M-H.; Che, C-M.; Sun, H. Controlled synthesis of high crystalline bismuth sulfide nanorods: using bismuth citrate as a precursor. *J. Mater. Chem.,* **2005**, *15*, 4540-4545. [http://dx.doi.org/10.1039/b510299e]

[56] Chen, R.; So, M-H.; Yang, J.; Deng, F.; Che, C-M.; Sun, H. Fabrication of bismuth subcarbonate nanotube arrays from bismuth citrate. *Chem. Commun. (Camb.),* **2006**, (21), 2265-2267. [http://dx.doi.org/10.1039/b601764a] [PMID: 16718324]

[57] Wallraff, G.M.; Hinsberg, W.D. Lithographic imaging techniques for the formation of nanoscopic features. *Chem. Rev.,* **1999**, *99*(7), 1801-1822. [http://dx.doi.org/10.1021/cr980003i] [PMID: 11849011]

[58] Swihart, M.T. Vapor-phase synthesis of nanoparticles. *Curr. Opin. Colloid Interface Sci.,* **2003**, *8*, 127-133. [http://dx.doi.org/10.1016/S1359-0294(03)00007-4]

[59] Wu, Y.; Yang, P. Direct observation of vapor−liquid−solid nanowire growth. *J. Am. Chem. Soc.,* **2001**, *123*, 3165-3166. [http://dx.doi.org/10.1021/ja0059084]

[60] Ge, J.; Li, Y. Controllable CVD route to CoS and MnS single-crystal nanowires. *Chem. Commun. (Camb.),* **2003**, (19), 2498-2499. [http://dx.doi.org/10.1039/b307452h] [PMID: 14587750]

[61] Ge, J.P.; Wang, J.; Zhang, H.X.; Li, Y.D. A general atmospheric pressure chemical vapor deposition synthesis and crystallographic study of transition-metal sulfide one-dimensional nanostructures. *Chemistry,* **2004**, *10*(14), 3525-3530. [http://dx.doi.org/10.1002/chem.200400008] [PMID: 15252798]

[62] Ge, J.P.; Li, Y.D. Selective atmospheric pressure chemical vapor deposition route to CdS arrays, nanowires, and nanocombs. *Adv. Funct. Mater.,* **2004**, *14*, 157-162. [http://dx.doi.org/10.1002/adfm.200305051]

[63] Li, X.L.; Ge, J.P.; Li, Y.D. Atmospheric pressure chemical vapor deposition: an alternative route to large-scale MoS2 and WS2 inorganic fullerene-like nanostructures and nanoflowers. *Chemistry,* **2004**, *10*(23), 6163-6171. [http://dx.doi.org/10.1002/chem.200400451] [PMID: 15515074]

[64] Ge, J.P.; Wang, J.; Zhang, H.X.; Wang, X.; Peng, Q.; Li, Y.D. Halide-transport chemical vapor deposition of luminescent ZnS:Mn2+ one-dimensional nanostructures. *Adv. Funct. Mater.,* **2005**, *15*, 303-308. [http://dx.doi.org/10.1002/adfm.200400078]

[65] Jin, R.; Cao, Y.; Mirkin, C.A.; Kelly, K.L.; Schatz, G.C.; Zheng, J.G. Photoinduced conversion of silver nanospheres to nanoprisms. *Science,* **2001**, *294*(5548), 1901-1903. [http://dx.doi.org/10.1126/science.1066541] [PMID: 11729310]

[66] Cushing, B.L.; Kolesnichenko, V.L.; O'Connor, C.J. Recent advances in the liquid-phase syntheses of inorganic nanoparticles. *Chem. Rev.,* **2004**, *104*(9), 3893-3946. [http://dx.doi.org/10.1021/cr030027b] [PMID: 15352782]

[67] Yin, Y.; Alivisatos, A.P. Colloidal nanocrystal synthesis and the organic-inorganic interface. *Nature,* **2005**, *437*(7059), 664-670. [http://dx.doi.org/10.1038/nature04165] [PMID: 16193041]

[68] Wang, X.; Zhuang, J.; Peng, Q.; Li, Y. A general strategy for nanocrystal synthesis. *Nature,* **2005**, *437*(7055), 121-124. [http://dx.doi.org/10.1038/nature03968] [PMID: 16136139]

[69] Murphy, C.J.; Sau, T.K.; Gole, A.M.; Orendorff, C.J.; Gao, J.; Gou, L.; Hunyadi, S.E.; Li, T. Anisotropic metal nanoparticles: Synthesis, assembly, and optical applications. *J. Phys. Chem. B,* **2005**, *109*(29), 13857-13870. [http://dx.doi.org/10.1021/jp0516846] [PMID: 16852739]

[70] Park, J.; Joo, J.; Kwon, S.G.; Jang, Y.; Hyeon, T. Synthesis of monodisperse spherical nanocrystals. *Angew. Chem. Int. Ed. Engl.,* **2007**, *46*(25), 4630-4660. [http://dx.doi.org/10.1002/anie.200603148] [PMID: 17525914]

[71] Wang, D.; Xie, T.; Li, Y. Nanocrystals: Solution-based synthesis and applications as nanocatalysts. *Nano Res.,* **2009**, *2*, 30-46. [http://dx.doi.org/10.1007/s12274-009-9007-x]

[72] Mokari, T.; Sztrum, C.G.; Salant, A.; Rabani, E.; Banin, U. Formation of asymmetric one-sided metal-tipped semiconductor nanocrystal dots and rods. *Nat. Mater.,* **2005**, *4*, 855-863. [http://dx.doi.org/10.1038/nmat1505]

[73] Park, J.; Privman, V.; Matijević, E. Model of formation of monodispersed colloids. *J. Phys. Chem. B,* **2001**, *105*, 11630-11635. [http://dx.doi.org/10.1021/jp011306a]

[74] Hunter, R.J. *Foundations of Colloid Science*; Oxford University Press: New York, **1987**.

[75] Napper, D.H. *Polymeric Stabilization of Colloidal Dispersions*; Academic Press: London, **1983**.

[76] Turkevich, J.; Stevenson, P.C.; Hillier, J. A study of the nucleation and growth processes in the synthesis of colloidal gold. *Discuss. Faraday Soc.,* **1951**, 55-75. [http://dx.doi.org/10.1039/df9511100055]

[77] Frens, G. Controlled nucleation for the regulation of the particle size in monodisperse gold suspensions. *Nat. Phys. Sci (Lond.),* **1973**, *241*, 20-22. [http://dx.doi.org/10.1038/physci241020a0]

[78] Furlong, D.N.; Launikonis, A.; Sasse, W.H.; Sanders, J.V. Colloidal platinum sols. Preparation, characterization and stability towards salt. *J. Chem. Soc., Faraday Trans. I,* **1984**, *80*, 571-588. [http://dx.doi.org/10.1039/f19848000571]

[79] Lee, P.C.; Meisel, D. Adsorption and surface-enhanced Raman of dyes on silver and gold sols. *J. Phys. Chem.,* **1982**, *86*, 3391-3395.
[http://dx.doi.org/10.1021/j100214a025]

[80] Rivas, L.; Sanchez-Cortes, S.; García-Ramos, J.V.; Morcillo, G. Growth of silver colloidal particles obtained by citrate reduction to increase the Raman enhancement factor. *Langmuir,* **2001**, *17*, 574-577.
[http://dx.doi.org/10.1021/la001038s]

[81] Pillai, Z.S.; Kamat, P.V. What factors control the size and shape of silver nanoparticles in the citrate ion reduction method? *J. Phys. Chem. B,* **2004**, *108*, 945-951.
[http://dx.doi.org/10.1021/jp037018r]

[82] Dong, X.; Ji, X.; Wu, H.; Zhao, L.; Li, J.; Yang, W. Shape control of silver nanoparticles by stepwise citrate reduction. *J. Phys. Chem. C,* **2009**, *113*, 6573-6576.
[http://dx.doi.org/10.1021/jp900775b]

[83] Tan, Y.; Dai, X.; Li, Y.; Zhu, D. Preparation of gold, platinum, palladium and silver nanoparticles by the reduction of their salts with a weak reductant–potassium bitartrate. *J. Mater. Chem.,* **2003**, *13*, 1069-1075.
[http://dx.doi.org/10.1039/b211386d]

[84] Sondi, I.; Goia, D.V.; Matijević, E. Preparation of highly concentrated stable dispersions of uniform silver nanoparticles. *J. Colloid Interface Sci.,* **2003**, *260*(1), 75-81.
[http://dx.doi.org/10.1016/S0021-9797(02)00205-9] [PMID: 12742036]

[85] Liu, J.; He, F.; Gunn, T.M.; Zhao, D.; Roberts, C.B. Precise seed-mediated growth and size-controlled synthesis of palladium nanoparticles using a green chemistry approach. *Langmuir,* **2009**, *25*(12), 7116-7128.
[http://dx.doi.org/10.1021/la900228d] [PMID: 19309120]

[86] Glavee, G.N.; Klabunde, K.J.; Sorensen, C.M.; Hadjipanayis, G.C. Borohydride reduction of cobalt ions in water. Chemistry leading to nanoscale metal, boride, or borate particles. *Langmuir,* **1993**, *9*, 162-169.
[http://dx.doi.org/10.1021/la00025a034]

[87] Gui, Z.; Fan, R.; Mo, W.; Chen, X.; Yang, L.; Hu, Y. Synthesis and characterization of reduced transition metal oxides and nanophase metals with hydrazine in aqueous solution. *Mater. Res. Bull.,* **2003**, *38*, 169-176.
[http://dx.doi.org/10.1016/S0025-5408(02)00983-2]

[88] Chen, D-H.; Hsieh, C-H. Synthesis of nickel nanoparticles in aqueous cationic surfactant solutions. *J. Mater. Chem.,* **2002**, *12*, 2412-2415.
[http://dx.doi.org/10.1039/b200603k]

[89] Hou, Y.; Gao, S. Monodisperse nickel nanoparticles prepared from a monosurfactant system and their magnetic properties. *J. Mater. Chem.,* **2003**, *13*, 1510-1512.
[http://dx.doi.org/10.1039/b303226d]

[90] Hirai, H.; Wakabayashi, H.; Komiyama, M. Polymer-protected copper colloids as catalysts for selective hydration of acrylonitrile. *Chem. Lett.,* **1983**, *12*, 1047-1050.

[http://dx.doi.org/10.1246/cl.1983.1047]

[91] Larpent, C.; Brisse-Le Menn, F.; Patin, H. New highly water-soluble surfactants stabilize colloidal rhodium(0) suspensions useful in biphasic catalysis. *J. Mol. Catal.,* **1991**, *65*, L35-L40. [http://dx.doi.org/10.1016/0304-5102(91)85061-6]

[92] Larpent, C.; Bernard, E.; Brisse-Le Menn, F.; Patin, H. Biphasic liquid-liquid hydrogenation catalysis by aqueous colloidal suspensions of rhodium: The choice of the protective-colloid agent and the role of interfacial phenomena. *J. Mol. Catal. Chem.,* **1997**, *116*, 277-288. [http://dx.doi.org/10.1016/S1381-1169(96)00189-6]

[93] Mayer, A.B.; Antonietti, M. Investigation of polymer-protected noble metal nanoparticles by transmission electron microscopy: control of particle morphology and shape. *Colloid Polym. Sci.,* **1998**, *276*, 769-779. [http://dx.doi.org/10.1007/s003960050309]

[94] Mayer, A.B.; Mark, J.E. Comparisons between cationic polyelectrolytes and nonionic polymers for the protection of palladium and platinum nanocatalysts. *J. Polym. Sci. A Polym. Chem.,* **1997**, *35*, 3151-3160. [http://dx.doi.org/10.1002/(SICI)1099-0518(19971115)35:15<3151::AID-POLA8>3.0.CO;2-Z]

[95] Mayer, A.B.; Mark, J.E. Transition metal nanoparticles protected by amphiphilic block copolymers as tailored catalyst systems. *Colloid Polym. Sci.,* **1997**, *275*, 333-340. [http://dx.doi.org/10.1007/s003960050090]

[96] Mayer, A.B.; Mark, J.E.; Morris, R.E. Palladium and platinum nanocatalysts protected by amphiphilic block copolymers. *Polym. J.,* **1998**, *30*, 197-205. [http://dx.doi.org/10.1295/polymj.30.197]

[97] Mayer, A.B.; Johnson, R.W.; Hausner, S.H.; Mark, J.E. Colloidal silver nanoparticles protected by water-soluble nonionic polymers and "soft" polyacids. *J. Macromol. Sci. Pure Appl. Chem.,* **1999**, *36*, 1427-1441.

[98] Yonezawa, T.; Onoue, S-Y.; Kimizuka, N. Preparation of highly positively charged silver nanoballs and their stability. *Langmuir,* **2000**, *16*, 5218-5220. [http://dx.doi.org/10.1021/la000186f]

[99] Mayer, A.B.; Hausner, S.H.; Mark, J.E. Colloidal silver nanoparticles generated in the presence of protective cationic polyelectrolytes. *Polym. J.,* **2000**, *32*, 15-22. [http://dx.doi.org/10.1295/polymj.32.15]

[100] Mayer, A.B.; Mark, J.E. Colloidal gold nanoparticles protected by water-soluble homopolymers and random copolymers. *Eur. Polym. J.,* **1998**, *34*, 103-108. [http://dx.doi.org/10.1016/S0014-3057(97)00074-8]

[101] Bönnemann, H.; Brijoux, W.; Joussen, T. The preparation of finely divided metal and alloy powders. *Angew. Chem. Int. Ed. Engl.,* **1990**, *29*, 273-275. [http://dx.doi.org/10.1002/anie.199002731]

[102] Bönnemann, H.; Brijoux, W.; Brinkmann, R.; Dinjus, E.; Jouben, T.; Korall, B. Formation of colloidal transition metals in organic phases and their application in catalysis. *Angew. Chem. Int. Ed. Engl.,* **1991**, *30*, 1312-1314.

[http://dx.doi.org/10.1002/anie.199113121]

[103] Bönnemann, H.; Bogdanovic, B.; Brinkmann, R.; Spliethoff, B.; He, D-W. The preparation of finely divided metal powders and transition metal complexes using "organically solvated" magnesium. *J. Organomet. Chem.,* **1993**, *451*, 23-31. [http://dx.doi.org/10.1016/0022-328X(93)83004-F]

[104] Viau, G.; Fievet-Vincent, F.; Fievet, F. Nucleation and growth of bimetallic CoNi and FeNi monodisperse particles prepared in polyols. *Solid State Ion.,* **1996**, *84*, 259-270. [http://dx.doi.org/10.1016/0167-2738(96)00005-7]

[105] Ayyappan, S.; Gopalan, R.S.; Subbanna, G.N.; Rao, C.N. Nanoparticles of Ag, Au, Pd, and Cu produced by alcohol reduction of the salts. *J. Mater. Res.,* **1997**, *12*, 398-401. [http://dx.doi.org/10.1557/JMR.1997.0057]

[106] Teranishi, T.; Hosoe, M.; Tanaka, T.; Miyake, M. Size control of monodispersed Pt nanoparticles and their 2D organization by electrophoretic deposition. *J. Phys. Chem. B,* **1999**, *103*, 3818-3827. [http://dx.doi.org/10.1021/jp983478m]

[107] Wang, Y.; Ren, J.; Deng, K.; Gui, L.; Tang, Y. Preparation of tractable platinum, rhodium, and ruthenium nanoclusters with small particle size in organic media. *Chem. Mater.,* **2000**, *12*, 1622-1627. [http://dx.doi.org/10.1021/cm0000853]

[108] Viau, G.; Brayner, R.; Poul, L.; Chakroune, N.; Lacaze, E.; Fievet-Vincent, F.; Fievet, F. Ruthenium nanoparticles: size, shape, and self-assemblies. *Chem. Mater.,* **2003**, *15*, 486-494. [http://dx.doi.org/10.1021/cm0212109]

[109] Wiley, B.; Herricks, T.; Sun, Y.; Xia, Y. Polyol synthesis of silver nanoparticles: use of chloride and oxygen to promote the formation of single-crystal, truncated cubes and tetrahedrons. *Nano Lett.,* **2004**, *4*, 1733-1739. [http://dx.doi.org/10.1021/nl048912c]

[110] Wiley, B.; Sun, Y.; Mayers, B.; Xia, Y. Shape-controlled synthesis of metal nanostructures: the case of silver. *Chemistry,* **2005**, *11*(2), 454-463. [http://dx.doi.org/10.1002/chem.200400927] [PMID: 15565727]

[111] Kim, K-S.; Choi, S.; Cha, J-H.; Yeon, S-H.; Lee, H. Facile one-pot synthesis of gold nanoparticles using alcohol ionic liquids. *J. Mater. Chem.,* **2006**, *16*, 1315-1317. [http://dx.doi.org/10.1039/b601478j]

[112] Zhang, Z.; Guo, L.; Wang, W. Synthesis and characterization of antimony oxide nanoparticles Synthesis and characterization of antimony oxide nanoparticles. *J. Mater. Res.,* **2001**, *16*, 803-805. [http://dx.doi.org/10.1557/JMR.2001.0096]

[113] Qian, Y.T.; Gu, Y.L.; Lu, J. *The chemistry of nanomaterials: synthesis, properties and applications, Rao, C.N.R.; Muller, A*; Cheetham, A.K., Ed.; Wiley-VCH: Weinheim, **2004**, Vol. 1, pp. 170-207.

[114] Demazeau, G. Solvothermal processes: a route to the stabilization of new materials. *J. Mater. Chem.,* **1999**, *9*, 15-18. [http://dx.doi.org/10.1039/a805536j]

[115] Cansell, F.; Chevalier, B.; Demourgues, A.; Etourneau, J.; Even, C.; Garrabos, Y.; Pessey, V.; Petit, S.; Tressaud, A.; Weill, F. Supercritical fluid processing: a new route for materials synthesis. *J. Mater.*

Chem., **1999**, *9*, 67-75.
[http://dx.doi.org/10.1039/a804964e]

[116] Li, J.; Chen, Z.; Wang, R-J.; Proserpio, D.M. Low temperature route towards new materials: solvothermal synthesis of metal chalcogenides in ethylenediamine. *Coord. Chem. Rev.,* **1999**, *190-192*, 707-735.
[http://dx.doi.org/10.1016/S0010-8545(99)00107-1]

[117] Rajamathi, M.; Seshadri, R. Oxide and chalcogenide nanoparticles from hydrothermal/solvothermal reactions. *Curr. Opin. Solid State Mater. Sci.,* **2002**, *6*, 337-345.
[http://dx.doi.org/10.1016/S1359-0286(02)00029-3]

[118] Xia, Y.; Yang, P.; Sun, Y.; Wu, Y.; Mayers, B.; Gates, B.; Yin, Y.; Kim, F.; Yan, H. One-dimensional nanostructures: synthesis, characterization, and applications. *Adv. Mater.,* **2003**, *15*, 353-389.
[http://dx.doi.org/10.1002/adma.200390087]

[119] Rao, C.N.; Deepak, F.L.; Gundiah, G.; Govindaraj, A. Inorganic nanowires. *Prog. Solid State Chem.,* **2003**, *31*, 5-147.
[http://dx.doi.org/10.1016/j.progsolidstchem.2003.08.001]

[120] Gao, S.; Zhang, J.; Zhu, Y-F.; Che, C-M. A convenient solvothermal route to ruthenium nanoparticles. *New J. Chem.,* **2000**, *24*, 739-740.
[http://dx.doi.org/10.1039/b004490n]

[121] Kittel, C. Theory of the structure of ferromagnetic domains in films and small particles. *Phys. Rev.,* **1946**, *70*, 965-971.
[http://dx.doi.org/10.1103/PhysRev.70.965]

[122] Bean, C.; Jacobs, I. Magnetic granulometry and superparamagnetism. *J. Appl. Phys.,* **1956**, *27*, 1448-1452.
[http://dx.doi.org/10.1063/1.1722287]

[123] Bean, C.; Livingston, J. Superparamagnetism. *J. Appl. Phys.,* **1959**, *30*, S120-S129.
[http://dx.doi.org/10.1063/1.2185850]

[124] Jacobs, I.; Bean, C. *Magnetism*; Rado, G. T.; Suhl, H., Eds.; Academic Press: New York, **1963**, Vol. III, pp. 271-350.

[125] Hyeon, T. Chemical synthesis of magnetic nanoparticles. *Chem. Commun. (Camb.),* **2003**, (8), 927-934.
[http://dx.doi.org/10.1039/b207789b] [PMID: 12744306]

[126] Gubin, S.P.; Koksharov, Y.A.; Khomutov, G.B.; Yurkov, G.Y. Magnetic nanoparticles: preparation, structure and properties. *Russ. Chem. Rev.,* **2005**, *74*, 489-520.
[http://dx.doi.org/10.1070/RC2005v074n06ABEH000897]

[127] Neveu, S.; Bee, A.; Robineau, M.; Talbot, D. Size-selective chemical synthesis of tartrate stabilized cobalt ferrite ionic magnetic fluid. *J. Colloid Interface Sci.,* **2002**, *255*(2), 293-298.
[http://dx.doi.org/10.1006/jcis.2002.8679] [PMID: 12505076]

[128] Grasset, F.; Labhsetwar, N.; Li, D.; Park, D.C.; Saito, N.; Haneda, H.; Cador, O.; Roisnel, T.; Mornet, S.; Duguet, E.; Portier, J.; Etourneau, J. Synthesis and magnetic characterization of zinc ferrite nanoparticles with different environments: powder, colloidal solution, and zinc ferrite–silica

core–shell nanoparticles. *Langmuir,* **2002**, *18*, 8209-8216. [http://dx.doi.org/10.1021/la020322b]

[129] Sun, S.; Zeng, H. Size-controlled synthesis of magnetite nanoparticles. *J. Am. Chem. Soc.,* **2002**, *124*(28), 8204-8205. [http://dx.doi.org/10.1021/ja026501x] [PMID: 12105897]

[130] Sun, S.; Murray, C.B.; Weller, D.; Folks, L.; Moser, A. Monodisperse FePt nanoparticles and ferromagnetic FePt nanocrystal superlattices. *Science,* **2000**, *287*(5460), 1989-1992. [http://dx.doi.org/10.1126/science.287.5460.1989] [PMID: 10720318]

[131] Shevchenko, E.V.; Talapin, D.V.; Rogach, A.L.; Kornowski, A.; Haase, M.; Weller, H. Colloidal synthesis and self-assembly of CoPt(3) nanocrystals. *J. Am. Chem. Soc.,* **2002**, *124*(38), 11480-11485. [http://dx.doi.org/10.1021/ja025976l] [PMID: 12236762]

[132] Chen, Q.; Rondinone, A.J.; Chakoumakos, B.C.; Zhang, Z.J. Synthesis of superparamagnetic MgFe2O4 nanoparticles by coprecipitation. *J. Magn. Magn. Mater.,* **1999**, *194*, 1-7. [http://dx.doi.org/10.1016/S0304-8853(98)00585-X]

[133] Park, J.; An, K.; Hwang, Y.; Park, J-G.; Noh, H-J.; Kim, J-Y.; Park, J-H.; Hwang, N-M.; Hyeon, T. Ultra-large-scale syntheses of monodisperse nanocrystals. *Nat. Mater.,* **2004**, *3*(12), 891-895. [http://dx.doi.org/10.1038/nmat1251] [PMID: 15568032]

[134] Park, S-J.; Kim, S.; Lee, S.; Khim, Z.; Char, K.; Hyeon, T. Synthesis and magnetic studies of uniform iron nanorods and nanospheres. *J. Am. Chem. Soc.,* **2000**, *122*, 8581-8582. [http://dx.doi.org/10.1021/ja001628c]

[135] Puntes, V.F.; Krishnan, K.M.; Alivisatos, A.P. Colloidal nanocrystal shape and size control: the case of cobalt. *Science,* **2001**, *291*(5511), 2115-2117. [http://dx.doi.org/10.1126/science.1057553] [PMID: 11251109]

[136] Lee, J.; Isobe, T.; Senna, M. Magnetic properties of ultrafine magnetite particles and their slurries prepared *via* in-situ precipitation. *Colloids Surf. A Physicochem. Eng. Asp.,* **1996**, *109*, 121-127. [http://dx.doi.org/10.1016/0927-7757(95)03479-X]

[137] Bee, A.; Massart, R.; Neveu, S. Synthesis of very fine maghemite particles. *J. Magn. Magn. Mater.,* **1995**, *149*, 6-9. [http://dx.doi.org/10.1016/0304-8853(95)00317-7]

[138] Ishikawa, T.; Kataoka, S.; Kandori, K. The influence of carboxylate ions on the growth of β-FeOOH particles. *J. Mater. Sci.,* **1993**, *28*, 2693-2698. [http://dx.doi.org/10.1007/BF00356205]

[139] Ishikawa, T.; Takeda, T.; Kandori, K. Effects of amines on the formation of β-ferric oxide hydroxide. *J. Mater. Sci.,* **1992**, *27*, 4531-4535. [http://dx.doi.org/10.1007/BF00541590]

[140] Kandori, K.; Kawashima, Y.; Ishikawa, T. Effects of citrate ions on the formation of monodispersed cubic hematite particles. *J. Colloid Interface Sci.,* **1992**, *152*, 284-288. [http://dx.doi.org/10.1016/0021-9797(92)90028-K]

[141] Cornell, R.M.; Schindler, P.W. Infrared study of the adsorption of hydroxycarboxylic acids onα-FeOOH and amorphous Fe (III)hydroxide. *Colloid Polym. Sci.,* **1980**, *258*, 1171-1175.

[http://dx.doi.org/10.1007/BF01382462]

[142] Willis, A.L.; Turro, N.; O'Brien, J.S. Spectroscopic characterization of the surface of iron oxide nanocrystals. *Chem. Mater.,* **2005**, *17*, 5970-5975. [http://dx.doi.org/10.1021/cm051370v]

[143] Cushing, B.L.; Kolesnichenko, V.L.; O'Connor, C.J. Recent advances in the liquid-phase syntheses of inorganic nanoparticles. *Chem. Rev.,* **2004**, *104*(9), 3893-3946. [http://dx.doi.org/10.1021/cr030027b] [PMID: 15352782]

[144] Sun, S.; Zeng, H.; Robinson, D.B.; Raoux, S.; Rice, P.M.; Wang, S.X.; Li, G. Monodisperse MFe2O4 (M = Fe, Co, Mn) nanoparticles. *J. Am. Chem. Soc.,* **2004**, *126*(1), 273-279. [http://dx.doi.org/10.1021/ja0380852] [PMID: 14709092]

[145] Redl, F.X.; Black, C.T.; Papaefthymiou, G.C.; Sandstrom, R.L.; Yin, M.; Zeng, H.; Murray, C.B.; O'Brien, S.P. Magnetic, electronic, and structural characterization of nonstoichiometric iron oxides at the nanoscale. *J. Am. Chem. Soc.,* **2004**, *126*(44), 14583-14599. [http://dx.doi.org/10.1021/ja046808r] [PMID: 15521779]

[146] Rockenberger, J.; Scher, E.C.; Alivisatos, A.P. A new nonhydrolytic single-precursor approach to surfactant-capped nanocrystals of transition metal oxides. *J. Am. Chem. Soc.,* **1999**, *121*, 11595-11596. [http://dx.doi.org/10.1021/ja993280v]

[147] Farrell, D.; Majetich, S.A.; Wilcoxon, J.P. Preparation and characterization of monodisperse Fe nanoparticles. *J. Phys. Chem. B,* **2003**, *107*, 11022-11030. [http://dx.doi.org/10.1021/jp0351831]

[148] Li, Y.; Afzaal, M.; O'Brien, P. The synthesis of amine-capped magnetic (Fe, Mn, Co, Ni) oxide nanocrystals and their surface modification for aqueous dispersibility. *J. Mater. Chem.,* **2006**, *16*, 2175-2180. [http://dx.doi.org/10.1039/b517351e]

[149] Samia, A.C.; Hyzer, K.; Schlueter, J.A.; Qin, C-J.; Jiang, J.S.; Bader, S.D.; Lin, X-M. Ligand effect on the growth and the digestion of Co nanocrystals. *J. Am. Chem. Soc.,* **2005**, *127*(12), 4126-4127. [http://dx.doi.org/10.1021/ja044419r] [PMID: 15783167]

[150] Jana, N.R.; Chen, Y.; Peng, X. Size- and shape-controlled magnetic (Cr, Mn, Fe, Co, Ni) oxide nanocrystals *via* a simple and general approach. *Chem. Mater.,* **2004**, *16*, 3931-3935. [http://dx.doi.org/10.1021/cm049221k]

[151] Suh, W.H.; Suslick, K.S.; Stucky, G.D.; Suh, Y.H. Nanotechnology, nanotoxicology, and neuroscience. *Prog. Neurobiol.,* **2009**, *87*(3), 133-170. [http://dx.doi.org/10.1016/j.pneurobio.2008.09.009] [PMID: 18926873]

[152] Kellar, K.E.; Fujii, D.K.; Gunther, W.H.; Briley-Saebo, K.; Bjornerud, A.; Spiller, M.; Koenig, S.H. Imaging of water and fat fractions in high-field MRI with multiple slice chemical shift-selective inversion recovery. *J. Magn. Reson. Imaging,* **2000**, *11*, 488-496. [http://dx.doi.org/10.1002/(SICI)1522-2586(200005)11:5<488::AID-JMRI4>3.0.CO;2-V] [PMID: 10813858]

[153] Zhao, X.; Harris, J.M. Novel degradable poly(ethylene glycol) hydrogels for controlled release of

protein. *J. Pharm. Sci.,* **1998**, *87*(11), 1450-1458.
[http://dx.doi.org/10.1021/js980065o] [PMID: 9811505]

[154] Jeong, Y.I.; Nah, J.W.; Na, H.K.; Na, K.; Kim, I.S.; Cho, C.S.; Kim, S.H. Self-assembling nanospheres of hydrophobized pullulans in water. *Drug Dev. Ind. Pharm.,* **1999**, *25*(8), 917-927.
[http://dx.doi.org/10.1081/DDC-100102252] [PMID: 10434135]

[155] Wu, W.; He, Q.; Jiang, C. Magnetic iron oxide nanoparticles: synthesis and surface functionalization strategies. *Nanoscale Res. Lett.,* **2008**, *3*(11), 397-415.
[http://dx.doi.org/10.1007/s11671-008-9174-9] [PMID: 21749733]

[156] Massia, S.P.; Stark, J.; Letbetter, D.S. Surface-immobilized dextran limits cell adhesion and spreading. *Biomaterials,* **2000**, *21*(22), 2253-2261.
[http://dx.doi.org/10.1016/S0142-9612(00)00151-4] [PMID: 11026631]

[157] De Palma, R.; Peeters, S.; Van Bael, M.J.; Van den Rul, H.; Bonroy, K.; Laureyn, W.; Mullens, J.B.; Orghs, G.; Maes, G. Silane ligand exchange to make hydrophobic superparamagnetic nanoparticles water-dispersible. *Chem. Mater.,* **2007**, *19*, 1821-1831.
[http://dx.doi.org/10.1021/cm0628000]

[158] Xu, C.; Xu, K.; Gu, H.; Zheng, R.; Liu, H.; Zhang, X.; Guo, Z.; Xu, B. Dopamine as a robust anchor to immobilize functional molecules on the iron oxide shell of magnetic nanoparticles. *J. Am. Chem. Soc.,* **2004**, *126*(32), 9938-9939.
[http://dx.doi.org/10.1021/ja0464802] [PMID: 15303865]

[159] Xie, J.; Xu, C.; Kohler, N.; Hou, Y.; Sun, S. Controlled PEGylation of monodisperse Fe3O4 nanoparticles for reduced non-specific uptake by macrophage cells. *Adv. Mater.,* **2007**, *19*, 3163-3166.
[http://dx.doi.org/10.1002/adma.200701975]

[160] Xie, J.; Xu, C.; Xu, Z.; Hou, Y.; Young, K.L.; Wang, S.; Pourmond, N.; Sun, S. Linking hydrophilic macromolecules to monodisperse magnetite (Fe3O4) nanoparticles *via* trichloro-s-triazine. *Chem. Mater.,* **2006**, *18*(23), 5401-5403.
[http://dx.doi.org/10.1021/cm061793c] [PMID: 18176627]

[161] Chen, Z.P.; Zhang, Y.; Zhang, S.; Xia, J.G.; Liu, J.W.; Xu, K.; Gu, N. Preparation and characterization of water-soluble monodisperse magnetic iron oxide nanoparticles *via* surface double-exchange with DMSA. *Colloids Surf. A Physicochem. Eng. Asp.,* **2008**, *316*, 210-216.
[http://dx.doi.org/10.1016/j.colsurfa.2007.09.017]

[162] Rossi, L.M.; Costa, N.J.; Silva, F.P.; Gonçalves, R.V. Magnetic nanocatalysts: supported metal nanoparticles for catalytic applications. *Nanotechnol. Rev.,* **2013**, *2*, 597-614.
[http://dx.doi.org/10.1515/ntrev-2013-0021]

[163] Laurent, S.; Forge, D.; Port, M.; Roch, A.; Robic, C.; Vander Elst, L.; Muller, R.N. Magnetic iron oxide nanoparticles: synthesis, stabilization, vectorization, physicochemical characterizations, and biological applications. *Chem. Rev.,* **2008**, *108*(6), 2064-2110.
[http://dx.doi.org/10.1021/cr068445e] [PMID: 18543879]

[164] Hao, R.; Xing, R.; Xu, Z.; Hou, Y.; Gao, S.; Sun, S. Synthesis, functionalization, and biomedical applications of multifunctional magnetic nanoparticles. *Adv. Mater.,* **2010**, *22*(25), 2729-2742.
[http://dx.doi.org/10.1002/adma.201000260] [PMID: 20473985]

[165] Lee, N.; Hyeon, T. Designed synthesis of uniformly sized iron oxide nanoparticles for efficient magnetic resonance imaging contrast agents. *Chem. Soc. Rev.,* **2012**, *41*(7), 2575-2589. [http://dx.doi.org/10.1039/C1CS15248C] [PMID: 22138852]

[166] Colombo, M.; Carregal-Romero, S.; Casula, M.F.; Gutiérrez, L.; Morales, M.P.; Böhm, I.B.; Heverhagen, J.T.; Prosperi, D.; Parak, W.J. Biological applications of magnetic nanoparticles. *Chem. Soc. Rev.,* **2012**, *41*(11), 4306-4334. [http://dx.doi.org/10.1039/c2cs15337h] [PMID: 22481569]

[167] Gallo, J.; Long, N.J.; Aboagye, E.O. Magnetic nanoparticles as contrast agents in the diagnosis and treatment of cancer. *Chem. Soc. Rev.,* **2013**, *42*(19), 7816-7833. [http://dx.doi.org/10.1039/c3cs60149h] [PMID: 23788179]

[168] Cohen, Y.; Shoushan, S.Y. Magnetic nanoparticles-based diagnostics and theranostics. *Curr. Opin. Biotechnol.,* **2013**, *24*(4), 672-681. [http://dx.doi.org/10.1016/j.copbio.2013.01.006] [PMID: 23415743]

[169] Kozissnik, B.; Bohorquez, A.C.; Dobson, J.; Rinaldi, C. Magnetic fluid hyperthermia: advances, challenges, and opportunity. *Int. J. Hyperthermia,* **2013**, *29*(8), 706-714. [http://dx.doi.org/10.3109/02656736.2013.837200] [PMID: 24106927]

[170] Thomas, R.; Park, I-K.; Jeong, Y.Y. Magnetic iron oxide nanoparticles for multimodal imaging and therapy of cancer. *Int. J. Mol. Sci.,* **2013**, *14*(8), 15910-15930. [http://dx.doi.org/10.3390/ijms140815910] [PMID: 23912234]

[171] Gawande, M.B.; Branco, P.S.; Varma, R.S. Nano-magnetite (Fe3O4) as a support for recyclable catalysts in the development of sustainable methodologies. *Chem. Soc. Rev.,* **2013**, *42*(8), 3371-3393. [http://dx.doi.org/10.1039/c3cs35480f] [PMID: 23420127]

[172] Chng, L.L.; Erathodiyil, N.; Ying, J.Y. Nanostructured catalysts for organic transformations. *Acc. Chem. Res.,* **2013**, *46*(8), 1825-1837. [http://dx.doi.org/10.1021/ar300197s] [PMID: 23350747]

[173] Govan, J.; Gun'ko, Y.K. Recent advances in the application of magnetic nanoparticles as a support for homogeneous catalysts. *Nanomater.,* **2014**, *4*, 222-241. [http://dx.doi.org/10.3390/nano4020222]

[174] Rossi, L.M.; Costa, N.J.; Silva, F.P.; Wojcieszak, R. Magnetic nanomaterials in catalysis: advanced catalysts for magnetic separation and beyond. *Green Chem.,* **2014**, *16*, 2906-2933. [http://dx.doi.org/10.1039/c4gc00164h]

[175] Cheng, T.; Zhang, D.; Li, H.; Liu, G. Magnetically recoverable nanoparticles as efficient catalysts for organic transformations in aqueous medium. *Green Chem.,* **2014**, *16*, 3401-3427. [http://dx.doi.org/10.1039/c4gc00458b]

[176] Tang, S.C.; Lo, I.M. Magnetic nanoparticles: essential factors for sustainable environmental applications. *Water Res.,* **2013**, *47*(8), 2613-2632. [http://dx.doi.org/10.1016/j.watres.2013.02.039] [PMID: 23515106]

[177] Gómez-Pastora, J.; Bringas, E.; Ortiz, I. Recent progress and future challenges on the use of high performance magnetic nano-adsorbents in environmental applications. *Chem. Eng. J.,* **2014**, *256*, 187-204.

[http://dx.doi.org/10.1016/j.cej.2014.06.119]

[178] Liu, Q.; Zhou, Q.; Jiang, G. Nanomaterials for analysis and monitoring of emerging chemical pollutants. *Trends Analyt. Chem.,* **2014**, *58*, 10-22. [http://dx.doi.org/10.1016/j.trac.2014.02.014]

[179] Saharan, P.; Chaudhary, G.R.; Mehta, S.K.; Umar, A. Removal of water contaminants by iron oxide nanomaterials. *J. Nanosci. Nanotechnol.,* **2014**, *14*(1), 627-643. [http://dx.doi.org/10.1166/jnn.2014.9053] [PMID: 24730287]

[180] Mattoussi, H.; Palui, G.; Na, H.B. Luminescent quantum dots as platforms for probing *in vitro* and *in vivo* biological processes. *Adv. Drug Deliv. Rev.,* **2012**, *64*(2), 138-166. [http://dx.doi.org/10.1016/j.addr.2011.09.011] [PMID: 21982955]

[181] Boxberg, F.; Tulkki, J. , **2004**.

[182] Scherer, A.; Craighead, H.G.; Beebe, E.D. Gallium-arsenide and aluminum gallium-arsenide reactive ion etching in boron-trichloride argon mixtures. *J. Vac. Sci. Technol. B,* **1987**, *5*, 1599-1605. [http://dx.doi.org/10.1116/1.583635]

[183] Tsutsui, K.; Hu, E.L.; Wilkinson, C.D. Reactive ion etched II-VI quantum dots–dependence of etched profile on pattern geometry. *Jpn. J. Appl. Phys. Part 1,* **1993**, *32*, 6233-6236.

[184] Chason, E.; Picraux, S.T.; Poate, J.M.; Borland, J.O.; Current, M.I.; delaRubia, T.D.; Eaglesham, D.J.; Holland, O.W.; Law, M.E.; Magee, C.W.; Mayer, J.W.; Melngailis, J.; Tasch, A.F. Ion beams in silicon processing and characterization. *J. Appl. Phys.,* **1997**, *81*, 6513-6561. [http://dx.doi.org/10.1063/1.365193]

[185] Leonard, D.; Krishnamurthy, M.; Reaves, C.M.; Denbaars, S.P.; Petroff, P.M. Direct formation of quantum-sized dots from uniform coherent islands of InGaAs on GaAs-surfaces. *Appl. Phys. Lett.,* **1993**, *63*, 3203-3205. [http://dx.doi.org/10.1063/1.110199]

[186] Xin, S.H.; Wang, P.D.; Yin, A.; Kim, C.; Dobrowolska, M.; Merz, J.L.; Furdyna, J.K. Formation of self-assembling CdSe quantum dots on ZnSe by molecular beam epitaxy. *Appl. Phys. Lett.,* **1996**, *69*, 3884-3886. [http://dx.doi.org/10.1063/1.117558]

[187] Leonardi, K.; Selke, H.; Heinke, H.; Ohkawa, K.; Hommel, D.; Gindele, F.; Woggon, U. Formation of self-assembling II-VI semiconductor nanostructures during migration enhanced epitaxy. *J. Cryst. Growth,* **1998**, *184*, 259-263. [http://dx.doi.org/10.1016/S0022-0248(98)80055-9]

[188] Kurtz, E.; Shen, J.; Schmidt, M.; Grun, M.; Hong, S.K.; Litvinov, D.; Gerthsen, D.; Oka, T.; Yao, T.; Klingshirn, C. Formation and properties of self-organized II-VI quantum islands. *Thin Solid Films,* **2000**, *367*, 68-74. [http://dx.doi.org/10.1016/S0040-6090(00)00665-9]

[189] Swihart, M.T. Vapor-phase synthesis of nanoparticles. *Curr. Opin. Colloid Interface Sci.,* **2003**, *8*, 127-133. [http://dx.doi.org/10.1016/S1359-0294(03)00007-4]

[190] Burda, C.; Chen, X.; Narayanan, R.; El-Sayed, M.A. Chemistry and properties of nanocrystals of

different shapes. *Chem. Rev.,* **2005**, *105*(4), 1025-1102.
[http://dx.doi.org/10.1021/cr030063a] [PMID: 15826010]

[191] Bang, J.; Yang, H.; Holloway, P.H. Enhanced and stable green emission of ZnO nanoparticles by surface segregation of Mg. *Nanotechnology,* **2006**, *17*(4), 973-978.
[http://dx.doi.org/10.1088/0957-4484/17/4/022] [PMID: 21727368]

[192] Bera, D.; Qian, L.; Sabui, S.; Santra, S.; Holloway, P.H. Photoluminescence of ZnO quantum dots produced by a sol-gel process. *Opt. Mater.,* **2008**, *30*, 1233-1239.
[http://dx.doi.org/10.1016/j.optmat.2007.06.001]

[193] Spanhel, L.; Anderson, M.A. Semiconductor clusters in the sol-gel process–quantized aggregation, gelation, and crystal-growth in concentrated ZnO colloids. *J. Am. Chem. Soc.,* **1991**, *113*, 2826-2833.
[http://dx.doi.org/10.1021/ja00008a004]

[194] Bera, D.; Qian, L.; Holloway, P.H. Time-evolution of photoluminescence properties of ZnO/MgO core/shell quantum dots. *J. Phys. D,* **2008**, *41*, 182002.
[http://dx.doi.org/10.1088/0022-3727/41/18/182002]

[195] Colvin, V.L.; Goldstein, A.N.; Alivisatos, A.P. Semiconductor nanocrystals covalently bound to metal-surfaces with self-assembled monolayers. *J. Am. Chem. Soc.,* **1992**, *114*, 5221-5230.
[http://dx.doi.org/10.1021/ja00039a038]

[196] Karanikolos, G.N.; Alexandridis, P.; Itskos, G.; Petrou, A.; Mountziaris, T.J. Synthesis and size control of luminescent ZnSe nanocrystals by a microemulsion-gas contacting technique. *Langmuir,* **2004**, *20*(3), 550-553.
[http://dx.doi.org/10.1021/la035397+] [PMID: 15773072]

[197] Ogawa, S.; Hu, K.; Fan, F.R.; Bard, A.J. Photoelectrochemistry of films of quantum size lead sulfide particles incorporated in self-assembled monolayers on gold. *J. Phys. Chem. B,* **1997**, *101*, 5707-5711.
[http://dx.doi.org/10.1021/jp970737j]

[198] Murray, C.B.; Norris, D.J.; Bawendi, M.G. Synthesis and characterization of nearly monodisperse CdE (E = S, Se, Te) semiconductor nanocrystallites. *J. Am. Chem. Soc.,* **1993**, *115*, 8706-8715.
[http://dx.doi.org/10.1021/ja00072a025]

[199] Qu, L.H.; Peng, Z.A.; Peng, X.G. Alternative routes toward high quality CdSe nanocrystals. *Nano Lett.,* **2001**, *1*, 333-337.
[http://dx.doi.org/10.1021/nl0155532]

[200] Zhu, J.J.; Koltypin, Y.; Gedanken, A. General sonochemical method far the preparation of nanophasic selenides: Synthesis of ZnSe nanoparticles. *Chem. Mater.,* **2000**, *12*, 73-78.
[http://dx.doi.org/10.1021/cm990380r]

[201] Qian, H.F.; Li, L.; Ren, J.C. One-step and rapid synthesis of high quality alloyed quantum dots (CdSe-CdS) in aqueous phase by microwave irradiation with controllable temperature. *Mater. Res. Bull.,* **2005**, *40*, 1726-1736.
[http://dx.doi.org/10.1016/j.materresbull.2005.05.022]

[202] Wang, L.C.; Chen, L.Y.; Luo, T.; Qian, Y.T. A hydrothermal method to prepare the spherical ZnS and flower-like CdS microcrystallites. *Mater. Lett.,* **2006**, *60*, 3627-3630.
[http://dx.doi.org/10.1016/j.matlet.2006.03.072]

[203] Yang, H.Q.; Yin, W.Y.; Zhao, H.; Yang, R.L.; Song, Y.Z. A complexant-assisted hydrothermal procedure for growing well-dispersed InP nanocrystals. *J. Phys. Chem. Solids,* **2008**, *69*, 1017-1022. [http://dx.doi.org/10.1016/j.jpcs.2007.11.017]

[204] Xie, Y.; Qian, Y.; Wang, W.; Zhang, S.; Zhang, Y. A benzene-thermal synthetic route to nanocrystalline. GaN. *Science,* **1996**, *272*(5270), 1926-1927. [http://dx.doi.org/10.1126/science.272.5270.1926] [PMID: 8662493]

[205] Dayal, S.; Burda, C. Surface effects on quantum dot-based energy transfer. *J. Am. Chem. Soc.,* **2007**, *129*(25), 7977-7981. [http://dx.doi.org/10.1021/ja071457c] [PMID: 17547406]

[206] Medintz, I.L.; Uyeda, H.T.; Goldman, E.R.; Mattoussi, H. Quantum dot bioconjugates for imaging, labelling and sensing. *Nat. Mater.,* **2005**, *4*(6), 435-446. [http://dx.doi.org/10.1038/nmat1390] [PMID: 15928695]

[207] Mittal, R.; Bruchez, M.P. Biotin-4-fluorescein based fluorescence quenching assay for determination of biotin binding capacity of streptavidin conjugated quantum dots. *Bioconjug. Chem.,* **2011**, *22*(3), 362-368. [http://dx.doi.org/10.1021/bc100321c] [PMID: 21314110]

[208] Shao, D.; Zeng, Q.; Fan, Z.; Li, J.; Zhang, M.; Zhang, Y.; Li, O.; Chen, L.; Kong, X.; Zhang, H. Monitoring HSV-TK/ganciclovir cancer suicide gene therapy using CdTe/CdS core/shell quantum dots. *Biomaterials,* **2012**, *33*(17), 4336-4344. [http://dx.doi.org/10.1016/j.biomaterials.2012.02.058] [PMID: 22440046]

[209] Schipper, M.L.; Iyer, G.; Koh, A.L.; Cheng, Z.; Ebenstein, Y.; Aharoni, A.; Keren, S.; Bentolila, L.A.; Li, J.; Rao, J.; Chen, X.; Banin, U.; Wu, A.M.; Sinclair, R.; Weiss, S.; Gambhir, S.S. Particle size, surface coating, and PEGylation influence the biodistribution of quantum dots in living mice. *Small,* **2009**, *5*(1), 126-134. [http://dx.doi.org/10.1002/smll.200800003] [PMID: 19051182]

[210] Dubois, F.; Mahler, B.; Dubertret, B.; Doris, E.; Mioskowski, C. A versatile strategy for quantum dot ligand exchange. *J. Am. Chem. Soc.,* **2007**, *129*(3), 482-483. [http://dx.doi.org/10.1021/ja067742y] [PMID: 17226998]

[211] Bonnemann, H.; Nagabhushana, K.S. Advantageous fuel cell catalysts from colloidal nanometals. *J. New Mater. Electrochem. Syst.,* **2004**, *7*, 93-108.

[212] Liu, H.S.; Song, C.J.; Zhang, L.; Zhang, J.J.; Wang, H.J.; Wilkinson, D.P. A review of anode catalysis in the direct methanol fuel cell. *J. Power Sources,* **2006**, *155*, 95-110. [http://dx.doi.org/10.1016/j.jpowsour.2006.01.030]

[213] Bing, Y.; Liu, H.; Zhang, L.; Ghosh, D.; Zhang, J. Nanostructured Pt-alloy electrocatalysts for PEM fuel cell oxygen reduction reaction. *Chem. Soc. Rev.,* **2010**, *39*(6), 2184-2202. [http://dx.doi.org/10.1039/b912552c] [PMID: 20502804]

[214] Guo, S.J.; Wang, E.K. Noble metal nanomaterials: Controllable synthesis and application in fuel cells and analytical sensors. *Nano Today,* **2011**, *6*, 240-264. [http://dx.doi.org/10.1016/j.nantod.2011.04.007]

[215] Cheng, F.; Chen, J. Metal-air batteries: from oxygen reduction electrochemistry to cathode catalysts.

Chem. Soc. Rev., **2012**, *41*(6), 2172-2192.
[http://dx.doi.org/10.1039/c1cs15228a] [PMID: 22254234]

[216] Singh, A.K.; Xu, Q. Synergistic catalysis over bimetallic alloy nanoparticles. *ChemCatChem,* **2013**, *5*, 652-676.
[http://dx.doi.org/10.1002/cctc.201200591]

[217] Valdes-Solis, T.; Fuertes, A.B. High-surface area inorganic compounds prepared by nanocasting techniques. *Mater. Res. Bull.,* **2006**, *41*, 2187-2197.
[http://dx.doi.org/10.1016/j.materresbull.2006.04.018]

[218] Orilall, M.C.; Wiesner, U. Block copolymer based composition and morphology control in nanostructured hybrid materials for energy conversion and storage: solar cells, batteries, and fuel cells. *Chem. Soc. Rev.,* **2011**, *40*(2), 520-535.
[http://dx.doi.org/10.1039/C0CS00034E] [PMID: 21152638]

[219] Jeffryes, C.; Campbell, J.; Li, H.Y.; Jiao, J.; Rorrer, G. The potential of diatom nanobiotechnology for applications in solar cells, batteries, and electroluminescent devices. *Energy Environ. Sci.,* **2011**, *4*, 3930-3941.
[http://dx.doi.org/10.1039/c0ee00306a]

[220] Poizot, P.; Laruelle, S.; Grugeon, S.; Dupont, L.; Tarascon, J.M. Nano-sized transition-metal oxides as negative-electrode materials for lithium-ion batteries. *Nature,* **2000**, *407*(6803), 496-499.
[http://dx.doi.org/10.1038/35035045] [PMID: 11028997]

[221] Rolison, D.R.; Hagans, P.L.; Swider, K.E.; Long, J.W. Role of hydrous ruthenium oxide in Pt-Ru direct methanol fuel cell anode electrocatalysts: The importance of mixed electron/proton conductivity. *Langmuir,* **1999**, *15*, 774-779.
[http://dx.doi.org/10.1021/la9807863]

[222] Zhang, X.; Chan, K.Y. Water-in-oil microemulsion synthesis of platinum-ruthenium nanoparticles, their characterization and electrocatalytic properties. *Chem. Mater.,* **2003**, *15*, 451-459.
[http://dx.doi.org/10.1021/cm0203868]

[223] Chan, K.Y.; Ding, J.; Ren, J.W.; Cheng, S.A.; Tsang, K.Y. Supported mixed metal nanoparticles as electrocatalysts in low temperature fuel cells. *J. Mater. Chem.,* **2004**, *14*, 505-516.
[http://dx.doi.org/10.1039/b314224h]

[224] Yang, H.; Vogel, W.; Lamy, C.; Alonso-Vante, N. Structure and electrocatalytic activity of carbon-supported Pt-Ni alloy nanoparticles toward the oxygen reduction reaction. *J. Phys. Chem. B,* **2004**, *108*, 11024-11034.
[http://dx.doi.org/10.1021/jp049034+]

[225] Lin, Y.; Cui, X.; Yen, C.H.; Wai, C.M. PtRu/carbon nanotube nanocomposite synthesized in supercritical fluid: a novel electrocatalyst for direct methanol fuel cells. *Langmuir,* **2005**, *21*(24), 11474-11479.
[http://dx.doi.org/10.1021/la051272o] [PMID: 16285828]

[226] Maillard, F.; Lu, G.Q.; Wieckowski, A.; Stimming, U. Ru-decorated Pt surfaces as model fuel cell electrocatalysts for CO electrooxidation. *J. Phys. Chem. B,* **2005**, *109*(34), 16230-16243.
[http://dx.doi.org/10.1021/jp052277x] [PMID: 16853064]

[227] Alayoglu, S.; Nilekar, A.U.; Mavrikakis, M.; Eichhorn, B. Ru-Pt core-shell nanoparticles for preferential oxidation of carbon monoxide in hydrogen. *Nat. Mater.,* **2008**, *7*(4), 333-338. [http://dx.doi.org/10.1038/nmat2156] [PMID: 18345004]

[228] Strasser, P.; Koh, S.; Anniyev, T.; Greeley, J.; More, K.; Yu, C.; Liu, Z.; Kaya, S.; Nordlund, D.; Ogasawara, H.; Toney, M.F.; Nilsson, A. Lattice-strain control of the activity in dealloyed core-shell fuel cell catalysts. *Nat. Chem.,* **2010**, *2*(6), 454-460. [http://dx.doi.org/10.1038/nchem.623] [PMID: 20489713]

[229] Crews, K.; Reeves, C.; Thomas, P.; Abugri, D.; Russell, A.; Curry, M.L. Heterogeneous catalysis of C–O bond cleavage for cellulose deconstruction: A potential pathway for ethanol production *ISRN Nanotechnol.,* **2014**, *2014*, 634679.

[230] Grugeon, S.; Laruelle, S.; Dupont, L.; Tarascon, J.M. An update on the reactivity of nanoparticles Co-based compounds towards Li. *Solid State Sci.,* **2003**, *5*, 895-904. [http://dx.doi.org/10.1016/S1293-2558(03)00114-6]

[231] Wang, J.; Polleux, J.; Lim, J.; Dunn, B. Pseudocapacitive contributions to electrochemical energy storage in TiO2 (anatase) nanoparticles. *J. Phys. Chem. C,* **2007**, *111*, 14925-14931. [http://dx.doi.org/10.1021/jp074464w]

[232] Sharma, Y.; Sharma, N.; Rao, G.V.; Chowdari, B.V. Nanophase ZnCo2O4 as a high performance anode material for Li-ion batteries. *Adv. Funct. Mater.,* **2007**, *17*, 2855-2861. [http://dx.doi.org/10.1002/adfm.200600997]

[233] Oumellal, Y.; Rougier, A.; Nazri, G.A.; Tarascon, J-M.; Aymard, L. Metal hydrides for lithium-ion batteries. *Nat. Mater.,* **2008**, *7*(11), 916-921. [http://dx.doi.org/10.1038/nmat2288] [PMID: 18849978]

[234] Chan, C.K.; Peng, H.; Liu, G.; McIlwrath, K.; Zhang, X.F.; Huggins, R.A.; Cui, Y. High-performance lithium battery anodes using silicon nanowires. *Nat. Nanotechnol.,* **2008**, *3*(1), 31-35. [http://dx.doi.org/10.1038/nnano.2007.411] [PMID: 18654447]

[235] Zhang, H.X.; Feng, C.; Zhai, Y.C.; Jiang, K.L.; Li, Q.Q.; Fan, S.S. Cross-stacked carbon nanotube sheets uniformly loaded with SnO2 nanoparticles: A novel binder-free and high-capacity anode material for lithium-ion batteries. *Adv. Mater.,* **2009**, *21*, 2299-2304. [http://dx.doi.org/10.1002/adma.200802290]

[236] Zhang, L.S.; Jiang, L.Y.; Yan, H.J.; Wang, W.D.; Wang, W.; Song, W.G.; Guo, Y.G.; Wan, L.J. Mono dispersed SnO2 nanoparticles on both sides of single layer graphene sheets as anode materials in Li-ion batteries. *J. Mater. Chem.,* **2010**, *20*, 5462-5467. [http://dx.doi.org/10.1039/c0jm00672f]

[237] Lu, Y.C.; Xu, Z.; Gasteiger, H.A.; Chen, S.; Hamad-Schifferli, K.; Shao-Horn, Y. Platinum-gold nanoparticles: a highly active bifunctional electrocatalyst for rechargeable lithium-air batteries. *J. Am. Chem. Soc.,* **2010**, *132*(35), 12170-12171. [http://dx.doi.org/10.1021/ja1036572] [PMID: 20527774]

[238] Wang, Z.; Zhou, L.; Lou, X.W. Metal oxide hollow nanostructures for lithium-ion batteries. *Adv. Mater.,* **2012**, *24*(14), 1903-1911. [http://dx.doi.org/10.1002/adma.201200469] [PMID: 22574316]

[239] Liu, N.; Lu, Z.; Zhao, J.; McDowell, M.T.; Lee, H-W.; Zhao, W.; Cui, Y. A pomegranate-inspired nanoscale design for large-volume-change lithium battery anodes. *Nat. Nanotechnol.,* **2014**, *9*(3), 187-192.
[http://dx.doi.org/10.1038/nnano.2014.6] [PMID: 24531496]

[240] Savage, N.; Diallo, M.S. Nanomaterials and water purification: Opportunities and challenges. *J. Nanopart. Res.,* **2005**, *7*, 331-342.
[http://dx.doi.org/10.1007/s11051-005-7523-5]

[241] Wigginton, N.S.; Haus, K.L.; Hochella, M.F., Jr Aquatic environmental nanoparticles. *J. Environ. Monit.,* **2007**, *9*(12), 1306-1316.
[http://dx.doi.org/10.1039/b712709j] [PMID: 18049768]

[242] Theron, J.; Walker, J.A.; Cloete, T.E. Nanotechnology and water treatment: applications and emerging opportunities. *Crit. Rev. Microbiol.,* **2008**, *34*(1), 43-69.
[http://dx.doi.org/10.1080/10408410701710442] [PMID: 18259980]

[243] Hua, M.; Zhang, S.; Pan, B.; Zhang, W.; Lv, L.; Zhang, Q. Heavy metal removal from water/wastewater by nanosized metal oxides: a review. *J. Hazard. Mater.,* **2012**, *211-212*, 317-331.
[http://dx.doi.org/10.1016/j.jhazmat.2011.10.016] [PMID: 22018872]

[244] Elliott, D.W.; Zhang, W.X. Field assessment of nanoscale bimetallic particles for groundwater treatment. *Environ. Sci. Technol.,* **2001**, *35*(24), 4922-4926.
[http://dx.doi.org/10.1021/es0108584] [PMID: 11775172]

[245] Xu, Y.; Zhao, D. Reductive immobilization of chromate in water and soil using stabilized iron nanoparticles. *Water Res.,* **2007**, *41*(10), 2101-2108.
[http://dx.doi.org/10.1016/j.watres.2007.02.037] [PMID: 17412389]

[246] Chen, H.M.; He, J.H. Facile synthesis of monodisperse manganese oxide nanostructures and their application in water treatment. *J. Phys. Chem. C,* **2008**, *112*, 17540-17545.
[http://dx.doi.org/10.1021/jp806160g]

[247] Hu, J.S.; Zhong, L.S.; Song, W.G.; Wan, L.J. Synthesis of hierarchically structured metal oxides and their application in heavy metal ion removal. *Adv. Mater.,* **2008**, *20*, 2977-2982.
[http://dx.doi.org/10.1002/adma.200800623]

[248] Zodrow, K.; Brunet, L.; Mahendra, S.; Li, D.; Zhang, A.; Li, Q.; Alvarez, P.J. Polysulfone ultrafiltration membranes impregnated with silver nanoparticles show improved biofouling resistance and virus removal. *Water Res.,* **2009**, *43*(3), 715-723.
[http://dx.doi.org/10.1016/j.watres.2008.11.014] [PMID: 19046755]

[249] Kiser, M.A.; Westerhoff, P.; Benn, T.; Wang, Y.; Pérez-Rivera, J.; Hristovski, K. Titanium nanomaterial removal and release from wastewater treatment plants. *Environ. Sci. Technol.,* **2009**, *43*(17), 6757-6763.
[http://dx.doi.org/10.1021/es901102n] [PMID: 19764246]

[250] Kaegi, R.; Voegelin, A.; Sinnet, B.; Zuleeg, S.; Hagendorfer, H.; Burkhardt, M.; Siegrist, H. Behavior of metallic silver nanoparticles in a pilot wastewater treatment plant. *Environ. Sci. Technol.,* **2011**, *45*(9), 3902-3908.
[http://dx.doi.org/10.1021/es1041892] [PMID: 21466186]

[251] Kapoor, V.; Wendell, D. Engineering bacterial efflux pumps for solar-powered bioremediation of surface waters. *Nano Lett.,* **2013**, *13*(5), 2189-2193.
[http://dx.doi.org/10.1021/nl400691d] [PMID: 23581993]

[252] Zhao, Z.; Arentz, J.; Pretzer, L.A.; Limpornpipat, P.; Clomburg, J.M.; Gonzalez, R.; Schweitzer, N.M.; Wu, T.; Millere, J.T.; Wong, M.S. Volcano-shape glycerol oxidation activity of palladium-decorated gold nanoparticles. *Chem. Sci. (Camb.),* **2014**, *5*, 3715-3728.
[http://dx.doi.org/10.1039/C4SC01001A]

[253] Shipway, A.N.; Katz, E.; Willner, I. Nanoparticle arrays on surfaces for electronic, optical, and sensor applications. *ChemPhysChem,* **2000**, *1*(1), 18-52.
[http://dx.doi.org/10.1002/1439-7641(20000804)1:1<18::AID-CPHC18>3.0.CO;2-L] [PMID: 23696260]

[254] Luo, X.L.; Morrin, A.; Killard, A.J.; Smyth, M.R. Application of nanoparticles in electrochemical sensors and biosensors. *Electroanal.,* **2006**, *18*, 319-326.
[http://dx.doi.org/10.1002/elan.200503415]

[255] Haick, H. Chemical sensors based on molecularly modified metallic nanoparticles. *J. Phys. D Appl. Phys.,* **2007**, *40*, 7173-7186.
[http://dx.doi.org/10.1088/0022-3727/40/23/S01]

[256] Anker, J.N.; Hall, W.P.; Lyandres, O.; Shah, N.C.; Zhao, J.; Van Duyne, R.P. Biosensing with plasmonic nanosensors. *Nat. Mater.,* **2008**, *7*(6), 442-453.
[http://dx.doi.org/10.1038/nmat2162] [PMID: 18497851]

[257] Hunt, H.K.; Armani, A.M. Label-free biological and chemical sensors. *Nanoscale,* **2010**, *2*(9), 1544-1559.
[http://dx.doi.org/10.1039/c0nr00201a] [PMID: 20820687]

[258] Lahav, M.; Shipway, A.N.; Willner, I. Au-nanoparticle-bis-bipyridinium cyclophane superstructures: assembly, characterization and sensoric applications. *J. Chem. Soc., Perkin Trans. 2,* **1999**, *9*, 1925-1931.
[http://dx.doi.org/10.1039/a902763g]

[259] Pavlov, V.; Xiao, Y.; Shlyahovsky, B.; Willner, I. Aptamer-functionalized Au nanoparticles for the amplified optical detection of thrombin. *J. Am. Chem. Soc.,* **2004**, *126*(38), 11768-11769.
[http://dx.doi.org/10.1021/ja046970u] [PMID: 15382892]

[260] Kolmakov, A.; Klenov, D.O.; Lilach, Y.; Stemmer, S.; Moskovits, M. Enhanced gas sensing by individual SnO2 nanowires and nanobelts functionalized with Pd catalyst particles. *Nano Lett.,* **2005**, *5*(4), 667-673.
[http://dx.doi.org/10.1021/nl050082v] [PMID: 15826106]

[261] Lee, K.S.; El-Sayed, M.A. Gold and silver nanoparticles in sensing and imaging: sensitivity of plasmon response to size, shape, and metal composition. *J. Phys. Chem. B,* **2006**, *110*(39), 19220-19225.
[http://dx.doi.org/10.1021/jp062536y] [PMID: 17004772]

[262] Huang, C.C.; Chang, H.T. Selective gold-nanoparticle-based "turn-on" fluorescent sensors for detection of mercury(II) in aqueous solution. *Anal. Chem.,* **2006**, *78*(24), 8332-8338.

[http://dx.doi.org/10.1021/ac061487i] [PMID: 17165824]

[263] Lee, J.S.; Han, M.S.; Mirkin, C.A. Colorimetric detection of mercuric ion (Hg2+) in aqueous media using DNA-functionalized gold nanoparticles. *Angew. Chem. Int. Ed. Engl.,* **2007**, *46*(22), 4093-4096. [http://dx.doi.org/10.1002/anie.200700269] [PMID: 17461429]

[264] Datar, R.; Kim, S.; Jeon, S. Hesketh, P.; Manalis, S.; Boisen, A.; Thundat, T. Cantilever sensors: nanomechanical tools for diagnostics. *MRS Bull.,* **2009**, *34*, 449-454. [http://dx.doi.org/10.1557/mrs2009.121]

[265] Galoppini, E. Linkers for anchoring sensitizers to semiconductor nanoparticles. *Coord. Chem. Rev.,* **2004**, *248*, 1283-1297. [http://dx.doi.org/10.1016/j.ccr.2004.03.016]

[266] Alstrum-Acevedo, J.H.; Brennaman, M.K.; Meyer, T.J. Chemical approaches to artificial photosynthesis. 2. *Inorg. Chem.,* **2005**, *44*(20), 6802-6827. [http://dx.doi.org/10.1021/ic050904r] [PMID: 16180838]

[267] Kamat, P.V. Meeting the clean energy demand: Nanostructure architectures for solar energy conversion. *J. Phys. Chem. C,* **2007**, *111*, 2834-2860. [http://dx.doi.org/10.1021/jp066952u]

[268] Jose, R.; Thavasi, V.; Ramakrishna, S. Metal oxides for dye-sensitized solar cells. *J. Am. Ceram. Soc.,* **2009**, *92*, 289-301. [http://dx.doi.org/10.1111/j.1551-2916.2008.02870.x]

[269] Hasobe, T.; Imahori, H.; Kamat, P.V.; Ahn, T.K.; Kim, S.K.; Kim, D.; Fujimoto, A.; Hirakawa, T.; Fukuzumi, S. Photovoltaic cells using composite nanoclusters of porphyrins and fullerenes with gold nanoparticles. *J. Am. Chem. Soc.,* **2005**, *127*(4), 1216-1228. [http://dx.doi.org/10.1021/ja047768u] [PMID: 15669861]

[270] Greene, L.E.; Yuhas, B.D.; Law, M.; Zitoun, D.; Yang, P. Solution-grown zinc oxide nanowires. *Inorg. Chem.,* **2006**, *45*(19), 7535-7543. [http://dx.doi.org/10.1021/ic0601900] [PMID: 16961338]

[271] Beek, W.J.; Wienk, M.M.; Janssen, R.A. Hybrid solar cells from regioregular polythiophene and ZnO nanoparticles. *Adv. Funct. Mater.,* **2006**, *16*, 1112-1116. [http://dx.doi.org/10.1002/adfm.200500573]

[272] Catchpole, K.R.; Polman, A. Design principles for particle plasmon enhanced solar cells. *Appl. Phys. Lett.,* **2008**, *93*, 191113. [http://dx.doi.org/10.1063/1.3021072]

[273] Catchpole, K.R.; Polman, A. Plasmonic solar cells. *Opt. Express,* **2008**, *16*(26), 21793-21800. [http://dx.doi.org/10.1364/OE.16.021793] [PMID: 19104612]

[274] Ham, M-H.; Choi, J.H.; Boghossian, A.A.; Jeng, E.S.; Graff, R.A.; Heller, D.A.; Chang, A.C.; Mattis, A.; Bayburt, T.H.; Grinkova, Y.V.; Zeiger, A.S.; Van Vliet, K.J.; Hobbie, E.K.; Sligar, S.G.; Wraight, C.A.; Strano, M.S. Photoelectrochemical complexes for solar energy conversion that chemically and autonomously regenerate. *Nat. Chem.,* **2010**, *2*(11), 929-936. [http://dx.doi.org/10.1038/nchem.822] [PMID: 20966948]

[275] Tang, J.; Huo, Z.; Brittman, S.; Gao, H.; Yang, P. Solution-processed core-shell nanowires for

efficient photovoltaic cells. *Nat. Nanotechnol.,* **2011**, *6*(9), 568-572.
[http://dx.doi.org/10.1038/nnano.2011.139] [PMID: 21857684]

[276] Linic, S.; Christopher, P.; Ingram, D.B. Plasmonic-metal nanostructures for efficient conversion of solar to chemical energy. *Nat. Mater.,* **2011**, *10*(12), 911-921.
[http://dx.doi.org/10.1038/nmat3151] [PMID: 22109608]

[277] Heurlin, M.; Magnusson, M.H.; Lindgren, D.; Ek, M.; Wallenberg, L.R.; Deppert, K.; Samuelson, L. Continuous gas-phase synthesis of nanowires with tunable properties. *Nature,* **2012**, *492*(7427), 90-94.
[http://dx.doi.org/10.1038/nature11652] [PMID: 23201685]

[278] Moreau, A.; Ciracì, C.; Mock, J.J.; Hill, R.T.; Wang, Q.; Wiley, B.J.; Chilkoti, A.; Smith, D.R. Controlled-reflectance surfaces with film-coupled colloidal nanoantennas. *Nature,* **2012**, *492*(7427), 86-89.
[http://dx.doi.org/10.1038/nature11615] [PMID: 23222613]

[279] Chen, C-C.; Dou, L.; Zhu, R.; Chung, C-H.; Song, T-B.; Zheng, Y.B.; Hawks, S.; Li, G.; Weiss, P.S.; Yang, Y. Visibly transparent polymer solar cells produced by solution processing. *ACS Nano,* **2012**, *6*(8), 7185-7190.
[http://dx.doi.org/10.1021/nn3029327] [PMID: 22789123]

[280] Krogstrup, P.; Jørgensen, H.I.; Heiss, M.; Demichel, O.; Holm, J.V.; Aagesen, M.; Nygard, J.; Fontcuberta i Morral, A. Single-nanowire solar cells beyond the Shockley–Queisser limit. *Nat. Photonics,* **2013**, *7*, 306-310.
[http://dx.doi.org/10.1038/nphoton.2013.32]

[281] Tuniz, A.; Kaltenecker, K.J.; Fischer, B.M.; Walther, M.; Fleming, S.C.; Argyros, A.; Kuhlmey, B.T. Metamaterial fibres for subdiffraction imaging and focusing at terahertz frequencies over optically long distances. *Nat. Commun.,* **2013**, *4*, 2706.
[http://dx.doi.org/10.1038/ncomms3706] [PMID: 24162458]

[282] Ferrari, M. Cancer nanotechnology: opportunities and challenges. *Nat. Rev. Cancer,* **2005**, *5*(3), 161-171.
[http://dx.doi.org/10.1038/nrc1566] [PMID: 15738981]

[283] Peer, D.; Karp, J.M.; Hong, S.; Farokhzad, O.C.; Margalit, R.; Langer, R. Nanocarriers as an emerging platform for cancer therapy. *Nat. Nanotechnol.,* **2007**, *2*(12), 751-760.
[http://dx.doi.org/10.1038/nnano.2007.387] [PMID: 18654426]

[284] Bhattacharya, R.; Mukherjee, P. Biological properties of "naked" metal nanoparticles. *Adv. Drug Deliv. Rev.,* **2008**, *60*(11), 1289-1306.
[http://dx.doi.org/10.1016/j.addr.2008.03.013] [PMID: 18501989]

[285] Jain, P.K.; Huang, X.; El-Sayed, I.H.; El-Sayed, M.A. Noble metals on the nanoscale: optical and photothermal properties and some applications in imaging, sensing, biology, and medicine. *Acc. Chem. Res.,* **2008**, *41*(12), 1578-1586.
[http://dx.doi.org/10.1021/ar7002804] [PMID: 18447366]

[286] Ghosh, P.; Han, G.; De, M.; Kim, C.K.; Rotello, V.M. Gold nanoparticles in delivery applications. *Adv. Drug Deliv. Rev.,* **2008**, *60*(11), 1307-1315.
[http://dx.doi.org/10.1016/j.addr.2008.03.016] [PMID: 18555555]

[287] De Jong, W.H.; Borm, P.J. Drug delivery and nanoparticles:applications and hazards. *Int. J. Nanomedicine,* **2008**, *3*(2), 133-149.
[http://dx.doi.org/10.2147/IJN.S596] [PMID: 18686775]

[288] Boisselier, E.; Astruc, D. Gold nanoparticles in nanomedicine: preparations, imaging, diagnostics, therapies and toxicity. *Chem. Soc. Rev.,* **2009**, *38*(6), 1759-1782.
[http://dx.doi.org/10.1039/b806051g] [PMID: 19587967]

[289] Khlebtsov, N.; Dykman, L. Biodistribution and toxicity of engineered gold nanoparticles: a review of *in vitro* and *in vivo* studies. *Chem. Soc. Rev.,* **2011**, *40*(3), 1647-1671.
[http://dx.doi.org/10.1039/C0CS00018C] [PMID: 21082078]

[290] Kumar, A.; Zhang, X.; Liang, X.J. Gold nanoparticles: emerging paradigm for targeted drug delivery system. *Biotechnol. Adv.,* **2013**, *31*(5), 593-606.
[http://dx.doi.org/10.1016/j.biotechadv.2012.10.002] [PMID: 23111203]

[291] Austin, L.A.; Mackey, M.A.; Dreaden, E.C.; El-Sayed, M.A. The optical, photothermal, and facile surface chemical properties of gold and silver nanoparticles in biodiagnostics, therapy, and drug delivery. *Arch. Toxicol.,* **2014**, *88*(7), 1391-1417.
[http://dx.doi.org/10.1007/s00204-014-1245-3] [PMID: 24894431]

[292] dos Santos, C.A.; Seckler, M.M.; Ingle, A.P.; Gupta, I.; Galdiero, S.; Galdiero, M.; Gade, A.; Rai, M. Silver nanoparticles: therapeutical uses, toxicity, and safety issues. *J. Pharm. Sci.,* **2014**, *103*(7), 1931-1944.
[http://dx.doi.org/10.1002/jps.24001] [PMID: 24824033]

[293] Caruso, F.; Caruso, R.A.; Mohwald, H. Nanoengineering of inorganic and hybrid hollow spheres by colloidal templating. *Science,* **1998**, *282*(5391), 1111-1114.
[http://dx.doi.org/10.1126/science.282.5391.1111] [PMID: 9804547]

[294] Gibson, J.D.; Khanal, B.P.; Zubarev, E.R. Paclitaxel-functionalized gold nanoparticles. *J. Am. Chem. Soc.,* **2007**, *129*(37), 11653-11661.
[http://dx.doi.org/10.1021/ja075181k] [PMID: 17718495]

[295] Dhar, S.; Gu, F.X.; Langer, R.; Farokhzad, O.C.; Lippard, S.J. Targeted delivery of cisplatin to prostate cancer cells by aptamer functionalized Pt(IV) prodrug-PLGA-PEG nanoparticles. *Proc. Natl. Acad. Sci. USA,* **2008**, *105*(45), 17356-17361.
[http://dx.doi.org/10.1073/pnas.0809154105] [PMID: 18978032]

[296] Cheng, Y.; C Samia, A.; Meyers, J.D.; Panagopoulos, I.; Fei, B.; Burda, C. Highly efficient drug delivery with gold nanoparticle vectors for *in vivo* photodynamic therapy of cancer. *J. Am. Chem. Soc.,* **2008**, *130*(32), 10643-10647.
[http://dx.doi.org/10.1021/ja801631c] [PMID: 18642918]

[297] Vivero-Escoto, J.L.; Slowing, I.I.; Wu, C.W.; Lin, V.S. Photoinduced intracellular controlled release drug delivery in human cells by gold-capped mesoporous silica nanosphere. *J. Am. Chem. Soc.,* **2009**, *131*(10), 3462-3463.
[http://dx.doi.org/10.1021/ja900025f] [PMID: 19275256]

[298] Wang, A.Z.; Yuet, K.; Zhang, L.; Gu, F.X.; Huynh-Le, M.; Radovic-Moreno, A.F.; Kantoff, P.W.; Bander, N.H.; Langer, R.; Farokhzad, O.C. ChemoRad nanoparticles: a novel multifunctional

nanoparticle platform for targeted delivery of concurrent chemoradiation. *Nanomedicine (Lond.),* **2010**, *5*(3), 361-368.
[http://dx.doi.org/10.2217/nnm.10.6] [PMID: 20394530]

[299] Valencia, P.M.; Pridgen, E.M.; Perea, B.; Gadde, S.; Sweeney, C.; Kantoff, P.W.; Bander, N.H.; Lippard, S.J.; Langer, R.; Karnik, R.; Farokhzad, O.C. Synergistic cytotoxicity of irinotecan and cisplatin in dual-drug targeted polymeric nanoparticles. *Nanomedicine (Lond.),* **2013**, *8*(5), 687-698.
[http://dx.doi.org/10.2217/nnm.12.134] [PMID: 23075285]

[300] Hu, C-M.; Fang, R.H.; Copp, J.; Luk, B.T.; Zhang, L. A biomimetic nanosponge that absorbs pore-forming toxins. *Nat. Nanotechnol.,* **2013**, *8*(5), 336-340.
[http://dx.doi.org/10.1038/nnano.2013.54] [PMID: 23584215]

[301] Gu, Z.; Dang, T.T.; Ma, M.; Tang, B.C.; Cheng, H.; Jiang, S.; Dong, Y.; Zhang, Y.; Anderson, D.G. Glucose-responsive microgels integrated with enzyme nanocapsules for closed-loop insulin delivery. *ACS Nano,* **2013**, *7*(8), 6758-6766.
[http://dx.doi.org/10.1021/nn401617u] [PMID: 23834678]

[302] Li, A.V.; Moon, J.J.; Abraham, W.; Suh, H.; Elkhader, J.; Seidman, M.A.; Yen, M.; Im, E.-J.; Foley, M.H.; Barouch, D.H.; Irvine, D.J. Generation of effect or memory T cell–based mucosal and systemic immunity with pulmonary nanoparticle vaccination. *Sci. Transl. Med.,* **2013**, *5*, 204-130.
[http://dx.doi.org/10.1186/1479-5876-11-204]

[303] Yao, L.; Daniels, J.; Moshnikova, A.; Kuznetsov, S.; Ahmed, A.; Engelman, D.M.; Reshetnyak, Y.K.; Andreev, O.A. pHLIP peptide targets nanogold particles to tumors. *Proc. Natl. Acad. Sci. USA,* **2013**, *110*(2), 465-470.
[http://dx.doi.org/10.1073/pnas.1219665110] [PMID: 23267062]

[304] Xu, X.; Xie, K.; Zhang, X.Q.; Pridgen, E.M.; Park, G.Y.; Cui, D.S.; Shi, J.; Wu, J.; Kantoff, P.W.; Lippard, S.J.; Langer, R.; Walker, G.C.; Farokhzad, O.C. Enhancing tumor cell response to chemotherapy through nanoparticle-mediated codelivery of siRNA and cisplatin prodrug. *Proc. Natl. Acad. Sci. USA,* **2013**, *110*(46), 18638-18643.
[http://dx.doi.org/10.1073/pnas.1303958110] [PMID: 24167294]

[305] Alqathami, M.; Blencowe, A.; Yeo, U.J.; Franich, R.; Doran, S.; Qiao, G.; Geso, M.J. Enhancement of radiation effects by bismuth oxide nanoparticles for kilovoltage x-ray beams: A dosimetric study using a novel multi-compartment 3D radiochromic dosimeter. *Phys.: Conf. Series,* **2013**, *444*, 012025.

[306] Stevens, D.M.; Gilmore, K.A.; Harth, E. An assessment of nanosponges for intravenous and oral drug delivery of BCS class IV drugs: Drug delivery kinetics and solubilization. *Polym. Chem.,* **2014**, *5*, 3551-3554.
[http://dx.doi.org/10.1039/c4py00207e]

[307] Jain, P.K.; El-Sayed, I.H.; El-Sayed, M.A. Au nanoparticles target cancer. *Nano Today,* **2007**, *2*, 18-29.
[http://dx.doi.org/10.1016/S1748-0132(07)70016-6]

[308] Huang, X.; Jain, P.K.; El-Sayed, I.H.; El-Sayed, M.A. Gold nanoparticles: interesting optical properties and recent applications in cancer diagnostics and therapy. *Nanomedicine (Lond.),* **2007**, *2*(5), 681-693.
[http://dx.doi.org/10.2217/17435889.2.5.681] [PMID: 17976030]

[309] Jain, P.K.; Huang, X.; El-Sayed, I.H.; El-Sayad, M.A. Review of some interesting surface plasmon resonance-enhanced properties of noble metal nanoparticles and their applications to biosystems. *Plasmon.,* **2007**, *2*, 107-118.
[http://dx.doi.org/10.1007/s11468-007-9031-1]

[310] Huang, X.; Jain, P.K.; El-Sayed, I.H.; El-Sayed, M.A. Plasmonic photothermal therapy (PPTT) using gold nanoparticles. *Lasers Med. Sci.,* **2008**, *23*(3), 217-228.
[http://dx.doi.org/10.1007/s10103-007-0470-x] [PMID: 17674122]

[311] Kennedy, L.C.; Bickford, L.R.; Lewinski, N.A.; Coughlin, A.J.; Hu, Y.; Day, E.S.; West, J.L.; Drezek, R.A. A new era for cancer treatment: gold-nanoparticle-mediated thermal therapies. *Small,* **2011**, *7*(2), 169-183.
[http://dx.doi.org/10.1002/smll.201000134] [PMID: 21213377]

[312] Doane, T.L.; Burda, C. The unique role of nanoparticles in nanomedicine: imaging, drug delivery and therapy. *Chem. Soc. Rev.,* **2012**, *41*(7), 2885-2911.
[http://dx.doi.org/10.1039/c2cs15260f] [PMID: 22286540]

[313] Maldonado, C.R.; Salassa, L.; Gomez-Blanco, N.; Mareque-Rivas, J.C. Nano-functionalization of metal complexes for molecular imaging and anticancer therapy. *Coord. Chem. Rev.,* **2013**, *257*, 2668-2688.
[http://dx.doi.org/10.1016/j.ccr.2013.04.014]

[314] Menon, J.U.; Jadeja, P.; Tambe, P.; Vu, K.; Yuan, B.; Nguyen, K.T. Nanomaterials for photo-based diagnostic and therapeutic applications. *Theranostics,* **2013**, *3*(3), 152-166.
[http://dx.doi.org/10.7150/thno.5327] [PMID: 23471164]

[315] Nam, J.; Won, N.; Bang, J.; Jin, H.; Park, J.; Jung, S.; Jung, S.; Park, Y.; Kim, S. Surface engineering of inorganic nanoparticles for imaging and therapy. *Adv. Drug Deliv. Rev.,* **2013**, *65*(5), 622-648.
[http://dx.doi.org/10.1016/j.addr.2012.08.015] [PMID: 22975010]

[316] Khan, M.S.; Vishakante, G.D.; Siddaramaiah H, Gold nanoparticles: a paradigm shift in biomedical applications. *Adv. Colloid Interface Sci.,* **2013**, *199-200*, 44-58.
[http://dx.doi.org/10.1016/j.cis.2013.06.003] [PMID: 23871224]

[317] Loo, C.; Lin, A.; Hirsch, L.; Lee, M.H.; Barton, J.; Halas, N.; West, J.; Drezek, R. Nanoshell-enabled photonics-based imaging and therapy of cancer. *Technol. Cancer Res. Treat.,* **2004**, *3*(1), 33-40.
[http://dx.doi.org/10.1177/153303460400300104] [PMID: 14750891]

[318] Chen, J.; Saeki, F.; Wiley, B.J.; Cang, H.; Cobb, M.J.; Li, Z.Y.; Au, L.; Zhang, H.; Kimmey, M.B.; Li, X.; Xia, Y. Gold nanocages: bioconjugation and their potential use as optical imaging contrast agents. *Nano Lett.,* **2005**, *5*(3), 473-477.
[http://dx.doi.org/10.1021/nl047950t] [PMID: 15755097]

[319] Farokhzad, O.C.; Cheng, J.; Teply, B.A.; Sherifi, I.; Jon, S.; Kantoff, P.W.; Richie, J.P.; Langer, R. Targeted nanoparticle-aptamer bioconjugates for cancer chemotherapy *in vivo. Proc. Natl. Acad. Sci. USA,* **2006**, *103*(16), 6315-6320.
[http://dx.doi.org/10.1073/pnas.0601755103] [PMID: 16606824]

[320] Huang, X.; El-Sayed, I.H.; Qian, W.; El-Sayed, M.A. Cancer cell imaging and photothermal therapy in the near-infrared region by using gold nanorods. *J. Am. Chem. Soc.,* **2006**, *128*(6), 2115-2120.

[http://dx.doi.org/10.1021/ja057254a] [PMID: 16464114]

[321] Dickerson, E.B.; Dreaden, E.C.; Huang, X.; El-Sayed, I.H.; Chu, H.; Pushpanketh, S.; McDonald, J.F.; El-Sayed, M.A. Gold nanorod assisted near-infrared plasmonic photothermal therapy (PPTT) of squamous cell carcinoma in mice. *Cancer Lett.,* **2008**, *269*(1), 57-66.
[http://dx.doi.org/10.1016/j.canlet.2008.04.026] [PMID: 18541363]

[322] Popovtzer, R.; Agrawal, A.; Kotov, N.A.; Popovtzer, A.; Balter, J.; Carey, T.E.; Kopelman, R. Targeted gold nanoparticles enable molecular CT imaging of cancer. *Nano Lett.,* **2008**, *8*(12), 4593-4596.
[http://dx.doi.org/10.1021/nl8029114] [PMID: 19367807]

[323] AshaRani, P.V.; Low Kah Mun, G.; Hande, M.P.; Valiyaveettil, S. Cytotoxicity and genotoxicity of silver nanoparticles in human cells. *ACS Nano,* **2009**, *3*(2), 279-290.
[http://dx.doi.org/10.1021/nn800596w] [PMID: 19236062]

[324] Kattumenu, R.; Lee, C.H.; Tian, L.; McConney, M.E.; Singamaneni, S. Nanorod decorated nanowires as highly efficient SERS-active hybrids. *J. Mater. Chem.,* **2011**, *21*, 15218.
[http://dx.doi.org/10.1039/c1jm12426a]

[325] Bitner, B.R.; Marcano, D.C.; Berlin, J.M.; Fabian, R.H.; Cherian, L.; Culver, J.C.; Dickinson, M.E.; Robertson, C.S.; Pautler, R.G.; Kent, T.A.; Tour, J.M. Antioxidant carbon particles improve cerebrovascular dysfunction following traumatic brain injury. *ACS Nano,* **2012**, *6*(9), 8007-8014.
[http://dx.doi.org/10.1021/nn302615f] [PMID: 22866916]

[326] Wang, H-N.; Fales, A.M.; Zaas, A.K.; Woods, C.W.; Burke, T.; Ginsburg, G.S.; Vo-Dinh, T. Surface-enhanced Raman scattering molecular sentinel nanoprobes for viral infection diagnostics. *Anal. Chim. Acta,* **2013**, *786*, 153-158.
[http://dx.doi.org/10.1016/j.aca.2013.05.017] [PMID: 23790305]

[327] Wang, Y.; Black, K.C.; Luehmann, H.; Li, W.; Zhang, Y.; Cai, X.; Wan, D.; Liu, S.Y.; Li, M.; Kim, P.; Li, Z-Y.; Wang, L.V.; Liu, Y.; Xia, Y. Comparison study of gold nanohexapods, nanorods, and nanocages for photothermal cancer treatment. *ACS Nano,* **2013**, *7*(3), 2068-2077.
[http://dx.doi.org/10.1021/nn304332s] [PMID: 23383982]

[328] Wason, M.S.; Colon, J.; Das, S.; Seal, S.; Turkson, J.; Zhao, J.; Baker, C.H. Sensitization of pancreatic cancer cells to radiation by cerium oxide nanoparticle-induced ROS production. *Nanomedicine (Lond.),* **2013**, *9*(4), 558-569.
[PMID: 23178284]

[329] Pan, L.; He, M.; Ma, J.; Tang, W.; Gao, G.; He, R.; Su, H.; Cui, D. Phase and size controllable synthesis of NaYbF4 nanocrystals in oleic acid/ionic liquid two-phase system for targeted fluorescent imaging of gastric cancer. *Theranostics,* **2013**, *3*(3), 210-222.
[http://dx.doi.org/10.7150/thno.5298] [PMID: 23471455]

[330] Wang, X.; Cai, X.; Hu, J.; Shao, N.; Wang, F.; Zhang, Q.; Xiao, J.; Cheng, Y. Glutathione-triggered "off-on" release of anticancer drugs from dendrimer-encapsulated gold nanoparticles. *J. Am. Chem. Soc.,* **2013**, *135*(26), 9805-9810.
[http://dx.doi.org/10.1021/ja402903h] [PMID: 23789713]

[331] Wang, C.; Daimon, H.; Onodera, T.; Koda, T.; Sun, S. A general approach to the size- and shape-controlled synthesis of platinum nanoparticles and their catalytic reduction of oxygen. *Angew. Chem.*

Int. Ed. Engl., **2008**, *47*(19), 3588-3591.
[http://dx.doi.org/10.1002/anie.200800073] [PMID: 18399516]

[332] Wang, C.; Daimon, H.; Lee, Y.; Kim, J.; Sun, S. Synthesis of monodisperse Pt nanocubes and their enhanced catalysis for oxygen reduction. *J. Am. Chem. Soc.,* **2007**, *129*(22), 6974-6975.
[http://dx.doi.org/10.1021/ja070440r] [PMID: 17500520]

[333] Tian, N.; Zhou, Z.Y.; Sun, S.G.; Ding, Y.; Wang, Z.L. Synthesis of tetrahexahedral platinum nanocrystals with high-index facets and high electro-oxidation activity. *Science,* **2007**, *316*(5825), 732-735.
[http://dx.doi.org/10.1126/science.1140484] [PMID: 17478717]

[334] Lim, B.; Lu, X.; Jiang, M.; Camargo, P.H.; Cho, E.C.; Lee, E.P.; Xia, Y. Facile synthesis of highly faceted multioctahedral Pt nanocrystals through controlled overgrowth. *Nano Lett.,* **2008**, *8*(11), 4043-4047.
[http://dx.doi.org/10.1021/nl802959b] [PMID: 18942882]

[335] Guo, S.; Dong, S.; Wang, E. A general method for the rapid synthesis of hollow metallic or bimetallic nanoelectrocatalysts with urchinlike morphology. *Chemistry,* **2008**, *14*(15), 4689-4695.
[http://dx.doi.org/10.1002/chem.200800077] [PMID: 18384027]

[336] Chen, Z.; Waje, M.; Li, W.; Yan, Y. Supportless Pt and PtPd nanotubes as electrocatalysts for oxygen-reduction reactions. *Angew. Chem. Int. Ed. Engl.,* **2007**, *46*(22), 4060-4063.
[http://dx.doi.org/10.1002/anie.200700894] [PMID: 17476642]

[337] Alia, S.; Zhang, G.; Kisailus, D.; Li, D.; Gu, S.; Jensen, K.; Yan, Y. Porous platinum nanotubes for oxygen reduction and methanol oxidation reactions. *Adv. Funct. Mater.,* **2010**, *20*, 3742-3746.
[http://dx.doi.org/10.1002/adfm.201001035]

[338] Sun, S.; Yang, D.; Villers, D.; Zhang, G.; Sacher, E.; Dodelet, J. Template- and surfactant-free room temperature synthesis of self-assembled 3D Pt nanoflowers from single-crystal nanowires. *Adv. Mater.,* **2008**, *20*, 571-574.
[http://dx.doi.org/10.1002/adma.200701408]

[339] Sun, S.; Jaouen, F.; Dodelet, J. Controlled growth of Pt nanowires on carbon nanospheres and their enhanced performance as electrocatalysts in PEM fuel cells. *Adv. Mater.,* **2008**, *20*, 3900-3904.
[http://dx.doi.org/10.1002/adma.200800491]

[340] Peng, Z.; You, H.; Wu, J.; Yang, H. Electrochemical synthesis and catalytic property of sub-10 nm platinum cubic nanoboxes. *Nano Lett.,* **2010**, *10*(4), 1492-1496.
[http://dx.doi.org/10.1021/nl100559y] [PMID: 20297767]

[341] Wang, L.; Yamauchi, Y. Facile synthesis of three-dimensional dendritic platinum nanoelectrocatalyst. *Chem. Mater.,* **2009**, *21*, 3562-3569.
[http://dx.doi.org/10.1021/cm901161g]

[342] Wang, L.; Guo, S.; Zhai, J.; Dong, S. Facile synthesis of platinum nanoelectrocatalyst with urchinlike morphology. *J. Phys. Chem. C,* **2008**, *112*, 13372-13377.
[http://dx.doi.org/10.1021/jp804484f]

[343] Jin, Y.; Shen, Y.; Dong, S. Electrochemical design of ultrathin platinum-coated gold nanoparticle monolayer films as a novel nanostructured electrocatalyst for oxygen reduction. *J. Phys. Chem. B,*

2004, *108*, 8142-8147.
[http://dx.doi.org/10.1021/jp0375517]

[344] Yoo, S.; Park, S. Platinum-coated, nanoporous gold nanorod arrays: synthesis and characterization. *Adv. Mater.,* **2007**, *19*, 1612-1615.
[http://dx.doi.org/10.1002/adma.200602551]

[345] Sasaki, K.; Naohara, H.; Cai, Y.; Choi, Y.M.; Liu, P.; Vukmirovic, M.B.; Wang, J.X.; Adzic, R.R. Core-protected platinum monolayer shell high-stability electrocatalysts for fuel-cell cathodes. *Angew. Chem. Int. Ed. Engl.,* **2010**, *49*(46), 8602-8607.
[http://dx.doi.org/10.1002/anie.201004287] [PMID: 20931587]

[346] Gong, K.; Su, D.; Adzic, R.R. Platinum-monolayer shell on AuNi(0.5)Fe nanoparticle core electrocatalyst with high activity and stability for the oxygen reduction reaction. *J. Am. Chem. Soc.,* **2010**, *132*(41), 14364-14366.
[http://dx.doi.org/10.1021/ja1063873] [PMID: 20873798]

[347] Lim, B.; Jiang, M.; Camargo, P.H.; Cho, E.C.; Tao, J.; Lu, X.; Zhu, Y.; Xia, Y. Pd-Pt bimetallic nanodendrites with high activity for oxygen reduction. *Science,* **2009**, *324*(5932), 1302-1305.
[http://dx.doi.org/10.1126/science.1170377] [PMID: 19443738]

[348] Guo, S.; Dong, S.; Wang, E. Ultralong Pt-on-Pd bimetallic nanowires with nanoporous surface: nanodendritic structure for enhanced electrocatalytic activity. *Chem. Commun. (Camb.),* **2010**, *46*(11), 1869-1871.
[http://dx.doi.org/10.1039/b922148d] [PMID: 20198236]

[349] Zhang, J.; Sasaki, K.; Sutter, E.; Adzic, R.R. Stabilization of platinum oxygen-reduction electrocatalysts using gold clusters. *Science,* **2007**, *315*(5809), 220-222.
[http://dx.doi.org/10.1126/science.1134569] [PMID: 17218522]

[350] Guo, S.J.; Li, J.; Dong, S.J.; Wang, E.K. Three-dimensional Pt-on-Au bimetallic dendritic nanoparticle: one-sep, high-yield synthesis and its bifunctional plasmonic and catalytic properties. *J. Phys. Chem. C,* **2010**, *114*, 15337-15342.
[http://dx.doi.org/10.1021/jp104942d]

[351] Gu, X.; Cong, X.; Ding, Y. Platinum-decorated Au porous nanotubes as highly efficient catalysts for formic acid electro-oxidation. *ChemPhysChem,* **2010**, *11*(4), 841-846.
[http://dx.doi.org/10.1002/cphc.200900927] [PMID: 20166117]

[352] Mazumder, V.; Lee, Y.; Sun, S. Recent development of active nanoparticle catalysts for fuel cell reactions. *Adv. Funct. Mater.,* **2010**, *20*, 1224-1231.
[http://dx.doi.org/10.1002/adfm.200902293]

[353] Chen, W.; Kim, J.; Sun, S.; Chen, S. Electrocatalytic reduction of oxygen by FePt alloy nanoparticles. *J. Phys. Chem. C,* **2008**, *112*, 3891-3898.
[http://dx.doi.org/10.1021/jp7110204]

[354] Xu, D.; Bliznakov, S.; Liu, Z.; Fang, J.; Dimitrov, N. Composition-dependent electrocatalytic activity of Pt-Cu nanocube catalysts for formic acid oxidation. *Angew. Chem. Int. Ed. Engl.,* **2010**, *49*(7), 1282-1285.
[http://dx.doi.org/10.1002/anie.200905248] [PMID: 20084649]

[355] Yang, H.; Zhang, J.; Sun, K.; Zou, S.; Fang, J. Enhancing by weakening: electrooxidation of methanol on Pt3Co and Pt nanocubes. *Angew. Chem. Int. Ed. Engl.,* **2010**, *49*(38), 6848-6851. [http://dx.doi.org/10.1002/anie.201002888] [PMID: 20715249]

[356] Kang, Y.; Murray, C.B. Synthesis and electrocatalytic properties of cubic Mn-Pt nanocrystals (nanocubes). *J. Am. Chem. Soc.,* **2010**, *132*(22), 7568-7569. [http://dx.doi.org/10.1021/ja100705j] [PMID: 20469855]

[357] Zhang, J.; Yang, H.; Fang, J.; Zou, S. Synthesis and oxygen reduction activity of shape-controlled Pt(3)Ni nanopolyhedra. *Nano Lett.,* **2010**, *10*(2), 638-644. [http://dx.doi.org/10.1021/nl903717z] [PMID: 20078068]

[358] Xu, C.; Wang, L.; Wang, R.; Wang, K.; Zhang, Y.; Tian, F.; Ding, Y. Nanotubular mesoporous bimetallic nanostructures with enhanced electrocatalytic performance. *Adv. Mater.,* **2009**, *21*, 2165-2169. [http://dx.doi.org/10.1002/adma.200702700]

[359] Xu, C.; Zhang, Y.; Wang, L.; Xu, L.; Bian, X.; Ma, H.; Ding, Y. Nanotubular mesoporous PdCu bimetallic electrocatalysts toward oxygen reduction reaction. *Chem. Mater.,* **2009**, *21*, 3110-3116. [http://dx.doi.org/10.1021/cm900244g]

[360] Mazumder, V.; Chi, M.; More, K.L.; Sun, S. Core/shell Pd/FePt nanoparticles as an active and durable catalyst for the oxygen reduction reaction. *J. Am. Chem. Soc.,* **2010**, *132*(23), 7848-7849. [http://dx.doi.org/10.1021/ja1024436] [PMID: 20496893]

[361] Wang, C.; van der Vliet, D.; More, K.L.; Zaluzec, N.J.; Peng, S.; Sun, S.; Daimon, H.; Wang, G.; Greeley, J.; Pearson, J.; Paulikas, A.P.; Karapetrov, G.; Strmcnik, D.; Markovic, N.M.; Stamenkovic, V.R. Multimetallic Au/FePt3 nanoparticles as highly durable electrocatalyst. *Nano Lett.,* **2011**, *11*(3), 919-926. [http://dx.doi.org/10.1021/nl102369k] [PMID: 20704335]

[362] Mazumder, V.; Chi, M.; More, K.L.; Sun, S. Synthesis and characterization of multimetallic Pd/Au and Pd/Au/FePt core/shell nanoparticles. *Angew. Chem. Int. Ed. Engl.,* **2010**, *49*(49), 9368-9372. [http://dx.doi.org/10.1002/anie.201003903] [PMID: 20853383]

[363] Khan, S.; Cao, Q.; Zheng, Y.M.; Huang, Y.Z.; Zhu, Y.G. Health risks of heavy metals in contaminated soils and food crops irrigated with wastewater in Beijing, China. *Environ. Pollut.,* **2008**, *152*(3), 686-692. [http://dx.doi.org/10.1016/j.envpol.2007.06.056] [PMID: 17720286]

[364] Singh, A.; Sharma, R.K.; Agrawal, M.; Marshall, F.M. Health risk assessment of heavy metals *via* dietary intake of foodstuffs from the wastewater irrigated site of a dry tropical area of India. *Food Chem. Toxicol.,* **2010**, *48*(2), 611-619. [http://dx.doi.org/10.1016/j.fct.2009.11.041] [PMID: 19941927]

[365] Fu, F.; Wang, Q. Removal of heavy metal ions from wastewaters: a review. *J. Environ. Manage.,* **2011**, *92*(3), 407-418. [http://dx.doi.org/10.1016/j.jenvman.2010.11.011] [PMID: 21138785]

[366] Kurniawan, T.A.; Chan, G.Y.; Lo, W.H.; Babel, S. Physico-chemical treatment techniques for wastewater laden with heavy metals. *Chem. Eng. J.,* **2006**, *118*, 83-98.

[http://dx.doi.org/10.1016/j.cej.2006.01.015]

[367] Obare, S.O.; Meyer, G.J. Nanostructured materials for environmental remediation of organic contaminants in water. *J Environ Sci Health A Tox Hazard Subst Environ Eng,* **2004**, *39*(10), 2549-2582.
[http://dx.doi.org/10.1081/ESE-200027010] [PMID: 15509009]

[368] Adesina, A.A. Industrial exploitation of photocatalysis progress, perspectives and prospects. *Catal. Surv. Asia,* **2004**, *8*, 265-273.
[http://dx.doi.org/10.1007/s10563-004-9117-0]

[369] Kabra, K.; Chaudhary, R.; Sawhney, R.L. Treatment of hazardous organic and inorganic compounds through aqueous-phase photocatalysis: A review. *Ind. Eng. Chem. Res.,* **2004**, *43*, 7683-7696.
[http://dx.doi.org/10.1021/ie0498551]

[370] Chitose, N.; Ueta, S.; Seino, S.; Yamamoto, T.A. Radiolysis of aqueous phenol solutions with nanoparticles. 1. Phenol degradation and TOC removal in solutions containing TiO2 induced by UV, gamma-ray and electron beams. *Chemosphere,* **2003**, *50*(8), 1007-1013.
[http://dx.doi.org/10.1016/S0045-6535(02)00642-2] [PMID: 12531706]

[371] Asahi, R.; Morikawa, T.; Ohwaki, T.; Aoki, K.; Taga, Y. Visible-light photocatalysis in nitrogen-doped titanium oxides. *Science,* **2001**, *293*(5528), 269-271.
[http://dx.doi.org/10.1126/science.1061051] [PMID: 11452117]

[372] Bae, E.; Choi, W. Highly enhanced photoreductive degradation of perchlorinated compounds on dye-sensitized metal/TiO2 under visible light. *Environ. Sci. Technol.,* **2003**, *37*(1), 147-152.
[http://dx.doi.org/10.1021/es025617q] [PMID: 12542303]

[373] Guzman, K.A.; Finnegan, M.P.; Banfield, J.F. Influence of surface potential on aggregation and transport of titania nanoparticles. *Environ. Sci. Technol.,* **2006**, *40*(24), 7688-7693.
[http://dx.doi.org/10.1021/es060847g] [PMID: 17256514]

[374] Nurmi, J.T.; Tratnyek, P.G.; Sarathy, V.; Baer, D.R.; Amonette, J.E.; Pecher, K.; Wang, C.; Linehan, J.C.; Matson, D.W.; Penn, R.L.; Driessen, M.D. Characterization and properties of metallic iron nanoparticles: spectroscopy, electrochemistry, and kinetics. *Environ. Sci. Technol.,* **2005**, *39*(5), 1221-1230.
[http://dx.doi.org/10.1021/es049190u] [PMID: 15787360]

[375] Schrick, B.; Blough, J.L.; Jones, A.D.; Mallouk, T.E. Hydrodechlorination of trichloroethylene to hydrocarbons using bimetallic nickel-iron nanoparticles. *Chem. Mater.,* **2002**, *14*, 5140-5147.
[http://dx.doi.org/10.1021/cm020737i]

[376] Zhang, W-X. Nanoscale iron particles for environmental remediation. *J. Nanopart. Res.,* **2003**, *5*, 323-332.
[http://dx.doi.org/10.1023/A:1025520116015]

[377] Kawashima, M.; Tainaka, Y.; Hori, T.; Koyama, M.; Takamatsu, T. Phosphate adsorption onto hydrous manganese (IV) oxide in the presence of divalent-cations. *Water Res.,* **1986**, *20*, 471-475.
[http://dx.doi.org/10.1016/0043-1354(86)90195-8]

[378] Takamatsu, T.; Kawashima, M.; Koyama, M. The role of Mn2+-rich hydrous manganese oxide in the accumulation of arsenic in lake-sediments. *Water Res.,* **1985**, *19*, 1029-1032.

[http://dx.doi.org/10.1016/0043-1354(85)90372-0]

[379] Mishra, S.P.; Dubey, S.S.; Tiwari, D. Inorganic particulates in removal of heavy metal toxic ions IX. Rapid and efficient removal of Hg(II) by hydrous manganese and tin oxides. *J. Colloid Interface Sci.,* **2004**, *279*(1), 61-67.
[PMID: 15380412]

[380] Wang, X.B.; Cai, W.P.; Lin, Y.X.; Wang, G.Z.; Liang, C.H. Mass production of micro/nanostructured porous ZnO plates and their strong structurally enhanced and selective adsorption performance for environmental remediation. *J. Mater. Chem.,* **2010**, *20*, 8582-8590.
[http://dx.doi.org/10.1039/c0jm01024c]

[381] Lee, J.H.; Kim, B.S.; Lee, J.C.; Park, S. Kim, H.S.; Park, S.-Y.; Hur, B.Y.; Lee, S.W. Removal of Cu^{++} ions from aqueous Cu-EDTA solution using ZnO nanopowder In: *Eco-Materials Processing & Design VI*; Trans Tech Publications: Korea, **2005**; pp. 510-513.

[382] Huang, J.; Wang, D.; Hou, H.; You, T. Electrospun palladium nanoparticle-loaded carbon nanofibers and their electrocatalytic activities towards hydrogen peroxide and NADH. *Adv. Funct. Mater.,* **2008**, *18*, 441-448.
[http://dx.doi.org/10.1002/adfm.200700729]

[383] Guo, S.; Dong, S.; Wang, E. Polyaniline/Pt hybrid nanofibers: high-efficiency nanoelectrocatalysts for electrochemical devices. *Small,* **2009**, *5*(16), 1869-1876.
[http://dx.doi.org/10.1002/smll.200900190] [PMID: 19384885]

[384] Guo, S.; Li, J.; Ren, W.; Wen, D.; Dong, S.; Wang, E. Carbon nanotube/silica coaxial nanocable as a three-dimensional support for loading diverse ultra-high-density metal nanostructures: facile preparation and use as enhanced materials for electrochemical devices and SERS. *Chem. Mater.,* **2009**, *21*, 2247.
[http://dx.doi.org/10.1021/cm900300v]

[385] Wen, D.; Guo, S.; Wang, Y.; Dong, S. Bifunctional nanocatalyst of bimetallic nanoparticle/TiO2 with enhanced performance in electrochemical and photoelectrochemical applications. *Langmuir,* **2010**, *26*(13), 11401-11406.
[http://dx.doi.org/10.1021/la100869r] [PMID: 20369896]

[386] Fang, Y.; Guo, S.; Zhu, C.; Zhai, Y.; Wang, E. Self-assembly of cationic polyelectrolyte-functionalized graphene nanosheets and gold nanoparticles: a two-dimensional heterostructure for hydrogen peroxide sensing. *Langmuir,* **2010**, *26*(13), 11277-11282.
[http://dx.doi.org/10.1021/la100575g] [PMID: 20232834]

[387] Cui, R.; Liu, C.; Shen, J.; Gao, D.; Zhu, J.; Chen, H. Gold nanoparticle–colloidal carbon nanosphere hybrid material: preparation, characterization, and application for an amplified electrochemical immunoassay. *Adv. Funct. Mater.,* **2008**, *18*, 2197-2204.
[http://dx.doi.org/10.1002/adfm.200701340]

[388] Mao, S.; Lu, G.; Yu, K.; Bo, Z.; Chen, J. Specific protein detection using thermally reduced graphene oxide sheet decorated with gold nanoparticle-antibody conjugates. *Adv. Mater.,* **2010**, *22*(32), 3521-3526.
[http://dx.doi.org/10.1002/adma.201000520] [PMID: 20665564]

[389] Rosi, N.L.; Mirkin, C.A. Nanostructures in biodiagnostics. *Chem. Rev.,* **2005**, *105*(4), 1547-1562.

[http://dx.doi.org/10.1021/cr030067f] [PMID: 15826019]

[390] Mirkin, C.A.; Letsinger, R.L.; Mucic, R.C.; Storhoff, J.J. A DNA-based method for rationally assembling nanoparticles into macroscopic materials. *Nature,* **1996**, *382*(6592), 607-609. [http://dx.doi.org/10.1038/382607a0] [PMID: 8757129]

[391] Elghanian, R.; Storhoff, J.J.; Mucic, R.C.; Letsinger, R.L.; Mirkin, C.A. Selective colorimetric detection of polynucleotides based on the distance-dependent optical properties of gold nanoparticles. *Science,* **1997**, *277*(5329), 1078-1081. [http://dx.doi.org/10.1126/science.277.5329.1078] [PMID: 9262471]

[392] Choi, Y.; Ho, N.H.; Tung, C.H. Sensing phosphatase activity by using gold nanoparticles. *Angew. Chem. Int. Ed. Engl.,* **2007**, *46*(5), 707-709. [http://dx.doi.org/10.1002/anie.200603735] [PMID: 17143915]

[393] Xu, X.; Han, M.S.; Mirkin, C.A. A gold-nanoparticle-based real-time colorimetric screening method for endonuclease activity and inhibition. *Angew. Chem. Int. Ed. Engl.,* **2007**, *46*(19), 3468-3470. [http://dx.doi.org/10.1002/anie.200605249] [PMID: 17385814]

[394] Taton, T.A.; Mirkin, C.A.; Letsinger, R.L. Scanometric DNA array detection with nanoparticle probes. *Science,* **2000**, *289*(5485), 1757-1760. [http://dx.doi.org/10.1126/science.289.5485.1757] [PMID: 10976070]

[395] McKenzie, F.; Faulds, K.; Graham, D. Sequence-specific DNA detection using high-affinity LNA-functionalized gold nanoparticles. *Small,* **2007**, *3*(11), 1866-1868. [http://dx.doi.org/10.1002/smll.200700225] [PMID: 17935063]

[396] Liu, J.; Cao, Z.; Lu, Y. Functional nucleic acid sensors. *Chem. Rev.,* **2009**, *109*(5), 1948-1998. [http://dx.doi.org/10.1021/cr030183i] [PMID: 19301873]

[397] He, S.; Li, D.; Zhu, C.; Song, S.; Wang, L.; Long, Y.; Fan, C. Design of a gold nanoprobe for rapid and portable mercury detection with the naked eye. *Chem. Commun. (Camb.),* **2008**, (40), 4885-4887. [http://dx.doi.org/10.1039/b811528a] [PMID: 18931727]

[398] Liu, J.; Lu, Y. Accelerated color change of gold nanoparticles assembled by DNAzymes for simple and fast colorimetric Pb2+ detection. *J. Am. Chem. Soc.,* **2004**, *126*(39), 12298-12305. [http://dx.doi.org/10.1021/ja046628h] [PMID: 15453763]

[399] Xue, X.; Wang, F.; Liu, X. One-step, room temperature, colorimetric detection of mercury (Hg2+) using DNA/nanoparticle conjugates. *J. Am. Chem. Soc.,* **2008**, *130*(11), 3244-3245. [http://dx.doi.org/10.1021/ja076716c] [PMID: 18293973]

[400] Han, M.S.; Lytton-Jean, A.K.; Mirkin, C.A. A gold nanoparticle based approach for screening triplex DNA binders. *J. Am. Chem. Soc.,* **2006**, *128*(15), 4954-4955. [http://dx.doi.org/10.1021/ja0606475] [PMID: 16608320]

[401] Li, H.; Rothberg, L. Colorimetric detection of DNA sequences based on electrostatic interactions with unmodified gold nanoparticles. *Proc. Natl. Acad. Sci. USA,* **2004**, *101*(39), 14036-14039. [http://dx.doi.org/10.1073/pnas.0406115101] [PMID: 15381774]

[402] Li, H.; Rothberg, L.J. Label-free colorimetric detection of specific sequences in genomic DNA amplified by the polymerase chain reaction. *J. Am. Chem. Soc.,* **2004**, *126*(35), 10958-10961. [http://dx.doi.org/10.1021/ja048749n] [PMID: 15339181]

[403] Wang, L.; Liu, X.; Hu, X.; Song, S.; Fan, C. Unmodified gold nanoparticles as a colorimetric probe for potassium DNA aptamers. *Chem. Commun. (Camb.),* **2006**, (36), 3780-3782. [http://dx.doi.org/10.1039/b607448k] [PMID: 16969455]

[404] Li, D.; Wieckowska, A.; Willner, I. Optical analysis of Hg2+ ions by oligonucleotide-gol--nanoparticle hybrids and DNA-based machines. *Angew. Chem. Int. Ed. Engl.,* **2008**, *47*(21), 3927-3931. [http://dx.doi.org/10.1002/anie.200705991] [PMID: 18404745]

[405] Wang, H.; Wang, Y.; Jin, J.; Yang, R. Gold nanoparticle-based colorimetric and "turn-on" fluorescent probe for mercury(II) ions in aqueous solution. *Anal. Chem.,* **2008**, *80*(23), 9021-9028. [http://dx.doi.org/10.1021/ac801382k] [PMID: 19551976]

[406] Wei, H.; Li, B.; Li, J.; Wang, E.; Dong, S. Simple and sensitive aptamer-based colorimetric sensing of protein using unmodified gold nanoparticle probes. *Chem. Commun. (Camb.),* **2007**, (36), 3735-3737. [http://dx.doi.org/10.1039/b707642h] [PMID: 17851611]

[407] Zhang, J.; Wang, L.; Pan, D.; Song, S.; Boey, F.Y.; Zhang, H.; Fan, C. Visual cocaine detection with gold nanoparticles and rationally engineered aptamer structures. *Small,* **2008**, *4*(8), 1196-1200. [http://dx.doi.org/10.1002/smll.200800057] [PMID: 18651718]

[408] Jiang, Y.; Zhao, H.; Lin, Y.; Zhu, N.; Ma, Y.; Mao, L. Colorimetric detection of glucose in rat brain using gold nanoparticles. *Angew. Chem. Int. Ed. Engl.,* **2010**, *49*(28), 4800-4804. [http://dx.doi.org/10.1002/anie.201001057] [PMID: 20533481]

[409] Jiang, Y.; Zhao, H.; Zhu, N.; Lin, Y.; Yu, P.; Mao, L. A simple assay for direct colorimetric visualization of trinitrotoluene at picomolar levels using gold nanoparticles. *Angew. Chem. Int. Ed. Engl.,* **2008**, *47*(45), 8601-8604. [http://dx.doi.org/10.1002/anie.200804066] [PMID: 18846526]

[410] Ai, K.; Liu, Y.; Lu, L. Hydrogen-bonding recognition-induced color change of gold nanoparticles for visual detection of melamine in raw milk and infant formula. *J. Am. Chem. Soc.,* **2009**, *131*(27), 9496-9497. [http://dx.doi.org/10.1021/ja9037017] [PMID: 19537721]

[411] Huang, C.C.; Chang, H.T. Parameters for selective colorimetric sensing of mercury(II) in aqueous solutions using mercaptopropionic acid-modified gold nanoparticles. *Chem. Commun. (Camb.),* **2007**, (12), 1215-1217. [http://dx.doi.org/10.1039/B615383F] [PMID: 17356760]

[412] Kim, S.; Park, J.W.; Kim, D.; Kim, D.; Lee, I.H.; Jon, S. Bioinspired colorimetric detection of calcium(II) ions in serum using calsequestrin-functionalized gold nanoparticles. *Angew. Chem. Int. Ed. Engl.,* **2009**, *48*(23), 4138-4141. [http://dx.doi.org/10.1002/anie.200900071] [PMID: 19425025]

[413] Yoosaf, K.; Ipe, B.; Suresh, C.; Thomas, K. In situ synthesis of metal nanoparticles and selective naked-eye detection of lead ions from aqueous media. *J. Phys. Chem. C,* **2007**, *111*, 12839. [http://dx.doi.org/10.1021/jp073923q]

[414] Lee, J.S.; Lytton-Jean, A.K.; Hurst, S.J.; Mirkin, C.A. Silver nanoparticle-oligonucleotide conjugates based on DNA with triple cyclic disulfide moieties. *Nano Lett.,* **2007**, *7*(7), 2112-2115.

[http://dx.doi.org/10.1021/nl071108g] [PMID: 17571909]

[415] Wei, H.; Chen, C.; Han, B.; Wang, E. Enzyme colorimetric assay using unmodified silver nanoparticles. *Anal. Chem.,* **2008**, *80*(18), 7051-7055. [http://dx.doi.org/10.1021/ac801144t] [PMID: 18662017]

[416] Xu, X.; Wang, J.; Yang, F.; Jiao, K.; Yang, X. Label-free colorimetric detection of small molecules utilizing DNA oligonucleotides and silver nanoparticles. *Small,* **2009**, *5*(23), 2669-2672. [http://dx.doi.org/10.1002/smll.200901164] [PMID: 19813222]

[417] Kanjanawarut, R.; Su, X. Colorimetric detection of DNA using unmodified metallic nanoparticles and peptide nucleic acid probes. *Anal. Chem.,* **2009**, *81*(15), 6122-6129. [http://dx.doi.org/10.1021/ac900525k] [PMID: 20337394]

[418] He, W.; Wu, X.; Liu, J.; Hu, X.; Zhang, K.; Hou, S.; Zhou, W.; Xie, S. Design of AgM bimetallic alloy nanostructures (M = Au, Pd, Pt) with tunable morphology and peroxidase-like activity. *Chem. Mater.,* **2010**, *22*, 2988-2994. [http://dx.doi.org/10.1021/cm100393v]

[419] Jv, Y.; Li, B.; Cao, R. Positively-charged gold nanoparticles as peroxidase mimic and their application in hydrogen peroxide and glucose detection. *Chem. Commun. (Camb.),* **2010**, *46*(42), 8017-8019. [http://dx.doi.org/10.1039/c0cc02698k] [PMID: 20871928]

[420] Guo, W.; Yuan, J.; Wang, E. Oligonucleotide-stabilized Ag nanoclusters as novel fluorescence probes for the highly selective and sensitive detection of the Hg2+ ion. *Chem. Commun. (Camb.),* **2009**, (23), 3395-3397. [http://dx.doi.org/10.1039/b821518a] [PMID: 19503882]

[421] Xie, J.; Zheng, Y.; Ying, J.Y. Highly selective and ultrasensitive detection of Hg(2+) based on fluorescence quenching of Au nanoclusters by Hg(2+)-Au(+) interactions. *Chem. Commun. (Camb.),* **2010**, *46*(6), 961-963. [http://dx.doi.org/10.1039/B920748A] [PMID: 20107664]

[422] Huang, C.C.; Yang, Z.; Lee, K.H.; Chang, H.T. Synthesis of highly fluorescent gold nanoparticles for sensing mercury(II). *Angew. Chem. Int. Ed. Engl.,* **2007**, *46*(36), 6824-6828. [http://dx.doi.org/10.1002/anie.200700803] [PMID: 17674391]

[423] Wei, H.; Wang, Z.; Yang, L.; Tian, S.; Hou, C.; Lu, Y. Lysozyme-stabilized gold fluorescent cluster: Synthesis and application as Hg(2+) sensor. *Analyst (Lond.),* **2010**, *135*(6), 1406-1410. [http://dx.doi.org/10.1039/c0an00046a] [PMID: 20411205]

[424] Shang, L.; Dong, S. Silver nanocluster-based fluorescent sensors for sensitive detection of Cu(II). *J. Mater. Chem.,* **2008**, *18*, 4636-4640. [http://dx.doi.org/10.1039/b810409c]

[425] Liu, Y.; Ai, K.; Cheng, X.; Huo, L.; Lu, L. Gold-nanocluster-based fluorescent sensors for highly sensitive and selective detection of cyanide in water. *Adv. Funct. Mater.,* **2010**, *20*, 951-956. [http://dx.doi.org/10.1002/adfm.200902062]

[426] Shang, L.; Dong, S. Sensitive detection of cysteine based on fluorescent silver clusters. *Biosens. Bioelectron.,* **2009**, *24*(6), 1569-1573. [http://dx.doi.org/10.1016/j.bios.2008.08.006] [PMID: 18823770]

[427] Guo, W.; Yuan, J.; Dong, Q.; Wang, E. Highly sequence-dependent formation of fluorescent silver nanoclusters in hybridized DNA duplexes for single nucleotide mutation identification. *J. Am. Chem. Soc.,* **2010**, *132*(3), 932-934.
[http://dx.doi.org/10.1021/ja907075s] [PMID: 20038102]

[428] Shang, L.; Yin, J.; Li, J.; Jin, L.; Dong, S. Gold nanoparticle-based near-infrared fluorescent detection of biological thiols in human plasma. *Biosens. Bioelectron.,* **2009**, *25*(2), 269-274.
[http://dx.doi.org/10.1016/j.bios.2009.06.021] [PMID: 19683912]

[429] Shang, L.; Qin, C.; Wang, T.; Wang, M.; Wang, L.; Dong, S. Fluorescent conjugated polymer-stabilized gold nanoparticles for sensitive and selective detection of cysteine. *J. Phys. Chem. C,* **2007**, *111*, 13414.
[http://dx.doi.org/10.1021/jp073913p]

[430] Shang, L.; Jin, L.; Dong, S. Sensitive turn-on fluorescent detection of cyanide based on the dissolution of fluorophore functionalized gold nanoparticles. *Chem. Commun. (Camb.),* **2009**, (21), 3077-3079.
[http://dx.doi.org/10.1039/b902216c] [PMID: 19462092]

[431] Shtoyko, T.; Matveeva, E.G.; Chang, I.F.; Gryczynski, Z.; Goldys, E.; Gryczynski, I. Enhanced fluorescent immunoassays on silver fractal-like structures. *Anal. Chem.,* **2008**, *80*(6), 1962-1966.
[http://dx.doi.org/10.1021/ac7019915] [PMID: 18288816]

[432] Steigerwald, M.L.; Brus, L.E. Semiconductor crystallites: A class of large molecules. *Acc. Chem. Res.,* **1990**, *23*, 183-188.
[http://dx.doi.org/10.1021/ar00174a003]

[433] Weller, H. Quantized semiconductor particles: A novel state of matter for materials science. *Adv. Mater.,* **1993**, *5*, 88-95.
[http://dx.doi.org/10.1002/adma.19930050204]

[434] Alivisatos, P. Perspectives on the physical chemistry of semiconductor nanocrystals. *J. Phys. Chem.,* **1996**, *100*, 13226-13239.
[http://dx.doi.org/10.1021/jp9535506]

[435] Kamat, P.V. Photophysical, photochemical and photocatalytic aspects of metal nanoparticles. *J. Phys. Chem. B,* **2002**, *106*, 7729-7744.
[http://dx.doi.org/10.1021/jp0209289]

[436] Adams, D.; Brus, L.; Chidsey, C.E.; Creager, S.; Cruetz, C.; Kagan, C.R.; Kamat, P.V.; Lieberman, M.; Lindsay, S.; Marcus, R.A.; Metzger, R.M.; Michel-Beyerle, M.E.; Miller, J.R.; Newton, M.D.; Rolison, D.R.; Sankey, O.; Schanze, K.S.; Yardley, J.; Zhu, X. Charge transfer on the nanoscale. *J. Phys. Chem. B,* **2003**, *107*, 6668-6697.
[http://dx.doi.org/10.1021/jp0268462]

[437] Ward, M.D. Photo-induced electron and energy transfer in noncovalently bonded supramolecular assemblies. *Chem. Soc. Rev.,* **1999**, 365-375.

[438] Templeton, A.C.; Wuelfing, W.P.; Murray, R.W. Monolayer-protected cluster molecules. *Acc. Chem. Res.,* **2000**, *33*(1), 27-36.
[http://dx.doi.org/10.1021/ar9602664] [PMID: 10639073]

[439] Schmid, G.; Corain, B. Nanoparticulated gold: syntheses, structures, electronics, and reactivities. *Eur.*

J. Inorg. Chem., **2003**, 3081-3098.
[http://dx.doi.org/10.1002/ejic.200300187]

[440] Imahori, H.; Arimura, M.; Hanada, T.; Nishimura, Y.; Yamazaki, I.; Sakata, Y.; Fukuzumi, S. Photoactive three-dimensional monolayers: porphyrin-alkanethiolate-stabilized gold clusters. *J. Am. Chem. Soc.,* **2001**, *123*(2), 335-336.
[http://dx.doi.org/10.1021/ja002838s] [PMID: 11456522]

[441] George Thomas, K.; Biju, V.; Guldi, D.M.; Kamat, P.V.; George, M.V. Photoinduced charge separation and stabilization in clusters of a fullerene-aniline dyad. *J. Phys. Chem. B,* **1999**, *103*, 8864-8869.
[http://dx.doi.org/10.1021/jp9915679]

[442] Nath, S.; Pal, H.; Palit, D.K.; Sapre, A.V.; Mittal, J.P. Aggregation of fullerene, C-60, in benzonitrile. *J. Phys. Chem. B,* **1998**, *102*, 10158-10164.
[http://dx.doi.org/10.1021/jp9824149]

[443] Kamat, P.V.; Barazzouk, S.; Hotchandani, S.; Thomas, K.G. Nanostructured thin films of C60-aniline dyad clusters: electrodeposition, charge separation, and photoelectrochemistry. *Chemistry,* **2000**, *6*(21), 3914-3921.
[http://dx.doi.org/10.1002/1521-3765(20001103)6:21<3914::AID-CHEM3914>3.3.CO;2-K] [PMID: 11126952]

[444] Hasobe, T.; Imahori, H.; Fukuzumi, S.; Kamat, P.V. Light energy harvesting using mixed molecular nanoclusters. Porphyrin and C60 cluster films for efficient photocurrent generation. *J. Phys. Chem. B,* **2003**, *107*, 12105-12112.
[http://dx.doi.org/10.1021/jp035854v]

[445] Hasobe, T.; Imahori, H.; Kamat, P.V.; Fukuzumi, S. Quaternary self-organization of porphyrin and fullerene units by clusterization with gold nanoparticles on SnO2 electrodes for organic solar cells. *J. Am. Chem. Soc.,* **2003**, *125*(49), 14962-14963.
[http://dx.doi.org/10.1021/ja0377192] [PMID: 14653711]

[446] Sudeep, P.K.; Ipe, B.I.; George Thomas, K.; George, M.V.; Barazzouk, S.; Hotchandani, S.; Kamat, P.V. Fullerene functionalized gold nanoparticles. A self-assembled photoactive antenna-metal nanocore assembly. *Nano Lett.,* **2002**, *2*, 29-35.
[http://dx.doi.org/10.1021/nl010073w]

[447] Biju, V.; Barazzouk, S.; Thomas, K.G.; George, M.V.; Kamat, P.V. Photoinduced electron transfer between 1,2,5-triphenylpyrrolidinofullerene cluster aggregates and electron donors. *Langmuir,* **2001**, *17*, 2930-2936.
[http://dx.doi.org/10.1021/la001644n]

[448] Ipe, B.I.; George Thomas, K.; Barazzouk, S.; Hotchandani, S.; Kamat, P.V. Photoinduced charge separation in a fluorophore-gold nanoassembly. *J. Phys. Chem. B,* **2002**, *106*, 18-21.
[http://dx.doi.org/10.1021/jp0134695]

[449] Solladié, N.; Hamel, A.; Gross, M. Synthesis of multiporphyrinic α-polypeptides: towards the study of the migration of an excited state for the mimicking of the natural light harvesting device. *Tetrahedron Lett.,* **2000**, *41*, 6075-6078.
[http://dx.doi.org/10.1016/S0040-4039(00)01049-2]

[450] Hasobe, T.; Kamat, P.V.; Troiani, V.; Solladié, N.; Ahn, T.K.; Kim, S.K.; Kim, D.; Kongkanand, A.; Kuwabata, S.; Fukuzumi, S. Enhancement of light-energy conversion efficiency by multi-porphyrin arrays of porphyrin-peptide oligomers with fullerene clusters. *J. Phys. Chem. B,* **2005**, *109*(1), 19-23. [http://dx.doi.org/10.1021/jp045246v] [PMID: 16850975]

[451] McIntosh, C.M.; Esposito, E.A., III; Boal, A.K.; Simard, J.M.; Martin, C.T.; Rotello, V.M. Inhibition of DNA transcription using cationic mixed monolayer protected gold clusters. *J. Am. Chem. Soc.,* **2001**, *123*(31), 7626-7629. [http://dx.doi.org/10.1021/ja015556g] [PMID: 11480984]

[452] Huang, X.; El-Sayed, I.H.; Qian, W.; El-Sayed, M.A. Cancer cell imaging and photothermal therapy in the near-infrared region by using gold nanorods. *J. Am. Chem. Soc.,* **2006**, *128*(6), 2115-2120. [http://dx.doi.org/10.1021/ja057254a] [PMID: 16464114]

[453] Huang, X.; Jain, P.K.; El-Sayed, I.H.; El-Sayed, M.A. Determination of the minimum temperature required for selective photothermal destruction of cancer cells with the use of immunotargeted gold nanoparticles. *Photochem. Photobiol.,* **2006**, *82*(2), 412-417. [http://dx.doi.org/10.1562/2005-12-14-RA-754] [PMID: 16613493]

[454] Huang, X.; El-Sayed, I.H.; Qian, W.; El-Sayed, M.A. Cancer cells assemble and align gold nanorods conjugated to antibodies to produce highly enhanced, sharp, and polarized surface Raman spectra: a potential cancer diagnostic marker. *Nano Lett.,* **2007**, *7*(6), 1591-1597. [http://dx.doi.org/10.1021/nl070472c] [PMID: 17474783]

[455] Yuan, Y.; Zhang, J.; Zhang, H.; Yang, X. Silver nanoparticle based label-free colorimetric immunosensor for rapid detection of neurogenin 1. *Analyst (Lond.),* **2012**, *137*(2), 496-501. [http://dx.doi.org/10.1039/C1AN15875A] [PMID: 22114758]

[456] Dubois, L.H.; Nuzzo, R.G. Synthesis, structure, and properties of model organic surfaces. *Annu. Rev. Phys. Chem.,* **1992**, *43*, 437-463. [http://dx.doi.org/10.1146/annurev.pc.43.100192.002253]

[457] Sapsford, K.E.; Algar, W.R.; Berti, L.; Gemmill, K.B.; Casey, B.J.; Oh, E.; Stewart, M.H.; Medintz, I.L. Functionalizing nanoparticles with biological molecules: developing chemistries that facilitate nanotechnology. *Chem. Rev.,* **2013**, *113*(3), 1904-2074. [http://dx.doi.org/10.1021/cr300143v] [PMID: 23432378]

[458] Amoozgar, Z.; Yeo, Y. Recent advances in stealth coating of nanoparticle drug delivery systems. *Wiley Interdiscip. Rev. Nanomed. Nanobiotechnol.,* **2012**, *4*(2), 219-233. [http://dx.doi.org/10.1002/wnan.1157] [PMID: 22231928]

[459] Wang, J.; Byrne, J.D.; Napier, M.E.; DeSimone, J.M. More effective nanomedicines through particle design. *Small,* **2011**, *7*(14), 1919-1931. [http://dx.doi.org/10.1002/smll.201100442] [PMID: 21695781]

[460] Alivisatos, A.P.; Johnsson, K.P.; Peng, X.; Wilson, T.E.; Loweth, C.J.; Bruchez, M.P., Jr; Schultz, P.G. Organization of 'nanocrystal molecules' using DNA. *Nature,* **1996**, *382*(6592), 609-611. [http://dx.doi.org/10.1038/382609a0] [PMID: 8757130]

[461] Glynou, K.; Ioannou, P.C.; Christopoulos, T.K.; Syriopoulou, V. Oligonucleotide-functionalized gold nanoparticles as probes in a dry-reagent strip biosensor for DNA analysis by hybridization. *Anal.*

Chem., **2003**, *75*(16), 4155-4160.
[http://dx.doi.org/10.1021/ac034256+] [PMID: 14632129]

[462] Rosi, N.L.; Giljohann, D.A.; Thaxton, C.S.; Lytton-Jean, A.K.; Han, M.S.; Mirkin, C.A. Oligonucleotide-modified gold nanoparticles for intracellular gene regulation. *Science,* **2006**, *312*(5776), 1027-1030.
[http://dx.doi.org/10.1126/science.1125559] [PMID: 16709779]

[463] Austin, L.A.; Kang, B.; Yen, C-W.; El-Sayed, M.A. Nuclear targeted silver nanospheres perturb the cancer cell cycle differently than those of nanogold. *Bioconjug. Chem.,* **2011**, *22*(11), 2324-2331.
[http://dx.doi.org/10.1021/bc200386m] [PMID: 22010874]

[464] Kang, B.; Mackey, M.A.; El-Sayed, M.A. Nuclear targeting of gold nanoparticles in cancer cells induces DNA damage, causing cytokinesis arrest and apoptosis. *J. Am. Chem. Soc.,* **2010**, *132*(5), 1517-1519.
[http://dx.doi.org/10.1021/ja9102698] [PMID: 20085324]

[465] Oyelere, A.K.; Chen, P.C.; Huang, X.; El-Sayed, I.H.; El-Sayed, M.A. Peptide-conjugated gold nanorods for nuclear targeting. *Bioconjug. Chem.,* **2007**, *18*(5), 1490-1497.
[http://dx.doi.org/10.1021/bc070132i] [PMID: 17630680]

[466] Liu, Z.; Kiessling, F.; Gatjens, J. *Advanced nanomaterials in multimodal imaging: design, functionalization, and biomedical applications.,* **2010**.
[http://dx.doi.org/10.1155/2010/894303]

[467] Prabaharan, M.; Grailer, J.J.; Pilla, S.; Steeber, D.A.; Gong, S. Gold nanoparticles with a monolayer of doxorubicin-conjugated amphiphilic block copolymer for tumor-targeted drug delivery. *Biomaterials,* **2009**, *30*(30), 6065-6075.
[http://dx.doi.org/10.1016/j.biomaterials.2009.07.048] [PMID: 19674777]

[468] Ting, B.P.; Zhang, J.; Gao, Z.; Ying, J.Y. A DNA biosensor based on the detection of doxorubicin-conjugated Ag nanoparticle labels using solid-state voltammetry. *Biosens. Bioelectron.,* **2009**, *25*(2), 282-287.
[http://dx.doi.org/10.1016/j.bios.2009.07.005] [PMID: 19665881]

[469] Fischler, M.; Sologubenko, A.; Mayer, J.; Clever, G.; Burley, G.; Gierlich, J.; Carell, T.; Simon, U. Chain-like assembly of gold nanoparticles on artificial DNA templates *via* 'click chemistry'. *Chem. Commun. (Camb.),* **2008**, (2), 169-171.
[http://dx.doi.org/10.1039/B715602B] [PMID: 18092076]

[470] Zhu, K.; Zhang, Y.; He, S.; Chen, W.; Shen, J.; Wang, Z.; Jiang, X. Quantification of proteins by functionalized gold nanoparticles using click chemistry. *Anal. Chem.,* **2012**, *84*(10), 4267-4270.
[http://dx.doi.org/10.1021/ac3010567] [PMID: 22540271]

[471] Paciotti, G.F.; Myer, L.; Weinreich, D.; Goia, D.; Pavel, N.; McLaughlin, R.E.; Tamarkin, L. Colloidal gold: a novel nanoparticle vector for tumor directed drug delivery. *Drug Deliv.,* **2004**, *11*(3), 169-183.
[http://dx.doi.org/10.1080/10717540490433895] [PMID: 15204636]

[472] Chen, Y.H.; Tsai, C.Y.; Huang, P.Y.; Chang, M.Y.; Cheng, P.C.; Chou, C.H.; Chen, D.H.; Wang, C.R.; Shiau, A.L.; Wu, C.L. Methotrexate conjugated to gold nanoparticles inhibits tumor growth in a syngeneic lung tumor model. *Mol. Pharm.,* **2007**, *4*(5), 713-722.
[http://dx.doi.org/10.1021/mp060132k] [PMID: 17708653]

[473] Dreaden, E.C.; Mwakwari, S.C.; Sodji, Q.H.; Oyelere, A.K.; El-Sayed, M.A. Tamoxifen-poly(ethylene glycol)-thiol gold nanoparticle conjugates: enhanced potency and selective delivery for breast cancer treatment. *Bioconjug. Chem.,* **2009**, *20*(12), 2247-2253. [http://dx.doi.org/10.1021/bc9002212] [PMID: 19919059]

[474] Dreaden, E.C.; Gryder, B.E.; Austin, L.A.; Tene Defo, B.A.; Hayden, S.C.; Pi, M.; Quarles, L.D.; Oyelere, A.K.; El-Sayed, M.A. Antiandrogen gold nanoparticles dual-target and overcome treatment resistance in hormone-insensitive prostate cancer cells. *Bioconjug. Chem.,* **2012**, *23*(8), 1507-1512. [http://dx.doi.org/10.1021/bc300158k] [PMID: 22768914]

[475] Dong, X.; Mumper, R.J. Nanomedicinal strategies to treat multidrug-resistant tumors: current progress. *Nanomedicine (Lond.),* **2010**, *5*(4), 597-615. [http://dx.doi.org/10.2217/nnm.10.35] [PMID: 20528455]

[476] Wang, F.; Wang, Y.C.; Dou, S.; Xiong, M.H.; Sun, T.M.; Wang, J. Doxorubicin-tethered responsive gold nanoparticles facilitate intracellular drug delivery for overcoming multidrug resistance in cancer cells. *ACS Nano,* **2011**, *5*(5), 3679-3692. [http://dx.doi.org/10.1021/nn200007z] [PMID: 21462992]

[477] El-Sayed, M.A. Some interesting properties of metals confined in time and nanometer space of different shapes. *Acc. Chem. Res.,* **2001**, *34*(4), 257-264. [http://dx.doi.org/10.1021/ar960016n] [PMID: 11308299]

[478] Link, S.; El-Sayed, M.A. Optical properties and ultrafast dynamics of metallic nanocrystals. *Annu. Rev. Phys. Chem.,* **2003**, *54*, 331-366. [http://dx.doi.org/10.1146/annurev.physchem.54.011002.103759] [PMID: 12626731]

[479] Nolsøe, C.P.; Torp-Pedersen, S.; Burcharth, F.; Horn, T.; Pedersen, S.; Christensen, N.E.; Olldag, E.S.; Andersen, P.H.; Karstrup, S.; Lorentzen, T. Interstitial hyperthermia of colorectal liver metastases with a US-guided Nd-YAG laser with a diffuser tip: a pilot clinical study. *Radiology,* **1993**, *187*(2), 333-337. [http://dx.doi.org/10.1148/radiology.187.2.8475269] [PMID: 8475269]

[480] Iyer, A.K.; Khaled, G.; Fang, J.; Maeda, H. Exploiting the enhanced permeability and retention effect for tumor targeting. *Drug Discov. Today,* **2006**, *11*(17-18), 812-818. [http://dx.doi.org/10.1016/j.drudis.2006.07.005] [PMID: 16935749]

[481] Hildebrandt, B.; Wust, P.; Ahlers, O.; Dieing, A.; Sreenivasa, G.; Kerner, T.; Felix, R.; Riess, H. The cellular and molecular basis of hyperthermia. *Crit. Rev. Oncol. Hematol.,* **2002**, *43*(1), 33-56. [http://dx.doi.org/10.1016/S1040-8428(01)00179-2] [PMID: 12098606]

[482] Pitsillides, C.M.; Joe, E.K.; Wei, X.; Anderson, R.R.; Lin, C.P. Selective cell targeting with light-absorbing microparticles and nanoparticles. *Biophys. J.,* **2003**, *84*(6), 4023-4032. [http://dx.doi.org/10.1016/S0006-3495(03)75128-5] [PMID: 12770906]

[483] Hirsch, L.R.; Stafford, R.J.; Bankson, J.A.; Sershen, S.R.; Rivera, B.; Price, R.E.; Hazle, J.D.; Halas, N.J.; West, J.L. Nanoshell-mediated near-infrared thermal therapy of tumors under magnetic resonance guidance. *Proc. Natl. Acad. Sci. USA,* **2003**, *100*(23), 13549-13554. [http://dx.doi.org/10.1073/pnas.2232479100] [PMID: 14597719]

[484] O'Neal, D.P.; Hirsch, L.R.; Halas, N.J.; Payne, J.D.; West, J.L. Photo-thermal tumor ablation in mice

using near infrared-absorbing nanoparticles. *Cancer Lett.,* **2004**, *209*(2), 171-176.
[http://dx.doi.org/10.1016/j.canlet.2004.02.004] [PMID: 15159019]

[485] Chen, J.; Wang, D.; Xi, J.; Au, L.; Siekkinen, A.; Warsen, A.; Li, Z.Y.; Zhang, H.; Xia, Y.; Li, X. Immuno gold nanocages with tailored optical properties for targeted photothermal destruction of cancer cells. *Nano Lett.,* **2007**, *7*(5), 1318-1322.
[http://dx.doi.org/10.1021/nl070345g] [PMID: 17430005]

[486] Lu, W.; Xiong, C.; Zhang, G.; Huang, Q.; Zhang, R.; Zhang, J.Z.; Li, C. Targeted photothermal ablation of murine melanomas with melanocyte-stimulating hormone analog-conjugated hollow gold nanospheres. *Clin. Cancer Res.,* **2009**, *15*(3), 876-886.
[http://dx.doi.org/10.1158/1078-0432.CCR-08-1480] [PMID: 19188158]

[487] Melancon, M.P.; Lu, W.; Yang, Z.; Zhang, R.; Cheng, Z.; Elliot, A.M.; Stafford, J.; Olson, T.; Zhang, J.Z.; Li, C. *In vitro* and *in vivo* targeting of hollow gold nanoshells directed at epidermal growth factor receptor for photothermal ablation therapy. *Mol. Cancer Ther.,* **2008**, *7*(6), 1730-1739.
[http://dx.doi.org/10.1158/1535-7163.MCT-08-0016] [PMID: 18566244]

[488] Van de Broek, B.; Devoogdt, N.; D'Hollander, A.; Gijs, H.L.; Jans, K.; Lagae, L.; Muyldermans, S.; Maes, G.; Borghs, G. Specific cell targeting with nanobody conjugated branched gold nanoparticles for photothermal therapy. *ACS Nano,* **2011**, *5*(6), 4319-4328.
[http://dx.doi.org/10.1021/nn1023363] [PMID: 21609027]

[489] Chaloupka, K.; Malam, Y.; Seifalian, A.M. Nanosilver as a new generation of nanoproduct in biomedical applications. *Trends Biotechnol.,* **2010**, *28*(11), 580-588.
[http://dx.doi.org/10.1016/j.tibtech.2010.07.006] [PMID: 20724010]

[490] Rai, M.; Yadav, A.; Gade, A. Silver nanoparticles as a new generation of antimicrobials. *Biotechnol. Adv.,* **2009**, *27*(1), 76-83.
[http://dx.doi.org/10.1016/j.biotechadv.2008.09.002] [PMID: 18854209]

[491] Sharma, V.K.; Yngard, R.A.; Lin, Y. Silver nanoparticles: green synthesis and their antimicrobial activities. *Adv. Colloid Interface Sci.,* **2009**, *145*(1-2), 83-96.
[http://dx.doi.org/10.1016/j.cis.2008.09.002] [PMID: 18945421]

[492] Sondi, I.; Salopek-Sondi, B. Silver nanoparticles as antimicrobial agent: a case study on *E . coli* as a model for Gram-negative bacteria. *J. Colloid Interface Sci.,* **2004**, *275*(1), 177-182.
[http://dx.doi.org/10.1016/j.jcis.2004.02.012] [PMID: 15158396]

[493] Morones, J.R.; Elechiguerra, J.L.; Camacho, A.; Holt, K.; Kouri, J.B.; Ramírez, J.T.; Yacaman, M.J. The bactericidal effect of silver nanoparticles. *Nanotechnology,* **2005**, *16*(10), 2346-2353.
[http://dx.doi.org/10.1088/0957-4484/16/10/059] [PMID: 20818017]

[494] Pal, S.; Tak, Y.K.; Song, J.M. Does the antibacterial activity of silver nanoparticles depend on the shape of the nanoparticle? A study of the Gram-negative bacterium *Escherichia coli. Appl. Environ. Microbiol.,* **2007**, *73*(6), 1712-1720.
[http://dx.doi.org/10.1128/AEM.02218-06] [PMID: 17261510]

[495] Fayaz, A.M.; Balaji, K.; Girilal, M.; Yadav, R.; Kalaichelvan, P.T.; Venketesan, R. Biogenic synthesis of silver nanoparticles and their synergistic effect with antibiotics: a study against gram-positive and gram-negative bacteria. *Nanomedicine (Lond.),* **2010**, *6*(1), 103-109.
[PMID: 19447203]

[496] Li, W.R.; Xie, X.B.; Shi, Q.S.; Zeng, H.Y.; Ou-Yang, Y.S.; Chen, Y.B. Antibacterial activity and mechanism of silver nanoparticles on Escherichia coli. *Appl. Microbiol. Biotechnol.,* **2010**, *85*(4), 1115-1122.
[http://dx.doi.org/10.1007/s00253-009-2159-5] [PMID: 19669753]

[497] Marambio-Jones, C.; Hoek, E.M. A review of the antibacterial effects of silver nanomaterials and potential implications for human health and the environment. *J. Nanopart. Res.,* **2010**, *12*, 1531-1551.
[http://dx.doi.org/10.1007/s11051-010-9900-y]

[498] Tian, J.; Wong, K.K.; Ho, C.M.; Lok, C.N.; Yu, W.Y.; Che, C.M.; Chiu, J.F.; Tam, P.K. Topical delivery of silver nanoparticles promotes wound healing. *ChemMedChem,* **2007**, *2*(1), 129-136.
[http://dx.doi.org/10.1002/cmdc.200600171] [PMID: 17075952]

[499] Liu, X.; Lee, P.Y.; Ho, C.M.; Lui, V.C.; Chen, Y.; Che, C.M.; Tam, P.K.; Wong, K.K. Silver nanoparticles mediate differential responses in keratinocytes and fibroblasts during skin wound healing. *ChemMedChem,* **2010**, *5*(3), 468-475.
[http://dx.doi.org/10.1002/cmdc.200900502] [PMID: 20112331]

[500] Kwan, K.H.; Liu, X.; To, M.K.; Yeung, K.W.; Ho, C-M.; Wong, K.K. Modulation of collagen alignment by silver nanoparticles results in better mechanical properties in wound healing. *Nanomedicine (Lond.),* **2011**, *7*(4), 497-504.
[PMID: 21272666]

[501] Chomoucka, J.; Drbohlavova, J.; Huska, D.; Adam, V.; Kizek, R.; Hubalek, J. Magnetic nanoparticles and targeted drug delivering. *Pharmacol. Res.,* **2010**, *62*(2), 144-149.
[http://dx.doi.org/10.1016/j.phrs.2010.01.014] [PMID: 20149874]

[502] Jennings, L.E.; Long, N.J. 'Two is better than one'--probes for dual-modality molecular imaging. *Chem. Commun. (Camb.),* **2009**, (24), 3511-3524.
[http://dx.doi.org/10.1039/b821903f] [PMID: 19521594]

[503] Bulte, J.W.; Kraitchman, D.L. Iron oxide MR contrast agents for molecular and cellular imaging. *NMR Biomed.,* **2004**, *17*(7), 484-499.
[http://dx.doi.org/10.1002/nbm.924] [PMID: 15526347]

[504] Jun, Y.W.; Lee, J.H.; Cheon, J. Chemical design of nanoparticle probes for high-performance magnetic resonance imaging. *Angew. Chem. Int. Ed. Engl.,* **2008**, *47*(28), 5122-5135.
[http://dx.doi.org/10.1002/anie.200701674] [PMID: 18574805]

[505] Chavanpatil, M.D.; Khdair, A.; Panyam, J. Nanoparticles for cellular drug delivery: mechanisms and factors influencing delivery. *J. Nanosci. Nanotechnol.,* **2006**, *6*(9-10), 2651-2663.
[http://dx.doi.org/10.1166/jnn.2006.443] [PMID: 17048473]

[506] Anzai, Y.; Piccoli, C.W.; Outwater, E.K.; Stanford, W.; Bluemke, D.A.; Nurenberg, P.; Saini, S.; Maravilla, K.R.; Feldman, D.E.; Schmiedl, U.P.; Brunberg, J.A.; Francis, I.R.; Harms, S.E.; Som, P.M.; Tempany, C.M. Evaluation of neck and body metastases to nodes with ferumoxtran 10-enhanced MR imaging: phase III safety and efficacy study. *Radiology,* **2003**, *228*(3), 777-788.
[http://dx.doi.org/10.1148/radiol.2283020872] [PMID: 12954896]

[507] Lee, J.H.; Huh, Y.M.; Jun, Y.W.; Seo, J.W.; Jang, J.T.; Song, H.T.; Kim, S.; Cho, E.J.; Yoon, H.G.; Suh, J.S.; Cheon, J. Artificially engineered magnetic nanoparticles for ultra-sensitive molecular

imaging. *Nat. Med.,* **2007**, *13*(1), 95-99.
[http://dx.doi.org/10.1038/nm1467] [PMID: 17187073]

[508] Cheng, F.Y.; Wang, S.P.; Su, C.H.; Tsai, T.L.; Wu, P.C.; Shieh, D.B.; Chen, J.H.; Hsieh, P.C.; Yeh, C.S. Stabilizer-free poly(lactide-co-glycolide) nanoparticles for multimodal biomedical probes. *Biomaterials,* **2008**, *29*(13), 2104-2112.
[http://dx.doi.org/10.1016/j.biomaterials.2008.01.010] [PMID: 18276001]

[509] Cao, A.H.; Shi, H.J.; Zhang, Y.; Teng, G.J. *In vivo* tracking of dual-labeled mesenchymal stem cells homing into the injured common carotid artery. *Anat. Rec. (Hoboken),* **2009**, *292*(10), 1677-1683.
[http://dx.doi.org/10.1002/ar.20951] [PMID: 19685506]

[510] Xie, J.; Chen, K.; Huang, J.; Lee, S.; Wang, J.; Gao, J.; Li, X.; Chen, X. PET/NIRF/MRI triple functional iron oxide nanoparticles. *Biomaterials,* **2010**, *31*(11), 3016-3022.
[http://dx.doi.org/10.1016/j.biomaterials.2010.01.010] [PMID: 20092887]

[511] Frullano, L.; Meade, T.J. Multimodal MRI contrast agents. *J. Biol. Inorg. Chem.,* **2007**, *12*(7), 939-949.
[http://dx.doi.org/10.1007/s00775-007-0265-3] [PMID: 17659368]

[512] Josephson, L.; Tung, C.H.; Moore, A.; Weissleder, R. High-efficiency intracellular magnetic labeling with novel superparamagnetic-Tat peptide conjugates. *Bioconjug. Chem.,* **1999**, *10*(2), 186-191.
[http://dx.doi.org/10.1021/bc980125h] [PMID: 10077466]

[513] Josephson, L.; Kircher, M.F.; Mahmood, U.; Tang, Y.; Weissleder, R. Near-infrared fluorescent nanoparticles as combined MR/optical imaging probes. *Bioconjug. Chem.,* **2002**, *13*(3), 554-560.
[http://dx.doi.org/10.1021/bc015555d] [PMID: 12009946]

[514] Nitin, N.; LaConte, L.E.; Zurkiya, O.; Hu, X.; Bao, G. Functionalization and peptide-based delivery of magnetic nanoparticles as an intracellular MRI contrast agent. *J. Biol. Inorg. Chem.,* **2004**, *9*(6), 706-712.
[http://dx.doi.org/10.1007/s00775-004-0560-1] [PMID: 15232722]

[515] Veiseh, O.; Sun, C.; Gunn, J.; Kohler, N.; Gabikian, P.; Lee, D.; Bhattarai, N.; Ellenbogen, R.; Sze, R.; Hallahan, A.; Olson, J.; Zhang, M. Optical and MRI multifunctional nanoprobe for targeting gliomas. *Nano Lett.,* **2005**, *5*(6), 1003-1008.
[http://dx.doi.org/10.1021/nl0502569] [PMID: 15943433]

[516] Klichko, Y.; Liong, M.; Choi, E.; Angelos, S.; Nel, A.E.; Stoddart, J.F.; Tamanoi, F.; Zink, J.I. Mesostructured silica for optical functionality, nanomachines, and Drug Delivery. *J. Am. Ceram. Soc.,* **2009**, *92*(s1), s2-s10.
[http://dx.doi.org/10.1111/j.1551-2916.2008.02722.x] [PMID: 19834571]

[517] Santra, S.; Kaittanis, C.; Grimm, J.; Perez, J.M. Drug/dye-loaded, multifunctional iron oxide nanoparticles for combined targeted cancer therapy and dual optical/magnetic resonance imaging. *Small,* **2009**, *5*(16), 1862-1868.
[http://dx.doi.org/10.1002/smll.200900389] [PMID: 19384879]

[518] West, C.M.; Jones, T.; Price, P. The potential of positron-emission tomography to study anticancer-drug resistance. *Nat. Rev. Cancer,* **2004**, *4*(6), 457-469.
[http://dx.doi.org/10.1038/nrc1368] [PMID: 15170448]

[519] Jarrett, B.R.; Gustafsson, B.; Kukis, D.L.; Louie, A.Y. Synthesis of 64Cu-labeled magnetic nanoparticles for multimodal imaging. *Bioconjug. Chem.,* **2008**, *19*(7), 1496-1504. [http://dx.doi.org/10.1021/bc800108v] [PMID: 18578485]

[520] Lee, H.Y.; Li, Z.; Chen, K.; Hsu, A.R.; Xu, C.; Xie, J.; Sun, S.; Chen, X. PET/MRI dual-modality tumor imaging using arginine-glycine-aspartic (RGD)-conjugated radiolabeled iron oxide nanoparticles. *J. Nucl. Med.,* **2008**, *49*(8), 1371-1379. [http://dx.doi.org/10.2967/jnumed.108.051243] [PMID: 18632815]

[521] Choi, J.S.; Park, J.C.; Nah, H.; Woo, S.; Oh, J.; Kim, K.M.; Cheon, G.J.; Chang, Y.; Yoo, J.; Cheon, J. A hybrid nanoparticle probe for dual-modality positron emission tomography and magnetic resonance imaging. *Angew. Chem. Int. Ed. Engl.,* **2008**, *47*(33), 6259-6262. [http://dx.doi.org/10.1002/anie.200801369] [PMID: 18613191]

[522] Rosengart, A.J.; Kaminski, M.D.; Chen, H.; Caviness, P.L.; Ebner, A.D.; Ritter, J.A. Magnetizable implants and functionalized magnetic carriers: A novel approach for noninvasive yet targeted drug delivery. *J. Magn. Magn. Mater.,* **2005**, *293*, 633-638. [http://dx.doi.org/10.1016/j.jmmm.2005.01.087]

[523] Iacob, G.; Rotariu, O.; Strachan, N.J.; Häfeli, U.O. Magnetizable needles and wires--modeling an efficient way to target magnetic microspheres *in vivo. Biorheology,* **2004**, *41*(5), 599-612. [PMID: 15477667]

[524] Berry, C.C.; Curtis, A.S. Functionalisation of magnetic nanoparticles for applications in biomedicine. *J. Phys. D Appl. Phys.,* **2003**, *36*, R198-R206. [http://dx.doi.org/10.1088/0022-3727/36/13/203]

[525] Banerjee, S.S.; Chen, D-H. Magnetic nanoparticles grafted with cyclodextrin for hydrophobic drug delivery. *Chem. Mater.,* **2007**, *19*, 6345-6349. [http://dx.doi.org/10.1021/cm702278u]

[526] Gu, H.; Xu, K.; Xu, C.; Xu, B. Biofunctional magnetic nanoparticles for protein separation and pathogen detection. *Chem. Commun. (Camb.),* **2006**, (9), 941-949. [http://dx.doi.org/10.1039/b514130c] [PMID: 16491171]

[527] Hultgren, A.; Tanase, M.; Chen, C.S.; Meyer, G.J.; Reich, D.H. Cell manipulation using magnetic nanowires. *J. Appl. Phys.,* **2003**, *93*, 7554-7556. [http://dx.doi.org/10.1063/1.1556204]

[528] Smith, J.E.; Wang, L.; Tan, W. Bioconjugated silica-coated nanoparticles for bioseparation and bioanalysis. *Trends Analyt. Chem.,* **2006**, *25*, 848-855. [http://dx.doi.org/10.1016/j.trac.2006.03.008]

[529] Sen, T.; Sebastianelli, A.; Bruce, I.J. Mesoporous silica-magnetite nanocomposite: fabrication and applications in magnetic bioseparations. *J. Am. Chem. Soc.,* **2006**, *128*(22), 7130-7131. [http://dx.doi.org/10.1021/ja061393q] [PMID: 16734444]

[530] Melzak, K.A.; Sherwood, C.S.; Turner, R.F.; Haynes, C.A. Driving forces for DNA adsorption to silica in perchlorate solutions. *J. Colloid Interface Sci.,* **1996**, *181*, 635-644. [http://dx.doi.org/10.1006/jcis.1996.0421]

[531] Yanase, M.; Shinkai, M.; Honda, H.; Wakabayashi, T.; Yoshida, J.; Kobayashi, T. Intracellular

hyperthermia for cancer using magnetite cationic liposomes: an *in vivo* study. *Jpn. J. Cancer Res.,* **1998**, *89*(4), 463-469.
[http://dx.doi.org/10.1111/j.1349-7006.1998.tb00586.x] [PMID: 9617354]

[532] Chan, D.C.; Kirpotin, D.B.; Bunn, P.A. Synthesis and evaluation of colloidal magnetic iron oxides for the site-specific radiofrequency-induced hyperthermia of cancer. *J. Magn. Magn. Mater.,* **1993**, *122*, 374-378.
[http://dx.doi.org/10.1016/0304-8853(93)91113-L]

[533] Ito, A.; Shinkai, M.; Honda, H.; Kobayashi, T. Medical application of functionalized magnetic nanoparticles. *J. Biosci. Bioeng.,* **2005**, *100*(1), 1-11.
[http://dx.doi.org/10.1263/jbb.100.1] [PMID: 16233845]

[534] Chantrell, R.W.; Lyberatos, A.; El-Hilo, M.; O'Grady, K. Models of slow relaxation in particulate and thin film materials. *J. Appl. Phys.,* **1994**, *76*, 6407-6412.
[http://dx.doi.org/10.1063/1.358281]

[535] Dormann, J.L.; Spinu, L.; Tronc, E.; Jolivet, J.P.; Lucari, F.; D'Orazio, F.; Fiorani, D. Effect of interparticle interactions on the dynamical properties of γ-Fe2O3 nanoparticles. *J. Magn. Magn. Mater.,* **1998**, *183*, L255-L260.
[http://dx.doi.org/10.1016/S0304-8853(97)01123-2]

[536] Gladysz, J.A. Introduction: recoverable catalysts and reagents-perspective and prospective. *Chem. Rev.,* **2002**, *102*(10), 3215-3216.
[http://dx.doi.org/10.1021/cr020068s] [PMID: 12371883]

[537] Abu-Reziq, R.; Alper, H.; Wang, D.; Post, M.L. Metal supported on dendronized magnetic nanoparticles: highly selective hydroformylation catalysts. *J. Am. Chem. Soc.,* **2006**, *128*(15), 5279-5282.
[http://dx.doi.org/10.1021/ja060140u] [PMID: 16608364]

[538] Guin, D.; Baruwati, B.; Manorama, S.V. Pd on amine-terminated ferrite nanoparticles: a complete magnetically recoverable facile catalyst for hydrogenation reactions. *Org. Lett.,* **2007**, *9*(7), 1419-1421.
[http://dx.doi.org/10.1021/ol070290p] [PMID: 17323965]

[539] Hu, A.; Yee, G.T.; Lin, W. Magnetically recoverable chiral catalysts immobilized on magnetite nanoparticles for asymmetric hydrogenation of aromatic ketones. *J. Am. Chem. Soc.,* **2005**, *127*(36), 12486-12487.
[http://dx.doi.org/10.1021/ja053881o] [PMID: 16144385]

[540] Ding, S.J.; Xing, Y.C.; Radosz, M.; Shen, Y.Q. Magnetic nanoparticle supported catalyst for atom transfer radical polymerization. *Macromolecules,* **2006**, *39*, 6399-6405.
[http://dx.doi.org/10.1021/ma061062y]

[541] Stevens, P.D.; Fan, J.; Gardimalla, H.M.; Yen, M.; Gao, Y. Superparamagnetic nanoparticle-supported catalysis of Suzuki cross-coupling reactions. *Org. Lett.,* **2005**, *7*(11), 2085-2088.
[http://dx.doi.org/10.1021/ol050218w] [PMID: 15901140]

[542] Duanmu, C.; Saha, I.; Zheng, Y.; Goodson, B.M.; Gao, Y. Dendron-functionalized superparamagnetic nanoparticles with switchable solubility in organic and aqueous media: matrices for homogeneous catalysis and potential MRI contrast agents. *Chem. Mater.,* **2006**, *18*, 5973-5981.

[http://dx.doi.org/10.1021/cm061782j]

[543] Gardimalla, H.M.; Mandal, D.; Stevens, P.D.; Yen, M.; Gao, Y. Superparamagnetic nanoparticle-supported enzymatic resolution of racemic carboxylates. *Chem. Commun. (Camb.),* **2005**, (35), 4432-4434.
[http://dx.doi.org/10.1039/b504128g] [PMID: 16136241]

[544] Zheng, Y.; Duanmu, C.; Gao, Y. A magnetic biomimetic nanocatalyst for cleaving phosphoester and carboxylic ester bonds under mild conditions. *Org. Lett.,* **2006**, *8*(15), 3215-3217.
[http://dx.doi.org/10.1021/ol061098a] [PMID: 16836369]

[545] Hu, J.; Chen, G.; Lo, I.M. Removal and recovery of Cr(VI) from wastewater by maghemite nanoparticles. *Water Res.,* **2005**, *39*(18), 4528-4536.
[http://dx.doi.org/10.1016/j.watres.2005.05.051] [PMID: 16146639]

[546] Zhou, L.; Gao, C.; Xu, W. Robust Fe3O4/SiO2-Pt/Au/Pd magnetic nanocatalysts with multifunctional hyperbranched polyglycerol amplifiers. *Langmuir,* **2010**, *26*(13), 11217-11225.
[http://dx.doi.org/10.1021/la100556p] [PMID: 20387849]

[547] Lin, F-H.; Doong, R-A. Bifunctional Au-Fe3O4 heterostructures for magnetically recyclable catalysis of nitrophenol reduction. *J. Phys. Chem. C,* **2011**, *115*, 6591-6598.
[http://dx.doi.org/10.1021/jp110956k]

[548] Ko, S.; Jang, J. A highly efficient palladium nanocatalyst anchored on a magnetically functionalized polymer-nanotube support. *Angew. Chem. Int. Ed. Engl.,* **2006**, *45*(45), 7564-7567.
[http://dx.doi.org/10.1002/anie.200602456] [PMID: 17042052]

[549] Yi, D.K.; Lee, S.S.; Ying, J.Y. Synthesis and applications of magnetic nanocomposite catalysts. *Chem. Mater.,* **2006**, *18*, 2459-2461.
[http://dx.doi.org/10.1021/cm052885p]

[550] Ranganath, K.V.; Kloesges, J.; Schäfer, A.H.; Glorius, F. Asymmetric nanocatalysis: N-heterocyclic carbenes as chiral modifiers of Fe3O4/Pd nanoparticles. *Angew. Chem. Int. Ed. Engl.,* **2010**, *49*(42), 7786-7789.
[http://dx.doi.org/10.1002/anie.201002782] [PMID: 20845343]

[551] Dálaigh, C.O.; Corr, S.A.; Gun'ko, Y.; Connon, S.J. *Angew. Chem. Int. Ed.,* **2007**, *46*, 4329-4332.
[http://dx.doi.org/10.1002/anie.200605216]

[552] Zhang, W-X. Nanoscale iron particles for environmental remediation: an overview. *J. Nanopart. Res.,* **2003**, *5*, 323-332.
[http://dx.doi.org/10.1023/A:1025520116015]

[553] Blowes, D.W.; Ptacek, C.J.; Benner, S.G.; McRae, W.T. Che, Bennett, T.A.; Puls, R.W. Treatment of inorganic contaminants using permeable reactive barriers. *J. Contam. Hydrol.,* **2000**, *45*, 123-137.
[http://dx.doi.org/10.1016/S0169-7722(00)00122-4]

[554] Li, L.; Fan, M.; Brown, R.C.; Leeuwen, J.V.; Wang, J.; Wang, W.; Song, Y.; Zhang, P. Synthesis, properties, and environmental applications of nanoscale iron-based materials: a review. *Crit. Rev. Environ. Sci. Technol.,* **2006**, *36*, 405-431.
[http://dx.doi.org/10.1080/10643380600620387]

[555] Huang, S-H.; Liao, M-H.; Chen, D-H. Fast and efficient recovery of lipase by polyacrylic acid-coated

magnetic nano-adsorbent with high activity retention. *Separ. Purif. Tech.,* **2006**, *51*, 113-117. [http://dx.doi.org/10.1016/j.seppur.2006.01.003]

[556] Mak, S-Y.; Chen, D-H. Fast adsorption of methylene blue on polyacrylic acid-bound iron oxide magnetic nanoparticles. *Dyes Pigm.,* **2004**, *61*, 93-98. [http://dx.doi.org/10.1016/j.dyepig.2003.10.008]

[557] Zargar, B.; Parham, H.; Hatamie, A. Fast removal and recovery of amaranth by modified iron oxide magnetic nanoparticles. *Chemosphere,* **2009**, *76*(4), 554-557. [http://dx.doi.org/10.1016/j.chemosphere.2009.02.065] [PMID: 19345980]

[558] Chang, Y-C.; Chen, D-H. Adsorption kinetics and thermodynamics of acid dyes on a carboxymethylated chitosan-conjugated magnetic nano-adsorbent. *Macromol. Biosci.,* **2005**, *5*(3), 254-261. [http://dx.doi.org/10.1002/mabi.200400153] [PMID: 15768445]

[559] Li, P.; Miser, D.E.; Rabiei, S.; Yadav, R.T.; Hajaligol, M.R. The removal of carbon monoxide by iron oxide nanoparticles. *Appl. Catal. B,* **2003**, *43*, 151-162. [http://dx.doi.org/10.1016/S0926-3373(02)00297-7]

[560] Chung, J.; Chun, J.; Lee, J.; Lee, S.H.; Lee, Y.J.; Hong, S.W. Sorption of Pb(II) and Cu(II) onto multi-amine grafted mesoporous silica embedded with nano-magnetite: effects of steric factors. *J. Hazard. Mater.,* **2012**, *239-240*, 183-191. [http://dx.doi.org/10.1016/j.jhazmat.2012.08.063] [PMID: 22985819]

[561] Tan, Y.; Chen, M.; Hao, Y. High efficient removal of Pb (II) by aminofunctionalized Fe3O4 magnetic nano-particles. *Chem. Eng. J.,* **2012**, *191*, 104-111. [http://dx.doi.org/10.1016/j.cej.2012.02.075]

[562] Wang, L.; Yang, Z.; Gao, J.; Xu, K.; Gu, H.; Zhang, B.; Zhang, X.; Xu, B. A biocompatible method of decorporation: bisphosphonate-modified magnetite nanoparticles to remove uranyl ions from blood. *J. Am. Chem. Soc.,* **2006**, *128*(41), 13358-13359. [http://dx.doi.org/10.1021/ja0651355] [PMID: 17031939]

[563] Lee, H.Y.; Bae, D.R.; Park, J.C.; Song, H.; Han, W.S.; Jung, J.H. A selective fluoroionophore based on BODIPY-functionalized magnetic silica nanoparticles: removal of Pb2+ from human blood. *Angew. Chem. Int. Ed. Engl.,* **2009**, *48*(7), 1239-1243. [http://dx.doi.org/10.1002/anie.200804714] [PMID: 19130522]

[564] Kim, J.; Piao, Y.; Hyeon, T. Multifunctional nanostructured materials for multimodal imaging, and simultaneous imaging and therapy. *Chem. Soc. Rev.,* **2009**, *38*(2), 372-390. [http://dx.doi.org/10.1039/B709883A] [PMID: 19169455]

[565] Shao, L.; Gao, Y.; Yan, F. Semiconductor quantum dots for biomedicial applications. *Sensors (Basel),* **2011**, *11*(12), 11736-11751. [http://dx.doi.org/10.3390/s111211736] [PMID: 22247690]

[566] James, M.L.; Gambhir, S.S. A molecular imaging primer: modalities, imaging agents, and applications. *Physiol. Rev.,* **2012**, *92*(2), 897-965. [http://dx.doi.org/10.1152/physrev.00049.2010] [PMID: 22535898]

[567] Kairdolf, B.A.; Smith, A.M.; Stokes, T.H.; Wang, M.D.; Young, A.N.; Nie, S. Semiconductor

quantum dots for bioimaging and biodiagnostic applications. *Annu. Rev. Anal. Chem. (Palo Alto, Calif.),* **2013**, *6*, 143-162.
[http://dx.doi.org/10.1146/annurev-anchem-060908-155136] [PMID: 23527547]

[568] Wang, Y.; Hu, R.; Lin, G.; Roy, I.; Yong, K-T. Functionalized quantum dots for biosensing and bioimaging and concerns on toxicity. *ACS Appl. Mater. Interfaces,* **2013**, *5*(8), 2786-2799.
[http://dx.doi.org/10.1021/am302030a] [PMID: 23394295]

[569] Mashinchian, O.; Johari-Ahar, M.; Ghaemi, B.; Rashidi, M.; Barar, J.; Omidi, Y. Impacts of quantum dots in molecular detection and bioimaging of cancer. *Bioimpacts,* **2014**, *4*(3), 149-166.
[http://dx.doi.org/10.15171/bi.2014.008] [PMID: 25337468]

[570] Pisanic, T.R., II; Zhang, Y.; Wang, T.H. Quantum dots in diagnostics and detection: principles and paradigms. *Analyst (Lond.),* **2014**, *139*(12), 2968-2981.
[http://dx.doi.org/10.1039/c4an00294f] [PMID: 24770716]

[571] Bhattacharya, P.; Ghosh, S.; Stiff-Roberts, A.D. Quantum dot opto-electronic devices. *Annu. Rev. Mater. Res.,* **2004**, *34*, 1-40.
[http://dx.doi.org/10.1146/annurev.matsci.34.040203.111535]

[572] Kim, J.Y.; Voznyy, O.; Zhitomirsky, D.; Sargent, E.H. 25th anniversary article: Colloidal quantum dot materials and devices: a quarter-century of advances. *Adv. Mater.,* **2013**, *25*(36), 4986-5010.
[http://dx.doi.org/10.1002/adma.201301947] [PMID: 24002864]

[573] Kershaw, S.V.; Susha, A.S.; Rogach, A.L. Narrow bandgap colloidal metal chalcogenide quantum dots: synthetic methods, heterostructures, assemblies, electronic and infrared optical properties. *Chem. Soc. Rev.,* **2013**, *42*(7), 3033-3087.
[http://dx.doi.org/10.1039/c2cs35331h] [PMID: 23361653]

[574] Erdem, T.; Demir, H.V. Color science of nanocrystal quantum dots for lighting and displays. *Nanophoton.,* **2013**, *2*, 57-81.
[http://dx.doi.org/10.1515/nanoph-2012-0031]

[575] Bisri, S.Z.; Piliego, C.; Gao, J.; Loi, M.A. Outlook and emerging semiconducting materials for ambipolar transistors. *Adv. Mater.,* **2014**, *26*(8), 1176-1199.
[http://dx.doi.org/10.1002/adma.201304280] [PMID: 24591008]

[576] Guzelturk, B.; Martinez, P.L.; Zhang, Q.; Xiong, Q.; Sun, H.; Sun, X.W.; Govorov, A.O.; Demir, H.V. Excitonics of semiconductor quantum dots and wires for lighting and displays. *Laser Photonics Rev.,* **2014**, *8*, 73-93.
[http://dx.doi.org/10.1002/lpor.201300024]

[577] Pichaandi, J.; van Veggel, F.C. Near-infrared emitting quantum dots: Recent progress on their synthesis and characterization. *Coord. Chem. Rev.,* **2014**, *263-264*, 138-150.
[http://dx.doi.org/10.1016/j.ccr.2013.10.011]

[578] Mukai, K. Semiconductor quantum dots for future optical applications. *J. Nanosci. Nanotechnol.,* **2014**, *14*(3), 2148-2156.
[http://dx.doi.org/10.1166/jnn.2014.8608] [PMID: 24745208]

[579] Kamat, P.V. Quantum dot solar cells. Semiconductor nanocrystals as light harvesters. *J. Phys. Chem. C,* **2008**, *112*, 18737-18753.

[http://dx.doi.org/10.1021/jp806791s]

[580] Tang, J.; Sargent, E.H. Infrared colloidal quantum dots for photovoltaics: fundamentals and recent progress. *Adv. Mater.,* **2011**, *23*(1), 12-29. [http://dx.doi.org/10.1002/adma.201001491] [PMID: 20842658]

[581] Huang, X.; Han, S.; Huang, W.; Liu, X. Enhancing solar cell efficiency: the search for luminescent materials as spectral converters. *Chem. Soc. Rev.,* **2013**, *42*(1), 173-201. [http://dx.doi.org/10.1039/C2CS35288E] [PMID: 23072924]

[582] Albero, J.; Clifford, J.N.; Palomares, E. Quantum dot based molecular solar cells. *Coord. Chem. Rev.,* **2014**, *263-264*, 53-64. [http://dx.doi.org/10.1016/j.ccr.2013.07.005]

[583] Kouhnavard, M.; Ikeda, S.; Ludin, N.A.; Khairudin, N.B.; Ghaffari, B.V.; Mat-Teridi, M.A.; Ibrahim, M.A.; Sepeai, S.; Sopian, K. A review of semiconductor materials as sensitizers for quantum dot-sensitized solar cells. *Renew. Sustain. Energy Rev.,* **2014**, *37*, 397-407. [http://dx.doi.org/10.1016/j.rser.2014.05.023]

[584] Liu, R. Hybrid organic/inorganic nanocomposites for photovoltaic. *Cells Mater.,* **2014**, *7*, 2747-2771. [http://dx.doi.org/10.3390/ma7042747]

[585] Wang, X.; Yang, L.; Chen, Z.G.; Shin, D.M. Application of nanotechnology in cancer therapy and imaging. *CA Cancer J. Clin.,* **2008**, *58*(2), 97-110. [http://dx.doi.org/10.3322/CA.2007.0003] [PMID: 18227410]

[586] Walling, M.A.; Novak, J.A.; Shepard, J.R. Quantum dots for live cell and *in vivo* imaging. *Int. J. Mol. Sci.,* **2009**, *10*(2), 441-491. [http://dx.doi.org/10.3390/ijms10020441] [PMID: 19333416]

[587] Gao, X.; Cui, Y.; Levenson, R.M.; Chung, L.W.; Nie, S. *In vivo* cancer targeting and imaging with semiconductor quantum dots. *Nat. Biotechnol.,* **2004**, *22*(8), 969-976. [http://dx.doi.org/10.1038/nbt994] [PMID: 15258594]

[588] Wu, X.; Liu, H.; Liu, J.; Haley, K.N.; Treadway, J.A.; Larson, J.P.; Ge, N.; Peale, F.; Bruchez, M.P. Immunofluorescent labeling of cancer marker Her2 and other cellular targets with semiconductor quantum dots. *Nat. Biotechnol.,* **2003**, *21*(1), 41-46. [http://dx.doi.org/10.1038/nbt764] [PMID: 12459735]

[589] Cai, W.; Shin, D.W.; Chen, K.; Gheysens, O.; Cao, Q.; Wang, S.X.; Gambhir, S.S.; Chen, X. Peptide-labeled near-infrared quantum dots for imaging tumor vasculature in living subjects. *Nano Lett.,* **2006**, *6*(4), 669-676. [http://dx.doi.org/10.1021/nl052405t] [PMID: 16608262]

[590] Tada, H.; Higuchi, H.; Wanatabe, T.M.; Ohuchi, N. *In vivo* real-time tracking of single quantum dots conjugated with monoclonal anti-HER2 antibody in tumors of mice. *Cancer Res.,* **2007**, *67*(3), 1138-1144. [http://dx.doi.org/10.1158/0008-5472.CAN-06-1185] [PMID: 17283148]

[591] Maysinger, D.; Behrendt, M.; Lalancette-Hébert, M.; Kriz, J. Real-time imaging of astrocyte response to quantum dots: *in vivo* screening model system for biocompatibility of nanoparticles. *Nano Lett.,* **2007**, *7*(8), 2513-2520.

[http://dx.doi.org/10.1021/nl071611t] [PMID: 17638392]

[592] Jiang, W.; Singhal, A.; Kim, B.Y.; Zheng, J.; Rutka, J.T.; Wang, C.; Chan, W.C. Assessing near-infrared quantum dots for deep tissue, organ, and animal imaging applications. *J. Assoc. Lab. Autom.,* **2008**, *13*, 6-12.
[http://dx.doi.org/10.1016/j.jala.2007.09.002]

[593] Zimmer, J.P.; Kim, S-W.; Ohnishi, S.; Tanaka, E.; Frangioni, J.V.; Bawendi, M.G. Size series of small indium arsenide-zinc selenide core-shell nanocrystals and their application to *in vivo* imaging. *J. Am. Chem. Soc.,* **2006**, *128*(8), 2526-2527.
[http://dx.doi.org/10.1021/ja0579816] [PMID: 16492023]

[594] Yang, H.S.; Santra, S.; Walter, G.A.; Holloway, P.H. Gd-III-functionalized fluorescent quantum dots as multimodal imaging probes. *Adv. Mater.,* **2006**, *18*, 2890-2894.
[http://dx.doi.org/10.1002/adma.200502665]

[595] Tan, W.B.; Zhang, Y. Multi-functional chitosan nanoparticles encapsulating quantum dots and Gd-DTPA as imaging probes for bio-applications. *J. Nanosci. Nanotechnol.,* **2007**, *7*(7), 2389-2393.
[http://dx.doi.org/10.1166/jnn.2007.415] [PMID: 17663257]

[596] Bakalova, R.; Zhelev, Z.; Aoki, I.; Ohba, H.; Imai, Y.; Kanno, I. Silica-shelled single quantum dot micelles as imaging probes with dual or multimodality. *Anal. Chem.,* **2006**, *78*(16), 5925-5932.
[http://dx.doi.org/10.1021/ac060412b] [PMID: 16906742]

[597] Yong, K.T.; Ding, H.; Roy, I.; Law, W.C.; Bergey, E.J.; Maitra, A.; Prasad, P.N. Imaging pancreatic cancer using bioconjugated InP quantum dots. *ACS Nano,* **2009**, *3*(3), 502-510.
[http://dx.doi.org/10.1021/nn8008933] [PMID: 19243145]

[598] Ducongé, F.; Pons, T.; Pestourie, C.; Hérin, L.; Thézé, B.; Gombert, K.; Mahler, B.; Hinnen, F.; Kühnast, B.; Dollé, F.; Dubertret, B.; Tavitian, B. Fluorine-18-labeled phospholipid quantum dot micelles for *in vivo* multimodal imaging from whole body to cellular scales. *Bioconjug. Chem.,* **2008**, *19*(9), 1921-1926.
[http://dx.doi.org/10.1021/bc800179j] [PMID: 18754572]

[599] Giepmans, B.N.; Deerinck, T.J.; Smarr, B.L.; Jones, Y.Z.; Ellisman, M.H. Correlated light and electron microscopic imaging of multiple endogenous proteins using Quantum dots. *Nat. Methods,* **2005**, *2*(10), 743-749.
[http://dx.doi.org/10.1038/nmeth791] [PMID: 16179920]

[600] Pons, T.; Mattoussi, H. Investigating biological processes at the single molecule level using luminescent quantum dots. *Ann. Biomed. Eng.,* **2009**, *37*(10), 1934-1959.
[http://dx.doi.org/10.1007/s10439-009-9715-0] [PMID: 19521775]

[601] Pinaud, F.; Clarke, S.; Sittner, A.; Dahan, M. Probing cellular events, one quantum dot at a time. *Nat. Methods,* **2010**, *7*(4), 275-285.
[http://dx.doi.org/10.1038/nmeth.1444] [PMID: 20354518]

[602] Cui, B.; Wu, C.; Chen, L.; Ramirez, A.; Bearer, E.L.; Li, W-P.; Mobley, W.C.; Chu, S. One at a time, live tracking of NGF axonal transport using quantum dots. *Proc. Natl. Acad. Sci. USA,* **2007**, *104*(34), 13666-13671.
[http://dx.doi.org/10.1073/pnas.0706192104] [PMID: 17698956]

[603] Dahan, M.; Lévi, S.; Luccardini, C.; Rostaing, P.; Riveau, B.; Triller, A. Diffusion dynamics of glycine receptors revealed by single-quantum dot tracking. *Science,* **2003**, *302*(5644), 442-445. [http://dx.doi.org/10.1126/science.1088525] [PMID: 14564008]

[604] Liu, W.; Howarth, M.; Greytak, A.B.; Zheng, Y.; Nocera, D.G.; Ting, A.Y.; Bawendi, M.G. Compact biocompatible quantum dots functionalized for cellular imaging. *J. Am. Chem. Soc.,* **2008**, *130*(4), 1274-1284. [http://dx.doi.org/10.1021/ja076069p] [PMID: 18177042]

[605] Hasegawa, U.; Nomura, S.M.; Kaul, S.C.; Hirano, T.; Akiyoshi, K. Nanogel-quantum dot hybrid nanoparticles for live cell imaging. *Biochem. Biophys. Res. Commun.,* **2005**, *331*(4), 917-921. [http://dx.doi.org/10.1016/j.bbrc.2005.03.228] [PMID: 15882965]

[606] Månsson, A.; Sundberg, M.; Balaz, M.; Bunk, R.; Nicholls, I.A.; Omling, P.; Tågerud, S.; Montelius, L. *In vitro* sliding of actin filaments labelled with single quantum dots. *Biochem. Biophys. Res. Commun.,* **2004**, *314*(2), 529-534. [http://dx.doi.org/10.1016/j.bbrc.2003.12.133] [PMID: 14733939]

[607] Lidke, D.S.; Nagy, P.; Heintzmann, R.; Arndt-Jovin, D.J.; Post, J.N.; Grecco, H.E.; Jares-Erijman, E.A.; Jovin, T.M. Quantum dot ligands provide new insights into erbB/HER receptor-mediated signal transduction. *Nat. Biotechnol.,* **2004**, *22*(2), 198-203. [http://dx.doi.org/10.1038/nbt929] [PMID: 14704683]

[608] Gaponik, N.; Talapin, D.V.; Rogach, A.L.; Hoppe, K.; Shevchenko, E.V.; Kornowski, A.; Eychmüller, A.; Weller, H. Thiol-capping of CDTe nanocrystals: An alternative to organometallic synthetic routes. *J. Phys. Chem. B,* **2002**, *106*, 7177-7185. [http://dx.doi.org/10.1021/jp025541k]

[609] Hauck, T.S.; Anderson, R.E.; Fischer, H.C.; Newbigging, S.; Chan, W.C. *In vivo* quantum-dot toxicity assessment. *Small,* **2010**, *6*(1), 138-144. [http://dx.doi.org/10.1002/smll.200900626] [PMID: 19743433]

[610] Wang, K.; Wang, K.; Shen, B.; Huang, T.; Sun, X.; Li, W.; Jin, G.; Li, L.; Bu, L.; Li, R.; Wang, D.; Chen, X. MR reporter gene imaging of endostatin expression and therapy. *Mol. Imaging Biol.,* **2010**, *12*(5), 520-529. [http://dx.doi.org/10.1007/s11307-009-0286-0] [PMID: 19957205]

[611] Chen, Z.; Chen, H.; Meng, H.; Xing, G.; Gao, X.; Sun, B.; Shi, X.; Yuan, H.; Zhang, C.; Liu, R.; Zhao, F.; Zhao, Y.; Fang, X. Bio-distribution and metabolic paths of silica coated CdSeS quantum dots. *Toxicol. Appl. Pharmacol.,* **2008**, *230*(3), 364-371. [http://dx.doi.org/10.1016/j.taap.2008.03.022] [PMID: 18495192]

[612] Biju, V.; Itoh, T.; Anas, A.; Sujith, A.; Ishikawa, M. Semiconductor quantum dots and metal nanoparticles: syntheses, optical properties, and biological applications. *Anal. Bioanal. Chem.,* **2008**, *391*(7), 2469-2495. [http://dx.doi.org/10.1007/s00216-008-2185-7] [PMID: 18548237]

[613] Frasco, M.F.; Chaniotakis, N. Bioconjugated quantum dots as fluorescent probes for bioanalytical applications. *Anal. Bioanal. Chem.,* **2010**, *396*(1), 229-240. [http://dx.doi.org/10.1007/s00216-009-3033-0] [PMID: 19714321]

[614] Gao, M.; Yang, Y.; Yang, B.; Bian, F.; Shen, J. Synthesis of PbS nanoparticles in polymer matrices. *J. Chem. Soc. Chem. Commun.,* **1994**, 2779-2780.
[http://dx.doi.org/10.1039/c39940002779]

[615] Qiao, Z-P.; Xie, G.; Huang, J-M.; Zhao, S-Y.; Chen, X-M. One-pot monomer self-assembly route to PbS/poly(methyl methacrylate) core/shell nanocomposite thin coatings. *J. Mater. Chem.,* **2002**, *12*, 611-613.
[http://dx.doi.org/10.1039/b105005m]

[616] Qiao, Z.; Xie, Y.; Zhu, Y.; Qian, Y. Synthesis of PbS/polyacrylonitrile nanocomposites at room temperature by γ-irradiation. *J. Mater. Chem.,* **1999**, *9*, 1001-1002.
[http://dx.doi.org/10.1039/a808965e]

[617] Qian, X-F.; Yin, J.; Feng, S.; Liu, S-H.; Zhu, Z-K. Preparation and characterization of polyvinylpyrrolidone films containing silver sulfide nanoparticles. *J. Mater. Chem.,* **2001**, *11*, 2504-2506.
[http://dx.doi.org/10.1039/b103708k]

[618] Sun, H.; Zhang, H.; Ju, J.; Zhang, J.; Qian, G.; Wang, C.; Yang, B.; Wang, Z.Y. One-step synthesis of high-quality gradient CdHgTe nanocrystals: A prerequisite to prepare CdHgTe-polymer bulk composites with intense near-infrared photoluminescence. *Chem. Mater.,* **2008**, *20*, 6764-6769.
[http://dx.doi.org/10.1021/cm8018296]

[619] Priyam, A.; Blumling, D.E.; Knappenberger, K.L., Jr Synthesis, characterization, and self-organization of dendrimer-encapsulated HgTe quantum dots. *Langmuir,* **2010**, *26*(13), 10636-10644.
[http://dx.doi.org/10.1021/la100866z] [PMID: 20481491]

[620] Decher, G. Fuzzy nanoassemblies: toward layered polymeric multicomposites. *Science,* **1997**, *277*, 1232-1237.
[http://dx.doi.org/10.1126/science.277.5330.1232]

[621] Rogach, A.L.; Koktysh, D.S.; Harrison, M.; Kotov, N.A. Layer-by-layer assembled films of HgTe nanocrystals with strong infrared emission. *Chem. Mater.,* **2000**, *12*, 1526-1528.
[http://dx.doi.org/10.1021/cm0000649]

[622] Rogach, A.L.; Kotov, N.A.; Koktysh, D.S.; Susha, A.S.; Caruso, F. II–VI semiconductor nanocrystals in thin films and colloidal crystals. *Colloids Surf. A Physicochem. Eng. Asp.,* **2002**, *202*, 135-144.
[http://dx.doi.org/10.1016/S0927-7757(01)01072-X]

[623] Wang, D.; Rogach, A.L.; Caruso, F. Composite photonic crystals from semiconductor nanocrystal/polyelectrolyte-coated colloidal spheres. *Chem. Mater.,* **2003**, *15*, 2724-2729.
[http://dx.doi.org/10.1021/cm031003o]

[624] Lee, B.; Podsiadlo, P.; Rupich, S.; Talapin, D.V.; Rajh, T.; Shevchenko, E.V. Comparison of structural behavior of nanocrystals in randomly packed films and long-range ordered superlattices by time-resolved small angle X-ray scattering. *J. Am. Chem. Soc.,* **2009**, *131*(45), 16386-16388.
[http://dx.doi.org/10.1021/ja906632b] [PMID: 19863066]

[625] Friedrich, H.; Gommes, C.J.; Overgaag, K.; Meeldijk, J.D.; Evers, W.H.; de Nijs, B.; Boneschanscher, M.P.; de Jongh, P.E.; Verkleij, A.J.; de Jong, K.P.; van Blaaderen, A.; Vanmaekelbergh, D. Quantitative structural analysis of binary nanocrystal superlattices by electron tomography. *Nano Lett.,*

2009, *9*(7), 2719-2724.
[http://dx.doi.org/10.1021/nl901212m] [PMID: 19505139]

[626] Kovalenko, M.V.; Scheele, M.; Talapin, D.V. Colloidal nanocrystals with molecular metal chalcogenide surface ligands. *Science,* **2009**, *324*(5933), 1417-1420.
[http://dx.doi.org/10.1126/science.1170524] [PMID: 19520953]

[627] Nag, A.; Kovalenko, M.V.; Lee, J-S.; Liu, W.; Spokoyny, B.; Talapin, D.V. Metal-free inorganic ligands for colloidal nanocrystals: S2-, HS-, Se2-, HSe-, Te2-, HTe-, TeS3(2-), OH-, and NH2- as surface ligands. *J. Am. Chem. Soc.,* **2011**, *133*(27), 10612-10620.
[http://dx.doi.org/10.1021/ja2029415] [PMID: 21682249]

[628] Lhuillier, E.; Keuleyan, S.; Rekemeyer, P.; Guyot-Sionnest, P. Thermal properties of mid-infrared colloidal quantum dot detectors. *J. Appl. Phys.,* **2011**, *110*, 033110.
[http://dx.doi.org/10.1063/1.3619857]

[629] Rauch, T.; Boberl, M.; Tedde, S.F.; Furst, J.; Kovalenko, M.V.; Hesser, G.; Lemmer, U.; Heiss, W.; Hayden, O. Near-infrared imaging with quantum-dot-sensitized organic photodiodes. *Nat. Photonics,* **2009**, *3*, 332-336.
[http://dx.doi.org/10.1038/nphoton.2009.72]

[630] Talapin, D.V.; Lee, J-S.; Kovalenko, M.V.; Shevchenko, E.V. Prospects of colloidal nanocrystals for electronic and optoelectronic applications. *Chem. Rev.,* **2010**, *110*(1), 389-458.
[http://dx.doi.org/10.1021/cr900137k] [PMID: 19958036]

[631] Ma, B.; Wang, L.; Dong, H.; Gao, R.; Geng, Y.; Zhu, Y.; Qiu, Y. Photocatalysis of PbS quantum dots in a quantum dot-sensitized solar cell: photovoltaic performance and characteristics. *Phys. Chem. Chem. Phys.,* **2011**, *13*(7), 2656-2658.
[http://dx.doi.org/10.1039/C0CP02415E] [PMID: 21165469]

[632] Lee, H.J.; Yum, J-H.; Leventis, H.C.; Zakeeruddin, S.M.; Haque, S.A.; Chen, P.; Seok, S.I.; Grätzel, M.; Nazeeruddin, M.K. CdSe quantum dot-sensitized solar cells exceeding efficiency 1% at full-sun intensity. *J. Phys. Chem. C,* **2008**, *112*, 11600-11608.
[http://dx.doi.org/10.1021/jp802572b]

[633] Lee, H.; Wang, M.; Chen, P.; Gamelin, D.R.; Zakeeruddin, S.M.; Grätzel, M.; Nazeeruddin, M.K. Efficient CdSe quantum dot-sensitized solar cells prepared by an improved successive ionic layer adsorption and reaction process. *Nano Lett.,* **2009**, *9*(12), 4221-4227.
[http://dx.doi.org/10.1021/nl902438d] [PMID: 19891465]

[634] Sambur, J.B.; Novet, T.; Parkinson, B.A. Multiple exciton collection in a sensitized photovoltaic system. *Science,* **2010**, *330*(6000), 63-66.
[http://dx.doi.org/10.1126/science.1191462] [PMID: 20929804]

[635] Hossain, M.A.; Koh, Z.Y.; Wang, Q. PbS/CdS-sensitized mesoscopic SnO_2 solar cells for enhanced infrared light harnessing. *Phys. Chem. Chem. Phys.,* **2012**, *14*(20), 7367-7374.
[http://dx.doi.org/10.1039/c2cp40551b] [PMID: 22531753]

[636] Li, T-L.; Lee, Y-L.; Teng, H. High-performance quantum dot-sensitized solar cells based on sensitization with CuInS2 quantum dots/CdS heterostructure. *Energy Environ. Sci.,* **2012**, *5*, 5315-5324.
[http://dx.doi.org/10.1039/C1EE02253A]

[637] Im, S.H.; Kim, H-J.; Kim, S.W.; Kim, S-W.; Seok, S.I. All solid state multiply layered PbS colloidal quantum-dot-sensitized photovoltaic cells. *Energy Environ. Sci.,* **2011**, *4*, 4181-4186. [http://dx.doi.org/10.1039/c1ee01774h]

[638] Etgar, L.; Zhang, W.; Gabriel, S.; Hickey, S.G.; Nazeeruddin, M.K.; Eychmüller, A.; Liu, B.; Grätzel, M. High efficiency quantum dot heterojunction solar cell using anatase (001) TiO2 nanosheets. *Adv. Mater.,* **2012**, *24*(16), 2202-2206. [http://dx.doi.org/10.1002/adma.201104497] [PMID: 22508497]

[639] Fu, H.; Tsang, S-W.; Zhang, Y.; Ouyang, J.; Lu, J.; Yu, K.; Tao, Y. Impact of the growth conditions of colloidal PbS nanocrystals on photovoltaic device performance. *Chem. Mater.,* **2011**, *23*, 1805-1810. [http://dx.doi.org/10.1021/cm200051j]

[640] Szendrei, K.; Gomulya, W.; Yarema, M.; Heiss, W.; Loi, M.A. PbS nanocrystal solar cells with high efficiency and fill factor. *Appl. Phys. Lett.,* **2010**, *97*, 203501-203503. [http://dx.doi.org/10.1063/1.3518067]

[641] Ma, W.; Swisher, S.L.; Ewers, T.; Engel, J.; Ferry, V.E.; Atwater, H.A.; Alivisatos, A.P. Photovoltaic performance of ultrasmall PbSe quantum dots. *ACS Nano,* **2011**, *5*(10), 8140-8147. [http://dx.doi.org/10.1021/nn202786g] [PMID: 21939281]

[642] Tang, J.; Kemp, K.W.; Hoogland, S.; Jeong, K.S.; Liu, H.; Levina, L.; Furukawa, M.; Wang, X.; Debnath, R.; Cha, D.; Chou, K.W.; Fischer, A.; Amassian, A.; Asbury, J.B.; Sargent, E.H. Colloidal-quantum-dot photovoltaics using atomic-ligand passivation. *Nat. Mater.,* **2011**, *10*(10), 765-771. [http://dx.doi.org/10.1038/nmat3118] [PMID: 21927006]

[643] Imahori, H.; Norieda, H.; Nishimura, Y.; Yamazaki, I.; Higuchi, K.; Kato, N.; Motohiro, T.; Yamada, H.; Tamaki, K.; Arimura, M.; Sakata, Y. Chain length effect on the structure and photoelectrochemical properties of self-assembled monolayers of porphyrins on gold electrodes. *J. Phys. Chem. B,* **2000**, *104*, 1253-1260. [http://dx.doi.org/10.1021/jp992768f]

[644] Boatman, E.M.; Lisensky, G.C.; Nordell, K.J. A safer, easier, faster synthesis for CdSe quantum dot nanocrystals. *J. Chem. Educ.,* **2005**, *82*, 1697-1699. [http://dx.doi.org/10.1021/ed082p1697]

[645] Yong, K-T.; Roy, I.; Law, W-C.; Hu, R. Synthesis of cRGD-peptide conjugated near-infrared CdTe/ZnSe core-shell quantum dots for *in vivo* cancer targeting and imaging. *Chem. Commun. (Camb.),* **2010**, *46*(38), 7136-7138. [http://dx.doi.org/10.1039/c0cc00667j] [PMID: 20820500]

[646] Cai, W.; Chen, X. Preparation of peptide-conjugated quantum dots for tumor vasculature-targeted imaging. *Nat. Protoc.,* **2008**, *3*(1), 89-96. [http://dx.doi.org/10.1038/nprot.2007.478] [PMID: 18193025]

CHAPTER 9

Future Prospects

Abstract: A property vs application analysis is herein carried out, using the materials focused in this book, recapitulating: phthalocyanines, polymethines, porphyrins, BODIPYs, dendrimers, carbon allotropes, organic frameworks and nanoparticles. Their own unique characteristics can be scrutinized, in order to, at least from a prospect point to view, anticipate the best property-application associations, with basis on their structure-property relationships for the most important applications.

Keywords: BODIPYs, Carbon allotropes, Dendrimers, Nanoparticles, Organic frameworks, Phthalocyanines, Polymethines, Porphyrins, Structure-property.

The establishment of highly accurate structure-property relationships is the one of the most important aspects when designing any new materials for any applications. Each application should, in an ideal model, match the best material, *i.e.*, one should predict, to the best possible approximation, which materials should chosen for any envisaged application.

This structure-property relationship can be perfected by several approaches, such as computational modeling, empirically, or by molecule screening. The computational modeling approach trusts the outcome of software for property prediction, usually based on existing or purposely created databases, relying on theoretical aspects of the structure, reactivity and properties. On the other hand, the empirical approach uses the observation outcome of experiences carried out on determining any property, related to a given application. Finally, the screening approach relies on the initial preparation of a set of molecules, closely related to each other, followed by application testing to obtain a certain number of "proof of concept molecules".

Nonetheless, the best approach is the one that manages to comprise parts of each of the above mentioned in a single approach. This way a researcher can extract the best outcomes, in a selective manner, where the theoretical knowledge can

Mário J. F. Calvete

symbiotically function with empiricism and screening. The best way is to flag a property or a set of properties for a certain application that requires them, in order to be distinguished as a useful application.

In the next paragraphs, a property *vs* application analysis will be carried out, using the materials focused in this book, recapitulating: phthalocyanines, polymethines, porphyrins, BODIPYs, dendrimers, carbon allotropes, organic frameworks and nanoparticles. Their own unique characteristics can be scrutinized, in order to, at least from a prospect point to view, anticipate the best property-application associations, with basis on their structure-property relationships for the most important applications.

CATALYSIS APPLICATIONS

For instance, application in catalysis requires compounds which fulfill certain requirements, such as inertness during application, structural features that allow effective reaction acceleration and recyclability. Moreover, the typically used homogeneous catalysis has turned into a more environmentally concerned heterogeneous catalysis, which is nowadays of paramount importance for a variety of research areas, from chemistry to energy, always thinking in an industrial prospect, as well. Thus, added to the aforementioned requirements, the best catalysts for any given catalytic application should also be easily recoverable.

Aiming on these premises, catalysts that are easier to immobilize onto/into any type of supports are the most promising ones. There are many possible supports, from organic, such as polymeric ones, or inorganic, such as based on silicates, among others, as well as several ways to immobilize the catalysts, from adsorption, covalent linkage or encapsulation.

Tetrapyrrolic macrocycles (porphyrin and phthalocyanine derivatives) are appealing molecules to be used in catalysis, since they can be straightforwardly fine-tuned. Both types of compound have already proven their worth in homogenous catalysis, but there is yet much to be established, regarding their use in heterogeneous catalysis. This means, as well, that there are plenty of possibilities ahead. For instance, they are very suitable from hosting metals in their cavities, which opens the possibility to use them in direct and asymmetric transition metal catalysis. Another possibility deals with their photosensitizing abilities, which makes them good candidates to be used in photocatalysis, mostly in oxidation reactions.

These macrocycles are also capable to be properly modulated to be linked to several supports, allowing a facilitated posterior separation from catalytic reaction mixture. As stated above, many supports are possible to be used. The main rule is that these supports stand inert during the course of reaction. The polymeric supports are quite useful, but since they are mostly from organic origin, one must prevent they reactivity, hence their stability during reaction. More stable, carbon allotropes are also quite promising supports for several types of reactions, provided that their functionalization process does not interfere afterwards with the catalytic reactivity, hence their inertness. On the other hand, supports of inorganic nature are more stable against harsh conditions, ensuring their inertness. As possible candidates, we can find double layered hydroxides, silicas, metal oxides, aluminium-silicate mesoporous materials, and nanoparticles, particularly magnetic nanoparticles. The last ones, without a doubt are probably the most promising, for their superior ease of recovery, specially the magnetic nanoparticles, which can easily be recovered by simple application of an external magnet.

Dendrimers are also good candidates as catalysts, since they mostly keep the advantages of homogeneous catalysts, when properly modulated to encapsulate appropriate catalyst synthons inside their branches or the branch endings. Their size makes them suitable for heterogeneous catalysis in the sense that, due to the simplicity in recovering the catalyst by ultra-filtration, after precipitation. Similarly, organic frameworks (metal or organic) are also pivotal candidates, both due to their size and ability to hold a variety of suitable metals for acting as catalysts, displaying the same straightforwardness in recovery and reuse. Finally, metal nanoparticles, as its name points, are of extreme use as catalysts, having features such as very large surfaces of contact, and chemical stability, two key properties in heterogeneous catalysis.

SENSING APPLICATIONS

Another important application is in sensing of any specific analytes. Since most of sensors focused in this book relate to chemical sensors or biosensors, the characteristics that sensors of this type usually require properties of entities that easily respond to concentrations of analytes, which selectively interact with the receptor molecules. Since the accuracy and efficiency of the sensor relies on its sensitivity, the most common sensors report their characteristic physical parameters in terms of optical signal. Thus, the best materials should be capable to use spectroscopic properties such as absorption and/or emission for recognition.

All compounds that possess a relatively high fluorescence quantum yield are potential candidates to be used as sensors, including phthalocyanines, porphyrins and BODIPYs. Modulation of these compounds' periphery with appropriate receptors might allow finding a suitable sensor for a vast array of analytes, upon alterations either in absorption or by increase/suppression of fluorescence after complexation.

Other molecules that can act sensors are the carbon allotropes, which are very sensitive chemiresistive sensors, but they are not as selective as is necessary for practical sensing applications. Therefore, modifying single-walled carbon nanotubes with suitable fragments can provide greater chemical selectivity.

Organic frameworks, predominantly the chemiluminescent ones, also might have the ability to perform in sensing applications, whose pores are perfect hosts for several analytes, particularly gases. Similarly, many metal nanoparticles can also act sensors, based on their luminescent and colorimetric properties.

ORGANIC PHOTOVOLTAIC APPLICATIONS

Organic photovoltaics, or as commonly called, dye sensitized solar cells (DSSC), have emerged as one of the most promising, and also studied, applications for organic materials, arising from the understanding how natural molecules, such as chlorophyll, take use of the solar irradiation to produce energy. Therefore, an ideal cell should possess a photosensitizer that can harvest most of the solar light, in order to be able to produce a measurable amount of energy. The photosensitizer should also display the smallest HOMO–LUMO energy gap possible, to permit a large photocurrent from the use of the long-wavelength region in the solar spectrum.

Compounds with strong molar extinction coefficients are suitable for use as sensitizers in DSSCs, including phthalocyanine, polymethine, porphyrin and BODIPY derivatives. Their UV-vis-NIR spectra should be their main "visit card", for which appropriate modulation can increase their suitability for photovoltaic applications. Modulation of the physical and optical properties of these compounds can be accomplished either by axial and/or peripheral substitution with a variety of ligands. Additionally, their redox properties are suitable both for sensitization and for dye regeneration by the electrolyte.

One major drawback on some of these materials (mostly for phthalocyanines and porphyrins) lies in their inherent tendency to form aggregates, which may result in efficiency decrease. Polymethine and BODIPY derivatives are less prone to such unfavorable phenomenon, due to their less planar structural features. Suppression of agglomeration issues can be lessened by appropriate axial coordination or introduction of bulky substituents, which in turn also increase material's solubility. Another way to avoid aggregation is by combinations with aggregation-immune materials, such as dendrimers, which have the advantage of providing additional number of linking sites, therefore enhancing the molar extinction coefficient.

Carbon-based materials, such as fullerene, nanotube and graphene derivatives are also good candidates to be used in organic photovoltaic applications. It is no secret that the most efficient bulk heterojunction organic solar cells reported so far are still based on fullerene-acceptor derivatives. The use of nanotubes should follow the outstanding work performed in earlier years with fullerenes, with the advantages of similar modulation proneness and economical saving. Furthermore, graphene, the youngest of the carbon allotropes, is still at its infancy, regarding practical uses in organic photovoltaics. To fully exploit their unique properties, graphene derivatives must be modulated in order to increase molar extinction coefficients and charge mobility.

Finally, nanoparticles have appeared as the next new building blocks to construct light energy harvesting assemblies, usually as starting synthons for larger material's production. A particular type of metal nanoparticles is the quantum dot, which offers high possibilities, once the stability problem is overcome. Thus, this remains as the main challenge, find suitable formulations that include quantum dots, which can even replace properly the so far used dyes in DSSCs.

PHOTOTHERAPY APPLICATIONS

Phototherapeutic modalities are nowadays one of the main areas of research, when dealing with malignancy's treatment, including of microbial origin. There are several types of possible therapeutics that can be efficiently used for disease eradication. Photodynamic therapy (PDT) is one of the most accepted types for this purpose, and it is based entirely on the capacity of a certain photosensitizer to act as a "cell killer", only active upon irradiation with a laser source that induce the photosensitizer to produce cell toxic entities (*e.g.* singlet oxygen). Other increasingly important modality is the photothermal therapy (PTT), which is considered an extension of PDT. Here the light activation excites the photosensitizer and then it releases vibrational energy (heat), killing the target cells. Therefore, the ideal photosensitizer should hold several important characteristics, such as low dark toxicity, good singlet oxygen producer (PDT), ultrafast radiantionless deactivations (PTT), ability to organ targeting and easy body excretion.

Phthalocyanine and porphyrin derivatives are very suitable candidates for use in phototherapeutic applications. In PDT, a vast array of such type of molecules has already been investigated and even approved for clinical use. The main characteristic these sensitizers should possess, besides the naturally necessary physiological solubility, which can be attained by proper modulation, is to absorb as much as possible in the therapeutic window, in the visible-NIR region, where body tissue absorbs much less. From that premise, the design of new molecules is almost free.

Similarly, polymethine and BODIPY dyes, which have been significantly less studied, are also very promising candidates, as they typically can be modulated to absorb in the NIR region and can be modulated in a similar fashion to the previously mentioned compounds.

In a different perspective, nanoparticles can also be used in disease therapeutics, as they are perfect host for a variety of therapeutic drugs and, since most metal nanoparticles undergo ultrafast radiantionless deactivations, they are suitable candidates for PTT.

MOLECULAR IMAGING APPLICATIONS

Molecular imaging is strongly related to the several types of therapy that use sensitizers for a certain treatment. Nevertheless, they can act as not only as disease killers, but also as disease "trackers". A sensitizer capable of correctly visualize any kind of cell alteration must hold several of the characteristics that are useful or any biological application, such as low toxicity to the organism, ability to organ targeting and easy body excretion. However, these molecules should hold other important properties, mainly related to each type of imaging modality, and have to be carefully chosen when envisaging any imaging modality. They should have at least one of several properties, such as be very good metal chelators, possess near infra red absorbance, have high fluorescence quantum yields or, in contrast, ultrafast radiantionless deactivations.

Molecules derived from phthalocyanines and porphyrins are quite suitable candidates for use in molecular imaging, once provided a suitable hydrophilic substitution pattern. These molecules, due to their extreme planarity are hydrophobic, and therefore unsuitable in the pristine form. However, there a considerable amount of substituents that can be attached, either axially or peripherally. They are quite known for their use in phototherapeutic treatment of several diseases, particularly in PDT, so their proven efficacy can only be an excellent starting point for the development of new materials for this purpose, with suitable absorption features, that can be modulated to the NIR, well inside the therapeutic window.

Similarly, polymethine and BODIPY dyes are excellent candidates, since they can be easily properly modified to be physiologically active. They can also have absorptions in the therapeutic window, on the NIR region, also possessing high fluorescence lifetimes.

Dendrimers, for their unique characteristics, can be modulated to host imaging probes, either inside their branches, by coordination, or at their periphery, though covalent linkages appropriately envisaged according to their ending functional groups, ensuring an excess in imaging responsible fragments and hydrophilicity, for solubility in physiologic media.

DRUG DELIVERY APPLICATIONS

Drug delivery technology is probably one of the most multidisciplinary areas of science nowadays, requiring integrated expertise of chemists, biologists, medicine professionals and engineers. The current efforts taken in optimizing drug delivery systems focus on the targeted delivery of drugs, reason why new delivery systems must contain a target oriented fragment, for instance for cancerous cell systems.

Dendrimers are incredibly suitable compounds for such purposes, due to their physical and chemical properties. The main principles for the use of dendrimers as delivery carriers involve the charge of the terminal groups that must be neutral or negative in order to avoid or minimize toxicity, the design of the molecular architecture to optimize the pharmacokinetics, the choice between dendritic encapsulation and covalent attachment to the branches, attachment of physiologically soluble groups for biodistribution and the use of targeting groups, this last the ultimate goal for the near future in using dendrimers as drug carriers.

ENERGY STORAGE APPLICATIONS

Energy storage devices are ever more important in nowadays life, for the need of storing big amounts of energy in increasingly more confined spaces. Materials that manage to provide a good basis for such application must possess high electrical conductivity, high surface areas and high chemical stability, managing to perform at a wide range of temperatures.

Carbon allotropic materials fulfill all the above premises, and are definitely the main materials to investigate in the near future, for which the main challenges are to suitably modulate the carbon nanostructures, particularly nanotubes, and mainly graphene, whose applications and their understanding is still in its infancy.

Also organic frameworks are very suitable for energy storage, mostly as gas storing superstructures. Their unique properties and relative ease of operation renders these nanomaterials the possibility of creation of new and more focused applications.

HYBRID MATERIALS

Multifunctionality may be the answer to your questions!

Among the vast possibilities for the materials mentioned in this book, already covered in previous chapters, it should be emphasized that research on any of the types of materials here focused is not circumscribed to themselves. Blending several materials is probably the best approach to achieve structures that encompass most of the desired properties. The next paragraphs will deal with applications that used hybrid materials for any give application.

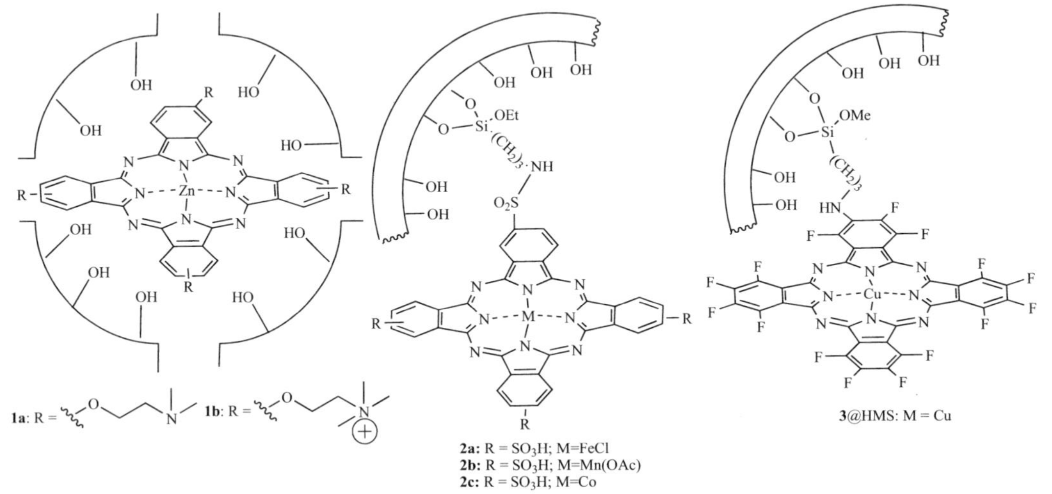

Fig. (1). Hybrid phthalocyanines for application in catalysis.

Hybrid materials containing phthalocyanines in their structure have appeared in literature for several applications. For instance, the use of phthalocyanine hybrids in heterogeneous catalysis is already documented. Immobilization by encapsulation (a method named ship-in-a-bottle synthesis) has been carried out for the immobilization of several phthalocyanine derivatives, like for instance tetrasubstituted zinc phthalocyanine derivatives **1a-d** [1] (Fig. **1**) and used for photocatalytic pesticide degradation. Other authors reported the covalent anchoring of iron(III), manganese(III) and cobalt(II) phthalocyanines (**2a-c**) into mesoporous silicas [2, 3]. Also copper hexadecafluorinated phthalocyanine **3** was encapsulated inside a mesoporous material (Fig. **1**) [4], through an innovative approach, which consisted in reacting hexadecafluorinated copper phthalocyanine with 3-aminopropyltrimethoxisylane, followed by the co-condensation with tetraethylorthosilicate in the presence of n-dodecylamine as micellar template forming an hexagonally ordered material, with a surface area of approximately 200 m^2g^{-1}.

In 2006, Torres and Blau [5] prepared hybrid material **4** for optical limiting (Fig. **2**), by combining a phthalocyanine and a fullerene in order to take advantage of the usually high nonlinear optical effects that both materials possess. Torres, in collaboration with Sessler [6], extended his research to the synthesis of BODIPY-Zn phthalocyanine-fullerene **5** (Fig. **2**), representing the first example in which a phthalocyanine core was connected to a BODIPY unit through a peripheral position, along with axial coordination with a fullerene unit. This hybrid material was used to evaluate its photophysical properties, particularly in terms of increased fluorescence and absorption, with a spectral panchromatic effect, for further use in photovoltaic applications.

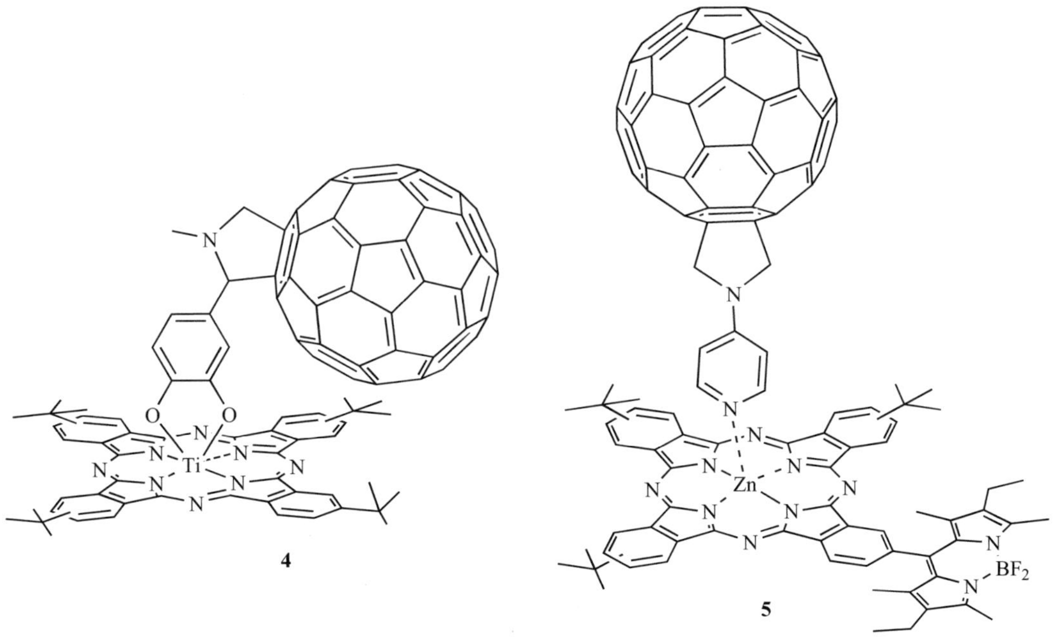

Fig. (2). Hybrid phthalocyanine-fullerene compounds.

Graphene was functionalized with phthalocyanines by covalent grafting of (2-aminoethoxy)(tri-*tert* -butyl)phthalocyanine zinc to exfoliated graphene sheets *via* direct nucleophilic addition of primary amines to obtain hybrid **6** (Fig. **3**), which was used for energy conversion studies [7]. Also a photosensitizer was formulated using dendrimer-phthalocyanine hybrid **7** (Fig. **3**) and tested for PDT, without showing skin phototoxicity [8].

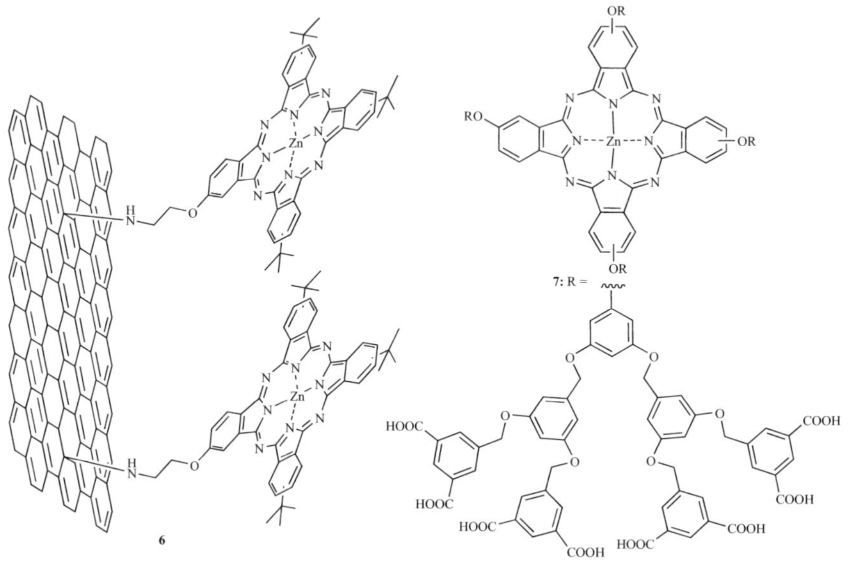

Fig. (3). Phthalocyanine based hybrid compounds.

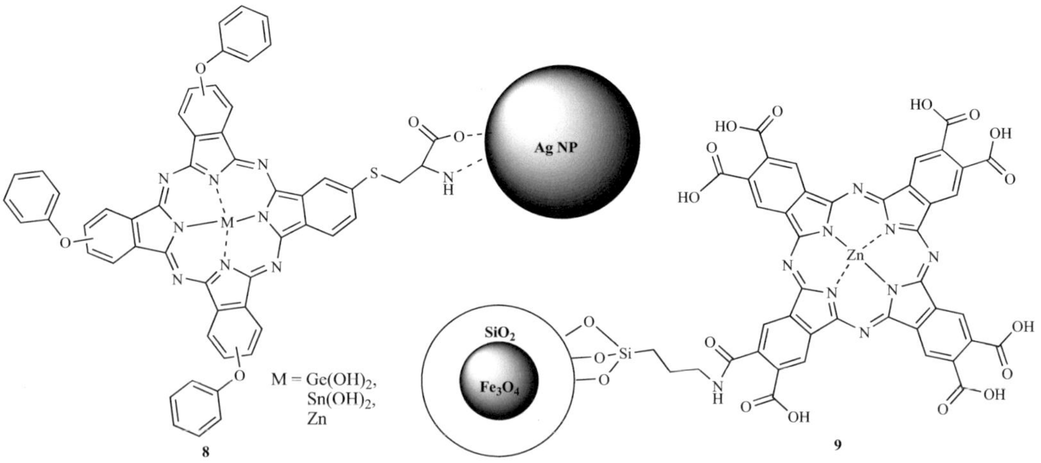

Fig. (4). Phthalocyanine based nanocomposites.

Other researchers reported the hybrid **8** (Fig. **4**) [9], using the phthalocyanine-nanoparticle material as photosensitizer for the antibacterial activity. The same group also reported phthalocyanine-magnetic nanoparticle hybrid **9** Fig. (**4**), for degradation of Orange-G [10].

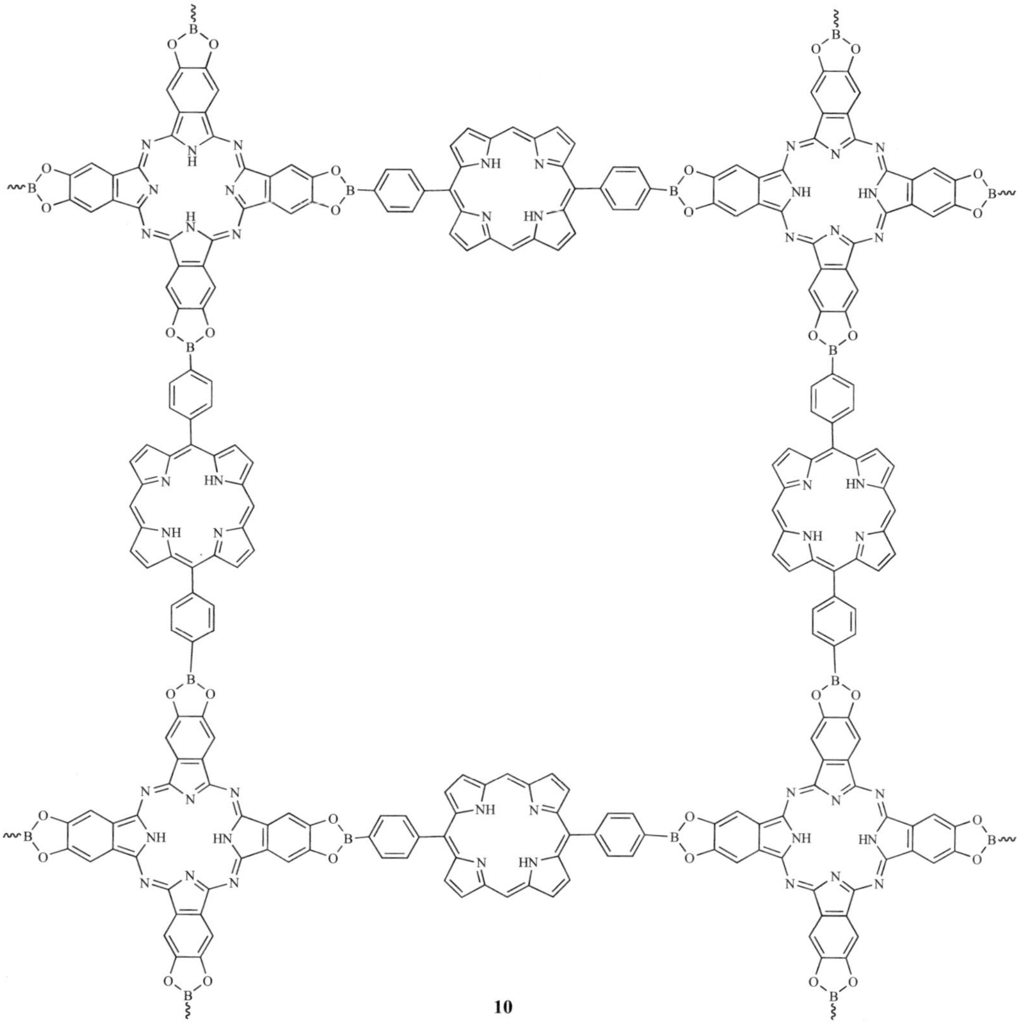

10

Fig. (5). Phthalocyanine-porphyrin based hybrid covalent organic framework.

Echegoyen [11] reported the synthesis of phthalocyanine-porphyrin based hybrid covalent organic framework **10** (Fig. **5**), having a high surface area (BET = 1315

m^2 g^{-1} and used it for evaluation in gas storage, managing to store up to 0.8 wt% of H_2 at 77 K and 1 bar and 0.6 wt% of CH_4 at 298 K and 1 bar.

Burgess [12] prepared bichromophoric energy-transfer hybrid **11** (Fig. **6**), based on a BODIPY donor linked *via* Sonogashira cross-coupling to cyanine acceptors.

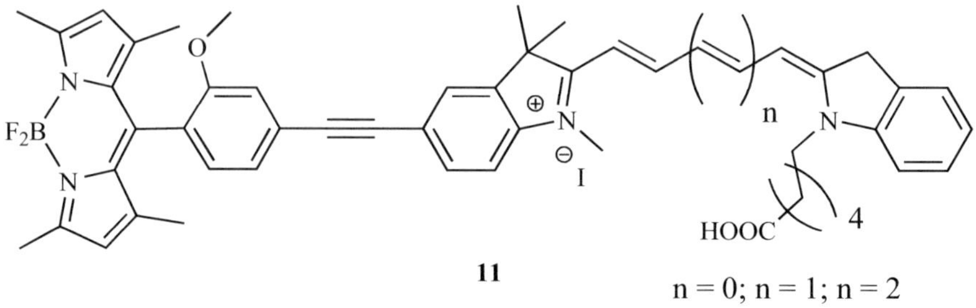

11

Fig. (6). Bichromophoric BODIPY-cyanine hybrid.

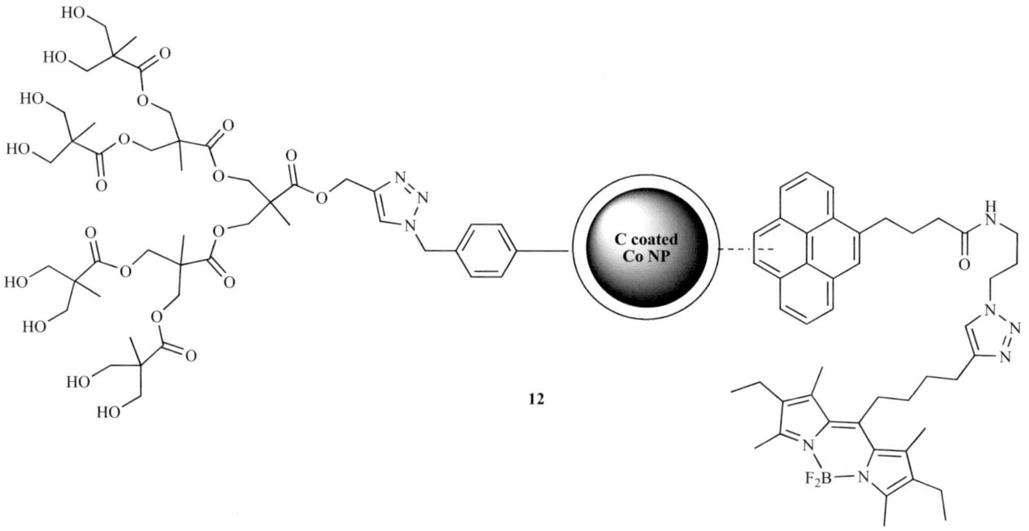

Fig. (7). Hybrid multifunctional nanomaterial.

Reiser [13] combined a noncovalent and a covalent strategy to prepare hybrid multifunctional nanomaterials **12** (Fig. **7**), for which highly magnetic carbon-coated cobalt nanoparticles were used as a scaffold for covalent functionalization by grafting dendrimers *via* "click chemistry", along with noncovalent functionalization of pyrene-tagged BODIPY fluorescent dye by π-π stacking interactions.

Rakovich [14] reported nanohybrid system **13** (Fig. **8**), by using electro-static assembly of complementary inorganic (quantum dots) and organic (cyanine dye), having highly efficient nonradiative energy transfer properties. Also Alberto and Caputo [15] prepared spherical hybrid cyanine-silica nanoparticles **14** (Fig. **8**), with 50 nm size, through reverse microemulsion method and studied its photoemission properties.

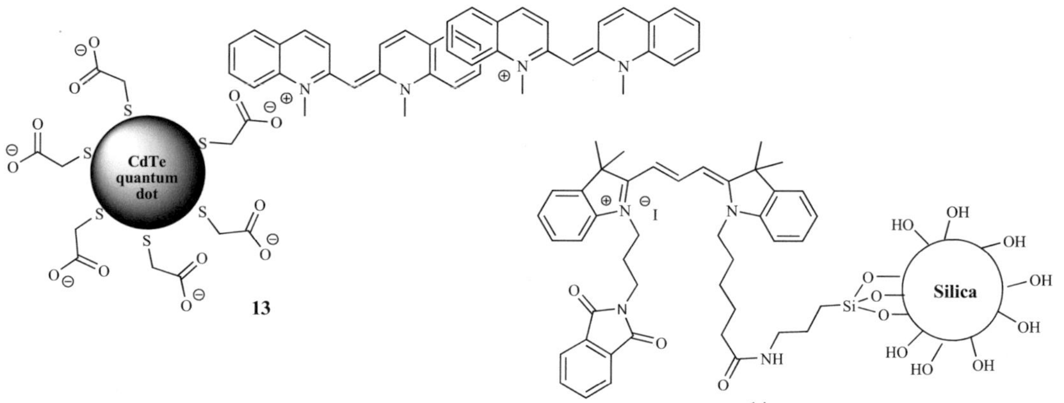

13

14

Fig. (8). Cyanine based hybrid materials.

Azenha and colleagues [16] evaluated the use of NaY zeolite and MCM-41 mesoporous molecular sieves as supports for immobilization of 5,10,15,20-tetra (2-fluorophenyl)porphyrin **15a** (Fig. **9**) *via* adsorption and ship-in-a-bottle routes, respectively, and used them for pesticide photodegradation with good results. Subsequently, 5,10,15,20-tetra (2,6-dichlorophenyl)-porphyrin, **16a** was immobilized also inside MCM-41 by ship-in-a-bottle method. In addition, the corresponding copper metal complexes, **15b** and **16b** were also prepared (Fig. **9**) [17].

Maldotti [18] reported the covalent linkage of chlorosulfonated (*meso*-tetra(2,-dichlorophenyl)porphyrin) **17** (Fig. **9**) to mesoporous material MCM-41 and amorphous SiO_2, comparing both materials on the oxidation of 1,4-pentanediol.

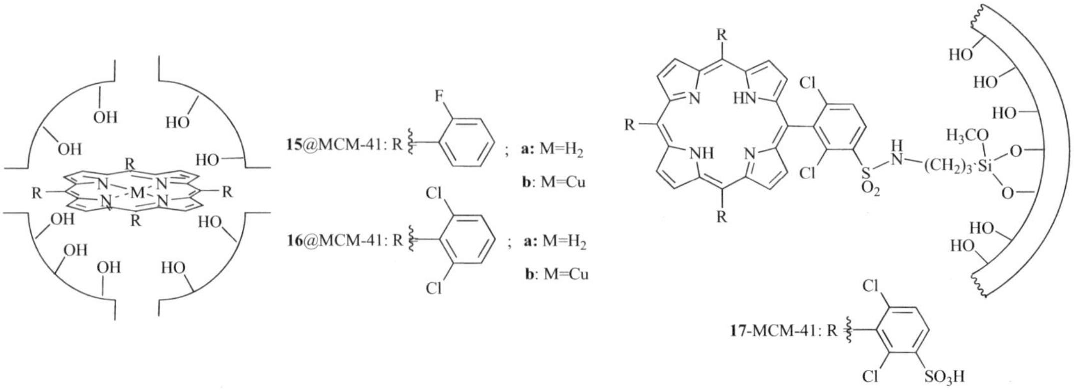

Fig. (9). Immobilized porphyrins for catalytic applications.

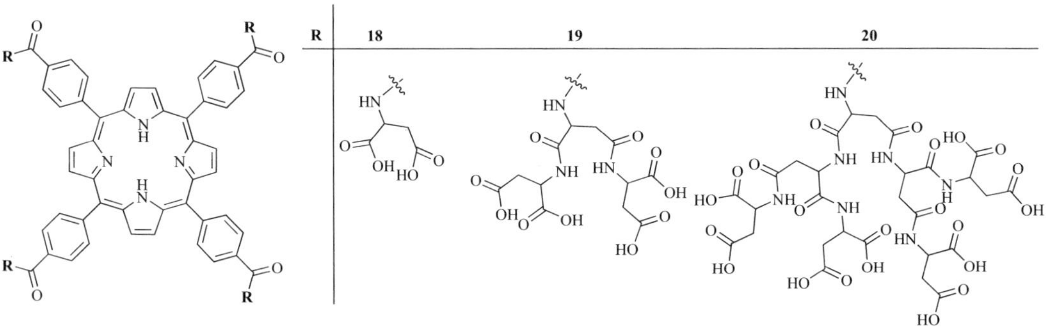

Fig. (10). Amino acid based porphyrin dendrimer.

Amino acid-based dendrimers offer unique properties including stereoselective sites for non-covalent interactions with photosensitizing molecules. Porphyrins have been conjugated with branched dendritic polyaspartic dendrimers **(18-20)** (Fig. **10**) and tested *in vitro* to evaluate their potential as photosensitizers [19]. With increasing dendron generation it was observed an increase in fluorescence and singlet oxygen quantum yields. *In vitro* studies performed in *HeLa* cells demonstrated that **18-20** conjugates were non-toxic in darkness and phototoxic after photoexcitation.

Fluorescence imaging is frequently based on fluorescence marking, using exogenous fluorophores, which selectively bind to specific targets in tissues. As an example, in 2005, Pandey [20] prepared porphyrin conjugate **21** (Fig. **11**), in which a chlorophyll analog was linked with a cyanine dye. The authors found the conjugate to be both an efficient tumor imaging as well as photosensitizing agent.

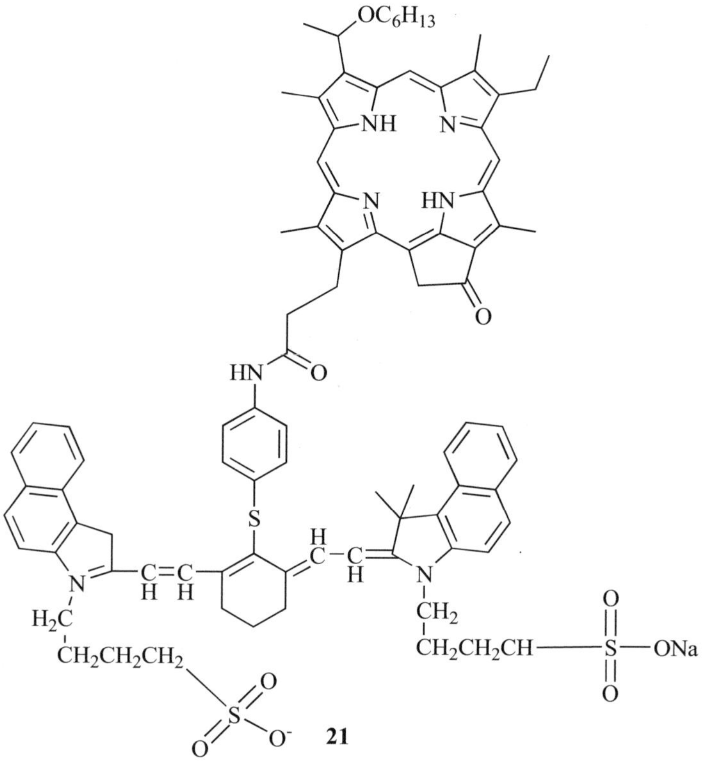

21

Fig. (11). Porphyrin conjugated derivatives for imaging disease diagnostics.

There are various studies on dendrimers that are constituted of electronic chromophores but have saturated linkages. Among them, propylene imine dendrimers **22** (Fig. **12**), containing with porphyrins at their periphery were prepared by a divergent route with the 2-aminoporphyrins coupled onto the surface groups of the preformed dendrimer using a diamide linking unit, to form a G3 dendrimer that had 16 porphyrin rings added to the surface [21].

The panchromatic trichromophoric system **23** (Fig. **13**) using antenna effect was also reported by Odobel for incorporation as a sensitizer in dye sensitized solar cells (DSSC) [22]. Compound **23** was synthesized by Heck alkynylation and presented an efficient intramolecular electron transfer between all the chromophoric units. The antenna effect was assessed by the photoaction spectrum, which featured all of a chromophore's absorption bands.

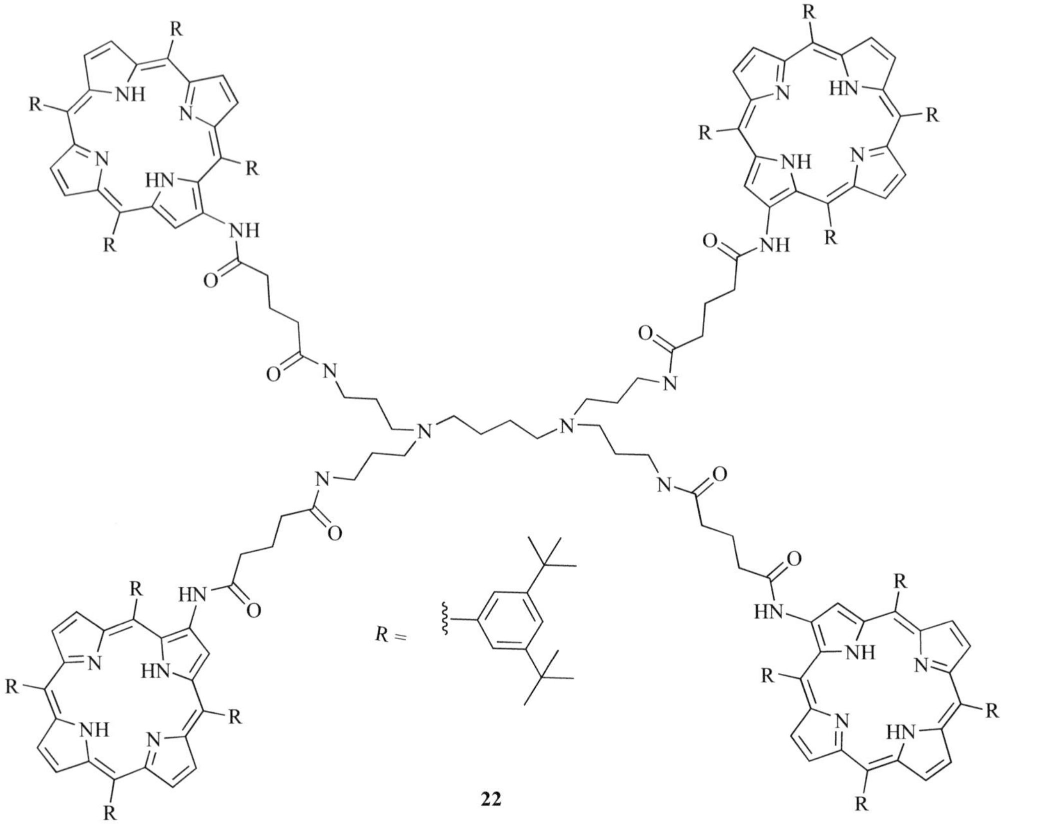

Fig. (12). Porphyrin substituted dendrimer for photovoltaics.

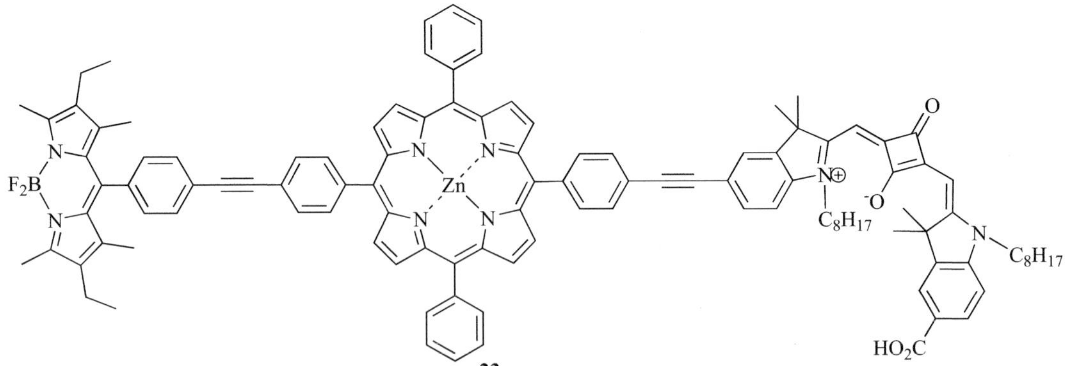

Fig. (13). BODIPY based triad for solar cell applications.

For example, Fukuzumi [23] reported hybrids of type **24**, constituted by spherical gold nanoclusters protected with porphyrin monolayers, and were used as light harvesting antennae (Fig. **14**), exhibiting efficient capability.

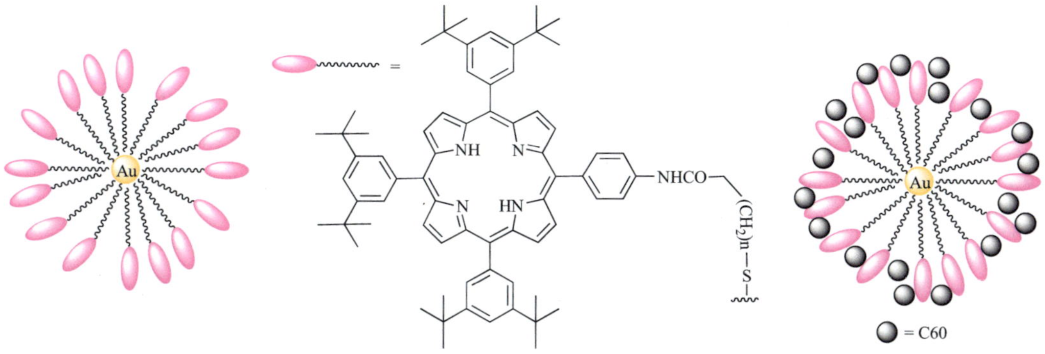

23a: n = 5
23b: n = 11
23c: n = 15

Fig. (14). Examples of gold nanoparticles functionalized with porphyrin 23.

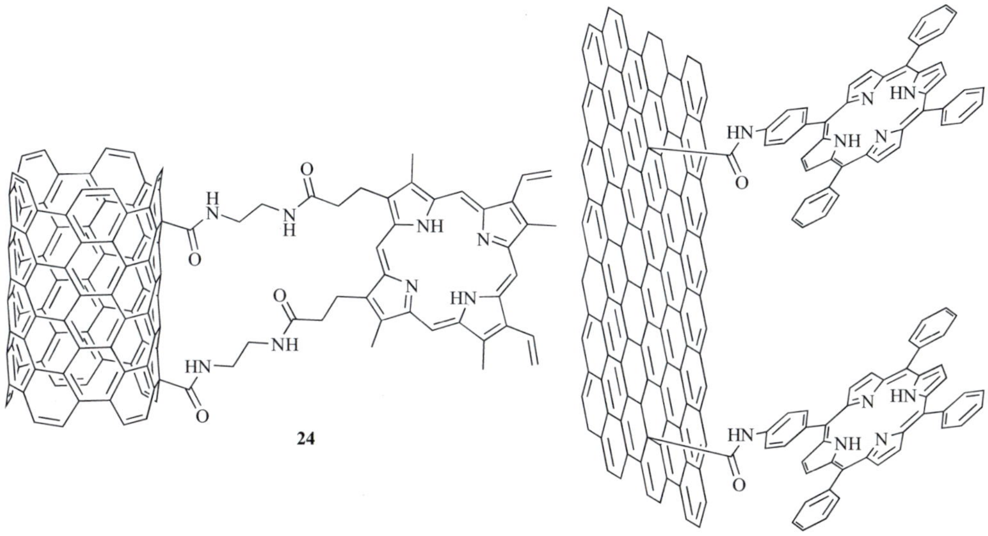

24

25

Fig. (15). Examples of carbon nanostructures functionalized with porphyrins.

Aiming to develop antimicrobial photosensitizers, Banerjee [24] reported the conjugation of protoporphyrin IX to a carbon nanotube to form nanocomposite **24**

(Fig. **15**), in ultrathin, flexible and transparent films. The authors used a carbodiimide mediated coupling to perform the attachment of the porphyrin moiety to the amine functionalized carbon nanotube.

Association of porphyrins with graphene was also reported, where the authors condensed an amino-substituted tetraphenylporphyrin (TPP) to the carboxylic groups present in graphene oxide (GO) [25] yielding GO–TPP hybrid structure 25 (Fig. **15**), whose luminescence quenching of TPP in the GO–TPP hybrid material was pointing to a strong electronic interaction between TPP in the excited state and GO.

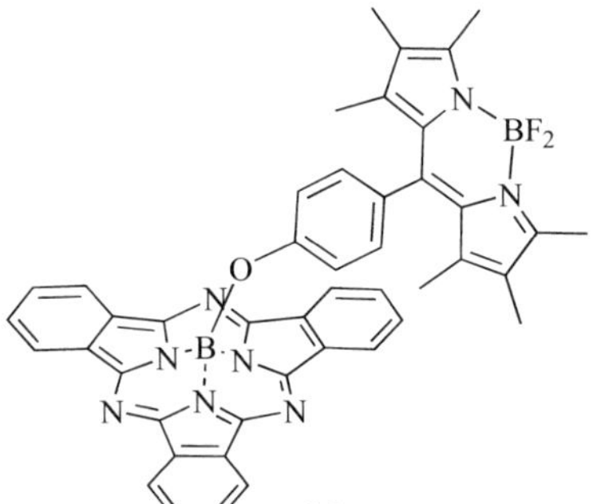

26

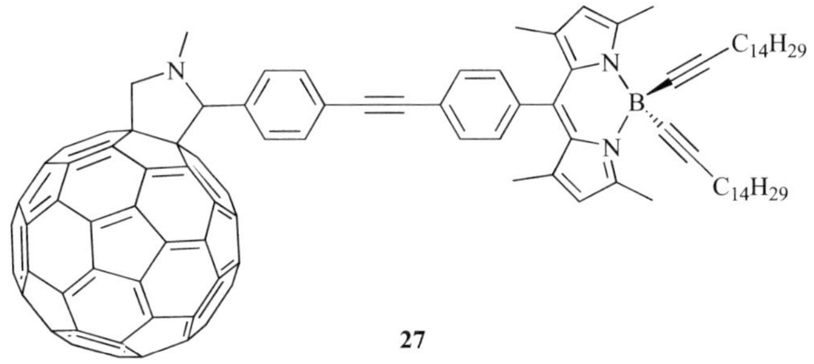

27

Fig. (16). Examples of hybrid BODIPY based composites.

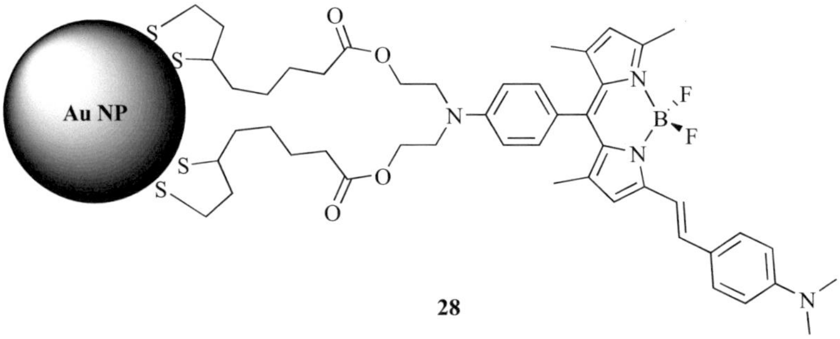

28

Fig. (17). Hybrid BODIPY-gold nanoparticle composites.

Ng prepared hybrid compound **26** (Fig. **16**) [26], constituted by a subphthalocyanine axially substituted with a BODIPY derivative, exhibiting highly efficient photoinduced energy transfer processes. Later, Harriman also reported fullerene-BODIPY dyad **27** (Fig. **16**) [27], which was used to study the transfer of energy and charge recombination between the two hybrid units.

The fluoro-chromogenic chemosensor based on BODIPY-functionalized gold nanoparticles **28** (Fig. **17**) was synthesized by Kang and Han [28]. The authors managed to observe an instantaneous change in color, along with a absorption band blue-shifting and emission spectrum quenching, upon addition of Cu^{2+} ions to aqueous solutions of **28**.

REFERENCES

[1] Silva, M.; Calvete, M.J.; Gonçalves, N.P.; Burrows, H.D.; Sarakha, M.; Fernandes, A.; Ribeiro, M.F.; Azenha, M.E.; Pereira, M.M. Zinc(II) phthalocyanines immobilized in mesoporous silica Al-MCM-41 and their applications in photocatalytic degradation of pesticides. *J. Hazard. Mater.,* **2012**, *233-234*, 79-88.
[http://dx.doi.org/10.1016/j.jhazmat.2012.06.059] [PMID: 22835771]

[2] Zalomaeva, O.V; Sorokin, A.B; GKholdeeva, O.A Clean catalytic oxidation of 8-hydroxyquinoline to quinoline-5,8-dione with tBuOOH in the presence of covalently bound FePcS–SiO2 catalysts. *Green Chem.,* **2010**, *12*, 1076-1082.
[http://dx.doi.org/10.1039/b923674k]

[3] Sirotin, S.V; Tolbin, A.Y; Moskovskaya, I.F; Abramchuk, S.S; Tomilova, L.G; Romanovsky, B.V Heterogenized Fe(III) phthalocyanine: Synthesis, characterization and application in liquid-phase oxidation of phenol. *J. Mol. Catal. Chem.,* **2010**, *319*, 39-45.
[http://dx.doi.org/10.1016/j.molcata.2009.11.017]

[4] DeOliveira, E.; Neri, C.R.; Ribeiro, A.O.; Garcia, V.S.; Costa, L.L.; Moura, A.O.; Prado, A.G.; Serra, O.A.; Iamamoto, Y. Hexagonal mesoporous silica modified with copper phthalocyanine as a photocatalyst for pesticide 2,4-dichlorophenoxiacetic acid degradation. *J. Colloid Interface Sci.,* **2008**, *323*(1), 98-104.
[http://dx.doi.org/10.1016/j.jcis.2008.04.025] [PMID: 18455176]

[5] Doyle, J.J; Ballesteros, B; de la Torre, G; McGovern, D.A; Kelly, J.M; Torres, T; Blau, W.J Combination of phthalocyanine and fullerene moieties for optical limiting. *Chem. Phys. Lett.,* **2006**, *428*, 307-311.

[6] Rio, Y.; Seitz, W.; Gouloumis, A.; Vázquez, P.; Sessler, J.L.; Guldi, D.M.; Torres, T. A panchromatic supramolecular fullerene-based donor-acceptor assembly derived from a peripherally substituted bodipy-zinc phthalocyanine dyad. *Chemistry,* **2010**, *16*(6), 1929-1940.
[http://dx.doi.org/10.1002/chem.200902507] [PMID: 20039345]

[7] Karousis, N.; Ortiz, J.; Ohkudo, K.; Hasobe, T.; Fukuzumi, S.; Sastre-Santos, A.; Tagmatarchis, N. Zinc phthalocyanine–graphene hybrid material for energy conversion: synthesis, characterization, photophysics, and photoelectrochemical cell preparation. *J. Phys. Chem. C,* **2012**, *116*, 20564-20573.
[http://dx.doi.org/10.1021/jp305783v]

[8] Nishiyama, N.; Nakagishi, Y.; Morimoto, Y.; Lai, P.S.; Miyazaki, K.; Urano, K.; Horie, S.; Kumagai,

M.; Fukushima, S.; Cheng, Y.; Jang, W.D.; Kikuchi, M.; Kataoka, K. Enhanced photodynamic cancer treatment by supramolecular nanocarriers charged with dendrimer phthalocyanine. *J. Control. Release,* **2009**, *133*(3), 245-251.
[http://dx.doi.org/10.1016/j.jconrel.2008.10.010] [PMID: 19000725]

[9] Masilela, N.; Nyokong, T. The interaction of silver nanoparticles with low symmetry cysteinyl metallophthalocyanines and their antimicrobial effect. *J. Photochem. Photobiol. B: Chem.,* **2013**, *255*, 1-9.
[http://dx.doi.org/10.1016/j.jphotochem.2013.01.009]

[10] Modisha, P.; Nyokong, T. Fabrication of phthalocyanine-magnetic nanoparticles hybrid nanofibers for degradation of Orange-G. *J. Mol. Catal. Chem.,* **2014**, *381*, 132-137.
[http://dx.doi.org/10.1016/j.molcata.2013.10.012]

[11] Pavan, V.S.; Neti, K.; Wu, X.; Deng, S.; Echegoyen, L. Synthesis of a phthalocyanine and porphyrin 2D covalent organic framework. *CrystEngComm,* **2013**, *15*, 6892-6895.
[http://dx.doi.org/10.1039/c3ce40706c]

[12] Ueno, Y.; Jose, J.; Loudet, A.; Pérez-Bolívar, C.; Anzenbacher, P., Jr; Burgess, K. Encapsulated energy-transfer cassettes with extremely well resolved fluorescent outputs. *J. Am. Chem. Soc.,* **2011**, *133*(1), 51-55.
[http://dx.doi.org/10.1021/ja107193j] [PMID: 21105708]

[13] Kainz, Q.M. ; Schaetz, A.; Zoepfl, A.; Stark, W.J.; Reiser, O. Combined covalent and noncovalent functionalization of nanomagnetic carbon surfaces with dendrimers and BODIPY fluorescent dye. *Chem. Mater.,* **2011**, *23*, 3606-3613.
[http://dx.doi.org/10.1021/cm200705d]

[14] Savateeva, D.; Melnikau, D.; Lesnyak, V.; Gaponik, N.; Rakovich, Y.P. Hybrid organic/inorganic semiconductor nanostructures with highly efficient energy transfer. *J. Mater. Chem.,* **2012**, *22*, 10816-10820.
[http://dx.doi.org/10.1039/c2jm30917c]

[15] Alberto, G. ; Miletto, I.; Viscardi, G.; Caputo, G.; Latterini, L.; Coluccia, S.; Martra, G. Hybrid cyanine-silica nanoparticles: homogeneous photoemission behavior of entrapped fluorophores and consequent high brightness enhancement. *J. Phys. Chem. C,* **2009**, *113*, 21048-21053.
[http://dx.doi.org/10.1021/jp907415q]

[16] Silva, M.; Azenha, M.E.; Pereira, M.M.; Burrows, H.D.; Sarakha, M.; Ribeiro, M.F.; Fernandes, A.; Monsanto, P.; Castanheira, F. Immobilization of 5,10,15,20-tetrakis-(2-fluorophenyl)porphyrin into MCM-41 and NaY: Routes toward photodegradation of pesticides. *Pure Appl. Chem.,* **2009**, *81*, 2025-2033.

[17] Silva, M. ; Azenha, M.E.; Pereira, M.M.; Burrows, H.D.; Sarakha, M.; Forano, C.; Ribeiro, M.F. Fernandes, Immobilization of halogenated porphyrins and their copper complexes in MCM-41: Environmentally friendly photocatalysts for the degradation of pesticides. *Appl. Catal. B,* **2010**, *100*, 1-9.
[http://dx.doi.org/10.1016/j.apcatb.2010.07.033]

[18] Molinari, A.; Maldotti, A.; Bratovcic, A.; Magnacca, G. Fe(III)-porphyrin heterogenized on MCM-41: Matrix effects on the oxidation of 1,4-pentanediol. *Catal. Today,* **2011**, *161*, 64-69.

[http://dx.doi.org/10.1016/j.cattod.2010.09.004]

[19] Gao, B.X.; Liu, Y.L.; Yin, H.J.; Li, Y.X.; Bai, Q.Q.; Zhang, L.C. Watersoluble dendritic polyaspartic porphyrins: potential photosensitizers for photodynamic therapy. *New J. Chem.,* **2012**, *36*, 28-31. [http://dx.doi.org/10.1039/C1NJ20733D]

[20] Chen, Y.; Gryshuk, A.; Achilefu, S.; Ohulchansky, T.; Potter, W.; Zhong, T.; Morgan, J.; Chance, B.; Prasad, P.N.; Henderson, B.W.; Oseroff, A.; Pandey, R.K. A novel approach to a bifunctional photosensitizer for tumor imaging and phototherapy. *Bioconjug. Chem.,* **2005**, *16*(5), 1264-1274. [http://dx.doi.org/10.1021/bc050177o] [PMID: 16173807]

[21] Li, W.S.; Jiang, D.L.; Suna, Y.; Aida, T. Cooperativity in chiroptical sensing with dendritic zinc porphyrins. *J. Am. Chem. Soc.,* **2005**, *127*(21), 7700-7702. [http://dx.doi.org/10.1021/ja0513335] [PMID: 15913359]

[22] Warnan, J.; Buchet, F.; Pellegrin, Y.; Blart, E.; Odobel, F. Panchromatic trichromophoric sensitizer for dye-sensitized solar cells using antenna effect. *Org. Lett.,* **2011**, *13*(15), 3944-3947. [http://dx.doi.org/10.1021/ol2014686] [PMID: 21721515]

[23] Imahori, H.; Arimura, M.; Hanada, T.; Nishimura, Y.; Yamazaki, I.; Sakata, Y.; Fukuzumi, S. Photoactive three-dimensional monolayers: porphyrin-alkanethiolate-stabilized gold clusters. *J. Am. Chem. Soc.,* **2001**, *123*(2), 335-336. [http://dx.doi.org/10.1021/ja002838s] [PMID: 11456522]

[24] Banerjee, I.; Mondal, D.; Martin, J.; Kane, R.S. Photoactivated antimicrobial activity of carbon nanotube-porphyrin conjugates. *Langmuir,* **2010**, *26*(22), 17369-17374. [http://dx.doi.org/10.1021/la103298e] [PMID: 20931992]

[25] Xu,, Y.; Liu, Z.; Zhang, X.; Wang, Y.; Tian, J.; Huang, Y.; Ma, Y.; Zhang, X.; Chen, Y. A graphene hybrid material covalently functionalized with porphyrin: synthesis and optical limiting property. *Adv. Mater.,* **2009**, *21*, 1275-1279. [http://dx.doi.org/10.1002/adma.200801617]

[26] Liu, J.Y.; Yeung, H.S.; Xu, W.; Li, X.; Ng, D.K. Highly efficient energy transfer in subphthalocyanine-BODIPY conjugates. *Org. Lett.,* **2008**, *10*(23), 5421-5424. [http://dx.doi.org/10.1021/ol8023677] [PMID: 19006394]

[27] Ziessel, R.; Allen, B.D.; Rewinska, D.B.; Harriman, A. Selective triplet-state formation during charge recombination in a fullerene/Bodipy molecular dyad (Bodipy=borondipyrromethene). *Chemistry,* **2009**, *15*(30), 7382-7393. [http://dx.doi.org/10.1002/chem.200900440] [PMID: 19551782]

[28] Lee, H.Y.; Son, H.; Lim, J.M.; Oh, J.; Kang, D.; Han, W.S.; Jung, J.H. BODIPY-functionalized gold nanoparticles as a selective fluoro-chromogenic chemosensor for imaging Cu2+ in living cells. *Analyst (Lond.),* **2010**, *135*, 2022-2027. [http://dx.doi.org/10.1039/c0an00129e]

SUBJECT INDEX

Mário J. F. Calvete

Made in the USA
Las Vegas, NV
03 November 2026

57425097R00343